BIOMEDICAL DEVICE TECHNOLOGY

ABOUT THE AUTHOR

Anthony Y. K. Chan graduated in Electrical Engineering (B.Sc. Hon.) from the University of Hong Kong and completed his M.Sc. in Engineering from the same university. He also completed a master's degree (M.Eng.) in Clinical Engineering and a Ph.D. in Biomedical Engineering from the University of British Columbia, Canada. Dr. Chan also holds a Certificate in Health Services Management from the Canadian Healthcare Association. Dr. Chan worked for a number of years as a project engineer in the field of electrical instrumentations, control, and systems, and was the director and manager of biomedical engineering in a number of Canadian acute care hospitals. He is currently the Program Head of the Biomedical Engineering Technology Program at the British Columbia Institute of Technology and is an Adjunct Professor of the School of Biomedical Engineering at the University of British Columbia. Dr. Chan is a Professional Engineer, a Chartered Engineer, and a Certified Clinical Engineer. He is a fellow member of CMBES, life senior member of IEEE, member of IET and HKIE.

Third Edition

BIOMEDICAL DEVICE TECHNOLOGY

Principles and Design

By

ANTHONY Y. K. CHAN, Ph.D., P.Eng., CCE

CHARLES C THOMAS • PUBLISHER • LTD.
Springfield • Illinois • U.S.A

Published and Distributed Throughout the World by

CHARLES C THOMAS • PUBLISHER, LTD.
2600 South First Street
Springfield, Illinois 62704

© 2023 by CHARLES C THOMAS • PUBLISHER, LTD.

ISBN 978-0-398-09392-1 (hard)
ISBN 978-0-398-09393-8 (ebook)

First Edition, 2008
Second Edition, 2016
Third Edition, 2023

Library of Congress Catalog Card Number: 2022036373 (print)
2022036374 (ebook)

With THOMAS BOOKS *careful attention is given to all details of manufacturing and design. It is the Publisher's desire to present books that are satisfactory as to their physical qualities and artistic possibilities and appropriate for their particular use.* THOMAS BOOKS *will be true to those laws of quality that assure a good name and good will.*

Printed in the United States of America
CM-C-1

Library of Congress Cataloging-in-Publication Data

Names: Chan, Anthony Y. K., author.
Title: Biomedical device technology: principles and design / by Anthony
   Y.K. Chan.
Description: Third edition. | Springfield, Illinois: Charles C Thomas,
   Publisher, Ltd., 2023. | Includes bibliographical references and index.
Identifiers: LCCN 2022036373 (print) | LCCN 2022036374 (ebook) | ISBN
   9780398093921 (hardback) | ISBN 9780398093938 (ebook)
Subjects: MESH: Biomedical Technology--instrumentation | Equipment and
   Supplies | Equipment Design | Equipment Safety
Classification: LCC R855.3 (print) | LCC R855.3 (ebook) | NLM W 26 | DDC
   610.285--dc23/eng/20221103
LC record available at https://lccn.loc.gov/2022036373
LC ebook record available at https://lccn.loc.gov/2022036374

This book is dedicated with love to
my wife Elaine,
my daughters Victoria and Tiffany,
and
in memory of
my brother David

PREFACE

For many years, the tools available to physicians were limited to a few simple handpieces such as stethoscopes, thermometers and syringes; medical professionals primarily relied on their senses and skills to perform diagnosis and disease mitigation. Today, diagnosis of medical problems is heavily dependent on the analysis of information made available by sophisticated medical machineries such as electrocardiographs, video endoscopic equipment and pulmonary analyzers. Patient treatments often involve specialized tools and systems such as cardiac pacemakers, electrosurgical units, and minimally invasive surgical instruments. Such biomedical devices play a critical and indispensable role in modern-day medicine.

In order to design, build, maintain, and effectively deploy medical devices, one needs to understand not only their use, design and construction but also how they interact with the human body. This book provides a comprehensive approach to studying biomedical devices and their applications. It is written for engineers and technologists who are interested in understanding the principles, design, and use of medical device technology. The book is also intended to be a textbook or reference for biomedical device technology courses in universities and colleges.

The most common reason for medical device obsolescence is changes in technology. For example, vacuum tubes in the 1960s, discrete semiconductors in the 1970s, integrated circuits in the 1980s, microprocessors in the 1990s and networked multiprocessor software-driven systems in today's devices. The average life span of medical devices has been diminishing; current medical devices have a life span of about 5 to 7 years. Some are even shorter. Therefore, it is unrealistic to write a book on medical devices and expect that the technology described will remain current and valid for years. On the other hand, the principles of medical device and their applications, the origins of physiological signals and their methods of acquisitions, and the concepts of signal analysis and processing will remain largely unchanged. This book focuses on the applications, functions and principles of medical devices (which are the invariant components) and uses specific designs and constructions to illustrate the concepts where appropriate.

The first part of this book discusses the fundamental building blocks of biomedical instrumentations. Starting from an introduction of the origins of biological signals, the essential functional building blocks of a typical medical device are studied. These functional blocks include electrodes and transducers, biopotential amplifiers, signal conditioners and processors, electrical safety and isolation, and output devices. The next section of the book covers a selection of biomedical devices. Their principles of operations, functional building blocks, special features, performance specifications are discussed. Architectural and schematic diagrams are used where appropriate to illustrate how specific device functions are being implemented. In addition, indications of use and clinical applications of each device are included. Common problems and hazards, and risk mitigation of each device are discussed. For those who would like to know more, a collection of relevant published papers and book references has been added to the end of each chapter.

Due to the vast variety of biomedical devices available in healthcare, it is impractical to include all of them in a single book. This book selectively covers diagnostic and therapeutic devices that are either commonly used or whose principles and design represent typical applications of the technology. To limit the scope, medical imaging equipment and laboratory instrumentations are excluded from this book.

Four appendices are included at the end of the book. These are appended for those who are not familiar with these concepts, yet an understanding in these areas will enhance the comprehension of the subject matters in the book. They are A1-A Primer on Fourier Analysis, A2-Overview of Medical Telemetry Development, A3-Medical Gas Supply Systems, and the newest addition A4-Concepts of Infection Control in Biomedical Device Technology.

In this third edition, many chapters have gone through revisions, some with significant updates and additions to keep up with new applications and advancements in medical technology. Based on requests, review questions are added for each chapter to help readers to assess their comprehension of the content material.

I am thankful to the readers, educators, and professionals who provided me with invaluable suggestions for this revision. I also would like to take the opportunity to thank Professor Euclid Seeram for inspiring me into book publishing, and Michael Thomas for his continuing support in publishing this new edition.

Anthony Y. K. Chan

CONTENTS

Page

Preface vii

Chapter

PART I–INTRODUCTION

1. Overview of Biomedical Instrumentation 5
2. Concepts in Signal Measurement, Processing, and Analysis 33

PART II–BIOMEDICAL TRANSDUCERS

3. Fundamentals of Biomedical Transducers 55
4. Pressure and Force Transducers 67
5. Temperature Transducers 80
6. Position and Motion Transducers 103
7. Flow Transducers 112
8. Optical Transducers 126
9. Electrochemical Transducers 151
10. Biopotential Electrodes 178

PART III–FUNDAMENTAL BUILDING BLOCKS
OF MEDICAL INSTRUMENTATION

11. Biopotential Amplifiers 191
12. Electrical Safety and Signal Isolation 216
13. Medical Waveform Display Systems 237

PART IV–MEDICAL DEVICES

14. Physiological Monitoring Systems 265
15. Electrocardiographs 283
16. Electroencephalographs 312
17. Electromyography and Evoked Potential Study Equipment 334
18. Invasive Blood Pressure Monitors 352
19. Noninvasive Blood Pressure Monitors 371
20. Cardiac Output Monitors 384
21. Cardiac Pacemakers 405
22. Cardiac Defibrillators 427
23. Infusion Devices 448
24. Electrosurgical Units 474
25. Pulmonary Function Analyzers 495
26. Mechanical Ventilators 513
27. Ultrasound Blood Flow Detectors 535
28. Fetal Monitors 545
29. Infant Incubators, Phototherapy Lights, Warmers and
    Resuscitators 552
30. Body Temperature Monitors 566
31. Pulse Oximeters, Oxygen Analyzers & Transcutaneous
    Oxygen Monitors 582
32. End-Tidal Carbon Dioxide Monitors 600
33. Anesthesia Machines 607
34. Dialysis Equipment 627
35. Surgical Lasers 654
36. Endoscopic Video Systems 679
37. Cardiopulmonary Bypass Units 701
38. Audiology Equipment 718

Appendices
A-1. *A Primer on Fourier Analysis* 753
A-2. *Overview of Medical Telemetry Development* 758
A-3. *Medical Gas Supply Systems* 762
A-4. *Concepts in Infection Control of Biomedical Device Technology* 765

Review Questions 773
Answers to Review Questions 848
Index 873

BIOMEDICAL DEVICE TECHNOLOGY

Part I

INTRODUCTION

Chapter 1

OVERVIEW OF BIOMEDICAL INSTRUMENTATION

OBJECTIVES

- Define medical device.
- Analyze biomedical instrumentation using a systems approach.
- Explain the origin and characteristics of biopotentials and common physiological signals.
- Introduce human factors engineering in medical device design.
- List common input, output, and control signals of medical devices.
- Discuss special constraints encountered in the design of biomedical devices.
- Define biocompatibility and list common implant materials.
- Explain tissue responses to foreign materials and state approaches to avoid adverse tissue reaction.
- Describe the basic functional building blocks of medical instrumentation.

CHAPTER CONTENTS

1. Introduction
2. Classification of Medical Devices
3. Systems Approach
4. Origins of Biopotentials
5. Physiological Signals
6. Human–Machine Interface
7. Input, Output, and Control Signals
8. Constraints in Biomedical Signal Measurements
9. Concepts on Biocompatibility
10. Functional Building Blocks of Medical Instrumentation

INTRODUCTION

Medical devices come with different designs and complexity. They can be as simple as a tongue depressor, as compact as an implantable pacemaker, or as sophisticated as a heart lung machine. Although most medical devices use similar technology as other consumer or industrial devices, there are many fundamental differences between devices used in medicine and devices used in other applications. This chapter will look at the definition of medical devices and the characteristics that differentiate a medical device from other household or consumer products.

According to the International Electrotechnical Commission (IEC), a medical device means:

> Any instrument, apparatus, implement, appliance, implant, in vitro reagent or calibrator, software, material or other similar or related article:
> a) intended by the manufacturer to be used, alone or in combination, for human beings for one or more of the specific purpose(s) of:
> • diagnosis, prevention, monitoring, treatment, or alleviation of disease,
> • diagnosis, monitoring, treatment, alleviation of, or compensation for an injury,
> • investigation, replacement, modification, or support of the anatomy or of a physiological process,
> • supporting or sustaining life,
> • control of conception,
> • disinfection of medical devices,
> • providing information for medical purposes by means of in vitro examination of specimens derived from the human body, and
> b) which does not achieve its primary intended action in or on the human body by pharmacological, immunological or metabolic means, but which can be assisted in its function by such means.

The United States Food and Drug Administration (FDA) defines a medical device as:

> An instrument, apparatus, implement, machine, contrivance, implant, in vitro reagent, or other similar or related article, including a component part, or accessory which is:
> • recognized in the official National Formulary, or the United States Pharmacopoeia, or any supplement to them,
> • intended for use in the diagnosis of disease or other conditions, or in the cure, mitigation, treatment, or prevention of disease, in man or other animals, or
> • intended to affect the structure or any function of the body of man or other animals, and which does not achieve any of its primary intended purposes

through chemical action within or on the body of man or other animals and which is not dependent upon being metabolized for the achievement of any of its primary intended purposes.

In the Canadian Food and Drugs Act, a medical device is similarly defined as:

Any article, instrument, apparatus or contrivance, including any component, part or accessory thereof, manufactured, sold or represented for use in:
(a) the diagnosis, treatment, mitigation or prevention of a disease, disorder or abnormal physical state, or the symptoms thereof, in humans or animals;
(b) restoring, correcting or modifying a body function, or the body structure of humans or animals;
(c) the diagnosis of pregnancy in humans or animals; or
(d) the care of humans or animals during pregnancy, and at, and after, birth of the offspring, including care of the offspring, and includes a contraceptive device but does not include a drug.

Apart from the obvious, it is clear from the above definitions that in vitro diagnostic products such as medical laboratory instruments are medical devices. Furthermore, accessories, reagents, or spare parts associated with a medical device are also considered to be medical devices. An obvious example of this is the electrodes of a heart monitor. Another example, which may not be as obvious, is the power adapter to a laryngoscope. Both of these accessories are considered as medical devices and are therefore regulated by the premarket and postmarket regulatory controls.

CLASSIFICATION OF MEDICAL DEVICES

There are many different ways to classify or group together medical devices. Devices can be grouped by their functions, their technologies, or their applications. A description of some common classification methods follows.

Classified by Functions

Grouping medical devices by their functions is by far the most common way to classify medical devices. Devices can be separated into two main categories: diagnostic and therapeutic.

Diagnostic devices are used for the analysis or detection of diseases, injuries, or other medical conditions. Ideally, a diagnostic device should not cause any change to the structure or function of the biological system. However, some diagnostic devices may disrupt the biological system due to their applications. For example, a real-time blood gas analyzer may require an invasive catheter (which puncture the skin into a blood vessel) to take dissolved carbon dioxide level (PCO^2) measurement. A computed tomography (CT) scanner will impose ionization radiation (transfer energy) on the human body in order to obtain diagnostic medical images.

Diagnostic devices whose functions are to detect changes of certain physiological parameters over a period of time are often referred to as monitoring devices. As the main purpose of this class of devices is trending, absolute accuracy may not be as important as their repeatability. Examples of monitoring devices are heart rate monitors used to track variation of heart rates during a course of drug therapy, and noninvasive blood pressure monitors to assess arterial blood pressure immediately after surgery.

Therapeutic devices are designed to create structural or functional changes that lead to improved function of the patient. Examples of such devices are electrosurgical units in surgery, radiotherapy linear accelerators in cancer treatment, and infusion devices in fluid management therapy. Assistive devices are a group of devices used to restore an existing function of the human body. They may be considered a subset of therapeutic devices. Examples of assistive devices are demand pacemakers to restore normal heart rhythm, hearing aids to improve hearing, and wheelchairs to enhance mobility of people with walking disability.

Based on the methods of application, these device classes can be further divided into invasive or noninvasive, automatic or manual sub-categories.

Classified by Physical Parameters

Medical devices can also be grouped by the physical parameters that they are measuring. For example, a blood pressure monitor is a pressure monitoring device, a respiration spirometer is a flow measurement device, and a tympanic thermometer is a temperature-sensing device.

Classified by Principles of Transduction

Some medical devices are grouped according to the types of transducers used at the patient-machine interface. Resistive, inductive, and ultrasonic devices are examples in this category.

Classified by Physiological Systems

Medical devices may also be grouped by their related human physiological systems. Examples of such grouping are cardiovascular devices (blood pressure monitors, electrocardiographs, etc.), and pulmonary devices (respirators, ventilators, etc.).

Classified by Clinical Medical Specialties

In another model, devices are grouped according to the medical specialties in which they are being used. For example, a fetal monitor is considered as an obstetric device, an x-ray machine as a radiological device.

Classified by Risk Classes

For biomedical engineers and regulatory personnel, medical devices are often referred to by their risk classes. Risk classes are created to differentiate devices by rating their level of risk on patients. A device risk classification determines the degree of scrutiny and regulatory control imposed by regulatory bodies on the manufacturers, suppliers and healthcare providers to ensure their safety and efficacy in clinical use. Table 1–1 shows examples of medical devices in each risk class under the Canadian Medical Device Regulations (MDR). Manufacturers of Classes II, III and IV medical devices require a Health Canada license before selling or advertising these devices. Similar risk classifications are used in the United States (Class I, II, III), and in Europe (Class I, IIa, IIb, III). Note that there are only 3 levels in the US risk classification with Class 3 devices having the highest risk as compared to 4 levels in the Canadian and European systems. Table 1-2 is the Canadian MDR and US FDA risk classifications of some devices covered in this book.

SYSTEMS APPROACH

In simple terms, a system is defined as a group of items, or parts, or processes working together under certain relationships. Collectively, the processes in the system transform a set of input entities to a set of output entities. Within a system there are aspects, variables, or parameters that mutually act on each other. A closed system is self-contained on a specific level and is separated from and not influenced by the environment, whereas an open system is influenced by the environmental conditions by which it is surrounded. Figure 1–1 shows an example of a system. The elements within

Table 1-1. Canada MDR Risk Classification

Four risk classes—from Class I (lowest risk) to Class IV (highest risk)	
Class I	conductive electrode gel, band-aids
Class II	latex gloves, contact lenses
Class III	IV bags, indwelling catheters
Class IV	heart valve implants, defibrillators

Table 1-2. US FDA & Health Canada Device Risk Classification Examples

Device	FDA Risk Class	Health Canada Risk Class
Electrocardiographs	2	2
Electroencephalographs	2	2
Electromyographs	2	2
Invasive Blood Pressure Monitors	2	3
Non-Invasive Blood Pressure Monitors	2	2
Cardiac Output Units, thermal dilution	2	3
Implantable Pacemakers	3	4
Cardiac Defibrillators	3	3
Infusion Pumps	2	3
Electrosurgical Units	2	3
Respiration Monitors	2	3
Mechanical Ventilators	2	3
Ultrasound Blood Flow Detectors	2	3
Fetal Scalp ECG Monitors	3	3
Infant Incubators	2	3
Body Temperature Monitors	2	2
Pulse Oximeters	2	3
Anesthesia Machines	2	3
Hemodialysis Machines	2	3
Neurosurgical Lasers	3	3
Flexible Endoscopes	2	2
Cardiac Pulmonary Bypass Machines	2	3
Audiometers	2	2
Hearing Aids	2	2
Acoustic Chamber (for hearing test)	1	1
Cochlear Implants	3	3

a system and their relationships as well as the environment can affect the performance of the system. A more complicated system may contain multiple numbers of subsystems (or simple systems).

In analyzing a large complex system, one can divide the system into several smaller subsystems with the output from one subsystem connected to the input of another. The simplest subsystem consists of an input, an output, and a process as shown in Figure 1–2. The process that takes the output and feeds it back to the input in order to modify the output is called a feedback process. A system with feedback is called a closed-loop system, whereas a system without any feedback is called an open-loop system. Most systems that we encounter contain feedback paths and hence are closed-loop systems.

Listening to a radio is an example of a simple closed-loop system. The

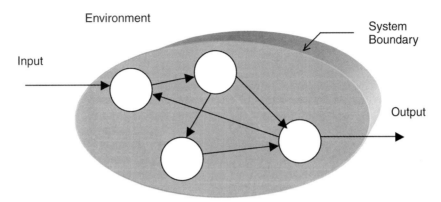

Figure 1-1. Typical System.

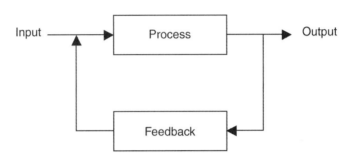

Figure 1-2. Basic Subsystem.

input to the system is the radio broadcast in the form of an electromagnetic wave that is received by the radio. The radio processes the received signal and produces the audible sound such as music. If the music (output) is not loud enough, the listener turns up the volume to increase the sound level. In doing this, the listener becomes the feedback process that analyzes the loudness of the music and invokes the action to turn up the volume.

The systems approach is basically a generalized technique to understand organized complexity. It provides a unified framework or a way of thinking about the systems and can be developed to handle specific problems. In order to solve a problem, one must look at all components within the system and analyze the input and output of each subsystem in view to isolate the problem and establish the relationships of the problem with respect to each component in the system.

Using block diagrams to analyze complex devices is an application of the systems approach. Figure 1–3 shows a music player system. The input to the player is the musical file either from a flash memory, radio broadcast, or via the internet, the output is sound (or music), and the feedback is the lis-

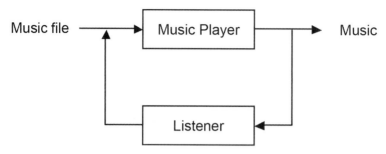

Figure 1-3. Music Player System.

tener who will switch to another file when it has finished playing or turn down the volume if it is too loud. If the player is not working properly, one may buy a new one and discard the malfunctioning unit.

The music player can be divided into its functional blocks as shown in Figure 1–4. One may be able to troubleshoot and isolate the problem to one of the functional blocks (or component). In this case, it will be cheaper just to replace the malfunctioning block. For example, if the speakers are not working, it may be more economical to get a pair of replacement speakers than to replace the entire music player.

Similarly, a complex biomedical device can be broken down into its functional building blocks. Figure 1-5 shows a block diagram of an electrocardiograph (ECG). The input to the device is the biopotential from the heart activities of the patient. The electrodes pick up the tiny electrical signals from the patient and send it to the amplifier block to increase the signal amplitude. The amplified ECG signal is then sent to the signal analysis block to extract information such as the heart rate. Finally, the ECG signal is sent to the output block such as a paper chart recorded to produce a hard copy of the ECG tracing. These functional blocks can be further subdivided, eventually down to the individual component level. Note that the cardiology technologist is also considered to be a part of the system. He or she serves

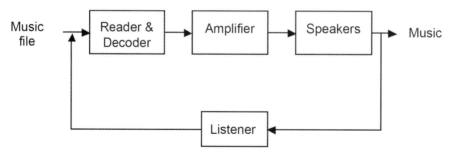

Figure 1-4. Music Player Functional Block.

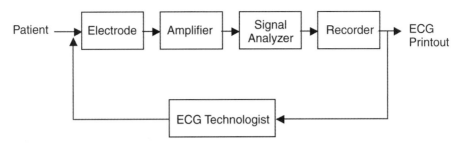

Figure 1-5. ECG Block Diagram.

as the feedback loop by monitoring the output and modifying the input.

When analyzing or troubleshooting a medical device, it is important to understand the functions of each building block, and what to expect from the output when a known input is applied to the block. Furthermore, medical devices are, in most cases, conceptualized, designed, and built from a combination of functional building blocks.

ORIGINS OF BIOPOTENTIALS

The source of electrical events in biological tissue is the ions in the electrolyte solution, as opposed to the electrons in electrical circuits. Biopotential is an electrical voltage caused by a flow of ions through biological tissues. It was first studied by Luigi Galvani, an Italian physiologist and physicist, in 1786. In living cells, there is an ongoing flow of ions (predominantly sodium [Na^+], potassium [K^+D] and chloride [Cl^-]) across the cell membrane. The cell membrane allows some ions to go through readily but resists others. Hence it is called a semipermeable membrane.

There are two fundamental causes of ion flow in the body: diffusion and drift. Fick's law states that if there is a high concentration of particles in one region and they are free to move, the particles will flow in a direction that equalizes the concentration; the force that results in the movement of charges is called diffusion force. The movement of charged particles (such as ions) that is due to the force of an electric field (static forces of attraction and repulsion) constitutes particle drift. Each cell in the body has a potential difference across the cell membrane known as the single-cell membrane potential.

Under equilibrium, the net flow of charges across the cell membrane is zero. However, due to an imbalance of positive and negative ions internal and external to the cell, the potential inside a living cell is about –50 mV to –100 mV with respect to the potential outside it (Figure 1–6). This membrane potential is the result of the diffusion and drift of ions across the high resistance but semipermeable cell membrane, predominantly sodium [Na^+]

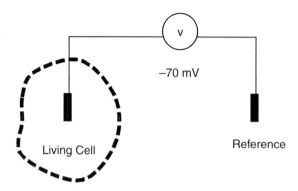

Figure 1-6. Cell Membrane Potential.

and potassium [K+] ions moving in and out of the cell. Because of the semi-permeable nature of the membrane, Na+ is partially restricted from passing into the cell. In addition, a process called the sodium-potassium pump moves sodium ions at two to five times the rate out of the cell than potassium ions into the cell. However, in the presence of diffusion and drift, an equilibrium point is established when the net flow of ions across the cell's membrane becomes zero. As there are more positive ions (Na+) moved outside the cells than positive ions (K+) moved into the cell, under equilibrium, the inside of the cell has less positive ions than the outside. Therefore, the inside of the cell is negative with respect to the outside. This is called the cell's resting potential, which is typically about –70 mV.

If the potential across the cell membrane is raised, for example by an external stimulation, to a level that exceeds the threshold; the permeability of the cell membrane will change, causing a flow of Na+ ions into the cell. This inrush of positive ions will create a positive change in the cell's membrane potential to about 20 mV to 40 mV more positive than the potential outside the cell. This action potential lasts for about 1 to 2 milliseconds. As long as the action potential exists, the cell is said to be depolarized. The membrane potential will drop eventually as the sodium-potassium pump repolarizes the cell to its resting state (–70 mV). This process is called repolarization and the time period is called the refractory period. During the refractory period, the cell is not responsive to any stimulation. The events of depolarization and repolarization are shown in Figure 1–7. The rise in the membrane potential from its resting stage (when stimulated) and return to the resting state is called the action potential. Cell potentials form the basis of all electrical activities in the body, including such activities as the electrocardiogram (ECG), electroencephalogram (EEG), electrooculogram (EOG), electroretinogram (ERG), and electromyogram (EMG).

When a cell is depolarized (during which the membrane potential

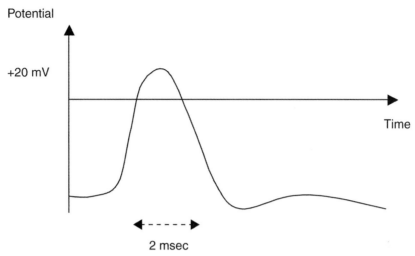

Figure 1-7. Single Cell Action Potential.

changes from negative to positive), the cells next to it may be triggered into depolarization. This disturbance is propagated either to adjacent cells, resulting in the entire tissue becoming depolarized (in an entire motor group), or along the length of the cell from one cell to the next (in a single motor unit or a nerve fiber).

In most biopotential signal measurements, unless one is using a needle electrode to measure the action potential of a single cell, the measured signal is the result of multiple action potentials from a group of cells or tissue. The amplitude and shape of the biopotential are largely dependent on the location of the measurement site and the signal sources. Furthermore, the biopotential signal will be altered as it propagates along the body tissue to the sensors. A typical example of biopotential measurement is measuring electrical heart activities using skin electrodes (electrocardiogram or ECG). Figure 1–8 shows a typical ECG waveform showing the electrical heart potential when a pair of electrodes is placed on the chest of the patient. This biopotential, which is the resultant of all action potentials from the heart tissue transmitted to the skin surface, is very different in amplitude and shape from the action potential from a single cell shown in Figure 1–7. In addition, placing the skin electrodes at different locations on the patient will produce very different looking ECG waveforms.

PHYSIOLOGICAL SIGNALS

Biopotential signals represent a substantial proportion of human physio-

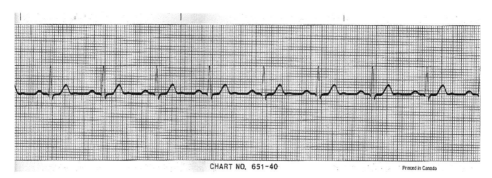

CHART NO. 651-40   Printed in Canada

Figure 1-8. Typical ECG obtained from Skin Electrodes.

logical signals. In addition, there are other forms of physiological signals, such as pressure and temperature, all of which contain information that reflects the well-being of an individual. Monitoring and analyzing such parameters is of interest to the medical professionals. Different physiological signals have different characteristics. Some physiological signals are very small compared with other background signals and noise; some change rapidly during the course of their measurement. Therefore, different transducers with matching characteristics are necessary in medical devices to accurately measure these signals. Table 1–3 shows some examples of physiological signals; their characteristics and examples of common transduction techniques used to capture these signals are also listed. The range and bandwidth quoted in the list are nominal values, which may not include some extreme cases. An example is severe hypothermia, in which the body temperature can become many degrees below 32°C.

An example of a physiological signal measurement is the electrocardiogram. When skin electrodes are placed on the surface of a patient's chest, they pick up a small electrical potential at the skin surface from the activities of the heart. If one plots this potential against time, this is called an electrocardiogram. An example of an electrocardiogram is shown in Figure 1–8. The spike is called the R- wave, which coincides with the contraction phase of the ventricles. The time interval between two adjacent R-waves represents one heart cycle. The amplitude and the shape of the ECG signal depend on the physiological state of the patient as well as the locations and the types of electrodes used. From Table 1–3, the amplitude of the R-wave may vary from 0.5 to 4 mV, and the ECG waveform has a frequency range or bandwidth from 0.01 to 150Hz.

There are many more physiological signals than those listed in Table 1-3. While some are common parameters in clinical settings (e.g., body temperature), others may only be measured sparingly (e.g., electroretinogram).

Table 1-3. Characteristics of Common Physiological Parameters

Physiological Parameters	Physical Units and Range of Measurement	Signal Frequency Range (Bandwidth)	Measurement Method or Transducer Used
Blood Flow	1 to 300 ml/s	0 to 20 Hz	Ultrasound Doppler flowmeter
Blood Pressure—Arterial	20 to 400 mmHg	0 to 50 Hz	Sphygmomanometer
Blood Pressure—Venous	0 to 50 mmHg	0 to 50 Hz	Semiconductor strain gauge
Blood pH	6.8 to 7.8	0 to 2 Hz	pH electrode
Cardiac Output	3 to 25 L/min	0 to 20 Hz	Thermistor (thermodilution)
Electrocardiography (ECG)	0.5 to 4 mV	0.01 to 150 Hz	Skin electrodes
Electroencephalography (EEG)—scalp	5 to 300 µV	0 to 150 Hz	Scalp electrodes
EEG—Brain surface or depth	10 to 5,000 µV	0 to 150 Hz	Cortical or depth electrodes
Electromyography (EMG)	0.1 to 5 mV	0 to 10,000 Hz	Needle electrodes
Nerve Potentials	0.01 to 3 mV	0 to 10,000 Hz	Needle electrodes
Oxygen Saturation— Arterial (noninvasive)	85 to 100%	0 to 50 Hz	Differential light absorption
Respiratory Rate	5 to 25 breath/min	0.1 to 10 Hz	Skin electrodes (impedance pneumography)
Tidal Volume	50 to 1,000 ml	0.1 to 10 Hz	Spirometer
Temperature—Body	32 to 40°C	0 to 0.1 Hz	Thermistor

HUMAN-MACHINE INTERFACE

A medical device is designed to assist clinicians to perform certain diagnostic or therapeutic functions. In fulfilling these functions, a device interfaces with the patients as well as the clinical users. Figure 1–9 shows the interfaces between a medical device, the patient, and the clinical staff. For a diagnostic device, the physiological signal from the patient is picked up and processed by the device; the processed information such as the heart rhythm from an ECG monitor or blood pressure waveform from an arterial blood line is displayed by the device and reviewed by the clinical staff. For a ther-

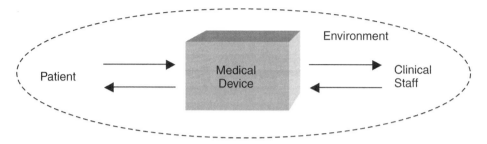

Figure 1-9. Human-Machine Interface.

apeutic device, the clinical staff will, using the device, apply certain actions on the patient. For example, a surgeon may activate the electrosurgical hand piece during a procedure to coagulate a blood vessel. In another case, a nurse may set up an intravenous infusion line to deliver fluid and medication to a patient.

These interfaces are important and often critical in the design of biomedical devices. An effective patient-machine interface is achieved through carefully choosing a transducer suitable for the application. For example, an implanted pH sensor must pick up the small changes in the hydrogen ion concentration in the blood; at the same time it also must withstand the corrosive body environment, maintain its sensitivity, and be nontoxic to the patient.

Other than safety and efficacy, human factor is another important consideration in designing medical devices. Despite the fact that human error is a major contributing factor toward clinical incidents involving medical devices, human factors are often overlooked in medical device design and in device acquisitions. The goal to achieve in user-interface design is to improve efficiency, reduce error, and prevent injury. Usability engineering, as defined in the international standard IEC 62366-1:2015 Medical devices – application of usability engineering to medical devices, is the application of knowledge about human behavior, abilities, limitations, and other characteristics related to the design of tools, devices, systems, tasks, jobs, and environments to achieve adequate usability. Usability or human factors engineering is a systematic, interactive design process that is critical to achieve an effective user-interface. It involves the use of various methods and tools throughout the design life cycle. Classical human factors engineering involves analysis of sensory limitations, perceptual and cognitive limitations, and effector limitations of the device users as well as the patients. Sensory limitation analysis evaluates the responses of the human visual, auditory, tactile, and olfactory systems. Perceptual and cognitive limitation analysis studies the nervous sys-

SUBJECTS   **LIMITATIONS**

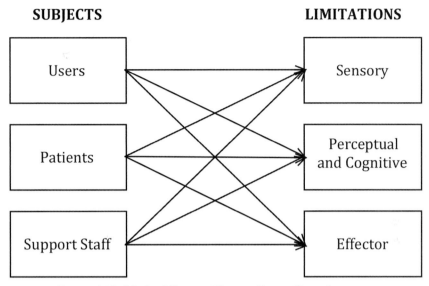

Figure 1-10. Medical Device Human Factor Considerations.

tem's response to the sensory information. Perception refers to how people identify and organize sensory input; cognition refers to higher-level mental phenomena such as abstract reasoning, formulating strategies, formation of hypothesis, et cetera. Effector limitation analysis evaluates the outputs or responses of the operators (e.g., the reaction time, force-exerting capability, etc.).

There are three subjects to be focused on in human factors design in medical devices: the user, the patient, and the support staff. The three areas of limitations described above must be considered in each case (Figure 1–10).

User Focus

For diagnostic devices, users rely on the information from the medical device to perform diagnosis. The display of information should be clear and unambiguous. It is especially important in clinical settings, where errors are often intolerable. In a situation in which visual alarms might be overlooked, loud audible alarms to alert one to critical events should be available. For therapeutic devices, ergonomic studies should be carried out in the design stage to ensure that the procedures could be performed in an effective and efficient manner. Critical devices should be intuitive and easy to set up. For example, a paramedic should be able to correctly perform a cardiac defibrillation without going through complicated initialization procedures since every second counts when a patient is in cardiac arrest.

A systems approach to analyze human interface related to users should consider the following:

- User characteristics
- Operating environment
- Human mental status
- Task priority
- Workflow

Human interface outputs may involve hand, finger, foot, head, eye, voice, et cetera. Each should be studied to identify the most appropriate choice for the application. A device should be ergonomically designed to minimize the strain and potential risk to the users, including long-term health hazards. For example, a heavy X-ray tube can create shoulder problems for radiology technologists who spend most of their working days maneuvering X-ray tubes over patients. Studies show that user fatigue is a major contributor to user errors. User fatigues include motor, visual, cognitive, and memory.

Human factors engineering is task-oriented. It examines and optimizes tasks to improve output quality, reduce time spent, and minimize the rate of error. Proactive human interface designers tend to be user-centered, who integrate the physical and mental states of the user into the design, including the level of fatigue and stress, as well as recruit emotional feedback. Ideally, a good human interface design will produce a device that is both user-intuitive and efficient. However, in most cases, there is a balance and trade-off between the two. An intuitive design is easy to use, that is, a user can learn to operate the device in a short time. However, the operation of such a device may not be efficient. An example of such a device is a PACS (picture archiving and communication system) using a standard computer mouse as the human-machine interface between the user and the PACS. The mouse is intuitive to most users. However, a radiologist may require going through a large number of moves and clicks to complete a single task. On the other hand, a specially designed, multi-button, task-oriented controller may be difficult to learn initially but will become more efficient once the radiologist has gotten used to it. Figure 1-11 shows the efficiency-time learn-ing curve of a device by a new user. The learning time for the intuitive device is shorter than the specially designed device, but the efficiency is much lower once the user becomes proficient with the specially designed device.

Patient Focus

Traditionally, in designing a medical device, much attention is given to the safety and efficacy of the system. However, it is also important to look at

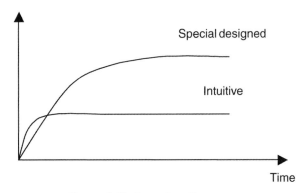

Figure 1-11. Learning Curve.

the design from the patient's perspectives. A good medical device design should be aesthetically pleasing to the eye and will not interfere with the normal routines of the patient. Some examples to illustrate the importance of human factors design related to patients are:

- A model of an infrared ear thermometer looks like a pistol with a trigger. The patient may feel threatened when the clinician points it into his or her ear and pulls the trigger.
- A motorized fan in an infant incubator is too noisy. It disturbs the sleep of the baby and may even inflict hearing damage.
- For a person who requires 24×7 mechanical ventilation, a tracheostomy tube that cannot be concealed properly may affect his (or her) social life.

Support Staff Focus

In designing a medical device, the ergonomics of maintenance tasks such as cleaning and servicing are often overlooked. Apart from its desired application, a medical device will be handled by many parties during its life span. A device that has difficult-to-assess hollow cavities will have problems in cleaning and sterilization and hence is not suitable for some medical procedures. Some devices are not service-friendly; many poorly designed devices require extensive dismantling in order to get access to replacement parts such as light bulbs and batteries. Other devices may not have taken into consideration the operating environment, which in most cases will result in expensive and labor-intensive maintenance. An example is a fan-cooled device used in a dusty environment. In addition, poor design of accessories may increase the chance of incorrect assembly, which can impose unnecessary risk on the patients.

Table 1-4. Medical Device Risk Mitigation

Degree of Effectiveness	Methods of Risk Mitigation	Examples of Methods to Mitigate IV Infusion Dose Error
1	Policies and Rules	Create a policy for users to double check flow rate setting on IV infusion pumps
2	Training and Education	Provide to all users initial and refresher training on infusion pumps
3	Checklists and Reminder Notes	Attach a checklist on each infusion pump to remind users to double check infusion setting
4	Standardization	Use the same make and model of infusion pumps in the hospital to avoid operation confusions
5	Constraints and Forcing Functions	Use "Smart" pump with built-in drug library to alert users when safety flow rate settings are violated
6	Automation	Use bar codes to identify patients and medications. Infusion pumps are automatically programmed with prescribed infusion rate.

Human Errors

Understanding human errors is essential in designing and analyzing medical devices. Human errors can be differentiated into three categories:

- Slips – slips are attention failures; the user has the right intention, but the action was incorrectly carried out. An example is putting salt instead of sugar into a cup of tea.
- Lapses – lapses are failure of memory; the operator has forgotten to carry out the intended action. An individual who forgot to turn off the stove before leaving home is an example.
- Mistakes – mistakes are choices of incorrect intention. A driver who took a wrong turn made a mistake in his driving route is an example. Mistakes can be further differentiated into rule-based and knowledge-based. Rule-

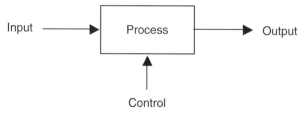

Figure 1-12. Medical Device System.

based mistakes may occur when a pre-programmed solution is used on a situation that is not anticipated, whereas knowledge-based mistakes will occur under a new situation that requires novel solution beyond the knowledge of the operator.

Human errors which occur in using a medical device may introduce risk and lead to serious injury to the patient or user. Manufacturers should aim to identify and remove risk during the design phase of medical device development, rather than fixing the problem through post-market recalls. In fact, most regulatory bodies require manufacturers to perform risk analysis as well as benefit-risk analysis as part of the design assurance process on high-risk devices. On the other side, users of medical devices should follow similar approaches in risk mitigation. The risk index of a hazardous situation is a measure of its risk severity based on its frequency of occurrence and the degree of harm it may cause. Once an unacceptable risk factor is identified, risk mitigation must be implemented such that the subsequent risk factor has become acceptable. While some methods of mitigation are simple, others can be complicated and costly. Nonetheless, it is important to understand the effectiveness of these migration methods. Column 2 of Table 1-4 lists some common methods of risk mitigation deployed in healthcare facilities. The order of effectiveness of these methods are listed in column 1 which is based on human factors consideration. Column 3 of the table provides examples of risk mitigation methods in minimizing dose error (wrong infusion administered to patient) in intravenous infusion (IV) devices.

Table 1-5. Examples of Medical Device Input/Output

Signal Input
    Electrical potential in ECG
    Pressure signal in blood pressure monitoring
    Heat in body temperature measurement
    Carbon dioxide partial pressure in end-tidal CO_2 monitoring
Device Output
    Printout in paper chart recorder
    Signal waveform in CRT display
    Alarm signal in audible tone
    Heat energy from a thermal blanket
    Grayscale image on an X-ray film
    Fluid flow from an infusion pump
Control Input
    Exposure technique settings on an X-ray machine
    Sensitivity setting on a medical display
    Total infusion volume setting on an infusion pump
    Alarm settings on an ECG monitor

INPUT, OUTPUT, AND CONTROL SIGNALS

A simple system has a single input and a single output. When we study a medical device using the systems approach, the first step is to analyze the input to the device. In most cases, input signal to biomedical devices are physiological signals. In order to study the characteristics of the output, one must understand the nature of the processes that the device applies to the input. In addition to the main input and output signals, most medical devices have one or more control inputs (Figure 1–12). These control inputs are used by the operator to select the functions and control the device. Table 1–5 lists some examples of input, output, and control signals in biomedical devices.

CONSTRAINTS IN BIOMEDICAL SIGNAL MEASUREMENTS

Medical devices in many respects are similar to devices we use in every-day life. In fact, most technologies used in health care are adapted from the same technologies used in military, industrial, and commercial applications. Since medical devices are used on humans, their reliability and safety requirements are usually more stringent than other devices. In addition, medical devices are often used in situations in which patients are vulnerable to even minor errors; therefore, special consideration in minimizing risk is necessary in designing medical devices. Listed next are some of the factors and constraints in designing medical devices.

Low Signal Level

The level of biological signal can be very small, for example, on the order of microvolts (μV) in EEG measurements. Therefore, very sensitive transducers as well as good noise rejection methods are required.

Inaccessible Measurement Site

Many signal sources are inside the human body and hidden by other anatomy. Biomedical measurements and procedures often require invasive means to access specific anatomy. For example, to access a nerve fiber for electrical activity measurement, the electrode must go through the skin, muscle, and other tissues.

Small Physical Size

Some measurement sites are very small. In order to measure the signal coming from these tiny sites while at the same time to avoid picking up the surrounding activities, special sensors that allow isolated measurement at the source are required. For example, in EMG measurement, needle electrodes with insulated stems are used to measure the electrical signal produced at a specific group of muscle fibers.

Difficult to Isolate Signal From Interfering Sources

As we cannot voluntarily turn ON or OFF, or remove tissue or organ to take a measurement, the measurand is subjected to much interference. As an example, in fetal monitoring, fetal heart activities are often masked by the stronger maternal heartbeats. It requires special techniques to extract information from these interfering signals.

Signal Varies With Time

Human physiological signals are seldom deterministic; they always change with time and with the activities of the body. It is therefore not an easy task to establish the norm of such signals. An example of such is in arterial blood pressure measurement; the blood pressure of a person is usually higher in the morning than at other times of day.

Signal Varies Among Healthy Individuals

Since every human being is different, the same physiological signal from one person is different from that of another. It is not a straightforward task to establish what is normal or abnormal and what is healthy or unhealthy when looking at some of these physiological parameters. For example, there is a huge difference in normal resting heart rate between an athlete and a person who seldom exercises. Nevertheless, there are generally recognized normal ranges. For example, systolic arterial pressure between 90 and 120 mmHg is considered acceptable.

Origin and Propagation of Signal Is Not Fully Understood

The human body is very complicated and nonhomogeneous. There are many signal paths within the body, and interrelationships between physiological events are often not fully understood. For example, the ECG obtained by surface electrodes looks very different from that obtained by invasive electrodes placed inside the heart chamber.

Difficult to Establish Safe Level of Applied Energy to the Tissue

Very often, electrical current from a medical device, whether by intention or by accident, will flow through the patient's body. Although such energy will impose risk on the patient, it is often difficult to establish the minimum safety limits of such signals.

Able to Withstand Harsh Clinical Environment

Medical devices and their accessories can be soiled, contaminated by chemicals and body fluid. They may be exposed to low or high temperature. They need to be cleaned by water, cleaning agents, and surface disinfectants. As well, some are required to be sterilized by heat or chemicals treatments.

Biocompatibility

The parts of a medical device that are in contact with patients must be nontoxic and must not trigger adverse reaction. In addition, they must be able to withstand the chemical corrosive environment of the human body.

CONCEPTS ON BIOCOMPATIBILITY

Definitions

Biocompatibility refers to the compatibility of nonliving materials with living tissues and organisms, whereas histocompatibility refers to the compatibility of different tissues in connection with immunological response. Histocompatibility is associated primarily with the human lymphocyte antigen system. Rejection of transplants may be prevented by matching tissues according to histocompatibility and by the use of immunosuppressive drugs. Biocompatibility entails mechanical, chemical, pharmacological, and surface compatibility. It is about the interactions that take place between the materials and the body fluid, tissues, and the physiological responses to these reactions.

Biocompatibility of metallic materials is controlled by the electrochemical interaction that results in the release of metal ions or insoluble particles into the tissues and the toxicity of these released substances. Biocompatibility of polymers is, to a large extent, dependent on how the surrounding fluids extract residual monomers, additives, and degradation products. Other than the chemistry, biocompatibility is also influenced by other factors such as mechanical stress imposed on the material.

Mechanism of Reaction

The adverse results of incompatibility include the production of toxic chemicals, as well as the corrosion and degradation of the biomaterials, which may affect the function or create failure of the device or implant. Protein absorption of the implant and tissue infection may lead to premature failure, resulting in removal and other complications. Compatibility between medical devices and the human body falls under the heading of biocompatibility.

Biocompatibility is especially important for implants or devices that for a considerable length of time are in contact with or inside the human body. Common implant materials include metal, polymers, ceramics, and products from other tissues or organisms.

Tissue Response to Implants

During an implant procedure, the process often requires injuring the tissue. Such injury will invoke reaction such as vasodilation, leakage of fluid into the extravascular space, and plugging of lymphatics. These reactions produce classic inflammatory signs such as redness, swelling, and heat, which often lead to local pain. Soon after injury of the soft tissues, the mesenchymal cells evolve into migratory fibroblasts that move into the injured site; together with the scaffolding formed from fibrinogen in the inflammatory exudates, collagen is deposited onto the wound. The collagen will dissolve and redeposit during the next 2 to 4 weeks for its molecules to polymerize in order to align and create cross-links to return the wound closer to that of normal tissue. This restructuring process can take more than 6 months.

The body always tries to remove foreign materials. Foreign material may be extruded from the body (as in the case of a wood splinter), walled off if it cannot be moved, or ingested by macrophages if it is in particulate or fluid form. These tissue responses are additional reactions to the healing processes described above.

A typical tissue response involves polymorphonuclear leukocytes appearing near the implant site followed by macrophages (foreign body giant cells). However, if the implant is chemically and physically inert to the tissue, only a thin layer of collagenous tissue is developed to encapsulate the implant. If the implant is either a chemical or a physical irritant to the surrounding tissue, then inflammation occurs at the implant site. The inflammation will delay normal healing and may cause necrosis of tissues by chemical, mechanical, and thermal trauma.

The degree of the tissue response varies according to both the physical and chemical nature of the implants. Pure metals (except the noble metals) tend to evoke a severe tissue reaction. Titanium has minimum tissue reaction of all the common metals used in implants as long as its oxide layer remains intact to prevent diffusion of metal ions and oxygen. Corrosion-resistant alloys such as cobalt-chromium and stainless steel have a similar effect on tissue once they are passivated. Most ceramic materials are oxides such as TiO_2 and Al_2O_3. These materials show minimal tissue reactions with only a thin layer of encapsulation. Polymers are quite inert toward tissue if there are no additives such as antioxidants, plasticizers, anti-discoloring agents, et cetera. On the other hand, monomers can evoke an adverse reaction since they are very reactive. Therefore, the degree of polymerization is related to the extent of tissue reaction. As 100% polymerization is not achievable, different sizes of polymer can leach out and cause severe tissue reaction.

A very important requirement for implant or materials in contact with blood is blood compatibility. Blood compatibility includes creating blood clot and damaging protein, enzymes, and blood elements. Damage to blood elements includes hemolysis (rupture of red blood cells) and triggering of platelet release. Factors affecting blood compatibility include surface roughness and surface wettability. A non-thrombogenic surface can be created by coating the surface with heparin, negatively charging the surface, or coating the surface with non-thrombogenic materials.

Systemic effect can be linked to some biodegradable sutures and surgical adhesives, as well as particles released from wear and corrosion of metals and other implants. In addition, there are some concerns about the possible carcinogenicity of some materials used in implantation.

Characteristics of Materials Affecting Biocompatibility

In addition to the above consideration, the following material characteristics should be analyzed to determine biocompatibility:

- Stress—the force exerted per unit area on the material; stress can be tensile, compressive, or shear.
- Strain—the percentage dimensional deformation of the material.
- Viscoelasticity—the time-dependent response between stress and strain.
- Thermal properties—include melting point, boiling point, specific heat capacity, heat capacity, thermal conductivity, and thermal expansion coefficient.
- Surface property—measure of surface tension and contact angle between liquid and solid surface.
- Heat treatment—for example, quenching of metal or surface compression of glass and ceramics to improve material strength.
- Electrical properties—determination of resistivity and piezoelectric properties.
- Optical properties—measurement of refractive index and spectral absorptivity.

- Density and porosity—include measurement of solid volume fraction.
- Acoustic properties—include acoustic impedance and attenuation coefficient.
- Diffusion properties—determination of permeability coefficients.

Each of these characteristics should be analyzed for its intended applications. It should be noted that the same material may have different degrees of biocompatibility under different environments and different applications.

In Vitro and In Vivo Tests for Biocompatibility

Tests for biocompatibility should include both material and host responses. The usual approach in testing a new product is to perform an in vitro screening test for quick rejection of incompatible materials. In vitro tests can be divided into two general classes: (1) tissue culture methods, and (2) blood contact methods.

Tissue culture refers collectively to the practice of maintaining portions of living tissues in a viable state, including cell culture, tissue culture, and organ culture. Blood contact methods are performed only for blood contact applications such as cardiovascular devices. Both static and dynamic (flow) tests should be performed.

After screening by in vitro techniques, the product is moved to in vivo testing. It is the practice to test new implant materials or existing materials in significantly different applications. In vivo tests are often done in extended-time whole animal tests before human clinical trials. In vivo tests in general are divided into two types: (1) nonfunctional, and (2) functional.

In nonfunctional tests, the product material can be of any shape and is embedded passively in the tissue site for a period of time (e.g., a few weeks to 24 months). Nonfunctional tests focus on the direct interactions between the material of the product and the chemical and biological species of the implant environment. In addition to being implanted, functional tests require that the product be placed in the functional mode as close as possible to the conditions of its intended applications. The purpose of functional tests is to study both the host and material responses such as tissue in growth into porous materials, material fatigue, and production of wear particles in load-bearing devices. Functional tests are obviously more involved and costly than nonfunctional tests.

FUNCTIONAL BUILDING BLOCKS OF MEDICAL INSTRUMENTATION

A typical diagnostic medical device acquires information from the patient, analyzes and processes the data, and presents the information to the

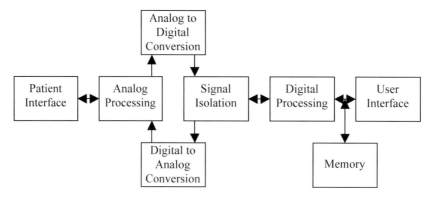

Figure 1-13. Functional Block Diagram of a Medical Instrument.

clinician. In a therapeutic device, it processes the input from the clinician and applies the therapeutic energy to the patient. Figure 1–13 shows the functional building blocks of a typical medical device. It includes the following functional building blocks:

- Patient interface
- Analog processing
- Analog to digital conversion and digital to analog conversion
- Signal isolation
- Digital processing
- Memory
- User interface

Interface

In diagnostic devices, the patient interface includes transducers or sensors to pick up and convert the physiological signal (e.g., blood pressure) to an electrical signal. In therapeutic devices, the patient interface contains transducers that generate and apply energy to the patient (e.g., ultrasound physiotherapy unit).

Analog Processing

The analog processing contains electrical circuits such as amplifiers (to increase signal level) and filters (to remove any unwanted frequency components such as high-frequency noise from the signal). The signal until this point is still in its analog format.

Analog to Digital Conversion

The function of the analog to digital converter (ADC) is to convert the analog signal to its digital format. The signal coming from the ADC is a string of binary numbers (1's and 0's).

Digital to Analog Conversion

If an analog output is necessary, a digital to analog converter (DAC) will be required to convert the digital signal from its "1" and "0" states back to its analog format. A DAC reverses the process of an ADC.

Signal Isolation

The primary function of signal isolation is for microshock prevention in patient electrical safety. The isolation barrier, usually an opto-coupler, provides a very high electrical impedance between the patient's applied parts and the power supply circuit to limit the amount of risk current flowing to or from the patient.

Digital Processing

After being digitized by the ADC, the signal is sent to the digital processing circuit. In a modern medical instrument, digital processing is done by one or more computers built into the system. The center of a digital computer is the central processing unit (CPU). Depending on the needs, the CPU may perform functions such as calculations, signal conditioning, pattern recognition, information extraction, et cetera.

Memory

Information such as waveforms or computed data is stored in its binary format in the memory module of the device. Signal stored in the memory can later be retrieved for display, analysis, or used to control other outputs.

Network & Communication Interface

Most medical devices today are able to communicate with their system components (such a patient monitor connected to its a central station), with other devices (e.g., a mechanical ventilator with a patient monitor), or with other systems (such as the electronic medical records). Device manufacturers often adhere to network and communication standards to facilitate compatibility.

User Interface

User interfaces can be output or input devices. Examples of output user interfaces are display monitor for physiological waveform and audio alarm annunciator. Examples of input devices are touch screen and voice command input.

BIBLIOGRAPHY

Andrade J. D., & Hlady, V. (1986). Protein absorption and materials biocompatibility: A tutorial review and suggested hypotheses. In *Advances in Polymer Science, Biopolymers/Non-Exclusion HPLC* (Vol. 79, pp 1-63). Berlin: Springer-Verlag.

Black J. (1992). *Biological Performance of Materials* (2nd ed.). New York: M. Dekker, Inc., 1992.

Byrne, J. H., & Schultz, S. G. (1988). *An Introduction to Membrane Transport and Bioelectricity.* New York: Raven Press.

Chen, Y. J., et al. (2018). A Comparative Study of Medical Device Regulations: US, Europe, Canada, and Taiwan. *Therapeutic Innovation & Regulatory Science.* 52(1), 62-69.

Clark, J. W., Jr. (2010). The origin of biopotentials. In J. G. Webster (Ed.), *Medical Instrumentation: Application and Design.* Hoboken, NJ: John Wiley & Sons.

Hall, J. E. (2010). *Guyton and Hall Textbook of Medical Physiology* (12th ed.). Philadelphia, PA: Saunders Elsevier.

International Association for Standardization. (2016). ISO 13485: 2016: *Quality Management System for Medical Devices.* Geneva, Switzerland. 2016.

International Association for Standardization. (2007). ISO 14971: 2007 (2019 update underway): *Medical Devices – Application of Risk Management to Medical Devices.* Geneva, Switzerland.

International Association for Standardization. (2007). ISO 62366: 2007: *Medical Devices – Application of Usability Engineering to Medical Devices.* Geneva, Switzerland.

Skyttner L. (2006). General Systems Theory: Problems, Perspective, Practice. Hackensack, NJ: World Scientific Publishing Company.

von Bertalanffy L. *General System theory: Foundations, Development, Applications* (Rev. ed.). New York, NY: George Braziller.

Vicente, K. J. (2003). *The Human Factor.* Toronto, Ontario: Knopf Canada.

Wickens, C. D., & Hollands, J. G. (2000). *Engineering psychology and human performance* (3rd ed.). New Jersey: Prentice Hall.

Williams, D. F. (2008). On the mechanisms of biocompatibility. *Biomaterials* 29(20), 2941-2953.

Chapter 2

CONCEPTS IN SIGNAL MEASUREMENT, PROCESSING, AND ANALYSIS

OBJECTIVES

- Explain device specifications and their significance in medical instrumentations.
- Define signal measurement parameters, including accuracy, error, precision, resolution, reproducibility, sensitivity, linearity, hysteresis, zero offset, and calibration.
- Describe steady state, transient, linear and nonlinear responses in transfer characteristics.
- Discuss and analyze time and frequency domain transformation and the effect of filters on biological signals.

CHAPTER CONTENTS

1. Introduction
2. Error of Measurements and Calibration
3. Device Specifications
4. Steady State Versus Transient Characteristics
5. Linear Versus Nonlinear Steady State Characteristics
6. Time and Frequency Domains
7. Signal Processing and Analysis

INTRODUCTION

Most medical devices involve measurement or sensing one or more physiological signals, enhancing the signals of interest, and extracting useful information from the signals. The concepts of signal measurement, processing, and analysis are fundamental in understanding scientific instrumentations, including medical devices. This chapter provides an introductory overview of these concepts.

ERROR OF MEASUREMENT AND CALIBRATION

Measurements are fundamental to science including measurement performed by instruments used in medicine. All measurements have a degree of uncertainty caused by two error factors, the limitation of the measuring instrument and the skill of the experimenter making the measurements. It is important that the results will not vary with a change of instrument, operator or measuring action. As users, we all expect a screw will fit into nuts of the same size even though they are from different manufacturers. In medical temperature measurement, a biomedical engineer trusts two J-type thermocouples will give the same output when they are exposed to the identical temperature. This confidence is based on international standard compliance and quality assurance in the measurement process. It is common practice to express the quality of a measurement by the measurement uncertainty associated with the result. The confidence in the value and the stated uncertainty relies on the traceability of measurement involving an unbroken and documented chain linking instrument used in the measurement to an internationally agreed measurement standard.

The error (ε) of a single measured quantity is the measured value (Q_m) minus the true value (Q_t), or

$$\varepsilon = Q_m - Q_t$$

There are three types of errors: gross error, systematic error, and random error. The total error is the sum of all three error quantities.

- Gross error is human error and arises from incorrect use of the instrument or misinterpretation of the measurement by the person taking the measurements (e.g., misread the scale of measurement, use a wrong constant in unit conversion).
- Systematic error results from bias in the instrument (e.g., improper calibration, defective or worn parts, adverse effect of the environment on the

equipment).
- Random error is fluctuations that cannot be directly established or corrected in each measurement (e.g., electronic noise, noise in photographic process). Random error can be reduced by taking the mean of repeated measurements.

Error may be expressed as absolute error or relative error.

- Absolute error ΔZ of a measured quantity Z is expressed in the specific unit of measurement and is always ≥ 0, for example, $Z = 15 \text{ W} \pm 1 \text{ W}$ (where $\Delta Z = 1 \text{ W}$ is the absolute error). The graphical representation is shown in Figure 2-1a. For a given input, the output will be equal to the corresponding output value within plus or minus the given error value; that is, within the dotted line in the graph.
- Relative error is expressed as a ratio of the measured quantity, such as, "$\pm$ 5% of the output (or measured value)" (Figure 2-1b).
- An alternative way to express absolute error is percentage of full scale, for example, "$\pm$ 5% of full-scale output" (Figure 2-1c).
- Error can also be a combination of the above, such as, "$\pm 1 \text{ W}$ or $\pm 5\%$ of output, whichever is greater" (Figure 2-1d).

In scientific measurements, the error of measurement (e.g., due to errors of the instruments) is often displayed as an error bar at the measurement point in the graph. The dimension of the error bar accounts for the sum of all errors in acquiring the measured value. An example is shown in Figure 2-1e where a straight line is best fitted to the 4 measured data points.

Quite often, the eventual output of an experiment (or a measurement process) is the combination of a number of intermediate measured quantities. Errors arising from each will propagate and contribute to the overall error of the final outcome. Below are the rules of error propagation:

1. A multiplication constant factor k does not affect the relative error

$Z = kA \rightarrow \Delta Z = k\Delta A$, therefore

$$\frac{\Delta Z}{Z} = \frac{\Delta A}{A}$$

2. Addition and subtraction with constants c, m

$Z = cA \pm mB \pm \cdots$

$$\Delta Z = \sqrt{(c\Delta A)^2 + (m\Delta B)^2 + \cdots}$$

(note that the error terms always add)

3. Multiplication and division with a constant factor k

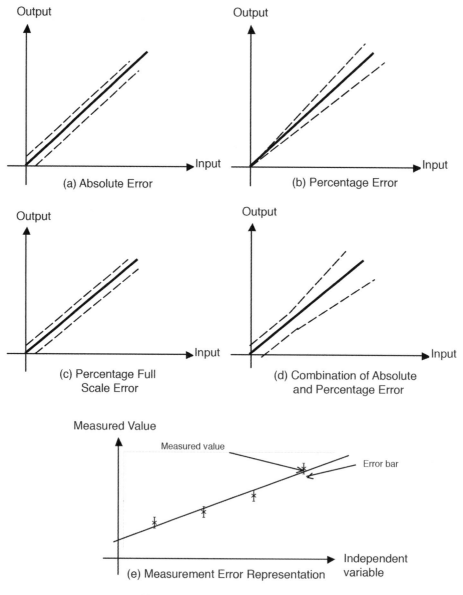

Figure 2-1. Error Representation.

$$z = \frac{kA \times B \times \ldots}{C \times D \times \ldots}$$

$$\frac{\Delta Z}{Z} = \sqrt{\left(\frac{\Delta A}{A}\right)^2 + \left(\frac{\Delta B}{B}\right)^2 + \left(\frac{\Delta C}{C}\right)^2 + \left(\frac{\Delta D}{D}\right)^2 + \cdots}$$

4. For functions of one variable, Z = F(A), if the quantity A is measured with error ΔA,

$$\Delta Z = \left(\frac{dF}{dA}\right) \Delta A$$

Example

Determine the error $\frac{\Delta Z}{Z}$ when Z is given by: $Z = xy^{-1}w^2 (1 - sin\theta)^{1/2}$

Solution

Let us rewrite the above equation to:

Z = A * B * C * D, where A = x, B = y^{-1}, C = w^2, D = $(1 - sin\theta)^{1/2}$

The relative error according to Rule#3 above becomes

$$\frac{\Delta Z}{Z} = \sqrt{\left(\frac{\Delta A}{A}\right)^2 + \left(\frac{\Delta B}{B}\right)^2 + \left(\frac{\Delta C}{C}\right)^2 + \left(\frac{\Delta D}{D}\right)^2}$$

Applying Rule#4

$$\frac{\Delta Z}{Z} = \sqrt{\left(\frac{\Delta x}{x}\right)^2 + \left(-\frac{\Delta y}{y}\right)^2 + \left(2\frac{\Delta w}{w}\right)^2 + \left(\frac{cos\theta \Delta\theta}{2 - 2sin\theta}\right)^2}$$

Calibration is the process of determining and recording the relationships between the values indicated by a measuring instrument and the true value of the measured quantity. Since the true value is usually difficult to obtain, the instrument is often calibrated against a device that is traceable to a national standard. The definition of "metrological traceability" or simply "traceability," according to the International Vocabulary of Metrology (VIM) is: "Property of a measurement result whereby the result can be related to a reference through a documented unbroken chain of calibrations, each contributing to the measurement uncertainty".

In calibration, metrological traceability is a fundamental consideration. If the calibration is not traceable, one cannot be sure if the measured value is correct or not. Table 2-1 shows a practical chain to achieve traceability.

In a hospital, there are many instruments (e.g., body temperature monitors) that are calibrated regularly by using some calibrators (e.g., an accurate thermometer). The calibrators is typically sent out to an external calibration laboratory or they are calibrated in-house by some reference standards (e.g., a precision temperature meter). The reference standards are sent out to an accredited external calibration laboratory to be calibrated. The external calibration laboratory will calibrate their references to assure traceability to the National Calibration laboratory (e.g. the NRC in Canada, or NIST in the US). The National Calibration laboratories work with International Level laboratories to assure that their calibrations are on the same level. The International Level laboratories based their measurements on international definitions and international comparisons. In addition to traceability, calibration needs to be properly scheduled, performed by trained professionals, and fully documented according to the user's quality system.

Table 2-1. Calibration Traceability Chain in Practice

	Traceability Chain	*Examples*
Hospital Instrument	Device to be measured	Temperature sensor in a body temperature monitor
Calibration equipment	Measuring instrument	An accurate hand held thermometer
Reference Standard	Working Standard	An in-house precision temperature meter
External Calibration Lab	Secondary Standard	High accuracy temperature source and meter
National Calibration Lab	National (primary) Standard	Fixed point cells (e.g. freezing point of ice)
International Level Lab	International Standard	International temperature scale (ITS90)

DEVICE SPECIFICATIONS

To understand the functions and performance of a medical device, one should start from reading the specifications of the device. Specifications of an instrument are the claims from the manufacturer on the characteristics and performance of the instrument. The specification document of a medical device should contain at the minimum the following information:

- List of device functions and intended use
- Input and output characteristics
- Performance statements
- Physical characteristics
- Environmental requirements
- Regulatory classifications
- User safety precautions where applicable

Instrument type: 12-channel, microcomputer-augmented, automatic electrocardiograph

Input channels: simultaneous acquisition of up to 12 channels

Frequency response: –3 dB @ 0.01 to 105 Hz

Sensitivities: 2.5, 5, 10, and 20 mm/mV, ±2%

Input impedance: >50 MΩ

Common mode rejection ratio: >106 dB

Recorder type: thermal digital dot array, 200 dots/in. vertical resolution

Recorder speed: 1, 5, 25, and 50 mm/s, ±3%

Digital sampling rate: 2000 samples/s/channel

ECG analysis frequency: 250 samples/s

Display formats: user-selectable 3, 4, 5, 6, and 12 channels with lead configurations

Dimensions: H × W × D = 90 cm × 42 cm × 75 cm

Weight: 30 kg

Power requirements: 90 VAC to 260 VAC, 50 or 60 Hz

Certification: UL 544 listed, meets ANSI/AAMI standards, complies with IEC 601 standards

---

Above is an example of a specification document of an electrocardiograph. Some common parameters found in medical device specifications are explained next.

Accuracy (A) is the error divided by the true value and is often expressed as a percentage.

$$A = \frac{Qm - Qt}{Qt} \times 100\%$$

Accuracy usually varies over the normal range of the quantity measured. It can be expressed as a percentage of the reading or a percentage of full scale. For example, for a speedometer with ± 5.0% accuracy, when it is reading 50 km/hr, the maximum error is ± 2.5 km/hr. If the speedometer is rated at ± 5.0% full-scale accuracy and the full scale reading is 200 km/hr, the maximum error of the measurement is ± 10 km/hr, irrespective of the reading.

The **precision** of a measurement expresses the number of distinguishable alternatives from which a given result is selected. For example, a meter that can measure a reading of three decimal places (e.g., 4.123 V) is more precise than one than can measure only two decimal places (e.g., 4.12 V). Another way to look at it is a precise measurement has a small variance under repeated measurements of the same quantity. Accuracy and precision in measurement are different but often related to each other. A digital clock with a display down to one tenth of a second may indicate time more precisely than a mechanical clock with only the hour and minute hands. However, the digital clock may not be as accurate as the mechanical clock is.

Resolution is the smallest incremental quantity that can be measured with certainty. If the readout of a digital thermometer jumped from 20°C to 22°C and then to 24°C when it is used to measure the temperature of a bath of water slowly being heated by an electric water heater, the resolution of the thermometer is 2°C. $S = \frac{\Delta Y}{\Delta X}$

Reproducibility (or repeatability) is the ability of an instrument to give the same output for equal inputs applied over some period of time.

Sensitivity is the ratio of the incremental output quantity to the incremental input quantity (   ). It is the slope or tangent of the output versus input curve. Note that the sensitivity of an instrument is a constant only if the output-input relationship is linear. For a nonlinear transfer function (as shown in Figure 2-2), the sensitivity is different at different points on the curve (S1 ≠ S2).

Zero offset is the output quantity measured when the input is zero. Input zero offset is the input value applied to obtain a zero output reading. Zero offsets can be positive or negative.

Zero drift has occurred when all output values increase or decrease by the same amount.

A **sensitivity drift** has occurred when the slope (sensitivity) of the input-output curve has changed over a period of time.

Perfect **linearity** of an instrument requires that the transfer function be a straight line. That is, a linear instrument has the characteristics

$$y = mx + c,$$

where x is the input, y is the output, and m and c are both constants.

Independent **nonlinearity** expresses the maximum deviation of points from the least-squares fitted line as either $\pm$ P% of the reading or $\pm$ Q% of full scale, whichever is greater. Percentage nonlinearity (Figure 2-3) is defined as the maximum deviation of the input (D_{max}) from the curve to the least square fit straight line divided by the full scale input range (I_{fs}). It is sometimes referred to as percent input nonlinearity (versus percent output nonlinearity).

$$\% \text{ nonlinearity} = \frac{D_{max}}{I_{fs}} \times 100\%$$

An instrument complies with the listed specifications (such as accuracy, % nonlinearity) only within the specified **input range(s)**. In other words, when the input of the instrument is beyond the specified input range, its characteristics may not be according to the labeled specifications.

Since transducers of a medical device often convert nonelectrical quan-

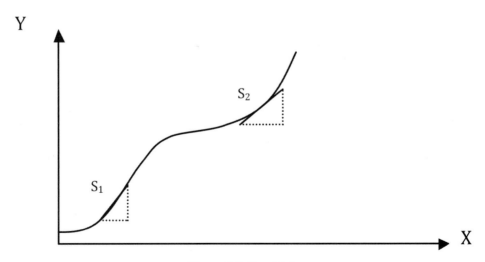

Figure 2-2. Sensitivity.

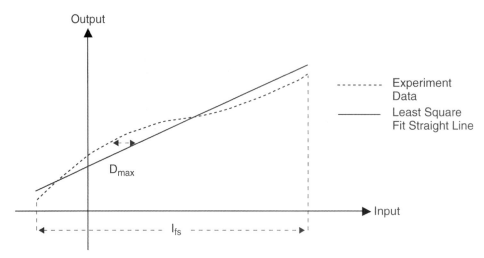

Figure 2-3. Percentage Nonlinearity.

tities to electrical quantities (voltage or current), their **input impedance** must be specified to evaluate the degree to which the device disturbs the quantity being measured. Input impedance is the ratio of the input voltage to the input current.

Hysteresis measures the capability of the output to follow the change of the input in either direction. Hysteresis often occurs when the process is lossy.

Response time is the time required for the output to change from its previous state to a final settled value given the tolerance (e.g., time to change to 90% of the steady state).

Statistical control ensures that random variations in measured quantities (result from all factors that influence the measurement process) are tolerable. If random variables make the output non-reproducible, statistical analysis must be used to determine the error variation. In fact, many medical devices rely on statistical means to determine their calibration accuracy.

STEADY STATE VERSUS TRANSIENT CHARACTERISTICS

For a typical instrument, the output will change following a change in the input. Figure 2-4 shows a typical output response when a step input (thin solid line) is applied to the system. Depending on the system characteristics, the output may experience a delay before it settles down (dotted line in Figure 2-4) or may get into oscillation right after the change of the input (thick solid line in Figure 2-4). However, in most instruments, this transient will eventually settle down to a steady state until the input is changed again.

The input-output characteristics when one ignores the initial transient period are called the steady state characteristics or static response of the system. When the input is a time-varying signal, one must take into consideration the transient characteristics of the system. For example, when the input is a fast-changing signal, the output may not be able to follow the input; that is, the output may not have enough time to reach its steady state before the input is changed again. In this case, the signal will suffer from distortion.

LINEAR VERSUS NONLINEAR STEADY STATE CHARACTERISTICS

A **linear** system response with a single input is one that follows a straight line relationship. In mathematical term:

$$y = mx + c,$$

where x is the input, y is the output, and m and c are both constants.

Figure 2-5a shows a linear characteristics with c = 0. However, in practice, system response often deviates from ideal linearity. Some common nonlinear characteristics are shown in Figure 2-5.

Saturation occurs when the input is increased to a point where the output cannot be increased further. A linear operational amplifier will become saturated when the input is close to the power supply voltage.

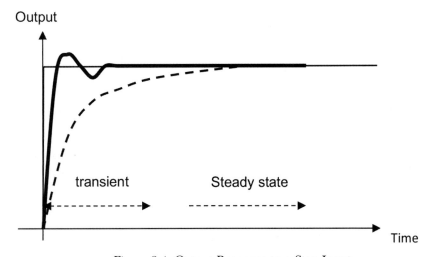

Figure 2-4. Output Response to a Step Input.

Breakdown is the phenomenon when the output abruptly starts to increase when the input changes slightly following a linear relationship. Some devices such as zener diodes have this type of nonlinear behavior.

Dead zone is a range of the input where the output remains constant. A worn-out gear system usually has some dead space (dead zone).

Bang-bang occurs when a minor reversal of the input creates an abrupt change in the output. This phenomenon can be observed in some thin metal diaphragm transducers. The diaphragm may flip from one side to another when the force applied to the center of the diaphragm changes direction.

Hysteresis is the phenomenon that the input-output characteristic follows different pathways depending on whether the input is increasing or decreasing. Hysteresis results when some of the energy applied during an increasing input is not recovered when the input is reversed. The magnetization characteristic of a transformer is a perfect example where some applied energy is loss in the eddy current in the iron core. Another example is the volume-pressure curve in patient respiration monitoring.

TIME AND FREQUENCY DOMAINS

A **time-varying signal** is a signal the amplitude of which changes with time. A periodical signal is a time-varying signal the wave shape of which repeats at regular time intervals. Mathematically, a periodical signal is given by:

$$G(t) = G(t + nT),$$

where n = any positive or negative integer, and
   T = fundamental period.

A sinusoidal signal is an example of a periodical signal. The mathematical expression of a sinusoidal signal is

$$G(t) = A \sin(\omega t + \phi)$$

where A = a constant,
   ω = angular velocity, and
   ϕ = phase angle.

Any periodical signal can be represented (through **Fourier-series** expansion) by a combination of sinusoidal signals

$$G(t) = a_0 + \sum_{n=1}^{\infty} [a_n \cos(n\,\omega_0 t) + b_n \sin(n\,\omega_0 t)],$$

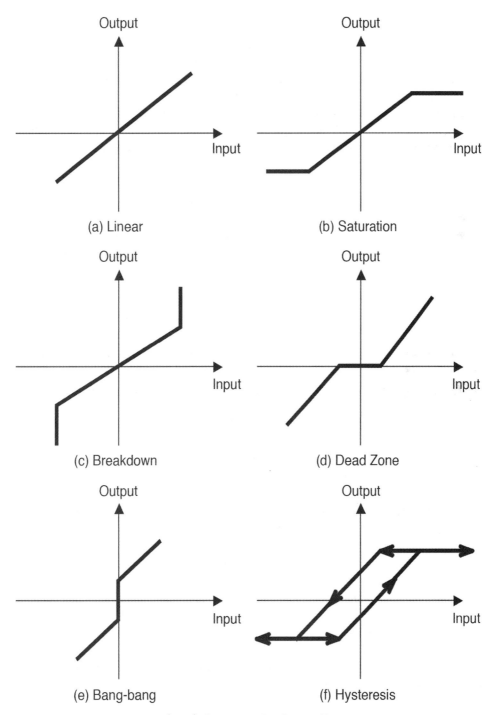

Figure 2-5 a) to f) Common Non-linear Characteristics

where a_o, a_n, and b_n are time-invariant values depending on the shape of $G(t)$, and $\omega_o = 2\pi f_o = 2\pi/T$.

Using the preceding equation, the Fourier series of a symmetrical square waveform (Figure 2-6a) with amplitude of ± 1 V and a period of 1 sec is:

$$V(t) = 4V/\pi \ [\sin(2\pi t) + 0.33\sin(6\pi t) + 0.20\sin(10\pi t) + 0.14\sin(14\pi t) + ...]$$

The frequency domain plot or frequency spectrum of the same signal is shown in Figure 2-6b. For a non-periodical signal, the frequency spectrum is continuous instead of discrete. An example of the theoretical frequency spectrum of an arterial blood pressure waveform is shown in Figure 2-7.

All biomedical signals are time-varying. Some signals may change very slowly with time (e.g., body temperature), while others may change more rapidly (e.g., blood pressure). Although some appear to be periodical, in

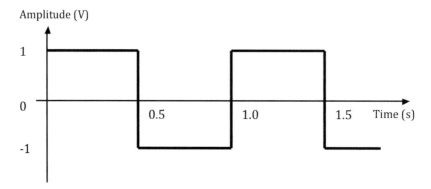

Figure 2-6a. Square Waveform in Time Domain.

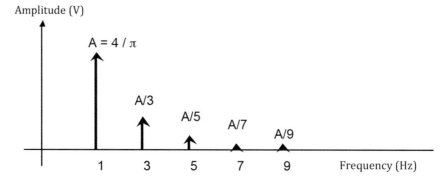

Figure 2-6b. Frequency Spectrum of the Square Wave in Figure a)

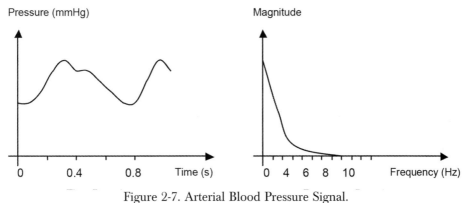

Figure 2-7. Arterial Blood Pressure Signal.

fact, each cycle of the signal differs from the others due to many factors. Physiological signals that appear to be periodical are considered pseudo-periodical. For example, each cycle of the invasive blood pressure waveform of a resting healthy person may look the same within a short period of time; however, the waveform and amplitude will be different when the person is engaged in different physical activities such as running. Furthermore, the waveform may be very different from cycle to cycle when the person has cardiovascular problems.

SIGNAL PROCESSING AND ANALYSIS

The purpose of signal processing and analysis in medical instrumentation is to condition the "raw" biological signal and extract useful information from it. For example, an ECG monitor can derive the rate of the heartbeats from biopotential signal of heart activities. It also can generate an alarm signal to alert the clinician should the heart rate fall outside a predetermined range (e.g., greater than 120 bpm or less than 50 bpm).

Transfer Function

Mathematically, an operation or process (Figure 2-8) can be represented by a transfer function *f(t)*. When a signal *x(t)* is processed by the transfer function, the output *y(t)* is equal to the time convolution between the input signal and the transfer function. That is,

$$y(t) = \int_0^t f(t - \lambda)x(\lambda)d\lambda$$

or simply denoted by *y(t) = f(t)* x(t)*.

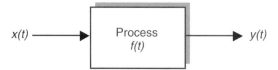

Figure 2-8. Time Domain Transfer Function.

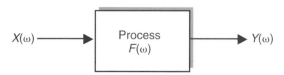

Figure 2-9. Frequency Domain Transfer Function.

In the frequency domain, the mathematical relationship between the output and the input signals when the input is processed by the transfer function F (Figure 2-9) is given by:

$$Y(\omega) = F(\omega)\, X(\omega),$$

where $X(\omega)$ is the input signal,
        $Y(\omega)$ is the output signal,
        $F(\omega)$ is the transfer function of the process, and
        ω is the angular velocity $= 2\pi f$.

Note that the output is simply equal to the input multiplied by the transfer function in the frequency domain. This is why signal analysis is often performed in the frequency domain.

Signal Filtering

A filter separates signals according to their frequencies. Most filters accomplish this by attenuating the part of the signal that is in one or more frequency regions. The transfer function of a filter is frequency dependent. A filter can be represented by a transfer function $F(\omega)$. Filters can be low pass, high pass, band pass, or band reject. The four types of filters with ideal characteristics are shown in Figure 2-10. The cutoff frequency (corner frequency) of a filter is usually measured at – 3dB from the mid-band amplitude (70.7% of the amplitude).

A low pass filter attenuates high frequencies above its cutoff frequency. High pass filters attenuate low frequencies and allow high-frequency signals to pass through. An example of such is the filter used to remove baseline

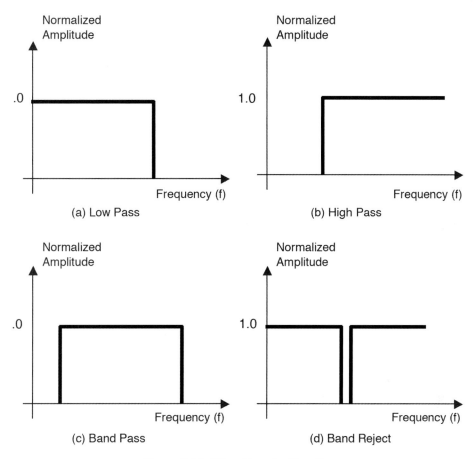

Figure 2-10. Filter Transfer Functions.

wandering signal in ECG monitoring; a 0.5-Hz high pass filter is switched into the signal path to remove the low-frequency component caused by the movement of the patient. Many biomedical devices have low pass filters with upper cutoff frequencies to remove unwanted high-frequency noise. A band pass filter is a combination of a high pass filter and a low pass filter; it eliminates unwanted low-and high-frequency signals while allowing the mid-frequency signals to go through. A band reject filter removes only a small range of frequency signal. A 60-Hz notch filter designed to remove 60-Hz power-induced noise is an example of a band reject filter. Filters can be inserted at any point in the signal pathway. Filters can be inherent (characteristic of the intrinsic or parasitic circuit components) or inserted to achieve a specific effect. For example, a low pass filter is inserted in the signal path way to remove high-frequency noise from the signal, which results in a "cleaner" waveform.

Figure 2-11 shows the effect of filters on an ECG waveform. Figure 2-11a is acquired using a bandwidth from 0.05 to 125 Hz. In Figure 2-11b, the upper cutoff frequency is reduced from 125 Hz to 25 Hz. The effect of eliminating the high-frequency components in the waveform is the attenuation of the fast-changing events (i.e., reduction of the amplitude of the R wave). Figure 2-11c shows the effect of increasing the lower cutoff frequency from 0.05 to 1.0 Hz. In this case, the low-frequency component of the signal is removed. Therefore, the waveform becomes more oscillatory.

Signal Amplification and Attenuation

An amplifier increases (amplifies) the signal amplitude; an attenuator decreases (attenuates) the signal amplitude. The transfer function of an amplifier is also called the amplification factor (A). A is expressed as the ratio of the output (Y) to the input quantity (X). That is,

$$A(\omega) = \frac{Y(\omega)}{X(\omega)}.$$

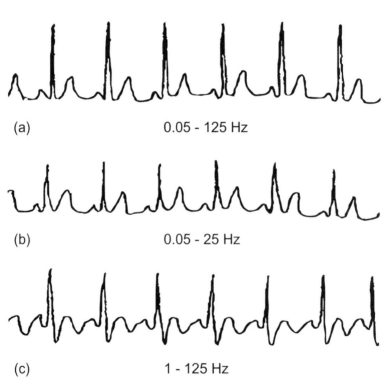

(a) 0.05 - 125 Hz

(b) 0.05 - 25 Hz

(c) 1 - 125 Hz

Figure 2-11. Effect of Filters on ECG Waveform.

The amplification factor may be frequency dependent (i.e., the value of A is different at different frequencies) or constant throughout the frequency range. The transfer function of an ideal amplifier or attenuator is independent of time and frequency (i.e., it has a constant magnitude at all frequencies).

Other Signal Processing Circuits

Other than filters and amplifiers, there are many other signal processors with different transfer function characteristics. Examples are integrators, differentiators, multipliers, adders, inverters, comparators, logarithm amplifiers, and so on. Readers who want to learn more about these signal processing circuits should refer to an integrated electronics textbook. Although many medical devices still use analog signal processing circuits, more and more of these signal processing functions are performed digitally by software in modern computer-based devices.

BIBLIOGRAPHY

BIPM, IEC, IFCC, ILAC, IUPAC, IUPAP, ISO, OIML (2008). *Evaluation of measurement data—guide for the expression of uncertainty in measurement.* Joint Committee for Guides in Metrology, JCGM 100:2008. www.bipm.org/en/publications/guides/gum.html

Coggan, D. A. (Ed.). (2005). *Fundamentals of Industrial Control: Practical Guides for Measurement and Control* (2nd ed.). Research Triangle Park, NC: ISA—The Instrumentation, Systems, and Automation Society.

Liptak, B. G. (Ed.). (2003). *Instrument Engineers' Handbook* (4th ed., vol. 1: Process Measurement and Analysis). Boca Raton, FL: CRC Press.

Part II

BIOMEDICAL TRANSDUCERS

Chapter 3

FUNDAMENTALS OF BIOMEDICAL TRANSDUCERS

OBJECTIVES

- Define the terms transducer, sensor, electrode, and actuator.
- Distinguish between the following modes of biological signal measurements: direct and indirect, intermittent and continuous, desired and interfering, invasive and noninvasive.
- Specify the three criteria for faithful reproduction of a transduction event.
- Evaluate the effect of non-ideal transducer characteristics on physiological signal measurements.
- Analyze Wheatstone bridge circuits in medical instrumentation applications.

CHAPTER CONTENTS

1. Introduction
2. Definitions
3. Types of Transducers
4. Transducer Characteristics
5. Signal Conditioning
6. Transducer Excitation
7. Common Physiological Signal Transducers

INTRODUCTION

Medical devices are designed either to measure physiological signals from the patient or to apply certain energy to the patient. To achieve that, a medical device must use a transducer to interface between the device and the patient. The transducer is usually the most critical component in a medical device because it must reliably and faithfully reproduce the signal taken from or applying to the patient. In addition, medical transducers are often in contact with or even implanted inside the human body. Transducers in such applications must be stable and nontoxic to the human body. Ideally, a transducer should respond only to the energy that is desired to be measured and exclude the others. Most of the medical device constraints discussed in Chapter 1 are also applicable to biomedical transducers.

DEFINITIONS

Generally speaking, a **transducer** is defined as a device to convert energy from one form to another. For example, the heating element on a kitchen stove is a transducer that converts electrical energy to heat energy for cooking. In instrumentation, a transducer is a device whose main function is to convert the measurand to a signal that is compatible with a measurement or control system. This compatible signal is often an electrical signal. For example, an optical transducer may convert light intensity to an electrical voltage. A **sensor** is a device that can sense changes of one physical quantity and transform them systematically into a different physical quantity; therefore, a sensor is a transducer. In instrumentation or measurement applications, sensors and transducers are often use synonymously. An **electrode** is a transducer that directly acquires an electrical signal without the need to convert it to another form; that is, both input and output are electrical signals. On the other hand, an **actuator** is a transducer that produces a force or motion. An electric motor is an example of an actuator that converts electricity to mechanical motion.

In biomedical applications, the transducer or sensor of the device often converts a physiological event to an electrical signal. With the event available as an electrical signal, it is easier to use modern computer technology to process the physiological event and display the output in a user-friendly format. Figure 3-1 shows a simple block diagram of a physiological monitor. In this simple block diagram, the patient signal is first acquired and convert-

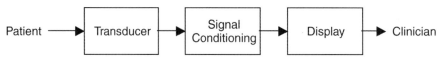

Figure 3-1. Physiological Monitor.

ed into an electrical signal by the transducer. It is then processed by the signal conditioning circuits, and the results are made available to clinicians on the visual display.

TYPES OF TRANSDUCERS

Passive Versus Active Transducers

Transducers can be passive or active. For a passive transducer, the input to the transducer produces change in a passive parameter such as resistance, capacitance, or inductance. On the other hand, an active transducer, such as a piezoelectric crystal or a thermocouple, acts as a generator, producing force, current, or voltage in response to the input.

Direct Versus Indirect Mode of Transducers

For a direct transducer, the measurand is interfaced to and measured directly by the transducer. Blood pressure may be measured directly by placing a pressure transducer inside a blood vessel. The electrical cardiac signal is directly picked up by a set of electrodes placed on the chest of a patient. Both are examples of direct mode of transduction. In indirect mode, the transducer measures another measurand that has a known relationship to the desired measurand. Indirect transducers are often used when the desired measurands are not readily accessible. An example of such indirect mode of transduction approach is in noninvasive blood pressure measurements, in which the systolic, mean, and diastolic pressures are estimated from the oscillatory characteristics of the pressure in a pneumatic cuff applied over the upper arm of the subject. Another example of indirect mode of transduction is an enzyme sensor which uses an oxygen sensor to determent glucose level in a sample of blood.

Intermittent Versus Continuous Measurement

In some cases, it is important to monitor physiological parameters in a continuous manner. Continuously monitoring the heart rate and blood oxygen level of a patient during general anesthesia is an example. In some applications, periodic measurement to track changes is sufficient. Charting the arterial blood pressure of a patient in the recovery room every 5 minutes is an example of periodic measurement. In other circumstances, a single measurement is sufficient to obtain a snapshot of the patient's condition. Measurement of oral temperature using a liquid-in-glass thermometer is an example of intermittent temperature measurement.

Desired Versus Interfering Input

Desired input to a transducer is the signal that the transducer is designed to pick up. Interfering input is any unwanted signal that affects or corrupts the output of the transducer. For example, maternal heart rate is the interfering input in fetal heart rate monitoring. Interfering input is sometimes referred to as noise in the system. In medical applications, interfering input is usually compensated for by adjusting the sensor location or through signal processing such as filtering.

Invasive Versus Noninvasive Method

A procedure that requires bypassing the skin of the patient is called an invasive procedure. Entering the body cavity such as through the mouth into the trachea is also considered to be invasive. In biomedical applications, measurement of a physiological signal often requires placing a transducer inside the patient's body. Using a needle electrode to measure myoelectric potential is an example of an invasive method of measurement. On the other hand, myoelectric measurements using skin (or surface) electrodes are noninvasive procedures.

TRANSDUCER CHARACTERISTICS

A transducer is often specified by the following:

- The quantity to be measured, or the measurand
- The principle of the conversion process
- The performance characteristics
- The physical characteristics

In biomedical measurements, common measurands are position, motion, velocity, acceleration, force, pressure, volume, flow, heat, temperature, humidity, light intensity, sound level, chemical composition, electric current, electrical voltage, and so on. Examples of their characteristics and method of measurements of some of these physiological parameters are tabulated in Table 1-2 in Chapter 1.

Many methods can be used to convert a physiological event to an electrical signal. The event can be made to modify, directly or indirectly, the electrical properties of the transducer, such as its resistance or inductance values. The primary functional component of a transducer or sensor is the transduction element. A variety of transduction elements are suitable for health care applications. Table 3-1 lists some common transducer categories and their

Table 3-1. Transducers and Their Operating Principles

Transduction Elements	Correlating Properties	Examples
Resistive	Resistance—temperature	Thermistor to measure temperature
	Resistance—displacement due to pressure	Resistive strain gauge to measure pressure
Capacitive	Capacitance—motion detection	Capacitive blanket in neonatal apnea
Inductive	Inductance—displacement due to pressure	Linear variable differential transformer in pressure measurement
Photoelectric	Electric current— light energy	Photomultiplier in scintillation counter, red and infrared LEDs and detectors in pulse oximeter
Piezoelectric	Electric potential—force	Ultrasound transducer in blood flow detector
Thermoelectric	Electric potential— thermal energy	Thermocouple junctions in temperature measurement
Chemical	Electric current—chemical concentration	Polarographic cell in oxygen analyzer

operating principles. Examples of transducers in each category are also listed in the table.

Ideally, a transducer should not introduce distortion to the original signal in the transduction process. Transducers should adhere to the following three criteria for faithful reproduction of an event:

1. Amplitude linearity—ability to produce an output signal such that it follows a linear relationship with the input. That is, the transfer function is a straight line $y = mx + c$ where x, y are the input and output signals respectively and m and c are both constants. .
2. Adequate frequency response—ability to follow both rapid and slow changes of the input signal. Ideally, the amplification (or attenuation) of the input signal caused by the transducer is a constant at all frequencies; that is m in the transfer function $y = mx + c$ is a constant at all input signal frequencies.
3. Free from phase distortion—ability to maintain the time difference between the output and input signal at all sinusoidal frequencies. That is, the phase shift introduced by the transducer is the same for all input signal frequencies.

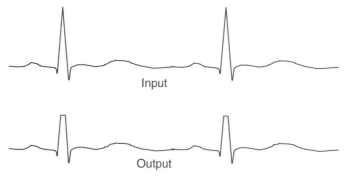

Figure 3-2. Effect of Nonlinearity (Saturation).

Amplitude Linearity

The output and input should follow a linear relationship within its operating range. The output will not resemble the input if the preceding is not true. A common example of a nonlinear input-output relationship is saturation of an operational amplifier when the input becomes too large. In Figure 3-2, when the input is within the linear region of the operational amplifier, any change of the input will produce a change of output proportional to the change of input. When the input becomes too large, it drives the amplifier into saturation with the effect that the output will not increase further with the input; the waveform is "clipped."

Adequate Frequency Response

In physiological signal measurements, the signals often change with time; body temperature changes slowly, whereas the heart potential (ECG) changes more rapidly. In order to accurately measure a changing signal, the transduc-

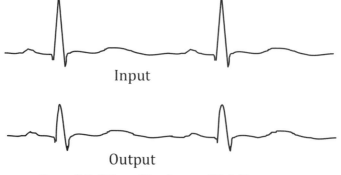

Figure 3-3. Effect of Inadequate High Frequency.

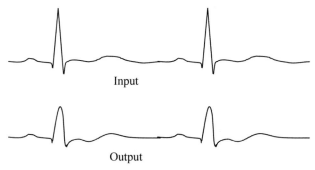

Figure 3-4. Effect of Phase Distortion.

er should be able to follow the changes of the input; that is, it must have a wide enough frequency response. Figure 3-3 shows the effect of inadequate frequency response. In this case, the high frequency of the ECG signal is attenuated or removed by the low pass filtering effect of the system. Note that in the output waveform, the amplitude of the R wave (the spike) is substantially reduced and sharp corners (which represent rapid signal change) are rounded.

Free From Phase Distortion

A system that creates different time delay at different signal frequency will create phase distortion. As we know, any time-varying signal can be represented by a number of sinusoidal signals of different frequencies and amplitudes; recombining or adding these signals will reproduce the original signal (see Appendix A-1, A Primer on Fourier Analysis). However, if the transducer in the measurement process creates different time delays on the sinusoidal signal components, recombining these signals at the output of the transducer will produce a distorted signal. Note that the phase distorted signal in Figure 3-4 has a different slew rate compared with the original signal. Phase distortion will prevent faithful reproduction of an event.

Any deviations from these three criteria will produce distorted output signals. Therefore, transducers must be carefully chosen to minimize distortion within the range of measurement (that is, amplitude and frequency). If signal distortion cannot be avoided due to non-ideal conditions, additional electronic circuits may be used to compensate for such distortions.

SIGNAL CONDITIONING

A transducer output may be directly coupled to a display device to be viewed by the user. Very often, the output of a transducer is coupled to a signal conditioning circuit. A signal conditioning circuit can be as simple as a passive filter or as complicated as a digital signal processor. A very common signal conditioning circuit for passive transducers is a Wheatstone bridge. Signal conditioning circuits commonly used in industrial electrical instrumentations are often used in medical devices. These signal conditioning functional circuits can be implemented using analog components or performed by computer software. In the latter case, the signal must be digitized and processed by a digital computer. Some common signal conditioning functions are amplifiers, filters, rectifiers, peak detectors, differentiators, integrators, and so on.

TRANSDUCER EXCITATION

Transducers that vary their electrical properties according to changes in their inputs are often used in biomedical applications. These transducers are usually coupled to operational amplifiers to increase their sensitivities and to reject noise. When the output of a transducer is a passive electrical parameter (e.g., resistance, capacitance, or inductance), it often requires an excitation to convert the passive output variable to a voltage signal. The excitation can be a constant voltage or a constant current source; it may be a direct current (DC) or an alternate current (AC) signal of any frequencies. A common transducer excitation method in biomedical application is the Wheatstone bridge.

A Wheatstone bridge is commonly used to couple a transducer to other electronic circuits. Figure 3-5 shows a simple or typical Wheatstone bridge with excitation voltage V_E and impedances Z_1, Z_2, Z_3, and Z_4 at each arm of the bridge. The output V_o of the bridge in the figure is:

$$V_o = V_a - V_b = V_E \left(\frac{Z_3}{Z_3 + Z_4} - \frac{Z_2}{Z_2 + Z_1} \right). \tag{3.1}$$

From the above equation, when the value inside the bracket is zero, i.e.,

$\left(\dfrac{Z_3}{Z_3 + Z_4} - \dfrac{Z_2}{Z_2 + Z_1} \right) = 0$, the bridge output voltage is zero ($V_o = 0$).

Such is called a balanced bridge condition.

In many transducer applications, one of the bridge arm impedances is replaced by a transducer whose impedance changes with the parameter being measured. Figure 3-6 shows an example of such an arrangement. The

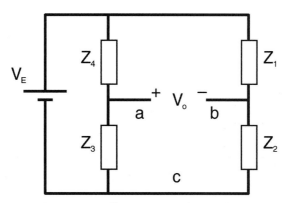

Figure 3-5. Wheatstone Bridge or Circuit.

transducer impedance can be written as $Z + \Delta Z$, where ΔZ is the impedance that changes with the quantity being measured, and Z is the invariable part of the transducer impedance. In the example shown in Figure 3-6, the impedances in the remaining arms are all equal to Z..

Substituting $Z_1 = Z + \Delta Z$ and $Z_2 = Z_3 = Z_4 = Z$ into Equation 3.1 gives

$$V_0 = V_E \frac{\Delta Z}{2(2Z + \Delta Z)}. \tag{3.2}$$

Equation 3.2 shows that V_0 and ΔZ has a non-linear relationship. However, if ΔZ is much smaller than Z ($\Delta Z \ll Z$), Equation 3.2 can be approximated by

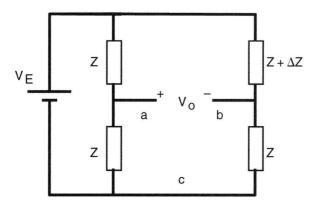

Figure 3-6. Wheatstone Bridge Transducer Circuit.

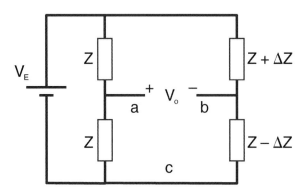

Figure 3-7. Half Bridge Transducer Circuit.

$$V_0 \approx \frac{V_E}{4Z} \Delta Z. \qquad (3.3)$$

Now consider the half bridge circuit in Figure 3-7 where there are 2 transducers with equal magnitude but opposite output change in respond to the change in the input. That is, one becomes $Z + \Delta Z$ and the other becomes $Z - \Delta Z$, each on one arm of the bridge.

Substituting $Z_0 = Z + \Delta Z$, $Z_2 = Z - \Delta Z$, and $Z_3 = Z_4 = Z$ in Equation 3.1 gives

$$V_0 = \frac{V_E}{2Z} \Delta Z. \qquad (3.4)$$

Notice that with the half bridge, the bridge output V_0 is proportional to the change in transducer impedance ΔZ with proportionality constant (or sensitivity) equal to $\frac{V_E}{2Z}$.

However, the half-bridge requires an additional transducer with matching characteristics (i.e., $Z_2 = Z - \Delta Z$) to achieve this improvement.

If we replace all the fixed impedances on the bridge arms with transducers as shown in Figure 3–8, the bridge circuit is called a full bridge.

Substituting $Z_1 = Z_3 = Z + \Delta Z$, $Z_2 = Z_4 = Z - \Delta Z$ into equation 3.1 gives

$$V_0 = \frac{V_E}{Z} \Delta Z. \qquad (3.5)$$

Similar to the half bridge circuit, the full bridge output V_0 is proportional to the change in transducer impedance ΔZ. However, the proportionality constant for a full bridge circuit is $\frac{V_E}{Z}$, which is two times the value of a half

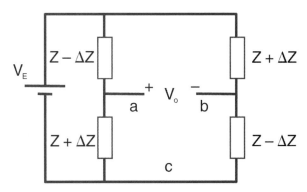

Figure 3-8. Full Bridge Transducer Circuit.

bridge circuit.

Among the three bridge circuits discussed, a full bridge transducer circuit produces a linear output voltage with respect to changes in the transducer impedance and it has the highest sensitivity. However, it requires two matching transducer pairs compared to one matching pair for a half bridge, and only a single transducer for a simple bridge circuit.

COMMON PHYSIOLOGICAL SIGNAL TRANSDUCERS

In biomedical measurement, the transducer is a component of the medical device that picks up the physiological signal from the patient. It is also the interface between the device and the human body. The transducer with its excitation circuit and its associated analog signal processing components is sometimes referred to as the front end of the medical device. There are many varieties of transducers in biomedical applications. Their characteristics, principles of operation, and design can be very different. The following major categories of biomedical transducers are covered in the next six chapters:

1. Pressure transducers
2. Temperature transducers
3. Motion transducers
4. Flow transducers
5. Optical transducers
6. Electrochemical transducers

Biomedical Device Technology

BIBLIOGRAPHY

Coggan, D. A. (Ed.) (2005). *Fundamental of Industrial Control: Practical Guides for Measurement and Control* (2nd ed.). Research Triangle Park, NC: ISA–The Instrumentation, Systems, and Automation Society.

Khazan, A. D. (1994). *Transducers and Their Elements: Design and Application.* Englewood Cliffs, NJ: Prentice Hall.

Liptak, B. G. (Ed.). (2003). *Instrument Engineers' Handbook* (4th ed., Vol. 1: *Process Measurement and Analysis*). Boca Raton, FL: CRC Press

Chapter 4

PRESSURE AND FORCE TRANSDUCERS

OBJECTIVES

- Recognize units of pressure measurement and conversion factors.
- Differentiate absolute and gauge pressure.
- Explain the principles and construction of barometers and manometers.
- Explain the principles of bourdon tube, bellow, and diaphragm pressure meters.
- Derive the gauge factor of a resistive stain gauge.
- Distinguish bonded and unbonded strain gauges, metal wire and diaphragm strain gauges.
- Apply strain gauges in pressure and force measurements.

CHAPTER CONTENTS

1. Introduction
2. Barometers and Manometers
3. Mechanical Pressure Gauges
4. Strain Gauges
5. Piezoelectric Pressure Transducers

INTRODUCTION

If a force F is acting uniformly on and perpendicular to a surface of area A, the pressure P on the surface is defined as:

$$P = \frac{F}{A}$$

From this definition, a force transducer may be used for pressure measurement and vice versa. Pressure transducers have many applications in biomedical instrumentations. Blood pressure measurement is one of the routine procedures performed in medicine. Several types of pressure-sensing elements are discussed in this chapter.

BAROMETERS AND MANOMETERS

One of the earliest transducers used in measuring atmospheric pressure was the barometer. A simple mercury barometer is constructed from inverting a mercury-filled glass tube in a bath of mercury (Figure 4-1). When the tube is tall enough, a vacuum is created on top of the mercury column. The atmospheric pressure P_{atm} is calculated from the height of the mercury column h by the equation:

$$P_{atm} = \rho g h, \tag{4.1}$$

where ρ is the density of mercury (equal to 13.6×10^3 kg/m^3),
   g is the acceleration due to gravity (equal to 9.8 m/s^2), and
   h is the height of the mercury column measured in meters.

The SI unit of pressure is Pascal (Pa). At standard atmospheric pressure (STP), one atmosphere is equal to 101 kilopascals (kPa) at temperature 273 kelvin (K) (0°C); the mercury column height h calculated from Equation 4.1 is 760 mm. Therefore, at STP, 1 atmospheric pressure corresponds to a mercury column height of 760 mm, or simply 760 mmHg. If one uses water instead of mercury in the column, using 1000 kg/m3 as the density of water,

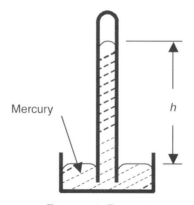

Figure 4-1. Barometer.

Table 4-1. Pressure Unit Conversion.

	Pa	*mmHg*	*cmH₂O*	*psi*	*milli Bar*
Pa	1	7.50×10^{-3}	1.02×10^{-2}	1.45×10^{-4}	1.00×10^{-2}
mmHg	133	1	1.36	1.93×10^{-2}	1.33
cmH2O	98.1	0.735	1	1.42×10^{-2}	0.981
psi	6.89×10^{3}	51.7	70.3	1	6.89×10^{1}
milli Bar	100	0.750	1.02	1.45×10^{-2}	1

the height of the water column is calculated to be 10.3 m using the same equation. In addition to the SI unit, the two most commonly used pressure units in biomedical measurement are mmHg (e.g., for blood pressure measurement) and cmH2O (e.g., for respiration pressure measurement). The conversion factors of some frequently used units in pressure measurements are tabulated in Table 4-1.

Example 4.1

a. What is 760 mmHg in kPa?
b. What is 2 cmH2O in psi?

Solution:

a. In the first column of Table 4-1, the multiplication factor to convert mmHg to Pa is 133. Therefore, 760 mmHg = 133 x 760 Pa = 101 080 Pa = 101.08 kPa.
b. In the fourth column of Table 4-1, the multiplication factor to convert cmH2O to psi is 1.42×10^{-2}. Therefore, 2 cmH2O = 2 x 1.42×10^{-2} = 2.84×10^{-2} psi.

In physiological pressure measurement, as the human body is exposed to the atmosphere, it is constantly under 1 atmospheric pressure. Physiological pressure measurements are expressed as pressure above atmospheric pressure *Patm* instead of absolute pressure *Pabs*. Absolute pressure is relative to perfect vacuum. The pressure above atmospheric pressure is called gauge pressure *Pgauge*. The relationship among absolute, atmospheric, and gauge pressures is given by the following equation:

$$P_{abs} = P_{gauge} + P_{atm} \tag{4.2}$$

Whereas a barometer is used to measure the atmospheric pressure, a manometer can be used to measure the pressure from any source. Figure 4-2 shows a manometer constructed from a liquid-filled U-shaped glass tube with one end connected to a known constant pressure source P_b and the other end connected to the source to be measured. If the source pressure is P_a, using the Pascal principle, the relationship between P_a, P_b, the difference in liquid column height h, and the liquid density ρ is given by

$$P_a = P_b + \rho g h \qquad\qquad (4.3)$$

In measuring gas pressure, one must remember that the pressure in a closed container changes with its temperature. The relationship between the gas pressure P, its volume V, and the temperature T is governed by the gas law, which states that:

$$PV = nRT,$$

where nR is a constant.

MECHANICAL PRESSURE GAUGES

Three types of mechanical pressure-sensing elements are often used in gas pressure measurement: bourdon tube, diaphragm, and bellow. A bourdon tube pressure transducer is shown in Figure 4-3; a bourdon tube is a hollow metal coil with an oval-shaped cross section. When the pressure inside the coil increases, the pressure creates a force to unwrap the coiled tube. A mechanical linkage translates this movement into a pressure readout scale.

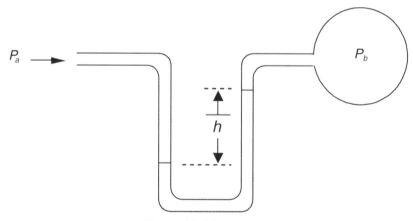

Figure 4-2. Manometer.

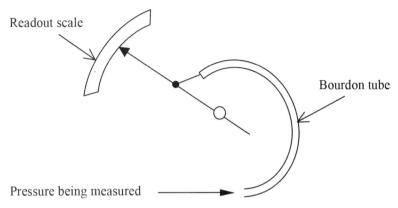

Figure 4-3. Bourdon Tube Pressure Transducer.

The scale is calibrated against known pressure sources. Bourdon tubes are often used in high pressure applications (e.g., measuring gas pressure in pressurized gas cylinders).

Figure 4-4 shows a diaphragm pressure transducer. When the pressure in the measurement chamber increases, the higher pressure on the measurement side pushes the diaphragm outward. The movement of the diaphragm is then converted to movement of a pointer needle to indicate the pressure to be measured. Diaphragm pressure gauges are often used in medium range pressures applications (e.g., measuring cuff pressure in a sphygmomanometer).

In Figure 4-5, a bellow is used as the sensing element. Similar to the diaphragm transducer, the bellow extends (or shortens) when the pressure inside the bellow becomes higher (or lower). Through a mechanical linkage,

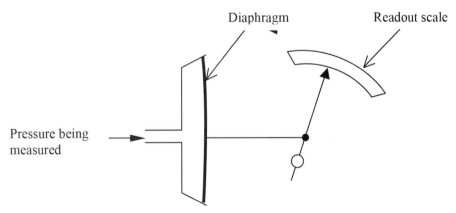

Figure 4-4. Diaphragm Pressure Gauge.

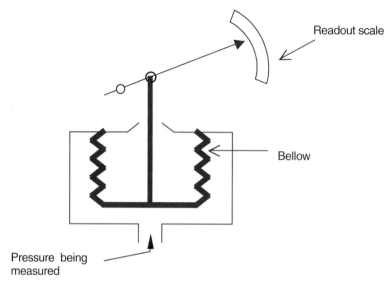

Figure 4-5. Bellow Gauge.

the bellow extension creates a motion on the pointer to indicate the pressure exerted on the bellow. Of the three types of mechanical pressure gauges, bellow pressure gauges has the highest sensitivity and is often used in low pressure applications (e.g., measuring atmospheric pressure changes).

STRAIN GAUGES

Passive resistive sensors are commonly used in physiological measurement. The resistance of a resistor wire is given by the equation:

$$R = \rho\frac{L}{A} \qquad (4.4)$$

where L is the length of the resistive wire,
$\qquad$ A is the cross-sectional area of the wire, and
$\qquad$ ρ is the resistivity.

A passive transducer can use one of the three parameters in Equation 4.4 (length, area, and resistivity) to change its resistance, and thereby transduce the physiological event. The two most important examples are the transduction of pressure and the transduction of temperature. For example, a device known as a strain gauge can be used to measure force or pressure. A strain gauge is either a length of thin conductor or a piece of semiconductor that is stretched or compressed in proportion to an applied force or pressure. The extension or contraction of the strain gauge element results in a change of resistance.

Consider a piece of conductor wire, the resistance of which is given by Equation 4.4. As the material is stretched, its length L will increase and its cross-sectional area A will decrease. Both of these changes will cause an increase in the resistance R of the conductor. Alternatively, if the material is compressed, the length will decrease and the area will increase and therefore will cause a reduction in the conductor's resistance. In fact, for most materials, the unit increase in length is proportional to the unit decrease in diameter. This proportional constant is called Poisson's ratio and is material dependent. Poisson's ratio v is defined as

$$v = -\frac{\Delta D/D}{\Delta L/L} \tag{4.5}$$

where $\Delta D/D$ is called the lateral strain and $\Delta L/L$ is the axial strain.

Both lateral and axial stain has no unit as their numerators and denominators have the same unit. However, the axial strain is often given a unit ε; and due to its small value, it is often expressed in $\mu\varepsilon$ (= 1×10^{-6}).

If one takes a partial derivative of R in equation 4.4, it becomes

$$\Delta R = \rho \frac{\Delta L}{A} - \rho L \frac{\Delta A}{A^2} + L \frac{\Delta \rho}{A} \tag{4.6}$$

Divide both sides of the equation by R (the right-hand side by R and the left-hand side by $\rho = \frac{L}{A} = R$) gives

$$\frac{\Delta R}{R} = \frac{\Delta L}{L} - \frac{\Delta A}{A} + \frac{\Delta \rho}{\rho}. \tag{4.7}$$

For a cylindrical resistant wire, $A = \frac{\pi D^2}{4}$, where D is the diameter of the wire. Taking the partial derivative of this equation and divide both sides by A gives

$$\frac{\Delta A}{A} = 2 \frac{\Delta D}{D}. \tag{4.8}$$

Substitute Equation 4.8 into 4.7 gives

$$\frac{\Delta R}{R} = \frac{\Delta L}{L} - 2\frac{\Delta D}{D} + \frac{\Delta \rho}{\rho} \qquad (4.9)$$

Substitute Equation 4.5 into 4.9 gives

$$\frac{\Delta R}{R} = \frac{\Delta L}{L} + 2v\frac{\Delta L}{L} + \frac{\Delta \rho}{\rho} = \frac{\Delta L}{L}(1 + 2v) + \frac{\Delta \rho}{\rho} \qquad (4.10)$$

Equation 4.10 shows that both the change in resistivity and the change in length of the wire affect the resistance. The unit change of resistance per unit change in length of a strain gauge is called the gauge factor (G.F.).

$$G.F. = \frac{\Delta R / R}{\Delta L / L} \qquad (4.11)$$

From Equations 4.10 and 4.11, the G.F. is equal to

$$G.F. = \frac{\Delta R / R}{\Delta L / L} = (1 + 2v) + \frac{\Delta \rho / \rho}{\Delta L / L} \qquad (4.12)$$

The G.F. value is different from one material to another. For a metal strain gauge, $\frac{\Delta \rho}{\rho}$ is 0 and the G.F. is between 2 and 4. For a semiconductor or piezoelectric strain gauge, $\frac{\Delta \rho}{\rho}$ is nonzero. The G.F. of a piezoelectric strain gauge can be several hundreds.

Example 4.2

For a metal wire resistive strain gauge with nominal resistance R_o = 120.0 Ω and G.F. = 2.045, find the change in resistance if an axial strain of 7320 με is applied to the strain gauge.

Solution:

From Equation 4.11

$$G.F. = \frac{\Delta R / R}{\Delta L / L}$$

$$\Rightarrow \quad \Delta R = R \times G.F. \times \frac{\Delta L}{L}$$

$$\Rightarrow \quad \Delta R = 120.0 \ \Omega \times 2.045 \times 7320 \times 10^{-6} = 1.796 \ \Omega.$$

A strain gauge may be either unbonded or bonded. An unbonded strain gauge uses posts to hold the ends of the gauge wires. The posts are attached to the structure for which movement is to be measured. A bonded strain gauge means that the wire or semiconductor material is attached to a flexible bonding material, such as plastic or paper. It is then cemented to the surface of the structure. Figure 4-6 shows unbonded and bonded metal wired strain gauges.

Other than metal wire strain gauges, a common transducer element in pressure measurement is the diaphragm piezoresistive strain gauge. It is fabricated using semiconductor technology in which a layer of piezoresistive material is deposited on a flexible diaphragm and lead wires are bonded to it (Figure 4-7). The gauge resistance changes as the strain element on the diaphragm is deformed by the applied pressure. Because the gauge factor of piezoresistive material is much higher than that of metal, it creates a much more sensitive pressure transducer than metal wire stain gauges do. Furthermore, by controlling the position and shape of the piezoresistive deposit on the diaphragm, it can produce a linear output response to the applied pressure.

Although the discussion so far has been on the use of the strain gauge for pressure measurements, it does have many other biomedical applications such as force and acceleration measurements. An interesting application in cardiology technology is the use of strain gauges to measure cardiac contractility. In this application, a bonded strain gauge is sutured directly to the ventricular wall of the heart. The contractile force of the muscle fibers causes a change in the strain gauge resistance, which is then measured, processed, and displayed. Another example is a load cell to measure the body weight

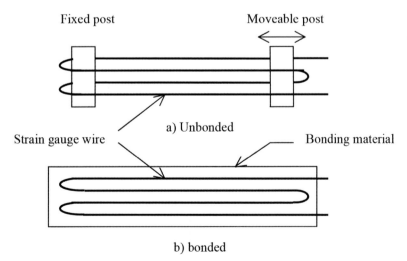

Figure 4-6. Unbonded and Bonded Wired Strain Gauges.

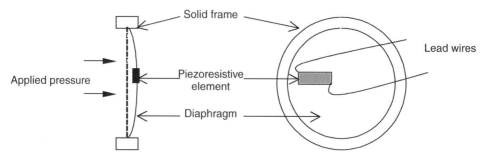

Figure 4-7. Diaphragm Piezoresistive Strain Gauge.

of a patient undergoing renal dialysis. A load cell is a transducer that converts a loading (or force acting on it) to an analog electrical signal. An example of a load cell is shown in Figure 4-8, where the force to be measured is converted to a change in resistance of the strain gauge. In this example, the force on the cantilever of the load cell creates unequal deformation on the two elements of the bonded strain gauge. The amount of deformation on the strain gauge elements as a result of the applied force depends on the dimensions and material properties of the cantilever.

PIEZOELECTRIC PRESSURE TRANSDUCERS

Some materials such as quartz crystal and certain ceramic materials undergo physical deformation when an electrical potential is applied. Also, these materials generate opposite electrical charges at their surfaces when subjected to mechanical strain. When an external force creates a strain on a piezoelectric

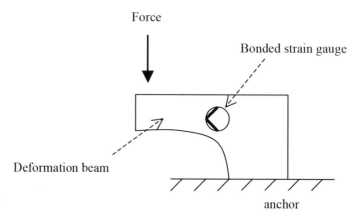

Figure 4-8. A Load Cell.

material, the charge Q produced is proportional to the applied force F. The proportionality constant is called the piezoelectric constant K. For the piezoelectric transducer in Figure 4-9, the charge Q created on the surface of the transducer is

$$Q = K \times F$$

Assuming the transducer is a thin circular disk with thickness d and area A, the disk can be considered as a parallel plate capacitor with charge residing on the surface. The capacitance C is given by $C = \varepsilon \times A/d$, where ε is the material permittivity. The voltage V across the capacitor is then given by

$$V = \frac{Q}{C} = \frac{K \times F}{C} = \frac{K \times F \times d}{\varepsilon \times A} = \frac{K \times d}{\varepsilon} \times \frac{F}{A} = \frac{K \times d}{\varepsilon} \times P \qquad (4.13)$$

Since K, d, and ε are constants, the equation shows that the voltage V generated at the surface of the transducer is proportional to the applied pressure P.

A biomedical application of such a transducer is the ultrasound transmitter and receiver. In an ultrasound receiver, the sound pressure to be detected produces mechanical strain on the piezoelectric element, which generates a potential difference whose amplitude varies according to the sound pressure. In an ultrasound transmitter, the electrical signal that imposes on the transducer produces deformation according to the amplitude and frequency of the signal. This physical deformation creates mechanical pressure waves in the medium in which the transducer element is submerged.

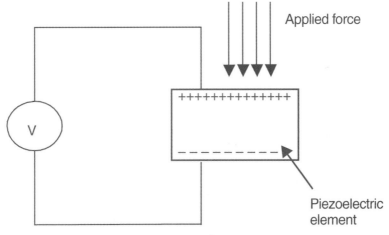

Figure 4-9. Circular Disc Piezoelectric Element.

Example 4.3

A pressure transducer with nominal resistance $R_0 = 100\ \Omega$ (at zero pressure) and sensitivity $S = 3.00\ \Omega$ per 100 mmHg is placed on one arm of a Wheatstone bridge as shown in the illustration below. If the resistors R_1, R_2, and R_3 are all 100 Ω and the excitation voltage $V_E = 5.00$ V, find the change in output voltage V_0 when the pressure is increased from 0 to 60.0 mmHg.

Solution:

When P = 0 mmHg, $R_x = R_1 = R_2 = R_3 = 100\ \Omega$, the bridge output V_0 = 0.0 V (balanced bridge).

When the applied pressure P = 60.0 mmHg, the change in resistance of the transducer $\Delta R_x = S \times \Delta P = 3.00\ \Omega/100$ mmHg $\times$ 60.0 mmHg = 1.80 Ω.

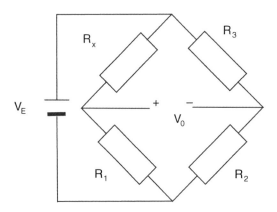

Therefore, $R_x = 100 + 1.80 = 101.8\ \Omega$, substituting into the bridge circuit gives

$$V_0 = V_E\left[\frac{R_1}{R_x + R_1} - \frac{R_2}{R_2 + R_3}\right] = 5.00\left[\frac{100}{101.8 + 100} - \frac{100}{100 + 100}\right] = -0.0223\,V = -22.3\,mV$$

BIBLIOGRAPHY

Beckwith, T. G., Roy, D. M., & John, H. L. (1993). *Measurement of Low Pressures. Mechanical Measurements* (5th ed., pp. 591–595). Reading, MA: Addison-Wesley. Benedict,

R. P. (1969). *Fundamentals of Temperature, Pressure and Flow Measurements.* New York:

John Wiley & Sons, Inc.

Coggan, D. A. (Ed.). (2005). *Fundamental of Industrial Control* (2nd ed.). Research Triangle Park, NC: ISA–The Instrumentation, Systems, and Automation Society.

Khazan, A. D. (1994). *Transducers and Their Elements: Design and Application.* Englewood Cliffs, NJ: Prentice Hall.

Liptak, B. G. (Ed.) (2003). Instrument Engineers' Handbook (4th ed., Vol. 1: *Process Measurement and Analysis*). CRC Press.

Chapter 5

TEMPERATURE TRANSDUCERS

OBJECTIVES

- Examine the International Practical Temperature Scale (IPTS).
- Perform temperature unit conversion.
- Review common temperature measurement terminology
- Describe the principles of nonelectrical temperature gauges.
- Analyze the characteristics of resistance temperature devices (RTDs), thermistors, thermocouples, and integrated circuit (IC) temperature sensors.
- Compute errors due to lead resistance and self-heating effect in temperature measurement applications.
- Explain methods to linearize thermistor characteristics.

CHAPTER CONTENTS

1. Introduction
2. Reference Temperature and Temperature Scale
3. Nonelectrical Temperature Transducers
4. Electrical Temperature Transducers
5. Resistance Temperature Devices
6. Lead Errors
7. Self-Heating Errors
8. Thermistors
9. Thermocouples
10. Integrated Circuit Temperature Sensors
11. Comparison of Temperature Sensors' Characteristics

INTRODUCTION

Temperature transducers have widespread applications in biomedical instrumentation. They are used to measure body temperature in patient assessment, to control heating or cooling in therapeutic procedures, and to monitor medical device performance to ensure patient safety. Temperature transducers that are designed to provide temperature readings are called thermometers. This chapter covers concepts in temperature measurement and the principles of some commonly used temperature transducers.

REFERENCE TEMPERATURE AND TEMPERATURE SCALE

Unlike most other physical quantities, one cannot add temperatures together as one would add lengths to measure distances. We must rely on physical phenomena to establish observable and consistent temperature references. For example, the freezing point and boiling point of water at one atmospheric pressure are assigned as $0°$ and $100°$, respectively, in the Celsius (or centigrade) scale and as $32°$ and $212°$ in the Fahrenheit scale. The entire temperature scale can be constructed by making interpolation of these fixed temperature references. Table 5-1 shows the reference temperatures on which the IPTS is based. Interpolation of these references is performed by temperature transducers.

Common units of temperature measurement are degree Celsius (°C) in the Celsius scale, degree Fahrenheit (°F) in the Fahrenheit scale, and kelvin (K) in the absolute scale.

For a temperature of A°F, the equivalent temperature reading in the

Table 5-1. IPTS-68 Reference Temperatures

Reference Point	K	°C
Triple Point of Hydrogen	13.81	−259.34
Liquid/Vapor Phase of Hydrogen at 25/76 std. atm.	17.042	−256.108
Boiling Point of Hydrogen	20.28	−252.87
Boiling Point of Neon	27.102	−246.048
Triple Point of Oxygen	54.361	−218.789
Boiling Point of Oxygen	90.188	−142.962
Triple Point of Water	273.16	0.01
Boiling Point of Water	373.15	100.00
Freezing Point of Zinc	692.73	419.58
Freezing Point of Silver	1235.08	961.93
Freezing Point of Gold	1337.58	1064.43

Celsius scale is B°C, where

$$B = \frac{5}{9}(A - 32), \text{ or in reverse,} \qquad (5.1)$$

$$A = \frac{9}{5}B + 32.$$

The same temperature in kelvin (C K) is given as

$$C = B + 273.15. \qquad (5.2)$$

Example 5.1

The body temperature of a patient is measured to be 37.0°C. What are the temperature readings in the Fahrenheit and absolute scales.

Solution:

From equation 5.1, $A = \frac{9}{5}B + 32 = \frac{9}{5} \times 37 + 32 = 98.6°F.$

From equation 5.2, $C = B + 273.15 = 37.0 + 273.15 = 310.2$ K.

NONELECTRICAL TEMPERATURE TRANSDUCERS

Fluid-in-glass thermometers have long been used to measure temperature. Until recently, it has been the most popular type of thermometer in body temperature measurements. Mercury-in-glass and petrolate-in-glass are the two common types in this category. A fluid-in-glass thermometer is shown in Figure 5-1. It consists of a glass tube (or stem) with a uniform lumen connected to a reservoir. The top of the tube is sealed after air is removed. When the temperature increases, the fluid in the reservoir expands and pushes up the fluid level in the stem. The position of the fluid meniscus indicates the temperature under measurement. Two temperature references (e.g., freezing and boiling points of water) are chosen to calibrate two points on the stem of the thermometer. Mercury-filled thermometers usually have a temperature measurement range from –40 to +900°C, whereas red-dyed petrolate can measure from –200 to +260°C. In medical applications, some glass thermometers are Teflon®, encapsulated to avoid shattering.

Bimetallic sensors are often used in temperature gauges as thermal switches. Two metals with different temperature expansion coefficients are

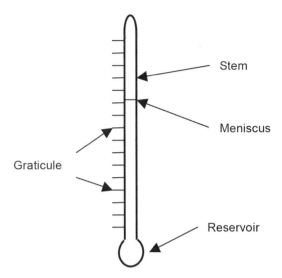

Figure 5-1. Fluid-in-Glass Thermometer.

bonded together to form a bimetallic strip. When the temperature changes, the differential expansion causes the strip to bend. The degree of bending can be calibrated and used in temperature measurement. Figure 5-2 shows a simple bimetallic strip. In this example, metal A has a higher coefficient of expansion than metal B has; the strip bends as shown at elevated temperature. Figure 5-3 shows a dial-type temperature gauge with the bimetallic strip formed into a helical coil to produce rotational motion with changes in temperature. Depending on the type of metals used, bimetallic gauges can measure temperature in the range of –40 to 1500°C.

Another nonelectrical temperature sensor is constructed by sandwiching a liquid crystal material between an adhesive backing and a transparent Mylar® film. The liquid crystal can be made to reflect different colors of light at different temperatures. Liquid crystal thermometers provide fast (1 to 2 seconds respond time) visual indication of temperature with a range from 0 to +60°C. Strip liquid crystal thermometers are available to measure

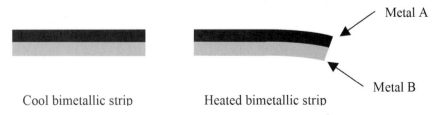

Figure 5-2. Bimetallic Temperature Sensor.

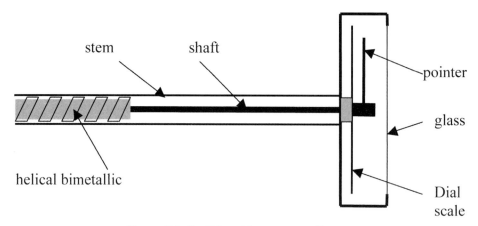

Figure 5-3. Dial-Type Temperature Gauge.

body temperature.

ELECTRICAL TEMPERATURE TRANSDUCERS

In general, electrical transducers can be divided into two types: passive and active. Resistance temperature devices and thermistors, both of which are passive transducers, are covered in this chapter. Thermocouples and integrated circuit sensors are covered as examples of active temperature transducers. Temperature sensors that function by detecting infrared energy emitting from heated sources are discussed in Chapter 8, Optical Transducers.

In addition to other instrumentation specification parameters covered in Chapter 2, some parameters that are commonly referred to in temperature measurement applications are

- Operating range—the temperature between two limits within which the characteristics conform to the specifications.
- Stability—the quality of the sensor to maintain a consistent output when a constant input is applied. The shelf life of a disposable transducer is an indication of the stability of the sensor.
- Dissipation constant—the power (in milliwatts [mW]) required to raise the sensor by 1°C above the surrounding temperature. For example, a thermistor may be listed to have a dissipation constant of 1.0 mW/°C in still air.
- Respond time—the time required to reach a certain percentage of the steady state output followed by a step input. The time to reach 63.2% is referred to as one time constant. For most temperature transducers, approximately five time constants are necessary to reach 99% of the steady

state output.

RESISTANCE TEMPERATURE DEVICES

For a metal conductor, the resistance increases with temperature. Therefore, one can measure the resistance of a metal wire to determine its temperature. Figure 5-4 shows the normalized resistance-temperature characteristics of metal RTDs, including nickel, copper, platinum, and tungsten. R_T is the resistance of the metal wire in ohm (Ω) at temperature T°C. By selecting different metals, RTD can measure temperature from –200 to over +800°C. As shown in Figure 5-4, despite its low sensitivity, the platinum RTD shows an almost linear resistance-temperature relationship and therefore can be approximated by

$$R_T = R_0(1 + \alpha\Delta T), \tag{5.3}$$

where $\Delta T = T - T_0$; T_0 is usually specified at 0°C;
and α is the fractional change in resistance per unit temperature.

The temperature coefficient is defined as (dR_T/dT), which is equal to αR_0 from Equation 5.3.

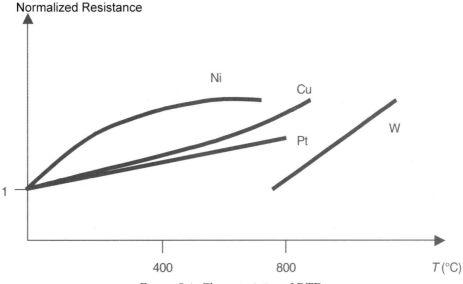

Figure 5-4. Characteristics of RTD.

Example 5.2

A piece of platinum RTD wire has a resistance of 100.0 Ω at 0°C. Find (a) its temperature coefficient and (b) its resistance at 189.0°C, given that α = 0.00385/°C for platinum RTD.

Solution:

(a) Using Equation 5.3, the temperature coefficient $= \alpha R_0 = 0.00385 \times 100 = 0.385$ Ω/°C.

(b) Using Equation 5.3, $R_{189} = 100 (1 + 0.00385 \times 189) = 171.8$ Ω.

Example 5.3

A platinum RTD is used in a bridge circuit as shown below. What is the bridge output V_0 at 10°C, given the excitation voltage $V_E = 2.0$ V, bridge resistance $R = R_0 = 100$ Ω at 0°C and α = 0.00385/°C.

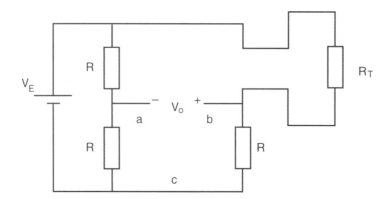

Solution:

Let $\Delta R = R_T - R_0$, at 10°C, $\Delta R = R_0 (1 + \alpha\Delta T) - R_0 = \alpha R_0\Delta T = 0.00852 \times 100 \times 10$ Ω $= 3.85$ Ω.

For the Wheatstone Bridge circuit,

$$V_0 = \left[\frac{R}{R + R_T} - \frac{R}{R + R}\right] V_E = \left[\frac{R}{R + R + \Delta R} - \frac{1}{2}\right] V_E = -\frac{\Delta R}{2R + \Delta R} \times \frac{V_E}{2}.$$

Substituting $R = 100$ Ω, $\Delta R = 3.85$ Ω, and $V_E = 2.0$ V into the above gives

$$V_0 = -18.9 \text{ mV}$$

The value of α quoted in this example (0.00385/°C) is according to the DIN Standard (European). In the United States, α is equal to 0.00392/°C for a standard platinum RTD. Usually, the RTD characteristics are specified in a lookup table of resistance against temperature supplied by the manufacturer.

LEAD ERRORS

It should be noted that RTDs in general have low sensitivities. That is, the output resistance changes only slightly for a relatively large change in temperature. For the platinum RTD in the previous example, the temperature sensitivity (or temperature coefficient) is equal to 0.385 Ω/°C. In temperature measurement using RTD, due to its low temperature coefficient, one must be aware of the errors introduced to the measurement from inadvertent additions and changes in the overall measured resistance.

Example 5.4

In a temperature measurement experiment using a platinum RTD, a pair of long lead wires are used to connect the RTD to an ohmmeter. If α = 0.00385/°C and each lead wire has an overall resistance of 0.5 Ω, find the lead error.

Solution:

The lead wires add a total resistance of 0.5 Ω + 0.5 Ω = 1.0 Ω to the RTD. The temperature error T_E due to the lead resistance (or simply called lead error) is

$$T_E = \frac{1.0}{0.385} = 2.6°\text{C},$$

which is a very significant error in medical applications.

Example 5.5

For the RTD application in Example 5.3, if each lead wire to the RTD has a resistance R_L of 0.50 Ω, what is the bridge output?

Solution:

With the lead resistance R_L, the effective resistance R_E across the bridge arm becomes $R_E = R_T + 2R_L$, substituting into the bridge equation gives

$$V_0 = \left[\frac{R}{R+R_E} - \frac{R}{R+R}\right]V_E = \left[\frac{R}{R+R+\Delta R+2R_L} - \frac{1}{2}\right]V_E = -\frac{\Delta R + 2R_L}{2R + \Delta R + 2R_L} \times \frac{V_E}{2}.$$

Substituting $R = 100\ \Omega$, $\Delta R = 3.85\ \Omega$ and $V_E = 2.0\ V$ into the above equation,

$$V_0 = -\frac{3.85 \times 2 \times 0.50}{2 \times 100 + 3.85 + 2 \times 0.50} \times \frac{2.0}{2}\ V = -23.7\ mV.$$

This result shows a −25% error compares to the bridge output without lead resistance (−18.9 mV in Example 5.3).

In any resistive transducer applications, lead resistance may introduce significant errors in the measurement. Lead errors can be accounted for if one knows the exact resistance of the lead wires. This may not be possible in some applications, however what follows are some methods used to minimize lead errors.

Three-Wire Bridge

One method to reduce the lead error is to use a three-wire bridge circuit. Instead of using a conventional two-wire RTD, a three-wire RTD is used (Figure 5-5). For the circuit in Figure 5-5, if we are using a high input impedance voltmeter to measure V_0, there will be no voltage drop across the middle lead of the 3-wire RTD because the middle lead does not carry any current. Given $R_T = R + \Delta R$, the bridge output becomes

$$V_0 = \left[\frac{R+R_L}{R+R_T+2R_L} - \frac{R}{R+R}\right]V_E = \left[\frac{R+R_L}{R+R+\Delta R+2R_L} - \frac{1}{2}\right]V_E = -\frac{\Delta R}{2R + \Delta R + 2R_L} \times \frac{V_E}{2}.$$

Example 5.6

For the three-wire RTD application in Figure 5.5, using the same circuit values as Example 5.5, that is, $R = 100\ \Omega$, $R_L = 0.05\ \Omega$, $\Delta R = 3.85\ \Omega$ (at 10°C) and $V_E = 2.0\ V$, what is the bridge output?

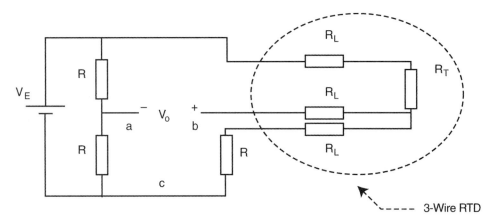

Figure 5-5. Three-Wire Wheatstone Bridge Circuit.

Solution:

Substituting R = 100 Ω, ΔR = 3.85 Ω, R_L = 3.85 Ω, and V_E = 2.0 V into the equation,

$$V_0 = -\frac{\Delta R}{2R + \Delta R + 2R_L} \times \frac{V_E}{2}, \text{ which gives}$$

$$V_0 = -\frac{3.85}{2 \times 100 + 3.85 + 2 \times 0.50} \times \frac{2.0}{2} V$$

$$= -0.0188 \text{ V} = -18.8 \text{ mV}.$$

This result shows a mere 0.5% error compared to the bridge output without lead resistance (-18.99 mV in Example 5.3). This is a significant improvement to the two-wire RTD bridge circuit with lead error (-25% error in Example 5.5).

To totally eliminate the error due to lead resistance, a four-wire RTD with a high impedance voltmeter and constant current source connected as shown in Figure 5-6 is needed. The RTD resistance (R_T) is simply calculated from dividing the voltmeter reading by the magnitude of the current from the constant current source.

SELF-HEATING ERRORS

An RTD is a resistor whose resistance changes with temperature. As we

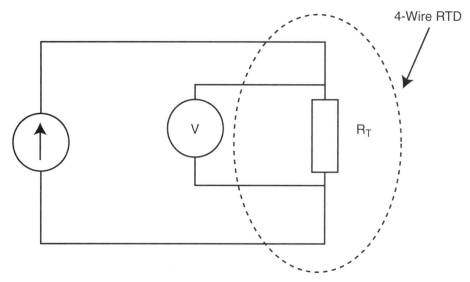

Figure 5-6. Four-Wire RTD Circuit.

all know, when a current I passes through a resistor R, it will produce heat at a rate equals to I^2R. This heat will raise the temperature of the RTD and therefore will create an error in the temperature measurement. Dissipation constant is a parameter specified for all resistive-type temperature transducers. The dissipation constant allows the user to determine the temperature error caused by heat generated in the transducer due to an externally applied current. The value of the dissipation constant depends on the type of transducer element, its packaging, as well as the medium in which it is being used.

Example 5.7

A 100-Ω platinum RTD in a bridge circuit is used to measure air temperature. The bridge excitation is 5.0 V and all the bridge resistors are 100 Ω. Calculate the error due to self-heating if the dissipation constant of the RTD in still air is 5.0 mW/°C (or 0.20°C/mW).

Solution:

When the bridge is balanced, in other words, R_T = R = 100 Ω, the voltage V_T across R_T is 0.5 times the excitation voltage V_E. The power P_T dissipated in the RTD is therefore given by

$$P_T = \frac{V_T^2}{R_T} = \frac{V_E^2}{4R} = \frac{5.0^2}{4 \times 100} = 0.0625 \text{ W} = 62.5 \text{ mW.}$$

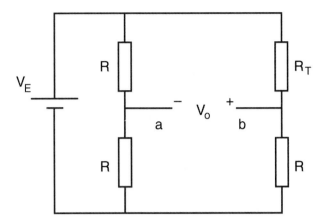

Therefore, self-heating error equals

$$\frac{62.5 \text{ mW}}{5.0 \text{ mW/°C}} = 12.5°C,$$

which means that the RTD will read 12.5°C higher than the correct temperature due to the self-heating effect. From this example, one can see that in order to reduce the self-heating error, one should reduce the excitation voltage, reduce the transducer current, or increase the bridge resistance.

THERMISTORS

Another common passive resistive temperature transducer is the thermistor. Thermistors are made from semiconductor materials such as oxide of nickel, manganese, iron, cobalt, or copper. Because a thermistor is a semiconductor, it has nonlinear resistance temperature characteristics. Thermistors can have a positive or negative temperature coefficient and an operating range from −80 to +150°C. Thermistors have higher nominal resistance than RTD (ranges from 1 kΩ to 1 MΩ) as well as higher sensitivity (from 3% to 5%/°C). Although they are less stable than RTDs, thermistors have decent long-term stability (less than 0.2% drift in resistance per year). The most common type of thermistor used in medical instruments is the YSI 400 series thermistor (YSI stands for Yellow Spring Instrument, a manufacturer of temperature transducers), which has a negative temperature coefficient (Figure 5-7). The characteristics of a thermistor are available in the manufacturer's lookup tables. In addition, most negative temperature coefficient thermistors can be approximated by

$$RT = Roe^{\beta\left(\frac{1}{T} - \frac{1}{To}\right)} \qquad (5.4)$$

where RT is the thermistor resistance in Ω at temperature T,
Ro is the thermistor resistance in Ω at temperature To,
T and To are absolute temperature in K (note that T is in °C in the RTD equation), and
β is the thermistor material constant in the range of 2500 to 5000 K.

The temperature coefficient (dRT/dT) can be obtained by differentiating Equation 5.4 with respect to temperature T, which gives

$$\left(\frac{dRT}{dT}\right) = Ro\frac{d\left[e^{\beta\left(\frac{1}{T}-\frac{1}{To}\right)}\right]}{dT} = Roe^{\beta\left(\frac{1}{T}-\frac{1}{To}\right)}\frac{d}{dT}\left(\frac{\beta}{T}\right) = RT\left(-\frac{\beta}{T^2}\right) = -\frac{RT\beta}{T^2}$$

Example 5.8

For a thermistor with β = 4000 K, find (a) the thermistor resistance and (b) the temperature coefficient at 37.0°C given the Ro = 7355 Ω at 0°C.

Solution:

(a) Using Equation (5.4)

$$RT = 7355e^{4000\left(\frac{1}{(37 + 273)} - \frac{1}{(0 + 273)}\right)} = 1280\ \Omega$$

(b) Temperature coefficient $= -\dfrac{RT\beta}{T^2} = \dfrac{-1280 \times 4000}{(37 + 273)^2} = $ -53.3 Ω/K.

One of the major drawbacks for thermistors is their highly nonlinear characteristics. Much research was done to produce thermistors with linear resistance temperature characteristics. A rather efficient method to produce a piecewise linear characteristic within a narrow temperature region is achieved by connecting a parallel resistor Rp to the thermistor. This combination produces an approximate linear region centered around the temperature of interest Tm. The parallel resistor value Rp is derived from finding the point of inflexion (d^2R/dT^2) of the resistor combination curve. Without going into the detailed derivation, Rp can be calculated from Equation 5.5.

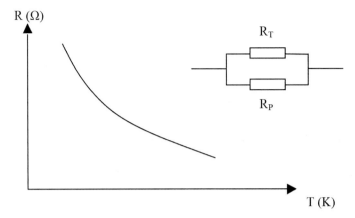

Figure 5-7. Thermistor Characteristics.

$$R_P = R_{Tm} \times \frac{\beta - 2T_m}{\beta + 2T_m} \tag{5.5}$$

The linearized characteristic using this method is shown in Figure 5-8. Note that this method reduces the overall sensitivity of the transducer.

Example 5.9

A YSI 400 thermistor is used for body temperature measurement. Choose a parallel resistor to the thermistor to linearize its resistance temperature characteristic around the temperature of interest.

Solution:

Normal body temperature is about 37°C (= 310 K). From the lookup table of the YSI 400 thermistor, R_{37} = 1355 Ω and β = 4000 K. Substitute into Equation 5.5 gives

$$R_P = 1355 \times \frac{4000 - 2 \times 310}{4000 + 2 \times 310} = 991 \; \Omega.$$

Another method to obtain a linear relationship is by using two thermistors. Figure 5-9a shows a YSI 700 series thermistor combination, and Figure 5-9b shows one example of using this thermistor to produce a linear resistance temperature curve (Figure 5-10). Note that there are three lead wires for a YSI 700 series thermistor (one more than in a 400 series). Other meth-

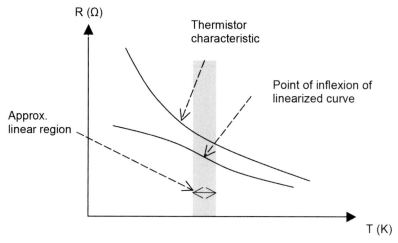

Figure 5-8. Thermistor Linearization.

ods are available to linearize thermistor characteristics. However, newer microprocessor-based devices with large memory to store the entire thermistor lookup table have made linearization less important. Other than the resistance-temperature characteristics, different sizes and shapes of thermistor probes are available for different applications.

THERMOCOUPLES

A thermocouple is an active temperature transducer that produces a small unique voltage according to temperature. A thermocouple consists of two dissimilar metals joined together. Thomas Seebeck discovered this property in 1821, and he named it the thermoelectric effect. He discovered that when two wires of dissimilar metals are joined at both ends and one end is at a higher temperature than the other, there is a continuous flow of current in the wire (Figure 5-11a). This current is called the thermoelectric current. When one junction is disconnected, a voltage called the Seebeck voltage can be measured across the two metal wires (Figure 5-11b).

For a small variation of the temperature difference between the hot and the cold junctions, the Seebeck voltage eAB is proportional to the temperature variation ΔT with a proportional constant α called the Seebeck coefficient. This relationship is represented by

$$eAB = \alpha \Delta T.$$

The Seebeck coefficient is different for different thermocouples. It is small

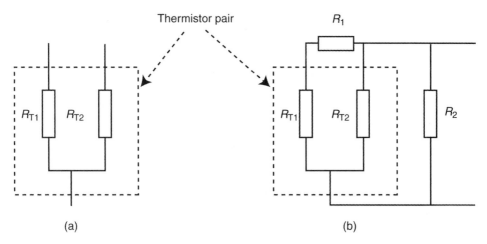

Figure 5-9. YSI 700 Series Thermistor and Application.

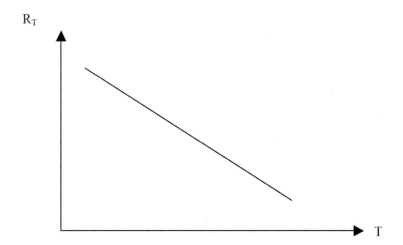

Figure 5-10. Linearized Characteristic of YSI 700 Series Circuit.

and varies with temperature. In addition, it has a nonlinear temperature characteristic. For example, the Seebeck coefficient for an E-type (chromel and constantan) thermocouple varies from about 25 μV/°C at 2200°C to 62 μV/°C at +800°C.

The thermoelectric sensitivity of a thermocouple material is usually given relative to a standard platinum reference. Since the Seebeck coefficient is temperature dependent, the sensitivity is also temperature dependent. Table 5-2 shows the sensitivities of some thermocouple metals referenced to platinum at 0°C. The sensitivity at 0°C of any combination of the materials can be obtained by taking the difference of the sensitivities of the materials

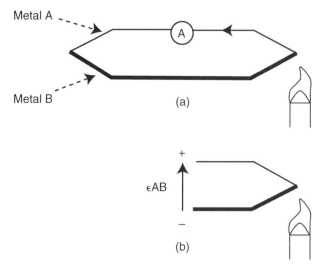

Figure 5-11. Seeback Effect.

forming the thermocouple. For example, the sensitivity of chromel and constantan (Type E thermocouple) is [+25 μV/°C − (−35 μV/°C)] = 60 μV/°C. In temperature measurement, one of the junctions is kept at a known (or reference) temperature and the other junction is at the temperature to be measured.

The reference temperature is often referred to as the cold temperature since thermocouples are usually used for high temperature measurements. Thermocouples can be used to measure temperature from −200 to +2000°C. In addition to its wide temperature range, it is rugged and accurate and can be made to be extremely small in size. The output characteristics of some common thermocouples are listed in Table 5-3 and shown in Figure 5-12.

Empirical Laws of Thermocouples

The following three empirical laws are important facts that enable thermocouples to be used as practical temperature measurement devices:

1. The law of homogeneous circuit
2. The law of intermediate metals
3. The law of intermediate temperature

The law of homogeneous circuit states that for a thermocouple of metals A and B, the output voltage is not affected by the temperature along metal A or B. It is determined by the temperatures at the two junctions (Figure 5-

Table 5-2. Thermoelectric Sensitivity of Thermocouple Materials

	Copper	*Chromel*	*Constantan*	*Iron*	*Platinum*
Composition	Cu	90 Ni/10 Cr 60	Cu/40 Ni	Fe	Pt
Sensitivity, μV/°C	+6.5	+25	-35	+18.5	0

Table 5-3. Characteristics of Thermocouples

Type	*Name*	*Temp Range*	*Output*, mV*
T	Copper/constantan	−200 to +300	4.25
J	Iron/constantan	−200 to +1100	5.28
E	Chromel/constantan	0 to +1100	6.30
K	Chromel/alumel	−200 to +1200	4.10
S	Platinum/Pt-rhodium	0 to +1450	0.64

*Output is measured at 100°C with reference at 0°C.

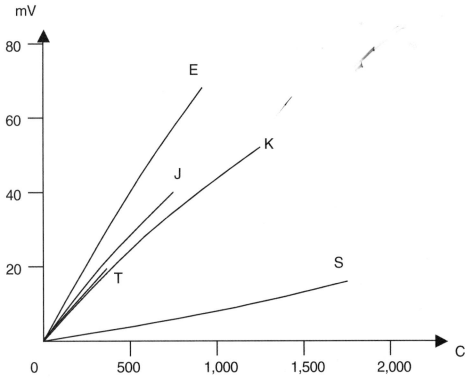

Figure 5-12. Thermocouple Output Characteristics.

13). This is required because the metals or lead wires between the thermo-couple junctions will be at temperatures different than either of the junctions.

The law of intermediate metals states that the output voltage is not affect-ed by the use of a third metal in the circuit provided that the new junctions are at the same temperature (Figure 5-14). This allows lead wires (usually copper wires) of a voltmeter to be connected to the thermocouple circuit to measure its output voltage.

The law of intermediate temperature states that for the same thermocou-ple, the output voltage V_{31} where the temperature at the junctions are T_3 and T_1, respectively, is equal to the sum of the output voltages of two separate measurements with junction temperatures T_3, T_2 and T_2, T_1 (i.e., $V_{31} = V_{32} + V_{21}$) (Figure 5-15). This allows using a standard thermocouple lookup table to calculate the junction temperature from the voltage measured.

In practice, the setup for temperature measurement using a thermocou-ple is as shown in Figure 5-16a. The empirical law of intermediate metal allows a third metal to connect the thermocouple to the voltmeter. It is im-

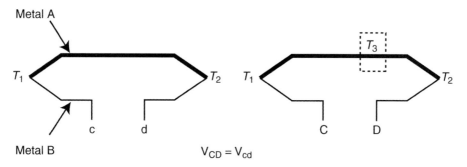

Figure 5-13. Law of Homogeneous Circuit.

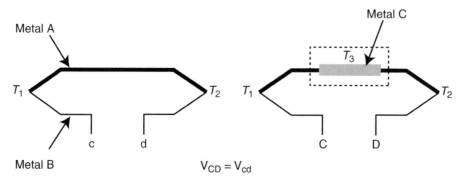

Figure 5-14. Law of Intermediate Metal.

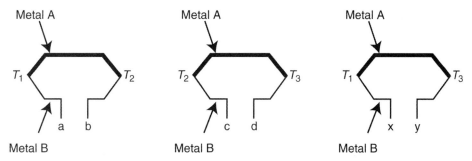

Figure 5-15. Law of Intermediate Temperature.

portant that the connections between the lead wires and the thermocouple metal wires be kept at the same temperature to avoid generating another voltage to upset the measurement. These connections are bonded to an isothermal block that has high thermal conductivity to ensure equal temperature at the connection points.

At the hot junction (temperature to be measured T_m), the junction voltage referenced to 0°C is equal to V_{m0}. At the isothermal block (at temperature T_r), the junction voltage referenced to 0°C is equal to V_{r0}.

Using the empirical law of intermediate temperature, we can write

$$V_{mr} + V_{r0} = V_{m0},$$

where $V_{mr} = V$ is the thermocouple output voltage, or

$$V = V_{m0} - V_{r0}. \tag{5.6}$$

If we know the reference temperature T_r and the output voltage, from the

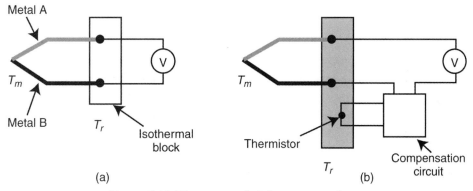

Figure 5-16. Thermocouple Measurement Setup.

thermocouple lookup table, we can use Equation 5.6 to find the hot junction temperature T_m.

Example 5.10

A J-type thermocouple is used to measure the temperature of an oven. If the temperature of the isothermal block is at 20.0°C and the output voltage measured is 9.210 mV, what is the temperature of the oven?

Solution:

STEP 1. Find the junction voltage at $T_r = 20°C$ (= V_{r0}). From the J-type thermocouple lookup table, the junction voltage V_{r0} of a J-type thermocouple at 20.0°C, referenced to 0°C is 1.019 mV.

STEP 2. Using Equation 5.6, $V_{m0} = V + V_{r0} = 9.210 + 1.019 = 10.229$ mV. Therefore, the junction voltage at the hot junction referenced to 0°C is 10.229 mV.

STEP 3. From the thermocouple lookup table, at $V_{m0} = 10.229$, $T_m = 190°C$. Therefore, the temperature of the oven is 190°C.

The setup in Figure 5-16b includes a temperature sensor (thermistor) to measure the reference temperature. The compensation circuit produces a compensation voltage according to the thermistor resistance. The output voltage then becomes the algebraic sum of the compensation voltage and the junction voltage, which can be converted to the junction temperature.

Thermocouples are available in a number of package configurations for different applications. The junction of the thermocouple can be exposed or protected, grounded or ungrounded. An exposed junction is recommended for temperature measurements in noncorrosive environments when fast response time is required. When a longer response time is acceptable, a thermocouple with a protective sheath to protect the junction against corrosion can be used. An ungrounded thermocouple probe is one in which the junction is electrically isolated from the protective sheath. An ungrounded junction with protective sheath is recommended for measurements in corrosive environments where electrical isolation is required. The sheath of an ungrounded thermocouple can be grounded to reduce interference when external electrical noise is present. The junction of a grounded thermocouple is welded to the protective sheath. Such configuration improves the response time of the thermocouple and allows it to be used in a corrosive environment. Because the output voltage magnitude of the thermocouple is usually very low, care must be taken to avoid errors due to electrical interference and voltaic effects.

INTEGRATED CIRCUIT TEMPERATURE SENSORS

The last temperature sensor to be discussed in this chapter is the IC temperature sensor. IC temperature sensors employ the temperature-dependent properties of the semiconductor P-N junction to measure temperature. IC temperature sensors are in general very accurate and linear. Due to the fact that they are constructed from semiconductor materials, however, IC temperature sensors are not suitable for extreme temperature conditions. The operating range of IC temperature sensors is from −55 to +150°C. IC temperature sensors can be current type or voltage type. In a current type sensor, the sensor will produce a current proportional to the temperature. In a voltage-type sensor, the output voltage is proportional to the temperature. An example of an IC temperature sensor is the National Semiconductor LM50, which has a temperature sensitivity of 10.0 mV/°C and ± 0.8% maximum nonlinearity and only requires a 4- to 10-V single power supply.

COMPARISON OF TEMPERATURE SENSORS' CHARACTERISTICS

Table 5-4 shows the differences and similarities of the four types of temperature sensors discussed earlier. Note that the quoted entries are approximate values. The choice of a sensor depends on matching its application and operational environment to the characteristics of the sensor.

Table 5-4. Comparison of Temperature Sensors Characteristics

	Thermistor	*Platinum RTD*	*Thermocouple*	*IC Sensor*
Sensitivity	50 to 100 Ω/°C*	0.38 Ω/°C‡	7 to 60 μV/°C	1 mV/°C
Linearity	Poor	Good	Fair	Very good
Stability	Stable	Very stable	Stable	Less stable
Power Required	Yes	Yes	Self powered	Yes
Min. Practical Span	1°C	25°C	100°C	25°C
Temp. Range, °C	−100 to +250	−200 to +750	−100 to +2000	−55 to +150
Reference Required	No	No	Cold junction	No
Ruggedness	Very rugged	Rugged	Very rugged	Rugged
Repeatability	Very good	Very good	Fair	Good
Hysteresis	Low	Low	High	Low

Note: * YSI 400 series thermistor, between 40 and 25°C; ‡ Platinum RTD with R_0 = 100 Ω

BIBLIOGRAPHY

Benedict, R. P. (1969). *Fundamentals of Temperature, Pressure and Flow Measurements.* New York, NY: John Wiley & Sons, Inc.

Coggan, D. A. (Ed.). (2005). *Fundamentals of Industrial Control: Practical Guides for Measurement and Control* (2nd ed.). Research Triangle Park, NC: ISA–The Instrumentation, Systems, and Automation Society.

Khazan, A. D. (1994). *Transducers and Their Elements: Design and Application.* Englewood Cliffs, NJ: PTR Prentice Hall.

Liptak, B. G. (Ed.). (2003). *Instrument Engineers' Handbook* (4th ed., Vol. 1: *Process Measurement and Analysis*). Boca Raton, FL: CRC Press.

Michalski, L., Eckersdorf, K., Kucharski, J., & McGhee, J. (2001). *Temperature Measurement* (2nd ed.). New York, NY: John Wiley & Sons, Inc.

Chapter 6

POSITION AND MOTION TRANSDUCERS

OBJECTIVES

- State the relationships among displacement, velocity, and acceleration.
- Derive the principles of resistive, capacitive, and inductive position and motion transducers.
- Describe application examples of resistive, capacitive, and inductive sensors in linear and angular displacement measurements.
- Explain the principles and applications of Hall effect sensors.
- Describe the principles accelerometers.

CHAPTER CONTENTS

1. Introduction
2. Resistive Displacement Transducers
3. Inductive Displacement Transducers
4. Capacitive Displacement Transducers
5. Hall Effect Sensors
6. Accelerometers

INTRODUCTION

The measurements of displacement, velocity, and acceleration are often performed in biomedical applications. Measuring motion of the heart wall during open heart surgery, quantifying the frequency and magnitude of hand tremors for a patient suffering from Parkinson's disease, and tracking the position of a surgical robotic handpiece are a few examples of the applications. The displacement x (in m), velocity v (in m/sec), and acceleration a

103

(m/sec^2) of an object are related by the following equations, where t is the time in seconds:

$$v = \frac{dx}{dt}$$

$$a = \frac{dv}{dt} = \frac{d^2x}{dt^2}$$

Therefore, if the displacement versus time of an object is recorded, its velocity and acceleration can be calculated. In a similar manner, if the angular change versus time is recorded, the angular velocity and angular acceleration can be computed. This chapter focuses on the measurement of displacement, including linear (or translational) and angular displacement.

RESISTIVE DISPLACEMENT TRANSDUCERS

A passive displacement transducer is a device such that a change in position causes a related change in a circuit parameter, such as change in resistance, inductance, or capacitance. For a resistive wire, the resistance R is given by

$$R = \frac{\rho L}{A} \tag{6.1}$$

An example of a linear displacement transducer using the preceding property is shown in Figure 6-1a, where the displacement to be measured is linked to the wiper of the potentiometer. In this example, the measured resistance is directly proportional to the displacement. A practical displacement transducer is shown in Figure 6-1b. Figure 6-2 shows an example of a resistive angular motion transducer using the same principle.

INDUCTIVE DISPLACEMENT TRANSDUCERS

By changing the physical dimension, an inductor can be used as a position transducer. The inductance *L* of a solenoid is given by

$$L = N^2\mu\frac{A}{l}, \tag{6.2}$$

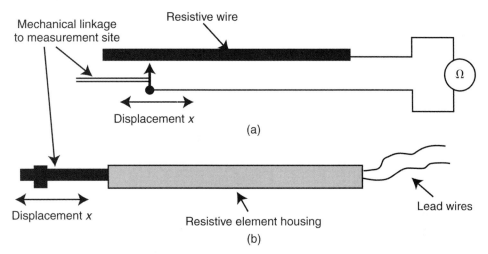

Figure 6-1. Linear Displacement Transducer.

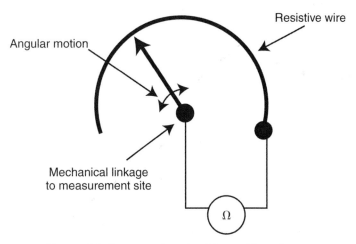

Figure 6-2. Resistive Angular Motion Transducer.

where N = the number of turns of the solenoid,
μ = the effective permeability of the material inside the solenoid core,
A = the cross-sectional area, and
l = the length of the solenoid.

A displacement transducer can be made by linking the change in displacement with one of the parameters in Equation 6.2. A simple example is shown in Figure 6-3a, where the ferromagnetic core of the solenoid is con-

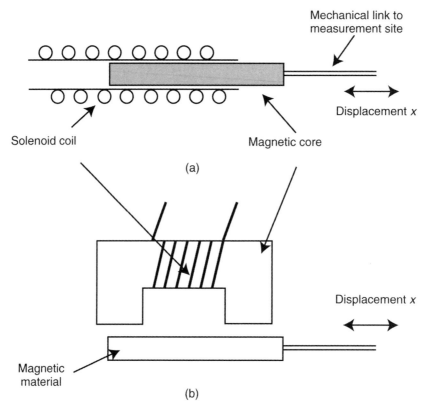

Figure 6-3. Inductance Displacement Transducer.

nected to the object of which the motion is to be measured. The effective permeability of the solenoid is changed by changing the position of the ferrous (or magnetic) core in the solenoid. The direction and magnitude of the displacement can thus be determined by measuring the inductance of the solenoid. Another example is shown in Figure 6-3b, where the displacement is measured by measuring the change in magnetic flux caused by the motion of the magnetic material.

Figure 6-4a shows a displacement transducer called the linear variable differential transformer (LVDT). It consists of a primary winding L1 and two equal but oppositely wound secondary windings L2 and L3. When L1 is connected to an AC excitation and the ferrous core is right at the middle position, the induced voltages of the two secondary windings are of equal magnitude and hence cancel each other. The output of the LVDT is therefore zero. When the core is off-centered, the induced voltage of winding L3 is different from that of L2. A nonzero output voltage Vo appears at the output. Vo is proportional to the distance of the core away from the center posi-

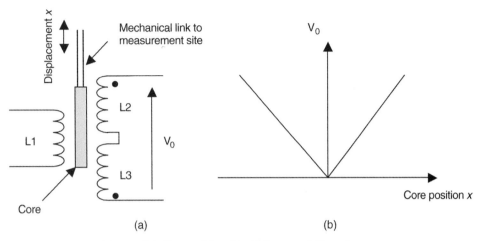

Figure 6-4. Linear Variable Differential Transformer.

tion. The input-output characteristic is as shown in Figure 6-4b. In addition to its linear characteristic, an LVDT theoretically has infinitely small resolution. Since *Vo* is a sinusoidal voltage, the output is always positive. A phase detector is therefore required to determine on which side of the center the magnetic core is located.

CAPACITIVE DISPLACEMENT TRANSDUCERS

For a parallel plate capacitor, the capacitance C is given by

$$C = \frac{\varepsilon A}{d}, \tag{6.3}$$

where ε = the permittivity of the dielectric,
A = the overlapping area of the plates, and
d = the distance between the two plates.

A capacitive displacement transducer element converts a change in the permittivity ε, the area A, or the plate separation d into a change in the value of the capacitance C. The change in capacitance will in turn be made to cause a change in an electrical signal. For example, one of the plates can be mechanically connected with the object to be measured; the setup will provide a signal according to the displacement of the object (Figure 6-5a). In Figure 6-5b, the permittivity ε is changed by the movement of the dielectric core linked to the object to be measured. Figure 6-5c shows how the dimen-

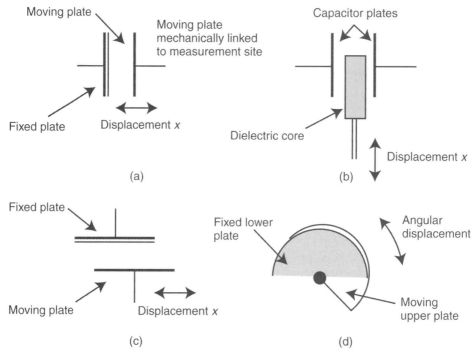

Figure 6-5. Capacitive Displacement Transducers.

sional change in the overlapping area A can be tied to measure displacement. A simple modification can be used for angular displacement measurement (Figure 6-5d).

HALL EFFECT SENSORS

Hall effect sensors are found in many medical devices. For example, they are used as a proximity sensor in the door switch of an infusion pump. Figure 6-6 illustrates the setup of a Hall effect sensor. The Hall element is a rectangular plate of metal or semiconductor material such as bismuth or tellurium. When the current-carrying plate is placed in a magnetic field perpendicular to the plate, a voltage (called the Hall voltage) is induced across the side faces of the plate in a direction perpendicular to the current in the plate. The Hall voltage E_h is given by the equation

$$E_h = \frac{KIB}{t},$$ (6.4)

Where, I = the current flowing through the Hall plate,
  B = the magnetic field perpendicular to the plate,
  t = the thickness of the plate, and
  K = the Hall coefficient, which depends on the material.

In practice, the current I is often held constant. From Equation 6.4, E_h is proportional to the magnetic field strength B perpendicular to the plate. Hall effect sensors are commonly used in magnetic field measurements with sensitivities of a few millivolts per kilo gauss (1 tesla = 10^4 gauss). A Hall effect sensor can easily be adapted to become a position or displacement sensor. Figure 6-7 shows several setups using Hall sensors as displacement transducers.

In Figure 6-7a, the Hall sensor is positioned between the two north poles of the magnets and supplied by a constant current. If the magnets are identical, the magnetic field strength is zero right at the middle and therefore the output of the sensor (Hall voltage) is zero. The magnetic field strength at both sides of the center varies with the distance from the center. By mechanically connecting the object to be measured to the sensor, the motion of the object can be monitored by measuring the Hall voltage.

In the setup as shown in Figure 6-7b, the Hall effect sensor is placed stationary between the poles of a magnet. The magnetic flux in the magnet, and therefore the magnetic field, at the Hall sensor varies with the air gap between the magnet and the ferromagnetic plate. Motion of the object can

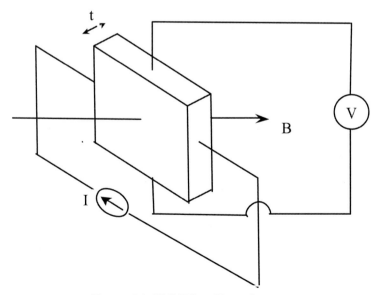

Figure 6-6. Hall Effect Trransducer.

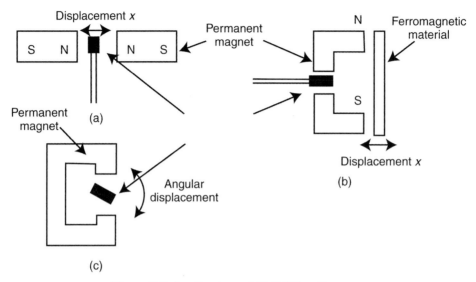

Figure 6-7. Applications of Hall Effect Devices.

thus be measured by the Hall sensor by connecting the ferromagnetic plate to the object. In Figure 6-7c, the magnetic field between the poles is constant. However, when the Hall effect sensor rotates, the magnetic field perpendicular to the surface of the sensor varies by a factor equal to sin Θ, where Θ is the angle between the magnetic axis and the surface of the sensor. In essence, a displacement sensor can be constructed by linking the object to a setup such that the Hall effect sensor element is exposed to changing magnetic field strength caused by the motion of the object.

ACCELEROMETERS

An accelerometer is a device to measure acceleration by measuring the force exerted on a test mass. Multiple elements can be configured to measure acceleration in more than one dimensions (that is, in x, y, z directions), or angular acceleration. A common type of accelerometer uses a piezoelectric force transducer attached to a small mass (m) to measure acceleration using the relationship

$$F = m \times A,$$

where, F = force exerts on the mass m,
A = acceleration of mass m.

Accelerometers can be used to measure direction and magnitude of acceleration. They are also commonly used in vibration and shock measurements. An application example is using an accelerometer to monitor the depth of compression during cardiopulmonary resuscitation (CPR). Sensors used to detect heel strikes in a pedometer are another application example.

BIBLIOGRAPHY

Bronzino, J. D. (Ed.). (2000). *The Biomedical Engineering Handbook* (2nd ed., Vol. 1: *Physical Measurements*). Boca Raton, FL: CRC Press & IEEE Press.

Coggan, D. A. (Ed.). (2005). *Fundamentals of Industrial Control: Practical Guides for Measurement and Control* (2nd ed.). Research Triangle Park, NC: ISA-The Instrumentation, Systems, and Automation Society.

Khazan, A. D. (1994). *Transducers and Their Elements: Design and Application.* Englewood Cliffs, NJ: PTR Prentice Hall.

Liptak, B. G. (Ed.). (2003). *Instrument Engineers' Handbook* (4th ed., Vol. 1: *Process Measurement and Analysis*). Boca Raton, FL: CRC Press.

Chapter 7

FLOW TRANSDUCERS

OBJECTIVES

- Explain the differences between laminar and turbulent flow.
- State Bernoulli's equation and Poiseuille's law.
- Define viscosity and Reynolds number.
- Derive the venturi tube equation and explain the principle of operation of venturi tube, orifice, pitot tube, and rotameter flowmeters.
- Explain the operation of turbine and paddle wheel flowmeters, hot wire anemometers, electromagnetic and ultrasound flow sensors.

CHAPTER CONTENTS

1. Introduction
2. Laminar and Turbulent Flow
3. Bernoulli's Equation and Poiseuille's law
4. Flow Transducers

INTRODUCTION

The center of the cardiovascular system is the heart, which generates pressure and flow to circulate blood through a network of blood vessels around the body. The blood pressure and flow velocity are different at different locations of the circulatory system. Chapter 4 covers pressure transducers that can be used to measure blood pressure; this chapter covers flow transducers to measure fluid flow velocity including blood flow. Major physiological parameters in the flow measurement category include, but are not limited to, blood flow, respiratory gas flow, and urine flow.

Flow measurement includes measuring the velocity v (m/sec), volume flow rate Q (L/sec), or mass flow rate F (kg/sec). The mass flow rate is defined as the mass m of the fluid that flows through a cross section of the vessel per unit of time t. Since mass is equal to volume V times density ρ, the mass flow rate is given by

$$F = \frac{m}{t} = \frac{\rho \times V}{t} = \frac{\rho \times A \times l}{t} = \rho \times A \times v = \rho \times Q, \tag{7.1}$$

where A = the cross-sectional area of the vessel, and
l = the distance of the fluid path.

LAMINAR AND TURBULENT FLOW

There are two types of fluid flow: laminar flow and turbulent flow. Laminar (or streamline) flow is smooth flow such that neighboring layers of the fluid slide by each other smoothly. It is characterized by the fact that each particle of the fluid follows a smooth path and the paths do not cross over one another; such a path is called a streamline. Laminar flow is usually slower flow. When the velocity of the flow increases, the flow eventually becomes turbulent. Turbulent flow is characterized by flow with eddies. Eddies absorb more energy and create large internal friction, which increases the fluid viscosity. Laminar and turbulent flows are shown in Figures 7-1a and 7-1b, respectively.

For an incompressible fluid continuously flowing through a vessel (Figure 7-2), the mass of the fluid flowing through section 1 per unit time is the same as that for section 2. Therefore,

$$\frac{m}{t} = \rho_1 \times A_1 \times v_1 = \rho_2 \times A_2 \times v_2.$$

Since $\rho_1 = \rho_2$, the equation becomes

$$A_1 \times v_1 = A_2 \times v_2, \tag{7.2}$$

which shows that the fluid velocity is higher when the cross section of the vessel is smaller.

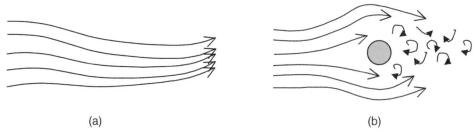

(a)   (b)

Figure 7-1. (a) Laminar Flow, (b) Turbulent Flow.

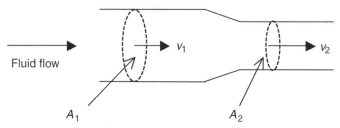

Figure 7-2. Fluid Flow Through Pipe With Varying Diameter.

BERNOULLI'S EQUATION AND POISEUILLE'S LAW

In the early eighteenth century, Daniel Bernoulli developed the Bernoulli's equation, which expresses the relationships among the velocity, pressure, and elevation of liquid flowing in a pipe with non-uniform cross section and varying height. Consider the setup shown in Figure 7-3. Assuming the flow is laminar, the fluid is incompressible, and the viscosity is negligible (i.e., there is no energy loss in the flow), equating the fluid energy at points 1 and 2 produces the following equation, which is called the Bernoulli's equation.

$$P_1 + \frac{1}{2}v_1{}^2 + \rho gh_1 = P_2 + \frac{1}{2}\rho v_2{}^2 + \rho gh_2, \qquad (7.3)$$

where ρ = the density of the fluid, and
   g = the acceleration due to gravity.

The equation shows that the quantity $P + 1/2\rho v^2 + \rho gh$ is constant at any point in a lossless fluid flow. Although any fluid flow in real life has certain energy loss, the Bernoulli's equation provides an estimate of the trade-off among the pressure, flow, and elevation of the fluid in a vessel.

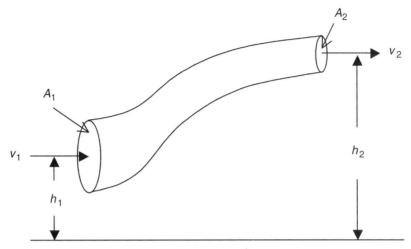

Figure 7-3. Fluid Flow in a Pipe.

Bernoulli's equation ignores the friction (called viscosity in fluid analysis) within a flowing fluid. This friction creates thermal energy, which is loss in the form of heat. This is why a pump is required (to input energy into the system) to move the fluid through a pipe.

Viscosity exists in both liquids and gases. In laminar flow, the fluid layer in immediate contact with the wall of the pipe is stationary due to the adhesive force between the molecules. The stationary layer slows down the flow of the layer in contact due to viscosity; this layer in turns slows down the next layer and so on. Therefore, the velocity of the fluid varies continuously from zero at the pipe wall to a maximum velocity at the center of the pipe, as shown in Figure 7-4. The viscosity of different fluids can be expressed quantitatively by the coefficient of viscosity η. The values of η for different fluids stated in Table 7-1 are at 20°C except for blood and blood plasma, which are measured at 37°C (body temperature).

Table 7-1. Coefficient of Viscosity for Different Fluids

Fluid	Coefficient of viscosity η ($\times 10^{-3}$ Pa.s)
Water	1.0
Ethyl Alcohol	1.2
Glycerine	1500
Air	0.018
Whole Blood	~4
Blood Plasma	~1.5

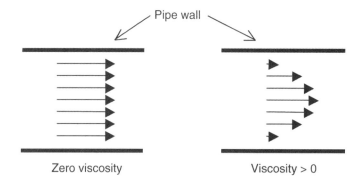

(length of arrow indicates the magnitude of flow velocity)

Figure 7-4. Velocity Flow Profile of Fluid in a Pipe.

For an incompressible fluid, the volume flow rate Q of the fluid undergoing laminar flow in a uniform cylindrical pipe is governed by the Poiseuille's law

$$Q = \frac{\pi r^4 (P_1 - P_2)}{8\eta L},\qquad(7.4)$$

where r = the radius of the pipe,
L = the length of the pipe section,
P_1 and P_2 = the pressure of the fluid at the beginning and the end of the section, respectively, and
η = the coefficient of viscosity.

Poiseuille's law states that the flow rate is directly proportional to the pressure gradient and to the fourth power of the radius but inversely proportional to the viscosity of the fluid. From Equation 7.4, one can easily note that if there is a stenosis in the blood vessel (decreased radius of the vessel), the blood pressure must increase substantially to maintain the same blood flow.

When the flow velocity is high, the flow will become turbulent and Poiseuille's law will become invalid. The onset of turbulence is often abrupt and is characterized by the Reynolds number R_e, where

$$Re = \frac{2\rho \bar{v} r}{\eta},\qquad(7.5)$$

where ρ = the density of the fluid,

$\bar{v}$ = the average velocity,

r = the radius of the pipe, and

η = the coefficient of viscosity.

Experiments have shown that when R_e exceeds about 2000, the flow becomes turbulent.

Example 7.1

The average blood velocity in a vessel of radius 3.0 mm is about 0.4 m/sec. Determine whether the blood flow is laminar or turbulent given that the density of blood is 1.05×10^3 kg/m^3 and the coefficient of viscosity of blood is 4.0×10^{-3} N.s/m^2.

Solution:

Using Equation 7.5, the Reynolds number is

$$\text{Re} = \frac{2 \times 1.05 \times 10^3 \times 0.4 \times 3.0 \times 10^{-3}}{4.0 \times 10^{-3}} = 1056,$$

which is less than 2000. The flow is therefore laminar.

FLOW TRANSDUCERS

There are many flow transducers with different characteristics. This section describes a number of common flow transducers that may be found in medical instrumentation.

Venturi Tube

Figure 7–5 shows the construction of a circular, horizontal venturi tube flowmeter. For an incompressible fluid flowing inside a pipe, assuming it is a frictionless flow, we can apply the Bernoulli's equation at points 1 and 2 of the fluid in the tube:

$$P_1 + \rho g h_1 + \frac{\rho v_1^2}{2} = P_2 + \rho g h_2 + \frac{\rho v_2^2}{2}.$$

Since the tube is horizontal, substituting $h_1 = h_2$ and rearranging the equation gives

$$P_1 - P_2 = \frac{\rho}{2}\left(v_2{}^2 - v_1{}^2\right). \tag{7.6}$$

Because the fluid is incompressible, the volume flow of fluid Q passing through section 1 and section 2 of the pipe is equal, in other words, $Q_1 = Q_2$. However, $Q = \left(\frac{\pi D^2}{4}\right)v$, where D is the diamter of the pipe and v is the fluid flow velocity. Therefore,

$$Q = \frac{\pi D_1{}^2}{4}\, v_1 = \frac{\pi D_2{}^2}{4}\, v_2$$

$$\Rightarrow v_1 = \frac{D_2{}^2}{D_1{}^2}\, v_2 \tag{7.7}$$

Substituting Equation 7.7 into Equation 7.6 yields

$$P_1 - P_2 = \frac{\rho}{2}\left[1 - \left(\frac{D_2}{D_1}\right)^4\right]v_{22} \tag{7.8}$$

$$\Rightarrow v_2 = \sqrt{\frac{P_1 - P_2}{[1 - (D_2/D_1)^4]}\times\frac{2}{\rho}}.$$

Therefore, the volume flow rate Q becomes

$$Q = \frac{\pi D_2{}^2}{4}\, v_2 = \frac{\pi D_2{}^2}{4}\sqrt{\frac{P_1 - P_2}{[1 - (D_2/D_1)^4]}\times\frac{2}{\rho}} = \frac{\pi D_2{}^2 D_1{}^2}{4}\sqrt{\frac{(P_1 - P_2)}{(D_1{}^4 - D_2{}^4)}\times\frac{2}{\rho}} \tag{7.9}$$

Equation 7.9 shows that the volume flow rate can be found by measuring the fluid pressures and pipe diameters at points 1 and 2 of the venturi tube.

Orifice Plate

A variation of the venturi tube is the orifice plate (Figure 7-6). The circular opening of the plate inside the pipe creates a reduction in the diameter of the fluid flow path downstream of the orifice opening. The differential pressure $\Delta P = P_1 - P_2$ is measured to determine the flow velocity v and the volume flow rate Q. Lookup tables of Q against ΔP are usually provided by

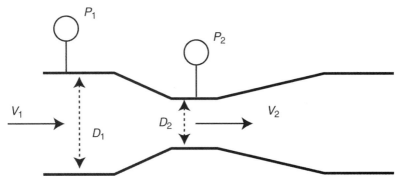

Figure 7-5. Venturi Tube.

the manufacturers.

Instead of an orifice, a wire mesh can also be used to restrict the fluid flow and create a differential pressure. An example of such a flow transducer is the pneumotachometer that is commonly used in respiratory gas flow measurements.

Pitot Tube

A pitot tube determines the velocity of the fluid by measuring the differences between the static pressure P_S in the flow and the impact pressure P_I. A pitot tube has two concentric pipes, as shown in Figure 7-7. The inner pipe has an opening at the outer wall on the side parallel to the direction of fluid flow and therefore measures the static pressure in the flow. The opening of the outer pipe is facing the flow and therefore detects the total pressure developed by the moving fluid.

By the Bernoulli's equation, the total pressure is given by $P_T = P + \rho g h$

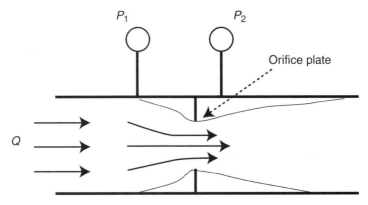

Figure 7-6. Orifice Plate Flowmeter.

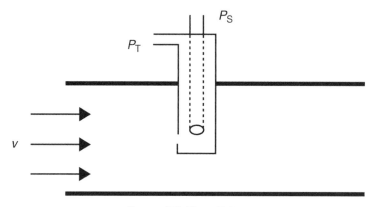

Figure 7-7. Pitot Tube.

+ $\rho v^2/2$, and the static pressure is $Ps = P + \rho gh$. The differential pressure is therefore given by

$$\Delta P = P_T - P_S = \frac{\rho v^2}{2}.$$

Compared with the orifice plate or the venturi tube, a pitot tube can easily be installed in the field by drilling a hole in the pipe and inserting the element through the hole.

Rotameter

A rotameter (or variable area flowmeter) consists of a float in a uniformly tapered tube as shown in Figure 7-8. An upward flow creates an upward force on the float. The float will go up or down until the upward force is equal to the weight of the float. The volume flow rate Q can be shown to be proportional to the area of the round gap between the float and the tube. The volume flow rate is usually calibrated and marked on the side of the tube. Rotameters are commonly used as reliable and maintenance-free indicators of gas flow, such as for oxygen or medical gases.

Turbine Flowmeters

A turbine (or vane or paddle wheel) installed in a pipe with flowing fluid will rotate at a speed proportional to the fluid flow velocity. A typical example of this class of flowmeter is a spirometer. Spirometers are used in respiratory gas flow measurements (Figure 7-9). The vane of a spirometer is made of very light material supported by bearings to reduce its resistance to the

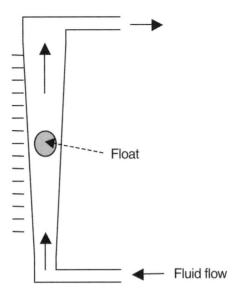

Figure 7-8. Rotameter.

gas flow. Gas flow causes the vane to rotate at a speed proportional to the flow velocity. Through a train of gears, the volume of the gas flow through the vane is indicated on the readout dial.

Electromagnetic Flowmeters

When a conductive fluid flows across a magnetic field, an electromotive force (emf) is induced between the two electrodes placed orthogonal to the magnetic field and the direction of fluid flow (Figure 7-10). The induced emf

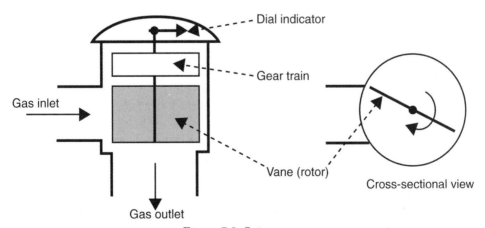

Figure 7-9. Spirometer.

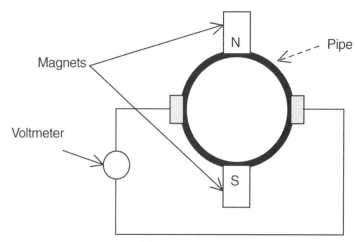

Figure 7-10. Electromagnetic Flowmeter.

is proportional to the average flow velocity of the conductive fluid in the tube. For a steady flow, factors such as fluid density, viscosity, turbulence, or Reynolds number, which normally affect other flow measurement methods, do not affect the output signal. In addition, since none of the components of the flowmeter is inside the fluid flow path, flow measurement without creating an obstruction to the flow is possible.

Ultrasound Flowmeters

An ultrasound flowmeter using the Doppler principle consists of an ultrasound transmitter and a receiver as shown in Figure 7-11a. Particles such as blood cells in the fluid reflect the sound signal from the transmitter to the receiver. The frequency shift of the ultrasound signal received by the receiver depends on the velocity of the reflecting particles traveling in the fluid.

An ultrasound flowmeter using the transit time principle consists of two ultrasound transmitter and receiver pairs placed in the fluid path and is shown in Figure 7-11b. Because the sound traveling upstream is slower than that traveling downstream, the time difference measured between the upstream and downstream transit times is used to derive the flow velocity of the fluid in the pipe. The principles and construction of ultrasound blood flow devices will be covered in Chapter 27.

Figure 7-12 shows the construction of an ultrasound vortex flowmeter. When the flowing fluid hits the wedge (vortex shedder) located at the center of the vessel, the viscosity-related effect creates vortices downstream. The vortex shedding frequency F is given by the equation $F = S \times v/d$, where v is the flow velocity, d is the shedder width, and S is a constant. At laminar

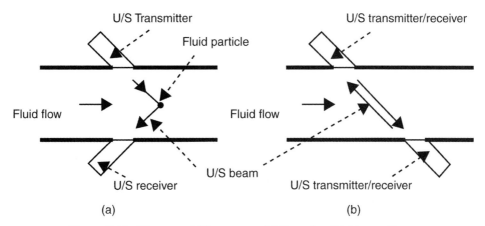

Figure 7-11. Ultrasound Flowmeters (a) Doppler, (b) Transit Time.

flow, the shedding frequency is directly proportional to the fluid velocity and is independent of the fluid density or viscosity. These vortices create turbulence in the fluid, which can be detected by an ultrasound transmitter and receiver pair setup as shown. Alternatively, piezoelectric force sensors can be mechanically connected to the vortex shedder to sense the vibration created by the vortices.

Thermal Flowmeters

A thermal flowmeter with a heating element and two temperature sen-

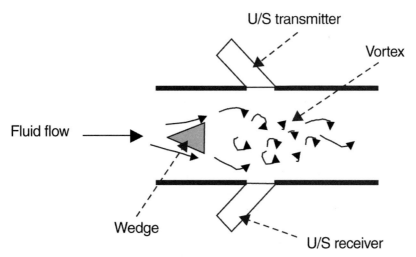

7-12. Ultrasound Vortex Flowmeter.

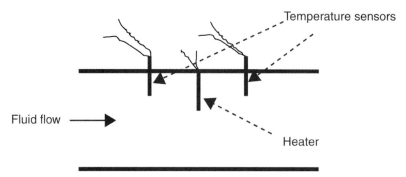

Figure 7-13. Thermal Flowmeter.

sors (e.g., thermistors) placed at equal distance to the element is shown in Figure 7-13. The heater heats the fluid in contact with the heating element. When there is no flow, the sensors are at the same temperature. The flow of the fluid causes a temperature imbalance. The flow rate, as well as the direction of flow, is determined by the temperature difference between the two temperature sensors.

Hot Wire Anemometer

Another type of thermal flowmeter is the anemometer. In a hot wire anemometer, a wire heated by an electrical current is placed in the fluid flow path. The fluid flows pass the wire cools it to a lower temperature. The rate of heat removal from the hot wire is a function of the fluid flow rate. Figure 7-14 shows a hot wire anemometer for respiratory gas flow measurement. When the gas flow rate becomes higher, it removes heat from the hot filament at a faster rate. To maintain a constant temperature on the filament, the flowmeter responds by increasing the filament heating current. The gas flow velocity is therefore a function of the filament current. To obtain the flow direction, another hot wire is placed behind a flow diverting pin at the

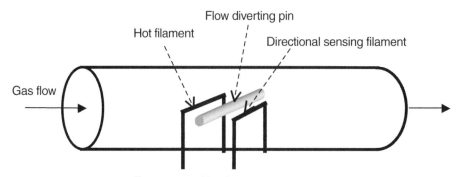

Figure 7-14. Hot Wire Anemometer.

same level as the first wire. The upstream wire will be cooled more rapidly than the downstream wire will.

A variation of the hot wire anemometer is the hot film anemometer, which replaces the heated wire with a heated metal film. The hot film provides a larger contact area between the gas and the heated element to increase the transducer's sensitivity.

BIBLIOGRAPHY

Benedict, R. P. (1969). *Fundamentals of Temperature, Pressure and Flow Measurements.* New York, NY: John Wiley & Sons, Inc.

Coggan, D. A. (Ed.). (2005). *Fundamentals of Industrial Control: Practical Guides for Measurement and Control* (2nd ed.). Research Triangle Park, NC: ISA–The Instrumentation, Systems, and Automation Society.

Khazan, A. D. (1994). *Transducers and Their Elements: Design and Application.* Englewood Cliffs, NJ: PTR Prentice Hall.

Liptak, B. G. (Ed.). (2003). *Instrument Engineers' Handbook* (4th ed., Vol. 1: *Process Measurement and Analysis*). Boca Raton, FL: CRC Press.

Miller, R. W. (1996). *Flow Measurement Engineering Handbook.* New York, NY: McGraw-Hill.

Chapter 8

OPTICAL TRANSDUCERS

OBJECTIVES

- Differentiate between photon detectors and thermal optical detectors.
- Relate radiometry and photometry; and list their parameters and units of measurement.
- Discuss black body, gray body, selective radiators, and color temperature.
- Analyze the principles and characteristics of common photo sensors.
- Explain the construction and operating principles of a charged-couple device image sensor.
- Describe the physical principles, constructions, and characteristics of fiber optic cables and sensors.

CHAPTER CONTENTS

1. Introduction
2. Quantum and Thermal Events
3. Definitions
4. Photo Sensing Elements
5. Fiber Optic Cables and Sensors
6. Application Examples

INTRODUCTION

With the rapid advancement of optical instrumentation and communication systems, optical transducers (also known as photo detectors) have been

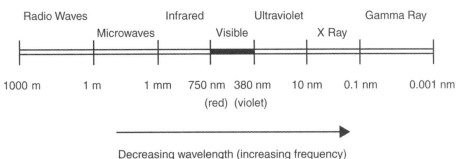

Figure 8-1. Electromagnetic Spectrum.

undergoing rapid development in the past decade. Many photo detectors have found applications in medical instrumentation. Photo detectors may be used to measure energy, flux, intensity, and so on of an electromagnetic radiation. Fiber optics are used in transmitting light, signals, and patient information including medical images. Electromagnetic radiation from low (e.g., 2.4 GHz or λ = 12.5 cm in telemetry) to high frequencies (e.g., 3 x 10^{22} Hz or λ = 10^{-14} m in gamma radiation) are used in various medical applications. Radiometry is the measurement of quantities associated with radiant energy, whereas photometry is the measurement of quantities associated with visible light. Visible light falls within a narrow band (λ = 380 to 780 nm) in the electromagnetic spectrum (Figure 8-1). This chapter covers the principles and construction of some common optical transducers used in medicine.

QUANTUM AND THERMAL EVENTS

There are two different mechanisms in radiant energy transduction: quantum (or photon) event and thermal event. In a quantum event, photons are absorbed by the sensing element, which generates an electrical output according to the amount and rate of absorption of the light quanta. Quantum event is governed by the equation

$$E = hf = hc, \tag{8.1}$$

Where h = the Planck's constant = 6.625 x 10^{-34}, in Js;
f = frequency of electromagnetic radiation, Hz;
A = wavelength of electromagnetic radiation, m;
c = speed of light = 3.00 x 10^8 msec^{-1} for all frequencies in a vacuum but lower in other media (e.g., 2.25 x 10^8 ms^{-1} in water for a 589 nm light source).

Because the electrons must gain sufficient energy in order to jump over the energy gap, a minimum photon energy E_{min} is required to initiate the event. Therefore, radiation with wavelengths longer than λ_{max} ($= hc/E_{min}$) will not produce a quantum event, λ_{max} is therefore defined as the maximum wavelength to produce a quantum event.

In a thermal event, radiant energy is first absorbed by the sensing element to produce heat energy; the heat in turn causes a change in an electrical value. Unlike quantum events, thermal events happen at all wavelengths. The sensitivity of a thermal photo sensor also depends on the wavelength of the incident radiation. The spectral response of a photo sensor is often provided in the device specifications by the manufacturer. In addition, filters can be used to modify the spectral response of a sensor.

DEFINITIONS

Thermal and Photon Detectors

Photo detectors can be divided into two classes: thermal detectors and photon detectors. The sensing element of a thermal detector first converts the electromagnetic radiation to heat and then to an electrical value. This change in electrical value is often proportional to the received energy. That is, they are using the thermal event of transduction. Photon detection is achieved by quantum events that stimulate the sensing element to produce an electrical signal proportional to the rate of absorption of light photons.

Radiometry and Photometry

Radiometry is the measurement of quantities associated with radiant energy. The units of measurement and definitions of some radiometric parameters are listed here:

- Radiant Energy—energy traveling in the form of electromagnetic waves.
- Radiant Density—radiant energy per unit volume.
- Radiant Flux—the time rate of flow of any parts of the radiant energy spectrum.
- Radiant Flux Density at a surface—the quotient of radiant flux to the area at that surface. Radiant emittance (M) refers to radiant flux density leaving a surface; irradiance (E) refers to radiant flux density incident on a surface.
- Radiant Intensity—the radiant flux from the source per unit solid angle.
- Radiance—the radiance in a direction at a point on a surface is the quotient of the radiant intensity leaving, passing through, or arriving at the

surface to the area of the orthogonal projection of the surface on a plane perpendicular to the given direction.

Photometry is the measurement of quantities within the visible spectrum. For the human eyes, this range is $\lambda = 380$ to 780 nm. Rods and cones are the two types of optical receptors in human eyes. Rods, which are more sensitive to light than are cones, are responsible for low light level vision. However, rods cannot discriminate between colors. Cones are responsible for color visions but are less sensitive to light. There are three types of cones, each of which is sensitive to a different spectral band. The human brain combines all signals from these light sensors to create visual perception. In photometry or lighting engineering applications when only the visible part of the electromagnetic spectrum is of interest, the radiometric quantities are weighted by a visual sensitivity (or luminosity) function corresponding to the color sensitivity of the human eye. The term "radiant" is changed to "luminous" for the measurement quantities. For example:

- Luminous Energy—radiant energy of the electromagnetic radiation in the visible spectrum (from 380 to 780 nm).
- Luminous Flux—the time rate of flow of the luminous part (380 to 780 nm) of the radiant energy spectrum.

The quantities, symbols, and SI units of these parameters are tabulated in Table 8-1.

Black Body, Gray Body, and Selective Radiators

The radiant energy from a practical source is often described by comparing it with a black body radiator. A black body is defined as an entity that absorbs all incident radiation, with no transmission or reflection. A black body also radiates the most power at any given wavelength than any other source operating at the same temperature. The ratio of the output power of a radiator at wavelength λ to that of a black body at the same temperature and the same wavelength is known as the spectral emissivity $\varepsilon(\lambda)$ of the radiator. When the spectral emissivity of a radiator is constant for all wavelengths, it is called a gray body. No known radiator has a constant spectral emissivity over the entire electromagnetic spectrum. However, some materials exhibit near gray body characteristics within a certain range of wavelengths (e.g., a carbon filament in the visible region). Non-gray body radiators are called selective radiators. The emissivity of a selective radiator is different at different wavelengths.

Color temperature is a way to describe the light appearance of a light

Table 8-1. Standard Units of Radiometry and Photometry.

Radiometric Parameter	SI unit	Photometric Parameter	SI Unit	Equation
Radiant Energy (Q)	J	Luminous Energy (Q)	lm.s	
Radiant Density (Ω)	J/m^3	Luminous Density (Ω)	lm.s/m^3	$\Omega = \dfrac{dQ}{dV}$
Radiant Flux (Φ)	W	Luminous Flux (Φ)	lm	$\phi = \dfrac{dQ}{dt}$
Radiant Flux Density at a Surface	W/m^2	Luminous Flux Density at a Surface	lm/m^2	$M = \dfrac{d\phi}{dA}$
• Radiant Emittance (M)		• Luminous Emittance (M)		$E = \dfrac{d\phi}{dA}$
• Irradiance (E)		• Illuminance (E)		
Radiant Intensity (I)	W/sr	Luminous Intensity (I)	cd	$I = \dfrac{d\phi}{d\omega}$
Radiance (L)	W/(sr.m^2)	Luminance (L)	nit	$L = \dfrac{d^2\phi}{d\omega(dA\cos\theta)}$

source. It is the temperature of a black body radiator at which the color of the light source and the color of the black body appear to be the same. In practice, the color temperature of a tungsten filament lamp is pretty much the same as the color emitting from a black body at the same temperature of the filament.

The radiant curve of a selective radiator will never be identical to that of a black body. The color temperature of a selective radiator is the temperature of a black body such that their profiles of color distribution have the closest match. (Figure 8-2). Note that color matching does not imply equal radiant output. The color temperature of a light source using this approximate matching method is referred to as the "correlated color temperature" of the light source. Although the match is never perfect, color temperature values are used to represent spectral distribution of light sources. For example, midday sunlight has a color temperature of about 6,500 K, and the output of a tungsten filament incandescent lamp has a color temperature of 3,000 K. Table 8-2 shows the visible color corresponding to the absolute temperature (K) of a black body radiator.

PHOTOSENSING ELEMENTS

Some examples of photosensors are discussed in this section. Photoresistive

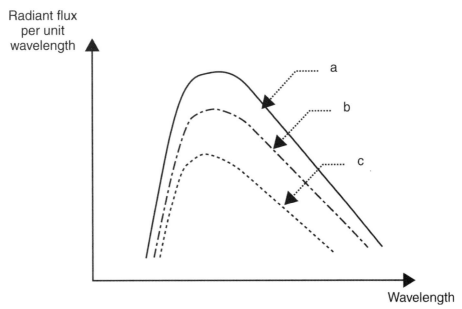

Figure 8-2. Radiant Curves for (a) Black Body, (b) Gray Body, and (c) Selective Radiator at the Same Color Temperature.

Table 8-2. Color Temperature of a Black Body Radiator

Temperature (K)	Color
800–900	Red
3000	Yellow
5000	White
8000–10000	Blue
60 000–100000	Brilliant sky blue

sensors, thermopiles, and pyroelectric sensors are thermal event detectors; photoconductors, photoemissive sensors, photodiodes, phototransistors, and charge-coupled devices (CCDs) are quantum event detectors.

Photoresistive Sensors

For a semiconductor material, the electron and hole mobility, and hence its resistivity, vary with temperature. This property of temperature dependent resistivity of semiconductor material can be used to detect thermal energy from a radiant source. The resistivity increases with temperature for lightly doped silicon but decreases at high doping level.

Thermopiles

A thermopile is made up of a number of thermocouples connected in series (Figure 8-3). Radiant energy is first converted to heat, creating a differential temperature between the hot and the cold junctions of the thermocouples. Each thermocouple will generate a small voltage (in the order of μVs) according to this temperature difference. The output of the thermopile is the summation of all the voltages from the thermocouples in the cell. Therefore, the sensitivity of the thermopile is proportional to the number of thermocouple elements in series. Miniature thermopiles, such as Si/Al thermopile, are fabricated from microelectronic technology.

Pyroelectric Sensors

In a pyroelectric sensor, a conductive material is deposited on the opposite surfaces of a slice of ferroelectric material, such as triglycine sulfate (TGS). The conductor on the sensing side is transparent to the light source to be measured. The ferroelectric material absorbs radiation and converts it to heat. The resulting rise in temperature changes the polarization of the material and creates charges. The current I flowing through the external resistor connected across the two conductive surfaces is proportional to the rate of change of temperature of the sensor. Pyroelectric sensors with cooling can detect radiant power down to 10^{-8} W with λ_{max} of up to 100 μm.

Photoconductive Sensors

In a semiconductor, electrons can be raised from the valence band to the conduction band by absorbing energy from light photons. The presence of

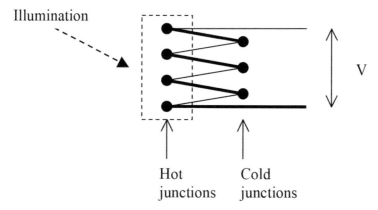

Figure 8-3. Thermopile Composed of Four Thermocouples.

these photon-induced electrons increases the conductivity of the semiconductor material. In order to raise the electron from the valence band to the conduction band, it must absorb enough energy from the photon to overcome the band gap energy E_G. Therefore, to create the photoconductive effect, the energy of the light photon should be greater than E_G and hence the wavelength of the radiation must be smaller than λ_{max} given by Equation 8.2:

$$\lambda_{max} \leq \frac{hc}{E_G}$$

(8.2)

For high-precision applications, the sensing element often must be cooled to reduce noise from thermionic electrons. These are free electrons generated in a material when its temperature is above absolute zero. Thermionic electrons decrease the signal-to-noise ratio (SNR) of the sensor.

Photoemissive Sensors

The phenomenon that electrons are liberated to the free space from the surface of a material when excited by light photons is called photoelectric effect. For high-efficiency conversion, the potential barrier or work function E_0 of the material must be much smaller than the photon energy. The efficiency or quantum yield of the sensor is defined as the ratio of the number of emitted electrons to the number of absorbed photons. Materials with low E_0, such as NaKCsSb or cesium oxide on GaAs substrate, have high quantum yields and therefore are good materials for photocathodes (photoemissive elements). In a simple photoelectric tube (Figure 8-4), under light illumination, electrons are liberated from the photocathode and conducted through the external circuit. If the wavelength is shorter than the work function, the photoelectric current produced in a vacuum photoelectric tube changes linearly with the level of illumination. In a gas-filled tube (e.g., a glass tube filled with low-pressure argon), collisions of the electrons with gas atoms may produce secondary electron emissions, resulting in higher sensitivity photo detection.

Photodiodes

A diode is constructed of an n-type semiconductor (e.g., phosphorus doped silicon) in contact with a p-type semiconductor (e.g., boron doped silicon). Under normal conditions, electrons readily break away from the impurity in the n-type material to become free electrons. In the p-type semiconductor, mobile holes are created instead. Despite the presence of mobile electrons in the n-type and mobile holes in the p-type semiconductor, the

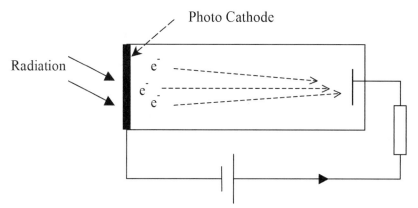

Figure 8-4. Photoelectric Tube.

individual semiconductors are both neutral in charge (i.e., the number of electrons is the same as the number of protons). When the two types of semiconductors are put together, however, the electrons from the n-type semiconductor will migrate across to fill the holes in the p-type semiconductor, forming a P-N junction (Figure 8-5). This migration of electrons also creates a net positive charge in the n-type semiconductor and a negative charge in the p-type. These charge layers create a net electric field and will eventually prevent further electrons from traveling across the junction. The electric field at the junction acts as a diode, creating a barrier for electrons to move from the n-type to p-type but not the other way around.

When light is incident upon the P-N junction of a semiconductor, a photon with sufficient energy will free a number of electrons from the atom. The number of free electrons (or accumulation of charges) can be found by measuring the voltage across the photodiode (photovoltaic mode) or by measuring the reverse bias current of the photodiode (photoconductive mode). The voltage-current characteristic of a typical photodiode and its equivalent circuit are shown in Figure 8-6, where v is the voltage across the diode and i is

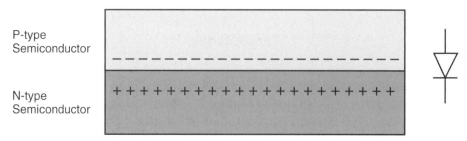

Figure 8-5. Semiconductor P-N Junction.

the diode current. The current source represents the current produced when the junction is exposed to radiation. Notice that when the output is short-circuited (i.e., $v = 0$), the short-circuit current is negative and its magnitude depends on the wavelength and the intensity of the incident illumination.

In the photoconductive mode of operation, a reverse bias voltage is applied to the photodiode as shown in Figure 8-7 (i.e., v is negative). The magnitude of this reverse bias current i will increase with the intensity I of the illumination. Although the reverse bias current is almost the same for a constant illumination, a higher reverse bias voltage will produce a faster response time. However, dark current (noise) increases with the magnitude of the applied bias voltage. Diodes in photoconductive modes operate in

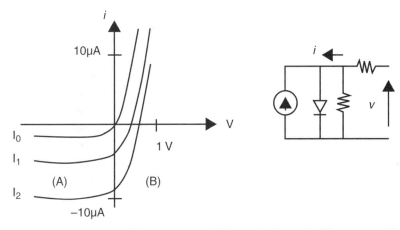

Figure 8-6. Current-Voltage Characteristic of Photodiode with Illumination Intensity $I_0 < I_1 < I_2$.

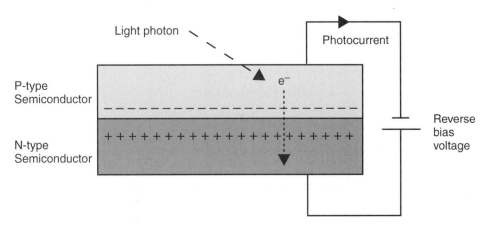

Figure 8-7. Photovoltaic Cell.

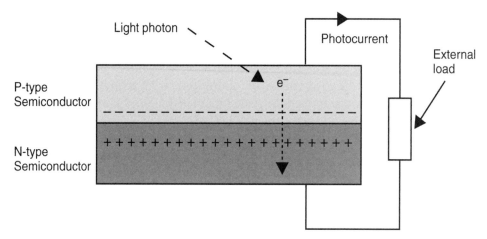

Figure 8-8. Photovoltaic Cell (Solar Cell).

quadrant A of the photodiode characteristic curve (Figure 8-6).

In a photovoltaic mode of operation, no bias voltage is required. The free electrons created in the p-type semiconductor, being moved by the positive electric field, will cross the junction into the n-type semiconductor. These electrons will accumulate at the junction and create a potential difference (photovoltaic) between the p-type and n-type semiconductors. A photocurrent will flow if an external path is established between the semiconductors (Figure 8-8). The magnitude of the photocurrent increases with the light intensity. Diodes in photovoltaic modes operate in quadrant B of the photodiode characteristic curve (Figure 8-6). Since dark current is a function of the bias magnitude, photovoltaic mode is ideal for low signal level detection. However, it is less suitable for high-frequency application because of its slow respond time. A diode operating in photoconductive mode is often referred to as a solar cell. Other than silicon, semiconductor materials such as gallium arsenide, and copper indium diselenide are common materials for photodiodes.

Phototransistors

A phototransistor can provide amplification of the photocurrent within the sensing element. The basic construction of the phototransistor is similar to that of a bipolar transistor except that the base normally has no connection and is exposed to the illumination being measured. For the phototransistor connection shown in Figure 8-9, the emitter current is given by

$$i_E = (i_\lambda + i_R)(hfe + 1),$$

where i_λ = light-induced base current,
  i_R = reverse leakage current, and
  h_{fe} = forward current transfer ratio.

If $i_\lambda >> i_R$ *and* $h_{fe} >> 1$, the above equation can be simplified to

$$iE \approx i_\lambda h_{fe} \tag{8.3}$$

Equation 8.3 shows a linear high gain characteristic that converts luminous intensity to a relatively large emitter current.

Charge-Coupled Device

Charge-coupled devices (CCDs) are often used in digital image acquisitions. The sensing elements of a CCD are metal oxide semiconductor (MOS) capacitors. Figure 8-10 illustrates the cross section of a buried channel MOS capacitor. A typical buried channel MOS capacitor has a n-type semiconductor layer (about 1 mm) above a p-type semiconductor substrate with an insulation formed by growing a thin oxide layer (about 0.1 mm) on top of the n-type layer. A conductive layer (metal or heavily doped semiconductor) is then deposited on top of the oxide to serve as the metal gate.

When light photons are allowed to incident on the P-N junction, electron-hole pairs are formed. Similar to a photodiode, the electrons will migrate to the n-type side of the junction and will be trapped in the "buried channel" (Figure 8-11). To create separation between adjacent pixels, a p-type stop region on each side of the metal gate is formed to confine the charges under the gate. The amount of trapped charge is proportional to the number of incident light photons.

A CCD consists of an array of these individual elements (pixels) built on

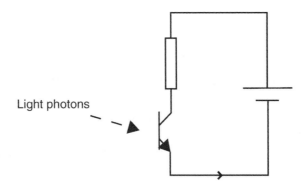

Figure 8-9. Phototransistor Circuit.

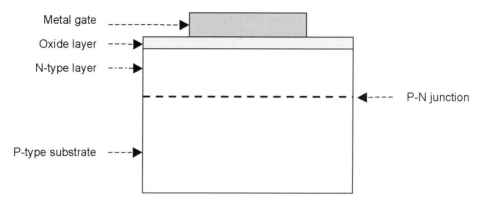

Figure 8-10. Buried Channel MOS Capacitor.

a single substrate. A 256 by 256 ($2^8 \times 2^8$) array CCD contains 2^{16} number of MOS capacitor elements. To understand the operation of CCDs, we can represent a MOS capacitor element by a bucket and the amount of trapped charges by the level of water in the bucket. When light of nonuniform intensity incidents on the CCD array for a predetermined duration of time (usually controlled by a timer-controlled shutter), the amount of charges created and trapped in each MOS capacitor is proportional to the number of light photons received by the capacitor. Figure 8-12 shows a 1 × 3 CCD linear array at the top and its water bucket analogy at the bottom. A larger amount of light photons creates a brighter CCD pixel, which is represented by a higher water level in the bucket. The process of reading the amount of charge in each pixel involves moving the charges from the site of collection

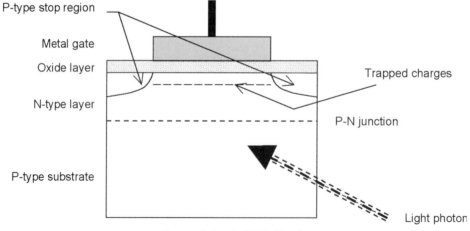

Figure 8-11. A CCD Pixel.

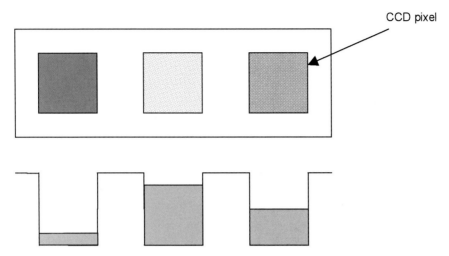

Figure 8-12. A 1 x 3 CCD Array and Its Water Bucket Analogy.

to a charge-detecting circuit located at one end of the linear array. Reading out the charges in a CCD array is a sequential process. This process is illustrated in Figure 8-13.

In general, each pixel has three gates; each gate is connected to a time controlled bias voltage V_1, V_2, and V_3 as shown (upper diagram of Fig. 8-13). Consider the charges stored in Pixel number 2. At time T_1, V_3 becomes positive. As a result, some charges below the V_2 electrode migrate to the region under V_3. At time T_2, the V_2 electrode is turned off; the rest of the charges under the V_2 electrode have now all migrated to the region under V_3. Similar charge shifts happen at time T_3, T_4, T_5, and T_6. After time T_6 (one shift cycle), the charges under Pixel 2 has been shifted to Pixel 3. Note that all charge packets in the array move (or shift) simultaneously one pixel to the right. For an N-pixel linear array, N shift cycles are required to read out the entire array.

Figure 8-14 shows the physical layout of a 3 × 2 (6 pixels) CCD array. The pixels are represented by the rectangular boxes. In this particular configuration, reading out the charges is achieved by shifting the charged in the vertical direction.

CCDs used in imaging are usually in square or rectangular arrays. An N x M array (Figure 8-15) can be considered to be made up of M linear arrays, each with N pixel elements. Reading out the array requires simultaneously shifting all rows of charge packets one pixel downward toward the serial register. The charge packets are then transferred from the serial register to the output amplifier one row at a time.

The unwanted charge accumulated in CCD pixels due to thermally gen-

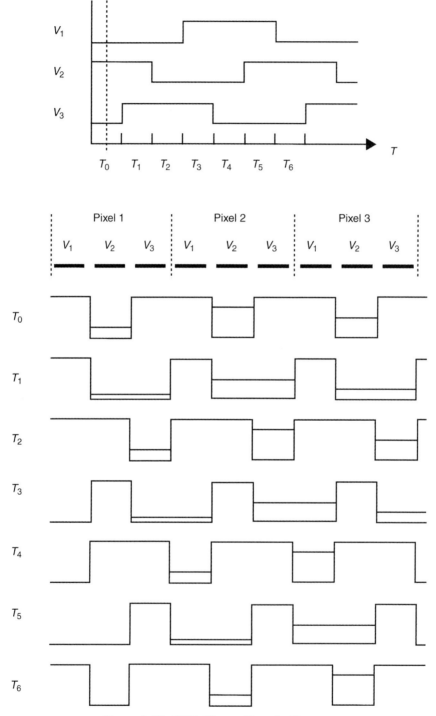

Figure 8-13. CCD Charge Transfer Process.

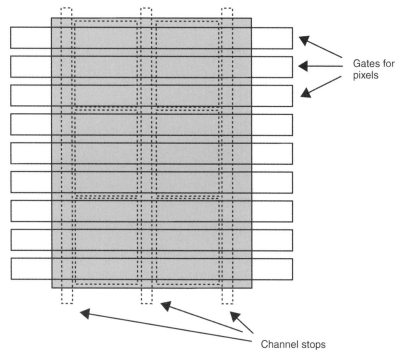

Figure 8-14. A 3 x 2 CCD Array.

erated electrons instead of light photons is called the dark current. Dark current contributes to noisy picture, especially when the input light level is low (i.e., low SNR). Since this noise is due to thermal effects, dark current can be reduced by cooling the CCD to a lower temperature. Quantum efficiency (Q_E) measures the efficiency of incident photon detection. It is defined as the ratio of the number of detected electrons N divided by the number of incident photons M times the expected number of electrons R generated from each photon.

$$Q_E = \frac{N}{M \times R} \times 100\%.$$

FIBER OPTIC CABLES AND SENSORS

Physics of Fiber Optics

In a homogenous medium, light travels in a straight line. When light hits the boundary of two media with different refractive indices, a portion of light

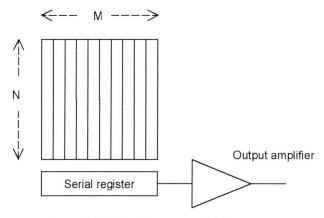

Figure 8-15. Two-Dimensional CCD Array.

will be reflected at the boundary and a portion will be transmitted. The transmitted light beam, according to Snell's Law, changes direction at the boundary of the two media. This phenomenon is called refraction. Figure 8-16 shows light travels from glass into air. As the incident angle θ increases, the transmission angle α becomes larger and approaches 90°. The incident angle θ is called the critical angle when α becomes 90°; light with incident angle larger than the critical angle cannot exit the glass-air interface and will be total internally reflected.

An optical fiber is used to transmit light from a source to a receiver. Fiber optic cables can be found in data communication, in signal transfer, in energy transport, and in image transmission. Some examples of optical fiber applications in medicine are laser surgeries and video endoscopic proce-

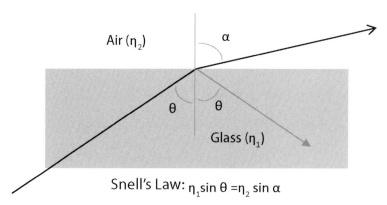

Snell's Law: $\eta_1 \sin \theta = \eta_2 \sin \alpha$

Figure 8-16. Snell's Law and Light Refraction.

dures. A fiber optic cable can be single core or bundled. The simplest fiber optic cable consists of an inner cylindrical core with an outer cladding. Light travels inside the fiber due to total internal reflection as the refractive index (η) of the core material is higher than its cladding. Fluorinated polymers (η = 1.3 to 1.4) is an example of the materials used for claddings. Glass (η = 1.45 to 1.6) is a common core material due to its low transmission loss. Although with higher loss than glass fibers, plastic optical fibers (e.g., acrylic fiber with core η = 1.50 to 1.53), are commonly used due to its lower cost and higher flexibility. When a light beam inside the fiber hits the core-cladding interface with an incident angle greater than the critical angle, the beam becomes totally reflected and stays inside the core of the fiber. In specifying fiber optic cables, in addition to stipulating the index of refraction of its core and cladding materials, the term numeric aperture (NA) is often provided in the specifications of optical fibers. NA is a measure of the light gathering ability of a fiber. It also indicates how easy it is to couple light into a fiber. Equation 8.5 is used to calculate the NA of a step-index fiber from the index of refraction (η) of its core and cladding. Acceptance angle (Θa) of the optical fiber is the half cone angle of light that can be sent into an optical fiber so that the entire light beam will be reflected internally. Acceptance angle is determined from the numeric aperture by equation 8.5. When the incident half cone angle of light into an optical fiber is larger than the acceptance angle, energy will be lost.

$$NA = \sqrt{(\eta core^2 - \eta cladding^2)} \tag{8.4}$$

$$\Theta a = \sin^{-1} NA \tag{8.5}$$

Light is attenuated as it travels inside an optical fiber due to scattering and absorption. The degree of attenuation depends on the fiber optic cable construction, material types, impurities and manufacturing tolerances. Fiber optic cable loss (α), which provides a measure of the light power reduction as the light travels along the fiber, is the logarithm of the ratio of output power to the input power (Equation 8.6), when the input power is 1 mW, α is expressed in dBm. For example, at 25 °C, light from a 660 nm LED light source traveling in a plastic optical fiber cable used in short distance signal transmission has an attenuation α = 0.22 dB/m. Each additional connection in the cable will add about 0.6 dB of loss to the system.

$$\alpha = log \frac{Po}{Pi} \tag{8.6}$$

The velocity v of light traveling inside the core of an optical fiber depends on the refractive index η of the core material. The relationship is given by the equation below where c is the velocity of light in vacuum. In addition, there is a slightly different refractive index for different wavelengths of light traveling in the core of an optical fiber.

$$v = \frac{c}{\eta} \tag{8.7}$$

Propagation delay (or latency) measures the time it takes for light to travel through a distance in the core of an optical fiber. In optical fiber specifications, propagation delay is often expressed in μs/km (e.g., 5 μs/km in a quartz single-mode fiber).

Modes and Types of Optical Fibers

Light traveling in an optical fiber can be single-mode or multi-mode. In a multi-mode fiber, the light photons travel in the core through multiple paths as they are reflected at different angles at the core-cladding interface, creating multi-modes or multiple paths of lights (Figure 8-17a). As light travels at different path for each mode in the fiber, higher modes travel a longer path than lower modes. In a long fiber, this path length variation has a lengthening effect on the light pulse and is called modal dispersion. In a single-mode fiber (Figure 8-17c), the entire beam of light travels in a single path contains entirely within the small central core. Single-mode fiber transmits only one mode of light, there is no spreading of the signal. Single-mode fibers have a small core (5 to 10 μm in diameter) as its core dimension needs to be close to the wavelength of its light carrier. Single-mode fibers can carry higher frequency signal over a longer distance. Multi-mode fibers have a relatively larger core (usually > 60 μm) and are used for shorter distance transmission.

There are three types of optical fibers, step-index, graded-index, and single-mode. Figure 8-17a shows a step-index fiber with the core covered by a layer of cladding. Step-index fibers produce multi-mode transmission. In a step-index fiber, higher-order modes travel longer distance than lower-modes. To compensate for the lengthening effect, a graded-index fiber (Figure 8-17b) can be considered as constructed of multiple layers of core materials with the lowest refractive index at outermost layer of the core. The lower refractive index core speed up the light photons, compensating for the longer distance traveled. Graded-index fiber produce multi-mode transmission but with less modal dispersion than step-index fiber. The construction of a single-mode fiber (Figure 8-17c) only allows a single-mode to pass efficiently. The diameter of the core of a single-mode fiber must be very close to the wavelength of the light to be transmitted. For example, a 1300 nm sin-

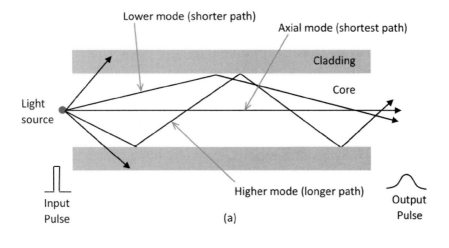

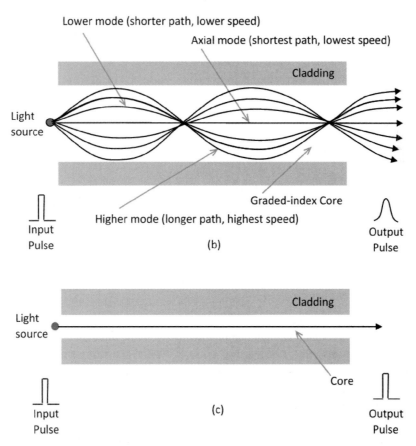

Figure 8-17. (a) Multi-mode Fiber, (b) Graded-index Fiber, (c) Single-mode Fiber.

gle-mode fiber will not be in single-mode when it is used with a 660 nm light source. The effect of modal dispersion of a light impulse caused by each type of optical fiber is illustrated by the output pulses in Figure 8-17.

Dispersion of Light in Optical Fiber

Chromatic dispersion is the term given to the phenomenon by which different spectral components of a light pulse travel at different velocities. Material dispersion is the main constituent in chromatic dispersion. It is due the wavelength dependence of the refractive index of common fiber core materials. As the propagation speed of light is inversely proportional to the wavelength, different wavelengths (colors) of light travel at different speeds in the optical fiber cable, resulting in spreading out of the signal according to the light spectrum. An effective way to decrease chromatic dispersion is to narrow the spectral width of the source transmitter, for example, by using a monochromatic laser to replace a wide spectral light source.

A light pulse entering the fiber will be broadened by modal and chromatic dispersion. The degree of dispersion limits the bandwidth of data transmission. A monochromatic laser used with a single-mode fiber will eliminate dispersion (both chromatic and modal). Such a combination supports high bandwidth transmission over a long distance. Note that in multimode fiber applications, the effect of chromatic dispersion is negligible when it is compared to that of modal dispersion.

Fiber Optic Cables and Connectors

A fiber optic cable may be single-core or multi-core. A multi-core fiber optic cable may contain a number of optical fibers encased by a common cable jacket each transmitting an independent signal. In some applications, multiple fibers are bundled together to serve a common function. A fiber bundle can be coherent (image transmission) or non-coherent (light source); bonded (rigid) or unbonded (flexible). A coherent bundle has fibers packed together in a fixed arrangement at both ends. A coherent fiber bundle can be used to transmit an image. A non-coherent bundle has fibers packed in a random manner which scrambles the image. In a rigid fiber bundle, all the fibers are fused together along the entire length. Rigid fiber bundles cannot be bent. A coherent flexible fiber bundle is made by bonding the fibers together at both ends while maintaining the alignments to accurately transmit the image. The middle region of the individual fibers of the bundle is free to move to allow flexibility. These fibers are usually encased inside a flexible sheath for mechanical and environmental protection. The image quality of the optical fiber bundle is determined by the number of fibers per unit area. An optical fiber is surrounded by cladding, therefore a fiber bundle is not 100% core. When an image or light source falls onto one end of the fiber, some light falls on the cladding and leads to transmission lost. Packing fraction is defined as the ratio of the total core area divided by the total surface area of the optical fiber bundle. A bundle with higher packing fraction has higher coupling efficiency than one with a lower packing fraction. In medicine, coherent optical fiber bundles are commonly used in endoscopy for transmission of images and illumination. For example, in a laparoscopic procedure, a flexible fiber cable (flexible non-coherent bundle) is used to transmit light from an external light source to illuminate the surgical field inside the abdominal cavity; a rigid endoscope (rigid coherent bundle) is used to transmit the image of the surgical field to be displayed real-time on a monitor to be viewed by the surgeon during the procedure.

Improper connection of two optical fibers will introduce significant loss in light transmission. Losses are caused by problems at the fiber connection such as numeric aperture mismatch, core diameter mismatch, end separation, concentricity misalignment, angular misalignment, and poorly polished surface. Two fibers can be connected by a connector or a splice. A good fiber optic connector is designed to be easily connected and disconnected repeatedly with low loss and without affecting its connection integrity. A splices is a permanent connection between two fibers. A splice is made by first cleaving the ends of both fibers and then either glueing or fusing the fiber ends together. Special splicing tools are required to assure low loss and physical durability. Splices in general have lower losses than connectors. For example, a fusion splice may introduce a loss of less than 0.1 dB to the system, whereas a connector may introduce 0.3 dB loss. Splitters or couplers are used to connect 3 or more optical fibers together.

Fiber Optic Sensors

In addition to be used for illumination and to transmit images, fiber optics can be configured as sensors. Optical fiber sensors can be separated into the following categories – interruption probes, intensity sensors, interferometric sensors, and polarization sensors. They are described below:

Interruption Probes – Interruption probes detects the presence or absence of light. A vey simple application is shown in Figure 8-18a where one fiber is a light source and the other connects to a light detector, the probe detects the object when it is interrupting the transmission path of light. Figure 8-18b shows a liquid level detector; the narrow light beam from one fiber is reflected from the liquid surface, the detecting fiber is aligned such that it will receive the light beam when the liquid surface has risen to a specified level.

Intensity Sensors – Environmental factors can affect the characteristics of optical fibers. Fiber optic intensity sensors utilize the environmental dependency of these characteristics. Although the effects may be small, the variation can be amplified by increasing the length of the fiber. An example is a temperature sensor fabricated from a plastic clad glass fiber. Due to change in refractive index of the cladding with temperature, the numeric aperture of the fiber decreases as temperature increases. By using a stable light source coupled to the optical fiber, the light intensity measured by a photodetector at the other end of the fiber will be a function of temperature.

Interferometric Sensor – An interferometric sensor uses a single-mode fiber to detect very small phase shift of light created by external factors such as temperature, pressure, etc. Figure 8-19a illustrates the principle of the interferometric sensor. The middle waveform Y is an inverted image (180° phase shifted) of the original waveform X at the top. When X & Y are combined, the output is zero. However, if X is slightly phase shifted (becomes X'), the output will not yield zero. In an interferometric sensor, the phase of the sensing arm is compared with the phase of a reference arm. In an optical fiber interferometric sensor, one fiber is the sensor and the other is the reference. The sensing fiber changes the phase of its optical signal by increasing or decreasing its optical path length. The change in path length can be a result of changes in refractive index in the core of the fiber, which is temperature dependent. Figure 8-19b shows a typical fiber optic interferometric

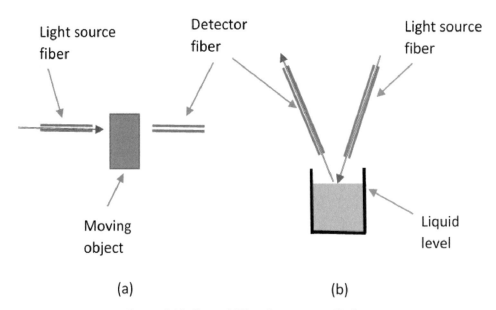

Figure 8-18. Optical Fiber Interruption Probes.

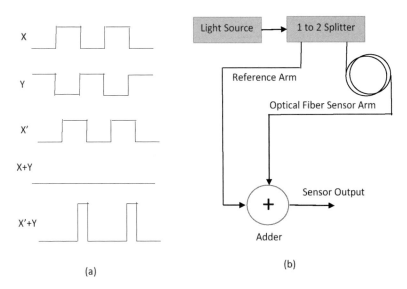

Figure 8-19. Fiber Optic Interferometric Sensor.

sensor.

Polarization Sensors – Special single-mode optical fiber can be fabricated to retain the polarization of light inside the fiber. External parameters can affect the polarization inside the fiber. Polarization sensors have been developed using polarization effect as a sensing mechanism. An example is a fiber optic polarization electric current sensor which measures the current flow by measuring the polarization of light inside the fiber. As magnetic field will rotate the polarization of light and electric current will generate magnetic field, the magnitude of the electric current can be measured by monitoring the rotation of polarization vectors of light at the exit of an optical fiber placed in proximity to the current carrying conductor.

APPLICATION EXAMPLES

There are a wide varieties of optical sensors used in medicine. A few examples follow:

- LED-photodiode pair in infusion pump fluid drop sensors
- Red and infrared LEDs and photo detector in pulse oximeters
- Thermopiles in tympanic thermometers
- Photo detectors in capnometers
- CCD cameras in medical video systems

- Optical fiber bundle light source in endoscopy
- Ambient light sensors in monitor screen dimmers
- Flat panel detectors in digital x-ray systems

BIBLIOGRAPHY

Boyle, W. S., & Smith, G. E. (1970). *Charge coupled semiconductor devices.* Bell System Technical Journal, 49(4), 587–593.

DiLaura, D., Houser, K., Mistrick, R., & Steffy, G. (Eds.). (2011). *The Lighting Handbook: Reference and Application* (10th ed.). New York, NY: Illuminating Engineering Society of North America.

Haus, J. (2010). *Optical Sensors: Basics and Applications.* Weinheim, Germany: WileyVCH.

Hecht, J. *Understanding Fiber Optics* (5th Ed.) (2005), Prentice Hall., NJ, USA.

Khazan, A. D. (1994). *Transducers and Their Elements: Design and Application.* Englewood Cliffs, NJ: PTR Prentice Hall.

Theuwissen, A. J. (1995). *Solid-State Imaging with Charge-Coupled Devices.* New York, NY: Springer.

Chapter 9

ELECTROCHEMICAL TRANSDUCERS

OBJECTIVES

- Differentiate galvanic and electrolytic electrochemical cells.
- Describe the construction of a galvanic cell and its electrochemical reactions.
- Define standard half-cell potential and compute cell potential under non-standard conditions.
- Explain how galvanic cells can be used as electrolyte analyzers.
- Analyze the constructions and the electrochemical reactions of reference electrodes, including hydrogen, calomel, and silver/silver chloride (Ag/AgCl) standard electrodes.
- Analyze the constructions and electrochemical reactions of ion selective electrodes including pK, pH, and pCO_2.
- Define amperometry and potentiometry.
- State the principles of enzyme biosensors.
- Describe the principles of batteries and fuel cells.

CHAPTER CONTENTS

1. Introduction
2. Electrochemistry
3. Reference Electrodes
4. Ion Selective Electrodes
5. Biosensors (Enzyme Sensors)
6. Batteries and Fuel Cells

INTRODUCTION

Electrochemistry plays an important role in understanding biopotential signals as well as their measurements. This chapter covers the principles of galvanic cells and electrode potentials. Reference electrodes including hydrogen, Ag/AgCl, and calomel electrodes are studied. Application examples such as K^+, pH, pO_2, and pCO_2 electrodes are discussed to illustrate the principles of ion selective electrodes. Enzyme sensors for glucose and cholesterol measurement are used to introduce this growing field of biosensors. The principles of energy storage cells and fuel cells are also discussed.

ELECTROCHEMISTRY

Electrochemistry is the study of the interconversion processes concerning the relationship between electrical energy and chemical changes. An electrochemical cell is a device that permits the interconversion of chemical and electrical energy. There are two types of electrochemical cells: galvanic and electrolytic cells. In a galvanic cell, also known as a voltaic cell, chemical energy is converted to electrical energy. The reverse happens in an electrolytic cell, where electrical energy is converted to chemical energy. In the context of medical instrumentation, we consider only galvanic cells in this chapter.

Galvanic (Voltaic) Cells

In a galvanic cell, the electrical energy produced is the result of a spontaneous redox reaction among the substances in the cell. Redox reaction is a chemical reaction involving oxidation (losing electrons) and reduction (gaining electrons). Consider the simple redox reaction caused by placing a zinc rod into a solution of copper sulfate. One would immediately notice a black dull deposit forming on the shiny zinc surface. The chemical equation of this spontaneous reaction is:

$$Zn(s) + Cu^{2+} \rightarrow Zn^{2+} + Cu(s)$$

In this chemical reaction, copper ions from the solution are reduced to fine particles of copper metal that grow to form a spongy layer on top of the zinc metal. The blue color of the copper sulfate solution gradually becomes paler (from its blue color), indicating that hydrated copper ions are used up in the reaction. The redox equation can be separated into an oxidation equation and a reduction equation as shown below. Each of these is called a half-reaction.

$$Zn(s) \rightarrow Zn^{2+} + 2e \qquad \text{(oxidation)}$$

$$Cu^{2+} + 2e^- \rightarrow Cu(s) \qquad \text{(reduction)}$$

The free-energy change of the overall redox reaction is –212 kJ when the reactants and products are in their standard states (i.e., pure metal and 1 M ionic concentration). This large, negative, free energy indicates a strong tendency for electrons to be transferred from the zinc metal to the copper ions. The free-energy change in a reaction depends only on the nature and state of the reactants and products and not on how the process takes place.

Consider the setup in Figure 9-1, in which the zinc rod is separated from the copper sulfate solution. In the setup, the zinc rod is immersed in a solution of zinc sulfate and a copper rod is immersed in a solution of copper sulfate. The two compartments of the cell are separated by a porous partition such as a piece of porcelain or clay. The zinc and copper rods become electrodes to provide surfaces at which oxidation and reduction half-cell reactions can take place.

If the zinc and copper electrodes are connected by an external conductor, electrons will leave the zinc metal and travel through this external conductor to the copper electrode. These electrons reduce the copper ions in the solution adjacent to the copper electrode to form atomic copper, which deposits on the surface of the copper electrode. In this galvanic cell, the

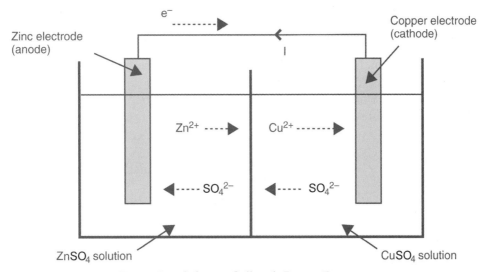

Figure 9-1. Galvanic Cell with Porous Partition.

anode (the electrode at which oxidation takes place) is the zinc electrode and the cathode (the electrode at which reduction takes place) is the copper electrode. The positive ions (in this case Zn^{2+}) which migrate toward the cathode, are called cations; the negative ions (SO_4^{2-}), which migrate toward the anode, are called anions.

When the external circuit is broken and a voltmeter of high-input impedance is connected across the cathode and anode, the voltmeter reading represents the electrical potential difference of the galvanic cell. The cell voltage depends on the type of metal and its electrolyte concentration in each of the two half-reaction compartments. The potential difference measured between the cathode and the anode for this zinc-copper cell at 1 M ionic concentration and 25°C is 1.10 V. (Note that the cathode is positive and the anode is negative.)

The purpose of the porous partition is to prevent direct transfer of electrons in the solution from the zinc metal to the copper ions. Without the porous partition, there will be no electron flow in the external circuit because the copper ions are able to migrate in the solution to the zinc electrode and capture the electrons directly from the anode. The porous partition in Figure 9-1 can be replaced by a "salt bridge" as shown in Figure 9-2. A salt bridge is a tube filled with a conductive solution such as potassium chloride (KCl). In the salt bridge, K^+ migrates toward the cathode and Cl^- toward the anode. A salt bridge provides physical separation between the galvanic cell compartments and establishes electrical continuity within the cell. In addition, it reduces the liquid junction potential. Liquid junction potential is a voltage produced when two dissimilar solutions are in contact and when the rates of migration of the cations and anions are not the same across the contact region (or the junction). The ions in the salt bridge are chosen such that cations and anions migrate across the junction at almost equal rates, thus minimizing the liquid junction potential. Calculation of the cell voltage is simplified if no liquid junction potential is present.

The Zn-Cu galvanic cell we have discussed so far in which electrons must travel through an external circuit before reaching the cathode compartment is called a Daniell cell. The galvanic cells described in Figures 9-1 and 9-2 are both Daniell cells. Galvanic cells can be depicted by a line notation called a cell diagram. The cell diagram of the Daniell cell is

$$Zn(s) \mid ZnSO_4(aq) \mid CuSO_4(aq) \mid Cu(s)$$

In this cell diagram, the anode is at the left and the vertical lines represent the junctions. When the salt bridge is present, the junction is represented by a double vertical line:

$$Zn(s) \mid ZnSO_4(aq) \parallel CuSO_4(aq) \mid Cu(s)$$

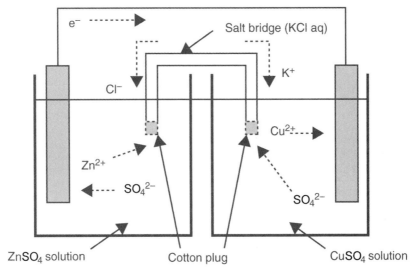

Figure 9-2. Galvanic Cell with Salt Bridge.

This can be further simplified by showing only the reacting ions in the solution phases:

$$Zn(s) \mid Zn^{2+} \parallel Cu^{2+} \mid Cu(s)$$

Standard Electrode Potentials

Consider the two half-reactions of the Daniell cell:

$$Zn(s) \rightarrow Zn^{2+} + 2e^- \qquad \text{(oxidation)}$$

$$Cu^{2+} + 2e^- \rightarrow Cu(s) \qquad \text{(reduction)}$$

If we reverse the first half-reaction, it becomes a reduction reaction:

$$Zn^{2+} + 2e \rightarrow Zn(s) \qquad \text{(reduction)}$$

It is not possible to measure the absolute potential of a single electrode (or half-cell) because all measuring devices can measure only the difference in potential. However, we can use a standard reference electrode to establish the half-cell potentials of different electrodes. The standard electrode potentials are measured against a hydrogen electrode under standard conditions; that is, all concentrations are 1 M, partial pressure of gases of 1 atm, and temperature at 25°C. The potential of this standard reference hydrogen elec-

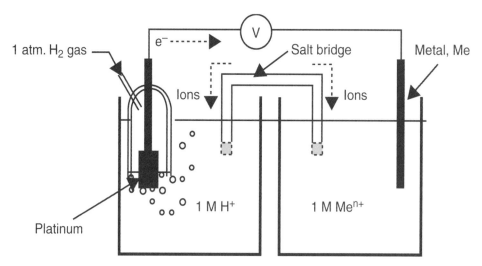

Figure 9-3. Measurement of Standard Electrode Potential.

trode is given a value of 0 V. Figure 9-3 shows the setup of a standard reference hydrogen electrode used to measure the standard electrode potential of a metal Me.

By convention, the half-cell reaction is written as a reduction reaction and the potential measured is denoted as the standard reduction potential E^0:

$$Me^{n+} + ne^- \rightarrow Me(s) \qquad \text{(reduction)}$$

Table 9-1 lists the standard electrode potentials of some half-cell reactions. Note that the sign of the standard potential may be positive or negative. A more negative standard potential implies higher reducing strength of the reaction. The standard oxidation potential has the same magnitude but opposite polarity to the standard reduction potential. For example, the oxidation potential of the copper half-reaction is –0.34 V:

$$Cu(s) \rightarrow Cu^{2+} + 2e^-$$

The potential of a galvanic cell E^0_{cell} is the sum of the standard reduction potential for the reaction at the cathode $E^0_{cathode}$ and the standard oxidation potential for the reaction at the anode $(-E^0_{cathode})$. Therefore, in the case of the Daniell cell,

$$E^0_{cell} = E^0_{cathode} + (-E^0_{anode}) = 0.34 \text{ V} + (+0.76 \text{ V}) = 1.10 \text{ V}$$

Table 9-1. Standard Electrode Potentials

Half-Reaction (Reduction)	E^0 (V)
$Li^+ + e^- \rightarrow Li(s)$	−3.05
$Na^+ + e^- \rightarrow Na(s)$	−2.71
$Al^{3+} + 3e^- \rightarrow Al(s)$	−1.66
$Zn^{2+} + 2e^- \rightarrow Zn(s)$	−0.76
$Fe^{2+} + 2e^- \rightarrow Fe(s)$	−0.44
$2H^+ + 2e^- \rightarrow H_2(g)$	0.00
$AgCl(s) + e^- \rightarrow Ag(s) + Cl^-$	+0.22
$Hg_2Cl_2(s) + 2e^- \rightarrow 2Hg(l) + 2Cl^-$	+0.27
$Cu^{2+} + 2e^- \rightarrow Cu(s)$	+0.34
$Fe^{3+} + e^- \rightarrow Fe^{2+}$	+0.77
$Ag^+ + e^- \rightarrow Ag(s)$	+0.80
$O_2(g) + 4H^+ \rightarrow H_2O(l)$	+1.23
$Cl_2(g) + 2e^- \rightarrow 2Cl^-$	+1.36

Example 9.1

Find the standard voltage produced by the cell

$$Ag(s) \mid AgCl \mid Cl^- \parallel Cu^{2+} \mid Cu(s)$$

Solution:

The separate half-reactions of the cell are:

$$Ag(s) + Cl^- \rightarrow AgCl(s) + e^- \qquad E^0_{ox} = -E^0 = -0.22 \text{ V}$$

$$Cu^{2+} + 2e^- \rightarrow Cu(s) \qquad E^0 = +0.34 \text{ V}$$

For the galvanic cell,

$$2Ag(s) + 2Cl^- + Cu^{2+} \rightarrow 2AgCl(s) + Cu(s)$$

$$E^0_{cell} = +0.34 \text{ V} - 0.22 \text{ V} = +0.12 \text{ V}$$

Example 9.2

Find the standard voltage produced by the cell

$$Pt(s) \mid Fe^{2+}, Fe^{3+} \parallel Cl^- \mid Cl^2 \text{ (g)} \mid Pt(s)$$

Solution:

The separate half-reactions of the cell are:

$$Fe^{2+} \rightarrow Fe^{3+} + e^- \qquad E^0_{ox} = -E^0 = -0.77 \text{ V}$$

$$Cl_2(g) + 2e^- \rightarrow 2Cl- \qquad E^0 = +1.36 \text{ V}$$

For the galvanic cell:

$$2Fe^2 + Cl_2(g) - 2Fe^{3+} + 2Cl^-$$

$$E^0_{cell} = +1.36 \text{ V} - 0.77 \text{ V} = +0.59 \text{ V}$$

Nernst Equation

Consider a hypothetical half-cell reaction:

$$aOx + ne^- \rightarrow bRed$$

When it is under nonstandard conditions (i.e., concentration other than 1 M and temperature is not equal to 25°C), the cell voltage can be quantitatively estimated by the Nernst equation:

$$E = E^0 - \frac{2.303RT}{nF} \log Q,$$

where R is the gas constant = 8.314 J K^{-1} mol^{-1};
T is the temperature in kelvin;
F is the Faraday constant = 96 485 C mol^{-1};
Q is the reaction quotient = $[Red]^b/[Ox]^a$ where concentrations are in mol^{-1}.

At 25°C, the Nernst equation can be simplified to

$$E = E^0 - \frac{0.0592}{n} \log \frac{[Red]^b}{[Ox]^a}$$

Note that the concentration for solid and liquid is given a value of 1 in the equation.

Example 9.3

(a) Find the half-cell potential of a copper electrode immersed in a 0.010-M Cu^{2+} solution at 25°C.
(b) If the above half-cell is connected to a standard zinc half-cell, what would be the cell potential?

Solution:

(a) The half-cell reaction is

$$Cu^{2+} + 2e^- \rightarrow Cu(s)$$

At standard condition, $E^0 = +0.34$ V.

$$E = 0.34 - \frac{0.0592}{2} \log \frac{1^1}{0.010^1} = 0.34 - \frac{0.0592}{2} \log 100$$

$$= 0.34 - 0.0592 = +0.28 \text{ V}$$

(b) Since $E_{anode} = E^0_{zinc} = -0.76$ V, therefore, the cell potential is

$$E_{cell} = E_{cathode} + (-E_{anode}) = +0.28 + 0.76 = +1.04 \text{ V}$$

The cell potential is +1.04 V (instead of +1.10 V under standard conditions).

Example 9.4

When the copper half-cell in Example 9.3 is connected to a nonstandard zinc half-cell at 25°C, the cell potential is measured to be +0.98 V. What is the Zn^{2+} concentration in the solution?

Solution:

The zinc half-cell reaction ($E^0 = -0.76$ V) is

$$Zn^{2+} + 2e^- - Zn(s)$$

Since $E_{cell} = E_{cathode} + (-E_{anode})$

$$E_{zinc} = E_{cathode} - E_{cell} = 0.28 - 0.98 = -0.70 \text{ V}$$

and

$$E^0_{zinc} = -0.76 \text{ V}$$

Using the Nernst equation,

$$E = E^0 - \frac{0.0592}{n} \log \frac{[\text{Re}d]^b}{[Ox]^a}$$

$$\Rightarrow -0.76 = -0.70 - \frac{0.0592}{2} \log \frac{1^1}{x^1}$$

$$\Rightarrow 2.027 = \log \frac{1}{x}$$

$$\Rightarrow x = 0.0094M$$

The concentration of Zn^{2+} in the solution is 0.0094 M.

We have shown in Example 9.4 that the concentration of a cation in a solution can be measured by using a reference electrode and a metal indicator electrode of the same cation. Consider the example of an Ag indicator electrode to find the concentration of Ag^+ in a solution. The half-reaction and the standard potential are

$$Ag^+ + e^- \rightarrow Ag(s) \qquad E^0 = +0.80 \text{ V}$$

Using the Nernst equation, the half-cell potential is

$$E = +0.80 - \frac{0.0592}{1} \log \frac{1}{[Ag^+]} = +0.80 - 0.0592\log \frac{1}{[Ag^+]}$$

From the previous equation, if the half-cell potential E is known, then the centration of Ag^+ can be calculated.

Note that if more than one ion is reduced (or oxidized) at the same time, such a metal indicator electrode will fail to provide the correct measurement. For example, a copper electrode will measure both Cu^{2+} and Ag^+ ions in the solution. Furthermore, many metals react with dissolved oxygen from the air and therefore a deaerated solution must be used.

A metal electrode can also be used to measure the concentration of an anion if the cations of the metal form a precipitate with the anions. Consider an Ag electrode in a saturated solution of AgCl.

$$AgCl(s) \leftrightarrow Ag^+(aq) + Cl^-(aq)$$

The solubility product K_{sp} is given by

$$K_{sp} = [Ag^+][Cl^-] \Rightarrow [Ag^+] = \frac{K_{sp}}{[Cl^-]}$$

Substituting $[Ag^+]$ into the above Nernst equation gives

$$E = +0.80 - \frac{0.0592}{1} \log \frac{1}{[Ag^+]} = +0.80 - 0.0592 \log \frac{[Cl^-]}{K_{sp}}$$

The concentration of Cl^- can therefore be calculated if the half-cell potential E is known.

REFERENCE ELECTRODES

Example 9.4 shows that one can use the Nernst equation to determine the concentration of an analyte (in this case Zn^{2+} ion concentration) in the solution of a half-cell by measuring the potential difference of the cell when the potential of the other half-cell is known. A typical cell used in potentiometric analysis is denoted by

Reference Electrode || Analyte | Indicator Electrode

The cell voltage is the potential difference between the indicator electrode and the reference electrode. An indicator electrode is a half-cell whose potential varies in a known way with the concentration of the analyte; a reference electrode is a half-cell with a known constant potential that is independent of the analyte. In addition to the standard hydrogen electrode mentioned earlier, two other commonly used reference electrodes are discussed in this section. They are the Ag/AgCl electrode and the calomel electrode.

Hydrogen Reference Electrode

The hydrogen electrode is a gas-ion electrode. A gas-ion electrode uses a gas in contact with its anion or cation in a solution. The gas is bubbled into the solution and electrical contact is made by means of a piece of inert metal, usually platinum. Figure 9-4 shows the construction of a hydrogen electrode. The cell diagram and electrode half-reaction of this hydrogen half-cell are

$$Pt(s) \mid H_2(g) \mid H^+ \parallel \qquad\qquad 2H^+ + 2e^- \rightarrow H_2(g)$$

Under standard condition (1 atm and 25°C), the hydrogen half-cell potential is assigned a value of 0 V.

Silver/Silver Chloride Reference Electrode

An Ag/AgCl electrode consists of a piece of Ag coated with AgCl immersed in a solution of KCl saturated with AgCl. Figure 9-5 shows a reference Ag/AgCl electrode (enclosed by dotted line) replacing the hydrogen electrode in Figure 9-3 to measure the standard electrode potential of a metal Me or the ionic concentration of the metal in the solution. The cell diagram of this half-cell is

$$Ag(s) \mid AgCl(sat'd) \mid Cl^- \parallel$$

The half-cell reaction and the standard reduction potential E^0, respectively, are

$$AgCl(s) + e^- \rightarrow Ag(s) + Cl^- \qquad E^0 = 0.22 \text{ V}$$

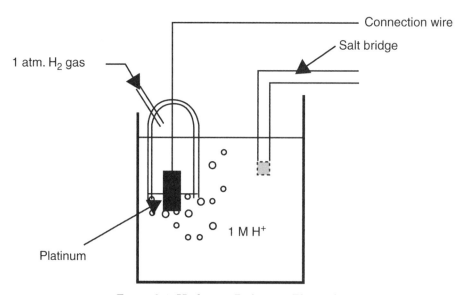

Figure 9-4. Hydrogen Reference Electrode.

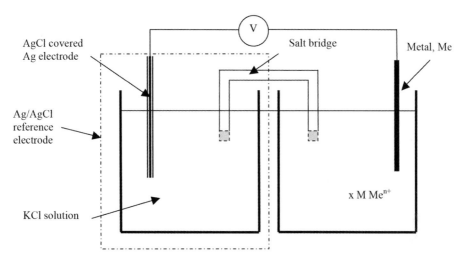

Figure 9-5. Measurement Using a Ag/AgCl Reference Electrode.

Applying the Nernst equation to this half-cell reaction at 25°C, the half-cell potential of the Ag/AgCl electrode is

$$E = E^0 - \frac{0.0592}{n} \log \frac{[Ag(s)][Cl^-]}{[AgCl(s)]} = E^0 - 0.059\log[Cl^-]$$

This potential depends on the concentration of the chloride ion in the KCl solution. Instead of using a standard 1-M solution, in order to maintain a constant concentration (to keep E constant), a saturated solution of KCl (approx. 4.6 M) is used. With a saturated KCl solution and at 25°C, the half-cell voltage becomes 0.20 V.

Figure 9-6 shows a typical commercial Ag/AgCl reference electrode. The salt bridge is replaced by a porous plug at the base of the electrode to allow passage of ions and completion of the electrical circuit. The air vent allows the electrolyte to drain slowly through the porous plug. The presence of some solid KCl and AgCl at the bottom maintains a solution saturated with KCl and AgCl.

Calomel Reference Electrode

A calomel electrode consists of a platinum electrode in contact with mercury, mercury chloride (also known as calomel), and a KCl solution of known concentration. The half-cell is denoted by this cell diagram:

$$Hg(l) \mid Hg_2Cl_2(s) \mid Cl^-(x\ M) \parallel$$

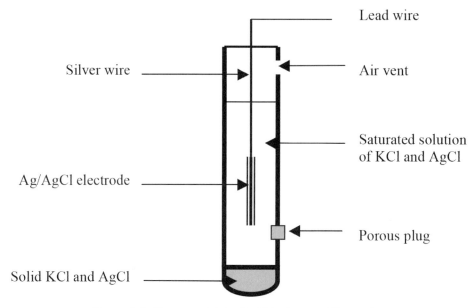

Figure 9-6. Commercial Ag/AgCl Reference Electrode.

where x = the molar concentration of KCl in the solution.

The half-reaction and standard reduction potential (1 M Cl⁻ ions) is given by

$$Hg_2Cl_2(s) + 2e^- \rightarrow 2Hg(l) + 2Cl^- \qquad E^0 = 0.27 \text{ V}$$

In practice, saturated KCl (4.6 M) is often used. The advantage is that the concentration, and therefore the potential, will not change even when some of the solution evaporates. At 25°C, a saturated calomel electrode (SCE) has a potential of 0.24 V. Figure 9-7 shows the construction of a laboratory calomel electrode. Similar to the Ag/AgCl reference electrode, the salt bridge is replaced by a porous plug in a commercial calomel reference electrode (Figure 9-8).

ION SELECTIVE ELECTRODES

An ion selective electrode (ISE) is an indicator electrode that produces a voltage when it is in contact with a solution of a particular ion. In general, this is achieved by using a membrane that is selective for the ion being analyzed. An ISE is also known as a membrane indicator electrode.

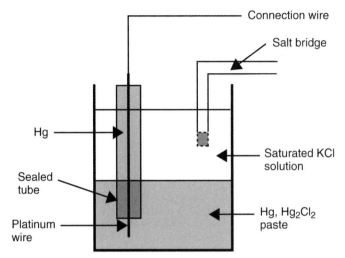

Figure 9-7. Laboratory Calomel Reference Electrode.

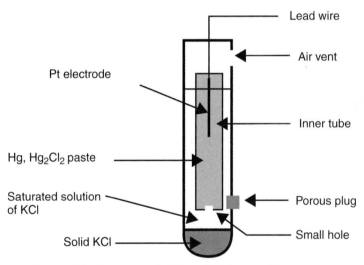

Figure 9-8. Commercial Calomel Reference Eleectrode.

pK Electrode

An example of an ISE is a K^+ electrode or pK electrode. The ion selective membrane of a pK electrode is a hydrophobic synthetic material containing ionophores such as valinomycin. The membrane is impermeable to H^+, OH^-, K^+, Cl^-, and so on. Ionophores are antibiotics produced by bacteria that are used to facilitate the movement of cations across the synthetic membrane. Valinomycin (an ionophore) can bind K^+ tightly but has a 1000 times lower affinity for Na^+. The valinomycin-K^+ complex can readily pass through the membrane from a solution of high K^+ concentration to a solution of low K^+ concentration. In the case of a solution of 0.1 M KCl and a solution of 0.01 M KCl separated by the membrane, potassium ions (K^+) are transported via the valinomycin-K^+ complex from the high concentration solution into the low concentration solution. Because Cl^- ions cannot pass through, however, a slight positive potential will develop on the low concentration side of the membrane with respect to the high concentration side. This potential eventually stops the net transfer of the K^+ ions across the membrane. The potential across the membrane is the difference in potential E_1 and E_2 at the surfaces of membrane due to different concentration of K^+ at both sides of the membrane. At 25°C this membrane potential is given by

$$E_{membrane} = E_1 - E_2 = \frac{0.0592}{n} \log \frac{[K^+_1]}{[K^+_2]} = 0.059 \log \frac{0.01}{0.1} = -0.059 \text{ V}$$

where n is the charge on the ion; for K^+, $n = 1$.

Figure 9-9 shows the galvanic cell setup to measure the concentration of the analyte K^+ in the solution. The indicator electrode consists of a reference Ag/AgCl electrode in a solution with a known KCl concentration (e.g., 0.1 M KCl). A calomel electrode is used as the other reference electrode. The cell diagram of this cell can be denoted as

$$Hg(l) \mid Hg_2Cl_2(s) \mid Cl^-(sat'd) \parallel K^+_{analyte} \mid membrane \mid K^+(0.1 \text{ M}), Cl^- \mid AgCl(s) \mid Ag(s)$$

The cell galvanic potential of this setup is

$$E_{cell} = E_{reference} + E_{indicator} = E_{calomel} + E_{Ag/AgCl} + E_{membrane}$$

Since both $E_{calomel}$ and $E_{Ag/AgCl}$ are known

$$E_{cell} = C + \frac{0.0592}{1} \log \frac{[K^+_{analyte}]}{[0.1]} = C' + 0.059 \log [K^+_{analyte}] = C' - 0.059 pK,$$

where $pK = -\log[K^+]$, and $E_{calomel} + E_{Ag/AgCl} = C$.

A plot of the cell potential E_{cell} versus pK at 25°C is a straight line with a negative slope of 0.059.

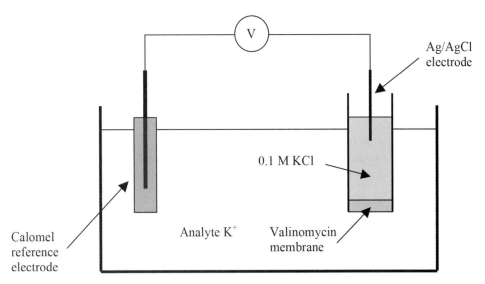

Figure 9-9. Measurement of K⁺ Concentration Using an ISE.

pH Electrode

Another example of an ISE is the glass membrane pH electrode. The membrane of this electrode is usually a thin (0.1 mm) sodium glass with a composition of 72% SiO_2, 22% Na_2O, and 6% CaO. The silicon and the oxygen in the glass membrane form a negatively charged structure with mobile positive ions to balance the charge. When the glass is soaked in an aqueous solution, the aqueous solution exchanges H^+ for Na^+ at the glass surface. The amount of Na^+–H^+ exchanged is proportional to the H^+ concentration in the solution.

When the two sides of the glass membrane are soaked in two solutions of different H^+ concentrations, a potential E_{glass} develops across the glass membrane. At 25°C,

$$E_{glass} = \frac{0.0592}{1} \log \frac{[H^+_2]}{[H^+_2]}$$

If one of the solutions has a known constant H^+ concentration, the concentration of H^+ in the other solution can be determined.

Figure 9-10 shows a setup to measure the pH of a solution. The cell diagram of this cell can be denoted as:

$Hg(l) \mid Hg_2Cl_2(s) \mid Cl^-(sat'd) \parallel H^+_{analyte} \mid glass \mid H^+(0.1\ M), Cl^- \mid AgCl(s) \mid Ag(s)$

Similar to pK electrode, the cell potential is

$$E_{cell} = C' + 0.059 \log [H^+_{analyte}] = C' - 0.059pH,$$

where *pH* = $-\log[H^+_{analyte}]$.

A plot of the cell potential E_{cell} versus *pH* at 25°C is a straight line with a negative slope of 0.059.

In a base solution, sodium glass starts to react with alkali metals such as Na^+ in addition to H^+. A sodium glass membrane electrode will start to produce noticeable error (referred to as alkaline error) at pH above 9. Therefore, to measure solution with pH greater than 9, a lithium glass membrane (72% SiO_2, 22% Li_2O, and 6% CaO) is used instead.

The reference electrode and the H^+ sensing glass membrane electrode in most industry models are combined into a single probe. Ag/AgCl electrodes may also be used (to replace the calomel electrode) as the reference electrode. A voltmeter connected to the electrode is usually calibrated to provide a direct readout of the measured pH. Because pH is affected by temperature, a pH meter is often calibrated by using two known buffers with known temperature characteristics.

pCO2 Electrodes

In pCO2 measurement, a membrane permeable to CO2 (e.g., silicon rubber) is used to separate the sample solution (e.g., blood) from a buffer

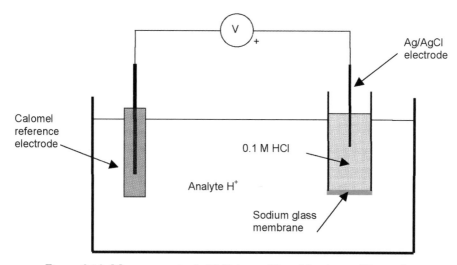

Figure 9-10. Measurement of pH Using a Glass Membrane Electrode.

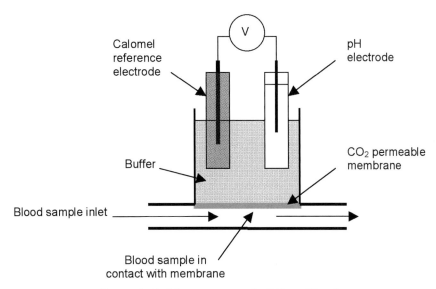

Figure 9-11. Measurement of pCO_2 in Blood.

(e.g., sodium bicarbonate). As the CO_2 diffuses from the sample into the buffer, the pH of the buffer is lowered. The change in pH in the buffer, measured by a pH electrode, correlates to the pCO_2 in the sample. A voltmeter is often connected across the pH electrode and a reference electrode (Figure 9-11) to produce a reading calibrated to read pCO_2.

pO_2 Electrodes

In a pO_2 electrode, a constant voltage source is applied across the pO_2 and reference electrodes and the current flowing through the electrode is measured. The magnitude of the current is a function of the pO_2 level in the sample solution. This electrochemical method is known as amperometry (measurement of current flowing through the electrodes with an applied voltage) in contrast to potentiometry (measurement of voltage between the electrodes with almost no current flow). An amperometric cell is also known as a polarographic cell.

Figure 9-12 shows a Clark polarographic oxygen electrode designed to measure pO_2 in a solution. The indicator electrode is a platinum cathode in a buffer solution of KCl. The anode is a Ag/AgCl electrode. An oxygen-permeable membrane (such as Teflon® or polypropylene) separates the buffer and the sample solution. When a voltage is applied across the anode and the cathode, oxygen molecules diffused through the membrane are reduced at the platinum cathode to form OH^- ions. The half-reactions at the electrodes are

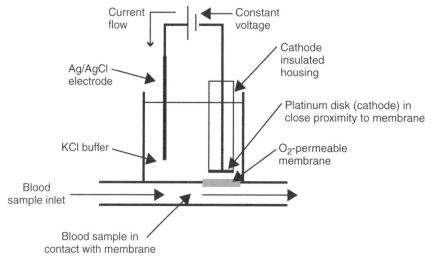

Figure 9-12. Measurement of pO$_2$ in Blood Serum.

$$Ag(s) + Cl^-(aq) \rightarrow AgCl(s) + e^- \qquad \text{Anode}$$

$$O_2 + 2H_2O + 4e^- \rightarrow 4OH^- \qquad \text{Cathode}$$

The magnitude of the current flowing out from the cathode is proportional to the dissolved O$_2$ level in the sample solution.

The cathode of a pO$_2$ electrode is usually made from an inert metal such as gold or platinum. To increase the respond time, oxygen must diffuse rapidly through the membrane and reach the cathode quickly. To achieve this, the membrane is made to be very thin (about 20 μm) and the cathode is in the form of a disk placed very close to the membrane (with separation about 10 μm).

A galvanic oxygen cell operates on the same principle, except that it does not have an external voltage source. Instead of measuring its current, the cell voltage is monitored. The cell potential is directly proportional to the concentration of oxygen for the gas outside the membrane.

pH, pCO$_2$, and pO$_2$ are the three analytes in blood gas analysis. In practice, the membranes of these electrodes must be kept clean and replaced regularly. Two point calibrations of the electrodes should be performed at 37°C and corrected to 37°C saturated vapor pressure.

BIOSENSORS (ENZYME SENSORS)

An oxygen sensor (e.g., Clark oxygen electrode) can be used indirectly to measure glucose or cholesterol concentration in blood. In serum glucose measurements, glucose in blood serum is first broken down by the enzyme glucose oxidase. This process consumes oxygen (see reactions below). The concentration of glucose is proportional to the amount of O_2 consumed, which can be measured using an oxygen electrode. The reaction is

$$Glucose + O_2 + H_2O \rightarrow gluconic\ acid + H_2O_2$$

In cholesterol measurement, the cholesterol ester in the blood serum is first broken down into cholesterol by the enzyme cholesterol esterase and then into cholest-4-ene-3-one and H_2O_2 by the enzyme cholesterol oxidase. The latter reaction consumes oxygen. The amount of O_2 consumed is proportional to the concentration of the blood serum cholesterol.

$$Cholesterol\ esters + H_2O \rightarrow cholesterol + fatty\ acid$$

$$Cholesterol + O_2 \rightarrow cholest-4-ene-3-one + H_2O_2$$

In addition to measuring the oxygen consumed, an alternative method is to measure the amount of hydrogen peroxide produced from the reaction, which is proportional to the concentration of glucose or cholesterol in the blood serum. The hydrogen peroxide is separated by an ion selective membrane (such as cellulose acetate membrane) and oxidized to give oxygen according to the reaction

$$H_2O_2 \rightarrow O_2 + 2H^+ + 2e^-$$

The amount of O_2, which is proportional to the concentration of glucose, is measured by an oxygen sensor.

In general, an enzyme sensor consists of three layers:

1. The first (outer) layer allows the analyte (e.g., glucose) to pass from the sample solution (e.g., blood serum) into the electrode.
2. The second layer contains an enzyme (e.g., glucose oxidase) to produce H_2O_2.
3. The inner layer collects and oxidizes H_2O_2 to form O_2. An O_2 sensor in this layer generates a current proportional to the concentration of the analyte.

Other analytes may be measured using this type of biosensor as long as an enzyme can be found that specifically oxidizes the analyte to produce hydrogen peroxide.

BATTERIES AND FUEL CELLS

Batteries are used in the industry and home as power sources. The basic element in a battery is the energy storage galvanic cell. A common galvanic cell is the Leclanché cell, also known as dry cell, or zinc-carbon cell. This galvanic cell consists of a zinc can, which serves as the anode; a central carbon rod is the cathode. The anode and the cathode are separated by a paste of manganese oxide, carbon, ammonium chloride, and zinc chloride moistened with water. At the anode, zinc is oxidized and at the cathode, MnO_2 is reduced. The half-reactions are:

$$Zn(s) \rightarrow Zn_2+ + 2e^- \qquad\qquad\qquad \text{Anode}$$

$$e^- + NH_4+ + MnO_2(s) \rightarrow MnO(OH)(s) + NH_3 \qquad \text{Cathode}$$

An external conductor connecting the anode and the cathode allows the flow of electrons from the anode to the cathode through the external load.

A lead acid cell is another example of an energy storage galvanic cell. The anode of the cell consists of a frame of lead filled with some sponge-like lead. When the lead is oxidized, Pb^{2+} ions immediately precipitate as $PbSO_4$ and deposit on the lead frame. The cathode is also a lead frame filled with PbO_2. The half-reactions are:

$$Pb(s) + HSO_4^- \rightarrow PbSO_4(s) + H^+ + 2e^- \qquad\qquad \text{Anode}$$

$$2e^- + PbO_2(s) + 2H^+ + HSO_4^- \rightarrow PbSO_4(s) + 2H_2O \qquad \text{Cathode}$$

When an external conductor is connected between the anode and the cathode, a current is drawn. The solid $PbSO_4$ is produced at both electrodes as the cell discharges; H^+ and HSO_4^- are removed from the solution at the same time.

A lead acid cell can be recharged by imposing a slightly larger reverse voltage on the cell. This reverse voltage forces the electrons to flow into the anode and out of the cathode to reverse the reactions, converting $PbSO_4$ back into Pb at the anode and into PbO_2 at the cathode. A lead acid cell can be recharged and reuse for a number of times. Cells that are specifically designed for reuse are called secondary cells, whereas those for single use are called primary cells. Some common primary and secondary cells and their nominal open circuit voltages are listed in Table 9–2. Multiple cells can be connected in series or parallel to create a battery for a specific applications. For example, the lead acid battery under the hood of a car is made up of 6 cells connected in series to produce a 12.6 V battery.

Key parameters used to describe the electrochemical functionality of a battery include the voltage, capacity and internal resistance. This capacity of a battery is measured in ampere-hours (Ah). A battery capacity can be viewed as being made up of 3 compartments

1. the empty compartment,
2. the available energy compartment, and
3. the dead compartment.

For an ideal battery, the available energy compartment should account for 100% of the entire battery compartment (a fresh and new primary battery is close to this). As the battery is used, the available energy decreases and the empty compartment increases until all available energy is consumed. In a

Table 9-2. Common Primary and Secondary Cells.

Negative electrode	Electrolyte	Positive electrode	Nominal voltage (V)	Common Name
Primary Cell				
Zinc	Ammonium chloride and Zinc chloride	Manganese dioxide	1.5	Zinc-carbon battery
Zinc	Alkali metal hydroxide	Manganese dioxide	1.5	Alkaline battery
Zinc	Alkali metal hydroxide	Oxygen	1.4	Zinc-air battery
Zinc	Alkali metal hydroxide	Silver oxide	1.55	Silver-oxide battery
Lithium	Organic electrolyte	Manganese dioxide	3.0	Lithium battery
Secondary Cell				
Lead	Sulfuric acid	Lead dioxide	2.1	Lead acid battery
Hydrogen absorbing alloy	Potassium hydroxide	Nickel oxide	1.2	Nickel-metal hydride battery
Cadmium	Potassium hydroxide	Nickel oxide	1.2	Nickel-cadmium battery
Lithium	Organic solvent such as DFSM2-doped electrolyte	Lithium oxide	3.6	Lithium-ion battery

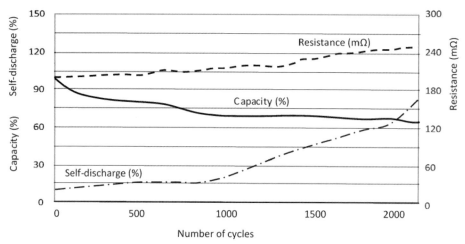

Figure 9-13. Relationship between number of cycles
and secondary battery characteristics.

secondary or rechargeable battery, the empty compartment can be recharged to become available energy.

A battery can only be charged to its active capacity (sum of the empty plus available energy compartments). Even when a battery is not used, the available energy decreases due to parasitic discharge paths within the cells, this is referred to as self-discharge of a battery. In addition, the size of the dead compartment increases as the secondary battery continues to be charged and recharged, leaving less room for the empty and available energy compartments. For some secondary batteries, the size of the dead compartment grows with time, leaving even less space for the active capacity. Both self-discharge and growth of the dead compartment result in decline in the available energy capacity of a battery. The rate of decline in active capacity depends on the electrochemistry of the battery, the charging/discharging practices (e.g., charge voltage, depth of discharge), as well as environmental conditions such as temperature. Figure 9-13 is an example of the relation between the number of cycles on the capacity, internal resistance, and the self-discharge rate of a secondary battery.

Secondary batteries can be useful for around 500 to 1000 cycles depending on the electrochemical system of the battery. The service life and life-cycle of a secondary battery depends on the requirements of its application. For examples, when the running voltage is lower than the minimum operating voltage of the device, the energy capacity becomes too low (say, less than 70% of the initial capacity for a cardiac defibrillator battery), or when the self-discharge rate becomes too high, it will need to be replaced.

The nominal voltage of a battery is the value assigned to power a circuit or system. The operating voltage of a battery can vary from the nominal voltage within a range that permits satisfactory operation of the system. The open circuit voltage of a battery (measured by a high impedance voltmeter) is higher than the nominal voltage due to the battery's internal resistance. Ideally, a battery should have no internal resistance. However, all batteries have some internal resistance from the electrodes and electrolytes. Increase in internal resistance lowers the output voltage and limits the maximum load current. Internal resistance of the battery increases with the cycles through its life time. As such, internal resistance provides useful information in detecting problems and signals when a battery needs to be replaced.

Self-discharge rate is the reduction of the deliverable charge capacity of a battery as a function of time, even when it is not being used. Self-discharge is associated with the parasitic chemical reactions within the battery. For a nickel-based batteries, the self-discharge rate is 10–15% in 24 hours right after fully charged, then 10-15% per month. For lead-acid batteries, it is about 5% per month.

A battery is not used immediately after it is manufactured, it may be stored prior to be shipped, or be sitting inside a device before it is being used. The shelf-life or maximum storage time of a battery is primarily dependent on the battery chemistry but some may also depend on their charge conditions. In general, primary alkaline and lithium batteries can be stored up to 10 years, nickel-based batteries up to five years, and secondary lithium ion up to three years when discharged to 30% to 50% capacity.

Battery performance is affected by temperature, both internal and environmental. Heat is produced inside a battery as it is used. If heat is not properly dissipated, it can lead to performance degradation or total failure. In the extreme case, thermal runaway of the lithium-ion battery due to rapid internal temperature rise initiates an unstoppable chain reaction to suddenly release energy stored in the battery. As such, thermal management features such as heat sink or temperature sensors are included to allow for safe operation of the battery and optimize the battery life.

A fuel cell is a galvanic cell in which the reactants are continuously fed into the cell to produce electricity. Figure 9–14 shows one type of hydrogen-oxygen fuel cell. The electrodes are made of porous carbon impregnated with platinum (as a catalyst). Hydrogen is oxidized at the anode and oxygen is reduced at the cathode. The half-reactions are:

$$H_2(g) + 2OH^- \rightarrow 2H_2O + 2e^- \qquad \text{Anode}$$

$$4e^- + O_2(g) + 2H_2O \rightarrow 4OH^- \qquad \text{Cathode}$$

Hydrogen and oxygen are combined to produce water (steam). The process produces an electron flow when an external load is connected between the anode and the cathode. The overall reaction is:

$$2H_2(g) + O_2(g) \rightarrow 2H_2O$$

For the hydrogen-oxygen fuel cell, if the water is in the form of liquid, then $E^0 = 1.23$ V; if it is a gas, then $E^0 = 1.18$ V. This is the voltage of a single cell. Practical fuel cells are built from a number of single cells in series or parallel to provide the desired voltage and power output.

There are many types of fuel cells using different electrolytes (e.g., alkaline, solid polymer, phosphoric acid) and different fuels (e.g., methane, carbon monoxide) "burned" at the anode. However, in most of these cells, oxygen from air is the oxidant at the cathode. Fuel cells are considered the energy conversion of choice for the future because they have no objectionable by-products and relatively high efficiencies (above 80% theoretical efficiency).

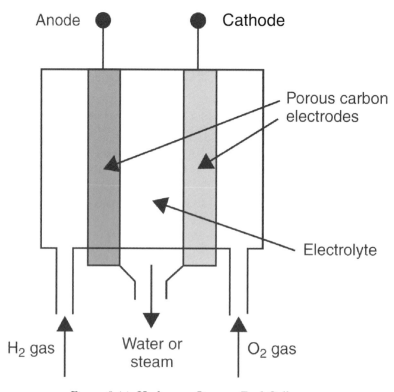

Figure 9-14. Hydrogen-Oxygen Fuel Cell.

BIBLIOGRAPHY

Andrews, D. H., & Kokes, R. J. (1962). *Fundamental Chemistry*. New York, NY: John Wiley & Sons, Inc.

Buchmann, Isidor. *Learn about Batteries*. Battery University. https://batteryuniversity.com/, accessed Oct. 20, 2020)

CSA Group. *Making Sense of Regulations for Medical Device Batteries*. CSA Group, www.csagroup.org/article/making-sense-regulations-medical-device-batteries/, accessed Oct. 20, 2020.

Donald, G. B. (1993). *Biosensors: Theory and Applications*. Lancaster, PA: Technomic Publishing Company, Inc.

Hoogers, G. (2013). *Fuel Cell Technology Handbook* (2nd ed.); Handbook Series for Mechanical Engineering). Boca Raton, FL: CRC Press.

Larminie, J., & Dicks, A. (2003). *Fuel Cell Systems Explained* (2nd ed.). New York, NY: John Wiley & Sons, Inc.

O'Hayre, R., Cha, S. K., Colella, W., & Prinz, F. B. (2009). *Fuel Cell Fundamentals*. New York, NY: John Wiley & Sons.

Reddy, T.B. (ed.), (2011). *Linden's Handbook of Batteries* (4th ed,). New York: McGraw Hill.

Chapter 10

BIOPOTENTIAL ELECTRODES

OBJECTIVES

- State the mechanism of ion flow in the body.
- Define biopotential and its origin.
- Explain the formation of cell membrane potential and action potential.
- List the characteristics of ideal biopotential electrodes.
- Explain half-cell potentials, offset potential, and their significance in biopotential measurements.
- Analyze the characteristics of a perfectly polarized and perfectly nonpolarized electrode.
- Sketch and analyze the electrical equivalent circuit of a Ag/AgCl skin electrode.

CHAPTER CONTENTS

1. Introduction
2. Origin of Biopotentials
3. Biopotential Electrodes

INTRODUCTION

Luigi Galvani, an Italian physiologist and physicist, was the first to explore electrical potentials of the body, now commonly called biopotentials. The element that produces electrical events in biological tissue is the ion in the electrolyte solution, as opposed to the electron in the electrical circuit. A biopotential, then, is an electrical voltage caused by a flow of ions through biological tissues. The devices that pick up these biopotentials are referred to as biopotential electrodes. Biopotential electrodes are a form of transducer.

ORIGIN OF BIOPOTENTIALS

Fick's law states that if there is a high concentration of particles in one region and they are free to move, they will flow in a direction that equalizes the concentration. The resulting movement of these charges is called diffusion.

The movement of charged particles (such as ions) that is due to the force of an electric field (the forces of attraction and repulsion) constitutes particle drift. Each cell in the body has a potential voltage across its membrane, known as the single-cell membrane potential Vm. The cell membrane separates the extracellular fluid and the cell contents. The membrane potential forms the basis for biopotentials of the body. Some of the biopotentials of interest include the ECG, EEG, EOG, ERG, and EMG.

The potential is the result of the diffusion and drift of ions across the high-resistance semipermeable cell membrane. The ions are predominantly sodium $[Na^+]$ ions moving into the cell, and potassium $[K^+]$ ions moving out of it (Figure 10-1). Because of the semipermeable nature of the membrane, Na^+ ions are partially restricted from passing into the cell. As a result, the concentration of Na^+ outside the cell is higher than that inside.

In addition, a process called the sodium-potassium pump keeps sodium largely outside the cell and potassium ions inside. In the process, potassium is pumped into the cell while sodium is pumped out. The rate of sodium pumping out of the cell is about two to five times that of potassium pumping into cells. In the presence of the offsetting effects of diffusion and drift and the sodium-potassium pump, the equilibrium concentration point is established when the net flow of ions is zero. Because there are more positive ions moved outside the cells (Na^+) than positive ions moved into the cells (K^+), the inside of the cell is less positive than the outside and more negative ions are present within the cell. Therefore, the cell is negative with respect to the outside; the cell becomes polarized. This potential difference between the inside and the outside of the cell at equilibrium is called the resting potential. The magnitude of the resting potential is −70 to −90 mV.

If, for any reason, the potential across the cell membrane is raised, say, by voluntary or involuntary muscle contractions, to a level above a stimulus threshold, the cell membrane resistance changes. Under this condition, the nature of the cell membrane changes and becomes permeable to sodium ions. The sodium ions will start to rush into the cell. The inrush of positively charged sodium ions caused by this change in cell membrane resistance

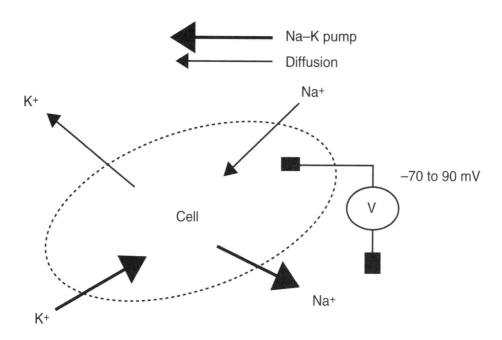

Figure 10-1. Mechanism of Cell Resting.

gives rise to a change in ion concentrations within and outside the cell. The result is a change in the membrane potential called the action potential (Figure 10-2). During this time, the potential inside the cell is 20 to 40 mV more positive than the potential outside. The action potential lasts for about 1 to 2 msec. As long as the action potential exists, the cell is said to be depolarized. Under certain conditions, this action potential disturbance is propagated from one cell to the next, causing the entire tissue to become depolarized. Eventually the cell equilibrium returns to its normal state (i.e., to its polarized state) and the –70 to –90 mV cell membrane potential is resumed. The time period when the cell is changing its polarization is called the refractory period. During this time, the cell is not responsive to any stimulation.

When cells are stimulated, they generate a small action potential. If a large group of cells is stimulated simultaneously, the resultant action potentials can be readily detected. For example, when the heart contracts and relaxes, the polarization and repolarization of the heart cells create a resultant action potential. This action potential can be monitored by external machines using electrodes placed on the surface of the body. This sequence of polarization and repolarization gives rise to a complete waveform known as the ECG.

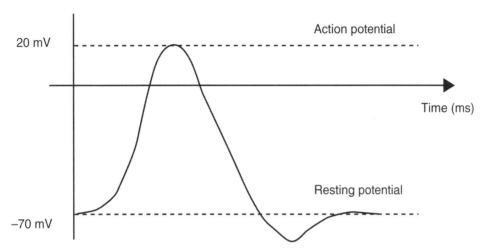

Figure 10-2. The Action Potential.

BIOPOTENTIAL ELECTRODES

Biopotential electrodes are transducers that pick up the body's electrical signals via body electrolytes. An electrode provides the interface that is needed between the ionic current flowing in the body and the electron current flowing in the machine. In other words, the electrode allows the machine to measure the electrical effects in tissue by making a transformation between ionic conduction to electronic conduction at the tissue-electrode interface. Ideally, the amount of current flowing through the electrode is zero; hence the purpose of the electrode is to measure the potential at the tissue due to the flow of ions across the biological tissue. Practically speaking, however, some small amount of current will flow through the electrode.

An ideal electrode is characterized by a number of features, including

- absence of distortion or electrical noise
- immunity to external interference
- inertness of the electrode in the presence of tissue and bodily fluid
- absence of interference with or influence on the tissue
- absence of interference with or influence on the movement of the subject
- ease of making contact with the biological source
- invariance of the contact even during long periods of time
- absence of discomfort to the subject
- repeatability of results
- low cost, durable, small size, and low weight

The Tissue-Electrode Interface

Given the preferred characteristics just listed, biopotential measurement using a pair of electrodes should not be considered as simply connecting two pieces of wire to two conductors and placing them onto a patient. Biopotential measurement at the tissue-electrode interface in the clinical setting is a rather complicated process. The electrode is the metal that makes contact with the electrolyte, and the electrolyte is interfaced to the biological tissue through which ions are free to flow. The tissue-electrode interface has a significant effect on the quality of the biopotential signal measured by the medical device.

Electrode Half-Cell Potential

In Chapter 9, it was discussed that if two dissimilar metals are submerged into a solution of electrolyte, a potential difference can be measured between the two metals. With the electrolyte each of these metals creates a half-cell potential. In biopotential measurements, each electrode-tissue interface therefore will create a half-cell potential.

In electrophysiology studies (study of the electrical activities within the human body) the difference in half-cell potentials that can be detected between two electrodes is referred to as the "offset potential" of the electrode pair. Under ideal conditions, when measuring biopotential using a pair of identical electrodes, the offset potential is zero since the half-cell electrode potential at each electrode-tissue interface will cancel out each other. In practice, however, the offset potential is never zero because the half-cell potentials at different locations are never identical. Offset potential on the order of ± 0.1 V is very common in body surface biopotential measurements.

Polarized and Nonpolarized Electrodes

As with any redox reaction, at any electrode/electrolyte interface, the electrode tends to discharge ions into the solution, and the ions in the electrolyte tend to combine with the electrode. That is,

Metal → electrons + metal ions   (oxidation reaction)

Electrons + metal ions → metal   (reduction reaction)

The net result of these reactions at the tissue-electrode interface is the creation of a charge gradient, the spatial arrangement of which is called the electrode double layer. Electrodes in which no ion transfer occurs across the metal-electrolyte interface are said to be perfectly polarized or perfectly non-

Perfectly Polarized Electrodes: Only one reaction occurs with ease.

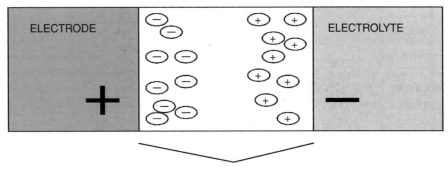

ELECTRODE DOUBLE LAYER FORMS

Potential exists between the electrode and electrolyte
due to the formation of the electrode double layer.

(a) METALLIC ELECTRODE

Perfectly Nonpolarized Electrodes: Both reactions occur with ease.

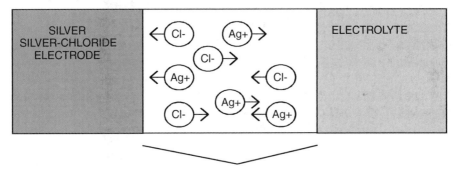

Silver chloride forms free silver ions (Ag+) and chloride ions
(Cl-) which prevent the formation of the electron double layer

(b) SILVER/SILVER CHLORIDE ELECTRODE

Figure 10-3. Perfectly Polarized and Perfectly Nonpolarized Electrode.

reversible electrodes; that is, only one of the two chemical reactions described above can occur. An example of this is shown in Figure 10-3a. Electrodes in which unhindered transfer of charge is possible are said to be perfectly nonpolarized or perfectly reversible; that is, both of the equations referred to above can occur with equal ease. An example of this is shown in Figure 10-3b. Perfectly non polarized electrodes are the ideal electrode of use since they allow for the best ion-electron interface. However, practical electrodes used in clinical situations have properties that lie between these ideal limits.

Silver/Silver Chloride Electrodes

The Ag/AgCl electrode is an important electrode in biopotential measurement because its interface characteristic is close to a perfectly nonpolarized electrode. Ag/AgCl electrodes may be manufactured in one of two ways: the electrode may consist of a solid Ag surface coated with a thin layer of solid AgCl, or it may consist of Ag powder and AgCl powder compressed into a solid pellet. In either case, the presence of the AgCl allows the electrode to behave as a near-perfect nonpolarizable or reversible electrode.

When the electrode is submerged in an electrolyte such as potassium chloride (KCl) containing chloride ions (Cl^-), dissociation of silver ions (Ag^+) from the electrode into the solution and association of the chloride ions (Cl^-) from the solution to the electrode will form free ionic and charge movement between the electrolyte and the electrode. This free ionic flow effect prohibits the formation of the electrode double layer. The net result is a low-impedance, low-offset potential interface between the Ag and the electrolyte. Another advantage of the Ag/AgCl electrode is, since AgCl is almost insoluble in a chloride-containing solution (note that bodily fluid contains chloride ions), very few free silver ions exist. Therefore, tissue damage as a result of silver ions is negligible.

Electrical Equivalent Circuit

A common form of electrodes used in biopotential measurement is the surface (or skin) electrode. An example of surface electrode is the skin electrode used in ECG acquisition. In such applications, the electrodes are placed on the surface of the skin over the chest of the patient; each pair of electrodes is placed at some specified locations and connected via lead wires to the input of a differential amplifier (Figure 10-4). The difference in potential at the two electrode sites is amplified and displayed.

An approximate electrical equivalent circuit of a tissue-electrode interface is shown in Figure 10-5. In this model, the impedance to ion flow between the tissue and the electrode through the skin is modeled as a 1000 Ω resistor, often referred to as the skin resistance (Rse). The double-layer impedance may be regarded as a 50 nF capacitor (Cd) in parallel with a reasonably high value resistor (Rd), typically 20 kΩ. Vhc represents the half-cell electrode potential: about 0.3 V for a typical Ag/AgCl electrode and KCl electrolyte.

Figure 10-4 shows the electrical equivalent circuit of a pair of electrodes attached to the patient's body and connected to a differential amplifier. The upper and lower branches represent the two electrode-tissue interfaces; a 2-MΩ resistance represents the input impedance of the differential amplifier;

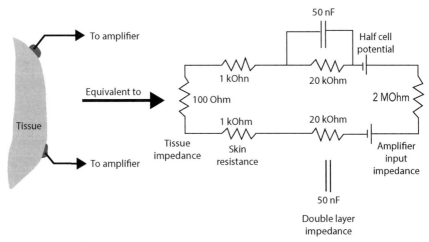

Figure 10-4. Electrical Equivalent Circuit of Biopotential Measurement.

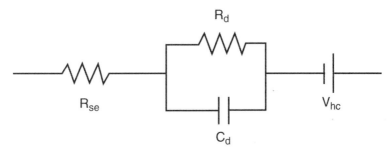

Figure 10-5. Electrical Equivalent Circuit of Single Electrode Tissue Interface.

the impedance to ion flow in the tissue between the electrodes is modeled by a 100-Ω resistance.

Note that in practice, the half-cell potential of the two electrodes will not be identical; therefore, a nonzero DC offset voltage will appear at the input of the differential amplifier and hence will appear as an amplified DC offset voltage at the output of the amplifier. In most applications, this DC offset is eliminated by using a high pass filter. The value of the double layer capacitance (C_d) is larger for a polarized electrode and smaller for a nonpolarized electrode (as there is little static charge accumulated at the electrode-electrolyte interface).

Floating Surface Electrode

At the electrode-tissue interface, any mechanical disturbances will create electrode noise. This is especially true of surface electrodes because the electrode double layer acts as a region of charge gradient. Figure 10-6a shows an Ag/AgCl electrode resting directly on the skin surface with a thin layer of electrolyte. Electrode gel, a pastelike jelly solution containing chloride ions, is often applied to the skin surface to provide a good electrical conduction interface between the electrode and the tissue. With this set up, any relative movement between the electrode and skin surface will create a disturbance to the charge gradient distribution in the gel. In the equivalent circuit, such disturbances will cause changes in capacitance (C_d) of the electrode double layer and therefore will affect the measured biopotential. The result of this disturbance is referred to as motion artifacts in electrophysiological measurements.

The electrical stability of the electrode may be considerably enhanced by mechanically stabilizing the electrode-electrolyte interface. This is achieved by using indirect-contact floating electrodes that interpose an electrolyte jelly paste or gel between the electrode and the tissue. This gel substantially reduces any electrical noise arising from mechanical disturbances in the double-layer charge gradient.

Figure 10-6b illustrates the stabilization of the double layer by using a gel-filled electrode housing. In this configuration, any movement between the electrode housing and the skin surface will cause little or no disturbance to the electrode double layer because it is at a distance away from the location of movement. A gel-filled foam between the electrode and tissue will have a similar effect. In the case of internal electrodes, which are usually constructed of stainless steel, the electrode-tissue interface is already enhanced and stabilized by the presence of extracellular fluid.

Other Biopotential Electrodes

The Ag/AgCl surface electrode described previously is only one of the

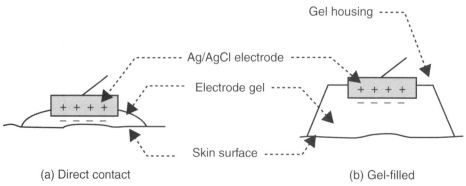

(a) Direct contact   (b) Gel-filled

Figure 10-6. Gel-Filled Electrode Housing to Minimize Motion Artifact.

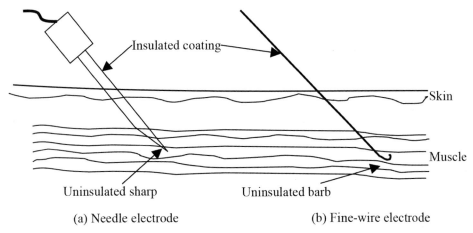

(a) Needle electrode (b) Fine-wire electrode

Figure 10-7. Invasive Biopotential Electrodes.

many different types of electrodes used in biopotential signal measurements. Metal (e.g., copper) electrodes are also used in measuring surface ECG. Other than surface electrodes, invasive electrodes are commonly used to measure signals deep inside the body or in a small localized region in the tissue. Figure 10-7a shows a needle electrode for measuring EMGs. It measures the localized electrical activities when inserted into a muscle fiber. Figure 10-7b is a fine-wire electrode for similar applications, but it can allow more movement by the subject because it is more flexible than a needle electrode is. Many specialized electrodes (for example, scalp electrodes of fetal ECG, and nasopharyngeal electrodes for EEGs) are available for different applications.

BIBLIOGRAPHY

Eggins, B. R. (1993). Skin contact electrodes for medical applications. *Analist., 188*, 439–442.

Fernández, M., & Pallás-Areny, R. (2000). Ag-AgCl electrode noise in high-resolution ECG measurements. *Biomedical Instrumentation & Technology, 34*, 125–130.

Ferris, C. D. (1972). *Introduction to Bioelectrodes.* New York, NY: Plenum Press.

Smith, D. C., & Wace, J. R. (1995). Surface electrodes for physiological measurement and stimulation. *European Journal of Anaesthesiology, 12*, 451–469.

Part III

FUNDAMENTAL BUILDING BLOCKS
OF MEDICAL INSTRUMENTATION

Chapter 11

BIOPOTENTIAL AMPLIFIERS

OBJECTIVES

- State the ideal characteristics of a biopotential amplifier.
- Define common mode and differential mode input.
- Review the characteristics of instrumentation amplifiers (IAs), including gain, common mode rejection ratio (CMRR), upper and lower cut-off frequencies, input and output impedances.
- Identify sources of external and internal noise in biopotential measurement.
- Analyze methods to reduce electric field-induced noise and interference.
- Explain interference due to magnetic induction and describe means to reduce such interferences.
- List sources of conductive interference and explain the principles of surge suppressors.

CHAPTER CONTENTS

1. Introduction
2. Instrumentation Amplifiers
3. Differential and Common Mode Signals
4. Noise in Biopotential Signal Measurements
5. Interference from External Electrical Field
6. Interference from External Magnetic Field
7. Conductive Interference

INTRODUCTION

A biopotential signal is often small in amplitude, mixed with other signals, and subjected to external interference. In addition to amplifying the signal, a biopotential amplifier is designed to extract the desired signal from interfering sources as well as to prevent external noise to corrupt the signal. This chapter studies the principles and design of biopotential amplifiers to achieve these objectives.

INSTRUMENTATION AMPLIFIERS

Figure 11-1 shows a simple operational amplifier equivalent circuit with gain = A, input voltage $V_i = V_2 - V_1$, output $V_0 = AV_i$, input impedance Z_i, and output impedance Z_0.

An ideal operational amplifier should have the following characteristics:

- Infinite open loop gain, or $A = \infty$
- Infinite input impedance, or $Z_i = \infty$
- Zero output impedance $Z_o = 0$
- Infinite bandwidth and no phase distortion

In reality, there is no ideal operational amplifier. However, in circuit analysis, when compared to other circuit parameters, a good operational amplifier may be considered to be ideal, which, to a great extent, simplifies the analysis process.

In biopotential measurements, a special amplifier called the IA is often used. An IA (Figure 11-2) is a closed-loop, differential input amplifier designed with the purpose of accurately amplifying the voltage difference applied to its inputs. Ideally, an IA responds only to the difference between

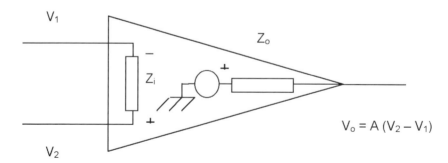

Figure 11-1. Operational Amplifier.

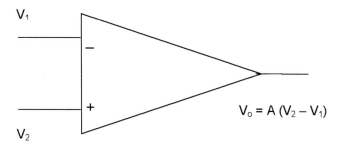

Figure 11-2. Instrumentation Amplifier.

the two input signals and has very high input impedance between its two input terminals and between each input to ground.

The characteristics of a good IA are as follows:

• Very large input impedance
• Very low output impedance
• Constant differential gain with zero nonlinearity
• High common mode rejection (that is, very small common mode gain)
• Very wide bandwidth with no phase distortion
• Low DC offset voltage or drift
• Low input bias current and offset current
• Low noise

Figure 11-3 shows a single Op-Amp differential amplifier.

The voltage V^+ at the noninverting input terminal of the amplifier is given by

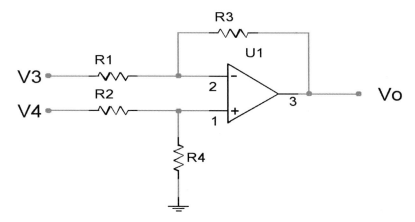

Figure 11-3. Differential Amp Stage of Instrumentation Amplifier.

$$V^+ = \frac{R_4}{R_2 + R_4} V_4. \tag{11.1}$$

For an Op-Amp in its active nonsaturated state, the voltages at the input terminals of the amplifier must be the same, in other words, $V^+ = V^-$. The current flowing through R_1 and R_3 are given by:

$$i_{R1} = \frac{V_3 - V^-}{R_1}$$

$$i_{R3} = \frac{V^- - V_o}{R_3}.$$

Since the input impedance of the Op-Amp is very large, there is no current flowing into the input of the amplifier; the currents flowing through R_1 and R_3 are the same, therefore:

$$\frac{V^- - V_o}{R_3} = \frac{V_3 - V^-}{R_1}. \tag{11.2}$$

Solving Equations (11.1 and 11.2) using $V^- = V^+$ yields

$$V_0 = \left[\frac{R_4}{R_1} \frac{(R_1 + R_3)}{(R_2 + R_4)} \right] V_4 - \left[\frac{R_3}{R_1} \right] V_3. \tag{11.3}$$

If $\dfrac{R_1}{R_3} = \dfrac{R_2}{R_4}$, Equation 11.3 can be simplified to

$$V_o = \frac{R_3}{R_1} (V_4 - V_3) = -\frac{R_3}{R_1} (V_3 - V_4). \tag{11.4}$$

Equation 11.4 is the characteristic of a true differential amplifier with differential gain DG equal to $-R_3/R_1$ and common mode gain CMG (when $V_3 = V_4$) equal to zero. The CMRR of this amplifier is therefore equal to infinity:

$$CMRR = \left| \frac{DG}{CMG} \right| = \frac{R_3/R_1}{0} = \infty.$$

The input impedance of this amplifier between the inverting and non-inverting input to ground is $R_1 + R_3$ and $R_2 + R_4$, respectively. Since these

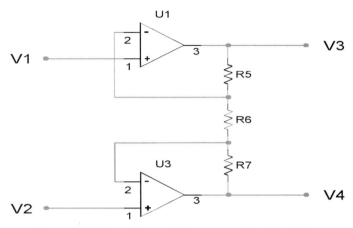

Figure 11-4. Input Stage of Instrumentation Amplifier.

resistances must be much smaller than the input impedance of the Op-Amp, typical input impedance of this differential amplifier stage is usually below 100 kΩ. To overcome this shortcoming, another amplifier stage shown in Figure 11-4 is required to increase the input impedance of the IA. Figure 11 below is the analysis of this impedance matching stage of the amplifier:

For an ideal Op-Amp, due to its large input impedance, no current flows into the input terminals; therefore, the currents flowing through R_5, R_6, and R_7 are identical and can be written as

$$i = \frac{V_3 - V_4}{R_5 + R_6 + R_7}. \tag{11.5}$$

Since the voltages at the two input terminals of an active unsaturated Op-Amp are identical, the current flowing through the resistor R_6 is equal to:

$$i = \frac{V_1 - V_2}{R_6}. \tag{11.6}$$

Therefore,

$$V_3 - V_4 = \frac{R_5 + R_6 + R_7}{R_6}(V_1 - V_2).$$

The differential gain of this stage is therefore $\left(\frac{R_5 + R_6 + R_7}{R_6}\right)$. From Equation 11.6, when $V_1 = V_2$, $i = 0$. Since there is no current flowing through the resistor R_5, V_3 is equal to the voltage at the inverting input of U1; therefore, $V_3 = V_1$. Similarly, we can show that $V_4 = V_2$. If the common mode voltage at the input is V_{cm}, the same common voltage will appear at the outputs V_2 and V_3 of the Op-Amp. Therefore, the common mode gain for this circuit is equal to unity. The CMRR for this stage is then given by

$$CMRR = \frac{|\,DG\,|}{|\,CMG\,|} = \frac{R_5 + R_6 + R_7}{R_6}.$$

Because the input signals are directly connected to the input terminals of the Op-Amp, the input impedance of the circuit is equal to the input impedance of the Op-Amp, which is usually on the order of 100 kΩ. Figure 11-5 shows the combination of these two amplifier stages forming a classic IA. The theoretical differential gain, common mode gain, CMRR, and input impedance of this classic IA are

$$DG = -\left(\frac{R_5 + R_6 + R_7}{R_6}\right)\frac{R_3}{R_1}.$$

$$CMG = \text{zero}$$
$$CMRR = \text{infinity, and}$$
$$Z_{in} = \text{infinity.}$$

However, these values are not achievable due to nonideal Op-Amp characteristics (such as nonzero input bias and offset current). In practice, a good IA can have CMRR > 100,000 and Z_{in} > 100 MΩ.

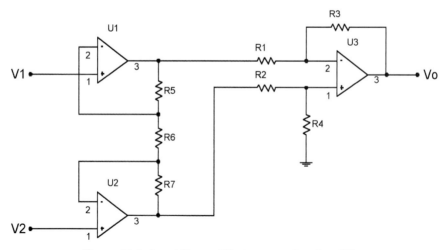

Figure 11-5. Input Stage of Instrumentation Amplifier.

DIFFERENTIAL AND COMMON MODE SIGNALS

Consider the IA shown in Figure 11-6a with voltage signals V_1 and V_2 connected to the input terminals of the amplifier. These input signals can be represented by their common mode and differential mode signals, V_c and V_d, respectively (Figure 11-6b). The mathematical relationships between these voltages are shown by the following equations:

$$V_c = \frac{V_1 + V_2}{2}$$

$$V_d = V_1 - V_2$$

or,

$$V_1 = V_c + \frac{1}{2}V_d$$

$$V_2 = V_c - \frac{1}{2}V_d.$$

If the amplifier is an ideal IA with a differential gain of A, the output is equal to AV_d. The common mode signal V_c will not appear at the output (since the common mode gain is zero).

Example 11.1

Referring to the amplifier in Figure 11-2, suppose the voltage measured at V_1 with respect to ground is 4.0 mVdc, and the voltage a V_2 with respect to ground is 2.5 mVdc. If the differential gain Ad of the differential

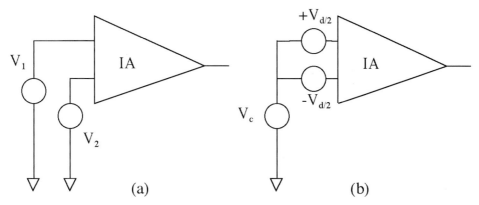

Figure 11-6. Common Mode and Differential Mode Inputs.

amplifier is 500, what is the output voltage V_{out} of the differential amplifier?

Solution:

V_{out} = A_d (V_2 – V_1) = 500 (4.0 – 2.5) mVdc = 500 x 1.5 mVdc = 750 mVdc.

Example 11.2

For an IA with differential gain A_d = 1000, common mode gain A_c = 0.001, what is the output voltage V_{out} if the differential input is a 1.5mV, 1.0-Hz sinusoidal signal and the common mode input is 2.0mV, 60-Hz noise?

Solution:

V_d = 1.5 sin(2πt) mV, V_c = 2.0 sin(120πt) mV

V_{out} = $A_d V_d$ + $A_c V_c$ = [1000 x 1.5 sin(2πt) + 0.001 x 2.0 sin(120πt)] mV
        = [1500 sin(2πt) + 0.002 sin(120πt)] mV
        = 1.5 sin(2πt) V + 2.0 sin(120πt) μV.

This example illustrates the function of the differential amplifier to amplify the differential input signal (desired) while suppressing the common mode signal (noise).

Example 11.3

In an experiment with the Op-Amp in Figure 11-6b, if V_{out} = 10 V when V_d = 1.0 mV and V_c = 0.0; V_{out} = 50 mV when V_d = 0.0 and V_c = 5.0 V, find the differential gain A_d, the common mode gain A_c, the CMRR and CMRdB.

Solution:

A_d = V_{out} / V_d = 10 / 0.001 = 10,000, A_c = V_{out} / V_c = 0.05 / 5 = 0.01

CMRR = A_d / A_c = 10,000 / 0.01 = 1,000,000

The CMRR expressed in dB (that is, CMRdB) is given as

CMRdB = 20 log(CMRR) = 20 log(1,000,000) = 120 dB.

NOISE IN BIOPOTENTIAL SIGNAL MEASUREMENTS

Biopotential signals are produced as a result of action potentials at the cellular level. In physiological monitoring, a biopotential signal is often the resultant electrical potentials from the activities of a group of tissues. We have learned in earlier chapters that a biopotential signal is usually small in magnitude and surrounded by noise. One of the many problems with the amplification of small signals is the concurrent amplification of noise or interfering signal. Noise is simply defined as any signal other than the desired signal. In physiological signal measurements, there are two different sources of noise and interference:

1. Artificial sources from the surrounding environment such as electromagnetic interference (EMI) or mechanical motion. For example, artifacts on an EEG recording caused by fluorescent lighting or unshielded power supply voltages are considered artificial interference.
2. Natural biological signal sources from the patient. For example, in ECG measurement, any signals other than ECG that arise from other biopotentials of the body are considered as natural noise. These include muscle artifact from the patient or electrical activity of the brain. Brain activity is noise when measuring ECG; an ECG signal is considered noise when brain waves (EEG) are measured.

One of the functions of a biopotential amplifier and its associated electronic circuitry is to amplify those biopotential signals of interest while rejecting or minimizing all other interfering signals. An example of a common form of interference in biopotential measurements is shown in Figure 11-7. The ECG signal is corrupted by 60-Hz (or 50-Hz) power line noise induced on the body of the patient (the power line frequency in North America is 60 Hz, whereas in Europe and some Asian countries it is 50 Hz). Noise from power line interference may have amplitude of several millivolts, which can be larger than the signal of interest. Fortunately, the power line-induced 60-Hz noise is on the entire body of the patient and therefore appears equally at both the inverting and noninverting input terminals of the biopotential amplifier. Such common mode signal can be substantially reduced by using a good IA with a large CMRR. Filters can also be used to remove the undesirable signal if the bandwidth of the interfering signal is not overlapping with that of the desired signal.

In order to obtain a good signal in a noisy environment, it is important to have a signal level much larger than the noise level. The ratio of signal to noise level is an important parameter in signal analysis and processing. SNR in decibels is defined as

Figure 11-7. 60Hz Power Line Interference on ECG Signal.

$$SNR(dB) = 20 \times \log \frac{V_s}{V_n},$$

where V_s and V_n are the signal and noise voltage respectively.

When dealing with medical instrumentations, the most common external noise source is from the power lines or electrical equipment in the patient care area. Interference from 60-Hz power can be induced by electric or magnetic fields. A 60-Hz electric field can induce current on lead wires as well as on the patient's body. A changing magnetic field (e.g., from 60-Hz power lines) can induce a voltage or current on a conductive loop. Other than 60-Hz power line interference, much equipment (e.g., switching regulators, electrosurgical units) emits electromagnetic noise into the surrounding area. These EMI can be of low or high frequencies (e.g., 500 kHz from an electrosurgical unit), which may create problems if it is not dealt with properly. EMI can be radiated as well as conducted through cables or conductor connections. For example, high frequency harmonics from switching power supplies can be transmitted through the power grid to other equipment in the vicinity. Switching transients, which may cause damage to electronic components, can be transferred in the same way.

In general, the design of the first stage of medical devices, which usually includes the patient interface and the IA, is critical to maintain a healthy SNR. The remainder of this chapter discusses the mechanism of interference and some practical noise suppression measures.

INTERFERENCE FROM EXTERNAL ELECTRICAL FIELD

A typical electromedical device measuring biopotential from the patient has conductor wires connecting the device to a patient. The patient and the

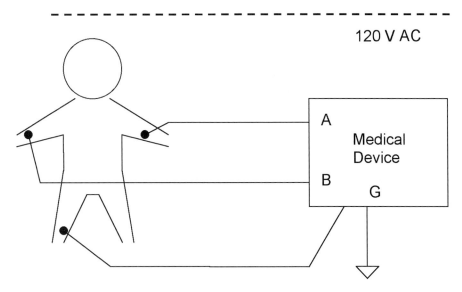

Figure 11-8. Connection of Medical Device to Patient.

device are usually working in an environment filled with an electric field produced by 120-V power lines and line-powered devices. Figure 11-8 shows such an arrangement. Using an ECG as an example, terminals A and B are the inputs of the IA and G is the ground-connecting terminal of the device.

Under typical operating conditions, through capacitive coupling, the electric field produced by the 120-V power sources will induce current flowing into the lead wires as well as into the patient's body to ground. The following sections examine the effects of such induced currents on the output and methods to reduce these interferences.

Currents Induced on Lead Wires

Consider that the medical device is an ECG with skin electrodes attached to the patient as shown in Figure 11-9. Z_1, Z_2, and Z_G represent the impedances of the skin electrode interfaces. C_1 and C_2 are the coupling capacitors between the 120-V power line and the lead wires. These capacitance values depend on the length of the conductors and their distance from the 120-V power sources. Due to these capacitive couplings, displacement currents (I_{d1} and I_{d2}) will flow in the wires to ground. Similarly, C_3 is the coupling capacitor between the power line and the chassis of the device. As the chassis is grounded, the displacement current I_{d3} will flow through the chassis to ground. C_b is the coupling capacitor between the power line and the body of the patient. A displacement current I_{db} will flow into the body of the

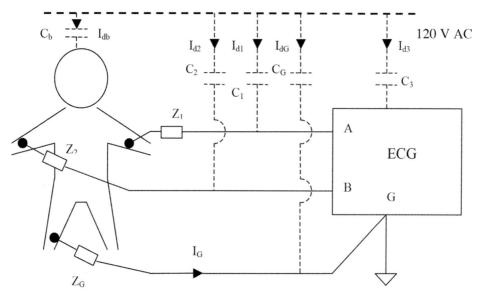

Figure 11-9. Interference From Power Line.

patient.

Assuming the input impedances of the IA are very large, the displacement currents I_{d1} and I_{d2}, which are 60-Hz induced currents, will flow through the skin-electrode interfaces into the patient's body and out through the skin-electrode interface (Z_G) to ground. This current path will create a voltage across the input terminals A and B of the IA given by

$$V_A - V_B = (I_{d1}Z_1 + I_GZ_G) - (I_{d2}Z_2 + I_GZ_G) = I_{d1}Z_1 - I_{d2}Z_2$$

(note the cancellation of the common mode voltage I_GZ_G by the differential amplifier).

If similar electrodes and lead wires are used and they are placed close together, one can simplify this expression by making $I_{d1} = I_{d2} = I_d$. In this case,

$$V_A - V_B = I_d (Z_1 - Z_2) \tag{11.7}$$

Example 11.4

In an ECG measurement using the setup in Figure 11-9, if the displacement current Id due to power line interference is 9 nA and the difference in skin-electrode impedances of the two limb electrodes are 20 kΩ, find the 60 Hz interference voltage across the input terminals of the ECG machine.

Solution:

Using the previous derived equation,

$$V_A - V_B = I_d (Z_1 - Z_2) = 9 \text{ nA} \times 20 \text{ k}\Omega = 180 \text{ }\mu V.$$

With a typical ECG signal amplitude of 1 mV, this represents 18% of the desired signal. This EMI creates a 0.18-mV, 60-Hz signal riding on the 1-mV amplitude ECG waveform (Figure 11-7).

Example 11.5

In Example 11.4, if the displacement current I_{db} through the patient's body is 0.2 μA, what is the common mode voltage at the input terminals of the ECG machine given that the skin-electrode impedance Z_G at the leg electrode is 50 kΩ and the body impedance Z_b is 500Ω?

Solution:

The common mode voltage V_{cm} at the input terminals of the ECG machine is due to the current flowing through the body impedance and the skin-electrode impedance.

$$V_{cm} = (I_{db} + I_{d1} + I_{d2}) (Z_G + Z_b).$$

Since I_{db} is much larger than I_{d1} and I_{d2} and Z_G is much larger than Z_b, we can write

$$V_{cm} = I_{db}Z_G = 0.2 \text{ }\mu A \times 50 \text{ k}\Omega = 10 \text{ mV.}$$

This common mode voltage is ten times the typical amplitude of an ECG signal. Fortunately, this 60-Hz common mode signal will not appear at the output due to the high CMRR of the IA.

From Equation 11.7, in order to reduce the interference signal (power line 60-Hz interference in this case), it is desirable to reduce the induced current ($I_{d1} = I_{d2} = I_d = 0$) or ensure that the skin-electrode impedances are the same ($Z_1 - Z_2 = 0$). The latter can be achieved by using identical electrodes and ensuring that proper skin preparation is done before the electrodes are applied. One attempt to reduce or even eliminate I_{d1} and I_{d2} is to use shielded lead wires as shown in Figure 11-10a. When the entire length of the lead wires is surrounded by a grounded sheath, the coupling capacitors between the power line and each lead wire are eliminated (i.e., $I_d = I_{d1} = I_{d2} = 0$). Therefore, from Equation 11.7, $V_A - V_B = 0$. However, the shield, which is in close proximity to the lead wires, creates coupling capacitances C_{s1} and C_{s2} with each of the wires. From the equivalent circuit shown in Figure 11-10b, one can show that the voltage across the input terminals of the ECG machine due to the nonzero common mode voltage on the patient body (*see* Example 11.5 for estimation of body common mode voltage) is equal to

$$V_A - V_B = \left(\frac{Z_{s1}}{Z_1 + Z_{s1}} - \frac{Z_{s2}}{Z_2 + Z_{s1}} \right) V_{cm}, \tag{11.8}$$

where Z_{s1} and Z_{s2} are the impedances due to capacitances C_{s1} and C_{s2}, respectively.

Equation 11.8 becomes zero only if $Z_1/Z_2 = Z_{s1}/Z_{s2}$. If this condition is not met, V_{cm} will appear at the output no matter how good the CMRR of the ECG is.

To prevent this, a second shield (called the guarding shield) is placed between the first shield and the lead wires, and the guarding shield is connected to the patient's body (e.g., the right leg) via an electrode. This setup is shown in Figure 11-11a, and its equivalent circuit is shown in Figure 11-11b. Note the potential of the guarding shield is at the same level as that of the patient body (i.e., at V_{cm}). Therefore, V_{cm} will not show up across the input terminals of the ECG because there is no current flow around the loops of Z_1–$Z_{s'1}$–Z_G and Z_1–$Z_{s'2}$–Z_G in the circuit. ($Z_{s'1}$ and $Z_{s'2}$ are the impedances of the coupling capacitors $C_{s'1}$ and $C_{s'2}$, respectively, between the guarding shield and the lead wires.)

By using the input guarding method, the induced lead current I_d and the common mode voltage V_{cm} will not appear across the input terminals of the IA of the ECG.

Right Leg-Driven Circuit

So far, we have been assuming that the ECG input stage is an ideal IA (i.e., with infinite input impedance and zero common mode gain). Under these ideal conditions, all common mode signals at the input terminals of the IA are rejected. Let us consider a more realistic situation when the input impedance of the IA has a finite value.

An ECG machine with a nonideal IA can be represented by an ideal IA coupled to finite input impedances Z_{in} to ground at each of its input terminals (Figure 11-12). The differential input voltage of this configuration is given by

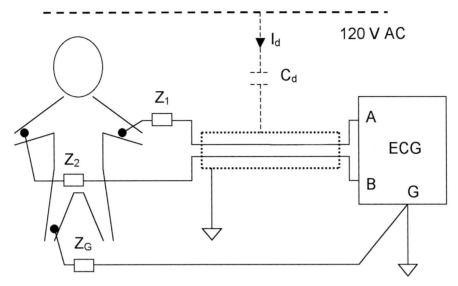

Figure 11-10a. Lead Shielding.

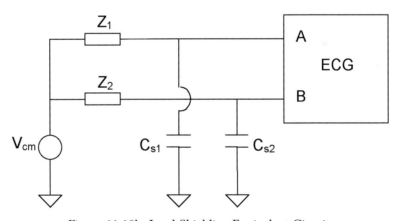

Figure 11-10b. Lead Shielding-Equivalent Circuit.

$$V_A - V_B = \left(\frac{Z_{in}}{Z_1 + Z_{in}} - \frac{Z_{in}}{Z_2 + Z_{in}} \right) V_{cm}.$$

If Z_{in} is much greater than Z_1 and Z_2, we can simplify the equation to

$$V_A - V_B = \frac{Z_2 - Z_1}{Z_{in}} V_{cm}. \tag{11.9}$$

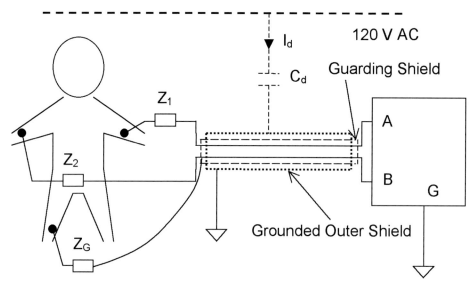

Figure 11-11a. Input Guarding.

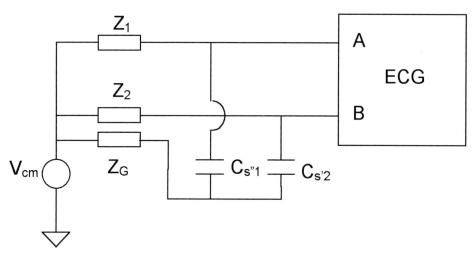

Figure 11-11b. Input Guarding-Equivalent Circuit.

Because the IA has a nonzero differential gain, this differential input voltage ($V_A - V_B$) to the IA due to the common mode voltage V_{cm} will be amplified no matter how large the CMRR is (or how small the common mode gain). To reduce this voltage, we can either choose an IA with very large Z_{in}, use perfectly matching electrodes with good skin preparation (i.e., make $Z_1 = Z_2$), or reduce V_{cm}.

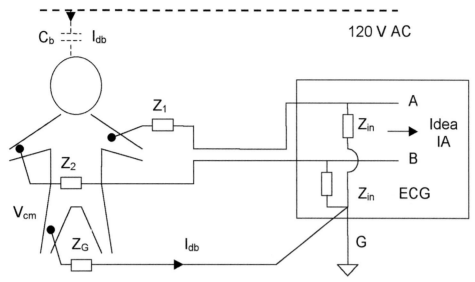

Figure 11-12. Effect of Common Mode Voltage on Finite Input Impedance.

Example 11.6

An IA with input impedances of 5 MΩ between each input terminal to ground is used as the first stage of an ECG machine. If the difference in the skin-electrode impedance is 20 kΩ and the common mode voltage induced from power lines on the patient's body is 10 mV, calculate the magnitude of the 60-Hz interference appearing across the input terminals of the IA.

Solution:

Substituting the value into Equation 11.9, the voltage across the input of the ideal IA is

$$V_A - V_B = \frac{20 \text{ k}\Omega}{5 \text{ M}\Omega} 10 \text{ mV} = 40 \text{ }\mu\text{V.}$$

In the preceding example, the power line voltage appearing across the input terminals of the IA is noticeable when it is compared with a typical ECG signal (amplitude = 1 mV). Since this is a differential input signal to the IA, this power frequency noise signal will be amplified together with the ECG signal and appear at the output of the ECG machine. A practical method using active cancellation to reduce Vcm is the right leg-driven (RLD) circuit shown in Figure 11-13.

The Op-Amps U1, U2, and U3 form a classic IA (*see* Figure 11-5 and its

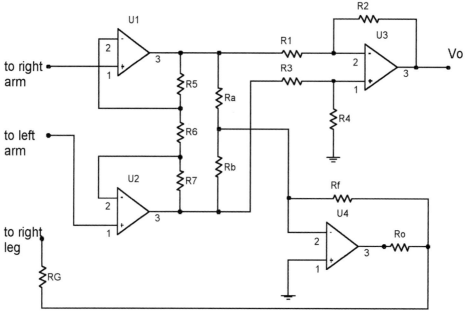

Figure 11-13. Instrumentation Amplifier with RLD Circuit.

corresponding analysis). The inputs are connected to the left and right arm electrodes of the patient. The output voltage V_o of U3 is the amplified biopotential signal between these two limb electrodes. V_o is coupled to the next stage of the ECG machine. U4 with R_f and R_o forms an inverting amplifier with input taken from the output of U1 and U2. This circuit extracts the common mode voltage from the patient's body, inverts it, and feeds it back to the patient via the right leg electrode. It creates an active cancellation effect on the common mode voltage induced on the patient's body and thereby reduces the magnitude of V_{cm} to a much smaller value. Figure 11-14 redraws the RLD circuit to facilitate quantitative analysis of this circuit.

For the RLD circuit in Figure 11-13, R_a is chosen to be equal to R_b. We have also proven earlier that the same common mode voltage V_{cm} at the patient's body appears at the output of U1 and U2. Therefore, we can represent this part of the circuit by a voltage source V_{cm} and a resistor with resistance equal to the parallel resistance of R_a and R_b (equal to $R_a/2$ since $R_a = R_b$). At the right leg electrode, the voltage at the electrode is also V_{cm} (electrode connecting to the patient's body), and there is an induced 60-Hz current I_{db} flowing from the patient into the ECG machine. This equivalent circuit is shown in Figure 11-14.

At the noninverting input terminal of U4, due to the large input impedance of the amplifier, there is no current flowing into or out of the Op-Amp.

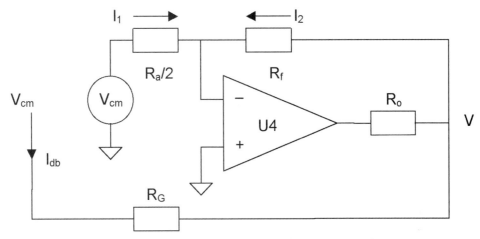

Figure 11-14. Equivalent Circuit of Right Leg-Driven Circuit.

Therefore, $I_1 + I_2 = 0$. But $I_2 = (V - 0)/R_f$ and $I_1 = (V_{cm} - 0)/R_a/2$; therefore,

$$\frac{V}{R_f} + \frac{2V_{cm}}{R_a} = 0 \Rightarrow V = -\frac{2R_f}{R_a} V_{cm}. \tag{11.10}$$

From the lower branch of the circuit,

$$V_{cm} = R_G I_{db} + V. \tag{11.11}$$

Combining Equations 11.10 and 11.11 gives

$$V_{cm} = \frac{R_G}{1 + 2R_f/R_a} I_{db}. \tag{11.12}$$

From Equation 11.12, we can see that if we want to have a small V_{cm}, we must make the denominator as large as possible. That is, the ratio of R_f/R_a should be very large. The function of R_o, usually of resistance equal to several mega ohms, is to limit the current flowing into the Op-Amp when there is a large V_{cm}. It is primarily added to protect the Op-Amp from damage by the high voltage on the patient body during cardiac defibrillation.

Example 11.7

For the RLD circuit in Figure 11-13, using the values in Example 11.5 (i.e., $I_{db} = 0.2~\mu A$, $RG = 50~k\Omega$,

(a) If R_f = 5 MΩ and R_a = 25 kΩ, find the common mode voltage V_{cm} on the patient's body.
(b) Using this new V_{cm} value from above, calculate the magnitude of the 60-Hz interference appearing across the input terminals of the IA in Example 11.6.

Solution:

(a) Substituting values into Equation 11.12 gives

$$V_{cm} = \frac{50 \text{ k}\Omega}{1 + (2 \times 5 \text{ M}\Omega\,/25 \text{ k}\Omega)}\, 0.2 \text{ μA} = 125 \text{ }\Omega \times 0.2 \text{ μA} = 25 \text{ μV}.$$

Using the RLD circuit, we have reduced the V_{cm} from 10 mV to 25 μV, a 400 times reduction.

(b) Substituting values into Equation 11.9, the voltage across the input of the ideal IA is

$$V_A - V_B = \frac{20 \text{ k}\Omega}{5 \text{ M}\Omega}\, 25 \text{ μV} = 0.1 \text{ μV}.$$

This magnitude of noise is negligible when compared to the 1-mV level of the ECG signal.

An alternative configuration of the RLD circuit is shown in Figure 11-15. Interested readers may go through a similar derivation to determine the common modes signal level using this feedback configuration.

INTERFERENCE FROM EXTERNAL MAGNETIC FIELD

Another source of interference is magnetic induction from changing magnetic fields. This can be from power lines or from devices that create magnetic fields, such as large electric motors or transformers. If a changing magnetic field passes through a conductor loop, it will induce a voltage Vi proportional to the rate of change of the magnetic flux Φ, that is

$$V_i \propto \frac{d\phi}{dt} \text{ or } V_i \propto \frac{d(B \times A)}{dt}, \tag{11.13}$$

where the magnetic flux Φ is the product of the area A of the conductor loop

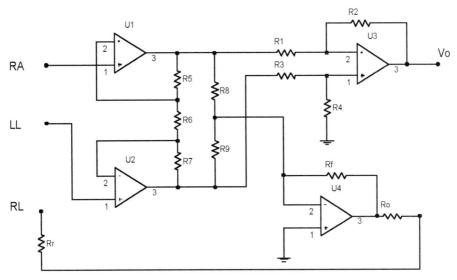

Figure 11-15. Instrumentation Amplifier with RLD (alternative).

and the magnetic field B perpendicular to A.

Figure 11-16a shows the magnetic field interference during an ECG measurement procedure. The conductor loop is formed by the lead wires and the patient's body. If the magnetic field is generated from the ballast of a fluorescent light fixture, a 60-Hz differential signal will appear across the input terminals of the ECG machine. From Equation 11.13, one can minimize the magnitude of interference by reducing loop area A. A simple approach is to place the lead wires closer together to reduce the magnetic induction area. Another method is to twist the wires together (as shown in Figure 11-16b) so that the fluxes cancel each other (flux generated in a loop is in opposite polarity to the flux generated in the adjacent loops). An unwanted magnetic field can also be shielded. However, it is relatively expensive to provide a magnetic shield to protect the device from magnetic field interference.

So far we have been using 60-Hz power line signals as examples of EMI. Other than power frequency interference, there are many other sources of interference that radiate EMI to the surrounding area. In medical settings, electrosurgical units that generate high frequency (e.g., 500 kHz) EMI and radio broadcasting and cellular phones that produce EMI in the GHz range are just a few of the many examples.

CONDUCTIVE INTERFERENCE

Other than EMI, which is often considered as radiated noise, interfer-

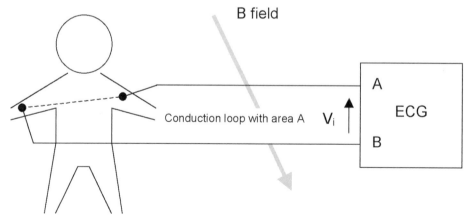

Figure 11-16a. Interference Due to Magnetic Field.

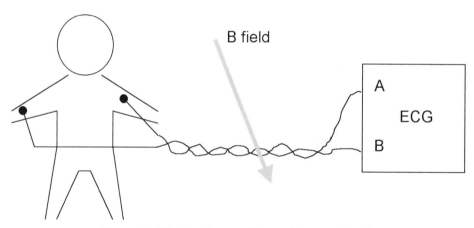

Figure 11-16b. Interference Due to Magnetic Field.

ence can also be caused by unwanted signals conducted to the device via lead wires, power cables, and so on. Some of these conductive interference sources are discussed in the following sections.

High Frequencies and Power Harmonics

This interference appears as high frequency riding on the 60 Hz power voltage. It is usually caused by poorly designed switch-mode regulators that return the high-frequency signal into the power lines. These harmonics and high frequency affect devices connected to the same power grid. They can be eliminated by placing power line filters at the input of the power supply of a device.

Switching Transients

Switching transients are produced when high voltage or high current is turned on and off by a switch or a circuit breaker. During the interruption of a switch or power breaker, arcing occurs across the contact of the switch. This arcing may generate an overvoltage with a short duration of high-frequency oscillation. Switching transients can damage sensitive electronic equipment if the device is not properly protected. Switching transient damage can be prevented by using power line filters and surge protectors.

Lightning Surges

When lightning strikes a conductive cable (such as an overhead power line, a telephone cable, or a network cable) connected to a device, the high voltage and high power surge will be conducted (through the power grid) into the medical device and cause component damage. Surge protection devices with adequate power capacities are required to protect electromedical devices from lightning damage.

Defibrillator Pulses

Medical devices are designed to be safe for patients and operators. Under normal operation, patients and users are not subjected to any electrical risk from medical devices. However, there are times a patient can cause electrical damage to a device. An example of such an occurrence is when a patient is undergoing cardiac defibrillation while an ECG monitor is still connected to the patient. In this case, a high voltage defibrillation pulse, which may be of several thousand volts, will be conducted via the lead wires connected to the patient's body into the ECG monitor. Special high voltage protection circuits (referred to as defibrillation protection circuits) are often built into devices that are subject to such risks. Figure 11-17 shows the defibrillator protection components of an ECG machine. In the case of a high voltage applied to the ECG lead wires, the voltage limiting device will limit the voltage to, say, 0.7 V to protect the IAs and other sensitive electronic components in the machine. The resistance R reduces the current flowing into the voltage limiting device. Without these current limiting resistors, excessively large current exceeding the rated capacity of these devices will flow through the voltage limiting devices to ground.

The voltage-limiting device can be two parallel diodes, two zener diodes in series, or a gas discharge tube as shown in Figure 11-18a. All these devices can limit the voltage level at the input of the ECG machine. The transfer function of the voltage limiting circuit (consisting of the voltage limiting device together with the current limiting resistor) is shown in Figure 11-18b.

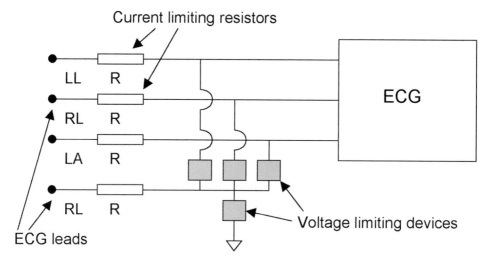

Figure 11-17. Defibrillator Protection Circuit.

For the defibrillator protection circuit shown in Figure 11-17, since the ECG signal is at the most a few millivolts, one can use a silicon diode (with turn on voltage = 0.7 V) as the voltage limiter. Under normal measurement conditions, the voltage at the input of the ECG machine will be equal to the ECG signal (about 1 mV amplitude). During defibrillation, although the voltage on the patient's body can be several thousand volts, the voltage at the input terminals of the ECG machine will be limited to 0.7 V, thereby protecting the electronic components in the ECG machine from being damaged by the high voltage pulses delivered to the patient's body by the defibrillator.

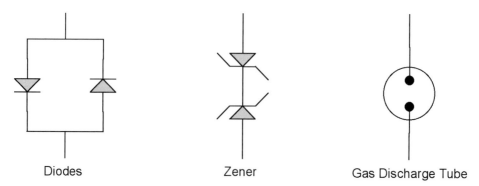

Figure 11-18a. Examples of Voltage Limiting Devices.

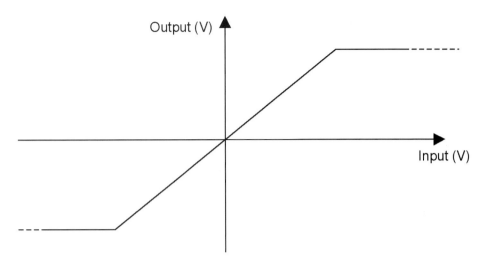

Figure 11-18b. Characteristics of Voltage Limiting Circuits.

BIBLIOGRAPHY

Huhta, J. C., & Webster, J. G. (1973). 60-Hz interference in electrocardiography. *IEEE Transactions on Biomedical Engineering, 20,* 91–101.

Nagel, J. H. (2000). Biopotential amplifers. In J. D. Bronzino (Ed.), *The Biomedical Engineering Hand Book* (2nd ed.). Boca Raton, FL: CRC Press.

Pallas-Areny, R., & Webster, J. G. (1990). Composite instrumentation amplifier for biopotentials. *Anals of Biomedical Engineering, 183*(3), 251–262.

Winter, B. B., & Webster, J. G. (1983). Driven-right-leg circuit design. *IEEE Transactions on Biomedical Engineering, 30,* 62–66.

Chapter 12

ELECTRICAL SAFETY AND SIGNAL ISOLATION

OBJECTIVES

- State the nature and causes of electrical shock hazards from medical devices.
- Explain the physiological and tissue effects of risk current.
- Differentiate microshocks and macroshocks.
- Define leakage current and identify its sources.
- List user precautions to minimize risk from electrical shock.
- Compare grounded and isolated power supply systems.
- Analyze the principles and shortfalls of grounded and isolated power systems in term of electrical safety.
- Explain the function of the line isolation transformer in an isolated power system.
- Explain the purpose of signal isolation and identify common isolation barriers.
- Describe other measures to enhance electrical safety.
- Evaluate the IEC601-1 leakage measurement device and its applications.

CHAPTER CONTENTS

1. Introduction
2. Electrical Shock Hazards
3. Macroshock and Microshock
4. Prevention of Electrical Shock
5. Grounded and Isolated Power Systems
6. Signal Isolation
7. Other Methods to Reduce Electrical Hazard
8. Measurement of Leakage Current

INTRODUCTION

Concerned with the increasing use of medical procedures that penetrate the skin barrier (e.g., catheterization), a number of studies published in the early 1970s suggested the potential occurrences of electrocution of patients from low-level electrical current passing directly through the heart (microshock). It was also demonstrated from animal studies that a 60-Hz current with a level as low as 20 μA directly flowing through the heart can cause ventricular fibrillation (VF).

Although the actual occurrence of death due to microshock in hospitals has never been documented, the potential of such an occurrence is believed to be present. In the interest of electrical safety and risk reduction, hospitals have implemented both infrastructure and procedural measures to prevent such electrical shock hazards. As a result, special considerations were given in designing medical devices to make them electrically safe. This chapter discusses these safety measures and device designs to prevent electrical shock to patients.

ELECTRICAL SHOCK HAZARDS

Electricity is a convenient form of energy. The deployment of electromedical devices in health care has advanced patient care, improved diagnosis, and enhanced the treatment of diseases. However, improper use of electricity may lead to electrical shock, fire, or even explosion, which may lead to patient and staff injuries. The heating and arcing from electricity may cause patient burn or ignite flammable materials such as cotton drapes, alcohol prep solutions, or even the body hair of patients. Together with enriched oxygen content in the surrounding atmosphere, fire and explosion hazards are eminent.

When an electrical current passes through a patient, it creates different effects on tissue. Tissue effects due to electrical current depend on the following factors:

- The magnitude and frequency of the electrical current
- The current path in the body
- The length of time that the current flows through the body
- The overall physical condition of the patient

Skin is a natural defense against electrical shock because the outermost layer of skin (the epithelium) has very low conductivity. The skin is a good insulator (relatively high resistance) surrounding the more susceptible internal organs. The resistance of 1 cm² of skin is about 15 kΩ to 1 MΩ (note that the resistance decreases with increasing contact area). However, skin conductivity can increase 100 to 1000 times (e.g., become 150 Ω) when it is wet.

Due to the nature of medical evaluations and treatments, patients are more susceptible to electrical shocks. Some reasons for the increased electrical hazards are listed below:

- In hospitals, skin resistance is often bypassed by conductive objects such as hypodermic needles, and fluid-filled catheters.
- Fluid inside the body is a good conductor for current flow.
- Surface electrodes (ECG, EEG, etc.) use electrolyte gel to reduce skin resistance.
- Patients are often in a compromised situation (e.g., under anesthesia). They may not be sensitive to or able to react to heat, pain, or other discomfort caused by electrical current.
- Clinicians do not necessarily have knowledge of electricity, nor understand how to maintain a safe electrical environment.

As an example, a patient in an intensive care unit may have several fluid-filled catheters connected to his or her heart to allow pressure measurements in the heart chambers. This same patient could be in an electric bed; be connected with ECG electrodes, temperature probes, respiration sensors, and IV lines; be covered by an electrical hypothermic blanket; and be connected to a ventilator. All of these connections have the potential to conduct a hazardous current to the patient.

An electrical current can create irreversible damage to tissue as well as stimulate muscle and nerve conduction. Table 12-1 shows the current level versus human physiological responses and tissue effects from a 1-second external contact with a 60 Hz electrical current.

Example 12.1

A patient is touching a medical device with one hand and grabbing the handrail of a grounded bed. If the ground wire of the medical device is broken and there is a fault in the medical device that shorted the chassis of the device to the live power conductor (120 V), what is the risk current passing through the patient? Assume that each skin contact has a resistance of 25 kΩ and the internal body resistance is 500 Ω.

Solution:

If the ground wire of the medical device is intact, a large fault current

Table 12-1. Potential Hazards from Electrical Current (External Contact)

Current Level	Physiological and Tissue Effect
1 mA	Threshold of perception. The person begins to sense the presence of the current.
5 mA	Maximum accepted safe current.
10 mA	Maximum current before involuntary muscle contraction. May cause the person's finger to clamp onto the current source ("let go" current).
50 mA	Perception of pain. Possible fainting, exhaustion, mechanical injury.
100–300 mA	Possible ventricular fibrillation (VF).
6A	Sustained myocardial contraction; temporary respiratory paralysis; may sustain tissue burns.
> 10 A	All of the above plus severe thermal burns.

will flow to ground and blow the fuse of the device or trip the circuit breaker of the power distribution circuit. If the ground of the device is open, a current will flow through the patient to ground. The total resistance R of the current path is

$$R = 25 \text{ k}\Omega + 25 \text{ k}\Omega + 500 \ \Omega = 50.5 \text{ k}\Omega.$$

The current I passing through the patient is therefore equal to

$$I = 120 \text{ V}/50.5 \text{ k}\Omega = 2.4 \text{ mA}.$$

According to Table 12-1, since this is above the threshold of perception, the patient should feel the presence of the current. Even though it is not large enough to blow the fuse or trip the circuit breaker, this level of electrical current is not high enough to endanger the patient.

Experimental work on dogs had shown that VF could be onset by a current as small as 20 µA (50–60 Hz) applied directly to the canine heart. Note that this current is 5000 times below the possible VF current (100 mA according to Table 12-1) applied externally. In addition, it was shown that the threshold current triggering these physiological effects increases with increasing frequency. For example, the "let go" current increases from 10 mA to 90 mA when the frequency is increased from 60 Hz to 10 kHz. Electrical current with very high frequency (e.g., 500 kHz current used in electrosurgical procedures) will not cause muscle contraction or nerve stimulation and therefore will not trigger VF. Moreover, skin burn and tissue damage can still occur at high current level with such a frequency.

MACROSHOCK AND MICROSHOCK

The physiological effects described in Table 12-1 are often referred to as macroshocks, whereas shocks that arise from current directly flowing through the heart are referred to as microshocks. Figures 12-1a and b show the differences between macroshocks and microshocks. In a macroshock, the electrical contacts are at the skin surface. The risk current is distributed through a large area across the patient's body. As shown in Figure 12-1a, only a portion of the risk current flows through the heart. In a microshock (Figure 12-1b), the entire risk current is directed through the heart by an indwelling catheter or conductor. Table 12-2 shows the current level versus potential hazard of microshock.

Example 12.2

If the patient in Example 12.1 has a heart catheter (a conductor connected directly to the heart) that is connected to ground, calculate the risk current.

Solution:

Since the catheter bypassed one skin-to-ground contact, the resistance of the current path is now reduced to 25 + 0.5 kΩ = 25.5 kΩ. The risk current therefore is equal to 120 V/25.5 kΩ = 4.7 mA.

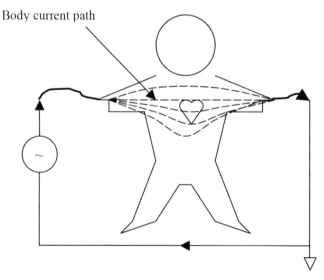

Body current path

Figure 12-1a. Macroshock.

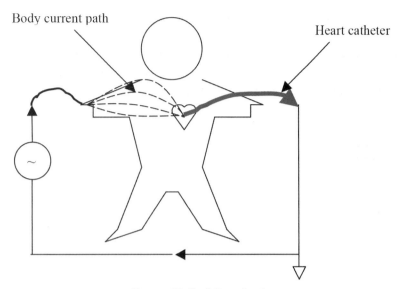

Body current path

Heart catheter

Figure 12-1b. Microshock.

Although this level of current is still considered safe for external contact (Table 12-1), it will trigger VF (>20 μA) under this situation because the current is directly flowing through the heart (see Table 12-2). It should be noted that the physiological effect of such a current is below the threshold of perception (Table 12-1), which suggests that the patient will not be able to feel the current even when it is sufficient to cause microshock. Both macroshock and microshock can cause serious injury or death to the patient. The characteristics of macroshock and microshock are summarized in Table 12-3.

The term leakage current is mentioned frequently in articles on patient safety. Leakage current is electric current that is not functional, but it is not a result of an electrical fault; it flows between any energized components and grounded parts of an electrical circuit. Leakage current generally has two components: one capacitive and the other resistive. Capacitive leakage current exists because any two conductors separated in space have a certain amount of capacitance between the conductors. This undesired capacitance is called stray

Table 12-2. Potential Hazards from Electrical Current (Cardiac Contact)

Current Level	Physiological Effect
0–10 μA	Safe for a normal heart
10–20 μA	VF may occur
20–800 μA	VF

Table 12-3. Characteristics of Macroshock and Microshock

Macroshock	*Microshock*
Requires two contact points with the electrical circuit at different potentials	Requires two contact points with electrical circuit at different potentials
High current passing through the body	Low current passing directly through the heart
Skin resistance is usually not bypassed, i.e., external skin contact	Skin resistance is bypassed
Usually due to equipment fault such as breakdown of insulation, exposure of live conductors, short circuit of hot line to case	Usually due to leakage current from stray capacitors

capacitance. When an alternating voltage is applied between two conductors, a measurable amount of current will flow (e.g., between the primary winding and the metal case of a power transformer, or between power conductors and the grounded chassis). The resistive component of leakage current is primarily due to imperfect insulation between conductors. Since no substance is a perfect insulator, some small amount of current will flow between a live conductor and ground. Because of its relatively small magnitude compared to the capacitive leakage current, however, resistive leakage current is often ignored.

When considering medical device electrical safety, leakage current can be divided into three categories: earth leakage current, touch current (or enclosure leakage current), and patient leakage current. Earth leakage current is current flowing from the mains part (live conductors) across the insulation into the protective earth conductor. Touch current is leakage current flowing from the enclosure or from parts accessible to the operator or patient in normal use, through an external path other than the protective earth conductor, to earth or to another part of the device enclosure. Touch current does not include current flowing from patient connections such as lead wires connecting to patient electrodes. Patient leakage current is leakage current flowing from the patient connections via the patient to earth.

In addition to electrical shocks and fire hazards described earlier, the loss of electricity in health care settings can also create problems and compromise the safety of patients. Table 12-4 summarizes different electrical hazards in the health care environment.

Example 12.3

The total stray capacitance between the live conductors and ground of a medical device powered by a 120 V, 60 Hz power supply is 0.22 nF. Find the total capacitive leakage current flowing to ground.

Solution:

Impedance due to the stray capacitance is

$$\frac{1}{2\pi fC} = \frac{1}{2\pi 60 \times 0.22 \times 10^{-9}} = 12 \ M\Omega.$$

Therefore, magnitude of the capacitive leakage current is

$$\frac{120\,V}{12M\Omega} = 10 \ \mu A.$$

PREVENTION OF ELECTRICAL SHOCK

Risk current is defined as any undesired current, including leakage current, that passes through the body of a patient. Although it cannot be avoided, it can be minimized by appropriate equipment deployment and proper use. Some simple user precautions that medical personnel can take to ensure patient and staff safety are as follows:

1. Medical personnel should ensure that all equipment is appropriate for the desired application. Medical equipment usually has an approval label

Table 12-4. Summary of Electrical Hazards

Type of Hazard	Nature of Hazard
Macroshock	• Burns, including external (skin), internal and cellular • Pain and muscle contraction, may cause physical injury • Ventricular fibrillation
Microshock	• Ventricular fibrillation
Fire and Explosion	• Damage and burn cause by heat or sparks from electrical short circuit or overload in the presence of fuel and enriched oxygen environment
Electrical Failure	• Loss of function of life supporting equipment • Disruption of service and treatment • Panic

on it that informs the operator of the risk level of the equipment. One example of such classification and its meaning is shown in Table 12-5. A patient leakage current is a current flowing from the patient-applied part through the patient to ground or from the patient through the applied part to ground.

2. Ensure that the medical equipment is properly connected to an electrical outlet that is part of a grounded electrical system. The power ground will provide a low-resistance path for the leakage current.

3. Users of medical devices should be cautioned about any damage to the equipment, including signs of physical damage, frayed power cords, and so forth.

4. Ensure that there is an equipment maintenance management program in place so that periodic inspections and quality assurance measures are performed by qualified individuals to ensure equipment performance and safety.

In addition to these user precautions, electrical systems and medical devices can incorporate designs to lower the risk of electrical hazards. The remainder of this chapter describes such designs.

GROUNDED AND ISOLATED POWER SYSTEMS

Figure 12-2 shows the line diagram of a common grounded electrical system. In a grounded system, there are three conductors: the hot, the neutral,

Table 12-5. Classification of Medical Electrical Equipment
(extracted from CAN/CSA–C22.2 No. 60601–1–2:16)

Equipment Type	Intended Application	Maximum Allowable Patient Leakage Current (μA) under	
		Normal Conditions	Single Fault Conditions
B	Equipment with casual patient contact, usually have no patient applied parts	100	500
BF	Equipment with patient applied parts	100	500
CF	Equipment with cardiac applied parts, i.e., connected to the heart or to great vessels leading to the heart	10	50

and the ground. These wires are colored black, white, and green, respectively, in a single phase 120-V North American power distribution system. The neutral wire is connected to the ground wire at the incoming substation or at the main distribution panel. In a three-wire grounded system, the voltages between the hot and the neutral wires and between the hot and the ground wires are both 120 V. Under normal conditions, there is no voltage between the neutral and the ground conductors.

An isolated power system is shown in Figure 12-3. In this distribution system, a line isolation transformer is placed between the power supply transformer and the power outlets. In an isolated power system, there is no neutral wire because both lines connected to the secondary of the transformer are not tied to ground. They are floating with respect to ground; that is, they have no conduction path to ground. The voltage between the lines is 120 V. The color coding for line 1, line 2, and ground conductor of an isolated power system is brown, orange, and green, respectively. Both line conductors are protected by circuit breakers that are mechanically linked together so that they are either open or close together.

In a grounded power distribution system, a short circuit between the hot conductor and the ground creates a large current to flow from the hot conductor to ground. The spark and heat produced can create a fire. In an enriched oxygen environment, an explosion may occur if flammable gas is present. However, in an isolated power system, a short circuit to ground will not produce any significant current since none of the line conductors are connected to ground (no return path exists for the fault current). As a result, fire and explosion hazards due to ground faults are eliminated. Isolated power using line isolation transformers was required in operating rooms by building codes to prevent explosion hazard when flammable anesthetic agents were used. Today, many jurisdictions have removed these requirements because flammable anesthetic agents are no longer used in health care facilities.

Consider a ground fault (a hot conductor is connected to the grounded chassis) occurring on a medical device plugged into a grounded power system (Figure 12-4). If the ground conductor is intact, very little current will flow through the patient; all current will be diverted through the ground conductor to ground even though the patient is touching the chassis. The circuit breaker in the hot wire will detect this excessive current and disconnect (or trip) the circuit.

An intact ground connection is an effective first line of defense against electrical shock. Normally, the circuit breaker will disconnect the device from power in a fraction of a second to minimize patient injury. A grounded system with a protective circuit breaker or fuse is relatively safe to prevent electrical shock to the patient when a ground fault happens. However, a

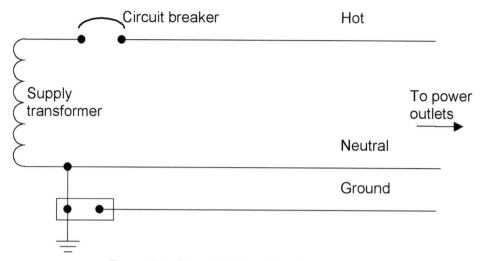

Figure 12-2. Grounded Three-Wire Power System.

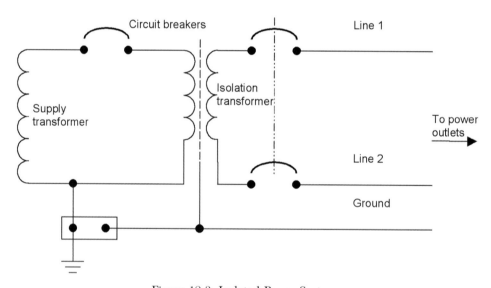

Figure 12-3. Isolated Power System.

spark may jump between the hot and ground conductor at the fault location. Sparks may create a fire or an explosion under the right situation.

Example 12.4

Consider the situation in which the chassis of the medical device in Figure 12-4 is not solidly grounded. If the ground fault current creates a 20

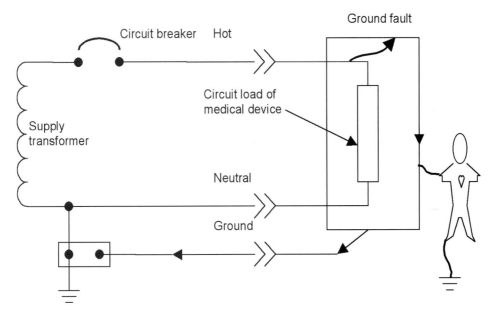

Figure 12-4. Ground Fault on Three-Wire Power System.

V potential difference between the chassis of the medical device to ground, calculate the risk current when

(a) The patient is touching the chassis and also touching a grounded object.
(b) The patient with a grounded heart catheter is touching the chassis.

Solution:

(a) Assuming the resistance of the current path is 50 kΩ, the risk current is then 20 V/50 kΩ = 0.4 mA. This current is harmless and is not even noticeable by the patient.
(b) Assuming the resistance of the current path is 25 kΩ when one skin contact resistance is bypassed by the catheter, the risk current is then 20 V/25 kΩ = 0.8 mA. Because the catheter directs this current to the heart, this micro shock current will trigger VF in the patient.

Example 12.5

The ground connection of a medical device has a resistance of 1.0 Ω. If the leakage current is 100 μA and a patient touching the grounded chassis has a resistance to ground = 25 kΩ, find the risk current flowing through the patient.

Solution:

The patient resistance is parallel to the ground connection resistance, the current flowing through the patient resistance is

$$\left| \frac{100 \ \mu A \times 1\Omega}{25 \ k\Omega + 1\Omega} \right| = 4.0 \ nA.$$

When the ground is intact, only 4 nA of current flows through the patient; with a broken ground, the full leakage current (100 μA) will flow through the patient.

For an isolated power system, a single ground fault between the line conductors and ground will not produce a noticeable fault current because there is no conduction path to complete the electrical circuit. If the fault is a short circuit between one of the line conductors to the chassis of the medical device, however, the chassis will become hot and therefore create a potential shock hazard to the patient. A line isolation monitor (LIM) is used in an isolated power system to detect this fault condition. A LIM is a device that monitors the impedance of the line conductors to ground by periodically (several times per second) connecting each of the line conductors to ground and measuring the ground current (Figure 12-5). A LIM is usually set to sound an alarm when the ground current exceeds 5 mA. Some alarms can be set from 1 to 10 mA. A LIM can detect a ground fault in an isolated power system as well as deterioration of the insulation between line conductors and ground. However, it cannot detect a broken ground conductor nor can it eliminate microshock hazards.

Example 12.6

An isolated power system with a LIM set to sound an alarm at 5 mA is supplying power to a patient location. The patient has a grounded heart catheter and is touching another medical device.

If the leakage impedance due to capacitive coupling of the windings of the line isolation transformer is 25 kΩ and there is an insulation failure between a line conductor and the chassis of the medical device (line shorted to metal chassis), what is the risk current to the patient assuming the patient impedance is 30 kΩ resistive?

Solution:

The current flowing through the patient is equal to the line voltage divided by the total impedance of the current path. The magnitude of the risk cur-

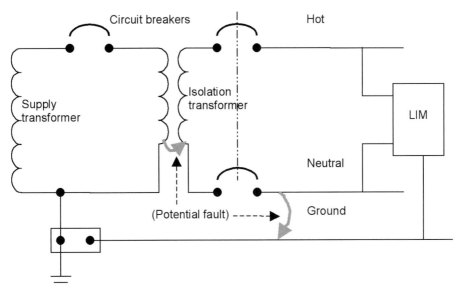

Figure 12-5. Isolated Power System with Line Isolation Transformer.

rent equals

$$\left| \frac{120 \ V}{30 \ k\Omega - j25 \ k\Omega} \right| = 3.0 \ \text{mA}.$$

The low level leakage current will not trigger the alarm of the LIM. However, this will be a fatal current if it is allowed to flow directly through the heart of the patient. In either case, the risk current will be prevented from flowing through the patient if the equipment enclosure is properly grounded.

SIGNAL ISOLATION

Example 12.6 shows that an isolated power system with a line isolation transformer is not sufficient to prevent microshock. In order to reduce microshock hazard, medical equipment with patient applied parts in contact with the heart or major blood vessels is designed such that the applied parts are electrically isolated from the power ground. Isolation is achieved by using an isolation barrier with electrical impedance of over 10 MΩ (such that the leakage current is lower than 10 μA). Figure 12-6 shows the block diagram of such a signal-isolation device. The components enclosed by the dotted line are electrically isolated from the power ground. Note that in order to achieve total isolation, the power supplies to the components before the signal isolation barrier will also need to be isolated. There are two different ground references in a patient applied part isolated medical device. The ground references for the nonisolated part of the circuit are connected to the power ground, whereas the ground references for the isolated components are connected to the ground reference of the isolated power supply. These two grounding references are not connected together and are distinguished by two different grounding symbols (Figure 12-6).

Signal isolation is achieved by the isolation barrier. Common isolation barriers used in medical instrumentation are transformer isolators and optical isolators (Figure 12-7). Isolation transformers used in signal isolation are much smaller in size than are those used in power system isolation because they do not need to transform high power. Furthermore, they have much higher isolation impedances. Optical isolators break the electrical conduction path by using light to transmit the signal through an optical path. Figure 12-7a shows a simple optical isolator using a light-emitting diode (LED) and a phototransistor. The signal applied to the LED turns the LED on at high voltage level and off at low voltage level. The phototransistor is turned on and off according to the light coming from the LED. Since low-frequency signals often suffer from distortion when passing through isolation barriers, a physiological signal is first modulated with a high-frequency carrier (e.g., 50 kHz) before being sent through the isolation barrier. A demodulator removes the carrier and restores the signal to its original form on the other side of the isolation barrier. The signal at the output of the isolation circuit should be the same as the signal at the input.

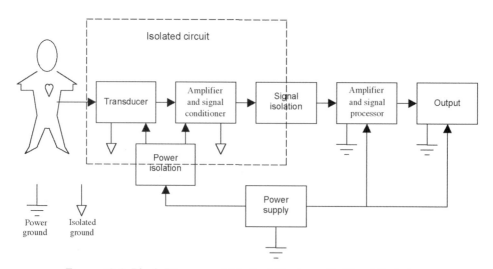

Figure 12-6. Block Diagram of Medical Device with Signal Isolation.

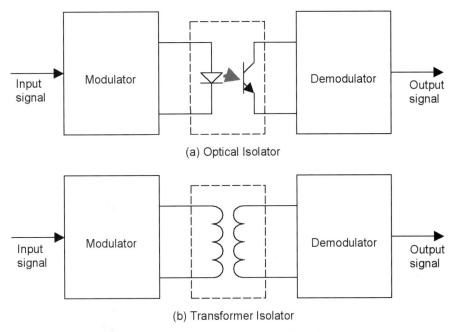

(a) Optical Isolator

(b) Transformer Isolator

Figure 12-7. Optical and Transformer Signal Isolation.

OTHER METHODS TO REDUCE ELECTRICAL HAZARD

Equipotential Grounding

For a patient with the skin impedance bypassed, a tiny voltage can create a microshock hazard. For example, with a current path impedance of 2 kΩ, a voltage difference of 20 mV is sufficient to cause a 10-μA patient risk current. For this reason, it is desirable to protect electrically susceptible patients by keeping all exposed conductive surfaces and receptacle grounds in the patient's environment at the same potential. This is achieved by connecting all ground conductors (equipment cases, bed rails, water pipes, medical gas outlets, etc.) in the patient's immediate environment together and making common ground distribution points in close proximity to patients.

Ground Fault Circuit Interrupters

For an electrical device, the current flowing in the hot conductor should be equal to the current in the neutral conductor. When there is a current flowing from the hot conductor to ground, such as the existence of leakage current or a ground fault, the current in the hot conductor is different from that in the neutral conductor. A ground fault circuit interrupter (GFCI) senses the difference between these two currents and interrupts the power when this difference, which must be flowing to ground, exceeds a fixed value (e.g., 6 mA). Figure 12-8 shows the principle of a GFCI. Under normal conditions, there is no magnetic flux in the sensing coil since the hot and neutral current are equal. When there is a large enough hot to ground current, the net magnetic flux will trip the circuit breaker.

If a person is touching a hot conductor with one hand and a grounded object with the other, a risk current will flow from the hot conductor through the patient to ground. The GFCI detects the current difference in the hot and neutral conductors and interrupts the power before it becomes lethal. This protects the person from macroshock. GFCIs are commonly used in wet locations where water increases the electrical shock hazard. However, a GFCI should not be used in critical patient care areas (OR, ICU, CCU) where life support equipment may be in use because it may be too sensitive to cause unnecessary power interruption.

Double Insulation

A device with double insulation has an additional protective layer of insulation to ensure that the outside casing has a very high value of resistance or impedance from ground. This additional layer of insulation prevents any conductive surface to be in contact with users or patients. Usually the outside casing of the equipment is made of nonconductive material such as plastic. Any exposed metal parts are separated from the conductive main body by the addition of a protective, reinforcing layer of insulation. All switch levers and control shafts must also be double insulated (e.g., using

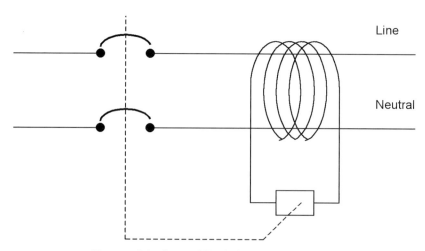

Figure 12-8. Ground Fault Circuit Interrupter.

plastic knobs). In order to be acceptable for medical equipment, the outer casing must be waterproof; that is, both layers of insulation should remain effective, even when there is spillage of conductive fluids. Double insulated equipment need not be grounded, so its supply cord does not have a ground pin.

Batteries and Extra Low-Voltage Power Supply

The higher the power supply voltage, the higher the current that will flow through a person in an electrical accident. To reduce the voltage, one can use a step-down transformer or a low-voltage battery. In general, a voltage not exceeding 30 Vrms is considered as extra low voltage. Such voltage is safe to touch for a healthy person. However, excessive heat from a direct short circuit (e.g., a battery short circuit) can create burn or explosion hazards. Battery-powered equipment is very common in health care settings because in addition to its lower electrical risk, it provides mobility to the equipment as well as to the patient. A common type of equipment using a battery as a source of power is the infusion pump.

Table 12-6 summarizes the effectiveness of different electrical safety protection measures discussed above toward microshock and macroshock.

MEASUREMENT OF LEAKAGE CURRENT

It was mentioned earlier that the levels of electrical shock threshold current increase with increasing frequency. Figure 12-9 shows the approximate relationships between the threshold current and the power frequency. The higher the frequency is, the higher the risk current threshold is to trigger physiological effects. To take into account the frequency-dependent characteristics of the human body's response to risk current, a measurement device to measure device leakage current is specified in the International Electrotechnical Commission standard (IEC601-1) on electrical safety testing of medical devices and systems. The measurement device consists of a passive network with a true root-mean-square (RMS) millivoltmeter, which essentially simulates the impedance of the human body as well as the frequency-dependent characteristics of the body to risk current.

Figure 12-10 shows the IEC601-1 leakage current measurement device. If the leakage current flowing into the measurement device is I_L, the current flowing through the capacitor I_1 is equal to

$$I_1 = I_L \left(\frac{1 \times 10^3}{1 \times 10^3 + 10 \times 10^3 + Z_C} \right), \text{ where } Z_C = \frac{1}{j2\pi f \times 0.0015 \times 10^{-6}} \ \Omega.$$

Table 12-6. Summary of Shock Prevention Methods

	Macroshock	*Microshock*
Proper Grounding	Yes	Yes
Double Insulation	Yes	Yes
Isolated Power System	Yes	No
Isolated Power with LIM	Yes	No
Isolated Patient Applied Parts	Yes	Yes
Equipotential Grounding	N/A	Yes
Ground Fault Circuit Interrupter	Yes	No
Battery Powered	Yes	Yes
Extra Low Voltage (AC)	Yes	No

The voltage across the capacitance as measured by the voltmeter is equal to

$$V_m = I_1 Z_C = I_L \left(\frac{1 \times 10^3}{1 \times 10^3 + 10 \times 10^3 + Z_C} \right) Z_C = I_L \left(\frac{1 \times 10^3}{11 \times 10^3 + Z_C} \right) Z_C$$

For a direct current, f = 0 and $Z_C = \infty$, therefore, $V_m = 10^3 I_L$.

From this relationship, if the voltmeter is measuring 1 mV, the leakage current is equal to 1 µA. Therefore, if the voltmeter is set to read in mV, the value displayed is the leakage current in µA.

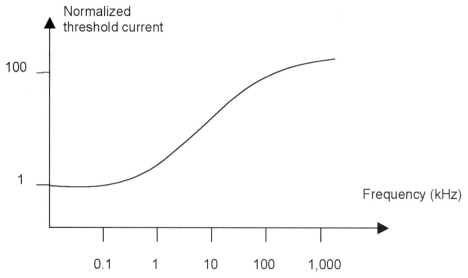

Figure 12-9. Threshold Electric Shock Current Versus Frequency.

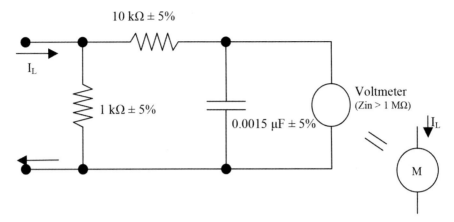

Figure 12-10. IEC601-1 Leakage Current Measurement Device.

For low-frequency current such as 60 Hz power frequency, $Z_C = -j177 \times 10^3\ \Omega$; therefore,

$$|V_m| = \left| I_L \frac{1 \times 10^3 \times (-j177 \times 10^3)}{11 \times 10^3 + (-j177 \times 10^3)} \right| = \frac{10^3 \times 177}{177.3} I_L \approx 10^3 I_L.$$

However, when f is very large, $Z_C \to 0$, therefore,

$$V_m = I_L\left(\frac{1 \times 10^3}{11 \times 10^3 + Z_C}\right) Z_C = 0.$$

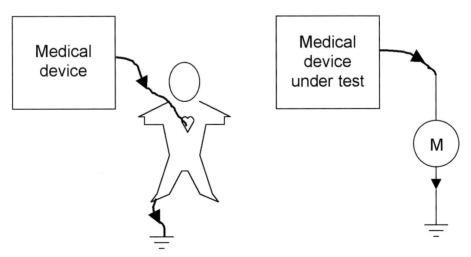

Figure 12-11. Use of IEC601-1 Measurement Device for Leakage Current.

The above analysis illustrates that the IEC601-1 leakage current measurement device has an inverse characteristic to the threshold electrical shock response (*see* Figure 12-9). Instead of having different threshold current levels for different leakage current frequencies (e.g., 10 μA at 60 *Hz* and 90 μA at 10 *kHz*), one can fix the threshold current limits and use the IEC601-1 measurement device to compensate for the frequency response (e.g., 10 μA maximum allowable leakage current for all frequencies). Figure 12-11 shows how the measurement device is used to measure the patient leakage current flowing from the device through a patient applied parts to ground. Interested readers should refer to the Standards (e.g., IEC601-1) for the leakage current limits and the details of how and what to measure.

BIBLIOGRAPHY

Geddes LA. Medical Device Accidents. CRC Press, Boca Raton, FL. 1998.

Daiziel CF. Reevaluation of Lethal Electric Currents. IEEE Transaction on Industry and General Applications. 1968 Sep;v.IGA-4(4):467-475.

Daiziel CF. Threshold 60-cycle fibrillating currents. AIEE Trans. (Power Apparatus and Systems) 1960 Oct;79:667-73

International Electrotechnical Commission. IEC 60601-1-2 (2014), Medical Electrical Equipment—Part 1-2: General Requirements for Basic Safety and Essential Performance. Geneva. 2014.

National Fire Protection Association, NFPA 99, Standard for Health Care Facilities. NFPA, Quincy, MA. 2002.

Standards Council of Canada. CAN/CSA-C22.2 No. 60601-1-2:16, Medical electrical equipment — Part 1-2: General requirements for basic safety and essential performance. Ottawa. 2016.

Chapter 13

MEDICAL WAVEFORM DISPLAY SYSTEMS

OBJECTIVES

- State the functions and characteristics of medical chart recorders and displays.
- Identify the basic building blocks of a paper chart recorder.
- Explain the construction of mechanical stylus and thermal dot array recorders.
- Explain the principles of operation of laser printers and inkjet printers.
- Explain the terms nonfade, waveform parade, and erase bar in medical displays.
- Describe common video signal interface
- Compare and explain the principles of different medical visual display technologies
- Analyze the performance characteristics of medical display systems.

CHAPTER CONTENTS

1. Introduction
2. Paper Chart Recorders
3. Visual Display Technology
4. Video Signal Interface
5. Performance Characteristics of Display Systems
6. Common Problems and Hazards

INTRODUCTION

For a medical device, the output device is the interface between the device

Table 13-1. Paper Chart Recorders.

Continuous Paper Feed	Mechanical stylus recorder; thermal dot array recorder
Single Page Feed	Inkjet printer; laser printer

and its users. Some common output devices found in medical equipment are listed in Table 1-3 of Chapter 1. They include paper records, audible alarms, visual displays, and so on. A video monitor that displays a medical waveform such as an ECG is a typical medical output device. The principles of paper chart recorders and video display monitors are discussed in this chapter.

PAPER CHART RECORDERS

The function of a paper chart recorder in a medical device system is to produce records of physiological waveforms and parameters. These records can be used

- as a snapshot of the patient's physiological condition for future reference, or
- to record an alarm condition so that it can be reviewed later by a medical professional.

Charts are often considered as medical records and therefore are required to be stored for a long period of time (e.g., 7 years for an adult patient in Canada). A paper chart recorder may use ink (e.g., ink stylus, inkjet) or heat (e.g., thermal stylus, thermal dot array) to produce the waveform on a piece of paper. In an ink recorder, ink from an ink reservoir is supplied to a writing mechanism to produce the waveform on a piece of paper. Recorders using thermal styli or thermal dot array print-heads require the use of heat-sensitive or thermal papers. Unlike ordinary paper, thermal paper is coated with a special chemical that turns dark when heat is applied. Thermal paper is more expensive than ordinary paper. However, the design and maintenance of thermal writers are simpler than those of ink writers.

Paper chart recorders can be divided into two categories: continuous paper feed recorders and single page recorders. Thermal dot array print-heads are commonly used in continuous paper feed recorders. Lasers printers are in general the recorder of choice for single page recorders. Examples of paper chart recorders are listed in Table 13-1.

Continuous Paper Feed Recorders

A continuous feed paper recorder draws paper from a paper roll or a stack of fan-folded paper. It can continuously record a waveform for an extended period of time. Two types of continuous paper feed chart recorders are found in medical devices: the mechanical stylus recorders and the thermal dot array recorders.

Mechanical Stylus Recorders

A mechanical stylus recorder consists of an electromechanical transducer to convert the analog electrical signal (e.g., amplified biopotential signal) to a mechanical motion. This motion is mechanically linked to a writing device such as an ink pen or a thermal stylus. The writing device leaves a trace on the chart paper as it moves across the paper. The paper is fed from a paper chart assembly that includes a paper supply mechanism and a writing table. The paper supply is driven by a motorized paper driving mechanism. Figure 13-1 shows a paper chart recorder setup using a galvanometer as the electromechanical transducer. A servomotor drive can replace the galvanometer to drive the mechanical stylus.

Thermal Dot Array Recorders

In a thermal dot array recorder, the electromechanical transducer is replaced by a thermal dot array print-head. The print-head of a thermal dot array recorder consists of a row of heater elements placed on top of the moving thermal paper as shown in Figure 13-2. Each of these heater elements can be independently activated to leave a black dot on the heat-sensitive paper. The analog electrical signal is first converted to a digital signal and then processed to heat the appropriate dots in the print-head. The paper sup-

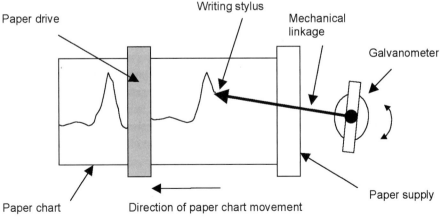

Figure 13–1. Mechanical Stylus Paper Chart Recorder.

ply and drive assembly is similar to that of the mechanical stylus recorder. Movement of the paper in conjunction with the appropriate addressing of the thermal elements on the print-head produces the image on the paper. Because there are only a finite number of thermal elements on the print-head, the trace recorded on the paper is not continuous like the trace produced by a mechanical stylus writer. The vertical (perpendicular to the paper movement) resolution of the recorder is limited by the number of thermal elements in the print-head. Paper chart recorders used in physiological monitors usually have resolution better than 200 dots per inch (dpi). Table 13-2 lists the major functional components of the mechanical stylus recorder and the thermal dot array recorder.

Mechanical stylus chart recorders are being replaced by thermal dot array recorders in medical devices because they have fewer mechanical moving parts. Mechanical stylus recorders require a higher level of maintenance due to wear and tear and misalignment problems.

Single Page Feed Recorders

Single page feed recorders use standard size paper (such as 8.5" by 11" paper) and therefore can record only a finite duration of the waveform on a single sheet of paper. Off the shelf laser printers or inkjet printers are commonly used. Figure 13-3 shows the construction of a laser printer.

Table 13-2. Main Building Blocks of Continuous Paper Feed Chart Recorders.

Mechanical Stylus Recorder	*Thermal Dot Array (Dot Matrix) Recorders*
An electromechanical device (galvanometric or servomotor drive) to convert electrical input signal to mechanical movement.	The analog electrical signal is converted to digital signal via an A/D converter.
An arm to transmit the mechanical movement to the stylus. print head.	The digital signal is being processed to heat the appropriate dots in the dot matrix
A stylus to leave a record of the signal on the chart paper as it moves across the paper. It can be an ink stylus or a thermal stylus.	The activated dot in the print head will leave a black dot on the heat-sensitive paper.

A chart paper assembly consisting of a paper supply mechanism and a paper writing table.

A paper drive mechanism to move the chart paper across the table.

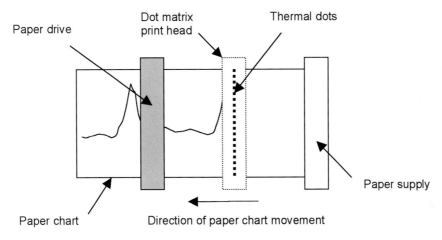

Figure 13-2. Thermal Dot Array Paper Chart Recorder.

Laser Printers

For a laser printer, the time varying signal, such as an ECG, is first converted to a digital signal by an analog to digital converter. A photosensitive drum in the printer rotates at a constant speed. The speed of rotation of the drum determines the paper speed of the chart. The rotational motion of the scanning mirror reflects the laser beam to move across the surface of the drum (Figure 13-3a). The laser diode is switched on and off by the print processor according to the ECG signal and the position of the scanning mirror. The section of the rotating photosensitive drum acquires a negative charge when it passes by the primary corona wire (Figure 13-3b). When the laser beam reflected by the scanning mirror strikes the spots on the drum where dots are to be printed, the spots on the surface of the negatively charged drum become electrically neutral. As the drum rotates, new rows of neutral spots form in response to the laser pulses.

The surface of the developing cylinder contains a weak electric field. As the developing cylinder rotates, it attracts a coating of dark resin particles (toner) that contain bits of negatively charged ferrite. When the resin particles on the developing cylinder move closer to the photosensitive drum, these particles, due to their negative charge, are repelled by the negatively charged area on the drum and moved to the neutral spots that were created earlier by the laser beam.

When the resin particles on the drum move toward the paper, they are being attracted to the paper by the positive charge on the paper created by the transfer corona wire. The image on the drum is therefore transferred onto the paper. To fix the image on the paper, the resin particles are heat

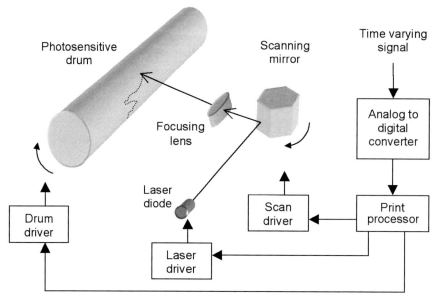

Figure 13-3a. Laser Printer Writer System.

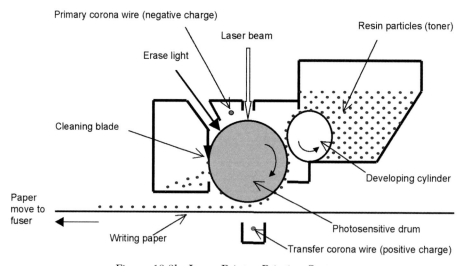

Figure 13-3b. Laser Printer Printing System.

fused onto the paper and the charge on the paper is neutralized by passing the paper over a grounded wire brush. Laser printers can have resolution of 600 dpi or higher.

As the drum continues to rotate, a cleaning blade removes all residue resin particles on the drum surface and an erase lamp introduces a fresh negative charge uniformly on the entire surface of the drum. This prepares the drum to receive the next part of the page information.

Inkjet Printers

An inkjet printer has a print head with multiple print elements as shown in Figure 13-4a. Each element consists of a tiny aperture with a heat transducer behind it. When activated, the heater boils the ink and forms a vapor bubble behind the aperture. The vapor pressure forces a minute drop of ink out toward the printing surface to create a single image dot.

The print head is coupled to an ink reservoir to form the print cartridge. The print cartridge is driven by a servomotor to move back and forth across the paper. Together with the translational motion of the paper, a time varying waveform can be recorded on the paper (Figure 13-4b). Instead of using heat to create the inkjet, some printers use the mechanical vibration force created by a piezoelectric crystal instead of a heat transducer to eject the ink onto the paper. Mechanical inkjet technology is also used in cell dispensing and tissue printing.

VISUAL DISPLAY TECHNOLOGY

In a physiological monitoring system, a visual display monitor provides real-time visual information for the medical professional to assess the condi-

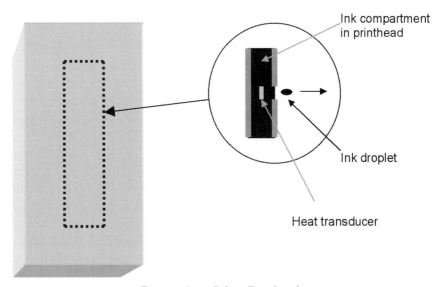

Figure 13-4a. Inkjet Printhead.

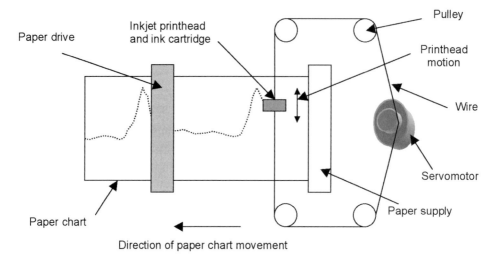

Figure 13-4b. Inkjet Paper Chart Recorder.

tion of the patient. Some common monitors used in medical applications are

- Cathode ray tube (CRT)
- Liquid crystal display (LCD)
- Plasma display
- Electroluminescent (EL) display
- Organic light-emitting diode (OLED) display

Cathode Ray Tube Displays

The CRT has been used for a long time as an efficient and reliable visual display monitor for medical devices. Figure 13-5a shows the main components of a CRT. It is also called an oscilloscope.

The electron gun provides a continuous supply of electrons by heating the cathode filament (thermal ionic effect). These electrons are attracted and accelerated toward the screen due to the high voltage (several kilovolts) applied across the anode and the cathode. When the high-velocity electron beam hits the phosphor screen, it emits visible light photons at the location of impact. To deflect the beam to a desired location on the screen, a voltage is applied across two pairs of deflection plates located above, below, and on each side of the electron beam. In normal time-base operations, a saw-tooth waveform voltage is applied across the horizontal deflection plates. If no signal is applied across the vertical deflection plates, this saw tooth voltage moves the electron beam horizontally back and forth across the tube. When the time varying waveform to be displayed is applied across the vertical deflection plates, it causes the electron beam to move in the vertical direction according to the voltage level of the signal. If the signal frequency is a multiple of the frequency of the saw tooth waveform, a steady waveform of the signal is displayed on the screen of the CRT (Figure 13-5b).

Adjusting the frequency of the saw tooth waveform so that a steady signal is displayed on the CRT screen is called triggering. Automatic triggering is done by feeding the signal to be displayed to a frequency detector and using this frequency to generate the frequency of the sawtooth horizontal deflection signal. The amplitude of the signal displayed on the screen can be adjusted by changing the amplification factor (sensitivity control) of the amplifier feeding the vertical deflection plates.

In order to control the brightness of the display and the convergence of the electron beam to a small dot when it reaches the phosphor screen, control voltages are applied to a set of control and focusing grids in front of the electron gun. Instead of using electrostatic deflection plates, some oscilloscopes use electromagnetic coils to provide horizontal and vertical deflections to the electron beam.

For a fast-moving repetitive waveform, the persistence of the phosphor and the response time of the human eye will make a triggered waveform appear to be continuous and stationary on the screen. For non-periodic waveforms, because each sweep (one cycle of the saw tooth waveform) produces a different trace of waveform on the screen, no stationary waveform will be seen. For slow time varying signals, since the phosphor can scintillate

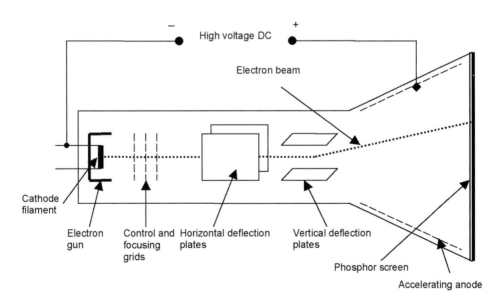

Figure 13-5a. Basic Structure of a CRT.

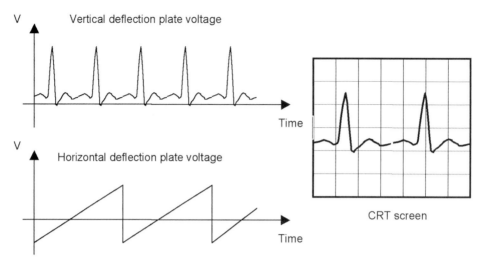

Figure 13-5b. Cathode Ray Tube Display.

only for a fraction of a second, the trace will appear as a dot moving up and down across the screen (this is why old physiological monitors were referred to as bouncing ball oscilloscopes). To produce a steady display even from non-periodic and slow varying signals, a storage oscilloscope is required. Figure 13- 6 shows the block diagram of a storage oscilloscope. Instead of sending the signal to be displayed directly to the vertical deflection plates, the signal is first converted to digital format by an ADC and stored in the memory. This slow varying signal is then reconstructed and swept across the screen many times faster than its original frequency so that it can be seen as a solid trace on the screen. This category of display is called nonfade displays.

There are two types of nonfade displays: one is "waveform parade" and the other is "erase bar." In a waveform parade nonfade display, the waveform appears to be moving across the screen with the newest data coming out from the right-hand side of the screen and the oldest data disappearing into the left-hand side (Figure 13-7a).

In an "erase bar" nonfade display, the data appear to be stationary. A cursor (or a line) sweeps across the screen from left to right (Figure 13-7b). The newest data emerges from the left-hand side of the cursor while the oldest data are erased as the cursor moves over them. When the cursor has reached the right edge of the screen, it disappears and then reappears from the left edge of the screen.

CRT displays are bright and have good contrast ratio (ratio of output light intensity between total bright and dark), high resolution, and high refreshing rate. However, they are heavy and bulky, and may have uneven

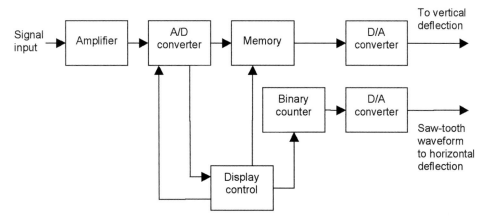

Figure 13-6. Storage Oscilloscope.

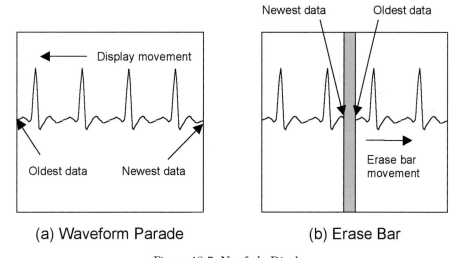

(a) Waveform Parade   (b) Erase Bar

Figure 13-7. Nonfade Display.

resolution across the screen.

Liquid Crystal Displays

LCDs have gained popularity to replace CRTs as the display of choice for medical devices in recent years. LCDs are lightweight, compact, and robust compared to CRTs. LCDs operate under the principle of light polarization. A typical liquid crystal cell is shown in Figure 13-8. Liquid crystal, which has the ability to rotate polarized light, is sandwiched between two transparent electrodes and two polarizers with axes aligned in the same direction to each other. The light from a light source at the back of the cell is polarized by the first polarizer before it passes through the liquid crystal and then to the other polarizer. When no voltage is applied across the electrodes, the axis of the polarized light is rotated (or twisted) 90° by the liquid crystals.

Because the axis of the analyzing polarizer is in line with that of the polarizing polarizer, the polarized light is blocked and therefore no light will exit from the other end. When a voltage (about 5 to 20 V) is applied across the electrodes, the twisting effect of the liquid crystal disappears. Since the axis of the polarized light is in line with the axis of the analyzing polarizer, the polarized light can exit through the other end with little attenuation. By switching the voltage across the electrodes, the liquid crystal cell can be turned on (bright) or off (dark). This is called a twisted nematic LCD.

To illustrate the operation of a two-dimensional display, a 5 pixel by 5 pixel LCD panel with column electrodes and a polarizer on one side plus row electrodes and a polarizer on the other side of the LCD crystal is shown in Figure 13-9. Each of the 25 pixels sandwiched between the horizontal and vertical addressing electrodes can be turned on and off by applying appropriate voltages to the rows and column electrodes. For example, if we want to turn on the pixel B-3, the column electrode B will be connected to a positive voltage and the row electrode 3 will be connected to a negative voltage. To display a time varying signal, a display driver converts the input signal to be displayed into column and row addressing sequences to turn the pixels on and off.

The brightness of a pixel is controlled by adjusting the "on" time or the

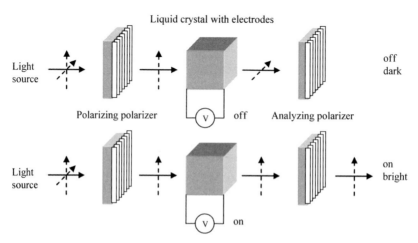

Figure 13-8. Principle of Operation of LCD.

applied voltage across the liquid crystal. The longer the duty cycle of the applied voltage, the brighter the pixel appears to be. To add color to the display, primary color filters (red, green, and blue) are overlaid on top of the pixel elements. Multiple colors are created by combining different intensities of these primary colors. Since it requires 3 pixels to form 1 color pixel, a color LCD display will require three times as many pixels to achieve the same resolution as a monochromatic monitor. A 640 by 480 color LCD panel requires 921,600 LCD pixel elements.

For this type of LCD display, the addressing frequency and hence the screen refreshing rate is limited by the capacitance formed by the addressing electrodes and the LCD crystal (two conductors separated by an insulator). Earlier LCD panels using passive addressing electrodes suffered from slow refreshing rate and often showed a "tail" following a fast-moving object. Employing thin film transistors (TFT) to form an active matrix (AM) has substantially increased the screen refreshing rate in modern LCD displays. However, since each pixel element requires one TFT, AMLCD panels are more expensive than passive LCD panels are. Figure 13-10 shows the schematic diagram of a single pixel element of an AMLCD. For a 640 by 480 color display, 921,600 TFT are required to be fabricated on a single substrate.

An LCD does not emit light photons. An external light source is therefore required to display images on a LCD. A backlit LCD has a light source located at the back of the LCD (Figure 13-8). The light source can be a row of fluorescent light or an array of LED. Liquid crystal displays using LED as

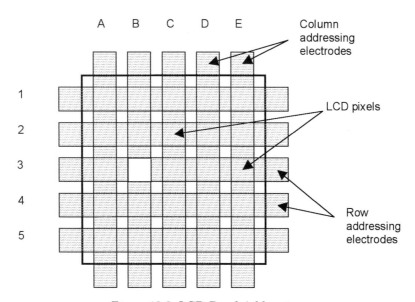

Figure 13-9. LCD Panel Addressing.

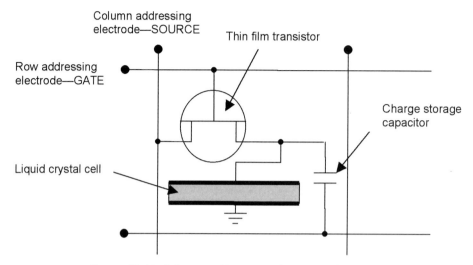

Figure 13-10. Schematic Diagram of an AMLCD Pixel.

light source are often referred to as LED displays. A reflective LCD uses a mirror at the back to reflect light coming from the front, such as daylight or ambient light. Other than its slower screen refreshing rate, a LCD has lower contrast ratio and narrower viewing angle (brightness of the LCD is lower when viewing from the sides and above or below the display) than a CRT has.

Instead of using red, green, blue filters over white light source, some display manufactures employ quantum dot layers to enhance the color output of the display. Quantum dots are nano-semiconductor particles. Depending on their sizes, photo-emissive quantum dots convert ultraviolet lights (from backlit LED of the display) into relatively pure blue, green and red light. It can also produce brighter picture as the color conversion is more efficient than a white light source on color filters.

Plasma Panel Displays

Plasma panels used in low-resolution and alphanumeric displays are established technology and have been proven to be rugged and reliable. High resolution large flat panel plasma displays are used in consumer products such as televisions as well as in medical applications. Plasma is a gas made up of free-flowing ions and electrons. When an electrical current is running through plasma, it creates a rapid flow of charged particles colliding with each other. These collisions excite the gas atoms, causing them to release energy in the form of photons. In xenon-neon plasma, most of these photons are in the ultraviolet region, which is invisible to the human eye.

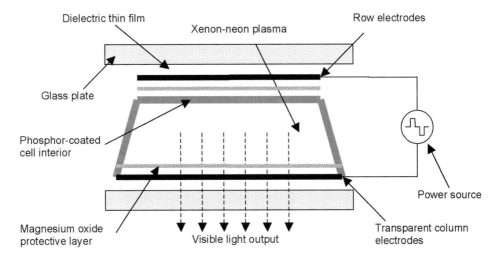

Figure 13-11. Basic Construction of a Plasma Display Cell.

However, these ultraviolet photons can interact with a phosphor material to produce visible light.

The cross-sectional view of a plasma display cell is shown in Figure 13-11. The cell is sandwiched between two glass plates. The addressing electrodes are surrounded by a dielectric insulating material covered by a magnesium oxide protective layer. The plasma is trapped in an enclosure coated with phosphor.

In a color plasma flat panel display, each pixel is made up of three separate plasma cells or subpixels with phosphors chosen to produce red, green, and blue light. Different colors can be produced by combining different intensities of these primary colors by varying the current pulses flowing through each of the three subpixels. Similar to an LCD, row- and column-addressing electrodes are used in plasma display to produce the image. Plasma panels are less bulky than a CRT, and they have a higher contrast ratio and higher response rate than an LCD. However, they require higher driving voltage (150–200 V) than for an LCD.

Electroluminescent Displays

Electroluminescent (EL) displays contain a substance (such as doped zinc sulfide) that produces visible light when an electric field is applied across it. Figure 13-12 shows an EL panel with the EL material between rows and columns of the addressing electrodes.

EL displays are thin and compac·t, with high respons rate and accept-

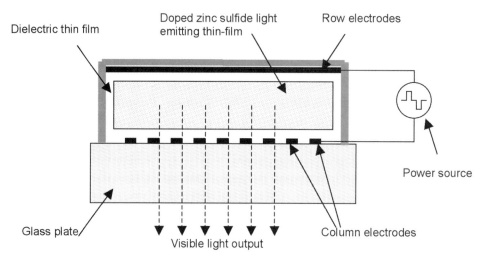

Figure 13-12. Basic Construction of EL Display.

able brightness. However, they are not as popular as LCDs in medical applications because they require relatively high driving voltage (170--200 V) and have lower power efficiency.

Organic Light-Emitting Diode Displays

Other than the display technologies discussed, organic LED (OLED) is becoming popular in many applications, including medical. An OLED consists of a layer of hydrocarbon (organic) compound sandwiched between rows and columns of transparent electrodes. Pixel-addressing in OLED is similar to LCD displays. However, no backlight is needed because the emissive EL material between the addressed electrodes (activated anode and cathode) will emit visible light. In contrast to passive addressing, active-matrix addressing OLED (AMOLED), with a TFT in each pixel, is used to increase the screen refreshing rate. The display panel of an OLED can be made to be flexible. Curved displays are available, and displays that can be rolled up and released are available. In addition, OLED displays have the advantage of high contrast ratio and are thinner and lighter compared to traditional LCD displays. They also have lower power consumption than CRT and plasma panels have. OLED displays are more expensive than flat panel displays using other technologies. Table 13-3 shows a general comparison of the common display technologies.

Table 13-3. Comparison Chart on Display Technologies

	CRT	*LCD*	*Plasma*	*EL*	*OLED*
BRIGHTNESS	High	Need light source	High	High	High
CONTRAST RATIO	High	Low	High	High	High
READABILITY UNDER BRIGHT DAYLIGHT	Washout	Yes	Washout	Washout	Washout
RESOLUTION	Good	Very good	Good	Good	Very good
VIEWING ANGLE	Wide	Narrow	Wide	Wide	Wide
REFRESHING RATE	High	Low (high with AMLCD)	High	High	High
WEIGHT	Heavy	Light	Medium	Medium	Light
SIZE	Bulky	Thin	Thin	Thin	Very Thin
DRIVING VOLTAGE	High	Low	Medium	Medium	Low
COST	Low	Medium	Medium	Medium	High

VIDEO SIGNAL INTERFACE

To display the image from a camera on a monitor, an interface is needed between the source and the receiver. The physical connection and the video signal format must be compatible with both the source and the receiving devices. The following analog video signal formats have traditionally been used:

- Composite—a one-wire format in which all signals are combined into one signal.
- S-Video (or Y/C)—a two-wire format that has two signal components. (Y is the luminance or brightness, and C is the chrominance or color.)
- YP_BP_R (or analog component video)—a three-wire format that has three signal components. (Y is the luminance or brightness, P_B carries the difference between blue and Y, and P_R carries the difference between red and Y; green color can be derived from these signals)
- RGB—a three-wire format with each video signal component carrying one

of the three colors (red, green, and blue). A fourth wire is typically included to carry video synchronized signal.

The preceding analog signals typically adhere to one of the following standards to display "standard definition" (SD) video or television on CRT displays:

- NTSC (National Television System Committee), a television format standard of 30 frames per second (fps) and 525 interlaced scan lines used mostly in American countries.
- PAL (Phase Alternation Line), a television format standard of 25 fps and 625 interlaced scan lines used in most European countries, except France.
- SECAM (Sequential Couleur Avec Memoire), a television format standard of 25 fps and 625 interlaced scan lines used in France, the former Soviet Union and Eastern-bloc countries, and parts of the Middle East and Africa.

In computer applications, analog signals (S-Video, RGB, YP_BP_R) are commonly transmitted via video graphic array (VGA) cables. A typical VGA cable has fifteen pin D-sub connectors on multicore cable. A set of BNC connectors on shielded wires can also be used. VGA interface is used on both CRT and flat panel displays. Analog video formats do not use signal compression. As a result, their picture resolution and color depth are limited by the transmission bandwidth. They are slowly being phased out by digital video signals.

The digital visual interface (DVI) was developed to transfer digital video content to connect a video source to a display. Video signal in the DVI standard is transmitted uncompressed and can be configured in three modes: DVI-A (analog), DVI-D (digital), and DVI-I (both digital and analog). DVI-A is backward compatible with VGA (via a passive adaptor). There is no audio channel in DVI. With three more pairs of cables, DVI-D(DL) and DVI-I(DL) support higher resolution (up to 2560 x 1600 at 60 Hz) video. DVI cables longer than 4.6 m will have signal degradation due to cable loss. Signal repeaters or boosters are available for longer cables. DVI supports a maximum video resolution of 2560×1600.

The high definition multimedia interface (HDMI) standard was developed to transfer digital video and audio signals. An HDMI cable can carry three-dimensional high definition video, up to eight channels of digital audio, and an Ethernet connection. The HDMI 2.1 specification can handle 4K and 8K video at up to 120 frames per second. HDMI and DVI have similar electrical specifications except HDMI is not compatible with VGA, and supports data packet transmission. A 1080p HDMI display can be driven directly via a passive adaptor by a DVI-D source. A DVI monitor may not display video signal from a HDMI source, however.

DisplayPort is a digital display interface developed with the intention of replacing DVI. The interface was designed to connect a digital video source to a display device. The cable can carry video and audio signals, USB, and other data streams all in digital package data format. The high bandwidth of a DisplayPort interface supports up to 8K resolution video. DisplayPort needs active adaptors to communicate with VGA, DVI, or HDMI devices. Dual-mode DisplayPort is compatible with DVI and HDMI with a passive adaptor. Mini DisplayPort is a small footprint version of DisplayPort. Thunderbolt is a successor of Mini DisplayPort with expanded connectivity.

PERFORMANCE CHARACTERISTICS OF DISPLAY SYSTEMS

The output of a display system is often used in diagnosis or to monitor the condition of a patient during medical treatment. The accuracy of the display can therefore affect the medical outcome. This section studies the performance characteristics of the paper chart recorders and the video display monitors discussed earlier in this chapter. Most of the medical device performance characteristics and parameters discussed in Chapter 2 apply to the display systems. Conventional visual display characteristics, such brightness, contrast ratio, and refreshing rates, apply to medical display systems. Some of the parameters critical to medical display systems are discussed in the following.

Sensitivity

The function of a paper chart recorder as well as a video display monitor is to convert the electrical input signal (usually a voltage signal) to a vertical deflection (distance). The sensitivity therefore is distance per unit voltage. For example, the vertical sensitivity of an ECG may be 5 mm/mV, 10 mm/mV, or 20 mm/mV. Many medical devices with an output display have an internal calibration signal to enable users to quickly verify the accuracy of the sensitivity. One example is the 1 mV internal calibration square pulse of an electrocardiograph. When invoked, the size of the square pulse will be shown according to the sensitivity setting. For example, when set at 10 mm/mV, a square pulse with 10 mm amplitude will be recorded or displayed. An external calibration signal can also be applied to the input to verify the accuracy of the display's sensitivity.

Paper Speed and Display Sweep Speed

A physiological signal is often a time varying signal. The distance on the horizontal axis of the display represents the elapsed time of the signal. For an ECG monitor, a common paper speed of the recorder and sweep speed of the monitor is 25 mm/sec. Slower speed of 12.5 mm/sec and higher speed of 50 mm/sec are also available. To check the accuracy of the paper speed or sweep speed, a repetitive signal of known frequency is applied to the input of the display device. The horizontal distance of one cycle of the output signal is measured. The sweep speed of the display is equal to this distance divided by the period of the applied signal.

Resolution

Resolution of a display is a measure of the smallest distinguishable dimension of an image on the display. For a paper chart recorder using a thermal dot array, the resolution is expressed in dots/mm. For example, 8 dots/mm in the vertical direction and 32 dots/mm in the horizontal direction. A high definition (HD) flat panel monitor has 1920 (horizontal) pixels and 1080 (vertical) pixels which can display 1920 vertical lines and 1080 horizontal lines. An ultra-high definition (UHD or 4k) monitor has 3842 × 2160 pixels while an 8K monitor has 7680 × 4320 pixels.

For a monitor with resolution 1280 × 800 pixels, if the dimensions of this monitor are 40 cm by 30 cm, the display resolution is 32 pixels/cm horizontally and 27 pixels/cm vertically. Display resolution can also be expressed in line pairs per mm (lp/mm). Because each line pair requires two lines, the resolution of this monitor has a horizontal resolution of 1.6 lp/mm and a vertical resolution of 1.3 lp/mm (3.2 pixels/mm by 2.7 pixels/mm). Note that radiographic films in medical imaging have a resolution of about 10 lp/mm.

Frequency Response

Like any transducers and functional components, the display of a medical device has a limited bandwidth. A typical frequency response of a display system is shown in Figure 13-13. An ideal display system should have a transfer function with the lower cutoff frequency (f_L) lower than that of the signal that is going to be displayed and an upper cutoff frequency (f_U) higher than that of the signal. The region of the transfer function between the upper and lower cutoff frequency should be constant or flat.

Example 13.1

An ECG is shown. If the sensitivity and paper speed settings are 10 mm/mV and 25 mm/sec, respectively, find the amplitude of the ECG signal and the patient's heart rate. (Note: One small square on the chart equals 1 mm).

Solution:

The amplitude of the R wave is 10 mm. Since the sensitivity setting is 10 mm/mV, the input ECG signal amplitude is 1 mV (*sensitivity = output/input*).

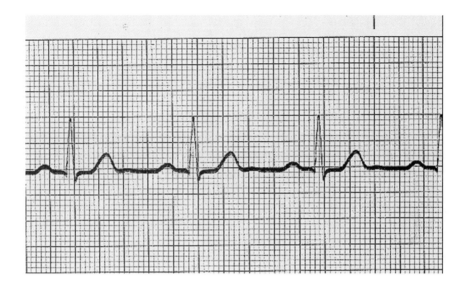

The distance between two QRS complexes is 25 mm (one cycle). With a paper speed of 25 mm/sec, this represents a period of 1 sec. Therefore, the heart rate is 60 beats per minute. (Heart rate in BPM = 60 × s/d, where s is paper speed in mm/sec, and d is the distance in mm between two adjacent R waves.)

The frequency response transfer function of the display system can be obtained by inputting a sinusoidal signal of known amplitude and frequency and measuring the amplitude of the output. By changing the input frequency and repeating the measurement, the frequency response transfer function can be plotted. Using this method to obtain the lower cutoff frequencies of a low frequency signal such as an electrocardiograph (f_L = 0.01 Hz) can be difficult and time consuming. In practice, the lower cutoff frequency of the display (or any system) can quickly be estimated by assuming that the display behaves like a first-order high pass filter.

To find the cutoff frequency of the high pass RC filter as shown in Figure 13-14, a step function is applied to the input to produce the step response from the output. From the step response, f_L can be obtained from the equation:

$$f_L = \frac{1}{2\pi RC}.$$

Where the exponential decay time constant RC can be obtained by using the equation:

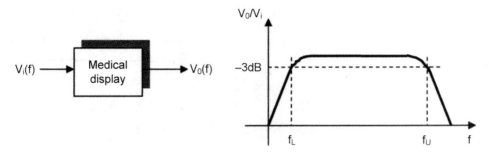

Figure 13-13. Frequency Response of a Medical Display System.

$$V_0 = V_i e^{-\frac{t}{RC}}$$

Example 13.2

Find the lower cutoff frequency of a paper chart recorder if the step response is as shown in Figure 13-14.

Solution:

Using the exponential decay equation

$$V_o = V_i e^{-\frac{t}{RC}},$$

at time t_1 and t_2,

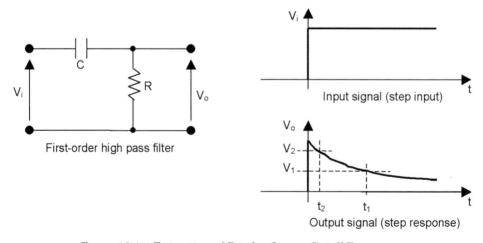

Figure 13-14. Estimation of Display Lower Cutoff Frequency.

$$V_1 = Vie^{-\frac{t_1}{RC}}, \text{ and } V_2 = Vie^{-\frac{t_2}{RC}},$$

dividing the second equation by the first gives

$$\frac{V_2}{V_1} = e^{-\frac{(t_2 - t_1)}{RC}} \Rightarrow RC = \frac{t_1 - t_2}{\ln \frac{V_2}{V_1}}. \text{ But } f_L. = \frac{1}{2\pi RC},$$

therefore, the lower cutoff frequency of the chart recorder can be calculated by looking up the voltage V_1 and V_2 at time t_1 and t_2 from the step response.

Alternatively, if we pick t_2 as the start of the step input (i.e., $t_2 = 0$) and t_1 is the time when the step response dropped from 100% to 50% of the initial value, from the above derived equation

$$RC = \frac{t_1 - t_2}{\ln \frac{V_2}{V_1}}, \; f_L = \frac{1}{2\pi RC} \Rightarrow f_L = \frac{1}{2\pi} \frac{\ln \frac{V_2}{V_1}}{t_1 - t_2} = \frac{1}{2\pi} \frac{\ln 2}{t_1} = \frac{0.11}{t_1},$$

Example 13.3

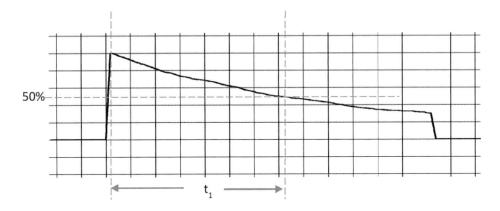

The following chart paper recording was made by an electrocardiograph in response to a step input. Estimate the lower cutoff frequency f_L of the unit if the paper speed is 25 mm/sec. The vertical axis of the chart is 1 mV/division and the horizontal axis is 5 mm/division.

Solution:

The 50% amplitude of the output is seven divisions from the beginning of the step input. Assuming the response is a first-order high pass filter. Using the equation $f_L = 0.11/t_1$ derived above,

$$t_1 = \frac{7 \times 5 \text{ mm}}{25 \text{ mm/s}} = 1.4\text{s. Therefore,}$$

$$f_L = \frac{0.11}{t_1} = \frac{0.11}{1.4} = 0.079 \text{ Hz.}$$

Figure 13-15 shows the effect of changing cut-off frequencies on a rectangular waveform. Note that the waveform becomes more distorted when the bandwidth becomes narrower. An inaccurate or inappropriately chosen medical display with too narrow bandwidth will introduce distortion to the waveforms, provide inaccurate information, introduce errors in diagnoses, and ultimately adversely affect the outcomes of medical treatment. Therefore, medical display systems must be inspected periodically to ensure their performance.

COMMON PROBLEMS AND HAZARDS

Other than electric shock hazards from line power, fluid spills on CRT displays pose high voltage shock hazards to users. Older thermal stylus printers may cause skin burns when users touch on the heated styli. Overheated

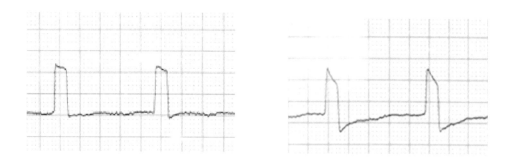

(0.1 то 100 Hz)   (1 to 40 Hz)

Figure 13-15. Effect of Filter Bandwidth on Waveform

styli on stagnant or jammed paper may create fire hazards. Excessive accumulation of ozone may occur when a laser printer is running in a confined space. Ultrafine particles from laser toner that escape to the environment may be inhaled by users and create health hazards.

BIBLIOGRAPHY

Boyes, W. (Ed.). (2003). *Instrumentation Reference Book* (3rd ed.). Burlington, MA: Elsevier Science.

Castellano, J.A. (1992). *Handbook of Display Technology.* Stanford Resources Inc., San Jose, California.

Chang, Y. L., & Lu, Z. H. (2013). *White organic light-emitting diodes for solid-state lighting.* Journal of Display Technology, 9(6), 459–468.

Gray, G. W., & Kelly, S. M. (1999). *Liquid crystals for twisted nematic display devices.* Journal of Materials Chemistry, 9(9), 2037–2050.

He, C., Morawska, L., & Taplin, L. (2007). Particle emission characteristics of office printers. *Environmental Science & Technology*, 41(17), 6039–6045.

Myers, R. L. (2002). *Display Interfaces: Fundamentals and Standards.* New York, NY: John Wiley and Sons.

Part IV

MEDICAL DEVICES

Chapter 14

PHYSIOLOGICAL MONITORING SYSTEMS

OBJECTIVES

- State the functions of a physiological monitor.
- Sketch the functional block diagram of a typical multiparameter patient monitor.
- List common physiological parameters being monitored in clinical settings.
- Identify the common features of a bedside, ambulatory, and central monitor.
- Explain the advantages and shortcomings of a telemetry patient monitoring system.
- Explain the two basic algorithms of ECG arrhythmia detection.
- List the desirable characteristics of a patient monitoring network.
- Differentiate ring, bus, and star network topologies.
- Differentiate host-terminal, client-server, and peer-to-peer networks.
- Describe the characteristics of Ethernet and Token Ring network protocols.
- State the characteristics of different types of network connections.
- List the functions of a network interface card, repeater, bridge, and router.

CHAPTER CONTENTS

1. Introduction
2. Functions of Physiological Monitors
3. Methods of Monitoring
4. Monitored Parameters
5. Characteristics of Patient Monitoring Systems
6. Telemetry

7. Arrhythmia Detection
8. Patient Monitoring Networks
9. Common Problems and Hazards

INTRODUCTION

Since the early 1960s, it has been recognized that some patients with myocardial infarction or those suffering from serious illnesses or recovering from major surgeries benefit from treatment in a specialized intensive care unit (ICU). In such units, cardiac patients thought to be susceptible to life-threatening arrhythmia could have their cardiovascular function continuously monitored and interpreted by specially trained clinicians.

Technological advances led to the ability to monitor other physiological parameters. Temperature transducers such as thermistors made it possible to continuously record patient temperature. Through impedance plethysmography, respiration can also be monitored from the same signals picked up by ECG electrodes. The development of accurate, sensitive pressure transducers gave clinicians the ability to continuously monitor venous and arterial blood pressures; improved catheters made it possible to monitor intracardiac pressures and provide relatively easy and safe methods of measuring cardiac output. Noninvasive means were developed to monitor parameters such as oxygen saturation level in blood.

As clinical knowledge continued to advance and become more sophisticated, special cardiac care units (CCUs) evolved to centralize cardiovascular monitoring. When patients in danger of cardiac arrest are grouped together with trained staff, resuscitation equipment, and vigilance and are combined with prompt responses to cardiac emergencies, lives can be saved. The concept of intensive specialized care assisted by continuous electronic monitoring of physiological parameters has been applied to specialties other than cardiology, resulting in the formation of other special care units such as pulmonary ICUs, neonatal ICUs, trauma ICUs, and burn units.

FUNCTIONS OF PHYSIOLOGICAL MONITORS

Patient monitors continuously or at prescribed intervals measure physiological parameters and store them for later review and thus can free up some clinician time for performing other tasks. In addition, monitors with built-in diagnostic capabilities can alert clinicians when abnormalities such as tachycardia (abnormally high heart rate) occur. Most physiological monitors carry out the following basic functions:

- **Sense**—pick up and transform the physiological signal into a more machine-friendly format (e.g., a pressure transducer to convert blood pressure waveform to an electrical signal)
- **Condition**—amplify, filter, level shift, and so on.
- **Analyze**—make measurements and interpretations of the signal (e.g., extract heart rates from ECG waveform)
- **Display**—show the output in visual (e.g., waveforms, numerical values) or audio formats
- **Alarm**—provide visual and audio warning when some limits are exceeded
- **Record**—store information on paper or electronic media
- **Communicate**—transmit and receive information such as electronic patient data to and from other medical devices, electronic medical records, and the hospital information system.

In a modern patient care ward such as an ICU, a number of physiological parameters of a patient are usually being monitored. Instead of having multiple single-parameter monitors clustered around a patient, a multiparameter monitor is used. A multiparameter monitor has the capability to capture several physiological signals simultaneously. In addition to saving space and providing a more organized scheme to connect the patient to the monitor, a multiparameter monitor is more economical because the modalities may share some common hardware components. Figure 14-1 shows a three-parameter monitor capable of measuring ECG, blood pressure, and temperature. Each of the three modules captures (senses and conditions) a physiological signal from the patient. Because the modules are sharing the remainder of the device components, the signals from the modules are multiplexed and sent to the central processor.

The processor analyzes the signals and extracts necessary information from them (such as heart rate from the ECG and systolic and diastolic pressure from the blood pressure waveform). Such information is then compared with preset parameters to trigger audio or visual alarms. The physiological waveforms as well as numerical information are displayed in the video monitor and hard copies are created by the paper chart recorder. The monitor may also be connected to a central monitor or to the hospital information system through a computer network. Health Level 7 (HL7) is the current standards to facilitate connectivity between different medical devices within the health information systems.

METHODS OF MONITORING

The simplest form of patient monitoring is shown in Figure 14-2a, where

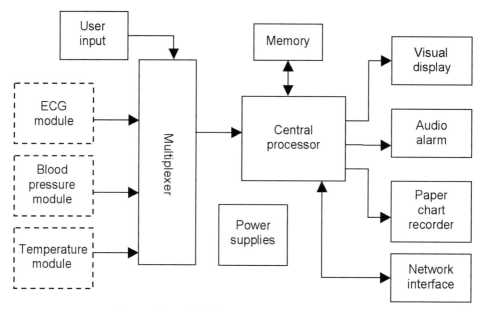

Figure 14-1. A Multiparameter Patient Monitor.

the patient's physiological parameters are displayed at bedside only. This assumes either a one-to-one nurse–patient ratio or a physical grouping of patients that permits viewing of the monitor by the nursing staff. Since alarms must be at the bedside in this type of system, nursing response must be immediate to prevent unnecessary psychological stress accompanying an alarm.

The need for patient privacy in the confines of a single room and the need to economize on nursing staff led to the use of both bedside and a remote monitoring station (commonly refers to central monitoring station). Extending the patient–nurse ratio from 1:1 to 2:1 or even 3:1 permits a more economical approach of surveillance (Figure 14-2b). Under such an arrangement, it is also possible to remove alarms from the patient's range of hearing. Furthermore, the central monitoring and physical layout allow nurses to maintain visual contact with patients and displayed parameters while being able to engage in other nursing duties in the work area.

The grouping of monitors permits the economical addition of components that can be selectively applied to any bed. For example, through suitable grouping, a single chart recorder to provide hard copy rhythm strips can be configured to serve multiple (e.g., four to six) patient beds. Such a central recorder can be programmed to print either on demand by the clinician or automatically in the event of an alarm.

MONITORED PARAMETERS

Some examples of physiological parameters being monitored are

- Electrocardiograph—heart rate, arrhythmia, ST segment level
- Hemodynamics—systolic, diastolic, and mean blood pressure; cardiac output
- Respiration
- Temperature
- End-tidal carbon dioxide level ($EtCO_2$)
- Percentage oxygen saturation ($\%SaO_2$)
- Blood gases (PO_2, PCO_2)
- Electroencephalography and bispectral (BIS) index.

Surface ECG, noninvasive blood pressure, and SpO_2 are basic parameters being monitored in most specialty areas, but additional monitoring needs are required in different areas of care. For example, $EtCO_2$ and BIS index (level of consciousness) monitoring on patients under general anesthesia are often required in the operating room.

CHARACTERISTICS OF PATIENT MONITORING SYSTEMS

The purpose of a patient monitoring system is to monitor vital physio-

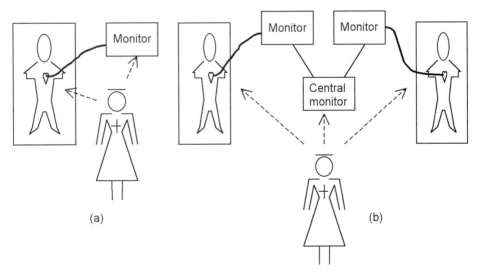

Figure 14-2. Methods of Monitoring.

logical parameters so that clinicians can be alerted to adverse changes in the patient's condition and provide appropriate interventions. A typical system often consists of a number of bedside monitors, a few ambulatory monitors, a central station, and sometimes a telemetry system. In a modern patient monitoring system, all of the preceding are connected via a patient monitoring network.

Bedside Monitors

A bedside monitor is positioned beside the patient bed location to acquire physiological signals from the patient. The monitor is either mounted on the wall or placed on a shelf beside the patient's bed. Catheters and leads physically connect the transducers or electrodes on the patient to the input modules of the monitor. Some of the common features of a bedside monitor are as follows:

- Multiple traces (or channels) display—able to display more than one trace of the same or different physiological parameters. For example, a four-channel monitor can be configured to display two channels of cascaded ECG, one arterial blood pressure, and $\%SaO_2$ waveforms.
- Alarms—provide visual and/or audio alerts when physiological variables are outside certain preset values; usually have silencing feature and are able to automatically reset into "ready" mode after powered off and on.
- Freeze capability—able to freeze the waveform displaying on the screen for more detailed analysis.
- Trending capability—receives input from any number of slowly changing physiological variables and plots a continuous record of this variable over a long period of time. For example, plotting the number of ectopic beats, heart rate, respiration rate, temperature, and blood pressure over time (e.g., 1, 8, or 24-hour basis).
- Recording—able to record a physiological waveform on a printing device. The printing device may be integrated with or networked to the beside monitor.

A bedside monitor can be preconfigured or modular. For a preconfigured monitor, all physiological parameters are built in as an integral part of the monitor at the factory (e.g., a monitor comes with one ECG, one temperature channel, and two blood pressure channels). For a modular bedside monitor, each physiological parameter is an individual module. A bedside monitor can be custom configured by the clinician at the bedside by selecting the modules to meet the monitoring needs of the patient. Modules can be inserted and removed easily by the user. Although the cost of a modular monitor is usually higher than that of a preconfigured monitor with the same features, a modular-design monitoring system is more economical and provides greater flexibility than that of a preconfigured design. For example, instead of having cardiac output measurement capability in all the monitors in a twelve-bed ICU, three cardiac output modules can be shared among twelve modular monitors because cardiac output measurements are done on an intermittent basis and not on all patients.

Ambulatory Monitors

Very often, a patient staying in a hospital has to be transported from one patient location to another or to another hospital. For example, a patient in the emergency department may need to be moved to radiology to have a CT scan. To facilitate patients who require uninterrupted monitoring, a smaller, battery-powered monitor that can be brought along with the patient is necessary. Ambulatory monitors are special monitors that can be transported with the patient. Some manufacturers have ambulatory monitors that use the same bedside monitor modules. Such a system can make disconnecting the patient cables and catheters from the bedside monitor and reconnecting them to the ambulatory monitor unnecessary. To prepare for transport, a user simply removes the modules from the bedside monitor (while still connected to the patient) and inserts them into the ambulatory monitor.

Central Station

The location relationship between the patient bed areas and the nursing work areas should be one in which the nurses are never far from their patients when they are carrying out their routine tasks away from the patients, and in which they can maintain visual observation of the patients. Similarly, the patients, frequently anxious, gain much reassurance by being able to keep the staff in view and by knowing that they are never far away should the patients need help. Therefore, a properly designed central station should maintain two-way visibility between the nurses on duty and each of their patients.

In practice, the central station is an extension of the bedside monitor and provides information from all patients at one location. Typically, one or more large multitrace central monitors and one or more chart recorders are located at the central station. By observing the central monitor, all patient activities can be observed at a glance. In addition, a chart can be printed manually or triggered to print by an alarm from the central recorder at the central station.

The central station monitor is usually a large multichannel instrument capable of displaying several waveform traces at the same time. A basic central monitor has the following capabilities:

- Display multiple traces per monitor
- Waveform selection and position controls for each of the traces on the central display
- Waveform freeze capability on all traces
- Display digital values indicated at the bedside along with alarm limit settings
- Selective trending of parameters

It may also include the following capabilities:

- Arrhythmia detection and ST segment analysis
- Record keeping (e.g., medication and drug interaction)
- Connection to the hospital information system for electronic charting, patient information downloading, electronic medical record retrieval, and billing

The central chart recorder is interfaced with the monitoring alarm system to instantly record the signal if an alarm occurs. It is usually a multichannel recorder capable of simultaneously printing a number of traces from the bedside monitors. The recorder can also be used manually to produce a printout of selected traces from any beside monitors upon demand or, when so equipped, can automatically provide a printout at predetermined time intervals.

TELEMETRY

A conventional ECG patient cable confines patients to the bedside monitors and limits their mobility. Exercise is considered beneficial for coronary patients; increase in mobility often increases the rate of recovery. This limitation can be removed by an ECG telemetry system. A telemetry system removes hardwired connections by replacing it with radio frequency links. ECG telemetry was developed in the 1950s for stress testing. It now replaces or supplements hardwired monitors in acute care units. A typical ECG telemetry system consists of

- A radio transmitter, about the size of a deck of cards, carried by the patient on a belt or in a pocket
- Surface (or skin) electrodes attached to the patient that through lead wires, feed the ECG signal into the transmitter

- A bedside or central station receiver that detects the radio wave and reconstructs the ECG waveform

An ECG telemetry system can be found in progressive care units and CCU units. It allows continuous monitoring of patients who require less care and need mobility. Typically, a CCU patient whose ECG rhythm has satisfactorily stabilized will be put on telemetry, where ECG monitoring continues to detect dangerous arrhythmia. ECG telemetry, however, has the following shortcomings:

- Increased system complexity and cost
- Increased ECG artifacts associated with increased mobility
- Decreased reliability (loss of signal, range limit, electrical interference)
- Delayed locating patients because the freedom afforded by telemetry encourages wandering

In a typical ECG telemetry system, the patient wears a transmitter that is connected to three or more skin electrodes that detect the ECG signal (Figure 14-3). The ECG signal is modulated and transmitted into the free space. The receiver at the receiving station demodulates the signal and displays the waveform and heart rate on the monitor. In some units, a LCD to show the real time ECG waveform and heart rate is integrated with the portable transmitter. Telemetry is also available for other physiological parameters such as pulse oximetry and noninvasive blood pressure measurement. "See Appendix A2 on application of wireless network communication in medical telemetry system."

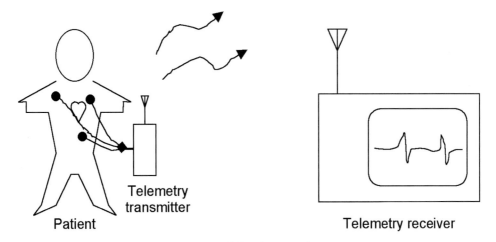

Figure 14-3 ECG Telemetry System.

ARRHYTHMIA DETECTION

ECG is an important physiological parameter in patient monitoring. When a patient has a heart problem such as myocardial infarction, the ECG waveform is different from that of a healthy individual. An arrhythmia is an abnormal heart rhythm. In an arrhythmia, the heart beat may be too fast (tachycardia), too low (bradycardia), irregular (e.g., fibrillation, reentry), or too early (premature ventricular contraction). A computerized arrhythmia detection system continuously analyzes the ECG waveforms and attempts to recognize arrhythmias and to alarm on certain ones. Most patient monitors with arrhythmia detection capability specifically identify and count different types of arrhythmia. The computer in the monitor measures QRS complexes, compares beat intervals, follows some algorithm to classify individual beats, and recognizes arrhythmias. The computer may use additional criteria, such as width, prematurity, heart rate, and compensatory pause, to further classify the type of arrhythmia. In general, there are two types of algorithms in arrhythmia detection:

1. Waveform feature extraction
2. Template matching (or cross-correlation variety)

Waveform feature extraction measures several QRS characteristics (e.g., width, prematurity, height, and area). This is often used in combination with a compression (data reduction) technique such as Amplitude-Zone Time-Epoch-Coding (AZTEC). AZTEC converts the sample ECG signal to a series of constant amplitude or sloped line segments for more efficient input and storage to the computer (Figure 14-4). The computer then identifies and quantifies the QRS complex.

These measurements are then compared to criteria stored in the computer system to differentiate between normal and abnormal beats. As the computer reviews each beat, it groups those with similar features (e.g., height, width). In most systems, the computer then classifies the beats as normal, paced, premature, and so on.

In template matching, the computer samples each beat at approximately 16 to 40 points. These values are then mathematically matched with the values from previously stored beats or a general set of templates. The beat is then classified as normal, abnormal, or questionable based on the number of points on the sampled beat that violate the template boundaries (Figure 14-5).

The template cross-correlation algorithm uses a calculated correlation coefficient (a number from −1 to +1) to mathematically determine how closely the beat matches one of a set of stored templates. A correlation coef-

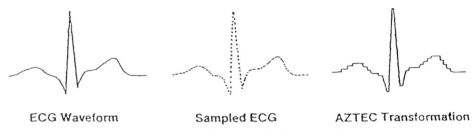

ECG Waveform   Sampled ECG   AZTEC Transformation

Figure 14-4. AZTEC ECG Compression.

ficient greater than or equal to a criterion (e.g., ≥0.9) constitutes a match between the sampled beat and the template.

With both algorithms, questionable beats are usually classified as noise or artifact if similar beats are not detected within a specified time period (generally under 1 minute) or within a certain number of beats (e.g., 500). Those that are seen again are classified as abnormal beats. All algorithms are subject to error, resulting in misclassification of noise and artifact as arrhythmia and incorrect categorization of arrhythmia. Since the "normal" QRS complex of a patient may change with time and condition, the system must "relearn" the patient's normal QRS from time to time to minimize false alarms.

PATIENT MONITORING NETWORKS

Patient monitors that can communicate with the electronic medical record and the hospital information systems can improve efficiency and

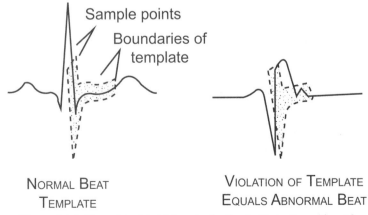

NORMAL BEAT
TEMPLATE

VIOLATION OF TEMPLATE
EQUALS ABNORMAL BEAT

Figure 14-5. Template Matching Arrhythmia Detection Algorithm.

reduce error. Automatic transferring of patient vital signs from patient monitors to electronic medical record with a synchronized time stamp can release nurses from manually filing paper patient charts. Using electronic means to confirm a patient ID with the hospital information system will reduce charting errors, allow up-to-date documentations, and facilitate admission and discharge accounting processes.

To connect bedside monitors to the central monitor, as well as to the hospital information system, a "Patient Monitoring Network" is required. A network is a group of computerized devices connected by one or more transmission media for communication or sharing of resources and data. The transmission links can be wired or wireless. Resources to be shared in a network can be hardware (such as hard disks, printers), software, or human resources (such as accessing remote clinical experts for consultation). Data to be shared can be any information, such as medical images, physiological waveform, or patient information. A network can be a local area network (LAN), which operates in a small area, or a wide area network (WAN), which connects a number of LANs over a large geographic area.

Some desirable characteristics of a patient monitoring network in health care applications are the following:

- Easily adaptable and configurable to all site-specific requirements (i.e., different number of monitored bedside systems, distances, etc.)
- Easy and fast to move information throughout the network
- Allows modifications or changes to system without loss of required function of the rest of the system
- Ability to disconnect and reconnect equipment to system without disruption
- Scalable and allowing variability of monitoring parameters
- Compatible with other network equipment and system (i.e., adheres to industry standards)

Network Topologies

Components in a network can be physically connected in different ways. Figure 14-6 shows some possible network topologies (physical ways of connections) for a patient physiological monitoring network system.

In a ring topology, data pass through each instrument on the way around the circle. In a bus (or tree) topology, the main data are on a pipeline or bus. Each computer receives the same information. In a star topology, the central controller connects to other computers like branches radiating out from the center. It may include satellites (branched out stars).

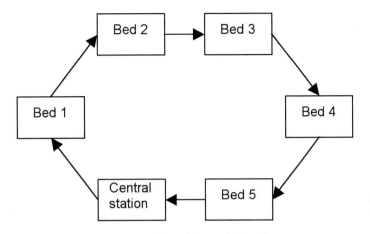

Figure 14-6a. Ring Network Topology.

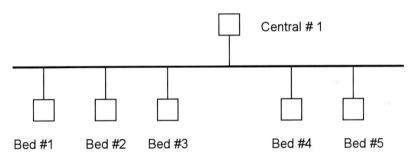

Figure 14-6b. Bus Network Topology.

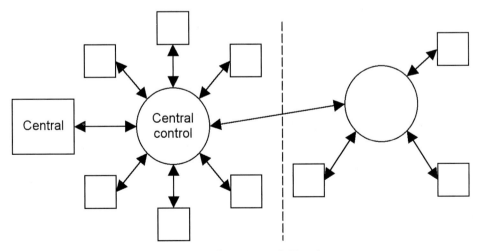

Figure 14-6c. Star Network Topology.

Network Protocols

The ARCnet network protocol, developed in the 1970s by Datapoint, was once a significant industrial standard to handle data links in networking. However, due to its low transmission rate (2.5 Mbps), it was slowly taken over by the Ethernet and Token Ring in the 1980s. Today, Ethernet is the dominant data link protocol used in computer networking, including physiological monitoring. The characteristics of the Ethernet and Token Ring protocols are as follows:

Token Ring

- Token Ring LANs were created by IBM and introduced as the IEEE 802.5 Standard
- Called a logical ring, physical star
- 4 or 16 Mbps bandwidth
- Uses Token-passing access methodology
- Guarantees no data collisions and ensures data delivery
- Sequential message delivery as opposed to Ethernet's broadcast delivery
- Contention is handled through a Token that circulates past all stations
- Token Ring LANs can be set up in a physical ring or a physical star
- The center of the star is called a multistation access unit

Ethernet

- Ethernet LANs were first developed by Xerox in the 1970s
- Adhere to the IEEE 802.3 Standard
- 10 Mbps (10 Base-2 Thinnet) to 100 Mbps (Cat. 5 UTP fast Ethernet) to gigabytes per second bandwidth
- Access methodology is carrier sensed with multiple access and collision detection
- All stations share the same bus
- A station ready to transmit will listen to make sure that the bus is not in use
- Upon a collision (two or more stations were transmitting simultaneously), each will wait for a random period of time and then retransmit
- Quite efficient for low traffic LANs. Data transmission rate suffers at high traffic due to increased frequency of collisions.

To handle networking and transportation of information, the transport control protocol and internet protocol (TCP/IP) is by far the most commonly used network protocol today to resolve addresses and route information and to ensure reliable data delivery.

Networks Models

There are three main network models, each of which is characterized by how it handles traffic and data.

Host-Terminal

- A host computer is connected to dump terminals
- The central host handles processing
- Terminals provide display and keyboard input

Client-Server

- Intelligent client workstations are connected to a server computer
- Application can be customized and processed at the workstation
- Considered to be a distributed processing network
- The server provides services such as file and printer access to the workstations
- Can have more than one server, for example, a file server and a printer server in a network

Peer-to-Peer

- Two or more computers are connected and running the same network software
- Each can do its own processing
- Good for small networks to share resources such as printers, storage, application software, and so on

Network Operating Systems

Most modern commercial network operating systems can work with Ethernet and Token Ring. Examples are the Novell Netware, Microsoft Windows NT and 2000 Servers, UNIX, Mac OS, and so on.

Network Connection Components

Transmission Links

The network hardware and software will determine the data transfer rates between networks and the components within a network. In LANs or WANs, the cables connecting the components are often one of the major factors affecting the data transfer rate. The type of connection and the distance will limit the maximum data transfer rate. Hardwired systems can use twisted copper wires (shielded STP or unshielded UTP), coaxial cables, or fiber-optic cables. Wireless links can use infrared, radio frequency, or microwave for data transmission. For rapid transmission of a large amount of data (e.g., for video conferencing), high-speed links are available through network service providers. Many organizations have installed such high-speed links as the backbone of their WAN.

Network Interconnection Devices

A number of interconnection devices can be found in a computer network system. Some of the common devices are as follows:

Network Interface Card (NIC)
- The interface between the computer and the physical network connection
- Responsible for sending and receiving binary signals according to the network standards
- Each NIC has a unique physical address, called media access control address or MAC address

Hubs
- A device to connect several computers
- Signals sent to the hub are broadcasted to all ports (where computers or network devices are connected) of the hub

Repeaters
- A device to extend network cabling segments over longer distances
- Basic functions are to receive, amplify, and retransmit signal

Switches
- A device to connect several computers
- Signals sent to the switch are broadcasted to selected ports based on the MAC addresses in the packages

Bridges
- An internetworking device to connect a small number of LANs together
- When a bridge receives a packet or a message, it reads the address and forwards the message if it is not local
- Reduce traffic congestion because it will not pass local packets to other LANs

Routers
- An internetworking device to direct traffic across multiple LANs or WANs

- Communicate to each other using a routing protocol that includes a routing table
- The routing table stores information about network accessibility and optimal routing routes

Health Care Network Standards

With the need to share information, LANs and WANs are installed in hospitals, clinics, and communities. To allow data transfer among terminal units connecting to these networks, standards are developed to enhance the connectivity of different medical devices and systems. Examples of standards for the exchange of patient and clinical data are Health Level 7 (HL7) for electronic data exchange in health and Digital Imaging and Communications in Medicine (DICOM) for vendor-independent digital medical image transfers between equipment.

DICOM standard sets the basis for interoperability among imaging devices that claim to support DICOM features. It facilitates communication among equipment from different manufacturers. DICOM 3 supports a subset of the OSI upper level service and is implemented in software.

HL7 is a standard for the exchange, management, and integration of data that support clinical patient care and the management, delivery, and evaluation of health care services. Level 7 refers to the highest level (the application layer) of the ISO OSI communication model. This level addresses definition of the data to be exchanged, the timing of interchange, and the communication of certain errors to the application. It supports such functions as security checks, participant identification, availability checks, exchange mechanism negotiations, and, most importantly, data exchange structuring.

COMMON PROBLEMS AND HAZARDS

Problems associated with patient monitors can be due to user errors (e.g., poor electrode attachment, alarm fatigue), accessory wear and tear (e.g., cable break, insulation breakdown), interface issues (e.g., component incompatibility, improper installation), software glitches (e.g., screen freeze, automatic reset), and external interferences (e.g., EMI from cell phones, lightning surges). These problems may lead to adverse consequences such as patient injury, misdiagnosis, delayed treatment, and loss of information. For monitors that are capable of automatic charting through network connections, accurate time synchronization and patient identification are critical.

Patient monitors may store patient information in the hardware memory.

This may create privacy or confidentiality risks through network connections or improper handling of the monitors. Network and access security must be in place to prevent unauthorized access. Policies and procedures must be established to remove such information when these units are sent out for repair or are being disposed of.

BIBLIOGRAPHY

Cox, J. R., Nolle, F. M., Fozzard, H. A., & Oliver, G. C. (1968). AZTEC, a preprocessing program for real-time ECG rhythm analysis. *IEEE Transactions on Biomedical Engineering, BME-15*(2), 128–129.

Dean, T. (2009). *Network operating systems. In Network+ Guide to Networks* (5th ed.) (pp. 421–483). Boston: Cengage Learning.

Graham, K. C., & Cvach, M. (2010). Monitor alarm fatigue: Standardizing use of physiological monitors and decreasing nuisance alarms. *American Journal of Critical Care, 19*(1), 28–34.

Krasteva, V., & Jekova, I. (2008). QRS template matching for recognition of ventricular ectopic beats. *Annals of Biomedical Engineering, 35*(12), 2065–2076.

Kurose, J. F., & Ross, K. W. (2005). *Computer Networking: A Top-Down Approach Featuring the Internet.* Upper Saddle River, NJ: Pearson Education.

Mandel, W. J. (Ed.). (1995). *Cardiac Arrhythmias: Their Mechanisms, Diagnosis, and Management* (3rd ed.). Philadelphia, PA: Lippincott Williams & Wilkins.

Pan, J., & Tompkins, W. J. (1985). A real-time QRS detection algorithm. *IEEE Transactions on Biomedical Engineering, BME-32*(3), 230–236.

Shaver, D. (2010). The HL7 Evolution: Comparing HL7 Versions 2 and 3. Corepoint Health. Available: http://www.corepointhealth.com/sites/default/files/white papers/hl7-v2-v3-evolution.pdf. Accessed February 16, 2014.

Webner, C. (2011). Applying evidence at the bedside: a journey to excellence in bedside cardiac monitoring. *Dimensions in Critical Care Nursing, 30*(1), 8–18.

Chapter 15

ELECTROCARDIOGRAPHS

OBJECTIVES

- Explain the origin of ECG signal and the relationships between the wave-form and cardiac activities.
- Explain projection and axis of the three-dimensional cardiac vector and analyze the relationships between the ECG leads.
- Define twelve-lead ECG, the electrode placements, connections, and their relationships.
- Differentiate between inpatient and outpatient ECG monitoring.
- Differentiate between diagnostic and monitoring ECG and explain the effects of changing bandwidth on the display waveform.
- Identify and analyze the functional building blocks of an ECG machine.
- List typical specifications of an electrocardiograph.
- Evaluate causes of poor ECG signal quality and suggest corrective solution.
- Describe common problems and hazards

CHAPTER CONTENTS

1. Introduction
2. Origin of the Cardiac Potential
3. The Electrocardiogram
4. Ambulatory ECG Monitors
5. ECG Lead Configurations and Twelve-Lead ECG
6. ECG Axis
7. Vectorcardiogram
8. Fundamental Building Blocks of an Electrocardiograph

9. Typical Specifications of Electrocardiographs
10. ECG Data Storage, Network, and Management
11. Common Problems and Hazards

INTRODUCTION

The class of medical instrumentation to acquire and analyze physiological parameters is called diagnostic devices. This chapter introduces an important diagnostic medical device to monitor and analyze the heart condition through collecting and evaluating electrical potential generated from cardiac activities. This medical instrument is called the electrocardiograph, and the record of the electrical cardiac potential as a function of time is called the electrocardiogram, both can be abbreviated as ECG. The first electrocardiograph came into clinical use in the 1920s using electron vacuum tube amplification, an oscilloscope for display, and a string galvanometer for recording. Electrocardiographs have since evolved into a group of highly sophisticated devices to acquire cardiac potentials, perform diagnostic analysis and interpretation, as well as provide information storage and communication.

ORIGIN OF THE CARDIAC POTENTIAL

The natural pacemaker of the heart is a small mass of specialized heart cells called the sinoatrial (SA) node. The SA node generates electrical impulses that travel through specialized conduction pathways in the atrium (Figure 15-1). As a result of this electrical activation, the atrial muscle contracts to push blood from the atria through the two atrioventricular valves into the ventricles.

While causing the atrial muscle to contract, this electrical impulse continues to travel and eventually reaches another specialized group of cells called the atrioventricular (AV) node. In the AV node, the electrical impulse is delayed by about 100 ms before it arrives at the bundle of His and its two major divisions, the right and left bundle branches. These branches then break into Purkinje's system, which conducts the electrical impulse to the inner wall of the ventricles, allowing synchronized contractions of its ventricles. The ventricular contraction pumps blood from the right ventricle into the lung and from the left ventricle to the rest of the body. The time delay of the electrical impulse in the AV node allows blood to be emptied from the atria to the ventricles before the ventricular contraction. This coordinated contraction of the atria and ventricles maximizes the throughput of the car-

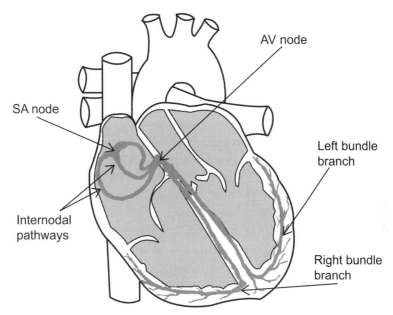

Figure 15-1. The Heart's Electrical Conduction Pathways.

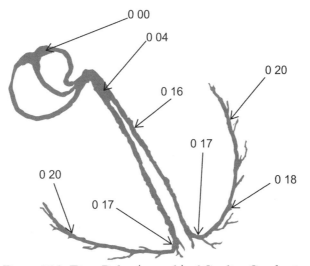

Figure 15-2. Time Delay (seconds) of Cardiac Conduction.

diac contraction. Figure 15-2 shows the time delay of the electrical stimulation reaching different locations of the heart conduction pathway.

The contraction and relaxation of the heart due to synchronized depolarization and repolarization of the cells in the cardiac muscles produce an electrical current that spreads from the heart to all parts of the body. The spreading of this current creates differences in potential at various locations on the body. Figure 15-3a shows a typical action potential plotted against time obtained from a pair of electrodes placed on a ventricular muscle fiber bundle under normal cardiac activity. It shows rapid depolarization (contraction) and then slow repolarization (relaxation) of the muscle fiber. Because there are many fiber bundles contracting and relaxing at slightly different times in a cardiac cycle, the combined result of these electrical potentials forms a cardiac vector of changing magnitude moving in three dimensions with time. The potential difference measured using a pair of electrodes placed on the surface of the body is the projection of the cardiac vector in the direction of the line joining the two electrodes. The waveform obtained by plotting this potential difference as a function of time is called the electrocardiogram.

THE ELECTROCARDIOGRAM

An ECG obtained from electrodes placed on the surface of the body (or skin) is called a surface ECG. A typical surface ECG is shown in Figure 15-3b. It consists of a series of waves (P, Q, R, S, and T) corresponding to different phases of the cardiac cycle. Roughly speaking, the P wave corresponds to the contraction of the atria, the QRS complex marks the beginning of the contraction of the ventricles, and the T wave corresponds to the relaxation of the ventricles. A smaller amplitude U wave is sometimes seen following the T wave; it is thought to represent the repolarization of the interventricular septum including the Purkinje's fibers. In a normal heart, relaxation of the atria occurs at the same time as the contraction of the ventricles. The voltage variation due to atrial relaxation is not visible because of the large amplitude of the QRS complex. The amplitude of the R wave for surface ECG is about 0.4 to 4 mV. Typical amplitude is 1 mV with a cycle time of 1 sec (60 beats per minute). Figure 15-4 shows the relationship between the surface ECG and the depolarization of the heart. In a normal cardiac cycle,

- The P wave represents the depolarization of the atria.
- The PQ (or PR) interval is a measure of the elapsed time from the onset of atrial depolarization to the beginning of ventricular depolarization. It reflects the atrial conduction. Normal PR interval is between 120 to 200 ms.
- The QRS complex marks the start of the depolarization of the ventricles. The normal duration of the QRS complex is between 80 and 100 ms.

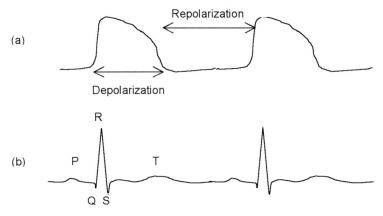

(a)

Repolarization

Depolarization

(b)

R

P

T

Q S

Figure 15-3. (a) Action Potential of a Cardiac Fiber Bundle and (b) Surface ECG.

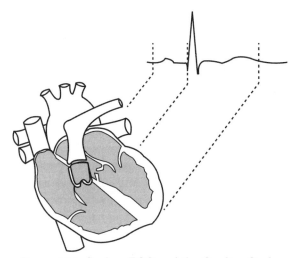

Figure 15-4. Surface ECG and the Cardiac Cycle.

- The QT interval marks the period of depolarization of the ventricle. The QT interval is dependent on the patient's heart rate. QTc is the corrected QT interval which is equal to the QT interval divided by the square-root of the duration between two consecutive R waves, or RR interval. QTc intervals are usually less than 400 ms in men and less than 450 ms in women.
- The T wave reflects ventricular repolarization.
- A small wave (U wave) may be seen in some ECG. It comes right after the T wave and may not always be observed due to its small size. The U wave is thought to represent repolarization of the Purkinje fibers.

Delay due to total interruption or non-responsiveness of some part of the pathway causes changes in the ECG. For example, if a large nonconductive area develops in the wall of the ventricle, the shape or duration of QRS will be altered. Any marked cardiac abnormality such as problems with the SA or AV nodes or in the ventricular conduction pathways will be reflected by changes in amplitude, shape, and interval of the ECG waveform. Surface ECG is an important diagnostic tool for clinicians to gain insight into different abnormal heart conditions.

An ECG can be used to diagnose physiological conditions of the heart (e.g., track heart rhythm and heart rate). Examples of some cardiac arrhythmias (abnormal heart rhythms) revealed in diagnostic ECG are shown in Figure 15-5. Figure 15-5b reviews a premature ventricular contraction caused by an ectopic focus from the ventricle. Figure 15-5c is the most severe consequence of ventricular condition; it occurs when groups of muscle fibers within the myocardium contract and relax at their own pace with no coordination. Under ventricular fibrillation (VF), the heart loses its ability to pump blood into the circulatory system.

An electrocardiogram is an important diagnostic and monitoring tool of the heart. Resting ECGs as well as stress test ECGs (ECG acquired during exercise) are often acquired using the standard twelve-lead ECG configura-

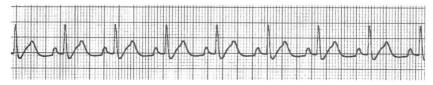

(a) Normal Heart Rhythm.

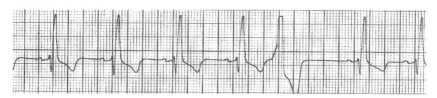

(b) Premature Ventricular Contraction (PVC).

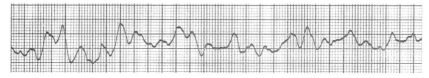

(c) Ventricular Fibrillation.

Figure 15-5. Normal and Arrhythmic ECG.

tion. Vectorcardiograms may be created from the same twelve-lead ECG configuration or from other special lead placements. Continuous inpatient ECG monitoring is a common practice on patient under surgery, in intensive care, and on patients who have heart conditions. Ambulatory ECG records patients' ECG during normal daily activities. Signal processing techniques such as waveform filtering are employed to remove noise from ECG recordings. Special signal processing algorithms are applied to extract specific diagnostic information from electrocardiograms. Application examples of special signal processing are automatic arrhythmia detection and signal averaging in high resolution ECG for ventricular late potential detection.

Resting twelve-lead and stress test ECGs are two examples of diagnostic ECG. When a patient's heart rhythm is monitored while staying in a hospital, it is called monitoring ECG. In general, diagnostic ECG contains more information than monitoring ECG does due to two major factors:

1. The bandwidth of diagnostic ECG is wider than that of monitoring ECG. The lower cutoff frequency of diagnostic ECG is lower and the high cutoff frequency is higher than those of monitoring ECG. (e.g., 0.05 Hz to 120 Hz versus 1 Hz to 40 Hz). Note that the values of the cutoff frequencies can vary between makes and models.
2. There are more leads (projections of the cardiac vector) taken simultaneously in diagnostic ECG than in monitoring ECG.

Another term for ECG with wide bandwidth to allow detailed diagnosis is full-disclosure ECG. The effects of machine bandwidth and lead configurations on ECG will be discussed in more detail later in this chapter.

In a critical care area in a hospital, monitoring of a patient's ECG provides the following information:

• Immediate detection of potentially fatal arrhythmia by means of alarms
• Early warning signs of more major arrhythmias that may follow
• Feedback on the effectiveness of a treatment intervention
• Correlation between cardiac rhythm and treatment variables
• Permanent record of ECG waveform on a routine basis

AMBULATORY ECG MONITORING

Ambulatory ECG monitoring is a procedure to acquire the cardiac rhythm of a patient during normal daily activities over an extended time period in contrast to capture a snap shot during a resting ECG acquisition.

This information is used to monitor patient's heart conditions, detect cardiac rhythm abnormalities, correlate symptoms with activities, or to adjust medications. During monitoring, the patient wears a small ECG module with a built-in data recorder. ECG skin electrodes are applied and the signal is stored in a semiconductor memory inside the module. Older systems use magnetic tapes as storage medium.

A Holter ECG is a special ECG monitor for outpatient use. It continuously records the patient's heart activities over an extended time period, typically for one to two days. Most units allow the patient to manually mark symptomatic events to correlate with the ECG recording. Skin electrodes (usually five electrodes) are applied and the signal is stored in a semiconductor memory inside the module. After the acquisition period, the patient returns the unit to a cardiology clinic. A technologist downloads the ECG recording from memory to a Holter workstation. Since it is extremely time consuming to browse through such a long signal, automatic signal analysis algorithm is built in to compute heart beats, detect abnormal rhythms, etc. However, the success of the automatic analysis is highly dependent on the signal quality. A cardiologist will review the machine interpretation and ECG waveform.

A cardiac event recorder, or loop recorder, records a short duration of ECG when it is activated by the user having a symptom. The system records the ECG events in memory. In addition to recording events, a loop recorder continuously records ECG into a buffer (configured with semiconductor memory) that is continuously updated and overwritten. The event duration recorded varies from a few seconds to a few minutes and is usually programmable. Event recorders capture the information before activation so that ECG data from before, during and after the symptom are captured. Some event recorders can detect abnormal heart rhythm and start to record automatically. During monitoring, the patient wears a small ECG module with a built-in data recorder. Skin electrodes are applied and the signal is stored in a semiconductor memory inside the module. For external event recorders, the patient transmits data over a telephone or wireless network to the designated monitoring station. Some systems employ wireless networks for data transmissions.

Some loop recorders are designed to be implanted inside the patient. An implantable loop recorder is of the size of a USB stick (e.g., 45×7×4 mm for the Reveal® LINQ™, Medtronic) and is placed subcutaneously on the left chest using a small incision with local anesthetic. After it is implanted, a loop recorder can last up to 3 years. As the electrodes are on the chassis of the implantable loop recorder, only one ECG lead can be captured. The recorded information is downloaded to an external receiver via telemetry or wireless network.

ECG LEAD CONFIGURATIONS AND 12-LEAD ECG

In an earlier discussion, we learned that the cardiac vector has a varying magnitude and pointing to different directions with time; also, an ECG is the potential difference measured against time from the projection of the cardiac vector into a direction according to the placement of the pair of electrodes. If ECG electrodes are connected to the right arm (RA), left arm (LA), and left leg (LL) of the patient, one projection of the cardiac vector can be obtained by connecting the LL and RA electrodes to the input terminals of a biopotential amplifier. A different projection of the same cardiac vector can be obtained from the LA and the RA electrodes, and similarly another

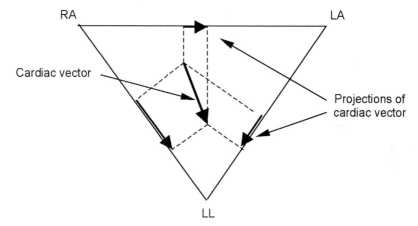

Figure 15-6. Projection of Cardiac Vector in the Frontal Plane.

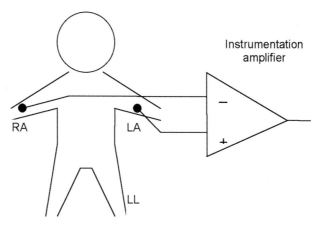

Figure 15-7. ECG Lead I Measurement.

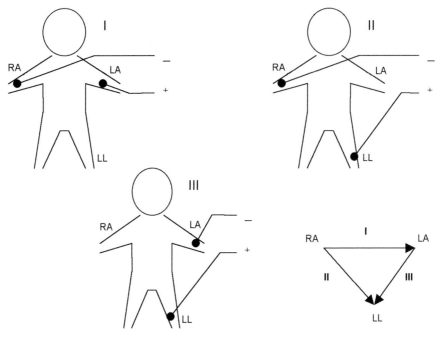

Figure 15-8. ECG Limb Leads.

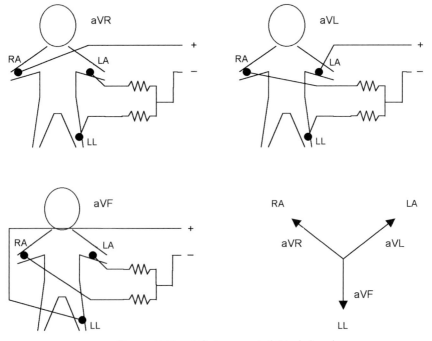

Figure 15-9. ECG Augmented Limb Leads.

projection from the LL and RA electrodes. These projections (or lead vectors) in the patient's frontal plane can be approximated by an equilateral triangle called the Einthoven's triangle. Figure 15-6 shows the projections of the cardiac vector at a certain time instant on the Einthoven's triangle.

The ECG obtained between the limb electrodes LA (+) and RA (–) is called lead I (Figure 15-7), between LL (+) and RA (–) is called lead II, and between LL (+) and LA (–) is called lead III. Figure 15-8 shows the configurations of these limb leads. Note the polarities of the electrodes.

The potential difference measured across a limb electrode and the average of the two other limb electrodes is called an augmented limb lead (e.g., aVR is obtained between RA and the average of LA and LL). There are three augmented limb leads; they are aVR, aVL, and aVF. (Note that R stands for right, L stands for left, and F stands for foot.) Figure 15-9 shows the configurations of the augmented limb leads. The average limb potential is obtained by connecting two identical value resistors to the two limb electrodes. It is then connected to the inverting input of the instrumentation amplifier. The limb leads (I, II, and III) and the augmented limb leads (aVR, aVL, and aVF) together are called the frontal plane leads.

The frontal plane leads represent the projection of the three-dimensional cardiac vector onto the two-dimensional frontal plane. In order to reconstruct the entire cardiac vector, the cardiac potential projected onto another plane is required. Figure 15-10 shows the position of the electrode placements on the chest of the patient to obtain the precordial leads (or the chest leads). The precordial leads represent the projection of the cardiac vector on the transverse plane of the patient. To measure the precordial leads, the potential of each of the chest electrodes is referenced to the average potential of the three limb electrodes (that is why precordial leads are also referred to as unipolar leads). There are six precordial leads. Figure 15-11 shows the connections to obtain the precordial leads. Note that all resistors to the limb

Table 15-1. Standard 12-Lead ECG Electrode Placement.

Lead	Electrode Placement	
	Positive Polarity	*Negative Polarity*
I	left arm (LA)	right arm (RA)
II	left leg (LL)	right arm (RA)
III	left leg (LL)	left arm (LA)
aVR	right arm (RA)	1/2 (LA + LL)
aVL	left arm (LA)	1/2 (RA + LL)
aVF	left leg (LL)	1/2 (RA + LA)
V1 through V6	chest positions	1/3 (LA + RA + LL)

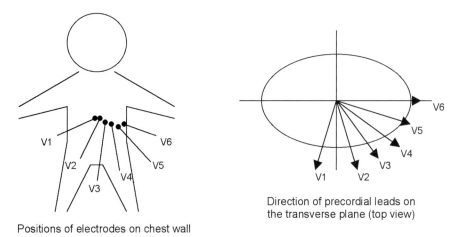

Positions of electrodes on chest wall

Direction of precordial leads on
the transverse plane (top view)

Figure 15-10. ECG Precordial Leads.

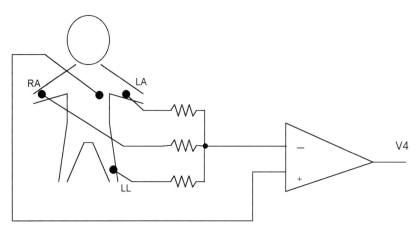

Figure 15-11. Connections for the Chest Leads.

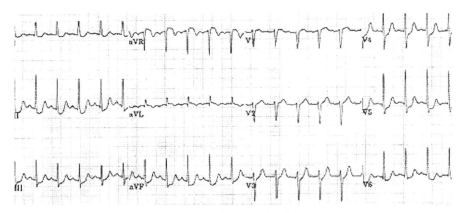

Figure 15-12. Standard 12-Lead ECG Waveform.

electrodes are of equal value. Which precordial lead is being measured depends on the position of the electrode on the chest of the patient (Figure 15-10). The six frontal plane leads (three limb leads plus three augmented limb leads) and the six precordial leads form the standard twelve-lead ECG configuration. A summary of the electrode positions for the standard twelve-lead ECG is shown in Table 15-1.

Note that altogether nine electrodes (three on the frontal plane and six on the transverse plane) are necessary to acquire the twelve-ECG leads simultaneously. In practice, a tenth electrode attached to the patient's right leg is used either as the reference (ground) or connected to the RLD circuit for common mode noise reduction (see Chapter 11). Figure 15-12 shows the characteristic ECG waveform from a standard twelve-lead measurement. Note that in this 3×4 format, each row contains four ECG lead recording segments. Row 1 records lead I, aVR, V1, and V4; row 2 records lead II, aVL, V2, and V5; row 3 records lead III, aVF, V3, and V6.

From the definition of the limb leads, lead I (or I) is the difference in potential between the electrodes attached to the LA and the RA. That is, lead I = $E_{LA} - E_{RA}$, similarly, lead II = $E_{LL} - E_{RA}$, and lead III = $E_{LL} - E_{LA}$. If we add lead I to lead III,

$$I + III = (E_{LA} - E_{RA}) + (E_{LL} - E_{LA}) = E_{LL} - E_{RA}$$

which is equal to lead II. Therefore, the sum of lead I and lead III equals lead II. This result agrees with the vector relationships between lead I, lead II, and lead III shown in Figure 15-8.

For the precordial lead V_n, where n = 1 to 6,

$$V_n = E_n - \frac{E_{RA} + E_{LA} + E_{LL}}{3}.$$

For the augmented limb leads,

$$aVR = E_{RA} - \frac{E_{LA} + E_{LL}}{2} = \frac{2E_{RA} - E_{LA} - E_{LL}}{2} = -\frac{(E_{LL} - E_{RA}) + (E_{LA} - E_{RA})}{2}.$$

Since Lead I (or I) = $E_{LA} - E_{RA}$ and Lead II (or II) = $E_{LL} - E_{RA}$,

$$aVR = -\frac{II + I}{2},$$

similarly, one can show that

$$aVL = E_{LA} - \frac{E_{RA} + E_{LL}}{2} = \frac{I - III}{2} \text{ and}$$

$$aVF = E_{LL} - \frac{E_{RA} + E_{LA}}{2} = \frac{II + III}{2}.$$

Furthermore, augmented leads can be obtained by subtracting the average potential of the three limb electrodes from one of the limb electrode potential:

$$E_{LA} - \frac{E_{RA} + E_{LA} + E_{LL}}{3} = \frac{2E_{LL}}{3} - \left(\frac{E_{RA}}{3} + \frac{E_{LA}}{3}\right) = \frac{2}{3}\left(E_{LL} - \frac{E_{RA} + E_{LA}}{2}\right) = \frac{2}{3}\,aVF.$$

Similarly, one can show that

$$E_{LA} - \frac{E_{RA} + E_{LA} + E_{LL}}{3} = \frac{2}{3}aVL, \text{ and}$$

$$E_{RA} - \frac{E_{RA} + E_{LA} + E_{LL}}{3} = \frac{2}{3}aVR.$$

Consider the Wilson network shown in Figure 15-13. If the corners of this triangular resistive network are connected to electrodes on the RA, LA, and the LL of the patient, V_-, V_{R-}, V_{L-} and V_{F-} are equal to

$$V_- = \frac{E_{RA} + E_{LA} + E_{LL}}{3}$$

$$V_{R-} = \frac{E_{LA} + E_{LL}}{2}$$

$$V_{L-} = \frac{E_{RA} + E_{LL}}{2}$$

$$V_{F-} = \frac{E_{RA} + E_{LA}}{2}.$$

These terminals on the network can therefore be used as the negative reference to measure the augmented and precordial ECG leads. The Wilson network allows using only one electrode at each patient location (LA, RA,

LL). It also avoids the need to remove and reconnect lead wires and electrodes during ECG measurement. Typical resistance values of R and R1 in the Wilson network (Figure 15-13) are 10 kΩ and 15 kΩ, respectively. Note that for this machine, if the IA is connected directly to a display or recorder, only one lead can be measured at a time. In order to simultaneously mea-

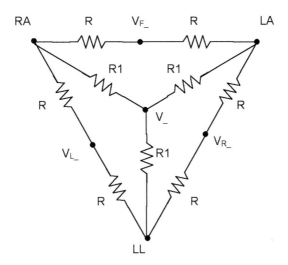

Figure 15-13. Wilson Network.

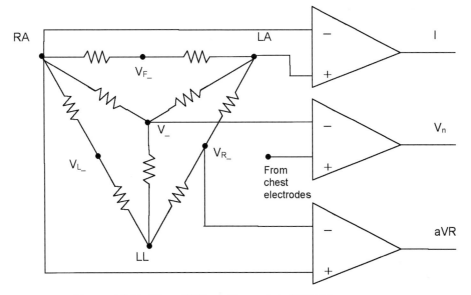

Figure 15-14. Use of Wilson Network in ECG Measurement.

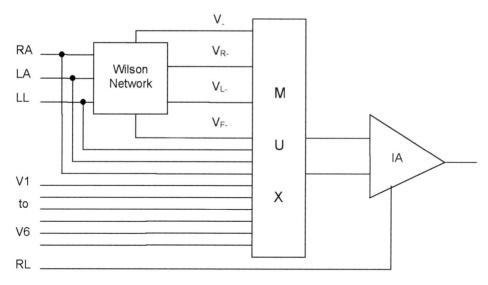

Figure 15-15. A Single Channel Twelve-Lead ECG Front End.

sure more than one ECG lead, more than one IA is usually required. Figure 15-14 shows a three-channel ECG machine measuring lead I, lead aVR, and one chest lead simultaneously. In general, to measure all twelve leads simultaneously, the electrocardiograph will need to have twelve sets of IAs as well as twelve display channels. In digital (computerized) electrocardiographs, some use sampling and time-division multiplexing techniques to avoid using multiple IAs to acquire simultaneous ECG leads.

Figure 15-15 shows the acquisition block (or patient interface module) of a single channel twelve-lead ECG machine. During operation, it uses a multiplexer or a number of mechanical switches to select which two input combinations of electrodes are connected to the IA. The Wilson resistor network may also be eliminated by using software algorithm to compute lead signals from individual electrode potentials using the relationships derived previously.

Other than the standard 12-lead ECG, other lead systems or lead locations are used in diagnostic electrocardiography. One such commonly used lead is the esophageal lead, which is obtained by swallowing an electrode into the esophagus so that the electrode is directly behind and close to the heart of the patient. The esophageal lead displays a higher amplitude P wave as it is closer to the atria than the ventricles. The esophageal electrode is often referenced to the average of the limb leads. Some models are offering more than 12-lead ECG recording, (e.g., adding posterior chest leads can better detect acute posterior and right ventricular myocardial infarctions).

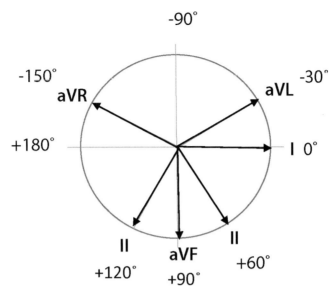

Figure 15-16. ECG Vector Diagram.

ECG Axis

An axis of an ECG is a direction of the electric cardiac vector. The limb leads (I, II & III) and the augmented limb leads (aVR, aVL, & aVF) are derived from the limb electrodes RA, LA and LL. The vector diagram formed by these six frontal plane leads is shown in Figure 15-16.

The axes of the QRS complex, P wave and T wave can be determined from these frontal plane leads. The vector direction of the frontal plan QRS complex can be used in cardiovascular diagnosis. Population data shows that a normal QRS axis is from −30° to +90°. For ECG interpretation, the QRS axis is divided into four segments – Normal Axis (QRS axis between +90° and -30°), Left Axis Deviation (LAD, QRS axis between -30° and -90°), Extreme Axis Deviation (EAD, QRS axis between -90° and -180°) also known as Northern Axis, and Right Axis Deviation (RAD, QRS axis between +180° and +105°). Figure 15-17 show the ECG axis diagram with QRS segments.

The QRS axis direction can be visually estimated by the polarity of the QRS of lead I and lead aVF. In the 12-lead ECG recording in Figure 15-12, the QRS of lead I and lead aVF are both positive. When they are transferred to the ECG vector diagram (dotted vectors in Figure 15-18), the resultant of the two vectors is in the Normal Axis segment. If the QRS of lead I is instead negative, and that of aVF is positive, the QRS axis of the patient would be

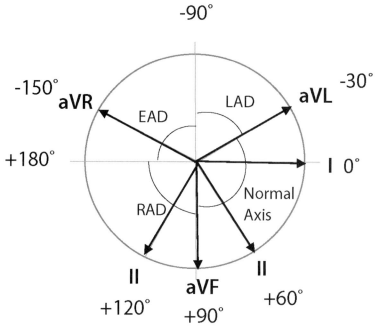

Figure 15-17. ECG Axis with QRS Segments.

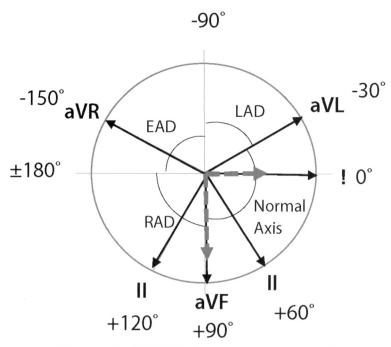

Figure 15-18. QRS ECG Axis from Lead 1 and aVF.

in the RAD segment. Deviation of the QRS axis from the Normal Axis segment to other segments is an indication of some heart conditions. Some possible causes leading to axis deviation are listed below:

- RAD - right ventricular hypertrophy, right ventricular strain, sodium-channel blockage
- LAD – left ventricular hypertrophy, inferior myocardial infarction, left bundle branch block
- EAD – severe right ventricular hypertrophy, ventricular ectopy, hyperkalaemia.

VECTORCARDIOGRAM

Another interpretation of the electrical cardiac activity is the vectorcardiogram. It was discussed earlier that the cardiac vector changes in both magnitude and direction (in three dimensions) as the electrical impulse spreads through the myocardium. A vectorcardiogram depicts the change in magnitude and direction of the cardiac vector as a function of time during the cardiac cycle.

Figure 15-19 shows the magnitude and direction of the cardiac vector projected onto the frontal plane at five different time intervals during the QRS complex (i.e., ventricular contraction). The vector at time t_1 is zero, which corresponds to the quiescent time before the ventricle starts to contract. When the current starts to flow toward the apex of the heart, causing

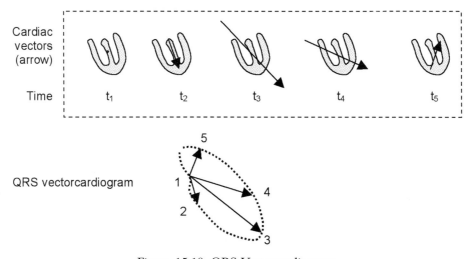

Figure 15-19. QRS Vectorcardiogram.

the ventricle to contract, the cardiac vector starts to grow in magnitude as well as change in direction. The vectors at time intervals t_2 to t_5 are shown in Figure 15-19. The elliptical figure (or loop) traced by the cardiac vector during the QRS interval using the quiescent point as reference is called the QRS vectorcardiogram. A smaller loop, referred to as the T vectorcardiogram, is also produced by the T wave about 0.25 sec after the QRS complex. As the magnitude of the T wave is about 0.2 to 0.3 mV, it is a much smaller loop that appears about 0.25 sec after the disappearance of the QRS vectorcardiogram. A still smaller P vectorcardiogram can be recorded during atrial depolarization. Like the conventional ECG, the vectorcardiogram can be used in the diagnosis of certain heart conditions.

FUNDAMENTAL BUILDING BLOCKS OF AN ELECTROCARDIOGRAPH

Figure 15-20 shows the functional block diagram of a typical electrocardiograph. The function of each block is described in what follows.

Defibrillator Protection

Because the ECG electrodes (or the defibrillator paddles which are also used to pick up ECG signals from the patient) are connected to the patient's chest, they will pick up the high-voltage impulses during cardiac defibrilla-

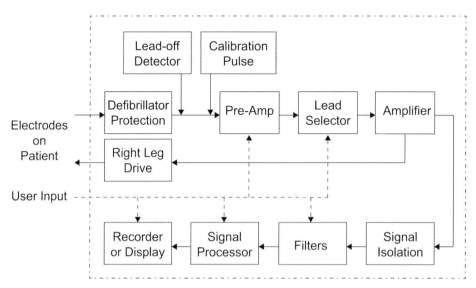

Figure 15-20. Functional Block Diagram of an Electrocardiograph.

tions. Gas discharge tubes and silicon diodes are used for defibrillator protection (see Chapter 11, Figures 11-17 and 11-18) to prevent the high-voltage defibrillation discharge from damaging sensitive electronic components in the ECG monitoring circuit.

Lead-Off Detector

When an electrode or lead wire is disconnected, the output of the ECG may display a flat baseline with noise. This may be misinterpreted as asystole. A lead-off (or lead fault) detector can prevent such misinterpretation. A simple lead-off detector is shown in Figure 15-21. In this design, a very large value resistor (>100 MW) is connected between the positive power supply and a lead wire to allow a small DC current to flow via the electrode through the patient to ground. Under normal conditions, due to the relatively small electrode/skin impedance, the DC voltages created at the input terminals of the operation amplifier are very small and almost of equal value. However, if an electrode or a lead wire comes off from the patient, the amplifier will be saturated since the voltage at one input of the amplifier will rise to the level of the power supply. In this case, the lead-off LED will turn on to alert the user to a lead fault.

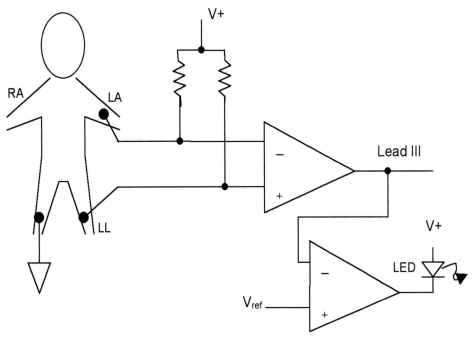

Figure 15-21. Lead-Off Detector.

Preamplifier

The magnitude of surface ECG is about 0.1 to 4 mV. A system, especially one with long unshielded lead wires, may pick up noise of up to several milli-volts through electromagnetic coupling. Therefore, it is important to amplify this small signal as close to the source as possible before it is corrupted by noise. Most ECG machines amplify the biopotential signals picked up by the electrodes in a preamplifier module or patient interface module located near the patient.

Lead Selector

The lead selector chooses the ECG lead to be displayed or recorded. In a multichannel machine, the lead selector also configures the sequence and format of the display or printout.

Amplifier

Typically, the magnitude of ECG biopotential at the surface of the body is about 1 mV, but this value may vary substantially from patient to patient. For example, the ECG of a critically ill patient may be as low as 0.1 mV or as high as 3 mV. Therefore, an electrocardiography must have some means of controlling the size of the ECG waveform. This is also called size, gain, or sensitivity adjustment. Typical sensitivity settings are 5, 10, or 20 mm/mV.

Right Leg-Driven Circuit

Electrical equipment and wiring near the electrocardiograph may induce common mode signal of several millivolts in magnitude on the patient's body. The RLD circuit is to suppress this common mode signal so that it will not over mask the ECG signal (see Chapter 11).

Calibration Pulse

For each ECG measurement, a built-in reference voltage of 1 mV is often applied to the input of the electrocardiograph. This reference signal is displayed on the screen and on the printout to inform the user that the machine is functioning properly and that it has the necessary gain to display the ECG signal coming from the patient.

Signal Isolation

The function of the signal isolation circuit is to reduce the leakage current to and from the patient through the electrode/lead connection for micro shock prevention. A module consisting of a FM modulator, an optical isolator, and a demodulator is commonly used to serve this purpose.

Filter

The frequency bandwidth for a diagnostic quality (or full-disclosure) electrocardiograph is from about 0.05 to 120 Hz. Such diagnostic mode bandwidth allows accurate presentation of the electrical activities of the patient's heart. In bed side monitoring, such wide bandwidth is often not necessary. Monitoring mode is used where a gross observation of the electrical activity of the patient's heart is necessary but requires little analysis or details. Interference and baseline drift can be reduced by a bandwidth less than that required for a diagnostic-quality ECG. For monitoring, a bandwidth of 1 to 40 Hz is reasonable and will allow recognition of common arrhythmias while providing reasonable rejection of artifacts and power frequency (60 or 50 Hz) interference. However, due to the reduced bandwidth, some distortion of the ECG will occur. Figure 15-22 show the effect of switching bandwidth on an ECG monitor. In Figure 15-22a, the ECG amplitude is slightly reduced (plus other slight distortions) when the monitor is switched from extended (0.1 to 100 Hz) to monitor (1 to 40 Hz) mode.

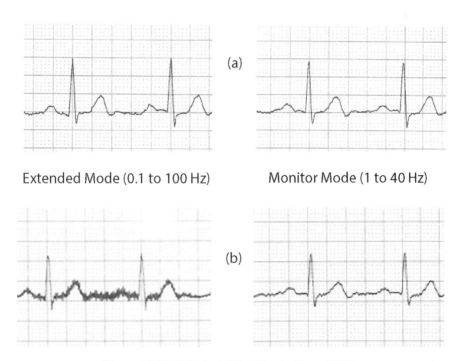

(a)

Extended Mode (0.1 to 100 Hz)   Monitor Mode (1 to 40 Hz)

(b)

Figure 15-22. Effect of Filter Bandwith on ECG.

Figure 15-22b shows removal of power frequency noise (60 Hz) when switched from extended to monitor mode. Most electrocardiographs have built-in upper and lower cutoff frequency selection to allow the user to choose the optimal bandwidth for the situation. A power frequency rejection filter (notch filter) can also be switched on by the user to minimize power frequency interference.

Signal Processor

Signal processing functions in ECG machines can range from simple heart rate detection to sophisticated waveform analysis and classification. Some common features for signal processing are

- Heart rate detection and alarm
- Pacemaker pulse detection and rejection
- Waveform duration measurement (e.g., PR interval, QRS duration, etc.)
- ECG axis calculation
- Arrhythmia analysis and classification (e.g., detection of PVC or premature ventricular contraction)
- Disease diagnosis and interpretation (note the machine interpretation displayed at the top of the 12-lead ECG recording on Figure 15-23).

Recorder or Display

The acquired waveform of diagnostic ECG can be viewed on a display monitor or printed out from a paper chart recorder. In either case, the speed of the waveform traveling across the screen of the monitor or the speed of the paper in the chart recorder can be adjusted. Typical speeds are 12.5, 25, and 50 mm/sec. For a multichannel ECG machine, the display format can be selected to display a combination of ECG leads. For example, a $3 \times 4 + 3R$ print format from a six-channel paper chart recorder is shown in Figure 15-23. In this format, the twelve ECG leads are displayed in three rows of four ECG leads. Each of the leads is displayed for 2.5 sec. In addition, three ECG leads selected by the user are displayed for the entire 10 sec in the lower three rows. Note that the 1 mV calibration pulse is printed at the beginning of every row to provide the user a quick reference of the performance of the machine.

TYPICAL SPECIFICATIONS OF ELECTROCARDIOGRAPHS

The specifications of a typical twelve-lead electrocardiograph are listed in the following:

- Input channels: simultaneous acquisition of up to twelve ECG leads
- Frequency response: –3 dB at 0.01 to 105 Hz
- CMRR: >110 dB
- Input impedance: >50 MΩ
- A/D conversion: 12 bits
- Sampling rate: 2000 samples per second per channel
- Writer type: thermal digital dot array with 200 dots per inch vertical resolution
- Writer speed: 1, 5, 25, and 50 mm/sec, user selectable
- Sensitivities: 2.5, 5, 10, and 20 mm/mV, user selectable
- Printout formats: three, four, five, six, and twelve channels, user selectable channel, and lead configurations
- Dimensions: 200 (H) × 40 (W) × 76 (D) cm
- Power requirements: 90 VAC to 260 VAC, 50 or 60 Hz
- Certifications: IEC 601

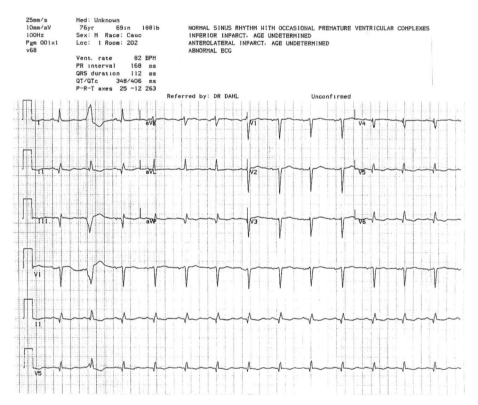

Figure 15-23. A 3 x 4 _ 3R Printout of a Twelve-Lead ECG.

ECG DATA STORAGE, NETWORK, AND MANAGEMENT

With the advancement in electronic data storage and computer network technologies, modern ECG machines are capable of electronically stored and shared information through computer networks. In a hospital, wireless ECG telemetry, diagnostic review stations, ECG machines, and electronic storage can be integrated into an "ECG data management system" via a LAN. Multiple hospitals, through a WAN, can also be configured to communicate and share resources, such as mass storage or archives. In a paper-less cardiology system, ECG data can be readily stored, retrieved, transferred, and viewed at any designated location.

COMMON PROBLEMS AND HAZARDS

ECG electrodes and lead wires in contact with the patient may become a source or drain of leakage current. However, designs such as defibrillator protection, signal isolation, and safety standards are well-established and adhered to by manufacturers to reduce risk and injury. The most common problem is artifacts or noise affecting ECG waveform. The causes of abnormal ECG waveform may be grouped into the following three categories:

1. Artifacts due to electrode problems, which may be caused by
   - Improper positioning of electrodes on the patient
   - Loose contact between the electrode and the patient
   - Dried-out electrode gel
   - Bad connection between the lead wire and the electrode
   - Failure to properly prepare (clean, shave, and abrade) the patient site for electrode attachment
2. Artifacts from the patient, which may be caused by
   - Skeletal muscle contraction
   - Breathing action
   - Patient movement
   - Involuntary muscle contraction (e.g., tremor)
3. Artifacts due to external interference, which may be caused by
   - Power frequency interference coupled to the lead wires or as common-mode voltage on the patient body (60 Hz interference in North America, 50 Hz in some other countries)
   - Radiated EMI from other equipment (e.g., 500 kHz interference from an electrosurgical unit)
   - Conductive interference from the power line or ground conductor (e.g., high-frequency noise from switching power supplies)

- Interfering signals from other equipment connected to the patient (e.g., pacemaker or neural stimulator pulses)
- Induced current or voltage from changing magnetic field
- Power interruption and supply voltage fluctuation

(1) Power Frequency (60 Hz) Interference.

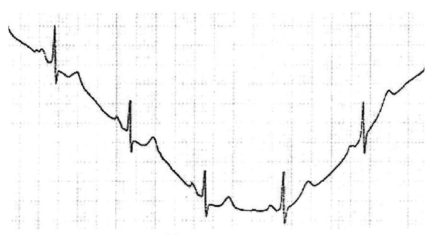

(b) Baseline Wander.

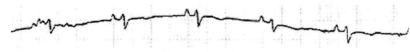

(c) Low Amplitude.

(d) Muscle Contraction.

Figure 15-24. Common ECG Artifacts.

Figure 15-24 shows some common artifacts in ECG acquisitions. Figure 15-24a shows a typical ECG waveform with power frequency interference. One can see sixty (or 50) even, regular spikes in a 1-sec interval if the timescale is expanded. Severe power frequency interference is often caused by improper patient or equipment grounding. It may also occur when the ECG lead wires are placed too close to power sources or unshielded electrical equipment (such as transformers, motors, cables). Turning on the built-in 60 (or 50) Hz notch filter (if available) can eliminate such interference. Grouping the lead wires together may reduce the interference amplitude. Figure 15-24b shows an ECG with wandering baseline. This can be caused by patient movement, or patient's respiratory action especially under the conditions of poor skin preparation, bad electrode contact, dried-out or expired electrode. Figure 15-24c shows an ECG waveform with abnormally small amplitude. Poor skin contact, improperly prepared skin, or dried-out electrode may be the cause. Figure 15-24d shows ECG artifacts due to skeletal muscle contraction. Muscle artifacts will usually disappear when the patient is relaxed and calmed down.

BIBLIOGRAPHY

Bailey, J. J., Berson, A. S., Garson, A., Horan, L. G., Macfarlane, P. W., Mortara, D. W., & Zywietz, C. (1990). Recommendations for standardization and specifications in automated electrocardio-graphy: Bandwidth and digital signal processing. A report for health professionals by an ad hoc writing group of the Committee on Electrocardiography and Cardiac Electrophysiology of the Council on Clinical Cardiology, American Heart Association. *Circulation, 81*(2), 730–739.

Baranchuk, A., Shaw, C., Alanazi, H., Campbell, D., Bally, K., Redfearn, D. P., Simpson, C. S., & Abdollah, H. (2009). Electrocardiography pitfalls and artifacts: The 10 commandments. *Critical Care Nursing, 29*(1), 67–73.

Braunwald, E. (Ed.). (1997). *Heart Disease: A Textbook of Cardiovascular Medicine* (5th ed.). Philadelphia, W. B. Saunders Co.

Conover, Mary Boudreau (2003-01-01). Understanding Electrocardiography. Elsevier Health Sciences. ISBN 978-0323019057.

Geselowitz, D. B. (1989). On the theory of the electrocardiogram. *Proceedings of the IEEE, 77*(6), 857–876.

Goldberger, A. L., Bhargava, V., Froelicher, V., & Covell, J. (1981). Effect of myocardial infarction on high-frequency QRS potentials. *Circulation, 64*(1), 34–42.

Holter, N. J. (1961). New method for heart studies: Continuous electrocardiography of active subjects over long periods is now practical. *Science, 134*(3486), 1214–1220.

Hurst, J. W. (1998). Naming of the waves in the ECG, with a brief account of their genesis. *Circulation, 98*(18), 1937–1942.

Krasteva, V., & Jekova, I. (2005). Assessment of ECG frequency and morphology parameters for automatic classification of life-threatening cardiac arrhythmias. *Physiological Measurement, 26*(5), 707–723.

Mittal, S., Movsowitz, C., et al. (2011). Ambulatory external electrocardiographic monitoring: focus on atrial fibrillation. *Journal of American College of Cardiology.* 58(17), 1741-1749.

Pérez Riera, A. R., Ferreira, C., Ferreira Filho, C., Ferreira, M., Meneghini, A., Uchida, A. H., Schapachnik, E., Dubner, S., & Zhang, L. (2008). The enigmatic sixth wave of the electrocardiogram: The U wave. *Cardiology Journal, 15*(5), 408–421.

Ripley, K. L., & Murray, A. (Eds.). (1980). *Introduction to Automated Arrhythmia Detection.* New York, NY: Institute of Electrical and Electronics Engineers, Inc.

Shouldice, R. B., & Bass, G. (2002). From bench to bedside—developments in electrocardiology. *The Engineers Journal, Institution of Engineers of Ireland, 56*(4), 47–49.

Simson, M. B. (1981). Use of signals in the terminal QRS complex to identify patients with ventricular tachycardia after myocardial infarction. *Circulation, 64,* 235–242.

Surawicz, Borys; Knillans, Timothy (2008). Chou's electrocardiography in clinical practice : adult and pediatric (6th ed.). Philadelphia, PA: Saunders/Elsevier. p. 12. ISBN 978-1416037743.

Wagner, G. S. (2007). *Marriott's Practical Electrocardiography* (11th ed.). Philadelphia, PA: Lippincott Williams & Wilkins

Yang, Y., Yin, D., & Freyer, R. (2002). Development of a digital signal processor-based new 12-lead synchronization electrocardiogram automatic analysis system. *Computer Methods and Programs in Biomedicine, 69*(1), 57–63.

Chapter 16

ELECTROENCEPHALOGRAPHS

OBJECTIVES

- Explain electroneurophysiology and the sources of signals.
- Outline the signal characteristics and clinical applications of EEGs.
- Compare different EEG electrode types and their applications.
- Explain the "10–20" electrode placements and montages.
- Identify the characteristics of EEG waveforms.
- Sketch and explain the functional block diagram of an EEG machine.
- Identify causes of poor EEG signal quality, and problems with EEG acquisition and suggest corrective solutions.

CHAPTER CONTENTS

1. Introduction
2. Anatomy of the Brain
3. Applications of EEG
4. Challenges in EEG Acquisition
5. EEG Electrodes and Placement
6. EEG Waveform Characteristics
7. Functional Building Blocks of EEG Machines
8. Common Problems and Hazards

INTRODUCTION

Electroneurophysiology refers to the study of electrical signals from the central and peripheral nervous systems for functional analysis and diagnosis.

These signals are recorded using extremely sensitive instruments to pick up tiny electrical signals produced by the system. There are four main areas in electroneurophysiology: EEG, evoked potential (EP) studies, polysomnography (PSG), and EMG.

EEG is a procedure in which small electrical signals produced by the brain are recorded. These signals are generated by the inhibitory and excitatory postsynaptic potentials of the cortical nerve cells. EEG includes the field of electrocorticography, which is a multichannel recording of biopotential signals from the exposed brain cortex. An electroencephalograph is a machine that captures these brain signals. The electrical potential from these signals plotted against time is called an electroencephalogram (also abbreviated as EEG). The first animal EEG and EP was published in 1912 by Vladimir Pravdich-Neminsky, a Russian physiologist. The first human EEG was recorded by Hans Berger, a German physiologist and psychiatrist, in 1924. Berger also invented the EEG machine. In the 1950s, William Grey Walter used a number of electrodes pasted on the scalp to create a map of the brain electrical activity. Using this, he demonstrated the use of delta waves to locate brain lesions responsible for epilepsy.

In EEG measurements, electrodes are generally placed on the skull of the patient; some procedures may use electrodes that penetrate the skin surface or electrodes that are placed directly on the surface of the cerebral cortex. The potential difference between a pair of electrodes is amplified and recorded. Before amplification, EEG signals measured directly at the surface of the brain or by a needle that penetrated the brain are typically of amplitude from 10 μV to 5 mV, whereas signals acquired on the surface of the scalp are typically from 5 to 500 μV. Figure 16-1 shows a general set up for EEG recording.

EP are performed to analyze the various nerve conduction pathways in the body. EP studies, for example, are useful in diagnosing problems in the visual and auditory pathways. During the procedure, stimulation such as sound or a flashing light is imposed on the subject to initiate a nerve signal transmission. If the signal is not getting through, a lesion in the particular nerve pathway may be present.

PSG is the study of sleep disorders by recording EEG, physiological parameters, and various muscle movements. PSG can be used in diagnosing and treating sleep disorders such as insomnia and sleep apnea. EMG is the study of the electrical activities of muscles and their peripheral nerves. It may be used to determine whether the muscles are functioning properly or if the nerve conduction pathway is healthy. This chapter focuses on EEG. EMG and EP studies are discussed in Chapter 17.

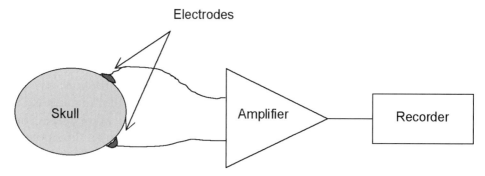

Figure 16-1. General Configuration of EEG Recording.

ANATOMY OF THE BRAIN

The brain is the enlarged portion and the major part of the central nervous system (CNS), protected by three protective membranes (the meninges) and enclosed in the cranial cavity of the skull. The brain and spinal cord are bathed in a special extracellular fluid called cerebrospinal fluid (CSF). The CNS consists of ascending sensory nerve tracts carrying information to the brain from different sensory transducers throughout the body. Information such as temperature, pain, fine touch, pressure, and so forth, is picked up by these sensors and delivered via the nerve tracts to be processed in the brain. The CNS also consists of descending motor nerve tracts, originating from the cerebrum and cerebellum and terminating on motor neurons in the ventral horn of the spinal column.

The three main parts of the brain are the cerebrum, the brainstem, and the cerebellum. The cerebrum consists of the right and left cerebral hemispheres, controlling the opposite side of the body. The surface layer of the hemisphere is called the cortex and is marked by ridges (gyri) and valleys (sulci); deeper sulci are known as fissures. The cortex receives sensory information from the skin, eyes, ears, and so on. The outer layer of the cerebrum, approximately 1.5 to 4.0 mm thick, is called the cerebral cortex. The layers beneath consist of axons and collections of cell bodies, which are called nuclei. The cerebrum is divided by the lateral fissure, central fissure (or central sulcus), and other landmarks into the temporal lobe (responsible for hearing), the occipital lobe (responsible for vision), the parietal lobe (containing the somatosensory cortex responsible for general sense receptors), and the frontal lobe (containing the primary motor and premotor cortex responsible for motor control).

The brainstem is an extension of the spinal cord, which serves three purposes:

1. Connecting link between the cerebral cortex, the spinal cord, and the cerebellum
2. Integration center for several visceral functions (e.g., heart and respiratory rates)
3. Integration center for various motor reflexes

The cerebellum receives information from the spinal cord regarding the position of the trunk and limbs in space, compares this with information received from the cortex, and sends out information to the spinal motor neurons.

APPLICATIONS OF EEG

A normal EEG usually consists of a range of possible waveforms from low frequency, near periodic waves with large delta components in deep sleep to high frequency, noncoherent beta waves measured on the frontal lobe channels during vigorous mental activity. Under a relaxed state, the EEG is characterized by alpha waves from the occipital lobe channels. Opening and closing the eyes results in an evoked response. EEG can be used as a clinical tool in diagnosing sleep disorders, epilepsy, multiple sclerosis, and so forth. It is used in BIS index monitoring as an indicator of the depth of anesthesia in surgery. EEG is an effective tool in ascertaining a patient's recovery from brain damage or to confirm brain damage. An EEG recording can be as short as 20 minutes or continue for a couple of days; the number of electrodes applied on the patient depends on the objective of the test and the required level of resolution. Some EEG applications are described in what follows.

Brain Death

Absence of EEG signals is a definition of clinical brain death.

Epilepsy and Partial Epilepsy

Epilepsy may be classified into the following categories:

- Generalized epilepsy—affects the entire brain
- Grand mal seizures—large electrical discharges from entire brain that last from a few seconds to several minutes. It is apparent on all EEG channels and may also be accompanied by skeletal muscle twitches and jerks.
- Petit mal seizures—a less severe form of epilepsy with strong delta waves

that last from 1 to 20 sec in only part of the brain and therefore appear in only a few EEG channels

Diagnosing Sleep Disorders

In normal sleep, the alpha rhythms are replaced by slower, larger delta waves. EEG monitoring can also determine if and when a subject is dreaming due to the presence of rapid, low-voltage interruptions indicating paradoxical sleep or rapid eye movement (REM) sleep.

CHALLENGES IN EEG ACQUISITION

Measurement of EEG signals using surface electrodes in general is more difficult than measuring ECG signals because

- The electrical potentials are conducted through a number of nonhomogeneous media before reaching the scalp. Table 16-1 lists the values of resistivity of different body tissues.
- Since tissues have higher resistivity than the CSF that overlies the brain, the CSF is acting as a region of high conductivity, having a shunting effect on electrical currents.
- Muscles over the temporal region and above the base of the skull provide pathways of high conductivity, allowing the shunting of local voltages well beneath the skin.
- Because of this spatial conductivity arrangement, the electrical potential difference measured actually shows the resultant field potential at a boundary of a large conducting medium surrounding an array of active elements (i.e., activities of the nuclei and some axons).
- In addition, utilization of any nervous functions would invoke or inhibit electrical activities of related parts of the brain, leading to a change in the electrical potentials on the scalp.

Table 16-1. Tissue's Electrical Resistivity.

Body Part	Resistivity (Ω-cm)
Blood	100–150
Heart muscle	300
Thoracic wall	400
Lung	1500
Dry skin	6,800,000

- Electrical activities will also be a function of age, state of consciousness, disease, drugs, and whether external stimulation is used.

EEG is then, in general, the scalp surface measurement of the total effects of all the electrical activities in the brain. It is assessed in conjunction with patient history, clinical symptoms, and other physiological parameters.

EEG ELECTRODES AND PLACEMENT

Types of EEG Electrodes

Depending on the nature of EEG studies, different types of electrodes are used. Surface electrodes, due to their noninvasive application, are the most commonly used electrodes. Needle, cortical, and depth electrodes are examples of invasive electrodes. Common materials for EEG electrodes are Ag/AgCl, gold plated Ag, stainless steel, and platinum. The constructions and placements of some are described next.

In an EEG measurement using surface (or scalp) electrodes, the electrodes are made to be in contact with the scalp of the patient. Electrodes may be in the form of a flat disk of 1 to 3 mm in diameter or a small cup with a hole at the center for injection of electrolyte gel. Materials such as platinum, gold, Ag, or Ag/AgCl are used for EEG surface electrodes. Earlobe electrodes and nasopharyngeal electrodes are some of the other noninvasive electrodes. In order to minimize noise and artifact problems, surface electrodes must be affixed to the scalp. One of two methods can be used:

1. Using collodion (a viscous and sticky fluid) to attach the electrode to the scalp. It is applied to the electrode site and dried using a jet of air. Electrolyte gel is then injected into the electrode through a hole in the center. Low melting point paraffin may be used as a substitute for collodion.
2. Adhesive conductive paste is placed directly on the desired location with the electrode pressed into the center of the paste.

Needle electrodes are sharp wires usually made of steel or platinum. They are inserted into the capillary bed between the skin and the skull bone. They can be applied quickly and provide slightly better signal quality than scalp electrodes do. Although it is relatively safe, EEG measurement using needle electrodes is an invasive procedure.

Cortical electrodes are used during neurosurgical procedures such as

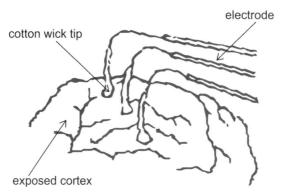

Figure 16-2. Cortical Electrodes.

excision of epileptogenic foci. They are applied directly onto the surface of the exposed cortex. A type of these electrodes consist of metal balls or wires with saline-soaked wicks. They may be held in place by swivel joints mounted on a bracket of a head frame for easy three-dimensional positioning. Figure 16-2 shows an example of the setup.

Subdural electrodes (another type of cortical electrodes) are used to localize epileptiform activity and to map cortical function. They consist of a number of disk electrodes mounted on a thin sheet of flexible translucent Silastic® rubber. The electrodes are made of platinum or stainless steel. Subdural electrodes are often configured as linear strips or rectangular grids with a number of electrode contact points. They are designed to be placed directly on the surface of the cortex. A single column strip can also be inserted into the intracranial cavity through a small burr hole opening. Subdural grids are placed over the cortical convexity in open cranial procedures to cover a large surface area. Figure 16-3 shows such electrodes.

Depth electrodes are fine, flexible plastic electrodes attached to wires that carry currents from deep and superficial brain structures. These currents are recorded through contact points mounted on the walls of the electrodes. Fine wires extending through the bores of the electrodes are inserted with stylets placed in the bores. Stereotactic depth electrodes are useful, for example, in determining the site of origin in temporal lobe epilepsy and as stimulating electrodes for the treatment of movement disorders. Either local or general anesthesia is applied when the electrodes are being inserted into the brain.

Surface (or Scalp) Electrode Placement

The international 10–20 system of electrode placement provides uniform coverage of the entire scalp. Based on the proven relationship between a measured electrode site and the underlying cortical structures and areas,

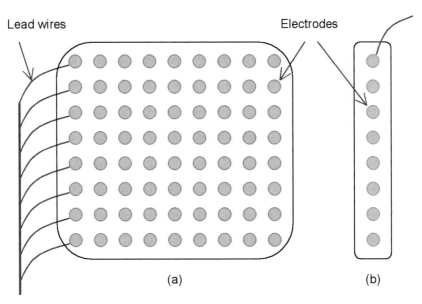

Figure 16-3. Subdural Electrodes (a) Grid, and (b) Single Column Strip.

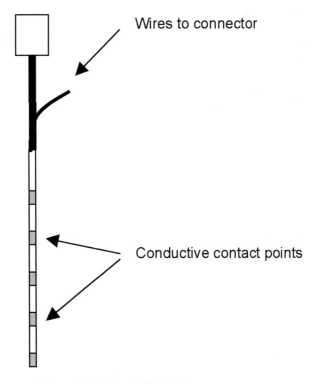

Figure 16-4. Depth Electrodes.

electrodes are symmetrically spaced on the scalp using identifiable skull anatomical landmarks as reference points. It is termed 10–20 because electrodes are spaced either 10% or 20% of the total distance between a given pair of skull landmarks. These landmarks are

- Nasion—the root of the nose
- Inion—ossification or bump on the occipital lobe
- Right auricular point—right ear
- Left auricular point—left ear

Figure 16-5 shows the locations of the electrodes in the 10–20 system and Table 16-2 lists the names of the electrode positions. The scalp is divided into five regions: (frontal (F), central (C), posterior (P), occipital (O), and temporal (T). The region letters are followed by numbers, with odd numbers on the left side and even numbers on right side of the patient's brain.

The use of the 10–20 system ensures reproducible electrode placement to allow more meaningful, more reliable comparison of EEGs from the same patient or different patients. Additional electrodes may be placed between a pair of adjacent electrodes to more accurately localize an event or abnormality. An example of electrode configuration to obtain higher resolution EEG is the standard international 10–10 system.

Table 16-2. Nomenclature for the 10–20 System.

Brain Area	Left Hemisphere	Midline	Right Hemisphere
Scalp Leads			
Frontal Pole	Fp1		Fp2
Frontal	F3		F4
Inferior Frontal	F7		F8
Midfrontal		Fz	
Midtemporal	T3		T4
Posterior Temporal	T5		T6
Central	C3		C4
Vertex or Midcentral		Cz	
Parietal	P3		P4
Midparietal		Pz	
Occipital	O1		O2
Nonscalp Leads (reference)			
Auricular	A1		A2
Nasopharyngeal*	Pg1		Pg2

Note: * optional leads

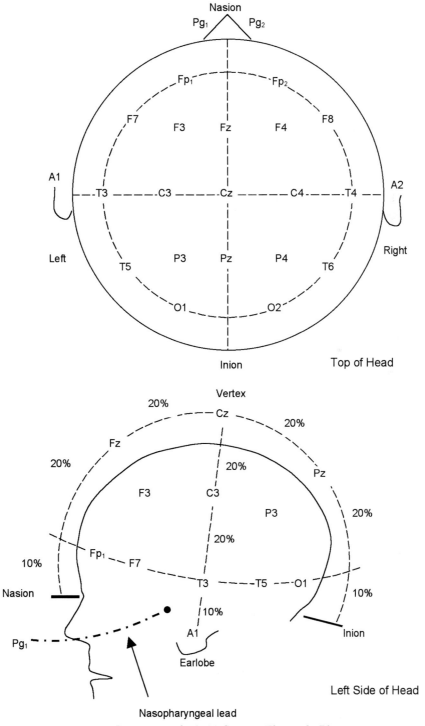

Figure 16-5. International 10–20 System Electrode Placement.

Scalp-Electrode Impedance

Because the EEG signal is of such low amplitude, the impedance of each electrode should be measured before every EEG recording. The impedance of each electrode should be between 100 Ω and 5 kΩ. Impedance below 100 Ω indicates a short circuit created by conductive gel between two electrodes. Impedance above 5 kΩ signals poor electrode skin contact. In practice, electrode impedance is usually measured using an ohmmeter by passing a small AC from one electrode through the scalp to all other connected electrodes. An AC of approximately 10 Hz is used to avoid electrode polarization and prevent measurement error due to DC offset potential. If only one pair of electrodes is used, the impedance should be between 200 Ω and 10 kΩ. In addition, minimizing the differences in impedance at different electrode sites can reduce EEG signal size variations.

EEG WAVEFORM CHARACTERISTICS

The peak to peak amplitude of EEG waveforms measured using scalp electrodes lies between 0 and 500 µV. A noticeable variation of EEG patterns can be seen in different persons of the same age. An even greater variation can be found in different age groups. An EEG is considered normal if there is no abnormal pattern known to be associated with clinical disorders. An EEG with no abnormal pattern does not guarantee the absence of problems because not all abnormalities of the brain produce abnormal EEG. Figure 16-6 shows an eight-channel EEG recording with normal rhythm followed by a run of epileptic events.

The frequency of EEG waveforms can be divided into four frequency ranges, they are beta, alpha, theta, and delta. The frequency bandwidths, general locations, and conditions of acquisitions of these four bands are listed in Table 16-3.

The EEG recording in Figure 16-6 shows the EEG in the time domain (amplitude against time). The same signal may be expressed in the frequency domain by Fourier transform. Representing EEG signals in the frequency domain provides a better visualization of rhythm (beta, alpha, theta, and delta) distribution. Rhythm distribution patterns over the brain have shown correlation with mental or physical activities, behaviors, and diseases. Figure 16-7 shows the amplitude-time signal and the frequency spectrum of an EEG recording. The dominant Alpha rhythm is highlighted in the frequency spectrum representation of the EEG recording. Figure 16-8 displays a sequence of EEG frequency spectra. Each frequency spectrum represents the frequency-domain representation of a segment of EEG signal recorded at the time

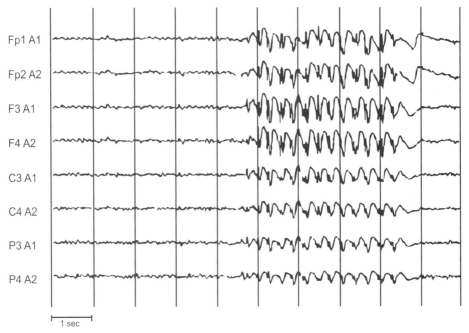

Figure 16-6. Normal and Abnormal (Epileptic) EEG.

indicated. In the compress spectral analysis (CSA) shown in Figure 16-8, the EEG was recorded from a patient during anesthetic induction before surgery. The CSA shows the shift of EEG rhythms at different level of anesthesia. CSA can also be used as a visual tool in analyzing drug response or functional response in electroneurophysiological studies.

CSA is a technique made possible by applying digital computer technology in EEG studies. Quantitative EEG (qEEG) is the analysis of the digitized EEG using different algorithms to quantify and localize the electrical signals. It is also referred to as "brain mapping." The qEEG is an extension of the analysis of the visual EEG interpretation.

An EEG application example is the BIS index used to assess the depth of sedation in anesthesia. The BIS index is a statistically based, empirically compiled parameter derived from a combination of EEG parameters to reduce the incidence of intraoperative awareness during general anesthesia. The BIS monitor captures the EEG signals from a number of electrodes attached to the forehead of the patient, processes the signal in real time, and provides a single number to indicate the depth of sedation. The BIS index ranges from 0 (equivalent to EEG silence) to 100 (fully awake). A range of BIS values (e.g., between 40 and 60) indicates status of deep sedation.

Table 16-3. Frequency of EEG Waveform.

Waveform	Frequency (Hz)	Remarks
Beta Rhythm	13–30	Frontoparietal leads Best when no alpha Prominent during mental activity
Alpha Rhythm	8–13	Parietooccipital Awake and relaxed subject Prominent with eyes closed Disappear completely in sleep
Theta Rhythm	4–8	Parietotemporal Children 2–5 years old Adults during stress or emotion
Delta Rhythm	0.5–4	Normal and deep sleep Children less than 1 year old Organic brain disease

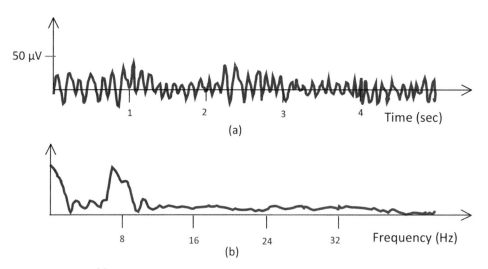

Figure 16-7. (a) EEG Waveform in Time-Domain. (b) Frequency Spectrum of (a).

FUNCTIONAL BUILDING BLOCKS OF EEG MACHINES

A very basic single-channel electroencephalograph is shown in Figure 16-1. In practice, an EEG machine for use in a diagnostic laboratory contains more functions and options. Figure 16-9 shows the functional block diagram of a typical EEG machine. Their functions are discussed in this section.

Electrode Connections and Head Box

A head box is used to interface electrodes from the skull to the switching system (or electrode selector). Each lead wire from the electrode applied to the skull is plugged into the corresponding location on the head box. A typical EEG referential head box for standard EEG application accepts twenty-three electrodes plus a few spares. The head box contains the first level of signal buffering and amplification to increase the signal level and provide a high input impedance to minimize common mode noise. Input impedance of a modern EEG machine is on the order of tens of mega ohms or higher.

Montage and Electrode Selector

Multichannel recordings are used to determine the distribution of electrical potential over the scalp. In order to gain insight into the activity at a

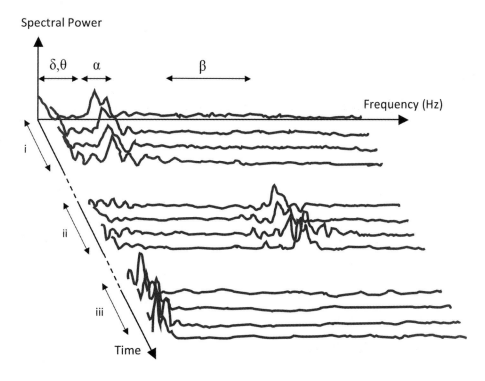

Figure 16-8. Typical CSA Anesthetic Induction Sequence: (i) initial stage—alpha activity dominates; mid stage, (ii) loss of alpha activity, replaced with beta activity; patient anesthetized, (iii) delta and theta activities dominate.

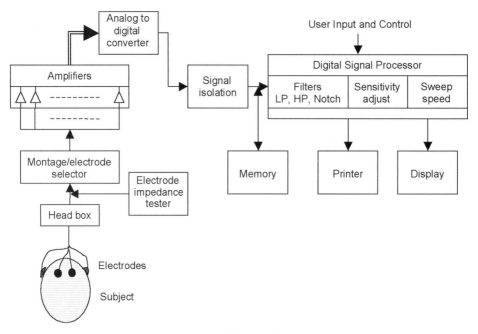

Figure 16-9. Functional Building Block of an EEG Machine.

given location, multiple differential signals from different combinations of electrode pairs are required. A montage consists of a distinct combination of differential signals of such multiple channel recordings.

Electrodes are attached in groups of eight (or ten) in a montage. Because of this, EEG machines usually have eight or sixteen (or ten or twenty) differential amplifiers. There are two types of amplifier input connections:

1. Bipolar connection—measurements taken between two electrodes
2. Unipolar connection—all measurements have a common reference point

Figure 16-10 shows a unipolar connection. A commonly used reference electrode for unipolar connection is the auricular electrodes (right auricular for the right cerebral electrodes and left for the left electrodes) The nasopharyngeal electrode is also used as a reference. To facilitate grouping of electrodes, an electrode selection circuit is available at the front end of the EEG machine. Figure 16-11 is a diagrammatic representation of a multichannel electrode selector matrix. Any two electrodes can be switched to the input of any of the differential amplifiers. To facilitate clinical diagnosis, certain electrode combinations are grouped together to form standard montages. Depending on the design, montage selection can be done digitally instead of

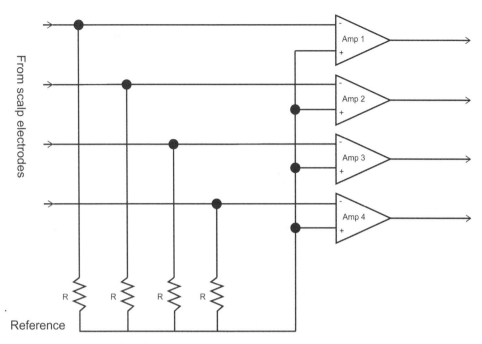

Figure 16-10. Unipolar Connection.

using analog switches. Standard montages are usually built into EEG electrode selection function and can be programmed or modified by the user. Figure 16-12 shows the standard "referential" and "transverse bipolar" montages.

Amplifiers

The amplifier increases the signal level to the desired amplitude for the analog to digital converter and the display. Together with the digital processing circuit, it allows the operator to select different levels of sensitivities. Most EEG machines have two ranges of sensitivities: mV/cm or μV/mm. A common sensitivity setting for general applications is 7 μV/mm. Each channel of an EEG machine consists of a high gain differential amplifier with a gain of approximately 10,000. Depending on the number of channels, an EEG machine typically has eight, ten, sixteen, or twenty differential amplifiers.

Analog to Digital Converter

The ADC samples and converts the analog EEG signals to a digital format so they can be processed by the digital computer. The analog signal from the amplifier is sampled before being fed to the ADC. A typical sam-

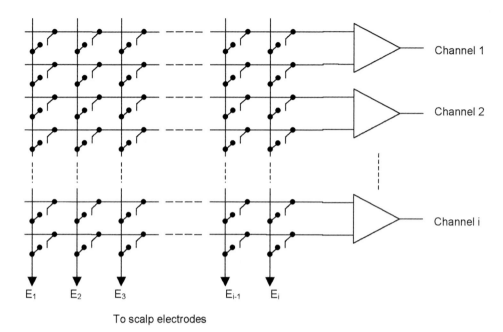

Figure 16-11. Electrode Selection Matrix.

pling rate is 1000 samples/sec with 12 bits (4096 vertical steps) resolution.

Signal Isolation

The patient-connected parts are isolated from the power ground via optical isolators. Signal isolation prevents electric shocks (microshocks and macroshocks) by reducing the amount of leakage current flowing to and from the patient.

Filters

The signal bandwidths are individually selectable through software digital filters (older machines use analog filters). The high pass filter (low filter) is usually adjustable in steps from 0.1 to 30 Hz and the low pass filter (high filter) from 15 to 100 Hz. In addition, a notch filter (60 Hz in North America and 50 Hz in Europe and Asia) can be selected to reduce power frequency noise from line interference.

Sensitivity Control

Sensitivity of each channel can be adjusted individually to match the input signal amplitude and the output display. Typical sensitivity range is 2 to 150 μV/mm or 2 to 150 mV/cm. A sensitivity equalizer control allows accurate verification of all channel sensitivities using a single input calibration signal.

Memory, Chart Speed, Display, and Recorder

Chart speeds of 10, 15, 30, and 60 mm/s are supported by most EEG machines. Mechanical paper chart recorders with ink styli on 11 by 17-inch fan-folded paper were used in older EEG machines. The huge volume of paper generated from each EEG study used to create storage problems in EEG departments. Today's digital technology allows EEG signals to be stored in electronic memories and viewed on flat panel displays instead of written on paper. To further reduce the storage requirement, neurologists may choose to remove non-pertinent EEG records and save only waveforms containing useful diagnostic information into patient records. To produce paper copies, laser printers are used for digital EEG machines.

Electrode Impedance Tester

A small 10-Hz AC is used to measure the impedance of the electrode skin contacts. Electrode impedance testing is available on demand in real-

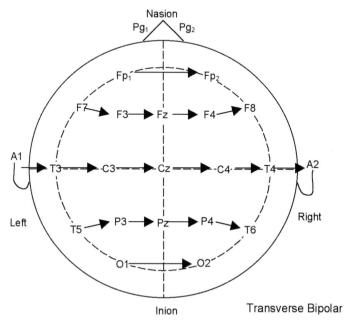

Figure 16-12. Transverse Bipolar and Referential Montages.

time modes on individual or all-channel basis. As discussed earlier, the impedance of a pair of electrodes should be between 200 Ω and 10 kΩ.

COMMON PROBLEMS AND HAZARDS

EEG Artifacts

In EEG measurements, recorded signals that are noncerebral in origin are considered artifacts. Artifacts can be either physiological or nonphysiological. Physiological artifacts arise from normal biopotential activities or movement activities of the patient. The primary sources of nonphysiological EEG artifacts include external EMI and problems with the recording electrodes. Device hardware malfunction may cause problems, but it is not a common source of EEG artifacts. Common sources of EEG artifacts are as follows:

Artifacts due to physiological interference, may result from

- The heart potential results from either patient touching metal and creating second ground or pulsatile blood flow in the brain
- Tongue and facial movement
- Eye movement
- Skeletal muscle movement (uncooperative patient or fine body tremors)
- Breathing
- High scalp impedance

The above may be mitigated by ensuring that the patient is calm and relaxed.

Artifacts due to electrode problems, may result from

- Improper electrode positioning
- Poor contact causing sharp irregular spikes or the pickup of 60 Hz noise
- Electrodes not secured properly
- Dried-out electrode gel
- Oozing of tissue fluids in needle electrodes
- Frayed connections
- Sweat resulting in changing skin resistance

The above artifacts may be minimized by reapplying electrodes on the scalp to ensure good electrode contacts (less than 10 kΩ between electrode pairs).

Artifacts due to EMI may result from

- Power frequency (60 or 50 Hz) common mode interference
- Radio frequency interference due to presence of electrical devices (e.g., an electrosurgical generator)
- Defibrillation
- Presence of pacemakers and neural stimulators

The above may be minimized by

- Proper grounding (no grounding or multiple ground loops) and shielding
- Removing sources of EMI
- Performing procedure in special EMI-shielded room

Troubleshooting an EEG Problem

Troubleshooting EEG problems is similar to troubleshooting other biopotential measuring devices using surface electrodes. Some common considerations are

- Common mode noise problems
- Problems compounded due to small signal levels (thousand times smaller than ECG)
- Problems with electrodes and leads (positioning, bad connections, etc.)
- Use internal or external calibration signals to check machine performance and distinguish them from electrode and lead problems
- Isolate problem to a single functional block; use a known input, if output is healthy, then problem is outside the functional block.
- For problems that happen with one channel only, can rule out common components such as power supply or display.

Hazards and Risks

Application of invasive EEG electrodes may lead to trauma or infection. Infection is considered the major risk of implanted EEG electrodes. Recent studies reported an infection rate of about 2% to 3%. Meticulous surgical techniques and procedures to prevent CF leakage keep the risk of infection low. Another risk from electrode placement is hemorrhage. Significant intracerebral hemorrhages have been reported, but the incidence is 1% or less. Direct brain injury due to passing of the depth electrodes has not been demonstrated because the electrodes are so thin that they normally dissect neural tissue without imposing much injury.

As with any other electromedical devices with patient applied parts, there is potential risk of electric shock. EEG acquisitions using noninvasive scalp electrodes are considered to be relatively safe medical procedures. However, EEG acquisition in conjunction with some other medical procedures may create hazards. For example, during magnetic resonance imaging (MRI) scan, the conductive loop created by EEG lead wires and the patient body may create tissue burn from current arising from electromagnetic induction.

Although it has become a common practice, the use of BIS monitoring as a reliable indicator of the level of sedation is still controversial. Studies indicated that different anesthetic agents affect the EEG differently. In addition, the same anesthetic agent used on different patients may produce different changes during the progression of anesthesia. When assessing a patient's condition, BIS should be used in conjunction with other patient information, such as vital signs from physiological monitoring systems.

BIBLIOGRAPHY

Avidan, M. S., Zhang, L., Burnside, B. A., Finkel, K. J., Searleman, A. C., Selvidge, J. A., . . . , & Evers, A. S. (2008). Anesthesia awareness and the bispectral index. *New England Journal of Medicine, 358*(11), 1097–1108.

Bickford, R. D. (1987). Electroencephalography. In G. Adelman (Ed.), *Encyclopedia of Neuroscience* (pp. 371–373). Basel, Switzerland: Birkhauser.

Borzova, V. V., & Smith, C. E. (2010). Monitoring and prevention of awareness in trauma anesthesia. *The Internet Journal of Anesthesiology, 23*(2), 8.

Bronzino, J. D. (1995). Principles of electroencephalography. In *The Biomedical Engineering Handbook* (pp. 201–212). Boca Raton, FL: CRC Press.

Collura, T. F. (1993). History and evolution of electroencephalographic instruments and techniques. *Journal of Clinical Neurophysiology, 10*(4), 476–504.

Ebersole, J. S., & Pedley, T. A. (2003). *Current Practice of Clinical Electroencephalography* (3rd ed.). Philadelphia, PA: Lippincott Williams & Wilkins.

Fisch, B. J. (1999). *Fisch and Spehlmann's EEG Primer: Basic Principles of Digital and Analog EEG* (3rd ed.). New York, NY: Elsevier.

Jasper, H.H. (1985). The ten-twenty electrode system of the International Federation. *EEG Clin. Neurophysiol.* 10, 371-335.

Kropotov, J. D. (2005). *Quantitative EEG, Event Related Potentials and Neurotherapy.* New York, NY: Elsevier.

Popa, L.L. et al. (2020). The role of Quantitative EEG in the Diagnosis of neuropsychiatric disorders. *Journal of Medicine and Life.* 1(1), 8-15.

Lemieux, L., Allen, P. J., Krakow, K., Symms, M. R., & Fish, D. R. (1999). Methodological issues in EEG-correlated functional MRI experiments. *International Journal of Bioelectromagnetism, 1*(1), 87–95.

Mirsattari, S. M., Lee, D. H., Jones, D., Bihari, F., & Ives, J. R. (2004). MRI compatible EEG electrode system for routine use in the epilepsy monitoring unit and intensive care unit. *Clinical Neurophysiology, 115*(9), 2175–2180.

Mullinger, K., Debener, S., Coxon, R., & Bowtell, R. (2008). Effects of simultaneous EEG recording on MRI data quality at 1.5, 3 and 7 tesla. *International Journal of Psychophysiology, 67*(3), 178–188.

Niedermeyer, E., & Lopes da Silva, F. (2005). Electroencephalography: Basic Principles, Clinical Applications, and Related Fields (5th ed.). Philadelphia, PA: Lippincott Williams & Wilkins.

Roche-Labarbe, N., Aarabi, A., Kongolo, G., Gondry -Jouet, C., Dumpelmann, M., Grebe, R., & Wallois, F. (2008). High-resolution electroencephalography and source localization in neonates. *Human Brain Mapping, 29*(2), 167–176.

Rosow, C., & Manberg, P. J. (2001). Bispectral index monitoring. *Anesthesiology Clinics of North America, 19*(4), 947–966.

Rush, S., & Driscoll, D. A. (1969). EEG electrode sensitivity—An application of reciprocity. *IEEE Transactions on Biomedical Engineering, BME-16*(1), 15–22.

Shellhaas, R. A., & Clancy, R. R. (2007). Characterization of neonatal seizures by conventional EEG and single-channel EEG. *Clinical Neurophysiology, 118*(10), 2156–2161.

Swartz, B. E., & Goldensohn, E. S. (1998). Timeline of the history of EEG and associated fields. *Electroencephalography and Clinical Neurophysiology, 106*(2), 173–176.

Waterhouse, E. (2003). New horizons in ambulatory electroencephalography. *IEEE Engineering in Medicine and Biology Magazine, 22*(3), 74–80.

Chapter 17

ELECTROMYOGRAPHY AND
EVOKED POTENTIAL STUDY EQUIPMENT

OBJECTIVES

- Explain the principles of EMG and EP studies.
- State clinical applications of EMG and EP studies.
- Outline typical functional components of an EMG/EP machine.
- Describe the constructions and applications of different types of surface and needle electrodes.
- Explain the foundation and characteristics of EMG signals.
- Differentiate between motor response and sensory nerve action potential.
- Illustrate common signal processing techniques used in EMG/EP studies.
- Explain the purpose of signal averaging and how it can reduce noise in EP studies.

CHAPTER CONTENTS

1. Introduction
2. Clinical Applications
3. Electrodes
4. EMG Signal Characteristics
5. Machine Settings
6. Signal Processing
7. Application Examples
8. Common Problems and Hazards

INTRODUCTION

In the previous chapter, the applications, signal acquisition, and functional building blocks of EEG were discussed. This chapter introduces other modalities in electroneurophysiology: EMG and EP studies. EMG studies the biopotentials from muscles and nerves that innervate the muscles; EP studies analyze the relationships between nerve stimulations and their responses.

CLINICAL APPLICATIONS

An EMG study may be used to establish the relationships between the signal morphology and the biomechanical variables. Comparing the biopotential signal frequency to muscle tension is an example. In an EP study, a nerve may be stimulated by an electrical signal at one end and the reaction measured somewhere along the nerve itself to determine the time-location relationship between the stimulus and the response. The stimulation may be visual, auditory, or somatosensory, and the responses may be detected in EEG and EMG signals. Parameters such as nerve conduction velocity can also be determined.

There are two main areas of applications of EMG and EP studies: one is in kinesiology and the other in diagnosis.

In kinesiology, the main areas of interest are

- Functional anatomy
- Force development
- Reflex connection of muscles

In electrodiagnosis, areas of analysis may involve

- Creation of strength-duration curves to assess nerve and muscle integrity
- Determination of nerve conduction velocity to diagnose nerve damage or compression
- Analyzing firing characteristics of motor neurons and motor units, including analysis of motor unit action potentials (MUAPs) to detect signs of pathology such as fibrillation potentials and positive sharp waves

Figure 17-1 shows a typical configuration of an EMG/EP study. The EMG or EP is picked up by a pair of electrodes, one being the sensing electrode and the other acting as the reference. The signal is amplified, processed, and sent to a chart recorder or a video display. Depending on the type of studies, a stimulation signal may be applied.

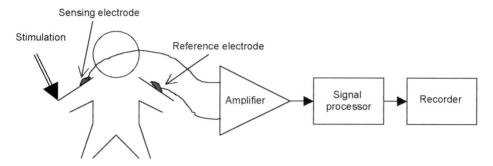

Figure 17-1. General Configuration of EMG/EP Recording.

ELECTRODES

Surface electrodes or intramuscular electrodes may be used in EMG/EP measurements. The basic electrodes, including grounding, stimulating, and recording, are described next.

Grounding Electrodes

As with all work involving measurement of biopotential signals, a ground electrode is needed. Grounding is essential for obtaining a response that is relatively free of artifact. In general, the ground electrode should be placed on the same extremity that is being investigated. The ground electrode in EP studies should be placed, if possible, halfway between the stimulating electrode and the active recording electrode. Usually the ground is a metal plate that is much larger than the recording electrodes and provides a large surface area of contact with the patient. Some clinicians may use a non-insulated needle inserted under the patient's skin for grounding. One should be careful not to apply more than one ground to the patient at any time. The presence of multiple grounds from different electrically powered devices can form "ground loops," which may create noise in the measurement.

Stimulating Electrodes

In most cases, a peripheral nerve can be easily stimulated by applying the stimulus near the nerve. Therefore, most nerve stimulation is done to segments of nerve that lie close to the skin surface. Because of the need for proximity, the number of nerves accessible to the stimulation and the locations of the stimulation of that nerve are limited.

The stimulating electrodes are normally two metal or felt pads placed about 1 to 3 cm apart (Figure 17-2). The electrodes are placed on the nerve

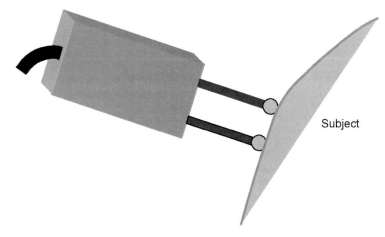

Figure 17-2. Stimulating Electrodes.

with the cathode toward the direction in which the nerve is to conduct. The stimulation amplitude is adjusted until a maximal response is obtained and then by 25% to 50% more to ensure that the response is truly maximal. One may use a needle electrode to stimulate nerves deep beneath the skin. Other than electrical stimulation, visual or audible stimulations may be used in EP studies.

Recording Electrodes

Positioning of recording electrodes depends on the type of response being studied. In motor response recording, the active electrode is placed over the belly of the muscle being activated. This placement should be over the motor point to give an initial clear negative deflection (note that in EMG/EP studies, an upward-going response is considered as negative) in the response. In testing of sensory nerve, the active electrode is placed over the nerve itself to record the nerve action potential. The reference electrode is placed distal from the active electrode and away from the stimulation.

In motor response recording, surface electrodes may be used. Surface electrodes can be made of pure metal or Ag/AgCl in the shape of a circular disk of 0.5 to 1 cm in diameter. Surface electrodes such as flat buttons, spring clips, or rings are frequently used in sensory recording. Some of the surface electrodes are shown in Figure 17-3. Other than surface electrodes, bare-tip insulated needle electrodes placed close to the nerves are also used by many investigators in motor response recording.

Figure 17-3. Surface Electrodes.

Needle Electrodes

Needle electrodes are commonly used in EMG/EP studies. They are used to evaluate individual motor units within a muscle to avoid picking up signals from other muscle units. The following paragraphs describe a few types of needle electrodes.

A monopolar needle electrode has a very finely sharpened point and is covered with Teflon or other insulating material over its entire length, except for a tiny (e.g., 0.5 mm) exposure at the tip (Figure 17-4). The needle serves as the active electrode, and a surface electrode placed on the skin close to it serves as a reference. The main advantage of monopolar needle electrodes is that they are of small diameter and the Teflon covering allows them to easily insert into and withdraw from the muscle. Moving the needle causes less discomfort to the patient. However, repeated use of this electrode changes the size of the bare tip, thereby limiting the number of examinations for which it can be used. Because the active needle electrode tip and the reference surface electrode are separated by some distance, it is easier to pick up background noise from remote muscle contractions.

A concentric needle electrode consists of a cannula with an insulated wire inserted down the middle (Figure 17-5). The active electrode is the small tip of the center wire, and the reference electrode is the outside cannula. Concentric needles may have two central wires (bipolar), in which case the active and reference electrodes are at the tip and the outside cannula acts as the ground. Because the active and reference electrodes are closer together, only local motor unit action potentials (MUAPs) are picked up by the electrode.

Another advantage of this electrode is that no reference surface electrode is needed. The main disadvantage of the concentric electrode is that it has a larger diameter relative to other needle electrodes. Large diameter needle electrodes tend to cause more pain and are uncomfortable to move around.

Single-fiber needle electrodes are used for special studies. A single-fiber needle consists of a thin (e.g., 0.5 mm) stainless steel cannula with a fine (e.g.,

Figure 17-4. Stimulating Electrodes.

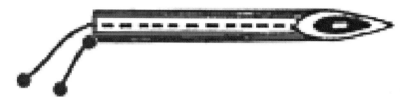

Figure 17-5. Concentric Needle.

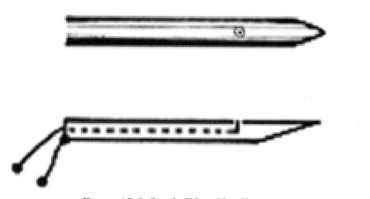

Figure 17-6. Single-Fiber Needle.

25 μm) platinum wire inside its hollow shaft. The cut end of the platinum wire is exposed from a side opening near its tip (Figure 17-6).

EMG SIGNAL CHARACTERISTICS

Most EMG signals have repetition frequencies in the range of 20 to 200 Hz. A single MUAP has amplitude on the order of 100 μV and duration of 1 msec. Typical EMG signals, which are the summation of multiple MUAPs, have amplitudes from about 50 μV to 20 mV. EMG signals can be used to diagnose neurogenic (e.g., denervation) or myogenic (e.g., muscular dystrophies) conditions. The following paragraphs describe EMGs obtained from needle electrode examinations and its electrical characteristics.

The Muscle at Rest

Insertion Activity

Insertion activity is the response of the muscle fibers to needle electrode insertion. It consists of a brief series of muscle action potentials in the form of spikes. It is caused by mechanical stimulation or injury of muscle fibers, which may disappear immediately or shortly after (a few seconds) stopping needle movements (Figure 17-7).

Spontaneous Activity

Any activity beyond insertion constitutes spontaneous activity. It can be due to normal end plate (neuromuscular junction) noise or to the presence of fasciculation (the random, spontaneous twitching of a group of muscle fibers or a motor unit).

Spontaneous activity due to end plate noise when the needle is in the vicinity of a motor end plate can be monophasic (end plate noise) or bipha-

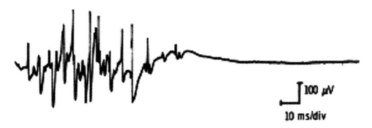

Figure 17-7. Needle Electrode Insertion Activity.

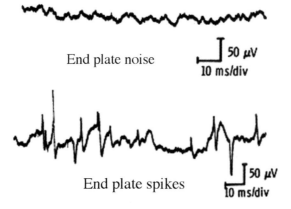

Figure 17-8. Needle Electrode End Plate Noise.

sic (end plate spikes) potentials (Figure 17-8). In EMG studies, a phase refers to the part of the wave between the departure and the return to the baseline. The monophasic potentials are of low amplitude and short duration. The biphasic activity consists of irregular, short, duration biphasic spikes with amplitude of 100 to 300 μv. A bandpass filter setting of 20 to 8000 Hz is often used to record insertion and spontaneous activities.

Muscle Voluntary Effort

Voluntary muscle effort involves recruitment of motor units. The strength of muscle contraction is controlled by the CNS and depends on:

1. the number of motor units activated (i.e., spatial recruitment)
2. the firing rate of individual motor unit (i.e., temporal recruitment)

Both mechanisms occur concurrently. For very low level muscle contractions, smaller motor units are recruited before larger motor units are, which provides a smooth gradual increase in contraction force. As the level continues to escalate, muscle strength is primarily increased by the addition of more motor units, but the firing rate of the initially recruited motor units also increases. When nearly all motor units are recruited, increase in firing frequency becomes the predominating mechanism to increase motor strength. At maximal voluntary muscle effort, the action potentials of individual motor units no longer can be distinguished from each other but are mixed together. This superimposition pattern of motor units at high voluntary muscular effort is called interference pattern. For a healthy individual, during a maximal voluntary muscle contraction, no individual MUAPs can be identified. An incomplete interference pattern may suggest neurogenic lesions or advanced stages of muscle disorder. A complete but low amplitude interference pattern may be an indication of myogenic conditions such as muscular dystrophy.

MUAPs are best studied with a similar filter setting used for insertion and spontaneous activity (i.e., 16–32 Hz low cutoff and 8000 Hz or more high cutoff frequency). The EMGs at different level of voluntary muscle efforts are described next.

Mild Effort

Only a few motor units are observed at this stage (Figure 17-9). These are the smaller motor units because they are the ones to be recruited first. Amplitude, duration, and number of phases of individual motor units are measured.

Moderate Effort

The frequency and recruitment of motor units are best assessed during this stage (Figure 17-10). Motor units seen at this stage are larger than those seen with mild effort. As muscle effort increases, motor unit firing rates are increased and new motor units are recruited.

Full Effort

At maximum contraction, it is difficult to distinguish individual motor units because the firing rates are so high and so many motor units are recruited that the motor units superimpose on each other (Figure 17-11). When all the motor units are recruited, a complete interference pattern is observed.

Motor Responses

A motor response is obtained by stimulating a nerve and recording from a muscle that it innervates. The muscle selected should have a fairly well-

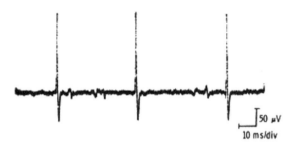

Figure 17-9. Mild Voluntary Effort.

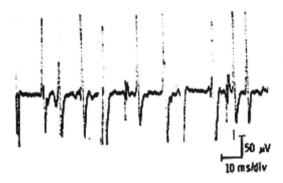

Figure 17-10. Moderate Voluntary Effort.

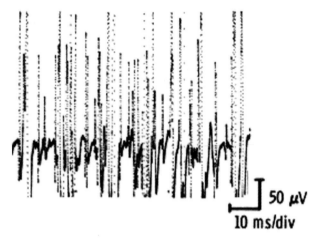

Figure 17-11. Full Voluntary Effort.

defined motor point and be isolated from other muscles innervated by the same nerve. The excitation of nearby muscles may alter the response and make it difficult to determine the exact onset of the desired motor response. Figure 17-12 shows a motor response recording. A motor response is characterized by its amplitude, duration, and wave shape. The amplitude is measured from the baseline to the top of the negative peak (upward) of the motor response. The latency is measured from the onset of the stimulus to the point of takeoff from the baseline. In motor response studies, it is important to ensure maximal motor response by using supramaximal stimulation of the nerve (i.e., using 15% to 20% more than the minimum level of stimulation). The number and size of muscle fibers being activated determine the

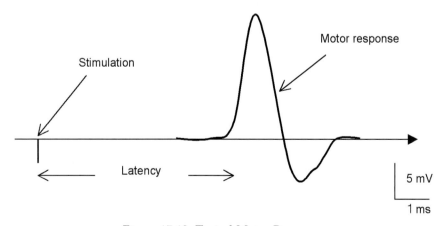

Figure 17-12. Typical Motor Resposne.

amplitude of the response.

Decrease in the number of motor units or muscle fibers responding to the stimulation will affect the amplitude of the motor response. The usual motor response has a fairly simple waveform. It may have one or two initial negative (up) peaks (the latter usually indicating two muscles being stimulated) and usually will be followed by a positive deflection (down) toward the end. If there is dispersion of the times when the motor units discharge, then the amplitude will be lowered and the response spread in time. The motor response also changes in relationship to the point of nerve stimulation. The more proximally the nerve is stimulated, the lower the amplitude and the longer the duration of responses.

Sensory Nerve Action Potentials

Sensory-nerve action potentials (SNAPs) are obtained by stimulating a nerve and recording directly from it or one of its branches. The recording site must be remote from muscles innervated by that same nerve because muscle responses will obscure the much smaller SNAP.

A typical SNAP is shown in Figure 17-13. A SNAP is characterized by its amplitude, duration, and wave shape. The amplitude of the SNAP is measured from the peak of the positive deflection to the peak of the negative deflection. The sensory distal latency is traditionally measured from the stimulus artifact to the takeoff or the peak of the negative deflection. To determine conduction velocities, the same takeoff of the proximal and distal responses are used to determine the latency. The amplitude depends on the number of axons being stimulated and the synchrony with which they transmit their impulses. If the axons transmit impulses at comparable velocities,

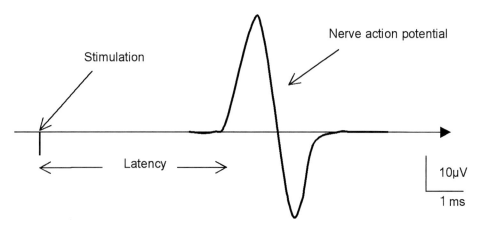

Figure 17-13. Typical Nerve Action Potential.

the response duration will be short and its amplitude high. However, if the axonal velocities are widely dispersed, the SNAP duration will be longer and its amplitude lower.

MACHINE SETTINGS

In studying sensory and motor responses, different filter, sweep speed, and sensitivity settings are used. Sensory studies are performed with the low frequency setting between 32 and 50 Hz and the high between 1.6 and 3 kHz. The sweep speed is set to 2 ms/div and the sensitivity at 10 to 20 μV/div. Motor studies are performed with the low frequency set to 16 to 32 Hz and the high frequencies to 8 to 10 kHz. Depending on the latency and duration of the response, the sweep speed can be set to anywhere between 2 and 5 ms/div and the sensitivity between 2 and 10 mV/div.

SIGNAL PROCESSING

Signal processing plays an important role in EMG/EP studies. Described next are some signal processing functions commonly found in EMG/EP machines.

Filtering

Filters are used to eliminate unwanted signals such as electrical noise and movement artifact. The frequency spectrum of muscle action potentials lies between 2 Hz and 10 kHz. In practice, a bandpass filter of 20 Hz to 8 kHz is often used because motion artifacts have frequencies less than 10 Hz and a high cutoff frequency is necessary to remove high-frequency noise.

Rectification and Integration

Because the raw signal is biphasic or polyphasic, a rectifier is sometime used to "flip" the signal's negative content across the zero axis, making the entire signal positive. Integration is also used to calculate the area under the curve for quantization and comparison.

Amplification

A single MUAP has an amplitude of about 100 μV; signals detected by surface electrodes are in the range of 0.1 to 1 mV; signals detected by indwelling electrodes are higher (up to 5 mV). All these signals must be amplified before they can be further processed. If a 1-V amplitude signal is required for the signal processor, an amplifier with a gain of 100 to 10,000 is necessary. Differential amplifiers with high common mode rejection ratio are used to minimize induced electrical noise, including 60-Hz power frequency noise, which is within the bandwidth of the signal. In addition, the impedance of the front-end amplifiers must be considerably higher than the impedance of the electrode/skin or electrode/muscle interfaces. Since indwelling electrodes have very high impedances (due to low surface area), very high amplifier input impedances (e.g., >100 MΩ) are necessary.

Spectral Analysis

Because EMG signal is actually a summation of MUAPs, some close and some at a distance from the recording electrodes, it is difficult to know which motor units contribute to the signal. In trying to differentiate normal from abnormal waveforms, some investigators, using spectral analysis, have tried to characterized the signals into its constituent frequencies.

Signal Averaging

Signal averaging is a technique used in EP studies to extract the low amplitude evoked response from noise. The amplitude of the evoked nerve response is on the order of microvolts, whereas noise can be on the order of millivolts. This technique assumes that noise is random and that the evoked responses at the same location from identical stimulations are the same. Instead of recording the nerve response from a single stimulus, multiple nerve responses are recorded from repeating the same stimulation periodically over a period of time. The response from each stimulus is stored and the average is computed by an analog or digital computer. Because all the nerve responses are the same, averaging will produce the same response. However, averaging random noise will reduce or eliminate the noise superimposing on the signal. In practice, an EP is acquired from averaging sixteen or more evoked responses.

APPLICATION EXAMPLES

Myoelectric Prostheses

To illustrate the previous signal processing concepts, an application of EMG in prosthetics is described. A myoelectric upper arm prosthesis is an assistive device to replace all or part of the functions of a lost arm. The pros-

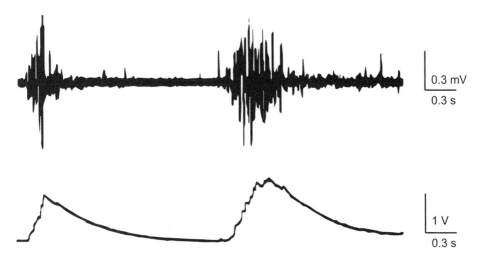

Figure 17-14. Surface EMG Signal (Top) and Myosignal (Bottom).

thesis is powered by batteries. Actuators (e.g., motors) to produce the desired function (e.g., hand grip) are controlled by myoelectric signals (myosignals) from voluntary contractions of muscle groups by the amputee.

Figure 17-14 shows the EMG of voluntary muscle contraction (top) and its processed myosignal (bottom). The EMG signal is acquired by applying surface electrodes on the belly of the flexor muscles on the forearm. The EMG is rectified, low pass filtered, and amplified to produce the myosignal. These myosignals are used to control prosthetic activation. For example, a high amplitude myosignal sent to a myoelectric hand will produce a strong grip force or two consecutive myosignals within 2 sec will switch the control from arm flexion to wrist rotation.

Nerve Conduction Velocity

Nerve conduction studies are used for evaluation of weakness of the arms and legs and paresthesias (numbness, tingling, burning). It is used to diagnose disorders such as carpal tunnel syndrome and Guillain-Barré syndrome.

Motor nerve conduction velocities (NCV) are determined by electrical stimulation of a peripheral nerve and recording from a muscle supplied by the nerve. The time it takes (latency) for the electrical impulse to travel from the location of stimulation to the recording electrode site is measured. The conduction velocity (v) expressed in meter per second (m/sec) is computed by measuring the distance (d) in millimeters (mm) between two recording points and dividing it by the difference in latency (ms) between the proximal (t_p) and distal recording points (t_d), as indicated in this equation:

$$v = \frac{d}{t_d - t_p}.$$

By applying stimulation at two different locations along the same nerve with one recording electrode, the NCV of the nerve segment between the two stimulation sites can be determined. Calculation is performed using the distance between the stimulating electrodes and the time difference between the two measured latencies. This method eliminates the neuromuscular transmission time and is used for most motor nerve studies. In sensory NCV studies, however, only one stimulation site is normally used. Figure 17-15 shows a nerve conduction study and the EP recordings at different locations along the nerve using surface electrodes.

Conduction velocities are different in different nerves due to different anatomical conditions. However, several general principles apply to nerve conduction studies:

• The more proximal the segment of nerve being evaluated, the faster the velocity will be.
• If the extremity being tested is cold, the velocity will be slowed and the amplitude increased. This effect occurs especially in cold weather, and some provisions for warming the patient and using a fairly constant room temperature should be made.
• The shorter the segment between the stimulation or recording points, the less reliable the calculated velocities will be, due to a greater effect on the margin of error by a shorter distance.
• Conduction velocities depend mostly on the integrity of the myelin sheath. In segmental demyelinating diseases, conduction velocities may drop to below 50% of normal values. Axonal loss will slow down the conduction velocity. Conduction velocity drop due to axonal loss is usually in the vicinity of 30% of normal values.

COMMON PROBLEMS AND HAZARDS

Due to small amplitude, EMG/EP signals are easily corrupted by noise. Excessive noise is often associated with electrode problems and may result from incorrect electrode placement, broken wires, and poor electrode site preparation. Patient movement and other muscle or nerve (including brain) activities will introduce artifacts leading to misinterpretation. Putting patients under sedation, especially in EP studies, can minimize such artifacts. Faulty equipment grounding, static electricity, conductive or radiated EMI, switching noise, and power surges from nearby equipment will overwhelm EMG/

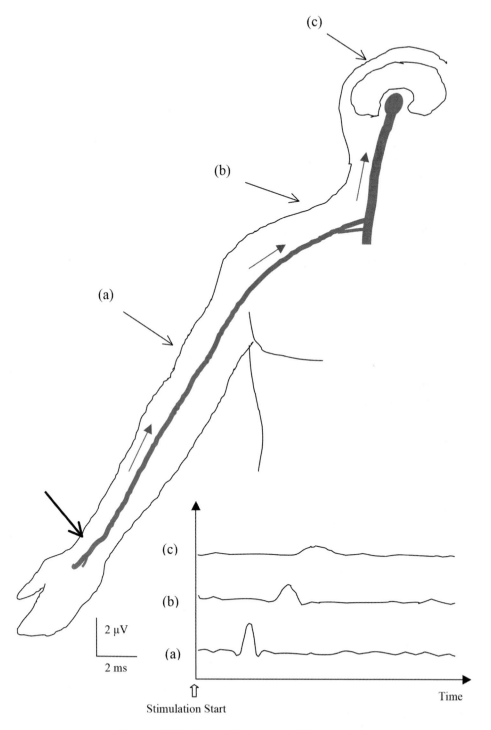

Figure 17-15. Nerve Conduction Velocity Study.

EP signals.

Recording electrodes with the same impedance produce more reliable signals. Differences in impedance can be caused by dislodged electrodes or poor electrode-tissue contact. Some machines have built in impedance testing to ensure electrode contact integrity. Impedance testing can also improve stimulating electrode performance. Constant-current stimulators allow the same current to be delivered to the tissue consistently and therefore enhance the reliability of nerve stimulation. Adjustment of sensitivity and filtering affect the amplitude and latency of EMG/EP waveforms. Therefore, machine settings such as sensitivity level and amount of filtering must remain constant throughout a recording session.

EMG/EP procedures are mildly invasive when needle electrodes are used. It may be painful during needle insertion and manipulation. Improper cleaning and handling may increase risk of infection. Use of surface electrodes and electrode gel may cause discomfort to or allergic reaction in some patients. Similar to other biopotential measurement devices, electric shock hazards due to leakage current and improper grounding are present.

BIBLIOGRAPHY

Chang, C. W., Shieh, S. F., Li, C. M., Wu, W. T., & Chang, K. F. (2006). Measurement of motor nerve conduction velocity of the sciatic nerve in patients with piriformis syndrome: A magnetic stimulation study. *Archives of Physical Medicine and Rehabilitation, 87*(10), 1371–1375.

Disselhorst-King, C., Schmitz-Rode, T., & Rau, G. (2009). Surface electromyography and muscle force: Limits in sEMG-force relationship and new approaches for applications. *Clinical Biomechanics, 24*(3), 225–235.

Dorfman, L. J., Howard, J. E., & McGill, K. C. (1989). Motor unit firing rates and firing rate variability in the detection of neuromuscular disorders. *Electroencephalography and Clinical Neurophysiology, 73*(3), 215–224.

Gordon, T., Thomas, C. K., Munson, J. B., & Stein, R. B. (2004). The resilience of the size principle in the organization of motor unit properties in normal and reinnervated adult skeletal muscles. *Canadian Journal of Physiology Pharmacology, 82*(8-9), 645–661.

Hodson-Tole, E. F., & Wakeling, J. M. (2009). Motor unit recruitment for dynamic tasks: Current understanding and future directions. *Journal of Comparative Physiology. B, Biochemical, System, and Environmental Physiology, 179*(1), 57–66.

Hoozemans, M. J., & van Dieen, J. H. (2005). Prediction of handgrip forces using surface EMG of forearm muscles. *Journal of Electromyography and Kinesiology: Official Journal of the International Society of Electrophysiological Kinesiology, 15*, 358–366.

Kothari, M. J., Heistand, M., & Rutkore, S. B. (1998). Three ulnar nerve conduction studies in patients with ulnar neuropathy at the elbow. *Archives of Physical*

Medicine and Rehabilitation, 79, 87–89.

McNerney, K. M., Lockwood, A. H., Coad, M. L., Wack, D. S., & Burkard, R. F. (2011). Use of 64-channel electroencephalography to study neural otolith-evoked responses. *Journal of the American Academy of Audiology, 22*(3), 143–155.

Misulis, K. E., & Fakhoury, T. (2001). *Spehlmann's Evoked Potential Primer* (3rd ed.). Boston, MA: Butterworth-Heinemann.

Navallas, J., Ariz, M., Villanueva, A., San Agustin, J., & Cabeza, R. (2011). Optimizing interoperability between video-oculographic and electromyographic systems. *Journal of Rehabilitation Research and Development, 48*(3), 253–266.

O'Shea, R. P., Roeber, U., & Bach, M. (2010). Evoked potentials: Vision. In E. B. Goldstein (Ed.), *Encyclopedia of Perception* (Vol. 1, pp. 399–400). Los Angeles, CA: Sage.

Pease, W. S., Lew, H. L., & Johnson, E. W. (2007). *Johnson's Practical Electromyography* (4th ed.). Philadelphia: Lippincott Williams & Wilkins.

Regan, D. (1966). Some characteristics of average steady-state and transient responses evoked by modulated light. *Electroencephalography and Clinical Neurophysiology, 20*(3), 238–248.

Regan, M. P., & Regan, D. (1988). A frequency domain technique for characterizing nonlinearities in biological systems. *Journal of Theoretical Biology, 133*(3), 293–317.

Sanders, D. B., Stalberg, E. V., & Nandedkar, S. D. (1996). Analysis of the electromyographic interference pattern. *Journal of Clinical Neurophysiology, 13*(5), 385–400.

Stalberg, E., Chu, J., Bril, V., Nandedkar, S., Stalberg, S., & Ericsson, M. (1983). Automatic analysis of the EMG interference pattern. *Electroencephalography and Clinical Neurophysiology, 56*(6), 672–681.

Willison, R. G. (1964). Analysis of electrical activity in healthy and dystrophic muscle in man. *Journal of Neurology, Neurosurgery, and Psychiatry, 27,* 386–394.

Chapter 18

INVASIVE BLOOD PRESSURE MONITORS

OBJECTIVES

- Explain the origin of blood pressure waveform.
- Analyze the relationships between blood pressure waveform and the cardiac cycle.
- Compare the magnitude and shape of blood pressure waveform at different locations in the cardiovascular system.
- Describe the clinical setup for invasive blood pressure (IBP) monitoring.
- Sketch the block diagram of a typical IBP monitor.
- Explain the construction and characteristics of a resistive strain gauge blood pressure transducer.
- Determine the systolic, diastolic, and mean blood pressure from the blood pressure waveform.
- List sources of errors in IBP measurement.
- Identify common problems and hazards.

CHAPTER CONTENTS

1. Introduction
2. Origin of Blood Pressure
3. Blood Pressure Waveforms
4. Arterial Blood Pressure Monitoring Setup
5. Functional Building Blocks of an Invasive Blood Pressure Monitor
6. Common Problems and Hazards

INTRODUCTION

The earliest attempt to record arterial blood pressure was performed in 1773 by Stephen Hales, a British scientist. Hales used an open-ended tube with one end inserted into a neck artery of a horse and the other end held at a high level (Figure 18-1). The blood from the horse rose to about 8 feet in the tube from the arterial insertion site and fluctuated by 2 to 3 inches between heartbeats. From the experiment, Hales was able to determine the blood pressure of the horse using the equation $P = \rho g h$ where ρ is the density of the blood, g the acceleration due to gravity, and h the height of the blood column. This chapter explores the principles and instrumentations of IBP measurements in clinical settings. Although only blood pressure measurement is discussed, the same principle and, in fact, similar instrumentations are used in other physiological pressure measurements such as bladder pressure and intracranial pressure.

ORIGIN OF BLOOD PRESSURE

In humans, circulation of blood is achieved by the pumping action of the heart. Atrial contraction pushes the blood from the right atrium through the tricuspid valve into the right ventricle and from the left atrium through the mitral valve into the left ventricle. The positive pressure created by the contraction of the ventricles forces blood to flow from the left ventricle through the aortic valve into the common aorta and from the right ventricle through the pulmonary valve into the pulmonary arteries (Figure 18-2). The blood

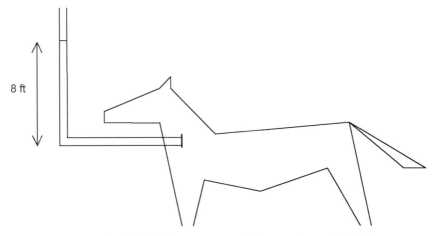

Figure 18-1. Hales's Experiment to Measure Arterial Blood.

Venous blood

To lung

From lung

Aterial blood

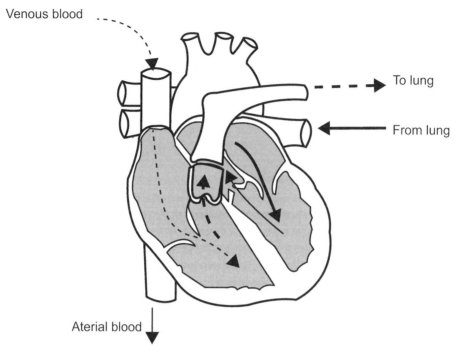

Figure 18-2. The Heart and Circulatory System.

from the aorta travels through the arteries and eventually reaches the capillaries, where oxygen and nutrients are delivered to the tissues and carbon dioxide and other metabolic wastes are diffused from the cells into the blood. This deoxygenated blood is collected by the veins and returned to the right atrium of the heart via the superior and inferior vena cavae. Contraction of the right atrium followed by the right ventricle delivers the deoxygenated blood to the lungs. Gaseous exchange takes place in the capillaries covering the alveoli of the lungs. Carbon dioxide is removed and oxygen is added to the blood. This oxygenated blood collected flows into the left atrium via the pulmonary veins and then into the left ventricle to start another round-trip in the cardiovascular system.

The heart is the center of the cardiovascular system, creating the pumping force. Every contraction of the heart produces an elevated pressure to push blood flow through the blood vessels. Relaxation of the heart allows blood to return to the heart chambers. Blood pressure within the cardiovascular system fluctuates in synchrony with the heart rhythm. The maximum pressure within a cardiac cycle is called systolic blood pressure; the lowest is called the diastolic blood pressure. Blood pressure measured in an artery is called arterial pressure, and pressure measured in a vein is called venous

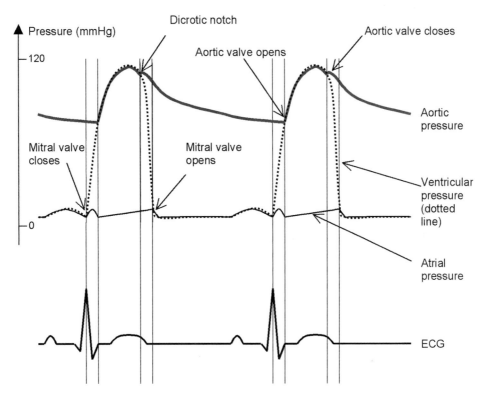

Figure 18-3. Events of a Cardiac Cycle and Blood Pressure Waveforms.

pressure. Although the SI unit of pressure is Pascal (Pa), the unit of blood pressure commonly used in North America is still in millimeters of mercury (mmHg). Figure 18-3 illustrates the timing of the cardiac cycle showing the blood pressures in the left ventricle, left atrium, and the common aorta. The pressure in the left ventricle starts to rise when the heart contracts (corresponding to the QRS complex of the ECG). When the blood pressure in the ventricle is above the pressure in the common aorta, the aortic valve opens, allowing blood to flow from the left ventricle into the aorta and then to the arteries. During the time when the aortic valve is open, the pressure in the common aorta is virtually the same as that in the ventricle. After the contraction, the heart relaxes, causing the ventricular pressure to drop rapidly. The pressure drop in the common aorta is slower than that in the ventricle due to the back pressure from downstream and the elasticity of the blood vessels. As the pressure in the left ventricle falls below the pressure in the common aorta, the aortic valve (which is a one-way valve) closes, hence separating the left ventricle from the common aorta. The blood pressure in the common aorta fluctuates between a low pressure and a high pressure.

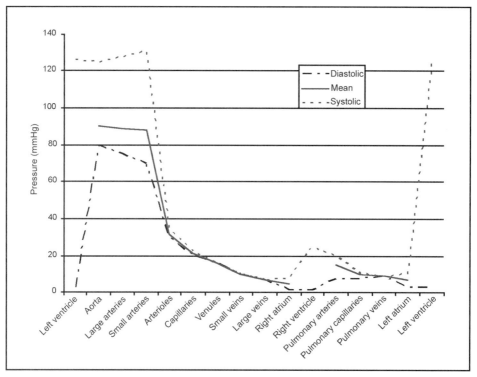

Figure 18-4. Typical Blood Pressure at Different Points of the Cardiovascular System.

BLOOD PRESSURE WAVEFORMS

As the arterial blood flows into smaller blood vessels, the average (mean) pressure as well as the magnitude of fluctuations (difference between systolic and diastolic pressure) drop due to reduction of vessel diameter, flow, and viscosity. Arterial blood pressure eventually reaches its lowest level in the capillaries. Venus blood pressure is the lowest just before it enters the right atrium. Figure 18-4 shows the values of typical mean, systolic, and diastolic blood pressure measured at different locations in the cardiovascular system. Since the left ventricle is the primary pumping device in the cardiovascular system, the blood pressure is elevated from the lowest level at the inlet of the left atrium to almost the highest as it leaves the left ventricle.

Figure 18-5 shows a typical blood pressure waveform. Note that the blood pressure is referenced to the atmospheric pressure and does not go negative. Each cycle of fluctuation corresponds to one cardiac cycle. The characteristic dicrotic notch is a result of the momentum of blood flow and the elasticity of the blood vessels. When the pressure inside the ventricle is lower than that in the common aorta, the aortic valve closes and suddenly stops blood from flowing out of the left ventricle. The blood flow continues in the aorta for a brief moment right after the valve closure due to the momentum of the blood velocity. This flow creates a transient pressure reduction in the aorta followed by a small pressure rebounce. This creates the characteristic dicrotic notch on the arterial blood pressure waveform. The dicrotic notch is less noticeable in smaller arteries and disappears altogether in the capillaries. Within a cardiac cycle, the blood pressure goes from a minimum to a maximum. The maximum pressure is called the systolic pressure (P_S), and the minimum is the diastolic pressure (P_D). The mean blood pressure (P_M) is determined by integrating the blood pressure waveform over one cycle and dividing the integral by the period (T):

$$P_M = \frac{1}{T} \int_0^T P(t)dt.$$

In some older blood pressure monitors, the mean blood presure is approximated by the equation

$$P_M \approx P_D + \frac{1}{3}(P_S - P_D).$$

Figure 18-6 shows typical blood pressure waveforms at different locations of the cardiovascular system.

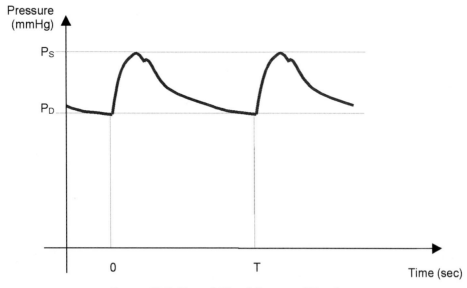

Figure 18-5. Typical Blood Pressure Waveform.

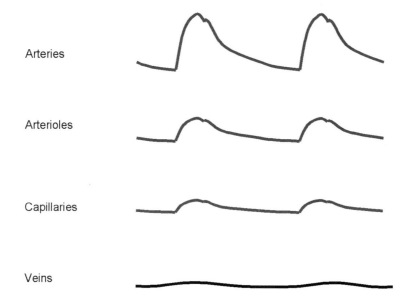

Arteries

Arterioles

Capillaries

Veins

Figure 18-6. Blood Pressure Waveforms at Different Points of the Cardiovascular System.

ARTERIAL BLOOD PRESSURE MONITORING SETUP

A typical arterial blood pressure monitoring setup is shown in Figure 18-7. Instead of inserting a pressure transducer into the blood vessel, it employs a less invasive approach by coupling a liquid column between an external pressure transducer and the blood in the vessel. In this commonly used IBP setup, a catheter is inserted into an artery (or a vein if venous pressure is monitored). Employing the Seldinger technique for catheter insertion, the blood vessel is first punctured with a hollow needle, a blunt guidewire is then advanced through the lumen of the needle, the needle is withdrawn, the catheter is then passed over the guidewire into the blood vessel. After the catheter has been inserted, an arterial pressure extension tube filled with saline is connected to the catheter. This setup is often referred to as the arterial line. The other end of the extension tube is connected to a pressure transducer. The transducer, which converts the pressure signal to an electrical signal, is connected to a pressure monitor to display the blood pressure waveform. From the signal, the systolic, mean, and diastolic pressure values are determined. To prevent a blood clot at the tip of the catheter inside the blood vessel (blood clot will block the pressure signal from reaching the transducer), a bag of heparinized saline is connected to the extension tube. The bag is pressurized (to about 300 mmHg) to above the blood pressure in the vessel. Together with the continuous flush valve at the transducer set, this setup produces a slow continuous flow of heparinized saline flushing the

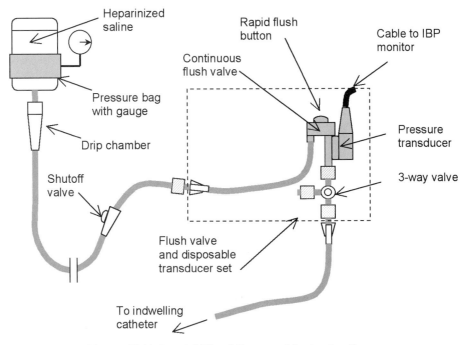

Figure 18-7. Arterial Blood Pressure Monitoring Setup.

catheter to prevent blood clot. The flow is very slow (less than 5 mL/hr) to avoid creating a pressure drop in the extension tube and catheter setup; otherwise it will affect the accuracy of blood pressure measurement. The rapid flush valve is used during initial setup to flush and fill the extension tube before it is connected to the indwelling catheter.

The equivalent hydraulic circuit of the arterial line setup is shown in Figure 18-8a. In the setup, the patient port (which is the location of the catheter) is h meters above the transducer port (the point where the liquid in the extension tube interfaces with the pressure transducer). Therefore, the pressure P_X, as seen by the transducer, is the sum of the pressure due to the liquid column and the pressure P_P at the patient port (blood pressure of the patient), in other words,

$$P_X = P_P + \rho gh, \tag{18.1}$$

where ρ is the density of the liquid in the extension tube and g is the acceleration due to gravity. This equation shows that the pressure as seen by the transducer is higher than the actual blood pressure by the product ρgh. Figure 18-8b shows the concept of introducing an offset P_0 to compensate for this overpressure phenomenon. The zeroing process performed during the initial setup is designed to allow the machine to determine the value of this offset.

From the compensation circuit and Equation 18.1,

$$P = P_X + P_0 = P_P + \rho gh + P_0. \qquad (18.2)$$

During the initial zeroing process, the patient port is open to the atmosphere so that P_P becomes zero. Equation 18.2 therefore becomes

$$P = \rho gh + P_0. \qquad (18.3)$$

While the patient port is still exposed to atmospheric pressure, the operation invokes the zeroing sequence of the blood pressure monitor, telling the monitor that this reading corresponds to zero pressure (i.e., P = atmospheric pressure = 0, or zero gauge pressure). Equation 18.3 now becomes

$$0 = \rho gh + P_0 \Rightarrow P_0 = -\rho gh.$$

The blood pressure monitor saves this value of P_0 in memory and exits the zeroing sequence. The operator then closes the patient port. The blood pressure monitor applies this offset to the transducer reading during subsequent pressure measurements. As long as the vertical height difference between the transducer port and patient port remains the same, the monitor will always display the true blood pressure of the patient. However, if the transducer is lowered (or the patient port is raised) after zeroing, the monitor will overread the pressure. For every 2.5-cm decrease in the height difference, the pressure is overread by about 2 mmHg. The zeroing process will also compensate for any other constant offsets in the system, including the

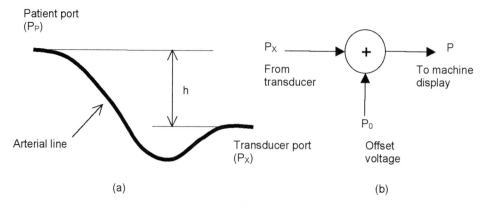

Figure 18-8. Zeroing of Blood Pressure Monitor.

offset from the pressure transducer.

Example 18.1

A patient is undergoing IBP monitoring. The arterial line was zeroed at setup. The patient's systolic pressure and diastolic pressure were 125 and 80 mmHg, respectively. If the patient bed is raised by 4.0 inches while the level of the transducer remains the same, what will the pressure readings be?

Solution:

Using the equation $P_0 = \rho gh$ and assuming the density of the saline in the extension tube is 1020 kg/m3, raising the patient by 4.0 inches (4.0 x 0.0254 = 0.10 m) will increase the offset pressure by

$$P_0 = 1020 \text{ kg/m}^3 \times 9.8 \text{ m/sec}^2 \times 0.10 \text{ m} = 1000 \text{ Pa} = 7.5 \text{ mmHg.}$$

The systolic and diastolic blood pressure reading therefore becomes 132.5 (125 + 7.5) mmHg and 87.5 (80 + 7.5) mmHg, respectively.

In most prepackaged disposable transducers, the zero ports are attached to the transducers. In order to properly zero the system, the setup procedure requires that the zero port be located at the same level as the patient's midaxillary line or the patient's heart. Instead of opening the patient port to the atmosphere during the zeroing process, the zero port attached to the transducer is open to the atmosphere. In this case, after correct zeroing, the monitor will display the blood pressure at the level of the patient's heart. This arrangement will make it easier for the clinicians to check (by simply verifying the transducer is at the heart level of the patient) if the transducer level is correctly positioned to ensure accurate blood pressure measurement. Alternatively, the transducer/zero port can be tapped to the patient body's phlebostatic axis which is the intersection of the line through the 4th intercostal space and the midauxillary line. Phlebostatic axis is regarded as the anatomical point that corresponds to the right atrium that most accurately reflects a patient's hemodynamic status.

FUNCTIONAL BUILDING BLOCKS OF AN INVASIVE BLOOD PRESSURE MONITOR

Figure 18-9 shows a typical functional block diagram of an IBP monitor. The following paragraphs describe the functions of each building block.

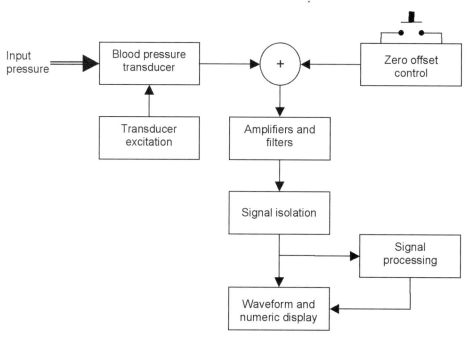

Figure 18-9. Functional Block Diagram of an Invasive Blood Pressure Monitor.

Transducer

The pressure transducer of an IBP monitor converts the pressure signal to an electrical signal. Ideally, the transducer should have a linear characteristic with adequate frequency response to handle the rate of pressure fluctuations.

Figure 18-10a shows the cross-sectional view of a four-wired resistive strain gauge pressure transducer. The central floating block is connected to the four pretensioned strain wires with the other ends of the wires connected to the stationary frame of the transducer. The diaphragm in contact with the fluid chamber (or pressure dome) is mechanically connected to the central floating block by a rigid connecting rod. The blood pressure to be measured is transmitted via saline in the extension tube to the fluid chamber, forcing the diaphragm to move according to the pressure fluctuation. As the diaphragm moves, the strain wires will stretch or relax according to the movement. The strain wires are connected in a bridge format with the electrical equivalent diagram as shown in Figure 18-10b.

For a higher applied pressure, strain wires 1 and 2 are extended while 3 and 4 are shortened. As a result, the resistance of the strain wires 1 and 2 (shown in the equivalent circuit) become higher while 3 and 4 become less.

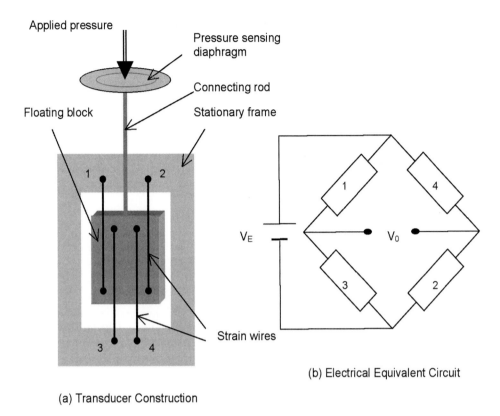

(a) Transducer Construction

(b) Electrical Equivalent Circuit

Figure 18-10. Resistive Strain Gauge Pressure Transducer.

If an excitation voltage V_E is applied to the bridge, the output voltage V_0 will vary according to the applied pressure. Although V_E is shown in the diagram as a DC voltage, AC excitation may be used. The sensitivity of a pressure transducer is expressed in output voltage per unit pressure (e.g., 10 mV/mmHg). To allow for designs using different choices of excitation, transducer manufacturers specify sensitivity in output voltage per unit excitation voltage per unit pressure (e.g., 2 mV/V/mmHg).

In older systems or systems using reusable pressure transducers, resistive strain gauge or piezoelectric element transducers are often used. Nowadays, disposable transducers using semiconductor piezoresistive strain gauges are commonly used. These transducers are mass produced using semiconductor fabrication technology, which yields consistent performance at low cost. Disposable IBP transducers are prepackaged with the extension tube, flush valves, and connectors for single patient use.

The sensitivity of a disposable blood pressure transducer as specified in the AAMI Standards BP22 and BP23 is 5 µV/V/mmHg ± 1% when an excitation voltage of 4 to 8 V, DC to 5 kHz is used. This allows interchangeability of blood pressure transducers across different manufacturers of compatible blood pressure monitors.

Example 18.2

A special-purpose reusable pressure transducer has an output sensitivity of 2.0 mV/V/mmHg. If an applied pressure of $P = 100$ mmHg is applied and the excitation V_E is a 5.0-V peak to peak 100-Hz sinusoidal voltage source, what is the output voltage of the transducer?

Solution:

Since V_0 = sensitivity x V_E x P, the transducer output voltage V_0 is calculated by

$$V_0 = 2.0 \text{ mV/V/mmHg} \times 5.0 \text{ V}_{p\text{-}p} \times 100 \text{ mmHg} = 1000 \text{ mV}_{p\text{-}p} \text{ or } 1.0 \text{ V}_{p\text{-}p}.$$

Due to the sinusoidal excitation, the output voltage is also a 100-Hz sinusoid signal.

Zero Offset

The zero offset functional block is used during the zeroing procedure to determine and store the zero offset value. During blood pressure monitoring, this stored value is used to compensate for the offset due to the static pressure of the setup and the offset of the transducer. For microprocessor-based machines, the zero offset value is stored digitally.

Amplifiers and Filters

For a systolic pressure of 120 mmHg, a standard disposable transducer (sensitivity = 5 μV/V/mmHg) with an excitation of 5 V produces an output of 3000 μV or 3 mV. Such small voltage must be amplified before it can be used by other parts of the monitor. IAs with high input impedance and high CMRR are used for this purpose.

The fundamental frequency of the blood pressure waveform is the same as the heart rate (which is about 1 Hz). However, spectral analysis of a typical arterial blood pressure waveform (Figure 18-11b) shows a bandwidth from DC (or zero frequency) to about 10 Hz (Figure 18-11b). In order to reduce higher frequency artifacts and interferences, a low pass filter with a high cutoff frequency of 20 to 50 Hz is often built into the front-end analog circuit of the monitor.

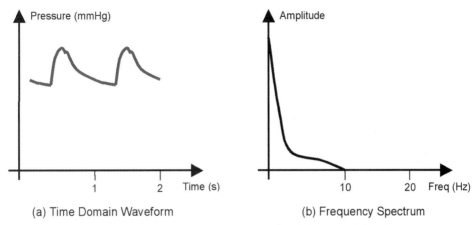

(a) Time Domain Waveform   (b) Frequency Spectrum

Figure 18-11. Frequency Spectrum of Blood Pressure Waveform.

Signal Isolation

IBP monitoring involves external access to major blood vessels. Both the saline solution in the arterial line and the blood in the artery conduct electricity. This setup forms a conduction path between the electromedical device and the patient's heart. The blood pressure monitor may become the source or sink of the risk current flowing through the heart. Signal isolation to break the conduction path is required to minimize the risk of electrical shock (both macroshocks and microshocks) to the patient.

Signal Processing

From the blood pressure waveform, the systolic, mean, and diastolic blood pressures are determined. In addition, the patient's heart rate can be derived because the frequency of the pressure cycle is the same as the cardiac cycle. The systolic blood pressure is obtained by using a peak detector circuit (Figure 18-12a). In order to track the fluctuating systolic pressure, a pair of peak detectors arranged in a sample and hold configuration are used. The diastolic pressure can be found by first inverting the pressure waveform and then finding the peak of this inverted waveform. Mean blood pressure is obtained using a low pass filter circuit (Figure 18-12b).

In modern monitors, the blood pressure waveform is sampled and converted to digital signals. Systolic, mean, and diastolic pressures are determined by software algorithms in the microprocessor.

Display

LCDs have replaced CRTs in as the display of choice for medical wave-

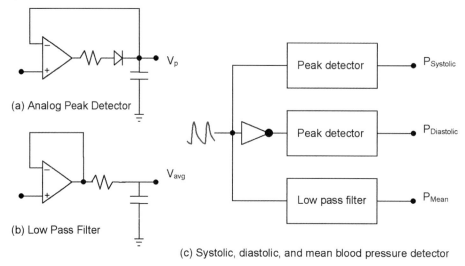

(a) Analog Peak Detector

(b) Low Pass Filter

(c) Systolic, diastolic, and mean blood pressure detector

Figure 18-12. Blood Pressure Detection from Pressure Waveform.

form displays. In addition to waveforms, numeric information is also displayed on the medical monitor.

COMMON PROBLEMS AND HAZARDS

Other than electric shock hazards described earlier (*see* Signal Isolation), a patient under IBP monitoring may develop infection due to the transducer setup. Although sterile procedures are followed in the handling and setting up of the arterial line, the invasive procedure will expose the patient to risk of infection.

Problems in an IBP measurement system may produce inaccurate pressure readings or distorted blood pressure waveform. These erroneous signals may cause improper diagnosis, leading to inappropriate medical intervention. Some common sources of errors are described next.

Setup Error

The most common problem in this category relates to the zeroing process. Incorrect zeroing procedure or change in vertical distance between the transducer and measurement site after initial setup produces a constant static pressure error in the measurement. It is important for the clinician to correctly perform the zeroing procedure, understand the principles, and be aware of the implications from setup variations.

Catheter Error

Although there is no active component in the catheter and it seems to be a very simple part of the blood pressure monitoring system, many artifacts and measurement errors may arise from the catheter. Some of the common problems are described next.

END PRESSURE, WHIPPING, AND IMPACT ARTIFACTS. The catheter in blood pressure monitoring is a small flexible tube inserted into a blood vessel with pulsating blood flow (Figure 18-13). If the blood is flowing in the same direction as the catheter, it creates a small negative pressure at the end of the catheter tip. In contrast, blood flowing toward the catheter tip will create a net positive pressure. Either of these will create an error in the blood pressure reading. The flow of blood may create turbulence and set the catheter tip into a whipping motion. Movement of the catheter may cause it to collide with the vessel wall or valves. Whipping motion and impact of the catheter will show up as distortion in the blood pressure waveform.

AIR BUBBLE, PINCHING, AND LEAK. Another area of pressure waveform distortions is caused by the reduction of the frequency response of the catheter extension tube. An air bubble in the fluid-filled catheter extension tube reduces the cutoff frequency of the low pass filter formed by the hydraulic circuit. Pinching the line or having a leak in the line has a similar effect. Such problems in the system will attenuate the high-frequency component of the blood pressure waveform. Figure 18-14 shows the effect of such problems on the frequency response of a catheter.

Blood Clot

The purpose of the pressured infusion bag is to prevent blood clots from occurring at the tip of the catheter in the blood vessel. Stagnant blood at the

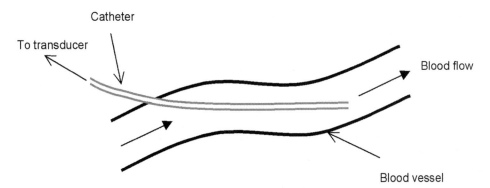

Figure 18-13. Catheter in Blood Vessel.

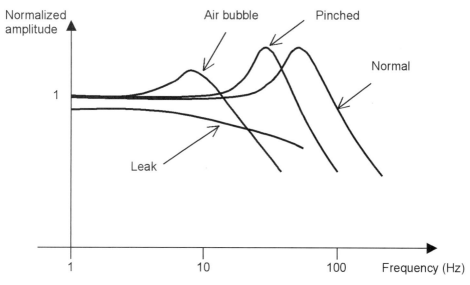

Figure 18-14. Frequency Response of Catheter Setup.

tip of the indwelling catheter will develop blood clots blocking the transmission of the pressure signal to the transducer. A partial block at the catheter tip will diminish the amplitude fluctuation (difference in systolic and diastolic pressure) and lower the high-frequency response of the setup. Periodic inspection of the drip chamber attached to the infusion bag to ensure a continuous flow of the heparinized saline will prevent clotting.

Transducer Calibration

Due to stringent manufacturing processes, there is no need to perform field verification of the accuracy of single-use disposable blood pressure transducers. However, blood pressure monitors must be checked periodically to ensure that they are functioning properly with amplification and frequency response according to manufacturers' specifications. In practice, a simulator is used to provide a known input to the monitor, and the output is measured and compared with the specifications.

For reusable pressure transducers, a known pressure source is used to determine the sensitivity of the transducer. Most pressure monitors have a calibration factor (F) adjustment to compensate for sensitivity drift of the transducers. A simple method to obtain the calibration factor of a particular transducer is as follows:

1. Apply a known pressure (Pi) to the transducer and read the pressure dis-

play (Pd) on the monitor.
2. Calculate the calibration factor by using the equation $F = Pi/Pd$.
3. Input the value of F into the calibration factor adjustment input of the monitor.
4. The monitor is now calibrated to use with this particular transducer.

Example 18.3

A 200 mmHg pressure source is used as input to determine the calibration factor of the monitor with a reusable pressure transducer. If the pressure reading of the monitor is 190 mmHg, what is the calibration factor?

Solution:

Using

$$F = \frac{P_i}{P_d}, \text{ the calibration factor is } F = \frac{200}{190} = 1.05.$$

Hardware Problems

As with all medical devices, there is always a possibility of component failure. It is important that users be able to differentiate between normal and abnormal performance of the monitoring system. Many monitors have built-in simple test procedures to allow the users to verify the function and performance of the system. In order to ensure that the monitor is functioning according to standards or manufacturers' specifications, periodic performance verification inspections by qualified professionals are required to detect nonobvious problems such as component parameter drifts.

BIBLIOGRAPHY

Booth, J. (1977). A short history of blood pressure measurement. *Proceedings of the Royal Society of Medicine, 70*(11), 793–799.

Eguchi, K., Yacoub, M., Jhalani, J., Gerin, W., Schwartz, J. E., & Pickering, T. G. (2007). Consistency of blood pressure differences between the left and right arms. *Archives of Internal Medicine, 167*(4), 388–393.

Guyton, A. C., & Hall, J. E. (2006). *Textbook of Medical Physiology* (11th ed.). Philadelphia, PA: Elsevier Saunders.

Klabunde, R. (2005). *Cardiovascular Physiology Concepts.* Philadelphia, PA: Lippincott Williams & Wilkins.

Lewis, O. (1994). Stephen Hales and the measurement of blood pressure. *Journal of Human Hypertension, 8*(12), 865–871.

Seldinger, S. I. (1953). Catheter replacement of the needle in percutaneous arteriography: A new technique. Acta *Radiologica, 39*(5), 368–376.

Shock N.W., Ogden E. (1939). The probable error of blood pressure measurements. *Quarterly Journ. Of Exp. Physiol.* 29:49-62.

Womersley, J. R. (1955). Method for the calculation of velocity, rate of flow and viscous drag in arteries when the pressure gradient is known. *Journal of Physiology, 127*(3), 553–563.

Chapter 19

NONINVASIVE BLOOD
PRESSURE MONITORS

OBJECTIVES

- Discuss the components of a sphygmomanometer and their functions.
- Describe the principles of operation and the limitations of using a manual auscultatory method to measure systolic and diastolic blood pressure.
- Differentiate between auscultatory and oscillometric methods employed in automatic noninvasive blood pressure (NIBP) measurement.
- Describe the principles of operation and limitations of NIBP measurement using the oscillometric method.
- Explain the functional building blocks of a typical oscillometric NIBP monitor.
- Discuss the principles of using Doppler ultrasound and tonometry in NIBP monitoring.
- Describe common problems and hazards in NIBP measurements.

CHAPTER CONTENTS

1. Introduction
2. Auscultatory Method
3. Oscillometric Method
4. Other Methods of NIBP Measurement
5. Common Problems and Hazards

INTRODUCTION

Blood pressure, an important physiological parameter, is measured routinely throughout the course of nearly all medical procedures and diagnosis. Although one may measure blood pressure more accurately using the invasive technique described in the previous chapter, being able to measure blood pressure noninvasively is a tremendous achievement made possible by the combination of creativity and technology.

In 1876, E. J. Marey, a French physiologist, performed the first experiment using counter-pressure to measure blood pressure. Marey was investigating the interaction between the arterial pulsatile pressure with an applied external pressure. He had his assistant place his hand into a jar filled with water; the jar was sealed at the wrist. The pressure in the jar was increased incrementally, and the pressure oscillation in the jar was measured at each pressure increment. Marey noticed that when the jar pressure was slightly above the systolic pressure, all the blood was driven out of the hand and no pressure oscillation was detected. He further noticed that the amplitude of oscillation began to rise as the jar pressure dropped below the systolic pressure, reached at maximum, and decreased as the pressure was further reduced. This method has since evolved into the oscillometric method of NIBP measurement—the most popular method to noninvasively measure blood pressure.

The auscultatory method was introduced in 1905 by N. S. Korotkoff, a Russian surgeon. By listening to the changes of sound from an artery when the pressure in the upstream occluding cuff was slowly released, Korotkoff was able to determine the systolic and diastolic blood pressure of a patient. This manual method using simple portable tools is considered as "the method" to measure blood pressure noninvasively, especially in physician offices.

Although results from NIBP measurement may not be as accurate as invasive methods, NIBP measurement is easy, nonhazardous, and inexpensive. It provides a safe and reliable method for measurements of a patient's blood pressure in clinical settings or at home. It is especially useful in trending. For example, the effectiveness of hypertension medication can be assessed by trending the systolic and diastolic blood pressure over the course of treatment. Today, NIBP measurement is performed in almost every medical examination. This chapter describes the principles and instrumentations of a number of common indirect and noninvasive methods in blood pressure measurement.

AUSCULTATORY METHOD

Indirect methods measure blood pressure without directly accessing the bloodstream. The most commonly used instrument is based on the auscultatory technique. The device used in this technique is called a sphygmomanometer, which is present at every hospital bedside and in every clinic and physician's office. In addition to the sphygmomanometer, a stethoscope is required to listen to the sounds in the artery during the measurement.

A sphygmomanometer (Figure 19-1) consists of

1. An inflatable rubber bladder enclosed in a fabric cover called the cuff
2. A rubber hand pump with valve assembly so that the pressure in the setup can be raised, and be released at a slow controlled rate
3. A pressure measurement device to display the air pressure inside the cuff. Mercury manometers were commonly used as the pressure measurement device. However, due to the health hazards of mercury exposure (to nervous, digestive and immune systems, and to lungs and kidneys), dial-type mechanical air pressure gauges have replaced mercury manometers in auscultatory indirect blood pressure measurements

During blood pressure measurement, the pressure cuff is wrapped around the upper arm of the subject and a stethoscope is placed on the inner elbow for the operator to listen to the sound produced by the blood flow in the brachial artery. While watching the pressure gauge, the operator manually squeezes the hand pump to raise the cuff pressure until it is above the systolic blood pressure (e.g., 150 mmHg or 20 kPa). At this pressure, the brachial artery is occluded. Since blood is not able to flow to the lower arm, no sound will be heard from the stethoscope. The cuff pressure is then slowly reduced, say, at a rate of approximately 3 mmHg (or 0.4 kPa) per second, by opening the pressure release valve. As the cuff pressure falls below the systolic pressure, the clinician will start to hear some clashing, snapping sounds from the stethoscope. This sound is caused by the jets of blood pushing through the occlusion. As the cuff pressure continues to decrease, the sound intensity will first increase and then turn into a murmur-like noise and become a loud thumping sound. The intensity and pitch of the sounds will change abruptly into a muffled tone when the cuff pressure is getting close to the diastolic pressure and will disappear completely when the pressure is below the diastolic pressure. These sounds are called Korotkoff sounds.

The cuff pressure at which the first Korotkoff sound appears corresponds to the systolic pressure; the disappearance of the sound corresponds to the diastolic pressure. Figure 19-2 shows the relationships between the arterial pressure and the cuff pressure during the course of measurement.

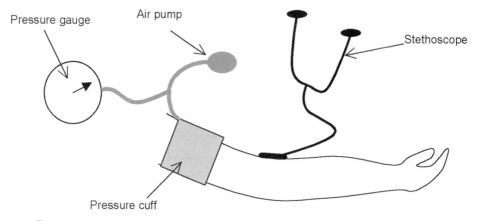

Figure 19-1. NIBP Measurement Setup Using Manual Auscultatory Method.

This method is suitable for most patients, including hypotensive and hypertensive patients. Other than applying the cuff over the upper arm, the cuff can be placed over the thigh or the calf of the patient. In each of these applications, the Korotkoff sounds should be detected downstream of the occlusions.

The limitation and accuracy of this method is affected by the following:

• The Korotkoff sounds are normally in the range of less than 200 Hz, where human hearing is normally less acute. Determination of the

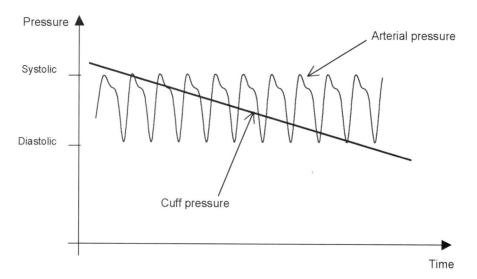

Figure 19-2. Relationships Between Arterial and Cuff Pressures.

Korotkoff sounds is affected by the hearing acuity of the operator, especially when it is used on hypotensive patients or infants, in whom the sound levels are low.

- The patient can affect the accuracy of blood pressure measurements. The American Heart Association (AHA) recommends that the patient be relaxed with legs uncrossed with back and arm supported. The middle of the cuff should be level with the right atrium; at least 2 readings be taken. The first reading is usually higher than the second. If the different is greater than 5 mmHg, additional readings should be taken.
- Inappropriate cuff size or incorrect placement can produce falsely high (undersized or loosely applied cuff) or falsely low (oversized cuff) readings. The AHA recommends that the ideal length of the bladder under the cuff be 80% of the circumference of the patient's limb and the width of the bladder is at least 40% (ideally 46%) of the circumference.
- Deflating the cuff too fast will produce an erroneous reading. The AHA recommends that the cuff should be deflated at a rate of 2 to 3 mmHg per second (or 0.4 kPa/s), higher rates can cause the systolic pressure to appear lower and the diastolic pressure to appear higher. Figure 19-3 shows an underestimation of the systolic pressure due to a deflation rate that is too fast.
- It is known that the Korotkoff sounds disappear early in some patients and then reappear as the cuff pressure is lowered toward the diastolic pressure. This phenomenon is referred to as the auscultatory gap.

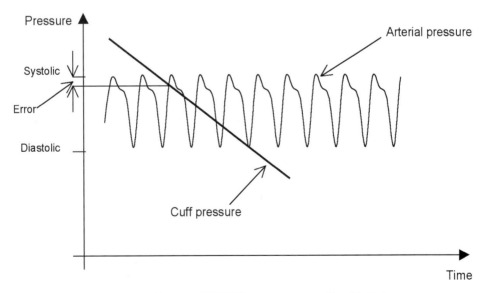

Figure 19-3. Error in NIBP Measurement on Fast Deflation.

• The auscultatory method can determine the systolic and diastolic pressures but not the mean blood pressure.

Although the sphygmomanometer is a relatively simple device, regular maintenance is still required which includes pressure gauge calibration; cleaning; and checking for leaks on tubing, cuff bladder, valves, and the hand pump.

An automatic NIBP monitor may use the same principle as the manual auscultatory method. Automation overcomes the hearing acuity limitation by employing a microphone inside the cuff to pick up the Korotkoff sounds instead of relying on the acuity of human hearing. It also replaces the manual pump with an automatic pump and uses an electronic pressure transducer instead of a mechanical pressure gauge. After the cuff is applied, the NIBP monitor automatically inflates the cuff to occlude the blood vessel. The bladder pressure is slowly released while the microphone listens for the Korotkoff sounds. These processes are automatically coordinated by the monitor. The systolic and diastolic pressures are determined by tracking the bladder pressure and correlating it to the different phases of the Korotkoff sounds picked up by the microphone.

OSCILLOMETRIC METHOD

Automatic NIBP monitors using the oscillometric method are similar to the auscultatory method except the oscillometric method detects the small fluctuations of pressure inside the cuff rather than listening to the Korotkoff sounds in the auscultatory method. When the cuff pressure falls below the systolic pressure, blood breaks through the occlusion, causing the blood vessel under the cuff to vibrate. This vibration of the vessel's wall causes fluctuation (or oscillation) of the cuff pressure. The onset of the vibration correlates well with the systolic pressure, while the maximum amplitude of oscillation corresponds to the mean arterial blood pressure. When the cuff pressure is at the mean arterial pressure, the net average pressure on the arterial wall is zero (both sides of the wall are of the same pressure), which allows the arterial wall to freely move in either direction. Under this condition, the amplitude of vibration of the arterial wall caused by blood pressure fluctuation in the artery is the highest. The diastolic pressure event on the oscillometric curve is somewhat less defined.

Figure 19-4a shows the relationships between the arterial blood pressure and the cuff pressure. The maximum amplitude of pressure oscillation is usually less than a few percentage of the pressure inside the cuff (e.g., 4 mmHg or 0.5 kPa amplitude fluctuation when the cuff pressure is 100

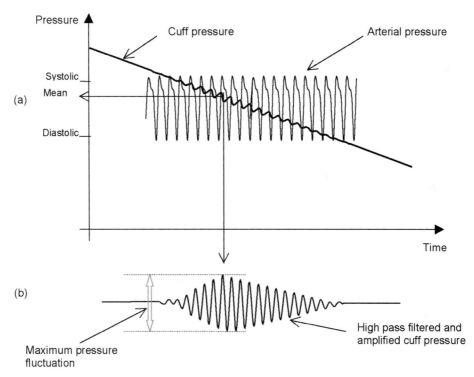

Figure 19-4. Relationships Between Arterial and Cuff Pressure in Oscillometric Method.

mmHg or 13 kPa). To extract only the oscillatory component from the pressure signal obtained by the pressure sensor, the low-frequency component of the signal (corresponding to the slowly deflating cuff pressure) is removed by a high pass filter. The remaining oscillatory component of the signal (shown amplified in Figure 19-4b) is then used to determine the mean, systolic, and diastolic blood pressures. The cuff pressure corresponding to the maximum oscillation amplitude is taken as the mean arterial pressure. The cuff pressure corresponding to the onset of oscillation is usually taken as the systolic pressure. Different manufacturers of NIBP monitors may use different algorithms to determine the diastolic pressures from this oscillometric signal. One commonly adopted approach to determine the diastolic pressure is to take the point where the amplitude of the oscillation has the highest rate of change; another approach estimates the diastolic pressure by locating the point where the cuff pressure corresponds to a fixed percentage (e.g., 20% used by one manufacturer) of the maximum oscillation amplitude.

Factors affecting the accuracy of automatic NIBP measurements using the oscillometric method are similar to those using the auscultatory method. Compared to the auscultatory method, NIBP measurements using the oscil-

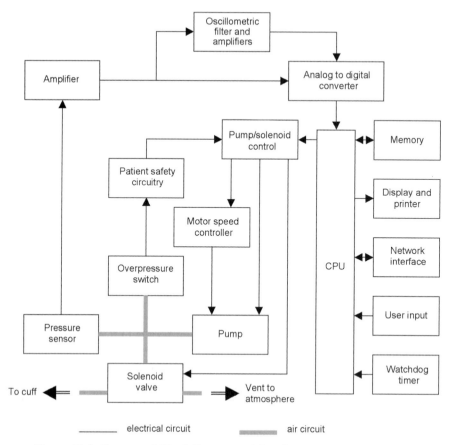

Figure 19-5. Functional Block Diagram of Oscillometric NIBP Monitor.

lometric method are not affected by audible noise and therefore can work in a noisy environment. On the other hand, because this method relies on detecting the amplitude of pressure fluctuation, any movement or vibration can lead to incorrect readings. Furthermore, in oscillometric NIBP monitors, the diastolic pressure is only an estimated quantity. In addition, the small pressure change at the onset of oscillation (which corresponds to the systolic pressure) is difficult to detect. Of the two automatic noninvasive methods, the oscillometric method is more commonly used than is the auscultatory method in automatic blood pressure monitors.

Figure 19-5 shows the functional building blocks of a NIBP monitor using the oscillometric method. The following descriptions explain the functions of the building blocks.

- **Pump and Solenoid Valve** - The motorized air pump inflates the cuff pressure to a predetermined pressure so that the artery is occluded under the cuff. The solenoid valve connects the air circuits between the pump and the cuff during inflation and connects the cuff to atmosphere during deflation. The rate of deflation can be controlled by pulsing the solenoid at a certain duty cycle.

- **Pressure Sensor** - Through the internal tubing connections, the cuff pressure is constantly monitored by the pressure sensor in the NIBP monitor.
- **Amplifiers and Oscillometric Filter** - The signal picked up by the pressure transducer is amplified (by about 100 times). This signal consists of two sets of information: the slowly decreasing cuff pressure and the oscillatory signal. The slow varying signal is separated from the oscillatory signal by a high pass filter (with cutoff frequency of about 1 Hz). Both signals are fed to the analog to digital converter.
- **Central Processing Unit (CPU) and Analog to Digital Convertor (ADC)** - The cuff pressure and oscillometric signal are digitized by the ADC and sent to the CPU to determine the mean, systolic, and diastolic pressures of the measurement. The heart rate can also be determined from the signals.
- **Display, Printer, Memory, and Network Interface** - The measured systolic, diastolic, and mean blood pressures are shown on a display (e.g., LCD). A hard copy may be printed for charting. These data may also be time stamped and saved in the memory of the monitor for trending or communicated via network connections to other devices or electronic medical record system.
- **Watchdog Timer and Overpressure Switch** - An independent overpressure safety switch activates the solenoid valve to release the cuff pressure down to atmospheric pressure should excessive pressure develop in the cuff. The solenoid will also open to the atmosphere if the cuff pressure remains high for a preset duration of time. Both features are in place to prevent compression damage of the tissues under the cuff.

OTHER METHODS OF NIBP MEASUREMENT

There are many other methods to measure or estimate blood pressure noninvasively. Compared to the auscultatory or oscillometric methods, these methods are either not as accurate or more complicated or not as easy to use in clinical settings. Two of the better methods are described next.

Doppler Ultrasound Blood Pressure Monitor

This class of device makes use of the Doppler effect to detect blood flow

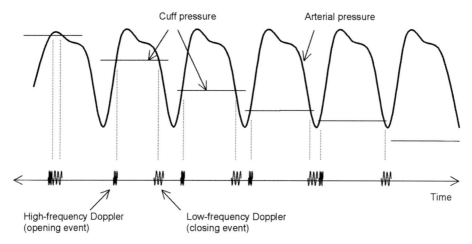

Figure 19-6. Doppler Events Due to Interaction of Cuff Pressure and Blood Pressure.

patterns in the artery of interest. An ultrasound transmitter and a receiver are used to replace the stethoscope. The monitor detects the Doppler shift when the incident sound wave is reflected from the blood flow in the subject. When the artery is occluded by the cuff, the Doppler shift is zero. When the cuff pressure is slowly reduced, the arterial pressure is able to overcome the cuff pressure occlusion, causing the occlusion to snap open. This jet of blood flowing through the cuff occlusion produces a Doppler shift. There are actually two Doppler events during each cardiac cycle — the opening and closing of the blood vessel under the cuff. When the arterial pressure exceeds the cuff pressure, the blood flowing through the opening of the occlusion produces a high-frequency Doppler shift (e.g., 200 to 500 Hz). When the arterial pressure recedes toward the diastolic pressure, the blood vessel will be re-occluded. This event produces a lower frequency Doppler shift (e.g., 15 to 100 Hz).

When the cuff pressure is allowed to be bled down at constant speed from the systolic pressure, the high- and low-frequency events appear and are next to each other. As the cuff pressure continues to drop, the two events become farther and farther apart. When the cuff pressure reaches the diastolic pressure, the low-frequency event will coincide with the high-frequency event of the next cardiac cycle (Figure 19-6). The frequency shift may be coupled to a loudspeaker to allow the operator to determine the systolic and diastolic pressures. Figure 19-6 shows the Doppler events due to the interaction of the cuff pressure and blood pressure.

Arterial Tonometry

None of the NIBP methods discussed earlier are able to measure the

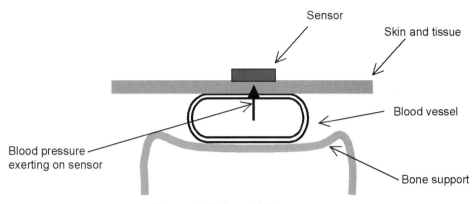

Figure 19-7. Arterial Tonometry.

blood pressure waveform. Arterial tonometry is a continuous pressure measurement technique that can noninvasively measure pressure in superficial arteries with sufficient bony support, such as the radial artery. A tonometer is a contact pressure sensor that is applied over a blood vessel. It is based on the principle that if the sensor is depressed onto the vessel wall of an artery such that the vessel wall is parallel to the face of the sensor, the arterial pressure is the only pressure perpendicular to the surface and is measured by the sensor (Figure 19-7). Theoretically, accurate real-time blood pressure waveform can be recorded using this noninvasive technique. However, experiments showed that although this method produces good-quality pulse waveform, it tends to underestimate the systolic and diastolic pressures.

To obtain good results, tonometry requires that the contact surface be stiff and the sensor be small relative to the diameter of the blood vessel. In addition, proper sensor application is critical because if the vessel is not flattened sufficiently (e.g., due to inadequate depression), the tonometer will measure forces due to arterial wall tension and bending of the vessel. However, too much depression force may occlude the blood vessel. In practice, blood pressure monitors using this method requires periodical calibration using blood pressure values obtained by invasive method.

COMMON PROBLEMS AND HAZARDS

As discussed earlier, NIBP monitors using the auscultatory method will be interfered with by audible noise. The oscillometric method is vulnerable to motion or vibration interference. In either method, it is important to select the correct size of cuff bladder. Cuff deflation too fast will produce errors. NIBP measurements, as explained in earlier sections, are not as accurate as invasive means discussed in the previous chapter. In addition, traditional NIBP measurements will not show the patient's blood pressure waveform which contains important diagnostic information. Too fast cuff deflation rate will lead to errors in pressure readings. Soft tissue injury including nerve compression damage can be caused by excessive and prolonged cuff pressure. NIBP has a maximum cuff pressure limit to reduce the risk of compression injury. Most clinical monitors, when first turned on, will inflate the cuff to an initial preset pressure (e.g., 160 mmHg or 8 kPa). In subsequent measurements, the monitor will assess the last systolic pressure, if it is less than the initial preset pressure, it will lower the inflation pressure to avoid unnecessary high cuff occlusion pressure. For neonatal monitors, the initial preset pressure as well as the maximum cuff pressure is programmed to be less for the adult version.

BIBLIOGRAPHY

Booth, J. (1977). A short history of blood pressure measurement. *Proceedings of the Royal Society of Medicine, 70*(11), 793–799.

Borow, K., & Newburger, J. (1982). Noninvasive estimation of central aortic pressure using the oscillometric method for analyzing systemic artery pulsatile blood flow: Comparative study of indirect systolic, diastolic, and mean brachial artery pressure with simultaneous direct ascending aortic pressure measurements. *American Heart Journal, 103*(5), 879–886.

Ernst, M. E., & Bergus, G. R. (2002). Noninvasive 24-hour ambulatory blood pressure monitoring: Overview of technology and clinical applications. *Pharmacotherapy, 22*(5), 597–612.

Giles, T. D., & Egan, P. (2008). Pay (adequately) for what works: The economic undervaluation of office and ambulatory blood pressure recordings. *Journal of Clinical Hypertension, 10*(4), 257–259.

Guyton, A. C., & Hall, J. E. (2006). *Textbook of Medical Physiology* (11th ed.). Philadelphia, PA: Elsevier Saunders.

Kjeldsen, S. E., Erdine, S., Farsang, C., Sleight, P., & Mancia, G. (2002). 1999 WHO/ISH hypertension guidelines—highlights & ESH update. *American Journal of Hypertension*, Jan; 20(1), 153–155.

Latman, N. S., & Latman, A. (1997). Evaluation of instruments for noninvasive blood pressure monitoring of the wrist. *Biomedical Instrumentation f3 Technology/Association for the Advancement of Medical Instrumentation, 31*(1), 63–68.

Livi, R., Teghini, L., Cagnoni, S., & Scarpelli, P. T. (1996). Simultaneous and sequential same-arm measurements in the validation studies of automated blood pressure measuring devices. *American Journal of Hypertension, 9*(12), 1228–1231.

Marey, E. J. (1876). *Pression et vitesse du sang.* Physiologie Experimentale, Masson, Paris.

Musso, N. R., Giacchè, M., Galbariggi, G., & Vergassola, C. (1996). Blood pressure evaluation by noninvasive and traditional methods: Consistencies and discrepancies among photoplethysmomanometry, office sphygmomanometry, and ambulatory monitoring. Effects of blood pressure measurement. *American Journal of Hypertension, 9*(4), 293–299.

Pesola, G. R., Pesola, H. R., Nelson, M. J., & Westfal, R. E. (2001). The normal difference in bilateral indirect blood pressure recordings in normotensive individuals. *American Journal of Emergency Medicine, 19*(1), 43–45.

Posey, J. A., Geddes, L. A., Williams, H., & Moore, A. G. (1969). The meaning of the point of maximum oscillations in the cuff pressure in the indirect measurement of blood pressure. *Cardiovascular Research Center Bulletin, 8*, 15–25.

Prasad, N., & Isles, C. (1996). Ambulatory blood pressure monitoring: A guide for general practitioners. *British Medical Journal, 313*(7071), 1535–1541.

Prisant, L. M. (1995). Ambulatory blood pressure monitoring in the diagnosis of hypertension. *Cardiology Clinics, 13*(4), 479–490.

Ramsey, M. (1979). Noninvasive automatic determination of mean arterial pressure. *Medical f3 Biological Engineering f3 Computing, 17*, 11–18.

Rutten, A., Ilsley, A., Skowronski, G., & Runcman, W. (1986). A comparative study of the measurement of mean arterial blood pressure using automatic oscillometers, arterial cannulation and auscultation. *Anaesthesia and Intensive Care, 14*(1), 58–65.

Sheps, S. G., Clement, D. L., Pickering, T. G., White, W. B., Messerli, F. H., Weber, M. A., & Perloff, D. (1994). Ambulatory blood pressure monitoring. Hypertensive Diseases Committee, American College of Cardiology. *Journal of the American College of Cardiology, 23*(6), 1511–1513.

Shevchenko, Y., & Tsitlik, J. (1996). 90th anniversary of the development by Nikolai S. Korotkoff of the auscultatory method of measuring blood pressure. *Circulation, 94*, 116–118.

Stergiou, G. S., Voutsa, A. V., Achimastos, A. D., & Mountokalakis, T. D. (1997). Home self-monitoring of blood pressure: Is fully automated oscillometric technique as good as conventional stethoscopic technique? *American Journal of Hypertension, 10*(4), 428–433.

Venus, B., Mathru, M., Smith, R., & Pham, C. (1985). Direct versus indirect blood pressure measurements in critically ill patients. *Heart & Lung: The Journal of Critical Care, 14*(3), 228–231.

Chapter 20

CARDIAC OUTPUT MONITORS

OBJECTIVES

- Define the terms cardiac output, stroke volume, and cardiac index.
- Apply the Fick principle in indicator dilution methods of cardiac output measurements.
- Describe how to measure cardiac output using oxygen and heat as the "tracer."
- Explain the principle and setup of the thermodilution method in cardiac output measurement.
- Sketch the functional block diagram of a cardiac output monitor using the pulmonary artery catheter thermal dilution method.
- Contrast transpulmonary thermodilution and pulmonary artery catheter thermodilution in cardiac output measurements.
- Explain the principle of continuous pulse contour cardiac output measurement
- Discuss potential sources of error in cardiac output measurements and methods to minimize errors.
- Discuss common problems and hazards in cardiac output measurements.

CHAPTER CONTENTS

1. Introduction
2. Definitions
3. Direct Fick Method
4. Indicator Dilution Method
5. Thermodilution Methods
6. Continuous Pulse Contour Analysis Method
7. Common Problems and Hazards

INTRODUCTION

Cardiac output (CO) is a measurement of the performance of the heart. It is also used to calculate many hemodynamic functions. The heart serves as a pump to circulate blood around the cardiovascular system. In fluid mechanics, the power produced by a pump is determined by its output pressure and volume flow rate. In the last two chapters, we have studied devices to measure blood pressure. In this chapter, we are going to study the cardiac output monitor—a medical device to measure blood flow of the heart.

Although there are many direct and indirect methods to measure cardiac output, since the introduction of the Swan-Ganz catheter in the 1970s, the thermodilution (TD) method using the pulmonary artery catheter (PAC) has become a standard procedure to measure cardiac output in critical care and surgical suites. The TD cardiac output method using a PAC has several advantages over other methods with respect to simplicity, accuracy, reproducibility, and ability to have repeated measurements at short intervals; in addition, there is no need for blood withdrawal. Other technologies, such as Doppler ultrasound and pulse contour cardiac output measurement show promise as noninvasive alternatives in measuring cardiac output.

The TD method in cardiac output measurement is an application of the indicator dilution method introduced by Stewart in 1897 based on the Fick principle. The Fick principle was proposed by Adolf Fick in 1870 and states that the rate Q of a substance delivered to an area with a moving fluid stream is equal to the product of the flow rate F of the fluid and the difference in concentration C of the substance at sites proximal and distal to the area. In equation format: $Q = F (C_p - C_d)$ or:

$$F = \frac{Q}{(C_d - C_p)}.$$ (20.1)

DEFINITIONS

For every contraction, the heart pumps a certain volume of blood into the common aorta. This volume of blood ejected in one cardiac contraction is defined as the stroke volume (SV). Therefore, the volume of blood pumped out from the heart per unit time is equal to the SV multiplied by the heart rate (HR). This product, which is the volume of blood pumped out by the heart per unit time is defined as the cardiac output. CO is commonly expressed in liters per minute (LPM). Expressed mathematically, CO = SV × HR, where SV is in liters, and HR is in beats per minute (BPM).

For a normal, young, heathy male adult, the average resting CO is about

5.6 L/min. For women, this value is 10% to 20% less. The CO of a healthy individual is usually proportional to the overall metabolism of the body. During exercise, both the HR and the SV become higher, resulting in a higher CO. The CO of the same individual may increase to several times than that at rest (e.g., from 5 to 20 L/min). The CO of a young healthy adult during extraneous exercise can increase to 20 to 25 L/min; for a young athlete, the CO can be as high as 30 to 35 L/min. As with all physiological signals, the resting CO varies from person to person and often depends on the body size. To facilitate comparison, CO is often normalized by dividing it by the weight or by the body surface area of the patient. The latter is called the cardiac index, which has a unit of L/min/m2. A typical resting cardiac index is 3.0 L/min/m2. One may wonder how body surface area is determined. In fact, lookup tables of body surface area based on the weights and heights of typical individuals are available. Alternatively, an empirical formula can be used to obtain the body surface area A after the weight and height of the individual are determined:

$$A = W^{0.425} \times H^{0.725} \times 0.007184 \qquad (20.2)$$

where A = total body surface area in m2,
W = body weight in kg, and
H = height in cm.

For example, the body surface area A of a 70-kg, 1.7-m tall patient is

$$A = 70^{0.425} \times 170^{0.725} \times 0.007184 = 1.73 \text{ m}^2.$$

DIRECT FICK METHOD

The direct Fick method is considered to be the "gold standard" in cardiac output measurement. It uses oxygen as the indicator and assumes that the left ventricular blood flow is equal to the blood flow through the lungs. This method involves measurement of the rate of oxygen uptake of the lungs and the oxygen content of the arterial blood and venous blood (Figure 20-1). Deoxygenated blood from the right ventricle, which has the lowest oxygen concentration in the cardiovascular system, enters the lungs and picks up oxygen to become oxygenated blood. The oxygenated blood then flows via the left atrium, left ventricle, and common aorta into the arteries. According to the Fick principle, the volume blood flow F through the lung (i.e., the cardiac output) can be obtained if the rate of oxygen uptake Q and the difference in oxygen concentration C_a of the arterial blood and oxygen

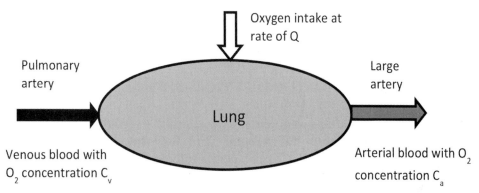

Figure 20-1. Direct Fick Method.

concentration C_v of the venous blood are known.

Using these quantities, Equation 20.1 then becomes

$$F = \frac{Q}{(C_a - C_v)}. \tag{20.3}$$

In practice, blood samples are drawn during measurement of oxygen consumption. The venous blood is drawn from the pulmonary artery and the arterial sample is taken from one of the main arteries (e.g., the brachial artery). Oxygen content of the venous and arterial blood is determined by laboratory analysis of these blood samples. The oxygen consumption Q is calculated from the rate of gas inhalation and the difference of the oxygen concentrations in the atmospheric air and the expired air from the patient. The rate of gas inhalation is measured using a spirometer and the expired gas oxygen concentration is measured using an oxygen analyzer. The blood flow rate F, or cardiac output, is then calculated. In this method, the subject must be in a steady state throughout the period of measurement (about 3 minutes) to avoid transient changes in blood flow or in the rate of ventilation.

Example 20.1

In a cardiac output measurement using the direct Fick method, the rate of oxygen consumption was found to be 300 ml/min. Blood sample analysis shows the arterial and mixed venous oxygen contents are 200 ml/L and 140 ml/L, respectively. Calculate the cardiac output.

Solution:

Using equation 20.3,

$$F = \frac{Q}{(C_a - C_v)} = \frac{300 \ ml/\text{min}}{(200 \ ml/L - 140 \ ml/L)} = \frac{300 \ ml/\text{min}}{60 \ ml/L} = 5 \ L/\text{min}.$$

INDICATOR DILUTION METHOD

The indicator dilution method is a variation of the Fick principle. The indicator dilution method to measure fluid flow rate is based on the upstream injection of a tracer (or detectable indicator) into a mixing chamber and measuring the concentration-time curve (or dilution curve) of the tracer downstream of the chamber (Figure 20-2a). To obtain accurate results, the tracer is required to thoroughly mix with the fluid in the mixing chamber. Theoretically, for the same flow and same tracer injection volume, the area under the dilution curve will be the same even though the shapes of the curves are different. Figure 20-2b shows the ideal indicator dilution curve obtained by immediate mixing of the tracer with the fluid after injection and having the same fluid velocity over the entire cross section of the tube. In the ideal case (Figure 20-2b), the fluid flow rate F is proportional to the amount of tracer m injected and inversely proportional to the concentration C of the tracer and the duration T of the concentration curve, or simply

$$F = \frac{m}{CT}.$$

Note that the product C and T is the area under the dilution curve.

Figure 20-2c shows a typical dilution curve in a realistic situation. Although it shows a rapid rise and an exponential fall in concentration, the area under the curve is still roughly the same as that of the idealistic curve (Figure 20-2b) as long as the fluid flow rate and the amount of tracer injected are the same.

$$F = \frac{m}{\varsigma\tau} = \frac{m}{A},$$

where A = area under the dilution curve.

Two types of indicators may be used in indicator dilution methods: diffusible and non-diffusible indicators. A non-diffusible indicator will remain in the system for a much longer period of time than a diffusible indicator does. For example, saline, which can be measured by a conductivity cell, is

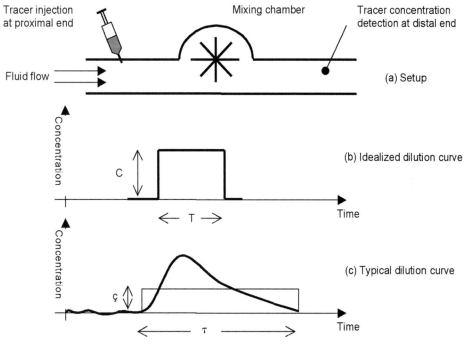

Figure 20-2. Indication Dilution Method.

a diffusible indicator in cardiac output measurement. It is estimated that over 15% of the salt will be removed from the blood in its first pass through the lung. Indocyanine green is a non-diffusible indicator that can be detected using optical sensors. Experiments showed that only about 50% of it will be lost in the first 10 minutes as it circulates around the cardiovascular system. Measurements using a diffusible indicator tend to overestimate the cardiac output, whereas recirculation of a non-diffusible indicator may result in lower cardiac output measurements. Recirculation is the effect of increased indicator concentration when the previous bolus of indicator returns to the measurement site during subsequent measurements. Figure 20-3 shows the dilution curve affected by recirculation. The dotted line shows the normal trace of the curve if no recirculation occurs.

Example 20.2

In an indicator dilution method to measure cardiac output, 10 mg of indicator is injected and the average concentration of the dilution curve is found to be 2.5 mg/L. If the indicator takes 60 sec to pass through the detector, what is the cardiac output?

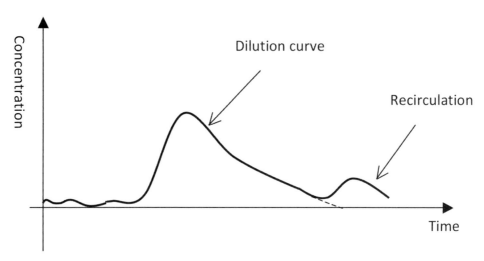

Figure 20-3. Effect of Recirculation in Indicator Dilution.

Solution:

Using the equation $F = \dfrac{m}{CT}$, the cardiac output is

$$\frac{m}{CT} = \frac{10 \text{ mg}}{2.5 \text{ mg/L} \times 60 \text{ s}} = \frac{4}{60}\frac{L}{s} = 4 \text{ L/min}.$$

THERMAL DILUTION METHOD

The TD method of cardiac output measurement is based on the indication dilution method, where heat is used as the indicator. In this method, a known volume of cold solution (0 °C, 5% dextrose or saline) is injected into the right atrium. This bolus of cold solution causes a decrease in the blood temperature when it mixes with the blood in the right ventricle. In cardiac output measurement by thermodilution method (TD) using a pulmonary artery catheter (PAC), the change of blood temperature in the pulmonary artery (downstream of the mixing chamber) is measured to obtain the TD curve as shown in Figure 20-4. It shows that the temperature of blood in the pulmonary artery drops when the indicator-blood mixture passes through the temperature sensor and gradually rises back to the normal body temperature. Note that the curve is inverted for easier reading since the cold solution causes a negative change in blood temperature.

From the TD curve, the heat loss dH from the blood over the time inter-

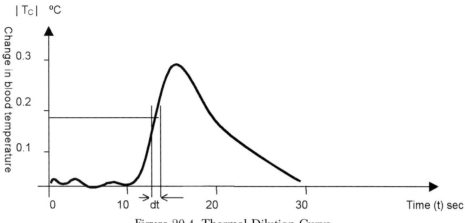

Figure 20-4. Thermal Dilution Curve.

val dt is

$$dH = C_B F \rho_B T_c dt, \qquad (20.4)$$

where C_B = specific heat capacity of blood,
    F = blood flow rate (cardiac output),
    ρ_B = density of blood, and
    T_c = temperature change (from body temperature) of the blood at time t.

Assuming the cardiac output F is a constant, the total heat loss of the blood to the injectate H is equal to the integral of Equation 20.4:

$$H = \int_0^\infty dH = \int_0^\infty C_B F \rho_B T_c dt = C_B F \rho_B \int_0^\infty T dt = C_B F \rho_B A, \qquad (20.5)$$

where A = $\int_0^\infty T_c dt$ is the area under the thermal dilution curve.

Theoretically, the total heat loss of the blood is equal to the heat gain of the injectate (to raise the injectate temperature to body temperature). The heat gain of the injectate H_I can be determined from the pre-injection condition of the injectate if we know the volume V_I, density ρ_I, specific heat capacity C_I and the initial temperature T_I of the injectate.

$$H_I = V_I C_I \rho_I (T_B - T_I). \qquad (20.6)$$

Since $H = H_I$, Equations 20.5 and 20.6 give

$$C_B F \rho B A = V_I C_I \rho_I (T_B - T_I)$$

$$\Rightarrow F = \frac{V_I C_I \rho_I (T_B - T_I)}{C_B \rho B A} \tag{20.7}$$

$$\Rightarrow CO = F = \frac{V_I K (T_B - T_I)}{A}, \tag{20.8}$$

where $K = \dfrac{C_I \rho_I}{C_B \rho B}$ is a constant for a particular indicator.

Equation 20.8 is known as the Stewart Hamilton equation. As heat (cold saline or dextrose) is a diffusible indicator, a correction factor K_1 (<1) is multiplied to Equation 20.8 to compensate for the warming effect of the indicator during measurement.

$$\Rightarrow CO = \frac{V_I K_1 K (T_B - T_I)}{A}. \tag{20.9}$$

Example 20.3

In a CO measurement using the PAC TD method, 5 ml of iced 5% dextrose is injected into the right atrium to obtain the TD curve. If the area under the curve is found to be 1.80 °C.s and a correction factor K_1 of 0.825 is used, find the cardiac output given that $\dfrac{C_I \rho B}{C_B \rho B} = 1.08$ for 5% dextrose and typical blood composition.

Solution:

Using 37 °C as the body temperature and 0 °C as the initial injectate temperature, from Equation 20.9, the cardiac output is calculated by:

CO = [5 ml x 0.825 x 1.08 x (37°C – 0°C)]/1.80 °C s = 91.6 ml/s = 5.5 L/min

In practice, a special catheter called the Swan-Ganz catheter is used to inject a known volume (e.g., 5 mL) of cold dextrose into the right atrium of the heart; the temperature of the blood in the pulmonary artery is measured by a temperature sensor embedded near the distal end of the catheter to obtain the TD curve. Mixing of indicator and blood occurs in the right ventricle. A Swan-Ganz catheter is a special multi-lumen catheter with a distal opening, a proximal opening, a thermistor temperature sensor, and an inflatable balloon. A typical catheter is about 110 cm long and 2 to 3 mm in diameter.

Figure 20-5 shows the construction of a four-lumen Swan-Ganz catheter, and Figure 20-6 shows the position of the catheter in the cardiovascular system during TD CO measurement. The catheter is positioned into the heart such that the injectate orifice is in the right atrium and the thermistor in the pulmonary artery. Such catheter with its placement in the pulmonary artery is referred to as a pulmonary artery catheter. During measurement, the injectate (cold saline or dextrose) delivered into the right atrium is mixed with blood as it travels through the right ventricle into the pulmonary artery. The temperature of blood flowing through the pulmonary artery is continuously measured by a thermistor located a few centimeters from the distal end of the catheter.

Listed below the procedures to perform PAC TD cardiac output measurement using a Swan-Ganz catheter

1. Create an intravascular access to a vein (subclavian, internal jugular, or basilic) using, for example, the Seldinger technique.
2. Insert the distal end of the catheter into the vein and slightly inflate the balloon.
3. Continue to insert the catheter into the vein. The balloon will be dragged by the blood flow to go through the right atrium, tricuspid valve, right ventricle, and pulmonary valve and into the pulmonary artery.
4. The position of the catheter can be estimated by the distant markings on the catheter and verified by
   a. using X-ray fluoroscopy (the tip of the catheter is radiopaque) or
   b. monitoring the characteristic changes in blood pressure waveform at thedistal lumen as it travels from the vein into the heart chambers and then into the pulmonary artery during the catheter insertion.
5. Deflate the balloon.
6. Connect the catheter to the cardiac output monitor and initialize the monitor.
7. Enter patient data.
8. Prepare injectate (saline or DW5–5% dextrose).
9. Measure injectate temperature (usually at 0 °C to 5 °C).
10. Inject a fixed volume of injectate (e.g., 5 ml) at a uniform rate (over a period of 2 to 4 sec) into the injectate port.
11. The cardiac output monitor will display the temperature change versus time curve (TD curve), measure the volume under the curve, and calcu-

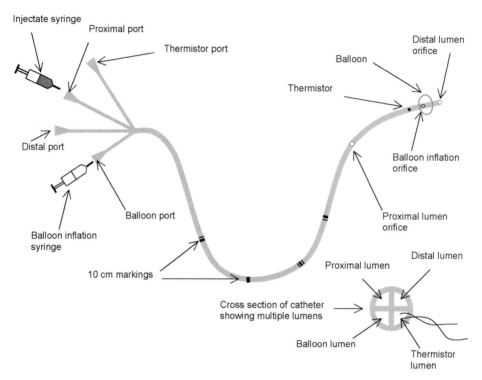

Figure 20-5. Construction of a Swan-Ganz Catheter.

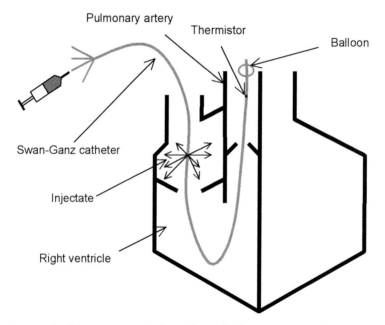

Figure 20-6. Positioning of Swan-Ganz Catheter in Heart Chambers.

late the cardiac output using Equation 20.9.

12. Obtain three measurements (and average the results) to obtain a reliable reading.

A Swan-Ganz catheter setup can also be used to monitor the pulmonary artery wedge pressure (PAWP). The PAWP, also termed the pulmonary capillary wedge (PCW) pressure, reflects the mean level of pressure in the left atrium. While the distal end of the catheter is in the pulmonary artery, the balloon is fully inflated to occlude the pulmonary artery. The PAWP is measured from the distal lumen by connecting a pressure transducer to the distal port of the catheter.

Note that the setup procedure to measure blood pressure using the Swan-Ganz catheter is the same as setting up an invasive arterial pressure line as discussed in Chapter 18. A cardiac output monitor usually consists of one or more blood pressure modules, one or more temperature modules, a computer, and interface components. The functional block diagram of a cardiac output monitor is shown in Figure 20-7. During a cardiac output measurement, a blood pressure transducer BP Tx1 is connected to the distal port to display the blood pressure waveform at the distal end of the catheter. This pressure waveform is a mean to allow the user to determine the position of the catheter during catheter insertion. These blood pressure modules may also be used to monitor intracardiac blood pressure. The thermistor temperature transducer Tx1 located near the distal end of the catheter is use to acquire the TD curve. A second thermistor, Temp Tx2, may be used to measure the injectate temperature. These pressure and temperature signals are digitized, multiplexed, and sent to the central processor to compute the cardiac output, cardiac index, blood pressures, and so on. The TD curve and the blood pressure waveform can be viewed from the integrated display of the monitor.

Instead of using cold solution (heat indicator), a non-diffusible dye (e.g., indocyanine green) may be used as the indicator. The dye dilution curve is obtained by measuring the dye concentration using a calibrated optical densitometer from a continuously drawn blood sample during the measurement period.

A modification to the TD cardiac output method was implemented to facilitate repeated measurements of cardiac output using the principle of thermal dilution. In this method, the Swan-Ganz catheter is modified by placing a length of heated filament (about 10 cm) near the usual injectate port. During cardiac output measurement, the heater is turned on to infuse a small amount of heat to the blood in the atrium. The distal blood temperature is recorded and cross-correlated with the heater output to produce a dilution curve to calculate the cardiac output. Without the need to prepare

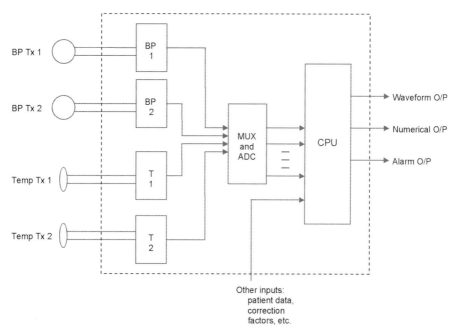

Figure 20-7. Functional Block Diagram of a Cardiac Output Monitor.

and handle the cold injectate, this measurement technique is considerably easier to use in clinical settings. Although the measurements are still intermittent, this method is termed continuous cardiac output method.

A modification to the PAC TD cardiac output method described above is used to facilitate repeated measurements of cardiac output using the principle of thermal dilution. In this method, the Swan-Ganz catheter is modified by placing a length of heated filament (about 10 cm) near the usual injectate port. During cardiac output measurement, the heater is turned on to infuse a small amount of heat to the blood in the atrium. The distal blood temperature is recorded and cross-correlated with the heater output to produce a dilution curve to calculate the cardiac output. Without the need to prepare and handle the cold injectate, this measurement technique is considerably easier to use in clinical settings. Although the measurements are still intermittent, this method is sometimes referred to as continuous cardiac output method.

Instead of using a pulmonary artery catheter in which the temperature transducer is located in a pulmonary artery, the temperature transducer can be placed in a large artery (e.g., femoral or brachial) supplied by blood from the common aorta. In this method, a bolus of cold saline is injected into a central vein (e.g., superior vena cava), venous blood is then mixed with the cold injectate inside the right heart chambers, passes through the lungs, back to the left heart chambers, and exit from the common aorta into the arteries. The thermodilution curve is obtained by plotting against time the decrease in blood temperature measured by the arterial temperature transducer. This transpulmonary thermodilution (TP TD) method to measure cardiac output is based on the same principle and mathematical concepts of the PAC TD method. Figure 20.8 compares the TD curves between the PAC and TP methods. With the same volume and temperature of the injectate (same heat input), the TP TD curve is broader and lower in magnitude than the PAC TD curve, but the areas under the curves of both are similar. As the reliability of the measurement requires a significant blood temperature drop, comparing to the PAC TD method, a larger volume and lower temperature injectate solution is needed (e.g., 15 ml, < 8 0C saline specified by one manufacturer) in TP TD CO measurements. Studies concluded that when due precautions are followed, accuracy of cardiac output measured by TP TD method is comparable to that of PAC TD method.

CONTINUOUS PULSE CONTOUR METHOD

While the direct Fick method is the gold standard in cardiac output measurement and the PAC TD method is considered as the "clinical gold standard", both methods can only provide intermittent measurement of the cardiac output. Continuous bedside monitoring of cardiac output by analyzing the arterial blood pressure waveform was made available in early 2000.

In 1904, Erlanger and Hooker hypothesized that cardiac output was proportional to the patient's arterial blood pressure. The pulse contour (PC)

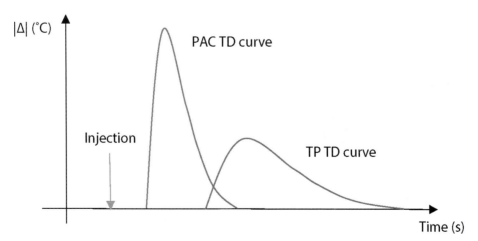

Figure 20-8. Thermodilution curves from the transpulmonary and pulmonary artery catheter methods.

method is based on analyzing the amplitude and shape of the arterial pressure curve to obtain beat by beat stroke volume. In general, pulse contour (PC) based devices estimate the arterial blood pressure waveform at the aorta by analyzing the geometry of the pressure waveform from a peripheral artery, and predict from it the stroke volume using proprietary algorithms.

Although CO measurements using PC method can be quite precise, the value derived from the algorithms is patient specific and may drift over time due to changes in the condition of the patient (such as blood pressure and arterial resistance). A PC CO monitor must first be calibrated using the patient's cardiac output obtained by a reliable mean such as the TP TD method. The monitor needs recalibration when there is significant change in the patient's condition (e.g., increase or decrease in blood pressure). In addition, periodic calibration (as often as once every hour) is needed to maintain accuracy. Figure 20-10 and Equation 20.10 depict the general concept of PC CO measurement used by manufacturers. The shaded area in Figure 20-9 is the area under the systolic portion of the pressure waveform which is supposed to be proportional to the beat-by-beat stroke volume. As the variation of blood vessel compliance affects the effect of volume blood flow due to blood pressure, Equation 20.10 incorporates the area under the systolic portion of the arterial waveform, the aortic compliance, a calibration factor specific to the patient's actual cardiac output (determined by a thermodilution method), and the shape of the pressure curve.

$$\text{PCCO} = k \times HR \times \int \left(\frac{P(t)}{SVR} + C(p) \times \frac{dP}{dt} \right) dt \qquad (20.10)$$

Where k = patient specific calibration factor

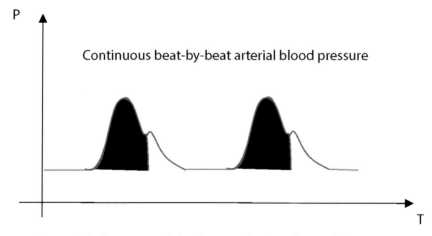

Figure 20-9. Continuous Pulse Contour Cardiac Output Measurement.

HR = patient heart rate

$\int \dfrac{P(t)}{AVR}\, dt$ = area (shaded) under the systolic portion of the pressure curve

$C(p)$ = aortic compliance

In addition to providing continuous beat-by-beat measurement of cardiac output, arterial pulse contour analysis together with TP TD measures hemodynamic variables such as stroke volume variation (SVV), intrathoracic blood volume (ITBV) and extravascular lung water (EVLW). These parameters are helpful to guide fluid therapy and to monitor effect of cardioavascular management in critically ill patients with circulatory compromise.

Although PC CO is less invasive comparing to PAC CO, it still needs to access a central vein and a peripheral artery. Much research has been underway to develop a non-invasive continuous bedside cardiac output monitor. One recent product employs an inflatable finger cuff and plethysmography technique to produce a continuous pressure waveform equivalent to the blood pressure inside the brachial artery. CO is then calculated by an algorithm similar to the pulse contour method described above. Although the finger cuff method provides a reasonable estimate of CO and blood pressure, it does not meet the criteria for clinical interchangeability with the currently used invasive methods.

There are other non-invasive methods able to measure cardiac output such as bioimpedance and transthoracic echocardiography. However, they are either not reliable or require extensive instrumentations which are not practical for bed side applications.

COMMON PROBLEMS AND HAZARDS

Pulmonary artery catheterization is an invasive procedure. Once inserted, a PAC will often stay for a period of time to allow repeated measurements and other hemodynamic assessment procedures. Complications include blood vessel perforation during insertion, bacterial infections, and higher rate of cardiac arrhythmias.

Cardiac output determination by TD, especially employing the TP method, may be unreliable or impossible in patients with low cardiac output and with tricuspid or pulmonary regurgitation. Baseline pulmonary artery temperature drift or the existence of intracardiac and extracardiac shunts can be observed from the TD curves.

Any parameter inaccuracy in Equation 20.9 will result in an incorrect cardiac output measurement. Most studies showed that PAC TD cardiac output measurements using Swan-Ganz catheters correlate well with the results from the direct Fick method. Although some manufacturers claimed to achieve accuracy of better than 5%, PAC TD cardiac output results are generally considered to have an error of 10% even under ideal circumstances. To avoid single measurement error, the usual practice is to average the results of three good measurements. Potential error sources of this technique are explained next.

Arterial compliance and impedance of patients with mitral and aortic insufficiency are different from patients without valve regurgitation. In severely haemodynamically compromised patients, TD method remains the best option for monitoring CO and cardiac pressures.

Catheter Dead Space and Injectate Warming

The TD cardiac output equation assumes that the entire volume of injectate enters the bloodstream, and heat transfer only occurs between the injectate and the blood in the right atrium. In practice, some injectate will remain in the catheter, and heat exchange between the injectate and the wall of the catheter occurs when the injectate travels along the catheter before reaching the heart. This will lead to a decrease in temperature change of the injectate-blood mixture (or lower amplitude thermal dilution curve). According to Equation 20.9, a smaller area (A) under the thermal dilution curve will lead to an overestimation of the cardiac output. To minimize such errors, the first reading should be discarded because the first injection contains warm fluid in the catheter dead space. A correction factor $(K_I < 1)$ is multiplied to Equation 20.8 to compensate for some of these errors.

Injection Timing

Respiratory action of the patient affects the blood pressure and flow; injections during TD cardiac output measurements should be timed at evenly spaced intervals of the ventilation cycle. When it is difficult to synchronize the injection with the respiratory cycle, using an average of three or more sequential measurements can minimize the variations.

Injection Rate

Erratic and long injection duration introduces errors in the TD curve and increases the injectate warming effect. Injection should be at a steady rate, and duration should be about 2 sec for a 5-mL solution and 4 seconds for a 10-mL solution.

Injectate Volume

The injectate volume should be measured accurately because this will affect the calculation. A small syringe (e.g., 10 mL) is usually included in the package of the catheter. TD cardiac output measurements using a small volume of injectate are more likely to be affected by errors such as injectate warming.

Injectate Temperature

A higher volume injectate at a lower temperature will increase the signal to noise ratio (SNR) in the measurements. Studies confirmed that cardiac output measurements using injectate at room temperature produced higher variability than using injectate at $0°C$ (iced injectate). It is recommended that for PAC TD measurement, if the volume of injectate is less than 10 mL, the solution should be iced. As well, clinicians should average more measurements if room temperature injectates are used. For TP TD CO measurements, injectate volume greater than 15 ml and temperature lower than $8°C$ should be used. Care should be taken to avoid warming the injectate during handling.

Thermistor Position

If the thermistor in the catheter is in contact with the wall of the pulmonary artery, the temperature reading will be higher, resulting in a lower value of A in Equation 20.9. There will be errors in the temperature measurement if the thermistor is positioned inside the ventricle instead of inside the pulmonary artery. Inside the ventricle, the injectate may not have been thoroughly mixed with the blood.

Frequency of Measurements and Recirculation

If consecutive measurements are made within a short period of time, the temperature of blood in the pulmonary artery may not have enough time to return to normal body temperature before the next bolus of injectate is introduced into the bloodstream. In theory, the cold bolus may be recirculated to the thermistor site during another measurement and introduce error in the TD curve. If too many measurements are done in a short period of time, it may lower the overall blood temperature of the patient. This latter problem is more significant in the dye dilution method because it takes longer for the kidneys and liver to remove the dye from the patient's bloodstream.

Intravenous Administration

Any intravenous fluid infusion will introduce below body temperature fluid into the bloodstream. This effect will create errors in the TD curve. Rapid intravenous fluid infusion should be discontinued during TD cardiac output measurements.

BIBLIOGRAPHY

Band, D. M., Linton, R. A., O'Brien, T. K., Jonas, M. M., & Linton, N. W. (1997). The shape of indicator dilution curves used for cardiac output measurement in man. *Journal of Physiology, 498,* 225–229.

Boyle, M., Steel, L., Flynn, G. M., Murgo, M., Nicholson, L., O'Brien, M., & Bihari, D. (2009). Assessment of the clinical utility of an ultrasonic monitor of cardiac output (the USCOM) and agreement with thermodilution measurement. *Critical Care and Resuscitation, 11*(3), 198–203.

Breukers, R. M., Sepehrkhouy, S., Spiegelenberg, S. R., & Groeneveld, A. B. (2007). Cardiac output measured by a new arterial pressure waveform analysis method without calibration compared with thermodilution after cardiac surgery. *Journal of Cardiothoracic and Vascular Anesthesia, 21,* 632–635.

Chittock, D. R., Dhingra, V. K., Ronco, J. J., Russell, J. A., Forrest, D. M., Tweeddale, M., & Fenwick, J. C. (2004). Severity of illness and risk of death associated with pulmonary artery catheter use. *Critical Care Medicine, 32*(4), 911–915.

Cholley, B. P., & Singer, M. (2003). Esophageal Doppler: Noninvasive cardiac output monitor. *Echocardiography, 20,* 763–769.

Costa, M. G., Della Rocca, G., Chiarandini, P., Mattelig, S., Pompei, L., Barriga, M. S., . . . , & Pietropaoli, P. (2008). Continuous and intermittent cardiac output measurement in hyperdynamic conditions: Pulmonary artery catheter vs. lithium dilution technique. *Intensive Care Medicine, 34,* 257–263.

Della Rocca, G., Costa, M. G., Pompei, L., Coccia, C., & Pietropaoli, P. (2002). Continuous and intermittent cardiac output measurement: Pulmonary artery catheter versus aortic transpulmonary technique. *British Journal of Anaesthesia, 88,* 350–356.

De Wilde, R. B., Geerts, B. F., Cui, J., van den Berg, P. C., & Jansen, J. R. (2009). Performance of three minimally invasive cardiac output monitoring systems. *Anaesthesia, 64,* 762–769.

Felbinger, T. W., Reuter, D. A., Eltzschig, H. K., Moerstedt, K., Goedje, O., & Goetz, A. E. (2002). Comparison of pulmonary arterial thermodilution and arterial pulse contour analysis: Evaluation of a new algorithm. *Journal of Clinical Anesthesia, 14,* 296–301.

Gardner, P. E., & Bridges, E. T. (1995). Hemodynamic monitoring. In S. L. Woods, E.S. Sivarajan-Froelicher, C. J. Halpenny, & S. U. Motzer (Eds.), *Cardiac Nursing* (3rd ed., pp. 424–458). Philadelphia, PA: Lippincott.

Hamilton W.F.R.R, Attyah A.M. (1948). Comparison of the Fick and dye injection methods of measuring the cardiac output in man. *Am J Physiol.* 153, 309–321.

Hamilton W.F.M.J., Kinsman J.I. (1932). Studies on the Circulation. IV. Further analysis of the injection method, and of changes in hemodynamics under physiological and pathological conditions. *Am J Physiol.* 99, 534–551.

Haryadi, D. G., Orr, J. A., Kuck, K., McJames, S., & Westenskow, D. R. (2000). Partial CO2 rebreathing indirect Fick technique for non-invasive measurement of cardiac output. *Journal of Clinical Monitoring and Computing, 16*(5–6), 361–374.

Hoel, B. L. (1978). Some aspects of the clinical use of thermodilution in measuring cardiac output. *Scandinavian Journal of Clinical and Laboratory Investigation, 38*, 383–388.

Hofhuizen C, Lansdorp B, van der Hoeven JG, et al. (2014). Validation of noninvasive pulse contour cardiac output using finger arterial pressure in cardiac surgery patients requiring fluid therapy. *J Crit Care, 29*, 161–165.

Jaeggi, P., Hofer, C. K., Klaghofer, R., Fodor, P., Genoni, M., & Zollinger, A. (2003). Measurement of cardiac output after cardiac surgery by a new transesophageal Doppler device. *Journal of Cardiothoracic and Vascular Anesthesia, 17*, 217–220.

Jain, S., Vafa, A., Margulies, D. R., Liu, W., Wilson, M. T., & Allins, A. D. (2008). Non-invasive Doppler ultrasonography for assessing cardiac function: Can it replace the Swan-Ganz catheter? *American Journal of Surgery, 196*(Dec), 961–968.

Jonas, M. M., & Tanser, S.J. (2002). Lithium dilution measurement of cardiac output and arterial pulse waveform analysis: An indicator dilution calibrated beat-by-beat system for continuous estimation of cardiac output. *Current Opinion in Critical Care, 8*, 257–261.

Kallasian, K. G., & Raffin, T. A. (1996). The technique of thermodilution cardiac output measurement. *Journal of Critical Illness, 11*, 249–256.

Lavdaniti, M. (2008). Invasive and non-invasive methods for cardiac output measurement. *International Journal of Caring Sciences, 1*(3), 112–117.

Linton, R. A., Band, D. M., & Haire, K. M. (1993). A new method of measuring cardiac output in man using lithium dilution. *British Journal of Anaesthesia, 71*, 262–266.

Mayer J, Boldt J, Schollhorn T, Rohm KD, Mengistu AM, Suttner S. (2007). Semi-invasive monitoring of cardiac output by a new device using arterial pressure waveform analysis: a comparison with intermittent pulmonary artery thermodilution in patients undergoing cardiac surgery. *Br J Anaesth.*, 98, 176-82.

Meier P.Z.K. (1954). On the theory of the indicator-dilution method for measurement of blood flow and volume. *J Appl Physiol.* 6, 731–744.

Monnet X, Teboul J.L. (2015). Minimally invasive monitoring. *Crit. Care Clin.* 31(1), 25-42.

Moore, F. A., Haenel, J. B., & Moore, E. E. (1991). Alternatives to Swan-Ganz cardiac output monitoring. *Surgical Clinics of North America, 71*, 699–721.

Moshkovitz, Y., Kaluski, E., Milo, O., Vered, Z., & Cotter, G. (2004). Recent developments in cardiac output determination by bioimpedance: Comparison with invasive cardiac output and potential cardiovascular applications. *Current Opinion in Cardiology, 19*(3), 229–237.

Nishikawa, T., & Dohi, S. (1993). Errors in the measurement of cardiac output by thermodilution. *Canadian Journal of Anaesthesia, 40*(2), 142–153.

Odenstedt, H., Stenqvist, O., & Lundin, S. (2002). Clinical evaluation of a partial

CO_2 rebreathing technique for cardiac output monitoring in critically ill patients. *Acta Anaesthesiologica Scandinavica, 46,* 152–159.

Phillips, R. A., Hood, S. G., Jacobson, B. M., West, M. J., Wan, L., & May, C. N. (2012). Pulmonary artery catheter (PAC) accuracy and efficacy compared with flow probe and transcutaneous Doppler (USCOM): An ovine cardiac output validation. *Critical Care Research and Practice, 2012,* 621496.

Schmid, E. R., Schmidlin, D., Tornic, M., & Seifert, B. (1999). Continuous thermodilution cardiac output: Clinical validation against a reference technique of known accuracy. *Intensive Care Medicine, 25,* 166–172.

Singh, A., Juneja, R., Mehta, Y., & Trehan, N. (2002). Comparison of continuous, stat, and intermittent cardiac output measurements in patients undergoing minimally invasive direct coronary artery bypass surgery. *Journal of Cardiothoracic and Vascular Anesthesia, 16,* 186–190.

Sommers, M. S., Woods, S. L., & Courtade, M. A. (1993). Issues in methods and measurement of thermodilution cardiac output. *Nursing Research, 42,* 228–233.

Stewart G.N. (1897). Researches on the circulation time and on influences which affect it. *J Physiol.* 22, 159–183.

Su, B. C., Tsai, Y. F., Chen, C. Y., Yu, H. P., Yang, M. W., Lee, W. C., Lin, C. C. (2012). Cardiac output derived from arterial pressure waveform analysis in patients undergoing liver transplantation: Validity of a third generation device. *Transplantation Proceedings, 44*(2), 424–428.

Swan, H. J., Ganz, W., Forrester, J., Marcus, H., Diamond, G., & Chonette, D. (1970). Catheterization of the heart in man with use of a flow-directed balloon-tipped catheter. *New England Journal of Medicine, 283,* 447–451.

Turner, M. A. (2003). Doppler-based hemodynamic monitoring: A minimally invasive alternative. *AACN Advanced Critical Care, 14*(2), 220–231.

Wesseling KH, Jansen JR, Settels JJ, et al. (1993). Computation of aortic flow from pressure in humans using a nonlinear, three-element model. *J Appl Physiol.* 74, 2566–2573.

Wesseling K H., Weber J.A.P. Smith N.T. (1983). A simple device for the continuous measurement of cardiac output. *Adv Cardiovasc Phys.* 5, 16–52.

Yelderman, M., Quinn, M. D., McKown, R. C., Eberhart, R. C., & Dollar, M. L. (1992). Continuous thermodilution cardiac output measurement in sheep. *Journal of Thoracic and Cardiovascular Surgery, 104*(2), 315–320.

Zisserman, D., Mantle, J.A., Smith, L.R., Rogers, W.J., Russell, R.O., Jr, Rackley, C.E. (1979). Clinical comparison of thermal dilution cardiac output to the Fick and angiographic methods [Abstract]. *Clinical Research, 27,* 736A.

Chapter 21

CARDIAC PACEMAKERS

OBJECTIVES

- Define arrhythmia and list indications for artificial cardiac pacemakers.
- Differentiate between endocardial and myocardial, bipolar and unipolar pacing leads.
- Discuss asynchronous, demand, and rate-modulated pacing.
- Describe the procedures of pacemaker implantation.
- Interpret the NBG (NASPE/BPEG) generic pacemaker codes
- Explain external cardiac pacing.
- Differentiate between invasive, transcutaneous external and non-invasive pacing.
- Describe the characteristics and design of implantable pacemakers.
- Analyze the block diagram of a demand pacemaker and describe its principles of operation.
- Discuss common problems and hazards in artificial pacing

CHAPTER CONTENTS

1. Introduction
2. Indication of Use
3. Types of Cardiac Pacemakers
4. Pacemaker Lead System
5. Implantation of Pacemaker
6. Pacing Mode Selection
7. Performance Characteristics

8. Functional Building Blocks of an Implantable Pacemaker
9. Temporary Pacing
10. Potential Problems and Hazards

INTRODUCTION

John Hopps, a Canadian electrical engineer, invented the first cardiac pacemaker in 1950. While working for the Canadian National Research Council, Hopps discovered that a heart that stopped beating due to cooling could be started again by artificial stimulation using mechanical or electrical means. This led to Hopps' invention of the world's first cardiac pacemaker in 1950. Hopps' device was too large to be implanted inside the human body; it was an external pacemaker powered from the AC mains. In 1952, Paul M. Zoll, working with engineers of the Electrodyne Company, developed a device that could stimulate the heart to contract through large electrodes placed on the chest wall of the patient. In 1957, C. Walton Lillehei and his colleagues, using an external pulse generator, successfully paced the heart by directly placing electrodes on the heart muscle. The first pacemaker clinically implanted into a human was in 1958, but it only lasted 3 hours. In 1958, Wilson Greatbatch developed a fully implantable pacemaker using an internal mercury battery. In 1960, the Medtronic Chardack-Greatbatch pacemaker, the first commercially produced pacemaker, was implanted into a patient. The first use of transvenous pacing in conjunction with an implanted pacemaker was by Parsonnet in 1962 in the USA; this method has since become the method of choice for pacemaker implantation. In 1986, the U.S. Food and Drug Administration gave market approval to a rate-responsive pacemaker made by Medtronic.

A pacemaker is a therapeutic device designed to rectify some heart problems arising from irregular heart rhythms. It performs its function by applying controlled electrical stimulations to the heart. A healthy heart is stimulated to contract by electrical impulses initiated from the SA node located in the atrium near the superior vena cava. The SA node is also called the natural pacemaker of the heart. An electrical impulse generated from the SA node is conducted through the atria, causing the atria to contract. The impulse eventually arrives at and depolarizes the AV node located in the septal wall of the right atrium. Through the bundle of His and the Purkinje fibers, the action potential is distributed to the myocardium, causing the ventricles to contract. The AV node creates a time delay between the atrial and ventricular contractions.

INDICATION OF USE

Normal sinus rhythm is a result of the continuous periodic and coordinated stimulation of the atria and ventricles of the heart. Failure of any part of this pathway will compromise the cardiac output. An arrhythmia is any disturbance in the rhythm of the heart with respect to rate, regularity, or propagation sequence of the depolarization wave. Depending on its nature, an arrhythmia can be mild or life threatening. There are two kinds of arrhythmias:

1. Arrhythmias caused by disturbances of conduction
2. Arrhythmias caused by disturbances of the origin of the stimulation

Disturbances of conduction includes

- Slowing of the spread of conduction of the electrical stimulation in one part of the conduction system so that part of the heart is activated significantly later than the rest, resulting in distortion of the ventricular contraction pattern
- Conduction through anomalous paths, which causes different parts of the heart muscle to contract in an uncoordinated manner
- Partial or complete blocks of the stimulation signal from the SA node to the ventricles. Heart blocks originate from some malfunctions of the heart's built-in electrical conduction system. The results of heart blocks include low heart rate, heart muscle not getting enough oxygen, and cardiac muscle becoming irritable and susceptible to irregular rhythm. A patient with heart blocks has inadequate body oxygen and low exercise tolerance and, in extreme cases, experiences loss of consciousness and convulsion due to lack of oxygen to the brain. There are three degrees of heart block:

  i. First degree—long delay in signal transmission, with prolonged PR intervals greater than 200 msec.
  ii. Second degree—intermittent complete blockage of transmission. There are two types of second degree heart block:
     - Mobitz I exhibits a progressive increase in PR intervals until a ventricular beat is dropped.
     - Mobitz II regularly drops ventricular beats.
  iii. Third degree—continuous complete blockage of transmission. No impulses travel from the atria to the ventricles. An ectopic foci (independent pacing site) usually appears to maintain a lower and irregular heart rate.

Disturbances of conduction includes

- erratic rate of SA node resulting in irregular heart beat
- more than one natural pacemaker site
- high heart rate, for example, sinus tachycardia
- low heart rate, for example, sinus bradycardia

Pacemakers are indicated to improve cardiac output, prevent symptoms, or protect against arrhythmias related to cardiac impulse formation and conduction disorders.

TYPES OF CARDIAC PACEMAKERS

A pacemaker has two physical parts, the pulse generator and the lead system (Figure 21-1). The pulse generator produces electrical stimulation pulses. Through the lead system, these pulses are delivered to the heart muscle, causing the heart to contract. There are three types of pacemakers: implantable, external invasive, and transcutaneous. For an implantable pacemaker, both the pulse generator and the leads are placed inside the patient's body without any exposed parts. For an external invasive pacemaker, the pulse generator is located outside the patient's body, and the lead wire connecting the pulse generator and the heart muscle is inserted through a vein into the right chamber of the heart. A transcutaneous pacemaker, sometimes called an external noninvasive pacemaker, has a pair of skin electrodes placed anterior and posterior to the surface of the chest. The electrical stim-

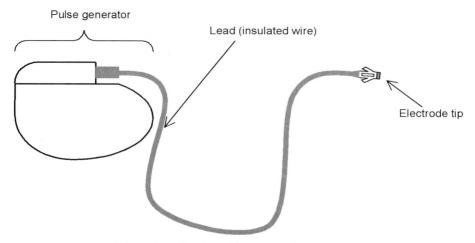

Figure 21-1. Implantable Pacemaker and Lead.

ulation from the external pulse generator is conducted across the externally placed electrodes through the heart, stimulating the heart muscles to contract.

According to how it regulates the pacing rate, a cardiac pacemaker has three modes of operation: asynchronous, demand, and rate modulated. A pacemaker in asynchronous mode can deliver only a fixed rate of stimulation. In demand mode, it senses the heart's activity to determine its pacing sequence. Rather than pacing at a fixed rate, a rate-modulated pacemaker can adjust its pacing rate based on the state of physical activity of the patient; therefore, it is able to adjust the patient's cardiac output to meet the body's demand. To determine the pacing rate, a rate-modulated pacemaker may detect body motion, blood temperature, ventilation rate, intracardiac pressure and volume, or other physiological parameters (such as blood pH, PO_2 and PCO_2), and use one or more of these parameters to adjust the pacing rate.

Some devices combine cardiac pacing with defibrillation. In addition to pacing the heart, an implantable cardioverter defibrillator (ICD) placed inside the body is capable to perform cardioversion and defibrillation. The device therefore provides first-line treatments of some life-threatening cardiac conditions such as ventricular fibrillation and ventricular tachycardia. Current devices can be programmed to detect abnormal heart rhythms and deliver anti-tachycardia therapy pacing in addition to low-energy and high-energy defibrillation shocks. Today's ICDs can last up to 8 years depending on the number of delivered pacing pulses and shocks. An ICD is slightly larger in size than a pacemaker due to the larger size of its battery; and it is implanted using similar methods as implantable cardiac pacemakers. Cardiac defibrillators are covered in Chapter 22.

Most modern pacemakers are programmable. Parameters that can be programmed include pacing rate, mode of pacing, pacing pulse amplitude, pulse duration, and sensitivity. In addition to the simple programmable features, multi-programmable pacemakers have built-in programmable diagnostic tests as well as the ability to log heart rhythms and pacing activities. To program or interrogate an implanted pacemaker, a pacemaker programmer transceiver (transmitter and receiver) is placed on the skin surface above the pacemaker. Programming commands and data are transmitted via telemetry (or electromagnetic coupling) between the programmer and the pacemaker.

PACEMAKER LEAD SYSTEM

The pacemaker lead system serves two functions. The first is to transmit

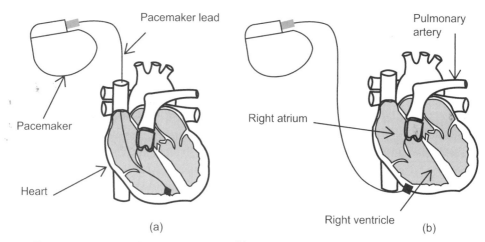

Figure 21-2. Pacemaker Lead System: (a) Endocardial Lead, (b) Myocardial Lead.

pacing pulses from the pulse generator to the heart. The second is to pick up electrical activities of the heart to modify the pacing sequence. The pacemaker lead is insulated with nonconductive material (e.g., silicon) except at the electrode tip and the connector to the pacemaker. The conductor is made of corrosion-resistive wire, which is coiled to increase its flexibility. The electrode (tip of the lead wire) may be attached to the surface of the heart or inserted through a vein into the chambers of the heart. The former is called the myocardial (or epicardial) lead system, and the latter is called the endocardial lead system. Figure 21-2 shows the two lead systems.

To complete the conduction path, the current produced from the pulse generator must return to the pulse generator after passing through the heart tissue. For a unipolar lead configuration, a single conductor lead is used. The conductor in the pacemaker lead carries the pacing current from the pulse generator circuit to the heart tissue. The metal housing of the pacemaker serves as the return electrode. Because there is only one conductor in the pacemaker lead, the return current must therefore return via the conductive body tissue of the patient to the metal housing and then back to the pulse generator circuit of the pacemaker. In a bipolar lead configuration, both the active and the return conductors are inside the insulated lead (a dual conductor lead). The pacing current (from the electrode tip) to the heart is picked up by the return electrode located near the tip electrode and returned to the pulse generator via the return conductor in the pacemaker lead wire. Figure 21-3 illustrates the two configurations. Bipolar leads, despite thicker, are usually preferred as they are less prone to problems.

A pacemaker with two leads (one in the atrium and the other in the ventricle) to allow pacing and/or sensing of both the ventricle and the atrium is

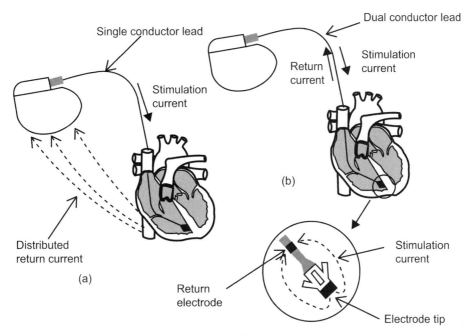

Figure 21-3. Lead Configuration: (a) Unipolar Lead, (b) Bipolar Lead.

referred to as a dual chamber pacemaker (Figure 21-4). A dual chamber pacemaker can more likely restore the natural contraction sequence of the heart. In conventional cardiac pacing, only the right atrium and the right ventricle are paced. During normal intrinsic heart contractions, both ventricles are activated at almost the same time. Under conventional (right heart only) cardiac pacing, contraction of the left ventricle is triggered through propagation of depolarization from the right ventricle. Such delay results in diminished cardiac output, which may cause significant problems for some patients. Cardiac resynchronization therapy refers to pacing of both ventricles simultaneously (biventricular pacing) to improve cardiac output. This is achieved by placing an additional lead in the lateral or posterolateral cardiac vein (located at the far side of the left ventricle). Special lead placement techniques are required because it is not easy to insert a lead into these cardiac veins.

Two electrode fixation mechanisms are used in endocardial lead system. In the active-fixation mechanism, a helical screw inserted into endocardium is used to fixate the electrode tip in position. In the passive fixation mechanism, tines on the electrode tip are caught onto muscular trabeculations at the surface of endocardium; scar tissue formation after implantation over the tines fixate them in position. Implantation of endocardial leads can cause inflammatory immune response of the cardiac tissue in contact with the elec-

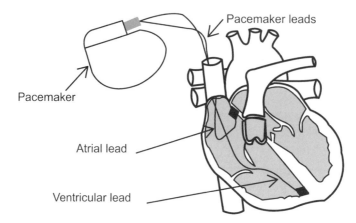

Figure 21-4. Dual Chamber Pacemaker.

trodes. Formation of fibrous tissue around the surface of the electrode increases contact impedance which causes rise in stimulation threshold over time. While active fixation allows more reliable electrode tissue contact than passive fixation, tissue trauma is known to temporarily raise stimulation threshold shortly after lead implantation. By applying steroids directly at the implantation site in both active and passive fixation, acute stimulation thresholds can be minimized and chronic thresholds are reduced. Dexamethasone acetate (DexA) and dexamethasone sodium phosphate (DexP) are commonly used on steroid eluted leads.

For modern pacemaker systems, to facilitate interchangeability, the connector end of the pacemaker lead is conformed to international standard IS-1, and it is sealed to prevent ingress of blood into the connector when the lead is inserted.

IMPLANTATION OF PACEMAKER

The procedures to implant myocardial and endocardial pacemaker leads are quite different. The following descriptions summarize the two procedures.

Myocardial Pacemaker Lead

Myocardial pacemaker lead implantation is performed under general anesthesia. The procedure starts with an incision between the ribs to expose the apex of the heart. An area of the apex free from coronary arteries is chosen. The electrode tip is inserted into the heart muscle (screw in for spiral electrode or stab-in for barb electrode) at the chosen location. Alternatively, a flat electrode is sutured on the surface of the heart. The other end of the lead is tunneled under the skin down toward the abdomen. A shallow incision is made in the abdomen so that the lead can be pulled out. Correct lead placement is confirmed by verifying the pacing and sensing thresholds (often by using a pacemaker analyzer). The lead is then plugged into the pulse generator. The pulse generator is pushed into the pocket made by the incision. The two incisions are closed to complete the procedure.

Endocardial Pacemaker Lead

Endocardial pacemaker lead (or transvenous lead) implantation is done under local anesthesia. The procedure starts with an incision over a vein (e.g., the right external jugular or subclavian). A cannula or trocar is inserted using the Seldinger technique. The pacemaker lead is passed through the cannula down within the vein, into the right atrium, through the right atrioventricular valve, and into the right ventricle. Correct placement is achieved when the tip of the lead is wedged firmly between the trabeculae at the apex of the right ventricle (or screwed into the endocardium) and the lead is observed to be fixed and immobile. Fluoroscopy is often used during the procedure to ensure proper lead placement. A shallow incision is then made in an appropriate area of the upper chest (e.g., under the clavicle). A tunnel is formed under the skin of both incisions so that the con-nector end of the lead wire is pulled through and is accessible at the chest incision. Correct lead placement is confirmed by verifying the pacing and sensing thresholds using a pacemaker simulator. The lead is then plugged into the pulse generator. The pulse generator is pushed into the pocket made by the incision. The two incisions are closed to complete the procedure.

After the procedure, the pacemaker parameters are programmed according to the patient's condition. Regular patient follow-up should be scheduled to monitor the condition of the pacemaker's battery and to confirm that the programmed parameter values are appropriate. Note that the pacing voltage threshold (minimum values to achieve pacing) will rise shortly after implantation and eventually become stabilized after about 3 to 4 months.

Figure 21-5a shows the strength duration curve of heart stimulation. In order to stimulate the heart, the stimulus must have its amplitude and duration above the curve. In the example shown, if the pulse amplitude is 1 V, the pulse width of the stimulation must be larger than 0.7 msec. A pulse duration of 1 msec or longer will be chosen. Figure 21-5b shows the variation in pacing voltage threshold after lead implantation. In this example, the initial pacing voltage amplitude should be set to over 3 V and subsequently reduced to about 2 V after 2 months. A lower pacing amplitude and narrower pacing pulse width will prolong the battery life.

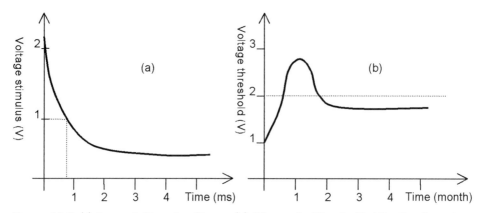

Figure 21-5. (a) Strength Duration Curve; (b) Change in Threshold After Implantation.

Pacemakers are pre-sterilized in the package. If sterility is compromised, re-sterilization should be performed (e.g., using ethylene oxide at 60 °C and 103 kPa). Most manufacturers specify not to re-sterilize pacemakers more than twice.

PACING MODE SELECTION

The pacemaker modes are defined in the NBG code. NBG stands for the North American Society of Pacing and Electrophysiology (NASPE) and the British Pacing and Electrophysiology Group (BPEG) Generic. It is a set of codes specifying the modes of operation of implantable pacemakers. The NBG code is intended for quick identification of the functionality of the pacemaker in case a pacemaker patient requires intervention. NBG code supersedes the older Intersociety Commission on Heart Disease (ICHD) code.

Each letter of the five-letter NBG code describes a specific type of operation. Table 21-1 describes the codes. Although there are five letters in the NBG code, some pacemakers may have only the first three or four letters imprinted on the pacemaker. The following examples illustrate how to interpret the NBG pacing code.

An AOO pacemaker will pace the atrium (first letter—A) at a fixed rate (i.e., at every sensor-indicated interval) irrespective of the intrinsic rate of the heart. Figure 21-6 shows the timing sequence of such a pacemaker. AP in the diagram stands for atrial paced. The dotted line indicates the timer, which keeps track of the pacing time intervals. When this timer reaches zero, a pacing pulse is generated.

A VVI pacemaker monitors or senses a ventricular contraction (second

Table 21-1. NBG Pacemaker Code

Position I Chamber Paced	O—None
	V—Ventricle
	A—Atrium
	D—Dual chamber (ventricle and atrium)
	S*—Single chamber (ventricle or atrium)
Position II Chamber Sensed	O—None
	V—Ventricle
	A—Atrium
	D—Dual chamber (ventricle and atrium)
	S*—Single chamber (ventricle or atrium)
Position III Mode of Response	O—None
	T—Triggered
	I—Inhibited
	D—Dual (triggered and inhibited)
Position IV Rate Modulation	O—None
	R—Rate modulated
Position V Multisite Pacing	O—None
	A—Atrium
	V—Ventricle
	D—Dual (ventricle and atrium)

*Note: Manufacturer's designation only.

letter—V). It will pace the ventricle (first letter—V) if it cannot sense a ventricular contraction (intrinsic rate slower than sensor-indicated interval). If a ventricular signal is sensed, it will inhibit (third letter—I) the pacing action. VVI pacemakers are often used to treat patients with second degree heart block. Figure 21-7 shows the timing sequence of such a pacemaker.

In the diagram, VP stands for ventricle paced and VS for ventricle sensed. After the first paced ventricular contraction, the timer started to count down. It was reset by the sensed intrinsic ventricular contraction. When the timer has reached the sensor-indicated interval without sensing any intrinsic ventricular activity, a pacing pulse is generated to trigger ventricular contraction.

A DDD pacemaker senses both atrial and ventricular activities (second letter—D). When it cannot detect atrial contraction (second letter—D), it will trigger the atrium (first letter—D, third letter—D). After an atrial paced or intrinsic contraction, if no ventricular contraction is detected (second letter—D) after the AV delay interval, the ventricle will be paced (first letter—D). If ventricular contraction is detected, the ventricular stimulation will be inhib-

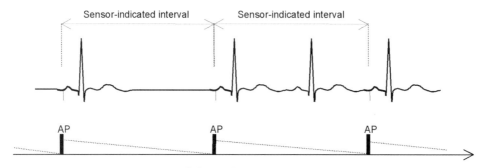

Figure 21-6. AOO Mode of Pacemaker Operation.

ited (third letter—D). A DDD pacemaker is a dual chamber pacemaker; that is, it has two leads, one in the right atrial chamber and the other in the right ventricular chamber. The timing sequence of a DDD pacemaker is shown in Figure 21-8. In the diagram, AP stands for atrial paced, AS for atrial sensed, VP for ventricular paced, and VS for ventricular sensed. If it is also a rate-modulated pacemaker, the pacing rate (or the sensor-indicated interval) will change with the patient's activities. Such a pacemaker will be labeled DDDR. The fourth letter indicates that it is a rate-responsive pacemaker.

Other than the modes described by the NBG code, most pacemakers can be switched to a magnet operation mode. Magnet mode is activated by placing a magnetized programming head or a permanent magnet over the pacemaker. During magnet operation, the pacemaker is paced asynchronously at a predetermined fixed rate. If it is also an ICD, defibrillation is inhibited. Magnet mode is usually combined with a threshold margin test and a self-diagnostic test to evaluate the integrity of the lead and pacemaker system.

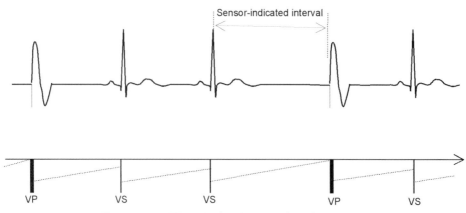

Figure 21-7. VVI Mode of Pacemaker Operation.

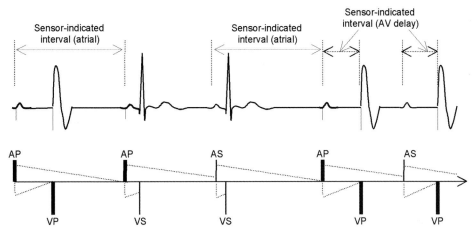

Figure 21-8. Example of DDDR Mode of Operation.

PERFORMANCE CHARACTERISTICS

Table 21-2 lists some common implantable pacemaker performance parameters and their nominal values.

- **Lower rate** is the programmed minimum pacing rate in the absence of sensor-driven pacing. The time between two consecutive pulses at the lower rate is called the escape interval.
- **Activities of daily living (ADL)** rate is the sensor-driven target rate that the patient's heart rate is expected to reach during moderate exercise.
- **Upper sensor rate** is the upper limit of the sensor-indicated rate during exercise.
- **Sensitivity** is the voltage level to which the pacemaker's sense amplifiers are responsive to electrical activities in the heart.
- **Refractory period** is the period of time following the onset of an action potential during which the heart tissue will respond to neither an intrinsic impulse nor an extrinsic impulse.
- **Hysteresis** is a pacing operation that allows a longer escape interval after a sensed intrinsic event. By waiting a bit longer after an intrinsic impulse, it gives the heart a greater opportunity to beat on its own.
- **Rate limit** is a nonprogrammable upper limit of the pacemaker. It is a built-in safeguard to limit the rate of the pulse generator.

In general, the acceptable lead impedance of the pacemaker is from about 200 to 1000 Ω. Very often, a resistor of 500 Ω is chosen as the patient

Table 21-2. Performance Parameters of an Implantable Pacemaker

Parameter	Capability	Nominal Setting
Lower rate	30 to 175 min^{-1} (± 2 min^{-1})	60 min^{-1}
ADL rate	10 to 189 min^{-1} (± 2 min^{-1})	95 min^{-1}
Upper sensor rate	80 to 180 min^{-1} (± 2 min^{-1})	120 min^{-1}
Amplitude	0.5 to 7.6 V ($\pm 10\%$)	3.5 V
Pulse width	0.12 to 1.5 msec (± 25 μs)	0.4 msec
Atrial sensitivity	0.25 to 4 mV ($\pm 40\%$)	2.8 mV
Ventricular sensitivity	5.6 to 11.2 mV ($\pm 40\%$)	8.5 mV
Refractory period	150 to 500 ms (± 9 msec)	330 msec
Single chamber hysteresis	40 to 60 min^{-1} (± 1 min^{-1})	Off
Rate limit	200 min^{-1} (± 20 min^{-1})	Nonprogrammable

load to verify pacemaker performance. A number of different output waveforms are used by different pacemaker manufacturers. Figure 21-9 shows different types of pacemaker waveforms. The simplest waveform is the square pulse while the biphasic is the most commonly used waveform in today's pacemakers.

The battery and electronic components of a pacemaker are usually placed inside a polypropylene container and then hermetically shielded by a titanium housing. The lead connectors are molded in and protected by an epoxy housing. A radiopaque identification code is placed inside the housing so that the pacemaker can be identified from a radiograph of the patient.

Mercury batteries were used in earlier implantable pacemakers, which suffered from low reliability and short life span. Lithium batteries (e.g., Lithium-iodine battery), with higher energy density, are now commonly used in implantable pacemakers. For example, the Medtronic Sigma 213 lithium-iodine battery is rated at 2.8 V with a capacity of 0.83 Ah. Lithium silver vanadium (Li/SVO) batteries with energy density of 2.0 kJ/cm^3 are often used in ICDs or pacemakers. In general, the lower the pacing rate, pulse amplitude, and pulse duration settings, the longer the battery will last. The

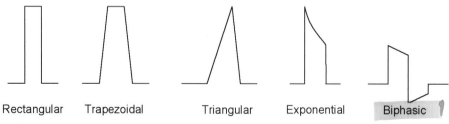

Rectangular   Trapezoidal   Triangular   Exponential   Biphasic

Figure 21-9. Example of Pacemaker Waveforms.

battery in most implantable pacemakers is designed to last for over 7 years under normal use.

Example 21.1

Estimate the battery life of an implantable pacemaker given that it is pacing at 100%, with a rectangular pulse at a rate of 70 beats per minute, a pulse width of 0.45 msec, and a pulse amplitude of 3.5 V, The useful capacity of the battery is 0.34 Ah at 2.8 V. Assume that the lead and/or tissue impedance is 500 Ω and that 80% of the battery energy is used to produce the impulses.

Solution:

The average output power *P* is the product of the output voltage, output current, and the duty cycle. Therefore,

$$P = \frac{V^2}{R}\frac{t_p}{T} = \frac{3.5^2 \times 0.45 \times 10^{-3}}{500 \times \frac{60}{70}} W = 13 \ \mu W.$$

The energy *E* of the battery is

$$E = V \times I \times t = 2.8 \times 0.83 \times 60 \times 60 = 3.4 \ kJ.$$

If 80% of the energy is used to produce the impulses, the longevity *t* of the battery is

$$t = \frac{E}{P} = \frac{0.8 \times 3.4 \times 10^3}{13 \times 10^{-6}} s = 0.21 \times 10^9 \ s = 6.6 \ years.$$

While a conventional implantable pacemaker delivers pacing energy to the heart tissue from a pulse generator positioned away from the heart via a long pacemaker lead, a miniature pacemaker system in the shape of a capsule (Micra by Medtronic) with integrated pulse generator and lead was developed such that the entire unit can be placed inside the right ventricle. This pacemaker, because of its small size (length = 26 mm, diameter = 6.7 mm), can be inserted through the femoral vein using a special catheter into the right ventricle. Tines at one end of the pacemaker secure it to the ventricular wall. The catheter is then withdrawn leaving the pacemaker inside the right ventricle. Due to its construction and only able to connect to one location of the heart, the initial version of this capsule pacemaker can only be used as a VVI pacemaker, mainly to correct bradycardia. A modified version utilizes a motion sensor to detect atrial contraction thus allowing it to support other modes of pacing.

Another type of implantable cardiac device that can be used to monitor the electrical activity of the heart is called an implantable loop recorder. A loop recorder is not a pacemaker but it allows continuous ambulatory recording of cardiac activities similar to a Holter ECG monitor. (Holter ECG is described in Chapter 15).

FUNCTIONAL BUILDING BLOCKS
OF AN IMPLANTABLE PACEMAKER

The hermetically sealed metal casing of an implantable pacemaker houses the battery power source and the electronic circuit for sensing, pacing, and communication. Pacemakers are not field serviceable; all malfunctioning pacemakers are sent back to the factory for analysis. They will not be repaired and reused. Figure 21-10 is a functional block diagram of a dual chamber pacemaker. The descriptions of the various functional blocks follow.

- **Battery Power Source**—The internal battery occupies most of the volume of the pulse generator. It has enough capacity to power the pacemaker for many years.
- **Battery Monitor**—This is a voltage level detector to monitor battery level. When the battery voltage drops below a certain limit (e.g., below 2.2 V for a 2.8-V lithium-iodine battery), the pacemaker will suspend most activities (e.g., cease to collect data) and fall back to a fixed rate to conserve power. A battery replacement message will be displayed when the programmer is engaged with the pacemaker.
- **Programming Signal Modulator/Demodulator**—The transmit/receive coil is the antenna for the programming signal. The signal from the external programmer is detected and demodulated to provide input signal to program the pacemaker parameters. Stored data from the pacemaker can be modulated and transmitted to the programmer via the transmit/receive coil.
- **Reed Switch**—This is a magnetic operated switch to connect and disconnect the pacemaker electronic components to the programming transmit/receive circuit. The switch is normally open. A strong magnet inside the programming module will close the switch when it is coupled to the pacemaker. Its purpose is to prevent external interference picked up by the programming transmit/receive coil getting into the pacemaker.

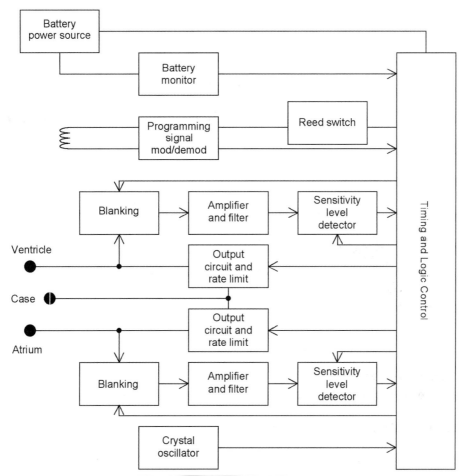

Figure 21-10.Functional Block Diagram of a Cardiac Pacemaker.

- **Atrial and Ventricular Sensing Amplifiers and Filters**–The input filter and amplifier function together provide bandpass and gain characteristics to identify the intrinsic atrial and ventricular activities. The sensitivity detector circuit senses a minimum amplitude level according to the programmed sensitivity setting. A blanking circuit disables the sensing circuit for a predetermined period of time after an impulse is delivered.
- **Output Circuit and Rate Limit**–The output circuit is designed to deliver an electrical pulse to the heart. The pulse amplitude and duration is controlled by the digital timing circuit. The rate limit circuit operates independent of other circuits in the pacemaker. It limits the pulse rate to a maximum value for patient safety (e.g., 185 min^{-1} for ventricular pacing).
- **Sensitivity Level Detector**–The ventricle (or atrium) sensing circuit

picks up ventricular (or atrial) contraction signal as well as noise (EMI or other biopotential signals). The adjustable level detector is to reject noise and allow the qualified signals to pass into the pacemaker's timing and logic control circuit.

- **Blanking**—This is to simulate the refractory period of the natural heart. After a ventricular (or atrial) pulse is sent to trigger the heart tissue, the blanking circuit will block any signal from entering the sensing circuit.

TEMPORARY PACING

Temporary cardiac pacing is achieved by employing external pacemakers. Temporary pacing is used under the following situations:

- After open-heart surgery until the heart reverts to a satisfactory condition
- For temporary pacing until a permanent pacemaker can be implanted
- When permanent pacing is not necessary such as in the case of someone recovering from a myocardial infarction
- During surgery to control the heart rate

There are two types of external pacemakers: invasive and noninvasive.

1. **External invasive pacemakers** have the pulse generator located outside the patient's body. The pacing lead is inserted into the heart chamber via a venous access (endocardial lead). A return lead is attached to the ground electrode placed on the patient's skin.
2. **External noninvasive pacemakers** (also known as transcutaneous or transthoracic pacemaker) have two large-surface, pre-gelled, adhesive, disposable electrodes that conduct the stimuli through the skin and skeletal muscle to the heart. One electrode supplies a current and the other collects the current from the body. They are applied to either the anteroposterior or anterior-anterior position of the patient's chest. External noninvasive pacing is preferred in some situations because it can be applied more quickly and easily and does not require the skills of a physician. However, each stimulating pulse causes contraction of skeletal muscle and can be painful to the patient. In demand mode, intrinsic heartbeats are generally picked up by an integrated ECG monitor using a separate set of leads and electrodes. Because the electrical current must overcome the impedance of the skin and underlying tissues and only a portion of the current flows through the heart, the pacing energy delivered by a transcutaneous pacemaker is substantially higher than that of an implantable or external invasive pacemaker (e.g., pacing

Table 21-3. Characteristics of Different Types of Pacemakers.

	Implantable	*External Invasive*	*Transcutaneous*
Pulse Generator Placement	Inside body	Outside body	Outside body
Lead Placement	Inside body; connected to heart via a blood vessel into the heart chamber	Inserted through skin via a blood vessel into the heart chamber	Outside body; on skin posterior and anterior to the chest
Pacing Energy Level	Low	Low	High

current up to 20 mA and pulse width up to 20 msec in transcutaneous pacing). Table 21-3 summarizes the differences of the three types of pacemakers.

POTENTIAL PROBLEMS AND HAZARDS

Although the electronic circuits of modern pacemakers are very reliable, the pacemaker and lead system may operate inappropriately due to the following problems:

- *Threshold Drift*—Threshold level may increase after initial implantation and may drift higher after prolonged usage. Increase in threshold may result in failure to capture.
- *Premature Battery Failure*—This may lead to shortened device life.
- *Lead Problems*—Broken lead wires, poor connections, and insulation failure may cause continuous or intermittent loss of capture, failure to sense properly, loss of sensing, cross talk between leads, and inhibition of pacing. Extraction of a pacemaker lead is a rather complicated procedure as fibrous tissue often enfolds the lead inside the blood vessel. Depending on the condition of the blood vessel, an additional lead may be implanted inside the same blood vessel. Special tools and procedures (such as using laser sheath catheter) have been develop to extract transvenous pacemaker leads.
- *Cross Talk*—This occurs in dual chamber pacemakers when a stimulus or intrinsic event from one chamber is sensed by the other chamber (e.g., the ventricular lead senses the pacing stimulation initiated in the atrium),

resulting in an inappropriate pacemaker response such as inhibition or resetting of the refractory period.

- *External Interference*–Pacemakers may be susceptible to certain sources of EMI. They include, but are not limited to, the following:

  o Communication equipment, such as cellular phones
  o Radio frequency transmitters such as RFID security systems, and electronic article surveillance devices
  o Magnetic resonance imaging (MRI) scanners
  o High-power electromagnetic fields from electrosurgical units
  o Therapeutic diathermy devices
  o Therapeutic ionization radiation equipment
  o Defibrillator voltage and current

  Pacemakers can be inhibited or reverted to asynchronous operation in the presence of EMI. Turning off the source or moving the pacemaker away from the source may return the pacemaker to normal operation. Extremely strong EMI, however, may reset the pacemaker to its default condition.
- *Muscle Sensing*–Electrical signals from skeletal muscle activities may be sensed and misinterpreted as heart activities. Muscle sensing may be corrected by increasing the sensitivity of the pacemaker.
- *Muscle Stimulation*–Refers to the stimulation of a muscle (other than the heart) by a pacing stimulus. Skeletal muscle stimulation only occurs with unipolar lead systems because it uses the body tissue as a return electrical path. Diaphragmatic stimulation can occur in either bipolar or unipolar systems, usually due to electrode placement that is too close to the diaphragm or phrenic nerve.
- *Mis-programming*–Due to incorrect setup, this can render the pacemaker less effective or reduce the battery life (e.g., too high pacing current).
- *Implant complications*–These include the following:

  o Infection at the incision or catheter site
  o Venous thrombosis caused by blood clot along the pacemaker leads
  o Hematoma
  o Pneumothorax if the lung is inadvertently punctured during the procedure
  o Pericarditis from myocardia lead placement
  o Damage to the vessel at the catheter insertion site
  o Lead dislodgment

- An implantable pacemaker may create problems in some medical procedures. For examples, it may create image artifacts in CT scans; image artifacts and patient heat injuries in MRI scans; and cardiac tissue burn at pacing electrode sites during electrosurgery.

BIBLIOGRAPHY

Ather, S., Bangalore, S., Vemuri, S., Cao, L. B., Bozkurt, B., & Messerli, F. H. (2011). Trials on the effect of cardiac resynchronization on arterial blood pressure in patients with heart failure. *American Journal of Cardiology, 107*(4), 561–568.

Bernstein, A. D., Daubert, J. C., Fletcher, R. D., Hayes, D. L., Lüderitz, B., Reynolds, D. W., . . . , & Sutton, R. (2002). The revised NASPE/BPEG generic code for antibradycardia, adaptive-rate, and multisite pacing. North American Society of Pacing and Electrophysiology/British Pacing and Electrophysiology Group. *Pacing and Clinical Electrophysiology: PACE, 25*(2), 260–264.

Bardy, G. H., Lee, K. L., Mark, D. B., Poole, J. E., Packer, D. L., Boineau, R., . . . , & Ip, J. H. (2005). Amiodarone or an implantable cardioverter–defibrillator for congestive heart failure. *New England Journal of Medicine, 352*(3), 225–237.

Byrd, C. L. (2000). Management of implant complications. In K. A. Ellenbogen, G. N. Kay, & B.L. Wilkoff (Eds.), *Clinical Cardiac Pacing and Defibrillation* (2nd ed.). Philadelphia, PA: W. B. Saunders.

Cleland, J. G., Daubert, J. C., Erdmann, E., Freemantle, N., Gras, D., Kappenberger, L., & Tavazzi, L. (2005). The effect of cardiac resynchronization on morbidity and mortality in heart failure. *New England Journal of Medicine, 352*(15), 1539–1549.

Ellenbogen, K. A., & Wood, M. A. (2008). *Cardiac Pacing and ICDs* (5th ed.). Hoboken, NJ: Blackwell Publishing.

Elshershari, H., Qeliker, A., Özer, S., & Özme, S. (2002). Influence of D-Net (EUROPEAN GSM-Standard) cellular telephones on implanted pacemakers in children. *Pacing and Clinical Electrophysiology: PACE, 25*(9), 1328–1330.

Ganjehei, L., Razavi, M., & Massumi, A. (2011). Cardiac resynchronization therapy: A decade of experience and the dilemma of nonresponders. *Texas Heart Institute Journal, 38*(4), 358–360.

Gimbel, J. R., & Kanal, E. (2004). Can patients with implantable pacemakers safely undergo magnetic resonance imaging? *Journal of the American College of Cardiology, 43*(7), 1325–1327.

Gupta, S., Annamalaisamy, R., & Coupe, M. (2010). Misplacement of temporary pacing wire into the left ventricle via an anomalous vein. *Hellenic Journal of Cardiology, 51*,175–177.

Hare, J. M. (2005). Cardiac resynchronization therapy for heart failure. *New England Journal of Medicine, 346*,1902–1905.

Hayes, D. L., & Friedman, P. A. (2008). *Cardiac Pacing, Defibrillation, and Resynchronization: A Clinical Approach* (2nd ed.). Hoboken, NJ: Wiley-Blackwell.

Hirose, M., Tachikawa, K., Ozaki, M., Umezawa, N., Shinbo, T., Kokubo, K., & Kobayashi, H. (2010). X-ray radiation causes electromagnetic interference in implantable cardiac pacemakers. *Pacing and Clinical Electrophysiology: PACE, 33*(10), 1174–1181.

Lee, K. L. (2010). In the wireless era: Leadless pacing. *Expert Review of Cardiovascular*

Therapy, 8(2), 171–174.

Liao, P.-C., Lai, L.-P., Lin, J.-L., & Huang, S. K. S. (2002). Inappropriate defibrillator discharges caused by an unusual interaction between an implantable cardioverter defibrillator and a pacemaker. *Journal of Cardiovascular Electrophysiology, 13*(11), 1178–1179.

Marco, D., Eisinger, G., & Hayes, D. L. (1992). Testing of work environments for electromagnetic interference. *Pacing and Clinical Electrophysiology: PACE, 15*(11), 2016–2022.

McWilliam, J. A. (1899). Electrical stimulation of the heart in man. *British Medical Journal, 1*, 348–350.

Mittal, S., Movsowitz, C., et al. (2011). Ambulatory external electrocardiographic monitoring: focus on atrial fibrillation. *Journal of American College of Cardiology. 58*(17), 1741-1749.

Mond, H., Sloman, J., & Edwards, R. (1982). The first pacemaker. *Pacing and Clinical Electrophysiology: PACE, 5*(2), 278–282.

Morita, H., Misawa, Y., Oki, S., & Saito, T. (2011). Infection of pacemaker lead by penicillin-resistant Streptococcus pneumoniae. *Annals of Thoracic and Cardiovascular Surgery, 17*(3), 313–315.

Parsonnet, V., Zucker, I. R., Gilbert, L., & Asa, M. M. (1962). An intracardiac bipolar electrode for interim treatment of complete heart block. *The American Journal of Cardiology, 10*(2), 261.

Reynolds D, Duray GZ, Omar R, et al. (2016). A Leadless Intracardiac transcatheter pacing system. N Engl J Med. February 11, 374(6), 533-541.

Seldinger, S. I. (1953). Catheter replacement of the needle in percutaneous arteriography: A new technique. *Acta Radiologica, 39*(5), 368–376.

Weirich, W. L., Gott, V. L., & Lillehei, C. W. (1957). The treatment of complete heart block by the combined use of a myocardial electrode and an artificial pacemaker. *Surgical Forum, 8*,360–363.

Wilkoff BL, Byrd CL, Love CJ, et al. (1999). Pacemaker lead extraction with the laser sheath: results of the pacing lead extraction with the excimer sheath (PLEXES) trial. *J Am Coll Cardiol 33*,1671–6.

Chapter 22

CARDIAC DEFIBRILLATORS

OBJECTIVES

- State the clinical applications of cardiac defibrillation.
- Sketch different defibrillation waveforms and analyze the basic circuits to generate such waveforms.
- Draw a block diagram of a cardiac defibrillator and explain the functions of each block.
- Describe built-in safety features, including isolated output and energy dump.
- Identify and explain the functions of critical components in a typical defibrillator.
- Explain synchronous cardioversion and its operating precautions.
- Identify common problems and hazards of cardiac defibrillators and the methods of mitigation.

CHAPTER CONTENTS

1. Introduction
2. Principles of Defibrillation
3. Defibrillation Waveforms
4. Waveform Shaping Circuits
5. Functional Building Blocks of Defibrillators
6. Output Isolation and Energy Dumping
7. Cardioversion
8. Automatic External Defibrillator
9. Defibrillation Protocols and Quality Assurance
10. Common Problems and Hazards

INTRODUCTION

Fibrillation is an arrhythmia. During fibrillation, the heart muscle quivers randomly and erratically as a result of individual groups of heart muscle contracting randomly instead of synchronously. It was suggested that fibrillation is caused by the interaction of the heart's electrical potential propagation with an obstacle in its pathway (e.g., a conduction block as a result from a myocardial infarction). Continuous fibrillation occurs when such abnormal wavefront propagates and circulates around the myocardium. This is often referred to as "establishment of re-entrant wavefront". If fibrillation occurs at the ventricles, it is called ventricular fibrillation (VF or VFib). This situation is life-threatening as it prevents effective pumping of blood to vital organs such as the brain, lungs, and the heart itself. If it occurs only at the atria, it is called atrial fibrillation (AF or AFib). AF is a less severe heart problem than VF and in most cases is not fatal. However, it compromises cardiac output and will likely lead to other more severe arrhythmias; in addition, uncorrected AF may cause thrombosis in the atria leading to strokes. Figure 22–1 shows a normal sinus rhythm, atrial fibrillation, and ventricular fibrillation waveforms.

The mechanism of defibrillation is not fully understood. It is thought that the extracellular potential gradient as a result of the electric shock from defibrillation alters the transmembrane potential, causing the cardiac cells to depolarize. This brings an abrupt halt to the reentrant wavefront. Prevost and Batelli in 1899 proved that an appropriately large, AC or DC electric shock could reverse VF. It was not until 1960 that open-chest defibrillation was replaced by the external defibrillation method. Today, the defibrillator is a critical life-saving medical device that is widely deployed in hospitals, clinics, ambulances, and even public areas to treat sudden cardiac arrest.

PRINCIPLES OF DEFIBRILLATION

Fibrillation can be caused by disruption of the electroconductive pathways in the myocardial muscle. It may also be triggered by an external electrical shock. Passing a very large momentary electrical current through the heart causes all musculature of the heart to be depolarized for a short period of time and to enter their refractory period together, disrupting all electrical potential propagation. This gives the SA node a chance to regain control and return to normal rhythm. Defibrillation can be external or internal. During defibrillation, an ECG monitor is necessary to detect VF and thereafter monitor the heart function until the patient can be placed in a critical care environment. When defibrillation is applied externally on the patient, a larg-

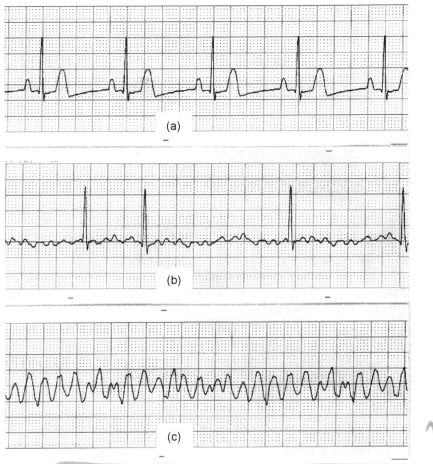

Figure 22-1. (a) Normal Sinus Rhythm, (b) Atrial Fibrillation, (c) Ventricular Fibrillation.

er voltage (and higher energy) is required to overcome the impedance of the body and to allow enough current to go through the heart. A typical discharge energy range is from 2 to 40 J for internal defibrillation and 50 to 400 J for external defibrillation. To account for the different loss of energy due to variation of the transthoracic impedance, some modern defibrillators have built-in sensing and compensation algorithms to deliver the desired dose of energy to the patient.

Defibrillation can also be used to treat certain atrial arrhythmias such as atrial flutter, atrial fibrillation, and ventricular tachycardia. However, an electrical shock applied to a nonfibrillating heart during the T wave of the heart rhythm may trigger VF. During atrial flutter or atrial fibrillation, the ventricles can still contract at regular intervals, thereby producing a certain amount of cardiac output (e.g., 70% of the normal value). It would be counterproductive and could create a life-threatening situation to the patient should this counter shock intended to correct atrial heart condition trigger VF.

In order to avoid discharging the energy during a T wave under cardioversion mode, the defibrillator should be synchronized to discharge its energy right after the patient's R wave and before the T wave. This is achieved by detecting the R wave with the help of an ECG monitor and electronically synchronizing the energy discharge with a short delay (e.g., 30 msec) from the R wave. This special mode of defibrillation is called synchronized cardioversion.

An ICD is a pacemaker with defibrillation capability. Upon sensing VF, an ICD will automatically produce a shock to the patient's heart. An ICD can be programmed to provide defibrillation, cardioversion, antitachycardia pacing, and antibradycardia pacing.

Since its initial conception, studies have indicated that many factors can affect the effectiveness of the defibrillation procedure. These include the waveform, energy, and amplitude of the electric shock; electrode position; electrode interface impedance (or transthoracic impedance); as well as the size and weight of the patient.

DEFIBRILLATION WAVEFORMS

Early experimental defibrillators used 60 Hz AC and a step-up transformer to create and increase the defibrillation voltage. Bursts of several hundred volts of sine wave were applied across the chest wall for a period of 0.25 to 1 second. The desire for portability led to the development of DC defibrillators. A DC defibrillator uses a battery as the power source so that connection to the AC outlet is not required during defibrillation. The battery may be replaced or recharged after use. It was later discovered that DC shocks were more effective than were AC shocks. Until recently, defibrillators have used one of the two types of waveforms: monophasic damped sinusoidal (MDS) and monophasic truncated exponential (MTE). The MDS waveform is also called the Lown waveforms. Figure 22-2 shows typical MDS and MTE defibrillation waveform. Note that a typical MDS waveform has a small negative component; therefore, strictly speaking, it is not truly monophasic.

Monophasic waveforms require a high energy level (up to 360 J) to defibrillate effectively. A MDS waveform requires a high peak voltage (e.g., 5000 V) to deliver such energy. The MTE waveform uses similar energy settings. However, it uses a lower voltage than the MDS waveform does. In order to deliver the same amount of energy, the MTE waveform requires longer time

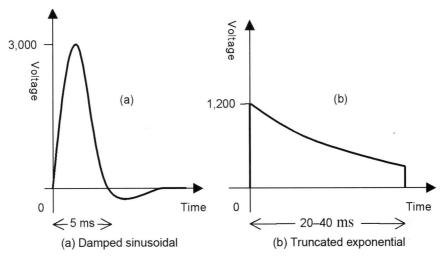

(a) Damped sinusoidal   (b) Truncated exponential

Figure 22-2. Monophasic Defibrillation Waveforms.

duration. Although studies have associated postdefibrillation myocardial damages with high peak voltages, long duration shocks have higher chances of refibrillation.

Studies in the early 1990s showed that biphasic defibrillation waveforms are more effective than monophasic waveforms. Biphasic waveforms have been the standard waveform for ICDs since they were introduced. With a biphasic waveform, the defibrillation current first passes through the heart in one direction and then in the reverse direction. A number of biphasic waveforms are used by different defibrillator manufacturers. Figure 22-3a shows one such waveform. It has been shown that defibrillations using biphasic waveforms not only defibrillate as effectively as traditional monophasic waveforms but also are associated with better postshock cardiac function, fewer postshock arrhythmias, and better neurological outcomes for survivors. In addition, biphasic defibrillators at lower energy settings produce the same results as traditional high energy monophasic defibrillators. Some manufacturers recommend that protocols to escalate energy shock used in traditional monophasic defibrillation are not required in biphasic defibrillation. Instead, a fixed energy setting (e.g., 150 J) is recommended for the first and all subsequent shocks. Recent studies showed that pulsed biphasic waveforms (PBW), which chop the continuous pulse at a high frequency with slightly higher peak voltage, can produce the same shock efficacy using lower energy (with lower average current) than ordinary biphasic waveforms can. An example of PBW is shown in Figure 22-3b. Triphasic defibrillation waveforms (biphasic waveform with a short duration of lower voltage "tail") have recently been proposed and suggested to have the potential to reduce postshock arrhythmias and cause less myocardial injury. An example is shown in Figure 22-3c.

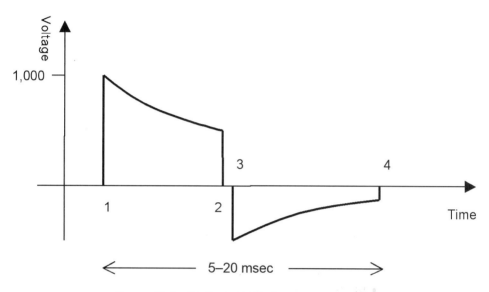

Figure 22-3a. Biphasic Defibrillation Waveforms.

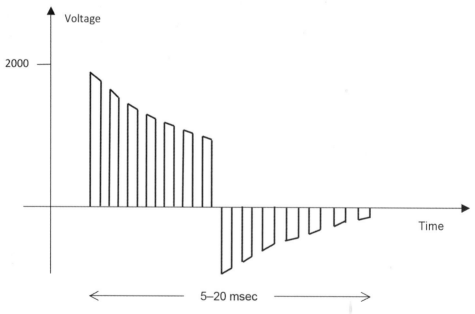

Figure 22-3b. Pulsed Biphasic Defibrillation Waveforms.

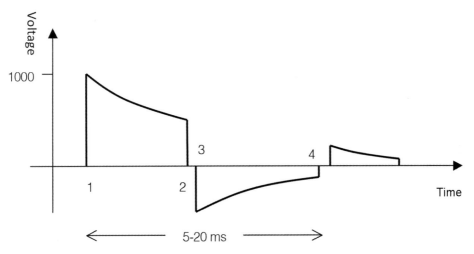

Figure 22-3c. Pulsed Biphasic Defibrillation Waveforms.

WAVEFORM SHAPING CIRCUITS

Figure 22-4 shows a simple functional block diagram of a DC cardiac defibrillator. The main component of a defibrillator is the energy storage capacitor. The capacitor is charged by the charging circuit, which is powered from the power supply. The charge control circuit monitors the amount of energy stored in the capacitor. It terminates the charging process when sufficient energy has been accumulated in the energy storage capacitor. The discharge control releases the stored energy from the capacitor when the user activates the discharge buttons. The waveform shaping circuit produces the particular type of waveform to effect defibrillation. Three common waveform shaping circuits are discussed in this section.

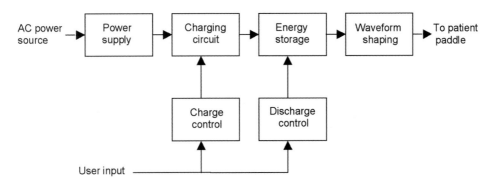

Figure 22-4. Block Diagram of a DC Defibrillator.

Monophasic Damped Sinusoidal Waveform

Figure 22-5 shows a simplified circuit to produce a damped sinusoidal defibrillation waveform. It consists of the energy storage capacity C, a step-up transformer, a rectifier, a charge relay, a discharge relay, and a wave-shaping inductor L. During charging, the charge relay is energized, and AC voltage from the power source is stepped up to the desired level by the step-up transformer. The charging circuit (a full wave rectifier in this example) converts the high voltage AC to a DC charging voltage. The capacitor C is being charged up by this high-level DC voltage until sufficient energy is stored in the capacitor. The energy Ec stored in the capacitor is related to the voltage across the capacitor according to the equation

$$E_C = \frac{1}{2} CV^2. \tag{22.1}$$

During the charging process, the voltage across the energy storage capacitor is monitored to determine the amount of energy stored. The charge relay is de-energized when sufficient energy is stored in the capacitor. When the operator pushes the discharge buttons, the discharge relay is energized. The energy stored in the capacitor flows through the inductor L into the patient. For ease of analysis, the patient load R is considered to be a resistive load of 50 Ω. The current discharging through this LRC circuit produces the damped sinusoidal waveform (Figure 22-2a).

Example 22.1

A cardiac defibrillator is designed to deliver up to 400 J of energy during discharge. If a capacitor of 16 μF is used as the energy storage capacitor, what is the minimum voltage across the capacitor at full charge?

Solution:

Using Equation 22.1,

$$V = \sqrt{\frac{2E}{C}} = \frac{2 \times 400}{16 \times 10^{-6}} = 7000 \text{ V}.$$

The minimum voltage across the capacitor is 7000 V.

In the preceding example, 7000 V is the minimum voltage across the energy storage capacity to deliver 400 J of discharge energy. In practice, not all energy stored in the capacitor is delivered to the patient during discharge.

Some energy, for example, is lost as heat in the discharge circuit. The energy E_D delivered to the patient is always less than the energy E_C stored in the capacitor. Assuming that the patient load is a constant resistive load, the energy delivered to the patient E_D for an MDS waveform defibrillator is given by

$$E_D = \int_{t=0}^{t=\infty} \frac{V^2}{R} \, dt = \frac{1}{R} \int_{t=0}^{t=\infty} V^2 \, dt.$$

Due to energy loss explained earlier, the energy stored must be higher than the discharge energy (or $E_D < E_C$).

A typical value of the waveform shaping inductor to produce an MDS waveform is 50 mH. The function of the resistor R_L is to limit the initial inrush current into the capacitor when the charge relay is first energized. Without R_L, the large inrush current may damage components in the charging circuit. A typical value of R_L is 3 kΩ, which will limit the initial inrush current to a worst case of 2.3 A (7000 V divided by 3 KΩ).

Monophasic Truncated Exponential Waveform

Figure 22-6 shows a simplified circuit of an MTE waveform defibrillator. This circuit is identical to the MDS circuit described earlier except that there is no waveform shaping inductor in the discharge circuit. A typical value of the energy storage capacitor is 200 µF. Without the inductor, the discharge

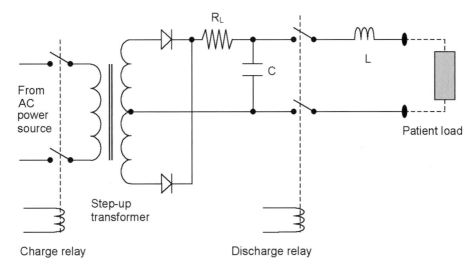

Figure 22-5. Simple MDS Defibrillator Circuit.

circuit is an RC instead of an LRC circuit, where *R* is the patient load. A RC discharge will produce an exponential decay curve (Figure 22-2b). Instead of allowing sufficient time to discharge all energy stored in the energy storage capacitor, a MTE defibrillator will terminate the discharge when enough energy is delivered to the patient. During defibrillation, the voltage across the paddles is monitored and the amount of energy discharged into the patient E_D is determined by

$$E_D = \int_{t=0}^{t=T} \frac{V^2}{R} \, dt = \frac{1}{R} \int_{t=0}^{t=T} V^2 \, dt,$$

where T is the time of discharge termination when sufficient energy has been delivered to the patient.

Biphasic Truncated Exponential Waveform

Figure 22-7 shows a simplified circuit of a biphasic truncated exponential waveform generator. To create the negative phase of the discharge, a bank of switches is added to the previously described MTE circuit. By closing and opening the biphasic switches S1 to S4, a biphasic waveform (Figure 22-3a) is produced. Table 22-1 shows the switching sequence of the switches and relays for the charging, discharging, and energy dumping functions. In the table, an "X" denotes switch closure. There are four phases in the discharge sequence: positive (P1), zero (P2), negative (P3), and discharge (P4).

During the charging period, the charge relay is energized, and switch SC

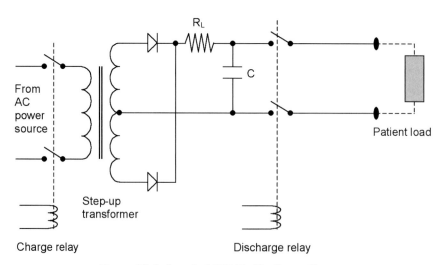

Figure 22-6. Simple MTE Defibrillator Circuit.

is closed so that the capacitor (e.g., a 200 µF metalized polypropylene capacitor) is charged by the charging circuit. SC will open when enough charge is stored in the capacitor. In the positive phase of the discharge sequence, S1, S4, and SD are closed. The flow path of the current from the capacitor is as follows: Capacitor positive terminal → R → S1 → SD-1 → patient → SD-2 → S4 → capacitor negative terminal. During the zero phase, only SD is closed. This zero phase provides a time separation to ensure that S1 and S4 are opened before S2 and S3 are closed. If all four switches are closed at the same time, a short circuit on the output circuit will result. In the negative phase, S2 and S3 are closed. The current flow path from the capacitor is R → S3 → SD-2 → patient → SD-1 → S2 and back to the capacitor. The direction of current flow in the patient load in the negative phase is opposite to that in the positive phase. After enough energy is discharged, SD is open and all biphasic switches (S1 to S4) are closed to remove the remaining charge stored in the energy storage capacitor. The previously described phases complete the discharge sequence. If the capacitor is charged but defibrillation is not necessary, the energy stored in the capacitor must be removed for safety reasons. All modern defibrillators have a programmed dumping sequence to remove the stored charge if defibrillation is not performed within a set time (e.g., after 60 seconds) and also when another energy level is selected (to avoid high residue energy stored in the capacitor). To dump the stored energy, all biphasic switches are closed to allow the energy to discharge through the resistor R.

SC and SD are mechanical switches. In fact, these are the contacts of

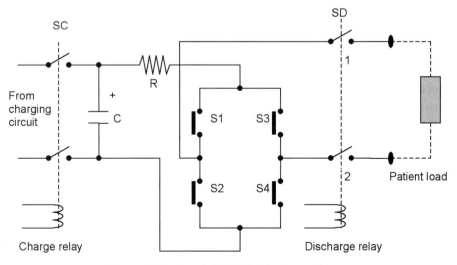

Figure 22-7. Simple BTE Defibrillator Circuit.

Table 22-1. Biphasic Waveform Generator Switching Sequence.

	SC	S1	S2	S3	S4	SD	
Charging	X						
P1 (1–2)		X			X	X	Positive
P2 (2–3)						X	Zero
P3 (3–4)			X	X		X	Negative
P4 (4–5)		X	X	X	X		Discharge
Energy dump		X	X	X	X		

high current relays or contactors. A relay is an electrical contact controlled by electromagnetic force (often created by a solenoid). The term contactor is used to denote a relay that carries a large current. The biphasic switches S1, S2, S3, and S4 are solid-state switches. S4 is usually a high-voltage, high-current switch (such as an insulated gate bipolar transistor) that is designed to interrupt the circuit at high current level. S1, S2, and S3 are usually silicon controlled rectifiers. The function of the resistor R is to limit current during capacitor discharge or energy dump. Some manufacturers may introduce an additional inductor to modify the shape of the waveform. For example, one manufacturer used a 5-Ω, 700-μH inductive resistor to serve both functions.

FUNCTIONAL BUILDING BLOCKS OF DEFIBRILLATORS

Figure 22-8 shows the functional block diagram of a DC defibrillator. After the user selects the energy setting and pushes the charge button, the charge control circuit energizes the charge relay. The voltage across the energy storage capacitor is monitored during charging. Using Equation 22.1, the charge relay is de-energized when the voltage across the capacitor is equal to the voltage corresponding to the selected defibrillation energy level. The charging sequence is then completed. The discharge relay is energized once the user pushes the discharge buttons on the defibrillator paddles (both buttons, one on each paddle, must be activated to prevent inadvertent discharge). The energy stored in the capacitor is then released through the waveform shaping circuit to the patient's chest to perform defibrillation. The energy being delivered to the patient is determined by the voltage and current monitors ($E_D = V \times I \times t$). When the total delivered energy has reached the user selected value, the discharge relay is de-energized.

Due to portability requirements, an internal rechargeable battery is used as the primary energy source of DC defibrillators. The capacity of a fully charged battery is usually sufficient to perform 20 to 80 defibrillation dis-

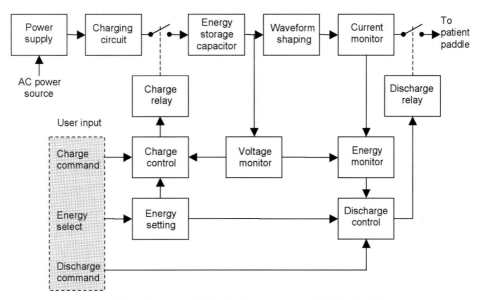

Figure 22-8. Functional Block Diagram of a DC Defibrillator.

charges. Defibrillators are always plugged into the AC mains on standby. AC voltage is rectified to charge the battery. When fully charged, the battery charge is maintained using a trickle charge system. During the charging phase (of the energy storage capacitor), the low-voltage DC from the battery is first converted to a high-frequency (e.g., 25 kHz) AC voltage by an inverter. This AC voltage is then stepped up to a higher voltage, say 5000 V, and rectified to charge the energy storage capacitor. The functional block diagram of the power supply and charging circuit is shown in Figure 22-9.

OUTPUT ISOLATION AND ENERGY DUMPING

A defibrillator produces an electrical shock on the patient to correct heart arrhythmias. If the operator inadvertently touches the discharge paddles or the patient during the delivery of an electrical shock, the shock may cause burns or trigger VF to the operator. Such injuries to the operator can be prevented by isolating the output of the defibrillator from ground. Figure 22-10a shows a nonisolated defibrillator output circuit. When an operator who has a ground connection is in contact with the output while the energy is discharged, a current will flow through the operator to ground and return to the energy storage capacitor. Figure 22-10b shows an isolated defibrillator output circuit. During shock delivery, the energy storage capacitor is not connected to ground. Theoretically, even when the operator is touching a defib-

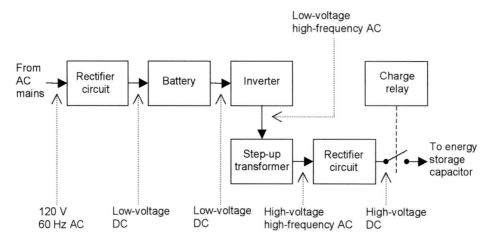

Figure 22-9. Defibrillator Power Supply and Charging Circuit.

rillator paddle, no current will flow through the operator because there is no return path for such current. In practice, however, a small current still flows through the operator.

Isolated output can also prevent secondary burn to the patient. For a grounded output circuit, a secondary burn occurs when the patient is in contact with any grounded object; such ground connection will provide an alternative return path for the discharge current. It will cause a burn at the ground contact site if sufficient current flows through this patient ground path.

After the energy storage capacitor is charged, if no defibrillation is necessary after a period of time, the charge in the capacitor will be dumped through a high power resistor. This is a safety feature to ensure that no hazardous high voltage is present in the unit for the safety of the users. In addition, if another energy level is selected, the charge stored in the capacitor will be released (or dumped) before it receives its new charge. This design is to prevent accumulation of charge in the capacitor, especially if the new selection is of lower energy.

CARDIOVERSION

Cardiac defibrillators are useful in correcting VF. However, the defibrillation shock may trigger VF in a healthy heart. During atrial fibrillation or atrial flutter, only the atrial muscle is contracting erratically. Studies have shown that a defibrillation pulse applied during the refractory period (T wave) of the ventricles may induce other more severe arrhythmias such as

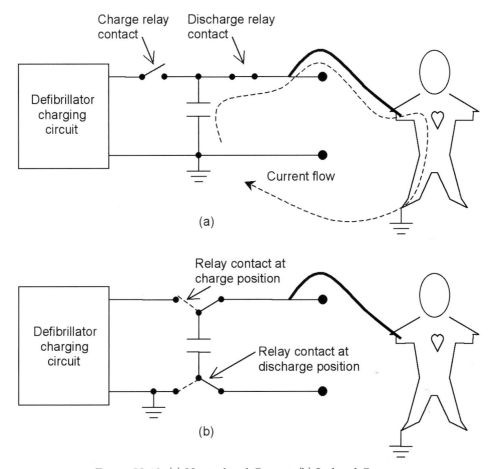

Figure 22-10. (a) Nonisolated Output, (b) Isolated Output.

VF; and the window of discharge for safe cardioversion is immediately after the QRS complex and before the T wave. Therefore, a synchronization circuit is required in cardioversion to avoid such complications.

A cardioversion synchronizing circuit consists of an R wave detector and a time delay circuit to synchronize the discharge within this safety window (Figure 22-11). An enabling signal is sent to the discharge control about 30 msec after detection of the R wave. When the synchronous cardioversion feature is selected, care should be taken to check whether the machine is able to lock onto the R wave (usually, successful detection of the R wave is highlighted on the ECG display of a cardioinverter). The user may have to increase the ECG sensitivity level to provide sufficient R wave amplitudes for the R wave detector to lock onto the signal.

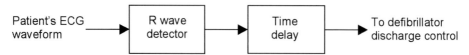

Figure 22-11. Cardioversion Synchronous Module.

AUTOMATIC EXTERNAL DEFIBRILLATOR

When a person suffers from VF, increasing the duration of VF reduces the chance of survival and increases the likelihood of postshock dysfunction. An in-hospital study showed that although almost all patients with VF survived when treated immediately, the survival rate dropped to 5% if treatment was delayed for more than 10 minutes. Because of the need for early intervention, The American Heart Association and the European Resuscitation Council have both endorsed public access defibrillation programs using automatic external defibrillators (AED).

An AED can be fully automatic or semiautomatic. For a fully automatic AED, after the user has applied the disposable electrodes on the patient and activated the unit, the AED analyzes the ECG waveform obtained from the electrodes and determines whether or not a defibrillation shock is needed. If it is, the device automatically charges and discharges a programmed energy shock to the patient.

Most AEDs are semiautomatic. A semiautomatic AED analyzes the patient's ECG, and when defibrillation is indicated, it charges automatically and prompts the operator to activate the defibrillation shock. The shock notification may be a visual message or voice instructions to step the operator through proper course of action.

Fully automatic AEDs are used in hospitals for cardiac monitoring of and automatically defibrillating patients who are deemed to be at high risk of sudden cardiac arrest. After setting up the device and attaching the disposable electrodes to the patient, no clinician intervention is needed to deliver the therapy. The AED continuously monitors the patient's cardiac rhythms through the disposable electrodes and responds by automatically delivering a defibrillation shock within seconds of detecting the corresponding arrhythmia. Semiautomatic AEDs are often used by paramedics and are deployed in public places such as airports and sports arenas.

DEFIBRILLATION PROTOCOLS AND QUALITY ASSURANCE

The procedures for safe operation of a manual cardiac defibrillation are as follows:

1. Apply electrolyte gel to the defibrillator paddles (to provide better electrical contact at the electrode-patient interface to allow unhindered energy flow and to reduce risk of burns).
2. Set energy level and press the charge button.
3. Allow capacitor to charge until a ready signal is given.
4. Press the paddles against the patient's chest.
5. Clear the patient area.
6. Press the defibrillator discharge buttons.
7. Check the patient's ECG waveform.
8. Repeat the above procedures if no sinus rhythm is detected (may need to select a higher energy level setting).

Guidelines are established by manufacturers and resuscitation professional organizations for initial and subsequent defibrillation shock energy level settings (e.g., 200 J, 200 J, and 360 J for monophasic waveform; 150 J, 150 J, and 150 J for biphasic waveform). Because a defibrillator delivers a high-voltage therapeutic pulse to critically ill patients, reliability of the device is crucial. A defibrillator should be tested regularly to ensure its performance. Most hospitals require daily testing by users. Depending on the hospital's protocol, testing includes functional checks and may include energy delivery on an internal dummy load. In addition, extensive performance verifications, including battery capacity and defibrillation energy delivery accuracy, are done periodically (e.g., every 6 months) by biomedical engineering personnel. A high power 50-Ω resistor is often used to simulate the patient load for energy delivery accuracy verifications. In some defibrillator testers, different patient loads (e.g., 25 to 200 Ω) can be selected to verify the energy delivery performance over a range of transthoracic impedance variations. When not in use, a defibrillator should always be plugged into the AC wall outlet to ensure that the internal battery is fully charged and ready for use.

COMMON PROBLEMS AND HAZARDS

Problems with defibrillators can be divided into two groups: common hardware problems and operational problems. They are described in the following.

Hardware Problems

- Batteries are considered high-maintenance components in a defibrillator. Failure of batteries prevents successful defibrillation. Common battery problems include a battery that is not fully charged, battery failure, and cell memory failure. Common batteries used in defibrillators are nickel-cadmium (NiCad), sealed lead acid, lithium-ion (Li-ion), lithium-manganese (Li-Mn), and nickel metal hydride (NiMH). The battery inside a defibrillator is sized to deliver at least 10 full energy discharges even though it is disconnected from the AC power source.

- Electronic components in general are quite reliable.
- Due to the need to deliver high-energy discharge pulses, relay failure is not uncommon in defibrillators. Relay contacts (especially the discharge relay) may be pitted from arcing, which creates high resistance at the contact, or fused due to excessive heat from high discharge current.
- Common problems with the energy storage capacitor are excessive leakage (which prevents the capacitor from maintaining the energy level), and short circuit due to insulation breakdown.

Hardware problems can be detected and disastrous breakdown may be prevented by periodic performance assurance inspections including battery analysis.

Operational Problems

- Users not familiar with the operation of the defibrillator
- Incorrect application of conductive gel, causing high paddle-skin resistance; high current density, or current shunt path, which may lead to unsuccessful defibrillation or patient injuries
- Incorrect paddle placement resulting in defibrillation current not passing through the heart
- Inappropriate defibrillation energy level selected
- Electrical shock to staff from gel spill, staff touching patient, or staff touching paddles
- A unit that is not picking up R wave properly in synchronous cardioversion will not discharge energy (missing or too low R wave level due to poor skin contact or too low sensitivity setting)

Most user errors can be prevented by proper in-service training, periodic practice, and equipment standardization.

Measures to Reduce Problems and Hazards

To prevent problems or hazards, the following must be done by the users regularly:

- Receive proper user in-service training.

- Perform operational check by charging and discharging into dummy load (e.g., weekly).
- Plug unit into wall outlet to maintain battery charge when not in use
- Clean unit, especially paddles, after every use to prevent dried gel and dirt from building up.
- Check quantities and expiration dates of all supplies (e.g., conductive gel, ECG electrodes) that are with the unit.
- Send unit periodically to biomedical engineering for complete functional and calibration check.

BIBLIOGRAPHY

Achleitner, U., Rheinberger, K., Furtner, B., Amann, A., & Baubin, M. (2001). Waveform analysis of biphasic external defibrillators. *Resuscitation, 50*(1), 61–70.

Alexander, S., Kleiger, R., & Lown, B. (1961). Use of external electric counter shock in the treatment of ventricular tachycardia. *JAMA, 177,* 916–918.

Berg, R. A., Hemphill, R., Abella, B. S., Aufderheide, T. P., Cave, D. M., Hazinski, M. F., . . . , & Swor, R. A. (2010). 2010 American Heart Association guidelines for cardiopulmonary resuscitation and emergency cardiovascular care Science. Part 5: Adult basic life support. *Circulation, 122*(18), S685–S705.

Charbonnier, F. M. (1996). External defibrillators and emergency external pacemakers. *Proceedings of IEEE, 84*(3), 487.

Cummins, R. O., Chesemore, K., & White, R. D. (1990). Defibrillator failures: Causes of problems and recommendations for improvement. *JAMA, 264*(8), 1019–1025.

Dillon, S. M., & Kwaku, K. F. (1998). Progressive depolarization: A unified hypothesis for defibrillation and fibrillation induction by shocks. *Journal of Cardiovascular Electrophysiology, 9,* 529–552.

ECC Guidelines. (2000). Part 6: Advanced cardiovascular life support. Section 2: Defibrillation. *Circulation, 102*(Suppl 1), I-90–I-94.

Ferris, L. P., King, B. H., & Spence, P. W. (1936). Effects of electric shock on the heart. *Electrical Engineering* (NY), *55,* 498.

Fuster, V., Ryden, L. E., Cannom, D. S., Crijns, H. J., Curtis, A. B., Ellenbogen, K. A., . . . , & Zamorano, J. L. (2006). ACC/AHA/ESC 2006 guidelines for the management of patients with atrial fibrillation—executive summary. *Circulation, 114,* 700–752.

Jones, J. L., & Jones, R. E. (1989). Improved safety factors for triphasic defibrillator waveforms. *Circulation Research, 64,* 1172–1177.

Jung, W., Manz, M., & Moosdorf, R. (1993). Comparative defibrillation efficacy of biphasic and triphasic waveforms. *New Trends in Arrhythmia, 9,* 765–769.

Lafuente-Lafuente, C., Mahé, I., & Extramiana, F. (2009). Management of atrial fibrillation. *British Medical Journal, 339,* b5216.

Lown, B., & Axelrod, P. (1972). Implanted standby defibrillators. *Circulation, 46*(4),

637–639.

Lown, B., Kleiger, R., & Wolff, G. (1964). The technique of cardioversion. *American Heart Journal, 67*(2), 282–284.

McWilliam, J. A. (1889). Electrical stimulation of the heart in man. *British Medical Journal, 1*(1468), 348–350.

Mittal, S., Ayati, S., Stein, K. M., Knight, B. P., Morady, F., Schwartzman, D., . . . , & Lerman, B. B. (1999). Comparison of a novel rectilinear biphasic waveform with a damped sine wave monophasic waveform for transthoracic ventricular defibrillation. *Journal of the American College of Cardiology, 34*(5), 1595–1601.

Mortensen, K., Risius, T., Schwemer, T. F., Aydin, M. A., Köster, R., Klemm, H. U., . . . , & Willems, S. (2008). Biphasic versus monophasic shock for external cardioversion of atrial flutter: A prospective, randomized trial. *Cardiology, 111*(1), 57–62.

Page, R. L., Kerber, R. E., Russell, J. K., Trouton, T., Waktare, J., Gallik, D., . . . , & Bardy, G. H. (2002). Biphasic versus monophasic shock waveform for conversion of atrial fibrillation: The results of an international randomized, double-blind multicenter trial. *Journal of the American College of Cardiology, 39*(12), 1956–1963.

Priori, S. G., Bossaert, L. L., Chamberlain, D. A., Napolitano, C., Arntz, H. R., Koster, R. W., . . . , & Wellens, H. H. (2004). ESC-ERC recommendations for the use of automated external defibrillators (AEDs) in Europe. *Resuscitation, 60*(3), 245–252.

Rho, R. W., & Page, R. L. (2007). The automated external defibrillator. *Journal of Cardiovascular Electrophysiology, 18*(8), 896–899.

Shea, J. B., & Maisel, W. H. (2002). Cardioversion. *Circulation, 106*(22), e176–e178.

Swedish Standards Institute. (2010). Medical electrical equipment—Part 2-4: Particular requirements for the basic safety and essential performance of cardiac defibrillators. International Electrotechnical Commission (IEC) 60601-2-4. Article number STD-570468.

Tang, W., Weil, M. H., Sun, S., Jorgenson, D., Morgan, C., Klouche, K., & Snyder, D. (2004). The effects of biphasic waveform design on post-resuscitation myocardial function. *Journal of the American College of Cardiology, 43*(7), 1228–1235.

Van Alem, A. P., Chapman, F. W., Lank, P., Hart, A. A., & Koster, R. W. (2003). A prospective randomised and blinded comparison of first shock success of monophasic and biphasic waveforms in out-of-hospital cardiac arrest. *Resuscitation, 58*(1), 17–24.

Waldo, A. L., & Witt, A. L. (2001). Mechanisms of cardiac arrhythmias and conduction disturbances. In V. Fuster, R. W. Alexander & R. A. O'Rourke (Eds.), *Hurst's the Heart* (10th ed., pp. 751–796). New York, NY: McGraw Hill.

Weaver, W. D., & Peberdy, M. A. (2002). Defibrillators in public places—one step closer to home. *New England Journal of Medicine, 347,* 1223–1224.

Wiggers, C. J. (1940). The mechanism and nature of ventricular fibrillation. *American Heart Journal, 20,* 399–412.

Wiggers, C. J., & Wégria, R. (1940). Ventricular fibrillation due to single, localized induction and condenser shocks applied during the vulnerable phase of ventricular systole. *American Journal of Physiology, 128,* 500–505.

Zoll, P. M., Linenthal, A. J., Gibson, W., Paul, M. H., & Norman, L. R. (1956). Termination of ventricular fibrillation in man by externally applied electric counter shock. *New England Journal of Medicine, 254,* 727–732.

Chapter 23

INFUSION DEVICES

OBJECTIVES

- Summarize the applications of intravenous infusion.
- Describe the setup and identify the components of a typical gravity flow manual infusion system.
- Discuss the limitations in manual gravity flow infusion.
- Differentiate between infusion pumps and infusion controllers.
- Analyze the pumping mechanisms of common infusion pumps.
- Evaluate operational and safety features of modern infusion pumps.
- Draw a functional block diagram of an infusion pump and describe its operation.
- Review performance verification procedures of infusion pumps.
- Discuss factors affecting the accuracy of infusion pumps.
- Describe common problems and hazards of infusion devices and methods of mitigation.

CHAPTER CONTENTS

1. Introduction
2. Purpose of Intravenous Infusion
3. Intravenous Access
4. Types of Infusion Devices
5. Manual Gravity Flow Infusion
6. Infusion Controllers
7. Infusion Pumps and Pumping Mechanisms
8. Typical Operational and Safety Features
9. Functional Block Diagram
10. Performance Evaluation

11. Factors Affecting Flow Accuracy
12. Common Problems and Hazards

INTRODUCTION

Infusion devices are employed to administer fluid into the body through either intravenous (IV), nasogastric, or epidural routes. Infusion devices for IV administration are commonly referred to as IV devices. Because the venous pressure is below 50 mmHg (6.7 kPa or about 0.6 mH2O), a one meter water column above the IV site is sufficient to allow gravity to overcome the venous blood pressure and drive the solution into the blood vessel. It is standard procedure to hang the IV bag more than 1 m above an adult patient's heart to ensure there is enough pressure to keep the IV running. Manual gravity flow IV infusion is used extensively in health care facilities for general-purpose infusion. To allow more controlled and accurate fluid delivery, more sophisticated devices have been developed. A variety of infusion devices are available for different applications. This chapter studies the principles and applications of a few of these devices.

PURPOSE OF INTRAVENOUS INFUSION

In general, four types of solutions are administered intravenously:

1. Water—Usually in the form of saline or dextrose, for rehydration or to prevent patient dehydration.
2. Medications and electrolytes—IV administration of drugs and electrolytes produces precise and fast-acting effects because it sends the drug directly into the bloodstream without going through the process of digestion and absorption. Examples include IV delivery of antibiotics, cardiovascular and chemotherapy drugs.
3. Nutrition—Although parenteral nutrition (PN) can be delivered through nasogastric enteral feeding, partial or total PN (TPN) is administered by IV infusion to patients when their normal diet cannot be ingested, absorbed, or tolerated for a significant period of time. A PN solution contains nutrients, including glucose, amino acids, lipids, vitamins, and dietary minerals.
4. Blood—Blood infusion (whole blood, blood products, or blood substitutes) may be performed by an IV infusion device. However, some may require special infusion sets to avoid problems associated with the relatively high viscosity of blood and potential hemolysis of blood cells.

IV infusion is a procedure commonly performed in health care facilities. Infusion devices can be found in most parts of a hospital, including the emergency department, medical imaging areas, and operating rooms.

INTRAVENOUS ACCESS

A venous access is needed in IV infusion. There are three types of vascular access: hypodermic needle, peripheral catheter (or cannula), and central line. A hypodermic needle is the most convenient way to gain access to the venous system. Veins on the back of the hand and the median cubital vein at the elbow are frequently used needle sites. A peripheral catheter is the most commonly used IV access in clinical settings. A short flexible tube is inserted through the skin into a peripheral vein using a needle (e.g., using the Seldinger technique). The end of the catheter outside the skin is connected to the IV line. A central IV line is used to deliver fluids and medications directly into the heart and distributed them quickly to the rest of the body. The tip of the central IV line is commonly placed in the superior or inferior vena cava or the right atrium of the heart.

TYPES OF INFUSION DEVICES

There are many different types of infusion devices on the market. Each has its own characteristics and serves some specific applications. In general, infusion devices can be divided into two main groups: gravity flow infusion devices and infusion pumps. A gravity flow infusion device relies on the gravitational force exerted by a liquid column to push the fluid via a venous access into the patient's bloodstream, whereas an infusion pump has a motorized pumping mechanism to generate the positive pressure to push the fluid into the blood vessel. Within the gravitation group are the manual gravity flow sets and the infusion controllers. There are two types of pumps in the infusion pump group: volumetric pumps and syringe pumps. Within the volumetric pump group are three different pumping mechanisms: piston cylinder, diaphragm, and peristaltic pumping mechanisms. Figure 23-1 shows the grouping of different types of infusion devices.

MANUAL GRAVITY FLOW INFUSION

The simplest infusion device is the manual gravity flow infusion set. Figure 23-2 shows a typical gravity flow infusion set. It consists of a long flex-

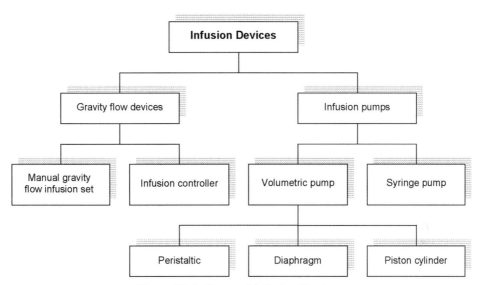

Figure 23-1. Types of Infusion Devices.

ible polyvinyl chloride (PVC) tubing with a solution bag spike at one end and a Luer-lock connector at the other. A number of components (described below) are coupled along the PVC tubing. In addition to these basic components shown in the figure, others such as in-line bacterial filters, one-way valves, and anti-siphon valve may be present. IV lines come in a pre-sterilized package.

The following sections describe the functional components of a gravity flow IV set.

IV Solution Bag

The solution bag contains the IV solution and often comes in a pre-sterilized package of different sizes (e.g., 500 cc, 1 liter, etc.). The bag is usually hung on an IV pole about 1.5 m above the infusion site to create enough hydrostatic pressure to overcome the venous pressure to sustain infusion. Reusable solution-filled glass bottles instead of disposable bags are used in some developing countries.

Solution Bag Spike

The solution bag spike is a rigid sharp-ended tubing connecting the IV set to the solution bag. This sharp spike is pushed through the seal of the solution bag to allow solution to flow from the bag into the line. A protective cap is covering the spike in a freshly opened IV set. To prevent contamination, the cap should not be removed until it is ready to be inserted into the solution bag.

Drip Chamber

The drip chamber is a clear transparent compartment that permits the clinician to see the solution drops coming down from the solution bag. The size of the drop nozzle is designed so that each drop produces a specific volume of the solution. A typical IV drop nozzle produces 0.05 ml per drop (or 20 drops per ml) of solution. A slower flow rate set (e.g., one used for neonatal infusion) may have a 60 drops/ml nozzle. By counting the number of drops within a known time interval, a clinician can determine the volume flow rate of the infusion.

Regulating Clamp

The regulating clamp is used to control the volume flow rate of infusion. It is also known as a roller clamp. By compressing the flexible PVC tubing, the position of the roller changes the inner lumen cross-sectional area of the IV line, thereby controlling the rate of infusion.

Y-Injection Site

The Y-connection (or Y-injection) site provides a point of access into the infusion line. Drugs or other solutions can be added to the infusion fluid by connecting another line to the site. For lines with Y-injection site, a needle is used to puncture the injection port. Needleless Y-connection ports are replacing Y-injection sites to minimize needle stick injuries. In a needleless Y-connection port, the needle injection port is replaced by a female Luer lock connector. To introduce a second solution into the IV line, a syringe or another line with a male Luer lock connector is connected to the Y-connection port of the IV line. To infuse a second solution (e.g., medication) when an infusion line has already been established, a setup called piggyback infusion is often used (Figure 23-3a). In piggyback infusion setup, since the secondary solution bag is located at a higher level (at least 15 cm higher) than the primary solution bag, only the solution from the secondary bag will flow downstream through the Y-site due to hydrostatic pressure.

Flow of the primary solution will resume automatically when the secondary solution bag becomes emptied. Since the secondary solution bag is placed higher than the primary solution bag, a one-way flow valve (or anti-reflux valve) in the primary line is needed to prevent the solution in the secondary bag from flowing into the primary solution bag.

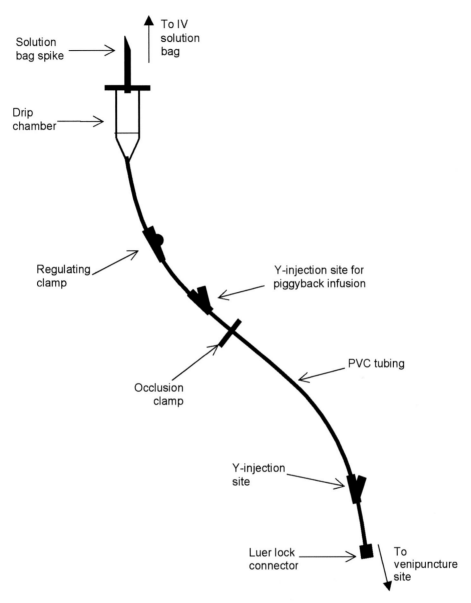

Figure 23-2. Gravity Flow Intravenous Infusion Set.

Occlusion Clamp

An occlusion clamp is used to totally occlude or shut down the infusion flow. Unlike the roller clamp, an occlusion clamp either fully opens the infusion line or totally occludes the line. A typical occlusion clamp is construct-

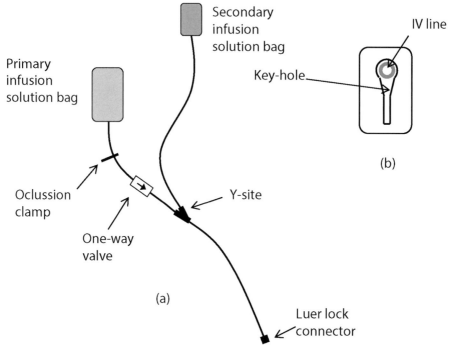

Figure 23-3. (a) Piggyback Infusion Setup, (b) Occlusion Clamp.

ed from a piece of thick plastic with the infusion line threaded through a key-hole-shaped opening in the middle (Figure 23-3b). The line is fully open when the PVC tubing is at the larger opening of the keyhole. If the line is pushed to the narrow end, the clamp will occlude the tubing and shut off the fluid flow.

One-way Flow Valve

A one-way flow valve (or anti-reflux valve) is built into some IV sets to prevent reverse flow of solution. Reverse flow can occur if the downstream pressure becomes higher than the pressure in the IV line, such as the case in piggyback infusion. Reverse flow may also happen when the venous pressure is higher than the hydrostatic pressure at the venipuncture site. In most cases, it happens when the solution bag is placed lower than one meter above the venipuncture site.

In-line Bacterial Filter

Some IV lines have an in-line bacterial filter to block bacteria from the solution to get into the patient. In line bacterial filter will introduce addition pressure drops in the IV line and is counter indicated when there are large molecules in the IV solution (such as blood products or PN infusion).

Luer-Lock Connector

A Luer-lock connector is a special twist lock mechanism to ensure a secure and liquid-tight connection in fluid lines. The luer connector design allows connection between different fluid delivery systems, such as intravenous, enteral feeding, and gas lines in patient care. Luer-lock fittings are securely joined by means of a tabbed hub on the female fitting which screws into threads in a sleeve on the male fitting.

To set up an IV infusion, the IV line is connected to a vascular access. A vascular access for IV can be a large bore needle or a catheter inserted into a vein using, for example, the Seldinger technique. The exposed end of the needle or catheter is a female Luer-lock connector. After the IV line is primed, the line is connected securely to the vascular access via the Luer-lock connection.

The procedures to set up and prime an IV line are described below:

1. Suspend the IV solution bag on the IV pole.
2. Open the bag containing the IV line.
3. Insert the solution bag spike of the line into the IV bag.
4. Open the roller clamp and the occlusion clamp to allow the solution to flush away all the air from the line.
5. Remove air bubbles trapped in the Y-connection site by inverting and gently tapping it with a finger.
6. Close the roller clamp and connect the Luer-lock at the end of the line to the Luer-lock at the indwelling catheter.
7. Squeeze and release the drip chamber compartment to fill about one third of the drip chamber with the IV solution.
8. Slowly open the roller clamp to set up the desired solution flow rate (by counting the drops using a stopwatch).

Example 23.1

A nurse is observing the drop rate in the drip chamber of a manual gravity flow infusion set in order to set the IV solution flow rate. How many drops per minute should be counted in the drip chamber if an infusion flow rate of 60 ml/hr is required? Assume that a 20 drops/ml nozzle is used in the drip chamber.

Solution:

At a flow rate of 60 ml/hr, 1200 drops (60 ml/hr × 20 drops/ml) will come down from the nozzle in one hour. Therefore, there will be 1200 ÷ 60 = 20 drops from the nozzle in one minute.

A major drawback of manual gravity flow infusion is the low flow rate accuracy. The mechanism to regulate the infusion flow rate (the roller clamp) relies on compressing the PVC tubing of the infusion line under the roller clamp. As the PVC tubing is not totally elastic, prolonged compression reduces the size of the inner lumen of the PVC tube under the roller, thereby reduces the fluid flow rate. In addition, the flow rate of manual gravity infusion is dependent on the gravitational force created from the height of the liquid column between the level of fluid in the solution bag and the vascular access location. As the liquid level drops in the solution bag, the flow rate decreases accordingly.

Figure 23-4 shows the change in flow rate after the initial setup when there is no user intervention. However, the decrease in flow is primarily caused by the collapse of the PVC tubing under the roller with time due to imperfect elastic nature of the PVC material. In normal use, the drop in flow rate due to fluid level reduction is often insignificant when compared to that due to collapse of the PVC tubing.

Experiments have shown that, in some IV lines, the flow rate can drop by almost 40% within a couple of hours after the initial setup (mainly due to deformation of the PVC tubing). To overcome such a problem, it is common nursing practice to recheck the flow rate from time to time after the initial

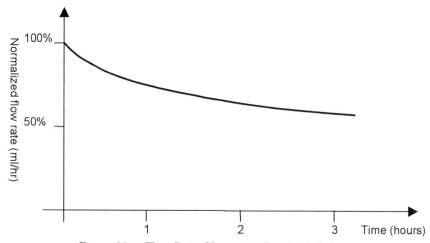

Figure 23-4. Flow Rate Change in Gravity Infusion.

setup and adjust the regulating clamp to reestablish the desired flow rate.

INFUSION CONTROLLERS

An infusion controller overcomes the problem of flow rate variation by automatically adjusting the regulating clamp. Figure 23-5 shows the setup of an infusion controller. An infusion controller monitors the flow rate by counting the drops in the drip chamber. A typical drop sensor consists of an infrared light emitting diode (LED) and a light sensor (e.g., a phototransistor), each located on the opposite side of the drip chamber. A fluid drop from the solution bag interrupts the optical path and produces an electrical pulse; the drop rate is computed by obtaining the time interval between drops. The flow rate is computed from the drop rate and the drop size. The calculated flow rate is then compared to the setting. If it is lower than the set-

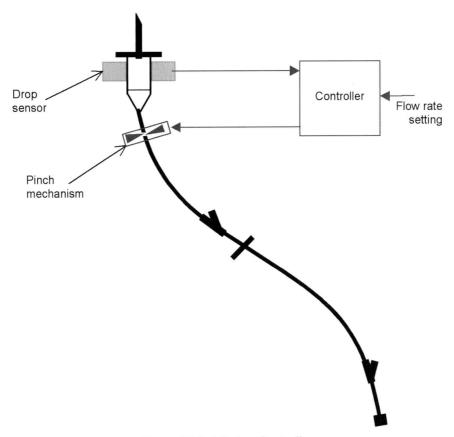

Figure 23-5. Infusion Controller.

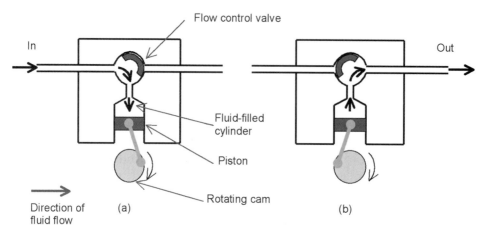

Figure 23-6. Piston Cylinder Infusion Mechanism.

ting, the pinching force of the pinch mechanism will be released to allow more fluid to flow through. If it is higher, it will increase the pinching force to reduce the flow. Such a feedback mechanism maintains a constant flow rate equal to the setting. Although it automatically regulates the fluid flow rate, an infusion controller still relies on gravitational force to generate the infusion. If there is too much resistance in the infusion line or when there is excessive backpressure, the gravity pressure created by the liquid column may not be sufficient to produce the desired flow rate.

INFUSION PUMPS AND PUMPING MECHANISMS

An infusion pump contains a motor-driven pumping mechanism to produce a net positive pressure on the fluid inside the infusion line. With the pumping mechanism, infusion pumps produce a more controlled and consistent flow than infusion controllers do. Infusion pumps can be divided into two types: volumetric pumps and syringe pumps.

Three common pumping mechanisms are used in volumetric infusion pumps. They are piston cylinder, piston diaphragm, and peristaltic. A syringe pump uses a screw and nut mechanism to drive the plunger of a syringe; it is also called a screw pump given that there is a long screw in the driving mechanism. The following sections describe these pumping mechanisms.

Piston Cylinder Pumps

The pumping mechanism of a piston cylinder pump consists of a cylinder, a piston, and valves that are mechanically linked to the piston motion. A stepper motor drives a cam to move the piston in and out of the cylinder in a reciprocal motion. Figure 23-6a shows that when the piston is moving downward, it creates a negative pressure inside the cylinder. The valve, which is linked to the cam, will be in such a position that the input port to the cylinder is opened and the output port is closed. Fluid from the IV bag will therefore be drawn into the cylinder. When the piston is moving upward (Figure 23-6b), the valve will close the input port and open the output port, allowing IV solution in the cylinder to exit through the output port. The stroke distance and the diameter of the piston determine the stroke volume, and the infusion flow rate is equal to the stroke volume times the cam's rotational speed.

Piston Diaphragm Pumps

The pumping mechanism of a piston diaphragm pump is similar to that of a piston cylinder pump except that the stroke motion is replaced by a moveable diaphragm and two coordinated valves. In the illustration shown (Figure 23- 7), when the diaphragm moves to the left, the intake valve is open to allow fluid to enter the fluid chamber. When it moves to the right, fluid is forced out of the chamber. Repeating the action provides a continuous flow of fluid.

Peristaltic Pumps

A peristaltic pump employs a protruding finger mechanism to occlude the flexible IV tubing. Its pumping action is similar to one using the thumb and index finger to squeeze on a plastic tubing filled with fluid and then running the fingers along the tube. This action will force the fluid to move along the direction of the finger motion. Repeating this action will produce a continuous fluid flow. Figure 23-8a shows the pumping mechanism of a rotary peristaltic infusion pump. In a rotary peristaltic pump (or roller pump), the rotor has several protruding rollers. The flexible IV tubing is placed inside a groove on the pumping mechanism housing with one side open to the rotor. The rollers on the rotating rotor push the tubing against the wall of the groove. The protruding rollers, while occluding the tubing, move in one direction along the IV tubing, creating a continuous fluid flow in the direction of motion of the rollers.

Instead of rotating the protruding rollers over the IV tubing, the protruding fingers in a linear peristaltic infusion pump sequentially occlude the IV tubing. Figure 23-8b shows the positions of the protruding fingers of a linear peristaltic pump at three sequential time instances. These coordinated

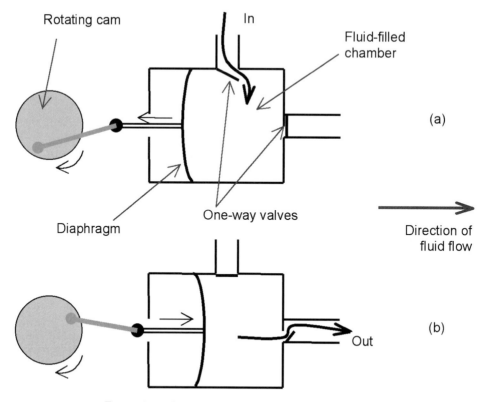

Figure 23-7. Piston Diaphragm Infusion Mechanism.

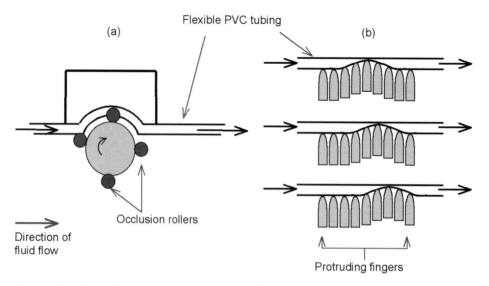

Figure 23-8. Peristaltic Infusion Mechanism. (a) Rotary Peristaltic; (b) Linear Peristaltic.

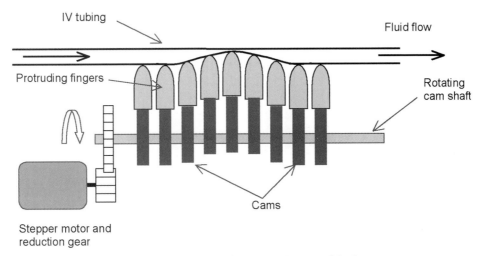

Figure 23-9. Linear Peristaltic Pump Driving Mechanism.

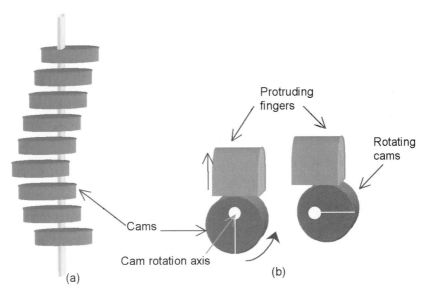

Figure 23-10. Linear Peristaltic Plunger Driving Mechanism.

motions of the protruding fingers produce a continuous flow of fluid in the direction shown. The driving mechanism of a linear peristaltic pump is shown in Figure 23-9. To create a linear peristaltic motion, cams with eccentric axes are attached to a rotating cam shaft (Figure 23-10a) such that when a shaft rotates, it moves the protruding finger up or down according to its eccentric angle of rotation (Figure 23-10b).

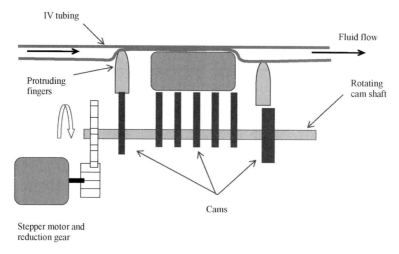

Figure 23-11. Modified Peristaltic Pumping Mechanism.

Figure 23-11 is a modified version of the peristaltic pumping mechanism. The principle of operation can be viewed as having the center cams all positioned in phase and combined together to produce a wide piston (or plunger) pressing on the IV tubing. During infusion, the left protruding finger occludes the IV tubing and the plunger moves slowly to compress the flexible PVC tubing. Fluid is therefore moved to the right at a flow rate according to the dimension of the tubing and the speed of the compression. At the lowest position of the plunger, the right protruding finger moves to occlude the tubing while the left finger releases the tubing; the plunger then retracts quickly, allowing fluid to fill the uncompressed portion of the IV tubing. The finger positions are then restored; another cycle of solution infusion will follow. This pumping mechanism may also be considered as a modified piston cylinder pumping mechanism. The volume flow rate of a peristaltic pump depends on the repetition speed of the protruding fingers and the inner diameter of the IV tubing.

Syringe Pumps

A syringe pump has a long screw mounted on the pump support. The screw is rotated by a stepper motor and gear combination. The screw is supported by two bearings to allow smooth operation. As the screw rotates, it moves a nut threaded onto the screw in the horizontal direction (Figure 23-12). The nut is attached to a pusher connecting to the plunger of a syringe, which is loaded with the solution to be infused. The flow rate of the fluid coming out of the syringe depends on the rotational speed of the screw, the

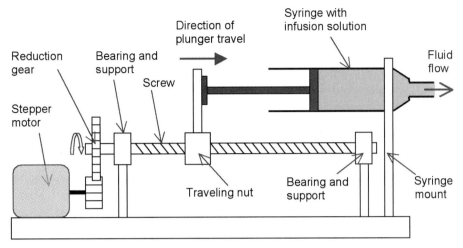

Figure 23-12. Syringe Pump.

screw pitch, and the inner cross-sectional area of the syringe body. In mathematical terms, the volume flow rate of a syringe pump is:

$$Q = R \times t \times A,$$

Where Q = volume flow rate in cubic centimeters per minute,
       R = rotational speed of the screw in revolutions per minute,
       t = screw pitch in centimeters, and
       A = cross-sectional area of the syringe plunger in square centimeters.

Syringe pumps are often used in high-accuracy, low volume (less than 60 ml), low-flow rate applications, and when a more uniform flow pattern is required. A full-term neonate typically receives IV fluids at rates of 10 to 20 ml/hr, whereas for extremely low birth weight neonates, infusion rates are often between 3 and 5 ml/hr. Rates below 2 ml/hr also occur in some clinical settings. When a syringe pump is used in neonates, the usual drug infusion duration ranges from short (3 to 5 minutes) with a slow push of a small dose to longer (30 minutes continuous infusion) durations. Syringe pumps are also used to infuse high-viscosity feeding solutions and for epidural infusion.

A syringe pump may also be configured as a patient-controlled analgesic (PCA) pump to allow patients to self-administer boluses of opioid analgesics such as hydromorphone for pain relief. A PCA pump allows on-demand pain medication to be controlled by the patient. To avoid over-dosing of

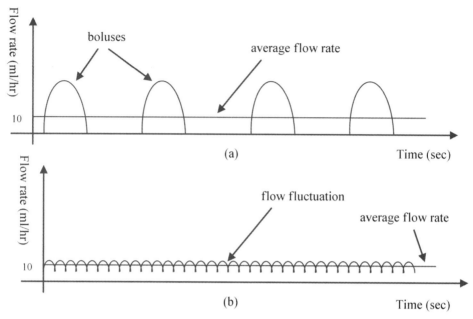

Figure 23-13. Low Flow Rate Infusion Flow Pattern.

medication, the size of the bolus as well as the minimum time interval between two sequential bolus deliveries can be selected. The maximum frequency of bolus delivery can be programmed into the unit. After a bolus has been delivered, the pump is locked-down to inhibit delivery of additional boluses until the pre-programed time interval (typically between a few minutes to an hour) has elapsed. For every patient bolus demand, irrespective of successful or locked-down status, the pump can be programmed to generate an audible beep to create a placebo effect to the patient. To prevent patients or others to temper with the narcotic loaded in the syringe, access to the syringe is protected by a locked and tempered-proof cover.

In general, piston cylinder infusion pumps and syringe pumps produce a more accurate flow output. During low flow rate settings, however, piston pumps (both piston cylinder and piston diaphragm) produce boluses of infusion rather than a smooth flow pattern. Figure 23-13a shows the theoretical flow pattern of a piston cylinder pump at a low flow rate setting (e.g., 10 ml/hr). A bolus in the diagram corresponds to the flow of one stroke of the pistons. The actual flow pattern, however, is not as abrupt due to the slight elastic nature of the tubing. A bolus type of infusion may not be suitable for some applications, such as administering medications to small infants. A peristaltic pump, with its multiple protruding finger pumping action, produces a flow pattern with less fluctuation than piston pumps do as shown in

Figure 23-13b. Under ideal conditions, a syringe pump will produce a uniform flow with little fluctuation. Flow rate accuracy as low as ±3% are claimed by some peristaltic pump manufacturers; accuracy of syringe pumps are usually better than peristaltic pumps.

TYPICAL FUNCTIONS AND FEATURES

Some common functions and features found in general-purpose infusion pumps are described in the following sections. Depending on the design application, an infusion pump may have other additional features.

Flow Rates

The flow rate of a general-purpose infusion pump can be set within the range of 1 to 999 ml/hr. In neonatal pumps, the range is 0.1 to 99 ml/hr.

Volume To Be Infused

A volume-to-be-infused (VTBI) of 1 to 1999 ml can be programmed such that the pump will stop after this volume has been delivered. Usually, when VTBI is reached, an audible tone will sound to alert the clinician. The pump will switch to its keep-vein-open (KVO) rate.

Keep Vein Open

When infusion has stopped, in order to prevent blood clot to blocking the catheter at the venipuncture site (due to stagnant blood in contact with the catheter), a slow infusion rate of about 1 to 5 ml/hr is maintained to flush the catheter to prevent blood clotting. The keep-vein-open or keep-vessel-open (KVO) feature can be disabled and the KVO rate can be selectable in some IV pumps.

Occlusion Pressure Alarm

A pressure sensor inside the pump positioned downstream of the pumping mechanism monitors the pressure of infusion. A high pressure indicates occlusion downstream of the pump. An alarm is set to notify the clinician to check the IV line. Downstream occlusion may be due to a clot in the IV catheter or pinching of the IV line (e.g., by the patient rolling over the line). In some infusion pumps, the occlusion pressure alarm may be adjusted to activate between 1 and 20 psi (7 to 140 kPa).

Fluid Depletion (or Upstream Occlusion) Alarm

When the IV bag is empty, a negative pressure will develop upstream of the pump. An alarm to indicate such a condition can alert the clinician about the empty solution bag and can prevent air from entering the IV line due to suction and subsequently being infused into the patient.

Infusion Runaway (or Free Flow) Prevention

Most modern-day pumps have a built-in mechanism to prevent the free flow of solution into the patient. Free flow can occur when the IV line is removed from the pump while the occlusion clamp and roller clamp are both open (both clamps need to be opened to allow normal pump operation.). The hydrostatic pressure from the solution bag creates the free flow of solution into the patient. When the pump is used to administer a potent drug to a patient, free flow can impose serious risk to the patient if a large dose of such medication is infused into the patient. In volumetric pumps, an interlock mechanism in the pump and an occlusion clamp on the IV line prevents free flow by shutting off the IV line when it is pulled out from the pump. In syringe pumps, free flow is prevented by inserting an anti-siphon valve in the IV line which requires a positive pressure to stay open. An anti-siphon valve prevents flow in the line due to gravity when the syringe is removed from the pump. During normal IV administration, the syringe pump pressure overcomes the pressure drop (about 150 mmHg or 20 kPa) across the anti-siphon valve.

Air-in-Line Detection

An air bubble of less than 30 µl may be harmlessly dissolved into the blood circulation. Although small air bubbles in the vein can leave the blood through the lungs, a larger volume of air in the bloodstream can cause life-threatening problems to the pulmonary circulation. An extremely large amount of air (3–8 µl per kg of body weight) can cause air lock and disable the heart's pumping mechanism. To prevent air embolism in patients, air-in-line detectors are built into infusion pumps to detect air bubbles in the IV line. Infusion will stop and an alarm will sound when a large air bubble (e.g., greater than 100 µl) in the line is detected.

Dose Error Reduction Systems

There are numerous reported incidents, including many fatal ones, due to dose errors in IV administration. The dose error reduction system is a software algorithm that checks programmed doses against preset limits specific to certain drugs and clinical profiles. It alerts clinicians if the programmed dose exceeds the preset limits. For example, drug X used in area A has a dose limit of 20 mcg/kg/hr. If the dose setting, based on the programmed flow rate and drug concentration, exceeded 20 mcg/kg/hr is entered, infusion will not start and an alarm will sound. This will force the clinician to check and rectify the setting to prevent erroneous drug infusion. A dose error reduction system is often associated with a barcode reading system to verify the patient ID and the prescription package to further avoid human errors in delivering drugs to the wrong patient.

Battery Operation

Most infusion pumps are powered by internal rechargeable batteries so that the patient can move around in the hospital or be transported while under infusion. A low battery detection circuit will alert the user to recharge the pump if the battery is running low.

FUNCTIONAL BLOCK DIAGRAM

Figure 23-14 shows a functional block diagram of a volumetric infusion pump. The user input/output interfaces are shown on the left-hand side of the diagram. User adjustable inputs include

- Infusion flow rate setting
- VTBI setting
- KVO enable/disable selection
- Pump start/stop control

The CPU, based on the input settings, controls the speed of the stepper motor driving the pumping mechanism to deliver the set infusion rate. The rotational speed of the pump driver is monitored by an LED and optical transistor pair slit detector. Based on this measured rotational speed, the volume of infusion is computed and compared to the VTBI setting. If KVO is enabled, the motor speed will be reduced to the KVO rate when VTBI has been reached.

The pressure in the IV line is monitored by a pressure sensor pressing on the IV tube inside the pump. When the pressure exceeds the occlusion pressure, the CPU will shut down the pump and sound an alarm.

Air bubbles in the IV line are detected by an ultrasound transmitter and receiver pair. The attenuation of ultrasound or acoustic impedance in air is higher than that in water. When an air bubble passes through the detector,

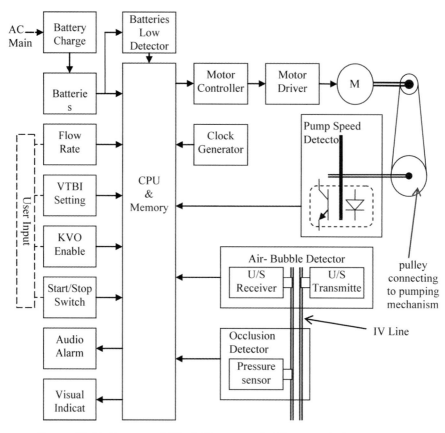

Figure 23-14. Functional Block Diagram of Volumetric Infusion Pump.

the intensity of ultrasound detected by the receiver will decrease. At a flow rate of q (in ml/hr), the duration t (in sec.) of this decreased signal corresponds to the size v (in ml) of the air bubble in the line according to the equation $v = (q \times t) / 3600$ ml. The CPU will stop the infusion and sound an alarm if a large air bubble (e.g., >100 µL) is detected.

PERFORMANCE EVALUATION

An important performance parameter of an infusion device is its flow rate accuracy. The flow rate of an infusion pump can be calculated by measuring the volume of solution delivered over a period of time. For example, a measuring cylinder can be used to collect the fluid infused over a period of, say 5 minutes, at a certain flow rate setting. Such a method is generally acceptable for most general purpose infusion devices. However, this method is not appropriate for measuring the accuracy of very low flow rate settings because it takes a very long time to collect enough solution to obtain an accurate volume measurement (e.g., it takes 2 hours to collect 10 ml of solution at 5 ml/hr setting). A container and weigh scale can replace a measurement cylinder if the density of the solution is known. Evaporation of fluid in the collection container will also affect the accuracy of a low-volume long-duration measurement. In addition, this method gives only the average flow rate over a period of time. No information regarding the flow pattern is obtained (flow fluctuation, bolus effect, etc.).

Another important performance parameter also relating to patient safety is to measure the downstream occlusion pressure alarm accuracy. This pressure is measured by connecting the IV line to a pressure meter. With the infusion pump running, the pressure inside the line will build up until it reaches the occlusion pressure alarm limit. It is important to leave an air buffer between the liquid line and the pressure meter should the meter response time is slow or not be able to measure wet fluid pressure.

Example 23.2

A measuring cylinder is used to collect fluid from an infusion pump during a flow rate performance evaluation test. During the test, 9.6 ml of fluid is collected over a period of 5 minutes. If the flow rate setting of the infusion pump is 120 ml/hr, what is the accuracy of the pump?

Solution:

From the test, 9.6 ml of fluid is infused in 5 minutes. Therefore, the calculated pump flow rate is 9.6 ml/5 min = 1.9 ml/min = 115 ml/hr. The percentage error of the infusion pump therefore is [(115 − 120)/120] x 100% = - 4.2%.

FACTORS AFFECTING FLOW ACCURACY

The following paragraphs describe common factors other than electronic component failures and mechanical wear and tear that affect the flow accuracy of infusion pumps.

Back pressure in the IV line that is too high can reduce the flow rate. Normal backpressure depends on the fluid flow rate, the diameter and length of the IV tubing, and the viscosity of the IV fluid. The smaller the inside lumen and the longer the tubing, the higher the backpressure; backpressure also increases with increased flow and fluid viscosity. In normal operation, the backpressure is low and will not affect the pump accuracy. A significant level of backpressure may be created when the IV line is kinked.

When the backpressure is too high, the pumping mechanism may not be able to overcome such pressure. For example, during high backpressure, if the occlusion created by the protruding fingers on the IV tubing is not high enough, fluid may leak backward at the location of the occlusion, causing a lower flow rate than its set value.

Another potential problem associated with high backpressure is bolus infusion. Because the flexible IV tubing is slightly elastic, its diameter will increase under high backpressure. Upon clearing the occlusion, the IV tubing will recoil to its original diameter thereby releasing the stored fluid along the length of the tubing. As a result, a large bolus of fluid may be infused into the patient.

For IV pumps using the peristaltic pumping mechanism, because the flow rate depends on the inner diameter of the IV tubing, variation of the inner diameter as well as the shape of the inner lumen (e.g., non-circular inner lumen) will affect the rate of infusion. It is therefore important to ensure that the inner diameter dimensions of IV lines used with peristaltic infusion pumps are manufactured within acceptable tolerances.

When an IV line is under prolonged use, because the segment of IV tubing under the protruding fingers is being compressed and because of the imperfect elastic property of the PVC tubing, the shape and therefore the inner cross-sectional area of the tubing will change. The deformation of the IV line with time (decrease in cross-sectional area) under the pumping mechanism will cause gradual reduction in the fluid flow rate. To overcome this problem, manufacturers often recommend users to periodically move the IV line so that a fresh segment of the PVC line is placed under the pumping mechanism. Other manufacturers insert a short segment of elastic silicon tubing (to be placed under the pumping mechanism) to minimize tubing deformation.

In general, among the different pumping mechanisms of the volumetric infusion category, the flow rate (averaged over a period of time) produced by the piston cylinder pump is the most accurate but it is the most expensive due to the special piston-cylinder cassette. In addition, a piston cylinder pump produces the highest bolus effect. The linear peristaltic pump, with its reasonably good accuracy and less bolus effect, is commonly used in general IV infusion since the IV lines without special cassette are less expensive. Some peristaltic pumps can even use ordinary gravity infusion sets. Given the high volume of infusions in a hospital, peristaltic pumps are the preferred choice of general infusion devices.

Theoretically, a syringe pump should produce an accurate and uniform flow pattern. In practice, however, under very low flow rate applications, the plunger in the syringe may stick to the wall of the cylinder until the pushing mechanism delivers enough force to overcome the static friction. Once the plunger is free, it will advance rapidly and stop after the force has been released, thereby pushing a bolus of solution into the patient. This sudden start and stop movement due to stiction is more prevalent and will repeat itself during low flow rate infusion. Use of low-stiction syringes will minimize this undesirable effect. As the flow rate of a syringe pump is proportional to the diameter of the syringe, any error that causes variation of the inner diameter of the syringe (tolerance in manufacturing, variation in diameter in different brands, use of wrong syringe size, etc.) will reduce flow rate accuracy.

COMMON PROBLEMS AND HAZARDS

IV infusion is invasive. Complications can be due to incorrect insertion, leading to blood leak, and infiltration to the surrounding tissue, leading to edema, pain, and tissue damage. An unprotected venipuncture site or improper site care may cause infection. Because the IV line is directly connected to the bloodstream, bacteria may get access to central circulation. To reduce the risk of infection, IV lines and indwelling catheters are often replaced periodically (e.g., every 72 hours)

Medication errors can be made at various stages throughout the medication process. Researchers have found that most errors occur during prescribing and administering of medications. Since the medication is being delivered into the patient, the most harmful errors occur during administration. There are many reported incidents due to IV dose errors leading to serious injuries and deaths Dose error reduction systems, described earlier, have substantially reduced infusion administration errors. In addition, human factor analysis has resulted in better user interface design, leading to safer use of infusion devices.

Occlusion pressure sensors to detect IV line occlusion, KVO infusion to prevent blood clots at the catheter tip, mechanisms to prevent free flow when the IV line is removed from the pump, and air-in-line detectors to prevent air embolism are now standard safety features in almost all infusion pumps.

Fluid overload and hypothermia from infusion flow rates that are too high and electrolyte imbalance from solutions that are too diluted or too concentrated are a few clinical related hazards of infusion.

BIBLIOGRAPHY

Ajmani, D. (2003). Evolution of infusion pumps—the search for safe and efficient medication delivery. MEEN. *Acuity Care Technology, Apr/May*, 38–44.

Breland, B. D. (2010). Continuous quality improvement using intelligent infusion pump data analysis. *American Journal of Health-System Pharmacy, 67*(17), 1446–1455.

Burdeu, G., Crawford, R., van de Vreede, M., & McCann, J. (2006). Taking aim at infusion confusion. *Journal of Nursing Care Quality, 21*(2), 151–159.

Center for Devices and Radiological Health, Food and Drug Administration, U.S. Department of Health and Human Services (2014), *Infusion Pumps Total Product Life Cycle-Guidance for Industry and FDA Staff.* https://www.fda.gov/media/78369/download (download on June 2020).

Chumbley, G., & Mountford, L. (2010). Patient-controlled analgesia infusion pumps for adults. *Nursing Standard, 25*(8), 35–40.

Cummings, K., & McGowan, R. (2011). "Smart" infusion pumps are selectively intelligent. *Nursing, 41*(3), 58–59.

Focus on five: Using infusion pumps safely. (2002). *Joint Commission Perspectives on Patient Safety, 2*(11), 11.

Gebhart, F. (2007, Aug. 20). Are smart pumps being used intelligently? [Online]. *Drug Topics.* Retrieved June 10, 2013 from http://drugtopics.modernmedicine.com

O'Grady, N. P., Alexander, M., Dellinger, E. P., Gerberding, J. L., Heard, S. O., Maki, D. G., . . . , & Weinstein, R. A. (2002). Guidelines for the prevention of intravascular catheter-related infections. *Morbidity and Mortality Weekly Report (MMWR: Recommendations and Reports), 51*(RR- 10), 1–29.

Guenter, P. (2001). Enteral feeding access devices. In P. Guenter & M. Silkroski, *Tube Feeding: Practical Guidelines and Nursing Protocols* (pp. 51–67). Gaithersburg, MD: Aspen Publishers.

Higgs, Z. C., Macafee, D. A., Braithwaite, B. D., & Maxwell-Armstrong, C. A. (2005). The Seldinger technique: 50 years on. *Lancet, 366*(9494), 1407–1409.

Kannan, S. (2001). Potential hazard with syringe infusion pump. *Anaesthesia, 56*(9), 913–914.

Leape, L. L., Bates, D. W., Cullen, D. J., Cooper, J., Demonaco, H. J., Gallivan, T., . . . , & Laffel, G. (1995). Systems analysis of adverse drug events. ADE Prevention Study Group. *The Journal of the American Medical Association, 274*(1), 35–43.

MacGillivray, N. (2009). Dr Thomas Latta: The father of intravenous infusion therapy. *Journal of Infection Prevention, 10* (Suppl 1), 3–6.

Neff, S. B., Neff, T. A., Gerber, S., & Weiss, M. M. (2007). Flow rate, syringe size and architecture are critical to start-up performance of syringe pumps. European Journal of *Anaesthesiology, 24*(7), 602–608.

Phelps, P. K. (2011). *Smart Infusion Pumps: Implementation, Management, and Drug Libraries* (2nd ed.). Bethesda, MD: American Society of Health-System Pharmacists.

Schmidt, N., Saez, C., Seri, I., & Maturana, A. (2010). Impact of syringe size on the performance of infusion pumps at low flow rates. *Pediatric Critical Care Medicine, 11*(2), 282–286.

Swedish Standards Institute. (2012). Medical electrical equipment–Part 2-24: *Particular requirements for the safety of infusion pumps and controllers* (2nd ed.).

International Electrotechnical Commission (IEC) 60601-2-24.

Wilkins, R. G., & Unverdorben, M. (2012). Accidental intravenous infusion of air: A concise review. *Journal of Infusion Nursing, 35*(6), 404–408.

Zhang, P., Wang, S. Y., Yu, C. Y., & Zhang, M. Y. (2009). Design of occlusion pressure testing system for infusion pump. *Journal of Biomedical Science and Engineering, 2*(6), 431–434.

Chapter 24

ELECTROSURGICAL UNITS

OBJECTIVES

- Explain the tissue response to electrosurgical current in terms of desiccation, fulguration, and cutting.
- Summarize the characteristics of the cut, coagulation, and blended electrosurgical waveforms.
- Differentiate the constructions and functions of active monopolar and bipolar electrodes, and the dispersive return electrode.
- Sketch the block diagram of an electrosurgical generator and explain the functions of each block.
- Analyze electrosurgical waveform generation circuits.
- Use quality assurance tests on electrosurgical unit generators.
- Describe common problems and potential hazards of electrosurgery.
- Discuss safety precautions during electrosurgical procedures.

CHAPTER CONTENTS

1. Introduction
2. Principle of Operation
3. Modes of Electrosurgery
4. Active Electrodes
5. Return Electrodes
6. Functional Building Blocks of electrosurgical Generators
7. Output Characteristics
8. Quality Assurance
9. Common Problems and Hazards

INTRODUCTION

An electrosurgical unit (ESU) delivers high-frequency electrical current through an active electrode to produce cutting and coagulation effects on tissues. The fundamental frequency of the output waveform for cutting is between 100 kHz and 5 MHz, which is within the radio frequency (RF) band. The heat pattern delivered to the tissue by the high frequency current creates the surgical effect. When appropriately modulated, the RF current can cut through tissues and cauterize bleeding blood vessels. The simultaneous cutting and hemostatic effect make an ESU useful for procedures on tissues with capillaries and on patients receiving a high dose of anticoagulant drugs.

The first electrosurgical device was invented by William T. Bovie using a spark-gap generator to produce the high frequency electrosurgical current. Its first clinical use was in 1926 to remove an intracranial tumor. Most ESUs today use solid-state technology with microprocessor-controlled output for better results and improved safety. Argon-enhanced ESU systems, using a jet of argon gas to cover the surgical site during electrosurgery, can provide rapid and uniform coagulation over a large bleeding surface and are therefore useful in surgeries on vascularized organs such as the liver, spleen, and lung. ESUs are also used in endoscopic procedures to perform ablation, desiccation, cauterization, and removal of tissues. Special ESU handpieces are designed for different applications.

The term cautery has been synonymously used with electrosurgery by many users. However, cauterization is only one of the functions of electrosurgery, and its principles are very different (cautery uses conductive heat to clot blood and destroy tissues). A Hyfrecator® is a lower power version of an ESU that is often found in medical offices to cauterize tissues. Hyfrecators do not employ return electrodes. They either operate in bipolar mode (current passes between the two tips of a forceps) or use the inherent capacitance of the patient as a current sink for the high frequency current.

PRINCIPLES OF OPERATION

An ESU is a RF current generator. Tissue damage at the surgical site is produced by the heating effect created by the RF current. Two electrodes, one called active and the other called passive, are used to apply the RF current from the ESU to the patient and to return the current back from the patient to the ESU. The passive electrode is also called the return or dispersive electrode. At the beginning of the procedure, the passive electrode is attached to the patient's body. The surgeon then applies the active electrode to the surgical site to achieve the surgical effect. The active electrode is usually a very small electrode; the return electrode has a large contact surface area with the patient. The high-frequency current passing through the tissue creates the surgical effect at the tip of the active electrode. The surgical effect is due to heat created by the RF current at the tissue-active electrode interface. The degree of heating in the tissue depends on the RF current density as well as the resistivity of the tissue. Figure 24-1 shows a typical setup of an electrosurgical procedure.

Different current densities create different effects on living tissues. Table 24-1 shows typical tissue effects at different levels of RF current density. In practice, a current density much higher than 400 mA/cm^2 at the surgical site is necessary to produce electrosurgical effect. However, to prevent tissue injury adjacent to the surgical site, the current density must be limited to below 50 mA/cm^2. This is achieved by attaching a large surface area electrode on the opposite side of the body of the patient to the active electrode so that the return current is dispersed over a larger area within the patient's body.

Figure 24-1 shows the active electrode applied to the surface of the tissue and the flow of current inside the body tissue when the return electrode is placed far from the active electrode. The density of the RF current flowing in the tissue closest to the active electrode is the highest, and it decreases rapidly (roughly inversely proportional to the square of the distance from the surgical site) at locations farther from the active electrode site.

Three different tissue effects can be created by an electrosurgical current at the active electrode site. They are desiccation, cut, and fulguration.

Desiccation

When a relatively small RF current flows through the tissue, it produces heat and raises the tissue temperature at the surgical site. At temperature above 60°C, protein will be denaturized. Heat will destroy and dry out the cells. This mechanism of tissue damage is called desiccation. Higher temperature may produce steam and bubbles and eventually turns the tissue into a

Table 24-1. Tissue Effect of RF Current Density.

Current Density	Tissue Effect
>50 mA/cm^2	Reddening of tissue
>80 mA/cm^2	Pain and blistering
>100 mA/cm^2	Intense pain
>400 mA/cm^2	Second-degree burn

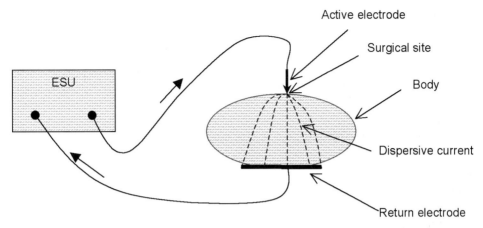

Figure 24-1. Electrosurgery Setup.

brownish colored mass. Desiccation is achieved by placing the active electrode in contact with the tissue and setting the ESU output to low power. Because desiccation is created by the heating (I^2R) effect, any current waveform may be used for desiccation. However, a cut waveform with a steady current will produce a more uniform effect.

Cut

By separating the active electrode by a small distance (about 1 mm) from the tissue and maintaining a few hundred volts or higher between the active and return electrodes, RF current will jump across the separation, producing sparks. Sparking creates intense heat, causing cells to explode. Such destruction of cells leaves behind a cavity. When the active electrode moves across the tissue, this continuous sparking creates an incision on the tissue to achieve the cutting effect. Note that it is not necessary for the surgeon to intentionally maintain a gap between the tip of the active electrode and the tissue because the steam created from the bursting cells creates the separation. In general, a high-frequency (e.g., 500 kHz) continuous sine wave is used to create the cutting effect. Cutting usually requires a high-output power setting.

Fulguration

To produce fulguration, the surgeon first touches the tissue with the energized active electrode and then withdraws it a few millimeters to create an air gap separation. As the active electrode moves away from the tissue, the high voltage creates an electric arc jumping across the active electrode to the tissue. This long arc burns and drives the current deep into the tissue.

Intermittent sparking does not produce enough continuous heating to explode cells, but it causes cell necrosis and tissue charring at the surgical site. Fulguration coagulates blood and seals lymphatic vessels. To achieve fulguration, most manufacturers use bursts of a short-duration damped sinusoidal waveform. The sinusoidal waveform is usually the same frequency used for cutting (e.g., 500 kHz), and the repetition frequency for the bursts is much lower (e.g., 30 kHz). Due to the large air gap, a higher voltage waveform is required to maintain the long sparks. Although the peak voltage is higher, fulguration requires less power than cutting does due to its low duty cycle. Table 24-2 summarizes the three mechanisms of electrosurgery.

MODES OF ELECTROSURGERY

The cut mode in electrosurgery applies a continuous RF waveform (sinusoidal or near sinusoidal) between the active and the return electrodes. The coagulation mode uses bursts of a higher voltage damped RF sinusoidal waveform (to create fulguration tissue effect). Instead of switching back and forth between cut and coagulation during a procedure, most ESUs have one or more blended modes that allow simultaneous cutting and coagulation.

A blended waveform has a lower voltage level but a higher duty cycle than the coagulation waveform has. Figure 24-2 shows an example of the cut,

Table 24-2. Mechanism of Electrosurgery.

	Tissue Effect	*Active Electrode*	*Power*
Desiccation	Heat dries up tissue, produces steam and bubbles. Turns tissue brown.	Monopolar or bipolar. In contact with tissue.	Low
Cut	Sparking produces intense heat, explodes cells leaving cavity. Incision on tissue caused by continuous sparking.	Monopolar. Electrode separated from tissue by a thin layer of steam.	High
Fulguration	Intermittent sparking does not produce continuous heating to explode cells. Heat causes necrosis to tissue. High voltage drives current deep into tissue, chars tissue	Monopolar. Electrode separated by an air gap.	Medium

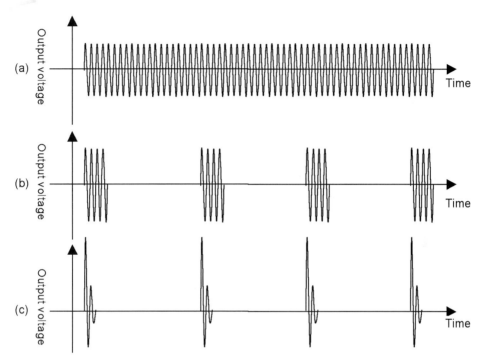

Figure 24-2. ESU Output Waveforms. (a) Cut, (b) Blended, (c) Coagulation.

blended, and coagulation output waveforms of an ESU. A blended mode with a higher duty cycle will have more cutting effect than one with a lower duty cycle.

The setup shown in Figure 24-1 with the active electrode and the large surface return electrode is called a monopolar operation. Instead of placing a separate return electrode away from the surgical site, a bipolar handpiece has both the active and return electrodes together (e.g., an ESU forceps). Bipolar electrodes are often used to perform localized desiccation on tissue. In Figure 24-3, the ESU is switched to bipolar mode to cauterize a section of a blood vessel before it is cut apart to avoid profuse bleeding.

In addition to the fundamental cut, coagulation, and blended modes of operation, some ESU manufacturers provide additional modes of operation by modifying the waveform characteristics of these fundamental modes. For example, one manufacturer added a fluid mode for urology procedures by providing a higher voltage at the onset to initialize the cutting effect in non-conductive fluid (such as glycine used in prostate transurethral resection procedures). Another manufacturer included a laparoscopic mode to limit the maximum ESU voltage (e.g., below 4000 V) for safety purpose.

Argon enhanced coagulation systems can provide rapid, uniform coagu-

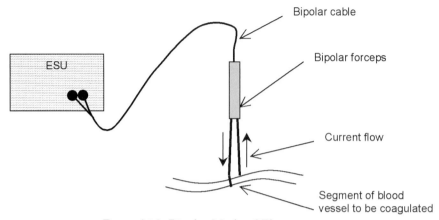

Figure 24-3. Bipolar Mode of Electrosurgery.

lation over large bleeding surfaces and promote better eschar formation. In argon-enhanced coagulation, the monopolar electrosurgical current ironizes the argon gas, creating an arc that flows over the electrode tip to the tissue surface. The handpiece of an argon-enhanced coagulator contains the argon gas supply tubing and the electrosurgical current conductor. During use, it is held at a distance (about 1 cm) from the tissue.

Table 24-3 lists the characteristics of ESU modes of operation. The crest factor (last column) represents the degree of hemostasis. It is defined as the peak voltage amplitude of the ESU waveform divided by its root mean square voltage. For a continuous sine wave, the crest factor is

$$\frac{V_P}{V_{rms}} = \frac{V_P}{V_P/\sqrt{2}} = \sqrt{2} = 1.41.$$

Since a pure sine wave has little or no hemostatic effect on tissues, most manufacturers use a lightly modulated sine wave to achieve a small degree of hemostatic effect in the cut mode. The crest factor of the coagulation waveform is the highest (about 9) since it has the largest peak voltage but the smallest duty cycle. In general, the higher the crest factor, the more hemostatic effect the ESU waveform will have on tissues.

ACTIVE ELECTRODES

ESU active electrodes for monopolar operations come in different forms and shapes. The most common active electrode is the flat blade electrode, which can be used to perform cutting and coagulation. Some of the other

Table 24-3. Characteristics of ESU Operation Modes.

	Effect	*Waveform*	*Voltage*	*Power*	*Crest Factor*
Monopolar					
Cut	Pure incision plus slight hemostatic effect	Continuous unmodulated sine wave to lightly modulated sine wave	Low	High	~1.41 to 2
Coagulation	Desiccation or fulguration	Burst of damped sine wave	High	Low	~9
Blended	Cut and coagulation	Burst of medium duty factor sine wave	Medium	Medium	Between cut and coagulation
Bipolar					
Coagulation	Desiccation	Continuous unmodulated sine wave	Lowest	Lowest	1.41

commonly used active electrodes are the needle, ball, and loop electrodes. Ball electrodes are usually used for desiccation (by pressing the electrode against the tissue and passing the RF current through the tissue). The loop electrode, with its conductive wire loop, is used to remove protruded tissues such as a nodule. The metal tips of the electrodes (Figure 24-4b) are single patient use disposable units. The electrode handles may be multiple use or single use. The handle part of the electrode may have one or more switches to activate the ESU cut or coagulation. A foot switch operated by the surgeon may be used instead of the hand-switched pencil. The combination of an ESU handle and an active electrode (Figure 24-4a) is often referred to as an ESU pencil or a hand-switched ESU pencil if a switch is located on the handle.

RETURN ELECTRODES

Although the function of the active electrode is to create the surgical effects, the return electrode (or passive electrode) in monopolar ESU operations provide the return path for the ESU current. As mentioned earlier, the maximum RF current density level to avoid causing any tissue injury is 50 mA/cm^2. A large surface area electrode is therefore required to limit the current density below this safe level in tissues away from the surgical site, including those in contact with the return electrode.

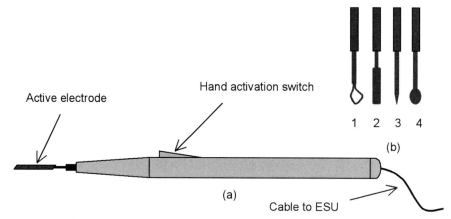

Figure 24-4. (a) Hand-Switched ESU Pencil with a Flat Blade Electrode; (b) Monopolar Tips: (1) Loop, (2) Flat Blade, (3) Needle, 4) Ball.

There are many types of return electrodes for ESU procedures. Bare metal plates placed under and in contact with the patient were used in early days. It was noted, however, that burns (primarily heat burns) and tissue damage occasionally occurred at the return electrode sites. Investigations revealed that the main cause of such patient injuries was poor electrode-skin contact (causing high electrode-skin contact resistance) or insufficient contact surface area between the electrode and the patient (causing high current density at electrode-skin interface). In addition, it was also noted that burns often appeared in the form of rings at the skin surface. Laboratory experiments showed that the current density at the skin–return electrode interface is highest around the rim of the electrode. Figure 24-5 shows the current density distribution of such an experiment measured just below the skin surface. This occurrence is due to the fact that electrons are negatively charged particles; when they are allowed to move freely in a conductive medium, they will repel each other while traveling toward the return electrode. Most will therefore be collected at the perimeter of the return electrode. This phenomenon is known as the "skin effect" in electrical engineering, where the current density of high-frequency current in a conductor is very much higher at the surface of the conductor than in its core. The higher current density (and hence higher temperature) at the electrode perimeter is the reason why return electrode burns often form a ring corresponding to the shape of skin contact at the return electrode site.

Today, flexible conductive gel pads are used for ESU return electrodes. A conductive gel pad electrode has a self-adhesive surface to avoid shift and falling off and is flexible to fit the contour of the patient's body. Return electrodes are designed so that, under normal use, no skin burn will occur at the

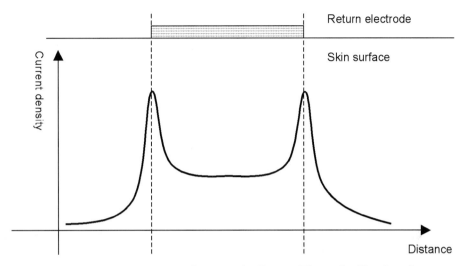

Figure 24-5. Current Density Crossing the Return Electrode–Skin Interface.

return electrode site. To ensure patient safety, technical standards are in place specifying the performance of return electrodes. For example, AAMI/IEC standards on ESU stipulate that the overall tissue-return electrode contact resistance shall be below 75 Ω. In addition, no part of the tissue in contact with the return electrode shall have more than a 6°C temperature increase when the ESU is activated continuously for up to 60 seconds with output current up to 700 mA.

Due to problems associated with burns, special monitoring devices are often built into ESUs to monitor the integrity of the return electrode path. If the integrity is breached, an alarm will sound and the ESU output will be disabled to prevent patient injury. Two levels of monitoring are often available for high output power ESU (e.g., output greater than 50 W). The first is return electrode monitoring and the second is return electrode quality monitoring.

Return Electrode Monitor

A return electrode monitor system monitors the return path of the electrode to the ESU. It detects the continuity of the return electrode cable from the electrode to the ESU. In a typical return electrode monitor system, a double conductor cable and a low-frequency, low-current (e.g., 140 kHz, 3 mA) isolated source from the ESU are used to measure the resistance of the return cables (Figure 24-6a). High resistance (e.g., > 20 Ω) due to a broken wire or poor connection between return electrode and the ESU, will trigger

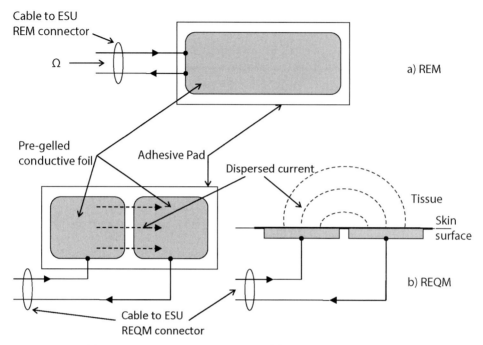

Figure 24-6. (a) Return Electrode Monitor, and (b) Return Electrode Quality Monitor.

the return electrode monitor alarm.

Return Electrode Quality Monitor

The return electrode monitor just described measures only the continuity of the return electrode cable, not the quality of contact between the electrode and the patient. A return electrode quality monitor (REQM) system monitors both cable continuity and electrode-skin contact quality. Figure 24-6b illustrates the principle of the REQM. In REQM, a dual conductive pad electrode is used. The right-hand side diagram in Figure 24-6b shows the cross-sectional view of the electrode–skin interface. The small monitoring current flows from the ESU REQM circuit to one of the conductive pads, passes through the two electrode–skin interfaces, and returns to the ESU via the second pad. Too high a REQM resistance (e.g., greater than 140 Ω) suggests poor electrode–skin contact or open circuit return cable; too low a resistance (e.g., less than 5 Ω) suggests a short circuit between the two conductive pads. In addition, some machines may sound an alarm if the REQM detects a large change in the resistance during use (e.g., resistance increase by more than 40% from the initial reference value).

For an ESU with REQM, both the REQM current and the electrosurgi-

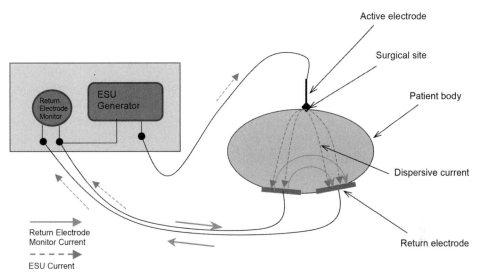

Figure 24-7. Functional Diagram of an ESU with REQM.

cal current will flow in the same return electrode cable although the frequency and amplitude of the REQM current are lower. Figure 24-7 shows the functional diagram of both circuits in the ESU generator.

The return electrodes described earlier are conductive electrodes. Since high frequency RF current is used in electrosurgery and capacitive impedance decreases with frequency, capacitive coupled return electrodes may be used. A typical capacitive coupled return electrode consists of a large sheet (e.g., 1.0 m × 0.5 m) of flexible conductor enclosed by a thin insulating material (e.g., urethane). The sheet forms a large electrode capacitively coupling the patient to the return path of the electrosurgical circuit. Unlike conductive electrodes, which are applied directly on the patient using adhesive, capacitive electrodes are not in direct contact with the patient. A capacitive electrode is often placed on the operating room table and covered with a protective cover sheet under the patient. The electrode is reusable; the cover sheet is replaced before a new procedure. Because it is not applied with adhesive directly onto the patient, it is used for patients with frail skin or extensive skin damage. Similar to conductive electrodes, skin burns may occur at the return electrode sites.

FUNCTIONAL BUILDING BLOCKS OF ESU GENERATORS

The spark-gap ESU generator developed in the 1920s consists of a step-up transformer T1, which increases the 60-Hz 120-V line voltage to above

2000 V (Figure 24-8). As the sinusoidal voltage at the secondary of T1 increases from zero, an electrical charge accumulates in the capacitor C1 and the gas inside the spark gap (a gas discharge tube) starts to ionize until an arc is formed between its electrodes. Arcing (or sparking) of the spark gap resembles the closing of a switch in the series resonance circuit formed by C1, L1, and the impedance of the spark gap. The fundamental frequency of the arcing current is approximately equal to the resonance frequency of L1/C1. The voltage amplitude of this high-frequency oscillation will decay until the arc is extinguished. Proper choice of L1 and C1 produces an RF-damped sinusoidal waveform that occurs twice within one period of the 60-Hz input signal. This RF damped sinusoidal waveform is coupled to the output circuit by induction between L1 and L2. The output level is selected by the taps selection on L2. The RF chokes L3 and L4 (or RF shunt capacitor C4) are used to block the RF signal from entering the power supply. Spark-gap generators were primarily used for coagulation or cauterization.

Spark-gap ESUs were commonly used until the early 1980s, when they began to be replaced by solid-state generators. In a solid-state ESU, the RF frequency (e.g., 500 kHz) and the burst repetition frequency (e.g., 30 kHz) are generated by solid-state oscillators. The shape of the ESU waveform (cut, blended, or coagulation) is created by combining the frequencies of these two oscillators. The waveform is then amplified by a power amplifier and the voltage is increased by a step-up transformer. The output of an ESU can go up to 1000 W, 9000 V (peak to peak open circuit voltage), and 10 A. Figure 24-9 shows the simplified functional block diagram of a solid-state ESU.

The output stage of an ESU is shown in Figure 24-10. In the circuit, the waveform created by the gating and wave-shaping circuit is fed into the base of a power amplifier Q1, and the output of the amplifier connects to the primary winding of the output transformer T1. The transformer and the power

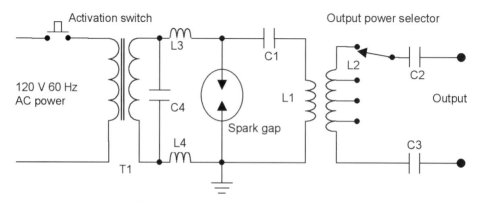

Figure 24-8. Spark-Gap ESU Generator.

amplifier circuit are connected to a 200 V DC power source. For high-power output ESUs, a number of power transistors connected in parallel form the output circuit. Each of these transistors shares a portion of the output power. The ESU waveform, after steps up by the output transformer to several thousand volts, is fed across the active and return electrodes via a pair of capacitors C1 and C2. These capacitors behave like a short circuit to RF but block low-frequency (60-Hz) leakage current to the patient.

The ESU output circuit shown in Figure 24-10 is considered an isolated output ESU because there is no connection from the patient circuit (the secondary of the output transformer) to the power ground. Theoretically, for an isolated output ESU, a person touching the active electrode but not the return electrode will not get a shock or burn when the ESU is energized. Due to the high frequency and nonzero leakage capacitance, however, if the person also touches a grounded object, some RF current will flow from the active electrode to the person and return to the ESU via this ground leakage path. High-frequency leakage current may be on the order of magnitude of a few tens of mA.

OUTPUT CHARACTERISTICS

Table 24-4 lists the output characteristics of a conventional ESU. Figure 24-11 illustrates the output characteristics of the ESU cut waveform at differ-

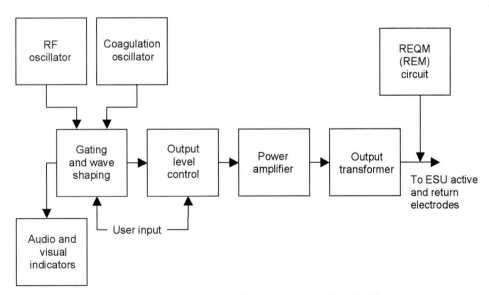

Figure 24-9. Functional Block Diagram of an ESU.

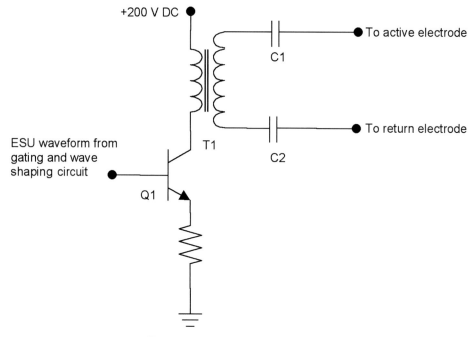

Figure 24-10. ESU Output Circuit.

ent values of patient load. Note that according to the output characteristics, the ESU is rated to produce 300 W of output only when the patient load is at 300 Ω. The output power (without dynamic control) is reduced to 180 W when the patient load becomes 800 Ω. According to the ESU output characteristics, the output power decreases as the patient load increases. Because the tissue impedance depends on the type of tissue as well as the condition of the tissue, this may create problems during the operation because the output power at a particular setting will fluctuate with the tissue impedance. To overcome this problem, some manufacturers have produced ESUs that can measure the tissue impedance and automatically restore the output power to the set value, this is termed dynamic control by some manufacturers. Figure 24-11 shows the ESU characteristics with and without dynamic control. Table 24-5 tabulates the approximate impedance of different tissues as seen across the active and return electrodes of an ESU.

In most electrosurgical procedures, the active electrode is energized only intermittently and each activation lasts for a short period of time (e.g., 15 sec for cutting in general surgery). Table 24-4 shows the peak to peak open circuit voltage of different modes of operation. When the current starts to flow (i.e., an arc has been established), however, the voltage across the active and return electrodes will drop substantially.

Table 24-4. ESU Output Characteristics

Mode	Waveform	Max. P-P* Open Circuit Voltage (V)	Rated Patient Load (Ω)	Output Power (at rated load) (W)
Cut	500 kHz sinusoidal	3000	300	300
Blend 1	500 kHz burst of sinusoidal at 50% duty cycle repeating at 30 kHz	3500	300	250
Blend 2	500 kHz burst of sinusoidal at 37.5% duty cycle repeating at 30 kHz	3700	300	200
Blend 3	500 kHz burst of sinusoidal at 25% duty cycle repeating at 30 kHz	4000	300	150
Coagulation	500 kHz burst of damped sinusoidal repeating at 30 kHz	7000	400	120
Bipolar	500 kHz sinusoidal	800	100	70

*Peak to Peak

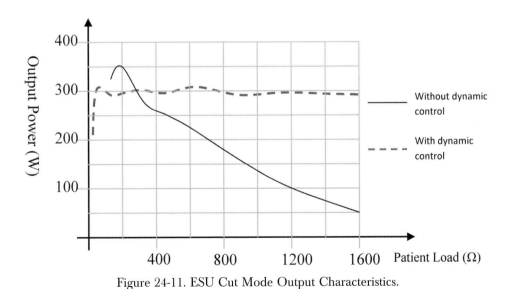

Figure 24-11. ESU Cut Mode Output Characteristics.

Table 24-5. ESU Tissue Impedance

Tissue	Approximate Impedance
Prostate in nonconductive solution (e.g., glycine)	20 to 1500 Ω
Muscle/liver	500 to 2000 Ω
Bowel	1000 to 2500 Ω
Gall bladder	1500 to 3000 Ω
Mesentery/omentum	2000 to 3500 Ω
Fat/scar/adhesions	3000 to 5000 Ω

QUALITY ASSURANCE

Since an ESU delivers high-energy therapeutic current, it is essential to ensure that the machine is safe and operating according to its designed specifications. Other than general electrical safety inspection, the following performance tests should be carried out periodically.

Output Power Verification Test

The output power of an ESU should be measured against the manufacturer's specifications. Figure 24-12 shows the setup to measure the ESU output power. The output waveform can be sampled across the sample resistor RS and displayed on the oscilloscope. The output voltage V0 of the ESU is calculated from the resistance values by the equation

$$V_0 = \frac{R + R_S}{R_S} \, V_S.$$

If the output voltage is a sine wave, the power output may be calculated from the equation

$$P = \frac{V_0{}^2}{R + R_S}.$$

Note that the load resistance R_L is equal to $(R + R_S)$ and both should be of sufficient power rating to withstand the ESU output. Since the inductive impedance is proportional to the product of the ESU frequency and the inductance, due to the high ESU frequency all resistors used in the testing circuit need to have very low inductance (or using non-inductive resistors).

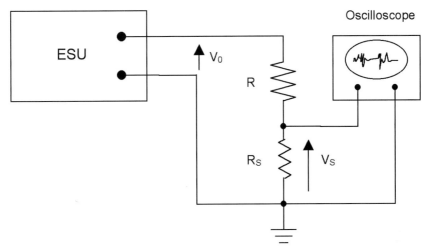

Figure 24-12. ESU Output Power Measurement.

High-Frequency Leakage Test

High-frequency leakage refers to the current flowing from either the active electrode to ground or the return electrode to ground when the ESU output is activated. Ideally, the amount of leakage current from an isolated-output ESU should be zero. However, due to the nature of the high fre-quency, a significant amount of capacitive leakage current will flow between the active electrode and the ground as well as between the return electrode and the ground. The high frequency leakage current may cause a secondary burn to the patient. For example, a patient will suffer a secondary burn to his/her hand when the patient's hand is touching a grounded object while the ESU is energized. Figure 24-13b shows the setup to measure the high-frequency leakage from the active electrode to ground. To measure the leak-age from the active electrode to ground, the load resistor R_L (e.g., 200 Ω) is connected to the active electrode connection of the ESU and the return elec-trode connection is left open. Alternatively, to measure the leakage from the return electrode to ground, the load resistor is connected to the return elec-trode connection of the ESU and the active electrode connection is left open.

Other than measuring the leakage current, the effectiveness of isolation can be found by measuring the power dissipated in the load resistance R_L. Percentage isolation is a common value to represent the degree of isolation. It is defined as

$$\% \text{ } Isolation = (1 - \frac{P_{isolation}}{P_{normal}}) \times 100\%.$$

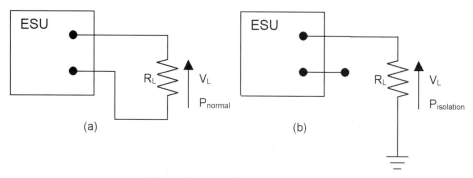

Figure 24-13. ESU Isolation Test.

Some manufacturers (and standards) call for the percentage isolation to be greater than 80% for a load resistance R_L within the range of 100 to 1000Ω. Special testers with built-in potential dividers, variable patient load, and switchable configurations are available to facilitate these measurements.

COMMON PROBLEMS AND HAZARDS

Electrosurgery is a potentially dangerous procedure. Users must understand its principles of operation and limitations and be fully aware of the hazards and safe operation. Hazards associated with electrosurgery may be grouped into four different categories: burns, fire, muscle/nerve stimulation, and EMI.

Burns

Burns may be inside the patient or on the surface of the skin or, occur on patient or the clinician in contact with the patient or the ESU accessories. Some common burn hazards follow:

- Skin burns at the return electrode site are one of the more common hazards for patients under electrosurgical procedures. The main causes of skin burns are poor electrode-skin contact, incorrect return electrode placement, inadequate site preparation, and pressure points (dents, creases, or bends on the electrode contact surface) that create a low resistance pathway for the ESU current.
- Electrosurgical injuries also occur at sites other than the return or active electrode. ESU current from damaged insulation of the active electrode and cable can cause burns to patients or operating room personnel.
- Internal tissue burns are caused by the concentration of ESU current

along a low resistance path such as a metal implant or a pacemaker lead wire near the active or return electrodes sites.

- For grounded ESUs or ESUs with isolation failure, RF current may flow through a secondary ground path on the patient (e.g., a patient's arm may receive a burn at the location where it is touching a grounded object).
- In endoscopic or laparoscopic procedures, an insulation failure on the shaft of the ESU handpiece will cause tissue burn when such failure creates a secondary conduction path between the active electrode and the tissue. Capacitive coupled leakage currents through electrode insulation (e.g., between the shaft of an ESU handpiece and the metal sheath of a laparoscopic trocar) with enough intensity will cause burns.
- Too high a power setting and too long an activation period (e.g., during a liver tumor ablation procedure) will cause burn on the return electrode site when an undersized return electrode was used or the return electrode was not properly applied.
- Patient or staff burns may be caused by an activated ESU pencil when it was inadvertently energized (e.g., someone accidentally stepped on the ESU foot activation switch) while touching the patient or a staff member.

Fire and Explosion

An electrosurgical procedure produces sparks and arcing. The sparks may ignite flammable materials such as body hair, cotton drapes, or a pool of alcohol used for disinfection. This situation is worsened under an enriched oxygen environment, which is commonly found in operating room areas.

There have been incident reports on cases of explosion inside the abdominal cavity when the ESU ignited flammable bowel gas inside the patient.

Muscle and Nerve Stimulation

The reason ESU frequency is above 100 kHz is to avoid muscle and nerve stimulation. Under normal circumstances, muscle and nerve fibers are not triggered by current higher than 100 kHz. However, studies have shown that arcing may produce current with lower frequency, which can stimulate nerve or muscle fibers. As a precaution, it is contraindicated to perform electrosurgery near major nerve fibers.

Electromagnetic Interference

An ESU is an RF source. Although the machine may be shielded to prevent radiation and conduction of EMI, the electrodes and cables act as antennae to broadcast the RF frequencies. Older medical devices, with lower electromagnetic immunity, can be adversely affected by the EMI from electrosurgery. Devices may reset, produce errors, or switch to another mode of operation under EMI influence. Improperly grounded devices are especially vulnerable to EMI.

Smoke Plume

Another problem associated with the use of ESUs is the hazardous smoke plumes formed by the arcing and vaporization of cells and tissues. Analysis of ESU smoke plume samples by electronic microscopy revealed irregular particles consistent with cellular components. The smoke plume (which may contain toxic chemicals, cellular material and viruses) released into the air in the operating room poses health risks to both the patient and the operating room staff when inhaled. As well, smoke plume will obscure visibility at the surgical site. Smoke evacuation systems (consisting of vacuum module, filters to remove submicron particles, tubing, and connectors) that capture and prevent the smoke plume from escaping to the surrounding area, are now standard configurations in both open and endoscopic electrosurgical procedures.

BIBLIOGRAPHY

Abu-Rafea, B., Vilos, G.A., Al-Obeed, O., AlSheikh, A., Vilos, A. G., & Al-Mandeel, H. (2011). Monopolar electrosurgery through single-port laparoscopy: A potential hidden hazard for bowel burns. *Journal of Minimally Invasive Gynecology, 18*(6), 734–740.

Al Sahaf, O. S., Vega-Carrascal, I., Cunningham, F. O., McGrath, J. P., & Bloomfield, F. J. (2007). Chemical composition of smoke produced by high-frequency electrosurgery. *Irish Journal of Medical Science, 176*(3), 229–232.

Boulay, B. R., & Carr-Locke, D. L. (2010). Current affairs: Electrosurgery in the endoscopy suite. *Gastrointestinal Endoscopy, 72*(5), 1044–1046.

Bovie, W. T., & Cushing, H. (1928). Electrosurgery as an aid to the removal of intracranial tumors with a preliminary note on a new surgical-current generator. *Surgical Gynecology and Obstetrics, 47*, 751–784.

Crossley, B. (2013). Video integration systems and electrosurgical units. *Biomedical Instrumentation & Technology, 47*(1), 81.

Gilbert, T. B., Shaffer, M., & Matthews, M. (1991). Electrical shock by dislodged spark gap in bipolar electrosurgical device. *Anesthesia and Analgesia, 73*(3), 355–357.

Lenz, L., Tafarel, J., Correia, L., Bonilha, D., Santos, M., Rodrigues, R., . . . , & Rohr, R. (2011). Comparative study of bipolar electrocoagulation versus argon plasma coagulation for rectal bleeding due to chronic radiation coloproctopathy. *Endoscopy, 43*(8), 697–701.

Martin, S. T., Heeney, A., Pierce, C., O'Connell, P. R., Hyland, J. M., & Winter, D. C. (2011). Use of an electrothermal bipolar sealing device in ligation of major mesenteric vessels during laparoscopic colorectal resection. *Techniques in Coloproctology, 15*(3), 285–289

Massarweh, N. N., Cosgriff, N., & Slakey, D. P. (2006). Electrosurgery: History, principles, and current and future issues. *Journal of the American College of Surgeons, 202*(3), 520–530.

Morris, M. L., Tucker, R. D., Baron, T. H., & Song, L. M. (2009). Electrosurgery in gastrointestinal endoscopy: Principles to practice. *American Journal of Gastroenterology, 104*(6), 1563–1574.

Pollack, S. V., Carruthers, A., & Grekin, R. C. (2000). The history of electrosurgery. *Dermatologic Surgery, 26*(10), 904–908.

Vellimana, A. K., Sciubba, D. M., Noggle, J. C., & Jallo, G. I. (2009). Current technological advances of bipolar coagulation. *Neurosurgery, 64*(3), 11–19.

Weld, K. J., Dryer, S., Ames, C. D., Cho, K., Hogan, C., Lee, M., . . . , & Landman, J. (2007). Analysis of surgical smoke produced by various energy-based instruments and effect on laparoscopic visibility. *Journal of Endourology/Endourological Society, 21*(3), 347–351.

Chapter 25

PULMONARY FUNCTION ANALYZERS

OBJECTIVES

- Explain the mechanics of breathing.
- Describe common respiration parameters.
- Explain the principle of operation and construction of medical spirometers.
- Define BTPS standardized gas volume.
- Explain the principles of bedside respiration monitoring using the impedance pneumography and thermistor methods.
- Sketch a block diagram of a respiration monitor using the method of impedance pneumography and explain the functions of each block.
- Examine factors affecting signal quality, accuracy, and patient safety in respiratory monitoring.

CHAPTER CONTENTS

1. Introduction
2. Mechanics of Breathing
3. Parameters of Respiration
4. Spirometers
5. Respiration Monitors
6. Common Problems and Hazards

INTRODUCTION

The primary function of the lungs is to exchange gases between the inspired air and the venous blood. Air is inhaled by voluntary or involuntary action and is presented to one side of the membrane of the alveoli, with venous blood on the other side. Gas exchange between air and blood occurs across this membrane. In the process, carbon dioxide is removed and oxygen is introduced into the bloodstream. For a normal adult, this blood-gas barrier is less than 1 µm thick and has a total surface area of about 100 m^2.

Disturbances of the respiratory system can be caused by a number of factors within the system or by other disorders. Diagnosis of respiratory disorders therefore can provide information about the well-being of the respiratory system as well as other organs or body functions. The rhythmic action of breathing is initiated in the respiration centers of the pons and medulla. The level and rate of respiration are controlled by the partial pressure of carbon dioxide and oxygen as well as the pH of the arterial blood. For example, a decrease in blood pH (e.g., due to increases in metabolism), accumulation of carbon dioxide in the arterial blood, and arterial hypoxemia will increase respiration.

This chapter introduces some methods of monitoring respiration patterns and parameters in pulmonary function laboratories and in the clinical environment.

MECHANICS OF BREATHING

The lung is elastic and will collapse if it is not held expanded. At the end of expiration or inspiration, the pressure inside the lung (or alveolar pressure) is the same as the atmospheric pressure, whereas the pressure outside the lung in the intrapleural space is below atmospheric pressure (approximately −5 cmH2O or −0.5 kPa). This negative pressure keeps the lung inflated. If air is introduced into the intrapleural space (e.g., punctured lung), the lung will collapse and the chest wall will move outward. This disorder is called pneu-mothorax.

The most important muscle for inspiration is the diaphragm. When it contracts, the abdominal contents are forced downward and forward. This action increases the vertical dimension of the chest cavity. In addition, the external intercostal muscles contract and pull the rib cage upward and forward, causing a widening of the transverse diameter of the thorax. In normal tidal breath (or passive breathing), the diaphragm descends by about 1 cm, but in forced breathing, a total descent of up to 10 cm may occur. Under active breathing (e.g., during heavy exercise), the abdominal muscles play an important role in expiration by pushing the diaphragm upward. The internal intercostal muscles assist active expiration by pulling the ribs downward and inward, thus further decreasing the volume of the thoracic cavity. Diseases that cause problems in these muscles or the nerves that innervate these muscles create disorders in breathing. In addition, the following are some of the diseases that compromise respiration:

- Reduced alveolar elasticity in a patient with emphysema
- Bronchoconstriction associated with asthma or chronic obstructive pulmonary disease (COPD)
- Inflammation in asthma, chronic bronchitis, COPD, and bronchiolitis
- Excess mucus production associated with asthma, chronic bronchitis, and cystic fibrosis

In all the preceding cases, there is an increase in airway resistance and a decrease in maximal expiratory flow. In restrictive lung disease, the volume of air entering the lungs is diminished. Patients with restrictive lung disease have difficulty fully expanding their lungs with air. In obstructive lung disease, airflow out of the lungs is diminished. Patients with obstructive lung disease have hard time breathing out. Both obstructive and restrictive lung disease patients share the same symptom, which is shortness of breath with exertion.

PARAMETERS OF RESPIRATION

Parameters of respiration include lung capacities, respiration rate, intrathoracic pressure, airway resistance, and lung compliance (or lung elasticity). In addition to these parameters, respiration waveform as well as end-tidal carbon dioxide concentration and its variations can all provide useful information in the assessment and disease diagnosis of the respiratory and related systems.

Figure 25-1 shows the volumes or capacities of respiration. The tidal volume (TV) measures the volume of inspired or expired gas during normal breathing. It is about 500 ml for a normal adult at rest. Inspiratory reserve volume (IRV) is the maximum amount of gas that can be inspired from the end-inspiratory level (or peak of the tidal volume). The sum of TV and IRV forms the inspiratory capacity (IC). The expiratory reserve volume (ERV) is the maximum amount of gas that can be exhaled from the end-expiratory level (or trough of the tidal volume). The sum of IC and ERV, which is the maximum volume of gas that the lung can expel or inhale, is the vital capacity (VC). The residual volume (RV) is the amount of gas remaining in the lung at the end of maximum expiration. This is the amount of gas that can-

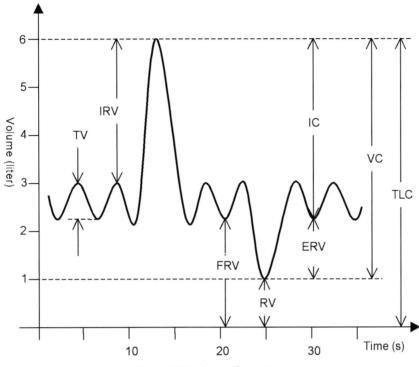

Figure 25-1. Lung Capacities.

not be squeezed out of the lung. The total lung capacity (TLC) is the sum of RV and VC. These parameters, as well as the ratios of some of them (e.g., RV/TLC), are used to assess the healthiness of the respiratory system.

Airway resistance measures the ease of airflow during inspiration and expiration through the bronchi and bronchioles. This is expressed as the pressure difference between the mouth and the alveoli per unit of airflow. The normal value is about 1 to 2 cmH2O per liter per second of flow at normal flow rate (e.g., 14 L/s). This resistance becomes higher at higher flow rates. Lung compliance measures the ability of the alveoli to expand and recoil to its original state during inspiration and expiration. The normal lung at rest expands by about 200 ml when the intrapleural pressure falls by 1 cmH2O. Therefore, the compliance of the lung is 200 ml/cmH2O. At high lung volumes, the lung is less easy to expand and thus its compliance falls. When the airway is restricted, the air resistance increases. When the lung becomes more fibrous, it loses its compliance.

To produce work to move the chest wall and force air along the airways, the respiratory muscle must consume oxygen. The total expenditure of energy necessary to accomplish the action of breathing is called the work of breathing (WOB). The oxygen cost of breathing is often used to indicate the work of breathing. WOB can be computed by multiplying the pulmonary pressure by the change in pulmonary volume (area enclosed by the pressure-volume loop). The oxygen consumption by the WOB accounts for about 5% of the total body oxygen consumption for a healthy individual. In patients with obstructive lung disease, however, the resistance to airflow becomes very high even at rest and therefore the work of breathing can be five or ten times its normal value. Under these conditions, the oxygen cost of breathing may become a significant fraction of the total oxygen consumption. Patients with a reduced compliance of the lung also have a higher work of breathing due to the stiffer structures. These patients tend to use shallower but more frequent breaths to reduce their oxygen cost of ventilation. However, the air exchange is not efficient in shallow breathing due to the fixed volume of air in the anatomic dead space in the lung, bronchi, and bronchioles.

One of the most useful tests in a pulmonary function laboratory is the analysis of a single forced expiration. The patient makes a full inspiration and then exhales as hard and as fast as possible into a spirometer (a flow and volume measurement device). The volume measured is called the forced vital capacity (FVC). FVC is usually less than the VC, which is obtained at slow expiration. The volume exhaled within the first one second is called the forced expiratory volume, or FEV_1. In obstructive lung disease (such as emphysema), due to high airway resistance, both FEV_1 and the ratio FEV_1/FVC are reduced. In restrictive lung disease (such as sarcoidosis), due to the limited lung expansion, FVC is low, but because the airway resistance is normal, the ratio FEV_1/FVC is high. Another index that can be derived from a forced expiration is the maximal midexpiratory flow ($FEF_{25-75\%}$), which is obtained by dividing the volume between 75% and 25% of the FVC by the corresponding elapsed time. This is a sensitive parameter to detect airway obstruction in early chronic obstructive lung disease. Figure 25-2 shows typical records of forced expiratory measurements from a normal individual, a patient with obstructive pulmonary condition, and with restrictive pulmonary condition. FEV_1 is a useful screening procedure to assess lung function and the efficacy of bronchodilator therapy and in following the progress of patients with asthma or chronic obstructive lung disease.

The flow-volume curve is a plot of the air flow against volume. Flow-volume curves of forced expiratory measurements from patients with different pulmonary conditions are shown in Figure 25-3. The positive portion of the curve is from expiration and the negative portion is from inspiration. From the curve, the peak expiratory flow (PEF) and peak inspiratory flow (PIF) are obtained. The characteristic shape of the flow-volume curve can be used to differentiate different pulmonary conditions.

The functional residual volume (FRV) is the volume of air in the lungs

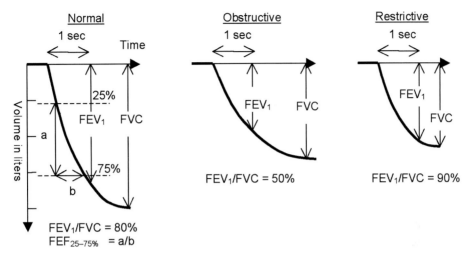

Figure 25-2. Forced Expiratory Volume Measurement.

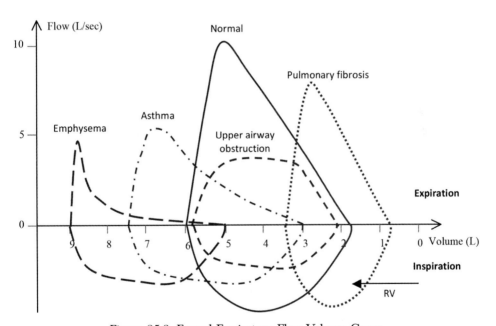

Figure 25-3. Forced Expiratory Flow Volume Curve.

at the end of expiration that is also the volume of air remaining in the lungs between breaths. It is an important lung function because it changes markedly in some pulmonary diseases. FRV is measured using an indirect method called the helium dilution method. In this method, a container of known volume (V) is filled with a mixture of air and helium of initial concentration C_{iHe}. The patient first breathes normally for a few cycles. At the end of the last expiration (the volume of gas inside the lungs is the FRV), the patient starts and continues to breathe from the container. After several breaths, the gas in the container is diluted and mixed thoroughly with the gas in the lung. FRV can be derived by equating the contents of helium gas before and after its inhalation. If the final helium concentration of the mixed gas is C_{fHe}, FRV can be calculated from the equation

$$FRV = \left(\frac{C_{iHe}}{C_{fHe}} - 1 \right) V.$$

The same method can be used to measure the residual volume (RV) of the lung. In RV measurement, the patient is asked to exhale as much gas as possible from the lung and then start breathing into the container of helium mixture. The residual volume (RV) is obtained by the same equation.

During inspiration, some of the air that a person breathes never reaches the alveoli for gas exchange to take place. During expiration, this volume of air expires first to the atmosphere before the air from the alveoli. This air that stays in the upper airway is called the anatomical dead space air. Anatomic dead space is the total volume of the conducting airways from the nose or mouth down to the level of the terminal bronchioles. It is about 150 ml on the average in humans. The volume of air that goes into non-functional alveoli (often caused by lack of pulmonary blood flow) are called functional dead space. When alveolar dead space is added to the anatomical dead space, it is called physiological dead space.

The nitrogen washout (or Fowler's) method is commonly used to measure the dead space volume. In this method, the patient first breathes normal air and inhales a breath of pure oxygen at the end of the exhalation. This intake of pure oxygen fills the entire dead space volume and some mixes with the alveolar air. The patient then expires through a nitrogen meter to produce a nitrogen concentration curve as shown in Figure 25-4. The initial expired air that comes from the dead space consists of pure oxygen (zero concentration of nitrogen). After a while, when the alveolar air reaches the nitrogen meter, the nitrogen concentration rises and then levels off. The concentration of nitrogen is plotted against the volume of expired air. The measurement terminates at the end of the expiration. The total volume of expired air V_E is also measured. The nitrogen concentration curve divides the graph into two regions with areas A_1 and A_2 as shown. The area covered by A_1 represents the dead space portion of the expired air; the area A_2 with nitrogen represents the alveolar portion of the expired air. Therefore, one can determine the volume of dead space air V_D from the equation

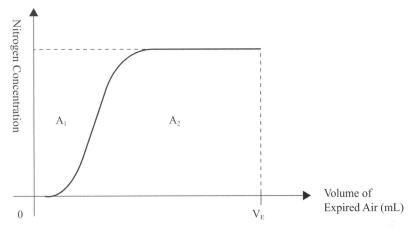

Figure 25-4. Nitrogen Dead Space Volume Measurement.

$$V_D = \left(\frac{A_1}{A_1 + A_2}\right) V_E ,$$

This method measures the anatomical dead space rather than the physiological dead space. Anatomical dead space can also be determined by whole-body plethysmography. In patients with compromised pulmonary function, the physiological dead space can be many times greater than the anatomical dead space.

Another important parameter in lung function assessment is the oxygen transfer from the alveolus into the red blood cell. According to the Fick's equation for gas diffusion:

$$D = \frac{k \times A \times \Delta P}{T}$$

where D = rate of diffusion = volume of gas transferred per unit time
k = diffusion coefficient of the gas
A = surface area for gas exchange
ΔP= partial pressure gradient of gas across the membrane
T = membrane thickness

According to the equation, the rate of diffusion is proportional to the surface area of the alveolar capillary membrane and the pressure gradient of oxygen between the alveolus and blood, and inversely proportional to the thickness of the alveolar capillary membrane. Decrease in the rate of diffusion may indicate diseases causing decrease in surface area of alveolar capillary membrane (such as pulmonary embolism or emphysema) and diseases that causing increase in the thickness of the alveolar capillary membrane (such as pulmonary fibrosis or pulmonary edema).

The diffusing capacity for carbon monoxide (DLCO), also known as transfer factor, is a measurement of oxygen transfer from the alveolus into the red blood cell. CO is used instead of oxygen because it binds more than 200 times faster and tighter to hemoglobin in the blood than oxygen. Therefore, the partial pressure of CO in blood remains very low, thus maintaining a diffusion gradient across the alveolar-capillary membrane. In the single-breath technique for measuring DLCO, a gas mixture of 0.3% CO and an inert gas (usually helium) is used. The patient first completely exhales to RV and then maximally inhales to TLC, the subject then holds his or her breath for 10 seconds to allow CO to diffuse into the blood. The concentration of CO in the exhaled gas mixture is measured. The uptake of CO (in ml/minute) is divided by the partial pressure gradient for CO (between alveolus and pulmonary capillary) to calculate DLCO (in ml/mmHg/minute or ml/kPa/minute).

Alveolar ventilation (VA), also known as alveolar minute ventilation, is less than minute ventilation due to the anatomical dead space. The volume of alveoli is equal to the difference between the tidal volume and anatomical dead space volume (DSV). Therefore VA is calculated as

$$\text{VA} = (\text{tidal volume} - \text{dead space volume}) \times \text{respiratory rate} = (\text{TV} - \text{DSV}) \times \text{RR}$$

VA is used to compensate for the inspired CO concentration which is diluted by the dead space air before reaching the alveoli. The parameter DLCO/VA (in ml/mmHg/minute/L or ml/kPa/minute/L) is computed to normalize the diffusion capacity according to the functional lung volume.

Maximal voluntary ventilation (MVV), also known as maximal breathing capacity (MBC), is another parameter in pulmonary function analysis. It is defined as the maximum minute volume of ventilation that the subject can maintain for 12 to 15 seconds. In a normal subject, MVV is about 15 to 20 times the resting minute volume. To measure MVV, the subject is asked to take deep breaths as fast as possible in 15 seconds, the total exhaled gas during the period is determined and converted to L/min.

SPIROMETERS

A spirometer is a device to measure the flow and volume of gas moving in and out of the lungs during inspiration and expiration. A critical characteristic of a spirometer is its low flow resistance. Flow resistance will create pressure

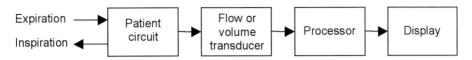

Figure 25-5. Block Diagram of a Spirometer.

drop across the spirometer. Due to the relatively low magnitude of the inhalation and exhalation pressures, additional pressure drop will disturb the intrinsic breathing characteristics of the patient. There are two categories of spirometers: one senses gas volume and the other senses gas flow. A volume-sensing spirometer has a container to measure the gas volumes; gas flow rate can be calculated from the volume-time information. A flow-sensing spirometer has a flow transducer placed in the gas flow pathway to measure the flow rate of gas. Gas volume can be derived from the flow-time information. Flow-sensing spirometers are usually smaller in dimension than the volume-sensing spirometers. Figure 25-5 shows the functional block diagram of a spirometer. The patient circuit allows the patient to breathe in and out of the spirometer; the transducer converts the volume or flow to an electrical signal. The processor computes the respiratory parameters from the collected information and displays them on the output device, such as a visual display or paper chart recorder.

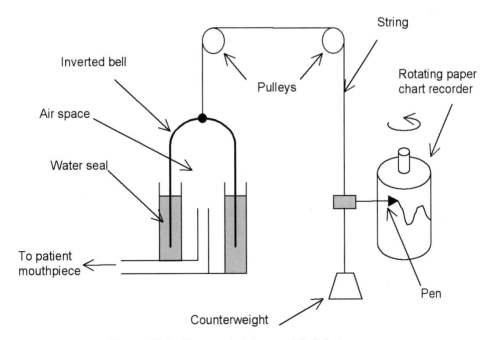

Figure 25-6. Water-Sealed Inverted Bell Spirometer.

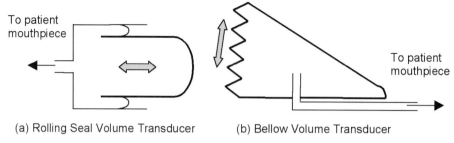

(a) Rolling Seal Volume Transducer   (b) Bellow Volume Transducer

Figure 25-7. Volume Transducers.

Volume Transducers

Three commonly used volume transducers for respiration measurement are shown in Figures 25-6 and 25-7. The water-sealed inverted bell spirometer (Figure 25-6) moves up and down according to the respiration of the patient. The low friction water seal and the counterweight attached to the inverted bell reduce the resistance and backpressure, thereby allowing accurate volume and flow measurements. A pen, which writes on a rotating drum, is mechanically linked to the inverted bell. A rolling seal with a horizontally mounted bell (Figure 25-7a) can also be used as the volume transducer in a spirometer. The horizontal mounting of the bell eliminates the need for a counterweight and therefore simplifies the construction of the spirometer. A third type of volume-measuring transducer is a bellow (Figure 25-7b). As gas moves in and out of the bellow, it inflates or deflates the bellow. The moving bellow can move a pen to record the changing volume on a paper chart.

Flow Transducers

Spirometers using flow transducers with no moving parts are commonly used to minimize errors due to mechanical wear and tear. They are usually smaller than volume spirometers. Figure 25-8 shows the block diagram of such a spirometer. Many different flow transducers can be used; examples include a hot air anemometer and a differential pressure flow transducer (see Chapter 7 for principles of flow transducers). As patients breathe directly into the spirometer, care must be taken to avoid contamination of the internal part of the spirometer. Some of the protective measures are using disposable mouthpiece and disposable patient breathing circuit, bacterial filter, and heated transducer chamber to prevent water condensation.

Modern spirometers are microprocessor-based and have built-in compensations for temperature and pressure fluctuations. The volume of gas in

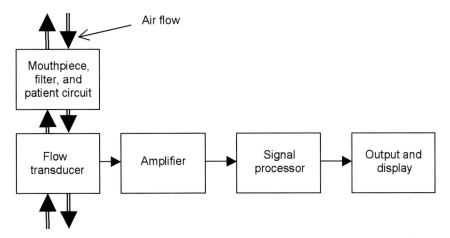

Figure 25-8. Flow-Sensing Spirometer.

the lungs is at body temperature, at atmospheric pressure, and is saturated with water vapor at body temperature (BTPS). In respiratory volume measurements, in order to relate the gas volumes measured outside the body to the condition inside the lungs, correction must be made to the gas volume obtained by a spirometer at ambient temperature and atmospheric pressure to gas volume under BTPS.

The following equation, derived from the gas laws, can be used for the correction of the volume to body temperature of 37°C and a saturated vapor pressure of 47 mmHg (or 6.3 kPa).

$$V \text{ (BTPS)} = V_t \times \left[\frac{273 + 37}{273 + t} \right] \times \left[\frac{P_B - P_{H_2O}}{P_B - 47} \right],$$

where t = temperature of the gas in the spirometer in °C,
V_t = volume collected at t °C,
P_B = barometric pressure (mmHg or kPa), and
P_{H_2O} = water vapor pressure (mmHg or kPa) of the gas in the spirometer at t °C.

In practice, the temperature and pressure corrections can be obtained from published tables.

RESPIRATION MONITORS

Clinical bedside monitoring of respiratory function is useful to assess the need for further respiratory intervention such as the introduction of mechanical ventilation. It is also a useful tool in evaluating the maturity of the regulatory functions of the respiratory system in neonatal development. The breathing rates as well as the waveform of breathing are the two parameters to be measured in bedside respiratory monitoring. In addition, the time elapsed of no breathing, or apnea, is often monitored. Respiration rate for a normal adult ranges from about 12 to 16 breaths per minute (bpm). Breathing rates for neonates are much higher (about 40 bpm). An apnea alarm is usually set at 20 sec.

There are a number of methods to obtain the respiratory waveform and determine the respiration rate. The impedance method, which measures the electrical impedance across the patient's chest, and the thermistor method, which detects the airflow in the patient's airway, are common methods in respiration monitoring.

Heated Thermistor Method

This method measures the change in temperature of a heated thermistor placed in the patient's breathing airway. A negative temperature coefficient thermistor is placed in the air path of the nostril as shown in Figure 25-9 or in a breathing circuit. A current source passes a constant heating current through the thermistor so that its temperature is above the ambient temperature but below the body temperature. When there is no air flowing across the thermistor, the voltage across the thermistor remains unchanged. During expiration, the warm air from the lung heats up the thermistor and decrease its resistance. The voltage across the thermistor will then decreases with its resistance. During inspiration, the colder outside air cools the thermistor, which causes the voltage across the thermistor to increase. The variation of voltage across the thermistor will vary with the airflow of respiration. This voltage is recorded and plotted against time.

Impedance Pneumographic Method

During inspiration, the volume of the thoracic cavity increases creating a negative pressure to pull air into the lung. The impedance across the chest therefore becomes higher. During expiration, the chest volume decreases and pushes air out of the lung. The impedance across the chest therefore becomes lower. In the impedance pneumography method, the monitor derives the respiration waveform and the breathing rate by measuring the change in impedance between a pair of electrodes applied on the chest of the patient.

To measure the impedance across the chest, a constant current I is applied across the chest through a pair of electrodes. The voltage V mea-

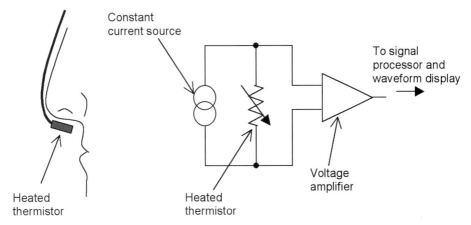

Figure 25-9. Heated Thermistor Respiration Monitor.

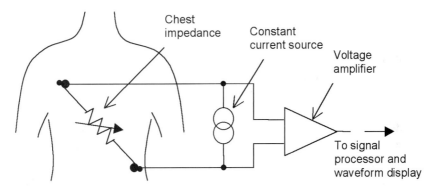

Figure 25-10. Impedance Pneumographic Respiration Monitor.

sured across the electrodes is therefore proportional to the chest impedance Z (V = I × Z). Figure 25-10 shows the setup to monitor respiration using the impedance pneumography method.

In order to prevent muscle and nerve stimulation and to prevent micro-shock (electrical safety), the injected current must be small and of high frequency. For this reason, respiration monitors use frequencies higher than 25 kHz and current amplitudes below 50 µA. The output of the amplifier in Figure 25-10 is an amplitude modulated signal with the carrier frequency equal to the frequency of the applied current. The amplitude variation of the modulated signal is proportional to the impedance change due to respiration. Figure 25-11 shows the respiration impedance waveform, the current source waveform, and the waveform of the detected voltage across the chest.

In most applications, respiration monitors using the impedance pneumography method employ the same sets of electrodes for ECG monitoring.

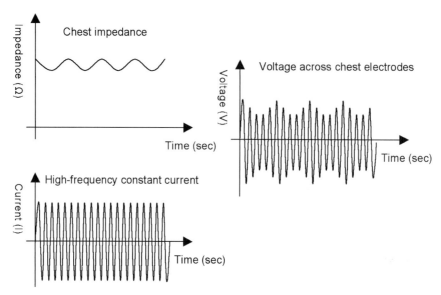

Figure 25-11. Impedance Pneumography Respiration Monitor Waveforms.

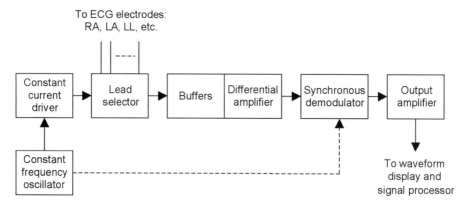

Figure 25-12. Block Diagram of Respiration Monitor Using Impedance Method.

Note that the impedance across the electrodes depends on the tissue imped-ance and the electrode–skin interface and is in the order of hundreds of ohms or kilo-ohms. The variation of the impedance due to respiration lies in the range of 0.1 to 4 ohms, however. As a result, the signal is very sensitive to electrode and body movement. In addition, since tissue impedance, which is not purely resistive, is frequency dependent, any fluctuation in frequency will create a change in the measured chest impedance. In order to minimize these errors, the applied current must have constant amplitude and be derived from a very stable frequency source.

Figure 25-12 shows the functional block diagram of a respiration monitor using the impedance method. This respiration monitor is part of the ECG/respiration monitor using the same set of ECG skin electrodes applied to the patient. The lead selector of the respiration monitor selects a pair of electrodes from the set of ECG electrodes. The high-frequency current (e.g., 50 kHz) flowing through the patient's chest from one electrode to another creates a voltage of the same frequency with amplitude equal to the product of the current and impedance across the chest. This voltage is captured to derive the respiration waveform and breathing rate. The voltage signal is first buffered so that it will not affect the ECG part of the monitor. The synchronous demodulator then removes the high frequency from the measured voltage and recovers the respiration waveform.

COMMON PROBLEMS AND HAZARDS

Hazards associated with pulmonary function analyzers and respiration monitors include electrical shocks and disease transmission. Electric shock hazard is mitigated with electric signal isolation design and using plastic hoses to electrically isolate the machine from the patient. Use of bacterial filters and single-use disposable patient mouthpieces and breathing circuits, as well as frequent cleaning and disinfection of internal parts, will decrease the risk of cross-contamination. However, special low resistance filter is required as it may increase airway resistance and therefore affect flow measurements.

Failures of modern spirometer components are not common; however, errors due to misuses or out-of-calibration devices affect the accuracy of measurements. The American Thoracic Society (ATS) and the European Respiratory Society (ERS) have published standards for spirometers and their methods of calibration to allow consistent, appropriate, and accurate measurement of flows and volumes. To reduce errors from environmental fluctuations, temperature and pressure compensations (to BTPS) are built into modern spirometers.

In respiration monitoring using the impedance pneumography method, the injection of an excitation current across the patient's chest creates electric shock hazard to the patient. Using high frequency and low amplitude current (e.g., 50 kHz and 100 μA) reduces such risk. The variation of chest impedance not related to breathing (such as patient movement) and fluctuation of frequency and amplitude of the excitation current will create artifacts to the respiration waveform. A constant current source and stable frequency oscillator are essential to minimize these possible errors.

BIBLIOGRAPHY

ATS Committee on Proficiency Standards for Clinical Pulmonary Function Laboratories. (2002). ATS statement: Guidelines for the six-minute walk test. *American Journal of Respiratory and Critical Care Medicine, 166*(1), 111–117.

Banner, M. J., Jaeger, M. J., & Kirby, R. R. (1994). Components of the work of breathing and implications for monitoring ventilator-dependent patients. *Critical Care Medicine, 22*(3), 515–523

Brochard, L. (1998). Respiratory pressure-volume curves. In M. J. Tobin (Ed.), *Principles and Practice of Intensive Care Monitoring* (pp. 597–616). New York, NY: McGraw-Hill.

Burgos, F., Torres, A., Gonzalez, J., Puig de la Bellacasa, J., Rodriguez-Roisin, R., & Roca, J. (1996). Bacterial colonization as a potential source of nosocomial respiratory infections in two types of spirometer. *European Respiratory Journal, 9*(12), 2612–2617.

Chu M.W., & Han J.K. (2008). Introduction to Pulmonary Function. *Otolaryngol Clin N Am., Apr;41*(2), 387-96.

Dondelinger, R. M. (2008). Pulmonary function analyzers. *Biomedical Instrumentation & Technology, 42*(5), 371–375.

Dubois, A. B., Botelho, S. Y., Bedell, G. N., Marshall, R., & Comroe, J. H., Jr. (1956). A rapid plethysmographic method for measuring thoracic gas volume: A comparison with a nitrogen washout method for measuring functional residual capacity in normal patients. *Journal of Clinical Investigation, 35*(3), 322–326.

Goldman, M. D., Smith, H. J., & Ulmer, W. T. (2005). Whole-body plethysmography. In R. Gosselink & H. Stam (Eds.), Lung Function Testing. *European Respiratory Society Monographs. Sheffield, UK: European Respiratory Society*, Vol. 31, pp. 26–54.

Graham, B. L., et al. (2018). 2017 ERS/ATS standards for single-breath carbon monoxide uptake in the lung. *Eur Respir J.*, Nov 22;52(5):1650016.

Grasso, S., Stripoli, T., De Michele, M., Bruno, F., Moschetta, M., Angelelli, G., & Fiore, T. (2007). ARDSnet ventilatory protocol and alveolar hyperinflation: Role of positive end-expiratory pressure. *American Journal of Respiratory and Critical Care Medicine, 176*(8), 761–767.

Grenvik, A., Ballou, S., McGinley, E., Millen, J. E., Cooley, W. L., & Safar, P. (1973). Impedance pneumography: Comparison between chest impedance changes and respiratory volumes in 11 healthy volunteers. *Chest, 62*(4):439–443.

Hathirat, S., Mitchell, M., & Renzetti, A. D. (1970). Measurement of the total lung capacity by helium dilution in a constant volume system. *American Review of Respiratory Disease, 102*(5), 760–770.

Hyatt, R. E., Scanlon, P. D., & Nakamura, M. (2008). *Interpretation of Pulmonary Function Tests: A Practical Guide* (3rd ed.). Philadelphia, PA: Lippincott Williams & Wilkins.

Johns, D. P., Ingram, C., Booth, H., Williams, T. J., & Walters, E. H. (1995). Effect of a microaerosol barrier filter on the measurement of lung function. *Chest, 107*(4), 1045–1048.

Kawamoto, H., Kimura, T., Kambe, M., Miyamura, I., & Kuraoka, T. (1999).

Significance of area under the flow volume curve–useful index of bronchial asthma. *Arerugi, 48*(7), 737–740.

Kelkar, S. P., Khambete, N. D., & Agashe, S. S. (2008). Development of movement artefacts free breathing monitor. *Journal of the Instrument Society of India, 38*(1), 34–43.

King, G. G. (2011). Cutting edge technologies in respiratory research: Lung function testing. *Respirology, 16*(6), 883–890.

Mason, R., Broaddus, V., Murray, J. F., & Nadel, J. A. (2010). Pulmonary function testing. In J. F. Murray & J. A. Nadel (Eds.), *Murray and Nadel's Textbook of Respiratory Medicine*. Philadelphia, PA: Saunders.

Miaskiewicz JJ. Chapter 103. Pulmonary Function Tests. In: Lawry GV, McKean SC, Matloff J, Ross JJ, Dressler DD, Brotman DJ, Ginsberg JS, eds. Principles and Practice of Hospital Medicine. New York: McGraw-Hill; 2012.

Miller, M. R., Crapo, R., Hankinson, J., Brusasco, V., Burgos, F., Casaburi, R., & Wanger, J. (2005). General considerations for lung function testing. *European Respiratory Journal, 26*(1), 153–161.

Miller, M. R., Hankinson, J., Brusasco, V., Burgos, F., Casaburi, R., Coates, A. & Wanger, J. (2005). Standardization of spirometry. *European Respiratory Journal, 26*(2), 319–338.

Mottram, C. (2013). *Ruppel's Manual of Pulmonary Function Testing* (10th ed.). Maryland Heights, MO: Mosby Elsevier.

Newth, C. J. L., Enright, P., & Johnson, R. L. (1997). Multiple-breath nitrogen washout techniques: Including measurements with patients on ventilators. *European Respiratory Journal, 10*(9), 2174–2185.

Patroniti, N., Bellani. G., Manfio, A., Maggioni, E., Giuffrida, A., Foti, G., & Pesenti, A. (2004). Lung volume in mechanically ventilated patients: Measurement by simplified helium dilution compared to quantitative CT scan. *Intensive Care Medicine, 30*(2), 282–289

Prutchi, D., & Norris, M. (2004). D*esign and Development of Medical Electronic Instrumentation: A Practical Perspective of the Design, Construction, and Test of Medical Devices*. Hoboken, NJ: Wiley Interscience.

Ranu, H., Wilde, M., & Madden, B. (2011). Pulmonary Function Tests. *Ulster Med J., May; 80*(2): 84–90.

Schlegelmilch, R. M., & Kramme, R. (2011). Pulmonary function testing. In R. Kramme, K-P. Hoffman & R. Pozos (Eds.), *Springer Handbook of Medical Technology* (pp. 95–117). Berlin, Germany: Springer-Verlag Berlin Heidelberg.

Stanojevic, S., Wade, A., Stocks, J., Hankinson, J., Coates, A. L., & Pan, H. (2008). Reference ranges for spirometry across all ages: A new approach. American *Journal of Respiratory and Critical Care Medicine, 177*(3), 253–260.

Tablan, O. C., Williams, W. W., & Martone, W. J. (1985). Infection control in pulmonary function laboratories. *Infection Control, 6*(11), 442–444.

Chapter 26

MECHANICAL VENTILATORS

OBJECTIVES

- Explain the applications of mechanical ventilation.
- Distinguish between positive and negative pressure ventilators.
- Classify ventilators based on the methods used to terminate inspiration.
- Contrast the types of breaths delivered by a ventilator, its modes and sub-modes of operation, and the ventilation parameters.
- List common user controls, alarm settings, and emergency modes in positive pressure ventilators.
- Sketch a block diagram of a positive pressure ventilator and explain the functions of each block.
- Analyze the gas delivery circuits and identify basic components of a positive pressure ventilator.
- Describe common problems and hazards

CHAPTER CONTENTS

1. Introduction
2. Indications for Mechanical Ventilation
3. Types of Ventilators
4. Modes of Ventilation
5. Ventilator Parameters and Controls
6. Basic Functional Building Blocks
7. Gas Delivery System Diagram
8. Safety Features
9. Special Ventilation Methods and Features
10. Common Problems and Hazards

INTRODUCTION

Patients may require respiratory intervention due to illness (e.g., asthma), injuries, congenital defects, postoperative conditions, or the influence of drugs (e.g., under general anesthesia). Oxygen therapy with a high flow nasal cannula is a common method to assist patients with compromised respiratory function such as pulmonary edema, pneumonia, chronic obstructive pulmonary disease (COPD), or asthma. Ventilators assist patients who cannot breathe on their own or who require assistance to maintain a sufficient level of ventilation. Patients with conditions such as bradypnea, apnea or acute lung injury are indicators for mechanical ventilation.

Philip Drinker of the United States developed the first mechanical ventilator in 1927. It was known as the "iron lung" for treating victims of poliomyelitis in the early 1950s. The iron lung is an airtight metal chamber enclosing the entire body of the patient except for the head, which is outside the chamber. The chamber isolates the patient's body from the outside atmosphere by an air seal around the neck of the patient. To create inspiration, the pressure inside the metal chamber is reduced to below atmospheric pressure. Because the outside pressure is higher than the chamber pressure, air is drawn into the patient's lungs through the patient's airway. Expiration is achieved by returning the chamber to atmospheric pressure. The iron lung is classified as a negative pressure ventilator because the inspiratory phase of the respiratory cycle is created by a negative pressure.

The biphasic cuirass ventilator (BCV) is a modified version of the iron lung. A BCV is a negative pressure ventilator; it uses a noninvasive cuirass or shell wrapped around the upper body of the patient. A power unit actively controls the inspiratory and expiratory phases of ventilation. To effect inspiration, air is pumped out of the cuirass, creating a negative pressure around the chest under the cuirass. Whereas most other types of ventilation depend on the passive recoil of the patient's chest, a BCV creates expiration by pumping air into the cuirass, creating a positive pressure around the chest, forcing air out from the lung into the atmosphere. An advantage of a BCV is its ability to achieve high tidal volume due to its active expiration mechanism. In addition, a BCV does not require patient airway circuits.

Today, positive pressure ventilators are used to avoid having to enclose the patient's body in a pressure chamber. A positive pressure ventilator uses a mask, an endotracheal tube or a tracheostomy tube to connect the machine to the patient's airway. The lungs are inflated by positive pressure during inspiration; expiration occurs upon release of the pressure. Using a tightly-fitted face mask in mechanical ventilation is considered as non-invasive ventilation, whereas the other two methods are invasive ventilation. Some patients with endotracheal tube may need sedation, analgesics or neuromuscular blocking drugs to lower stress and anxiety. This chapter describes the principles of operation, design, construction, common problems of positive pressure mechanical ventilators.

INDICATIONS FOR MECHANICAL VENTILATION

Breathing depends on rhythmic contraction of the inspiratory muscles. During normal respiration, a spontaneous breath begins with an electrical impulse generated by the respiratory centers in the brain. This impulse travels along the phrenic nerve, stimulates the diaphragm muscle to contract, causes descent or flattening of the diaphragm dome. It also stimulates contraction of the external intercostal muscles, causes elevation the rib cage. The action results in a volume increase of the thoracic cavity. As a result, the pressure in the airway drops, causing an inflow of air into the lungs. Due to differences in concentrations, inhaled atmospheric oxygen molecules move across the alveolar membrane into the bloodstream and carbon dioxide molecules offload from the bloodstream into the alveolar air. Relaxation of the muscles after the inspiratory phase (or at the beginning of the expiratory phase) of breathing reduces the volume of the thoracic cavity, causes pressure increase to push the alveolar air out to the atmosphere. The expiratory phase of breathing ends at the start of the next inspiratory phase. Forced or active breathing (e.g., during exercise) recruits additional muscles to enhance expansion and contraction the thoracic cavity volume.

Mechanical ventilation is indicated when the patient's spontaneous ventilation is inadequate to sustain life or when respiration control is needed in critically ill patients. Physiological indications include respiratory insufficiency and ineffective gas exchange. Below are some common indications for mechanical ventilation:

- Lung injury or acute respiratory distress syndrome (ARDS)
- Respiratory muscle injury or fatigue, and neuromuscular disease
- Coma or obtundation
- Bradypnea (low respiration rate < 12 bpm or respiratory arrest)
- Tachypnea (respiratory rate > 25 bpm)
- Low vital capacity (< 15 ml/kg)
- Low arterial oxygen saturation (S_aO_2 < 90%) even under elevated level of breathing oxygen
- High arterial carbon dioxide partial pressure ($PaCO_2$ > 50 mmHg or 6.7 kPa)

Mechanical ventilation is a therapy to allow the patient to recover from underlying problems. It is not a cure and is not without potentially harmful effects. Clinical judgment must be made based on the symptoms before starting mechanical ventilation. In addition to being used in acute care hospitals, there is a significant increase in ventilator use in homes and other nonclinical settings in the last couple of decades. This is attributed to the technological advancement and innovation in producing portable, reliable and user-friendly ventilators for use by patients with support of nonmedical caregivers, as well as the prevalence of chronic respiratory disorders and demands from the aging population.

TYPES OF VENTILATORS

Intermittent positive-pressure breathing (IPPB) therapy is a method to assist spontaneous breathing by applying a positive pressure to create inspiration and by removal of the inflated pressure to allow passive expiration. It is believed by some that IPPB can reduce the work of breathing, promote bronchopulmonary drainage, and provide more efficient delivery of bronchodilation drugs. In addition to the preset positive airway pressure during inspiration, a bilevel positive airway pressure (also known as BiPhasic or BiPAP) ventilator maintains a positive airway pressure during expiration.

Mechanical ventilators overcome respiration deficiencies by assisting spontaneous breathing or completely taking over the breathing of the patient. A positive pressure ventilator inflates the lungs by elevating the pressure in the airway to push air into the lungs to improve alveolar gas exchange. Expiration is usually achieved by opening the airway to the atmosphere so that the gas in the lungs is passively released to the atmosphere. Portable ventilators are used in step-down units, extended care facilities, and at homes. Home-use portable ventilators are often used 24 hours a day and are set up and operated by the patient or by home caregivers. Portable ventilators should be user friendly and are much simpler in operation than critical care ventilators are. For some patients who cannot tolerate the high inspiratory pressure and rapid airflow, high-frequency ventilators are used. These ventilators cycle at rates much higher than normal spontaneous respiration but deliver lower inspiratory volume for each breath.

Mechanical ventilators are classified according to how they terminate the inspiration phase of the ventilation cycle. Four parameters can be monitored and used to terminate inspiration; they are pressure, volume, time, and flow.

1. A pressure-cycled ventilator monitors the pressure in the airway and ends the inspiration when a certain preset pressure is reached.
2. A volume-cycled ventilator measures the volume of gas delivered to the

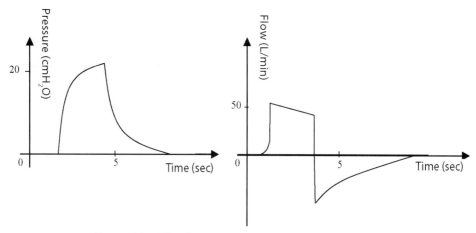

Figure 26-1. Ventilation Pressure and Flow Waveforms.

patient and ends the inspiration when a preset volume has been delivered.

3. A time-cycled ventilator tracks the time of inspiration and ends the inspiration when a preset time is reached.

4. A flow-cycled ventilator senses the flow in the airway and ends the inspiration when the inspiration flow falls to a preset level.

A mechanical ventilator typically provides pressure-controlled breaths or volume-controlled breaths in conjunction with a selected ventilation mode to provide ventilation to a patient. In pressure-controlled ventilation, the pressure of the delivered breath is controlled while the volume varies. In volume-controlled ventilation, the volume (by varying the flow and time) of the delivered breath are controlled. Ventilator modes of operations are discussed in the next section.

Figure 26-1 shows typical pressure and flow waveforms of positive pressure mechanical ventilation measured in the airway of a patient. At the start of the inspiratory phase, the airway and lung pressure increases rapidly until sufficient gas has entered the lungs. The gas flow rate is highest at the beginning of inspiration but decreases as the pressure builds up. The tidal volume (TV) is determined by integrating the flow with respective to time (area under the flow-time curve) during the inspiratory phase. During the expiratory phase, the airway pressure is lowered (e.g., opened to the atmosphere), the alveolar pressure drops and gas is pushed out of the lungs. Some ventilators allow the users to select different pressure or flow waveforms.

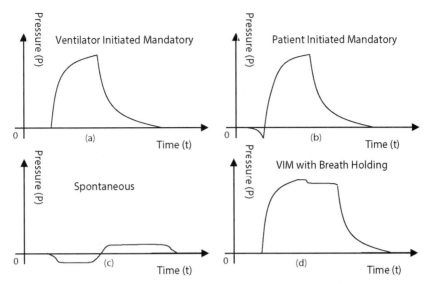

Figure 26-2. Mandatory and Spontaneous Breaths.

MODES OF VENTILATION

When a patient is undergoing positive pressure mechanical ventilation, there are two types of allowed breath: spontaneous breath and mandatory breath. In a spontaneous breath, the breathing parameters are determined by the patient's condition. However, in a mandatory breath, breathing parameters are determined by the machine. A ventilator-initiated mandatory (VIM) breath is initiated by the ventilator timing circuit, whereas a patient-initiated mandatory (PIM) breath is initiated by the patient. In a PIM breath, the patient attempts to breathe, and the breathing action creates a small negative pressure in the lungs and airway. Upon detecting this small negative pressure change, the machine will deliver a positive pressure breath to help the patient breathe. The waveform of a PIM breath is distinguished from that of a VIM breath by the slight momentary negative pressure before the onset of the positive inspiratory pressure (Figure 26-2a-b). The pressure waveform for a spontaneous breath is characterized by the small negative inspiratory pressure and small positive expiratory pressure (Figure 26-2c). When the pressure is maintained for a period of time after inspiration to allow longer time for oxygen diffusion (Figure 26-2d), it is called breath holding. The waveforms in Figure 26-2a to Figure 26-2c have no breath holding between the end of inspiration and beginning of expiration.

The modes of ventilation specify the characteristics of breaths delivered by a positive pressure ventilator in response to the patient's breathing attempts. However, there is neither a standard to define the modes of ventilation nor a universal vocabulary of ventilation. Manufacturers may use different names and abbreviations for the same mode of ventilation. These terminologies may not be consistent or they may even contradict each other. Below is an attempt to describe some of the common modes of ventilation using typical vocabulary. The pressure waveforms described are shown in Figure 26- 3.

- Controlled mandatory ventilation (CV)–Consists of VIM breaths at prescribed fixed time intervals. The CV mode is for patients who cannot breathe by themselves; breathing action is entirely controlled by the ventilator with no input from the patient.
- Intermittent mandatory ventilation (IMV)–Consists of VIM breaths at prescribed time intervals but allows the patient to breathe spontaneously between the controlled breaths.
- Continuous mandatory ventilation (CMV)–Consists of a mix of VIM breaths at prescribed time intervals or mandatory breaths initiated by the patient (PIM). The VIM breaths are synchronized with the patient's inspiration effort to generate PIM breaths. This is also called the assist control (AC) mode of ventilation by some manufacturers.
- Synchronized intermittent mandatory ventilation (SIMV)–produces VIM breaths at prescribed time intervals if there is no patient breathing initiation. The VIM breaths are synchronized with the patient's inspiration effort to generate PIM breaths. The patient is allowed to breathe spontaneously, within an s-phase window, right after a PIM breath. If the patient initiates a breath after the s-phase, the ventilator will generate a PIM breath. By gradually decreasing the ventilation rate, carbon dioxide will build up in the patient's blood, stimulating the respiratory control centers, and will trigger the patient to initiate mandatory (PIM) and spontaneous breaths. This method, call weaning, is used to gradually return the work of breathing to the patient.
- Continuous positive-airway pressure (CPAP) ventilation–in the CPAP mode, breathing is spontaneous. A continuous preset positive pressure is maintained throughout the breathing cycle. This elevated baseline pressure helps to open up alveoli and enhances oxygen diffusion from the lungs in to the alveolar capillaries.

In addition to the ventilation modes just described, the following submodes are commonly found in critical care ventilators.

- Positive end-expiratory pressure (PEEP)–PEEP is used with mandatory breaths; it maintains the lungs at a positive pressure at the end of expira-

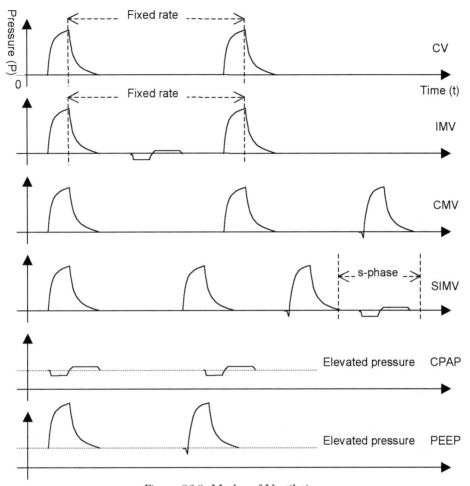

Figure 26-3. Modes of Ventilation.

tion (Figure 26-3). It is often used to increase the patient's arterial oxygen saturation without increasing the inspired O2 percentage.

- Apnea—Apnea is the cessation of breathing when the patient is under spontaneous breathing. The ventilator will measure the time duration of no breathing. On detection of an apnea, the ventilator will automatically go into a preset ventilation pattern stored under the apnea submode.
- Pressure support—pressure support creates an elevated pressure during a spontaneous inspiration. By elevating the pressure during inspiration, the patient does not have to create the entire pressure gradient to obtain a meaningful TV. Thus, this mode will reduce the patient's inspiratory work-of-breathing (WOB) while still allowing the patient to control many other breathing parameters. In pressure support mode, the end-expiratory pres-

sure is at atmospheric pressure, and the patient controls the respiration rate.
- Sigh—a sigh is a breath delivered by the ventilator that differs in duration and pressure from a normal breath. The profile (frequency, inspiratory volume, etc.) of a sigh is often user programmable.

VENTILATION PARAMETERS AND CONTROLS

The basic parameters of ventilation are pressure, flow, volume, and time. These parameters are interrelated during mechanical ventilation.

- Pressure—Pressure is the driving force against the resistance of the patient circuit and airway to produce flow, the unit of measurement is in centimeters of water (cmH2O).
- Flow—Flow is the rate of gas at which the TV is delivered. A higher flow setting will require a higher pressure to overcome the air resistance in the breathing circuit and patient's airway. A higher flow will take less time to deliver the required TV. The unit of flow measurement is liters per minute (LPM).
- Volume—Volume measures the quantity of gas delivered into the lungs of the patient in liters (l). Tidal volume (TV) setting prescribes the volume of inspired gas (same at the volume of expired gas when there is no air leak in the system) in each ventilation mandatory breath. The total inspired gas volume is also displayed on a ventilator as minute ventilation or minute volume. Minute volume, with unit in l/min, is the product of TV (in l) and respiration rate (in bpm).
- Time—Time is associated with the duration of the inspiratory and expiratory phases of the breathing cycle. The number of breathing cycles per unit time is the respiration rate. Inspiration time starts from the beginning of inspiration and ends at the start of expiration. Expiration time starts from the beginning of expiration and ends at the start of the next inspiration. The unit of measurement of breathing rate is bpm. The I:E ratio is a decimal value used to indicate the ratio of inspiratory time to expiratory time in a breathing cycle.

As ventilators are drawing and mixing gases from hospital gas supplies or from the atmosphere, ventilation volumes are measured outside the patient's body under different conditions than those inside the lungs of the patient. It is therefore important to convert gas volume (and flow) to the conditions inside the patient's lungs. Body temperature, sea level pressure, and gas saturated with water vapor (BTPS) are used as reference parameters for the conversion. BTPS volume is used to harmonize the volume of gas delivered to the patient by the ventilator.

Operator control parameters (with typical setting ranges) that modify the function of a mechanical ventilator include:

- Ventilation mode (CV, IMV, SIMV, etc.)
- Inspiratory flow waveform (square, descending ramp, etc.)
- I:E ratio (0.2–1)
- TV (0.1–2.5 L)
- Peak flow (120–180 LPM)
- Respiratory rate (0.5–70 bpm)
- Sensitivity (0.5–20 cmH_2O below baseline or PEEP)—for sensing patient-initiated breathing effort
- Inspired air O_2 concentration (21–100%)
- Manual breath (or sigh) (0.1–2.5 l not exceeding twice TV, 1–3 sighs per hour)
- PEEP/CPAP (0–45 cmH_2O)

Some common safety alarm features and settings are:

- High pressure limit (10–120 cmH_2O)
- Low inspiratory pressure (3–99 cmH_2O)
- Low PEEP/CPAP (0–45 cmH_2O)
- Low exhaled TV (0–2.5 l)
- Low exhaled minute volume (0–60 l)
- High respiratory rate (0–70 bpm)
- Low oxygen/air inlet pressure (35 psig or 241 kPa gauge)
- Apnea interval (10–60 sec)
- I:E ratio (< 1)

Some emergency ventilator operations are

- Apnea ventilation—Delivers preset ventilation when apnea is detected.
- Backup ventilation—Should the ventilator fail to provide the ventilation to the patient, backup ventilation function will be activated. For example, activation of backup air compressor to take over failed medical gas supplies.
- Safety valve open—If the ventilator fails, the patient breathing circuit is open to the atmosphere to allow spontaneous breathing and manual bagging.

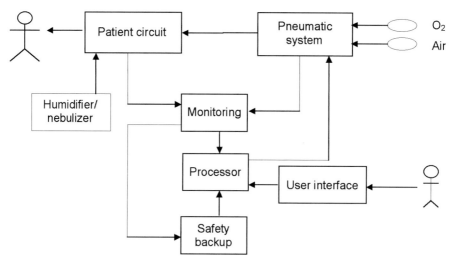

Figure 26-4. Functional Block Diagram of a Mechanical Ventilator.

BASIC FUNCTIONAL BUILDING BLOCKS

Figure 26-4 shows the block diagram of a positive pressure mechanical ventilator. The following paragraphs describe the functions of these building blocks.

- Medical air/oxygen supplies—Medical gas from wall outlets provides the air and oxygen necessary to produce the breathing gas mixture delivered to the patient. For portable ventilators, build in gas compressors replace wall medical gas supply. Oxygen gas cylinders attached to portable ventilators provide elevated level of oxygen to patients. Instead of using cylinders of oxygen gas, oxygen concentrators that continuously extract oxygen from atmospheric air can be used to provide elevated level of oxygen to patients. A typical portable home oxygen concentrator is able to produce a flow greater than 2 l/min at 90% oxygen concentration.
- Pneumatic system—The pneumatic system regulates the gas pressure, blends the air and oxygen to desired proportion, and controls the ventilation flow profile according to the control settings.
- Patient circuit—The patient circuit physically connects the pneumatic system to the patient. It supplies the inspired gas to the patient and removes the expired gas from the patient. It has one or more check valves to separate the inspired and expired gas flow and is fitted with bacteria filters to prevent contamination. Patient airway assess is part of the patient circuit. Airway access can be a face mask, an endotracheal tube, or a tracheostomy tube. A face mask covers the mouth and nose of a patient requires a

good air seal between the mask and the face of the patient to maintain the possible pressure during the inspiration phase of ventilation. An endotracheal tube is a flexible plastic tube that is placed through the mouth into the trachea. A small balloon near the distal end of the tube is inflated to seal off the airway from the atmosphere after the intubation tube is positioned. A tracheostomy is an opening created at the front of the neck so a tube can be inserted into the trachea, a balloon to seal off the airway from the atmosphere is also present. Face masks are used for temporary short duration ventilation, endotracheal tubes are used for prolonged procedures, tracheostomy access is for patients requiring continuous ventilation. Ventilation using face masks are considered non-invasive ventilation, while the other two are invasive ventilation.

- Processor—According to the user input and the information from the sensors, the processor produces control signals to the pneumatic circuit to produce breaths with desired characteristics.
- Monitoring—It measures the performance of the pneumatic system and feeds information back to the processor. Pressure and flow sensors at different locations of the pneumatic circuit are used to monitor and control ventilation parameters. Oxygen sensors are used to monitor the correct air/oxygen mixture being delivered to the patient.
- User interface—It allows users to set up ventilation parameters and displays system and patient information.
- Safety/backup—This system protects the patient under ventilation. It alerts the operator when preset conditions are violated and may initiate backup responses preset by the operator. In case of extreme circumstances, such as a loss of a gas source, the safety/backup system may take control of the pneumatic system and override settings previously selected by the operator.
- Humidifier (optional but often required)—Humidifiers are used to increase the water moisture content in the breathing gas before it is delivered to the patient. During normal breathing, the inspired gas is warmed and moisturized as it passes through the natural airway. During mechanical ventilation, prolonged inhalation of dry gas will cause patient discomfort and may damage the airway tissues. When a heated humidifier is used, the inspired gas in the patient circuit is bubbled through a reservoir of warm water to pick up moisture before entering the patient's airway. To prevent heat damage to the airway tissues, the temperature of the inspired gas must be monitored (by a temperature sensor) to ensure that it is below 42°C.
- Nebulizers—Nebulizers are used to deliver medication into the patient's airway during ventilation. The size of the vapor droplets determines the site of deposition. Larger droplets deposit in the upper airway; tiny

droplets (<1 μm) are deposited in the alveoli. Jet, ultrasonic or mesh nebulizers are devices used to produce tiny droplets of water in the inspired gas. In a jet nebulizer, pressurized air is directed through a small venturi nozzle to create a high flow velocity jet stream. By the Bernoulli Principle, the jet stream creates a negative pressure at the nozzle which draws the solution up a capillary tube from a reservoir containing the medication and turn it into an aerosol. In an ultrasonic nebulizer, a piezoelectric transducer element creates high frequency vibration in the solution causing aerosol to be produced at the liquid surface. Similar to ultrasonic nebulizers, a mesh nebulizer uses a piezoelectric transducer element to vibrate a mesh (with thousands of consistent micron-sized holes) in contact with solution. The vibration of the mesh pushes the liquid through the holes in the mesh to create the aerosol. The size of the holes determines the size of droplets in the aerosol. Users have to exercise care in using nebulizers because they have the potential to deliver too much water and overhydrate the patient.

• Air compressor (optional)—It is used as a backup to the medical air supply to allow the patient to breathe in case of medical air supply failure. The compressor will cut in automatically when it detects a low pressure in the medical air supply line.

GAS DELIVERY SYSTEM DIAGRAM

Figure 26-5 shows a typical gas delivery system diagram of a positive pressure mechanical ventilator including the pneumatic circuit, the patient breathing circuit and the medical gas supplies.

Medical air from the piped gas wall outlet is connected to the ventilator via a water trap and coarse filter to remove water condensation and particulates from the hospital's gas supply system. The inlet to the ventilator consists of another filter and a pressure sensor. The pressure sensor will switch off the air supply line and sound an alarm if the pressure becomes too low (e.g., < 35 psig or 241 kPa gauge). A check valve allows the gas to flow in only one direction, thereby preventing any contamination of the gas supply due to reverse flow. Medical air, which is usually about 50 psig (or 345 kPa gauge) from the wall outlet, is reduced to a lower pressure (e.g., 10 psig or 69 kPa guage) by the air regulator before being mixed with oxygen in the oxygen/air blender. In addition to creating the desired breathing gas mixture, the flow control within the blender generates the flow pattern and controls the ventilation rate. A flow sensor (e.g., hot air anemometer) monitors the volume flow rate of the air supply. (The oxygen supply line before the blender is identical to that of the air supply line.)

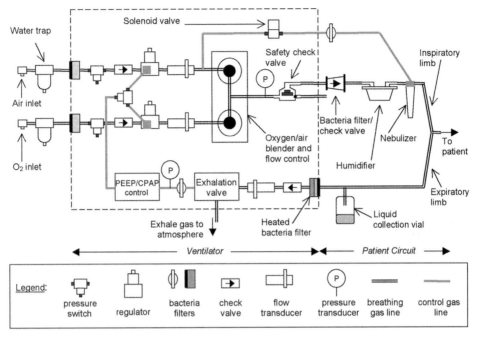

Figure 26-5. Pneumatic System Diagram of Mechanical Ventilator.

The patient circuit consists of an inspiratory limb and an expiratory limb connected at a Y-connection. The breathing gas, which contains the desired proportion of air and oxygen, exits the ventilator via a bacteria filter and check valve into the inspiratory limb of the breathing circuit. The gas mixture picks up moisture from the humidifier and medication (if needed) from the nebulizer before entering the patient's lungs. During the inspiration phase, the exhalation valve is closed to allow the inspired gas to inflate the lungs. During the expiratory phase, the flow control valve stops the supply gas flow, the exhalation valve opens to the atmosphere, and the thoracic cavity collapses and forces the gas to exhale from the lungs through the expiratory limb of the patient breathing circuit.

Because the exhaled gas from the lungs is at body temperature and saturated with water vapor, water will condense from the gas as its temperature becomes lower in the expiratory limb of the breathing circuit. Some ventilators fit a heating wire along the expiratory limb of the breathing circuit to reduce water condensation. The bacteria filter prevents contamination of the ventilator circuit components by the exhaled gas from the patient. However, a wet filter has a much lower efficiency for removing bacteria from the expired air. To prevent water condensation at the bacteria filter, a heater is required to warm the expired gas in the filter chamber. A collection vial is also connected to the inlet of the expiration compartment of the ventilator to remove water condensation and sputum from the patient's exhaled gas before reaching the filter. A check valve is in the expiratory limb to prevent reverse gas flow. A flow sensor measures the exhaled gas flow before it is vented to the atmosphere.

An elevated baseline pressure can be created by imposing a pressure on the main expiratory gas flow in the expiratory limb of the patient breathing circuit. Under CPAP or PEEP mode, the CPAP/PEEP controller exerts a pressure on the valve seat of the exhalation valve. Only exhaled gas with pressure higher than the CPAP/PEEP can be vented to the atmosphere.

SAFETY FEATURES

A mechanical ventilator is a critical life-supporting device for a patient who cannot breathe by himself or herself. Malfunctioning of any part in the system can threaten the life of the patient. Many safety features are built into the system. Some of the common safety features are:

- Disconnection alarm—A disconnection alarm detects disconnection of the breathing circuit from the ventilator or the patient. Pressure sensors detect a sudden drop in pressure in the patient circuit. To avoid a false alarm (e.g., due to pressure fluctuation), there is a tendency for caregivers to decrease the alarm pressure limit. Care must be taken not to lower the pressure limit too much to defeat the capability of disconnection detection. On the other hand, too high a setting may not be sensitive enough to trigger an alarm if the disconnection is at the distal end of the circuit.
- Air leak alarm—Excessive air leak in the system (especially in the patient breathing circuit) will compromise ventilation. Most ventilators utilize flow sensors to compare the inspired and expired gas flow volume. For a leak-free system, the volume of inspired gas over a period of time is equal to that of the expired gas. In the system shown in Figure 26-5, the inspired gas flow is measured by the flow sensors in the air and oxygen supply lines. The sum of these two flow sensors should be the same as the flow measured by the flow sensor in the expiration circuit. Note that all volumes must be adjusted to BTPS before making the comparison.
- High pressure alarm—A high pressure alarm can alert the user to a kink or obstruction in the patient breathing circuit. It can also prevent lung damage from inadvertent high pressure being developed in the breathing lines.
- Loss of power—Although breathing gas is derived from the medical gas wall outlets, most critical care ventilators employ electronic or micro-

processor circuits for control and alarm. In case of a power loss, the ventilator should sound an alarm to alert the clinicians. The ventilator and the patient circuit should be designed such that manual ventilation can be performed on the patient in a power failure situation. The unit should have a battery backed-up memory to store all the machine settings and data so that it can be powered up immediately without having to undergo lengthy initialization and programming after power has been restored.

- Loss of gas supplies—Ventilators should be designed to allow automatic switch-over of oxygen to medial air if there is no supply of oxygen (and vice versa). Similar to a loss of power, the ventilator should allow manual ventilation when all gases are lost. Some ventilators have a built-in electrical air compressor to supply compressed air to the gas lines in case the hospital gas supply has failed or is not available.

- Power-up self-test—Many critical care ventilators have a power-up self-test to check most operational conditions, including electronic diagnostic and leakage test of the pneumatic circuit. The compliance and flow resistance of the system, including the patient breathing circuit, are measured during the test to ensure accurate calculation of all breathing parameters.

OTHER VENTILATION MODES AND SPECIAL FEATURES

Some ventilators incorporate more than one method of control. For example, in the pressure-regulated volume-control (PRVC) method, the clinician presets a desired TV, and the ventilator delivers a controlled pressure breath until that preset TV is achieved. The inspiratory pressure of each breath is automatically adjusted according to the lung compliance and airway resistance of the patient to deliver the preset TV. The ventilator monitors each breath and compares the delivered TV with the set TV. If the delivered volume is too low, it increases the inspiratory pressure on the next breath. On the contrary, if the delivered volume is too high, it decreases the inspiratory pressure on the next breath. This adjustment gives the patient the lowest peak inspiratory pressure needed to achieve a preset TV.

A ventilator with mandatory minute volume (MMV) mode delivers mandatory breaths when the patient's spontaneous effort fail to meet the target minute volume set by the clinician. In MMV mode, the clinician sets the target minute volume plus either the TV or respiration rate.

In airway pressure release ventilation (APRV), the patient breathes spontaneously at a high pressure level for several breaths (CPAP for a set period of time), then followed by a very short period of lower pressure level. This creates the effect of inverting the I:E ratio under a bilevel ventilation mode. The long "inspiratory time" recruits alveoli and optimize gas exchange.

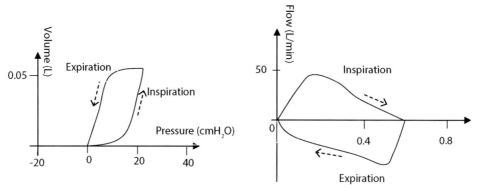

Figure 26-6. Ventilator Volume-Pressure and Flow-Volume Loops.

Pressure-controlled inverse-ratio ventilation (PC-IRV) is the name given by another manufacturer for this mode of ventilation.

A ventilator with volume support (VS) maintains a target tidal volume by varying the inspiratory pressure based on the patient's lung compliance and the airway resistance. Automatic tube compensation (ATC) or airway resistance compensation (ARC) is a feature that provides additional pressure to overcome the airway resistance during spontaneous breathing.

In addition to providing a graphic display of the pressure-time and flow-time breathing waveforms (see Figure 26-1), some ventilators can display volume-pressure and flow-volume breathing loops (Figure 26-6). The area enclosed by the pressure-volume loop represents the WOB created by the ventilator.

Conventional mechanical ventilation delivers inspiratory volume greater than the anatomical dead space to push air into the alveoli. Such ventilation may not be effective or may inflict injury in certain patient populations. High frequency ventilation is mechanical ventilation with higher respiration rate and smaller TV. It is used on some critically ill infants or patients with acute respiratory distress to minimize ventilation-related lung injuries. There are different variations of high frequency ventilation. A high frequency passive (or jet) ventilator uses an endotracheal tube to deliver a high frequency (e.g., 8 Hz), high pressure (e.g., 25 cmH_2O), short duration (e.g., 0.02 sec), low TV (e.g., 1 ml/kg of patient's body weight) jet of gas flows into the airway. Exhalation is passive with PEEP (e.g., 8 cmH_2O) to maintain alveolar inflation. PEEP is adjusted to establish appropriate arterial oxygen level. High frequency active (or oscillatory) ventilation is similar to high frequency passive ventilation except that a negative end expiratory pressure is used to actively remove gas from the lungs during the expiratory phase of ventilation. High frequency active ventilation is often used on patient with oxygenation diffusion issues, acute lung injury, or severe acute respiratory distress syndrome.

Neurally adjusted ventilatory assist (NAVA) is a mode of mechanical ventilation that uses the electrical activity of the diaphragm (EAdi) to trigger inspiratory assistance. The signal that excites the diaphragm arises from the respiratory center in the brain and controls the depth and cycling of breathing. EAdi is captured using a special catheter with an array of electrodes placed in the esophagus like an enteral feeding tube. It is used to deliver ventilation in synchronous with the patient's breathing effort and in proportion to the diaphragm activity. As ventilation employing NAVA uses the same stimulation signal to the diaphragm, NAVA potentially improves patient–ventilator interaction, reduces risks of over and under assistance of the patient, and limits the occurrence of asynchronies.

Pressure spikes due to coughs can create high fluctuation in airway pressure, trigger nuisance alarms, and disrupt patient-ventilator synchrony. A patient-responsive valve that can open and close rapidly in response to pulmonary pressure fluctuation will keep a stable pressure to maintain steady flow in the patient circuit.

COMMON PROBLEMS AND HAZARDS

Mechanical ventilators are critical life-support medical devices because patients often depend entirely on the ventilators for breathing. Ventilators should be inspected and performance operation be verified before every use by trained professionals. Patient's complications may arise from endotracheal intubation, mechanical ventilation, and prolonged immobility and inability to eat normally. Correct operation and proper maintenance is critical to prevent adverse effects from operating errors and machine failures. Below is a list of potential problems due to mechanical ventilation; many of them are life threatening.

The presence of endotracheal tube in the airway may cause tracheal stenosis, vocal cord injury and ventilator associated pneumonia. Trauma in the airway may occur during placement of an endotracheal tube. Inadvertent placement of an intubation tube into the esophagus will prevent proper ventilation. A prolonged period of mechanical ventilation increases the risk that the patient could acquire ventilator-associated pneumonia. To minimize risk of pneumonia, it is important to follow proper infection control procedures in breathing circuit and airway management. Prolonged mechanical ventilation also increases the risk of complications leading to acute respiratory failure.

The alveolar epithelium is at risk from both barotrauma and volutrauma. Barotrauma refers to the rupture of the alveolus with subsequent entry of air into the pleural space (pneumothorax). Large TVs and elevated peak inspiratory and plateau pressures are risk factors for barotrauma. Volutrauma refers to the local over distention of normal alveoli that may lead to local inflammation. Loosening and disconnection of the airway circuit to the patient's artificial airway will prevent the patient from being adequately ventilated. A ventilator should signal an audible and visual alarm whenever a leak or disconnection is detected. A leak in the breathing circuit may prevent the ventilator from functioning properly. It may cause inaccuracies in delivering gas flow and volume to the patient and fail to maintain desired pressure. Current ventilators have leak detection and compensation algorithms to help alleviate these problems. The recommended setting for low pressure alarms intended to detect breathing circuit disconnections is usually 5 to 7 cmH2O below the peak inspiratory pressure. However, they may be inappropriately adjusted below the detection threshold.

A patient breathing effort may fail to trigger the ventilator, or the ventilator may mistakenly interpret an airway pressure fluctuation (e.g., due to movement) as a patient's effort and deliver breaths at an inappropriate time. Improper setting of trigger sensitivity is a common cause of this patient-ventilator dyssynchrony that may lead to ineffective pulmonary gas exchange and respiratory distress.

Mechanical ventilation, especially using parameters deviate significantly from normal values of the patient, contributes to atrophy of the diaphragm muscles. Recent research showed about 6% reduction of diaphragm muscle per day under mechanical ventilation. Diaphragm weakness is a leading cause of difficulty in weaning patients from mechanical ventilation. Positive-pressure ventilation may reduce mucociliary motility in the airways and affect bronchial mucus transport, leading to retention of secretions.

Intrinsic PEEP is a complication that often occurs in patients with COPD or asthma who have high airway resistance leading to a prolonged expiratory phase of respiration. These patients may not have time to totally exhale the ventilator-delivered TV before the next machine breath is delivered. This breath stacking is the result of a portion of each subsequent TV being retained in the patient's lungs. If undetected, the patient's peak airway pressure may increase to a level that results in barotrauma and volutrauma and may lead to more serious injury. This may be alleviated by reducing the I:E ratio.

Positive-pressure ventilation can decrease cardiac preload and stroke volume, leading to lower cardiac output. Positive-pressure ventilation also affects renal blood flow, resulting in gradual fluid retention. Positive pressure maintained in the chest may decrease venous return from the head, increasing intracranial pressure.

BIBLIOGRAPHY

Baum, M., Benzer, H., Putensen, C., Koller, W., & Putz, G. (1989). Biphasic positive airway pressure (BIPAP)–A new form of augmented ventilation. *Anaesthesist, 38*(9), 452–458.

Bollen, C. W., Uiterwaal, C. S., & van Vught, A. J. (2006). Systematic review of determinants of mortality in high frequency oscillatory ventilation in acute respiratory distress syndrome. *Critical Care, 10*(1), R34.

Briscoe, W. A., Forster, R. E., & Comroe, J. H. (1954). Alveolar ventilation at very low tidal volumes. *Journal of Applied Physiology, 7*(1), 27–30.

Cereda, M., Foti, G., Marcora, B., Gili, M., Giacomini, M., Sparacino, M. E., & Pesenti, A. (2000). Pressure support ventilation in patients with acute lung injury. *Critical Care Medicine, 28*, 1269–1275.

Colice, G. L. (2006). Historical perspective on the development of mechanical ventilation. In M. J. Tobin (Ed.), *Principles and Practice of Mechanical Ventilation* (2nd ed., pp. 3–42). New York, NY: McGraw-Hill.

Downs, J. B., Klein, E. F., Jr., Desautels, D., Modell, J. H., & Kirby, R. R. (1973). Intermittent mandatory ventilation: A new approach to weaning patients from mechanical ventilators. *Chest, 64*, 331–335.

Dreyfuss, D., & Saumon, G. (1998). Ventilator-induced lung injury: Lessons from experimental studies. *American Journal of Respiratory and Critical Care Medicine, 157*(1), 294–323.

Epstein, S. K. (2011). How often does patient-ventilator asynchrony occur and what are the consequences? *Respiratory Care, 56*(1), 25–35.

Geddes, L. A. (2007). The history of artificial respiration. *IEEE Engineering in Medicine and Biology Magazine, 26*(6), 38–41.

Goligher, E. C. et al, (2018). Mechanical Ventilation–induced Diaphragm Atrophy Strongly Impacts Clinical Outcomes. *American Journal of Respiratory and Critical Care Medicine Volume*, 197(2). 204-217.

Henzler, D., Dembinski, R., Bensberg, R., Hochhausen, N., Rossaint, R., & Kuhlen, R. (2004). Ventilation with biphasic positive airway pressure in experimental lung injury: Influence of transpulmonary pressure on gas exchange and haemodynamics. *Intensive Care Medicine, 30*, 935–943.

Hess, D. R. (2011). Approaches to conventional mechanical ventilation of the patient with acute respiratory distress syndrome. *Respiratory Care, 56*(10), 1555–1572.

Kacmarek, R. M. (2011). The mechanical ventilator: Past, present, and future. *Respiratory Care, 56*(8), 1170–1180.

Kirby, R. R., Perry, J. C., Calderwood, H. W., Ruiz, B. C., & Lederman, D. S. (1975). Cardiorespiratory effects of high positive end-expiratory pressure. *Anesthesiology, 43*(5), 533–539.

Kleinstreuer, C., Zhang, Z., & Donohue J. F. (2008). Targeted drug-aerosol delivery in the human respiratory system. *Annu. Rev. Biomed. Eng.*, 10:195-220.

Konrad, F., Schreiber, T., Brecht-Kraus, D., & Georgieff, M. (1994). Mucociliary transport in ICU patients. *Chest, 105*(1), 237–241.

Krishnan, J. A., & Brower, R. G. (2000). High-frequency ventilation for acute lung injury and ARDS. *Chest, 118*(3), 795–807.

Levine, S., Nguyen, T., Taylor, N., Friscia, M. E., Budak, M. T., Rothenberg, P., & Shrager, J. B. (2008). Rapid disuse atrophy of diaphragm fibers in mechanically ventilated humans. *New England Journal of Medicine, 358*(13), 1327–1335.

Lunkenheimer, P. P., Rafflenebell, W., Keller, H., Frank, I., Dichut, H. H., & Fuhrmann, C. (1972). Application of transtracheal pressure oscillations as a modification of "diffusion respiration." *British Journal of Anaesthesia, 44*(6), 627.

Parker, J. C., Hernandez, L. A., & Peevy, K. J. (1993). Mechanisms of ventilator-induced lung injury. *Critical Care Medicine, 21*(1), 131–143.

Pritchard, J. N., Hatley, R., Denyer, J & von Hollen, D. (2018). Mesh nebulizers have become the first choice for new nebulized pharmaceutical drug developments. *Ther. Deliv., 9*(2), 121–136.

Putensen, C., Mutz, N. J., Putensen-Himmer, G., & Zinserling, J. (1999). Spontaneous breathing during ventilatory support improves ventilation-perfusion distributions in patients with acute respiratory distress syndrome. *American Journal of Respiratory and Critical Care Medicine, 159*(4 Pt 1), 1241–1248

Putensen, C., & Wrigge, H. (2004). Clinical review: Biphasic positive airway pressure and airway pressure release ventilation. *Critical Care, 8*(6), 492–497.

Ranu, H., Wilde, M., & Madden, B. (2011). Pulmonary Function Tests. *Ulster Med J.,* May; 80(2): 84–90.

Räsänen, J., Downs, J. B., & Stock, M. C. (1988). Cardiovascular effects of conventional positive pressure ventilation and airway pressure release ventilation. *Chest, 93*(5), 911–915.

Rathgeber, J., Schorn, B., Falk, V., Kazmaier, S., Spiegel, T., & Burchardi, H. (1997). The influence of controlled mandatory ventilation (CMV), intermittent mandatory ventilation (IMV) and biphasic intermittent positive airway pressure (BIPAP) on duration of intubation and consumption of analgesics and sedatives. A prospective analysis in 596 patients following adult cardiac surgery. *European Journal of Anaesthesiology, 14*(6), 576–582.

Schuster, D. P., Klain, M., & Snyder, J. V. (1982). Comparison of high frequency jet ventilation to conventional ventilation during severe acute respiratory failure in humans. *Critical Care Medicine, 10*(10), 625–630.

Sinderby C., et al. (1999). Neural control of mechanical ventilation in respiratory failure. *Nature Medicine, 5*, 1433–1436.

Standiford, T. J., & Morganroth, M. L. (1989). High-frequency ventilation. *Chest, 96*(6), 1380–1389.

Staudinger, T., Kordova, H., Roggla, M., Tesinsky, P., Locker, G. J., Laczika, K., . . ., & Frass, M. (1998). Comparison of oxygen cost of breathing with pressure-support ventilation and biphasic intermittent positive airway pressure ventilation. *Critical Care Medicine, 26*(9), 1518–1522.

Steinhoff, H., Falke, K., & Schwarzhoff, W. (1982). Enhanced renal function associated with intermittent mandatory ventilation in acute respiratory failure. *Intensive Care Medicine, 8*, 69–74.

Tobin, M. J. (Ed.). (2006). *Principles and Practice of Mechanical Ventilation* (2nd ed). New York, NY: McGraw-Hill.

Vockley, M. (2014). Clearing the air: Innovations and complications with ventilator technology. *Biomedical Instrumentation & Technology, 48*(4), 246–256.

Wilkins, R. L., & Stoller, J. K. (2003). *Egan's Fundamentals of Respiratory Care* (8th ed.) St. Louis, MO: CV Mosby.

Yang, K. L., & Tobin, M. J. (1991). A prospective study of indexes predicting the outcome of trials of weaning from mechanical ventilation. *New England Journal of Medicine, 324*(21), 1445–1450.

Chapter 27

ULTRASOUND BLOOD FLOW DETECTORS

OBJECTIVES

- Describe the properties of ultrasound.
- State the equations of sound propagation and Doppler effect in ultrasound.
- Explain the principles and derive equations of the Doppler and transit time blood flow measurement techniques.
- Sketch a block diagram of a Doppler blood flowmeter and explain the functions of each block.

CHAPTER CONTENTS

1. Introduction
2. Ultrasound Physics
3. Transit Time Flowmeter
4. Doppler Flowmeter
5. Functional Block Diagram of a Doppler Blood Flowmeter
6. Common Problems and Hazards

INTRODUCTION

Blood flowmeters and detectors are used to measure and evaluate the flow of blood in blood vessels. An ultrasound blood flow detector can be used noninvasively to detect blood vessel blockage or to locate and assess the degree of vascular restriction. For example, an ultrasound blood flow detector can be used to perform postoperative assessment after vascular surgery, in noninvasive diagnosis of deep venous thrombosis, or in detecting carotid artery occlusion by examining the pattern of dominant periorbital collaterals. In addition to detecting flow, some ultrasound blood flowmeters can noninvasively quantify the velocity and the volume of blood flow in blood vessels.

ULTRASOUND PHYSICS

Sound is a mechanical longitudinal (or compression) wave in which particles move back and forth parallel to the direction of wave travel. Ultrasound is sound beyond the upper audible frequency limit of human beings (i.e., of frequency 20 kHz or higher). Low-intensity (e.g., <0.1 W/cm^2) ultrasound is absorbed by human tissue without known damage. However, high-intensity ultrasound (e.g., 500 W/cm^2) can cause tissue injury due to heating effects. In addition, with appropriate frequency and setup, ultrasound can create shock wave and cavitation in tissues.

The wavelength λ of ultrasound is equal to the velocity c of sound in the medium divided by its frequency f, or

$$\lambda = \frac{c}{f}. \qquad (27.1)$$

Listed in Table 27-1 are the propagation speeds of ultrasound in different media. In diagnostic ultrasound, the average velocity of sound in soft tissue is 1540 m/sec or 1.54 mm/μs. This average velocity is used in distance calculation and in assessing the propagation of sound in body tissue. The distance of sound travel is equal to the sound velocity times the time of travel, or d = v × t.

When an ultrasound source and a receiver are moving at velocities of V_s and V_r, respectively, as shown in Figure 27-1, the apparent frequency f_r of

Table 27-1. Propagation Speed of Ultrasound

Medium	Propagation Speed (m/sec)
Air	300
Water	1480
Soft tissue	1440 to 1640
Fat	1450
Bone	2700 to 4100

Figure 27-1. Doppler Effect.

the ultrasound signal detected by the receiver is different from the source frequency f. The difference, called the Doppler shift, depends on the source frequency as well as the velocities of the source and the receiver. This is known as the Doppler effect.

In the case of the arrangement in Figure 27-1, the frequency of the received ultrasound is

$$f_r = \left(\frac{C - V_r}{C - V_s}\right) f, \tag{27.2}$$

where V_r = velocity of the receiver moving away from the source,
$\quad\quad V_s$ = velocity of the source in the same direction as V_r,
$\quad\quad f$ = frequency of the source, and
$\quad\quad C$ = speed of sound (in air = 330 m/sec).

The Doppler shift is defined as the change in frequency when the source and receiver are moving relative to each other. The Doppler shift is

$$f_D = f_r - f = \left[\left(\frac{C - V_r}{C - V_s}\right) - 1\right] f. \tag{27.3}$$

Example 27.1

What is the Doppler shift when the source is stationary and the receiver is moving toward the source at 100 m/sec in air? (C = 330 m/sec).

Solution:

Using Equation 27.3, substituting C = 330 m/sec, V_r = −100 m/sec, and V_s = 0.0,

$$f_D = \left(\frac{330 + 100}{330 - 0.0}\right) f - f = (1.3 - 1) f = 0.3 f. \tag{27.3}$$

TRANSIT TIME FLOWMETER

A transit time flowmeter computes the flow velocity by measuring the time difference between the sound traveling upstream and downstream of the flow. Figure 27-2 illustrates the principles of an ultrasound transit time flowmeter to measure blood flow in a blood vessel. An ultrasound transmitter and a receiver are positioned at an angle Θ external to the blood vessel. To start the measurement, the ultrasound transmitter A emits a short pulse. The time for the sound to reach receiver B is measured. In the next phase, ultrasound transmitter B emits a sound pulse. The time it takes to reach receiver A is again recorded. The velocity of blood flow in the vessel depends on the difference between the two recorded times. Below is the derivation. In the downstream transmission, the time for the sound to travel from point A to B is

$$T_{AB} = \frac{D}{C + v\cos\Theta} \text{ (downstream)},$$

where D = the distance between the ultrasound transducers,
C = the velocity of sound in the medium,
Θ = the angle between the direction of sound travel with the direction of blood flow, and
v = the velocity of blood.

In the upstream transmission, the time for the sound to transmit from point B to A is

$$T_{BA} = \frac{D}{C - v\cos\Theta} \text{ (upstream)},$$

The time difference ΔT between the upstream and downstream transmission is

$$\Delta T = T_{BA} - T_{AB} = \frac{D}{C - v\cos\Theta} - \frac{D}{C + v\cos\Theta} = \frac{2Dv\cos\Theta}{C^2 - v^2\cos^2\Theta}$$

$$\text{if } C \gg v, \Delta T = \frac{2Dv\cos\Theta}{C^2}$$

$$\Rightarrow v = \frac{C^2 \Delta T}{2D\cos\Theta}. \tag{27.4}$$

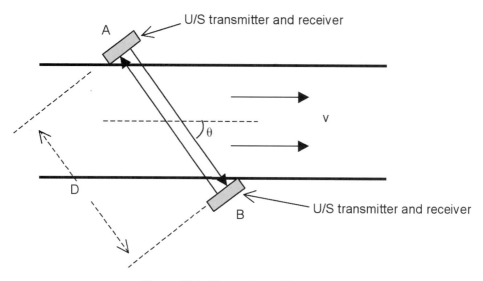

Figure 27-2. Transit Time Flowmeter.

DOPPLER FLOWMETER

Doppler flowmeters make use of the Doppler effect to determine the velocity of flow. For the setup shown in Figure 27-3, instead of a moving source and a moving receiver as shown in Figure 27-1, both transmitter and receiver are stationary. The sound wave is reflected from a moving reflector traveling at the same speed as the fluid flow. In blood flow measurement, the fluid is blood and the reflector is a red blood cell.

From the Doppler shift equation (Equation 27.3).

$$f_D = \left(\frac{C - VCos\Theta}{C + VCos\Phi} - 1 \right) f_s$$

$$= - \frac{(VCos\Theta + Cos\Phi)}{C + VCos\Phi} f_s.$$

Note that the negative sign indicates a decrease in frequency.

$$\text{if} \quad C \gg Vcos\Phi, \; f_D = - \frac{V(Cos\Theta + Cos\Phi)}{C} f_s. \qquad (27.5)$$

$$\text{if} \quad \Theta \text{ and } \Phi \text{ are both zero}, \; f_D = - 2\frac{V}{C} f_s.$$

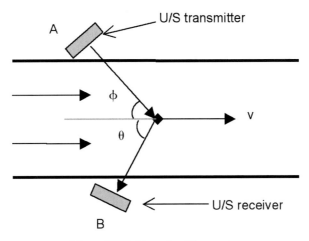

Figure 27-3. Doppler Flowmeter.

Example 27.2

For the ultrasound Doppler blood flowmeter as shown in Figure 27-3, if $\Theta = \Phi = 60°$, v = 100 cm/sec, f_s = 5 MHz, and C = 1.5 x 10^5 cm/sec, what is the Doppler shift?

Solution:

Using Equation 27.5,

$$f_D = 5 \times 10^6 \times \frac{100}{1.5 \times 10^5} \times (\text{Cos } 60° + \text{Cos } 60°) \text{ Hz} = -3.3 \text{ kHz.}$$

Note: The above results show a single frequency shift within the audible frequency range. In a real situation, as blood cells travel at different velocities, the backscattered ultrasound received will be of a broad frequency range.

In practice, an ultrasound Doppler blood flowmeter has the transmitter and receiver together so that the probe (containing both the transmitter and the receiver) can be placed on the surface of the skin or on top of a blood vessel during blood flow measurements (Figure 27-4). The Doppler shift in this case becomes

$$f_D = -2 \frac{V \cos\Phi}{C} f_s. \qquad (27.6)$$

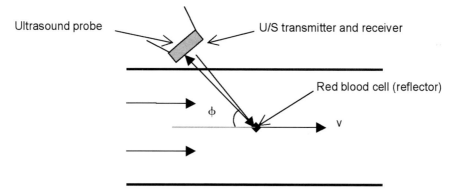

Ultrasound probe — U/S transmitter and receiver

Red blood cell (reflector)

φ   v

Figure 27-4. Doppler Blood Flowmeter.

FUNCTIONAL BLOCK DIAGRAM OF
A DOPPLER BLOOD FLOWMETER

Figure 27-5 shows a functional block diagram of an ultrasound Doppler blood flowmeter.

The RF oscillator generates the RF (e.g., 5 MHz) excitation signal to the ultrasound transmitter. The receiver detects the ultrasound reflected from the moving red blood cells in the blood vessel. The RF and the Doppler angle can be chosen such that the Doppler shifts due to the traveling blood cells are in the audio frequency range (*see* example 27.2). If all the blood cells are moving at one constant velocity, the received signal will have only one frequency that is equal to the transmitter frequency plus the Doppler shift ($f_s + f_D$). However, because blood flow is pulsatile and blood flow velocity is not the same across the blood vessel, f_D is not a single value and will occupy a range of frequencies. The signal received is frequency-modulated, with the Doppler shift proportional to the blood flow velocity. The detector is a frequency demodulator that removes the transmitter frequency f_s from the signal. In most cases, the output also contains a large amplitude low frequency wave caused by the motion of the blood vessel wall. This vessel wall motion artifact can easily be removed by a high pass filter. An audio frequency amplifier intensifies this signal and sends it to an audio speaker. Figure 27-6a shows the output from the detector and filter. Because the Doppler shift is in the audio frequency range, the clinician can hear the flow pattern of the blood in the blood vessel. A high pitch (large Doppler shift) corresponds to fast-moving blood and a low pitch corresponds to low blood flow. The Doppler shift (which is proportional to the blood flow velocity) can be converted to an analog flow velocity signal by passing it through a zero-crossing detector and a low pass filter (or integrator). The output from the zero-cross-

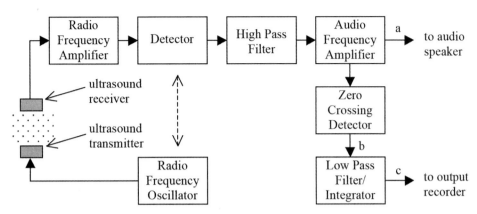

Figure 27-5. Doppler Blood Flowmeter Block Diagram.

ing detector is shown in Figures 27-6b; and the output from the low pass filter, which represents the blood flow velocity, is shown in 27-6c.

COMMON PROBLEMS AND HAZARDS

Ultrasonography is generally considered a safe imaging modality. High power ultrasound energy may cause tissue heating from mechanical vibration and cell damage from cavitation. The low power ultrasound from the transducer of a Doppler blood flow detector, however, will not cause any adverse effect to the patient. The physical compression by the ultrasound probe from the procedure (e.g., excessive by compressing the carotid artery) may cause tissue damage and restrict flow in the blood vessel.

The frequency of the sound received by the detector is modified by the movement of the blood cells or any reflected objects. Any moving object within the path of the sound beam will contribute to the Doppler shift. Noise can be introduced from movement of the detector probe, motion of the blood vessel or other tissues. The pulsatile blood pressure will cause the vessel wall to expand and contract, producing a large amplitude signal at the frequency of the cardiac cycle. Fortunately, most of these motion-related noises are of much lower frequency than the signal due to blood flow and therefore can be removed by simply using a high pass filter. The blood cells are not traveling at the same velocity all the time inside the blood vessel due to turbulence and higher friction near the wall of the blood vessel; the Doppler shift is not a single frequency at any instant of time.

From Equation 27.6, the Doppler shift is also proportional to $\cos\Phi$. When $\Phi = 90°$, $\cos\Phi$ is zero. This happens when the probe is held perpendicular to the blood vessel. In order to obtain a large frequency change from

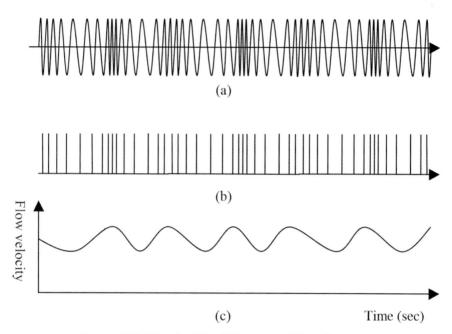

(a)

(b)

(c)　　　　　　　　　　　　　　　Time (sec)

Figure 27-6. Doppler Blood Flowmeter Block Diagram.

the change in blood flow velocity, the probe angle should be held constant and be as small as possible. In order to reduce loss of ultrasound intensity when it travels from the probe into the patient, ultrasound gel is applied between the patient's skin and the ultrasound probe. Aqueous gel is used to eliminate the air gap between the ultrasound probe and patient's skin to reduce the reflection loss due to the large acoustic impedance difference between the probe and the air (as well as the air and skin) interface. Without the ultrasound gel, the intensity of the reflected sound will become too small to provide a useful signal.

BIBLIOGRAPHY

Beldi, G., Bosshard, A., Hess, O., Althaus, U., & Walpoth, B. H. (2000). Transit time flow measurement: Experimental validation and comparison of three different systems. *Annals of Thoracic Surgery, 70*(1), 212–217.

Cobbold, R. S. C. (2007). *Foundations of Biomedical Ultrasound.* New York, NY: Oxford University Press.

Fish, P. (2003). *Physics and Instrumentation of Diagnostic Medical Ultrasound* (2nd ed.). Chichester, UK: John Wiley & Sons.

Hatle, L., & Angelsen, B. (1993). *Doppler Ultrasound in Cardiology: Physical Principles and Clinical Applications* (3rd ed.). Philadelphia, PA: Lea & Febiger.

Kearon, C., Julian, J. A., Newman, T. E., & Ginsberg, J. S. (1998). Noninvasive diagnosis of deep venous thrombosis. *Annals of Internal Medicine, 128*(8), 663–677.

Laustsen, J., Pedersen, E. M., Terp, K., Steinbrüchel, D., Kure, H. H., Paulsen, P. K., ... & Paaske, W. P. (1996). Validation of a new transit time ultrasound flowmeter in man. *European Journal of Vascular and Endovascular Surgery, 12*(1), 91–96.

Liptak, B. G. (Ed.). (2003). *Instrument Engineers' Handbook* (4th ed., Vol. 1: Process Measurement and Analysis). Boca Raton, FL: CRC Press.

Merritt, C. R. (1989). Ultrasound safety: What are the issues? *Radiology, 173*(2), 304–306.

Pagana, K. D., & Pagana, T. J. (2010). *Mosby's Manual of Diagnostic and Laboratory Tests* (4th ed.). St. Louis, MO: Mosby Elsevier.

Pinkney, N. (2005). *Ultrasound Physics, Imaging, Instrumentation and Doppler* (3rd ed.). West Babylon, NY: Sonicor, Inc.

Chapter 28

FETAL MONITORS

OBJECTIVES

- Describe the clinical significance of monitoring fetal heart rate (FHR) and maternal uterine activities (UAs) during labor.
- Describe and contrast different methods of monitoring FHRs, including direct, ultrasonic, maternal abdominal, and phono methods.
- Describe and compare external and intrauterine methods of monitoring maternal UAs.
- Explain the construction and principles of transducers and sensors used in fetal monitoring.
- Sketch a simple block diagram of a fetal monitor.

CHAPTER CONTENTS

1. Introduction
2. Monitoring Parameters
3. Methods of Monitoring Fetal Heart Rate
4. Methods of Monitoring Uterine Activities
5. Common Problems and Hazards

INTRODUCTION

Electronic fetal monitoring or cardiotocography provides graphic and numerical information to assist the clinician to assess the well-being of the fetus and the stage of labor. During labor, the FHR often accelerates and decelerates in response to the uterine contractions and fetal movements. Characteristics of these patterns may reveal labor problems, such as fetal hypoxia or decreased placental blood flow. Examining these patterns may indicate alternative courses of labor (e.g., cesarean section, suction or forceps delivery) or drug therapy (e.g., administering labor-inducing or labor-prohibiting drugs).

Antepartum (before birth) monitoring is used to monitor the development of the fetus in the uterus. Intrapartum monitoring includes monitoring the status of the mother and fetus as well as the progress of labor. Maternal monitoring includes measurements of the mother's vital signs such as heart rate, respiratory rate, blood pressure, temperature, oxygen saturation level, and UA. Fetal monitoring refers to the monitoring of the FHR and the maternal UA during labor and delivery. Electronic fetal monitors were first available in the late 1960s. Today, fetal monitoring is used in more than 60% of deliveries in North America.

MONITORING PARAMETERS

The two primary parameters in fetal monitoring are FHR and UAs. Other parameters that may be monitored are the maternal ECG and %SaO$_2$. FHR may reveal the conditions of the fetus during labor and delivery. Interpretation of FHR traces includes quantitative and qualitative analysis of its baseline, variability, change of patterns over time, accelerations, and decelerations. Normal FHRs fall within the range of 120 to 160 bpm during the third trimester of pregnancy and fluctuate from the baseline rate during contractions. Figure 28-1 shows a typical recording of FHR. Some abnormal FHR conditions and their indications are

- **Tachycardia (high heart rates)**–may be caused by maternal fever, fetal hypoxia, immaturity of fetus, anemia, or hypotension
- **Bradycardia (low heart rates)**–may be caused by congenital heart

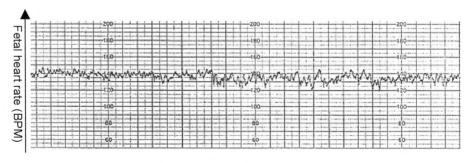

Figure 28-1. Fetal Heart Rate.

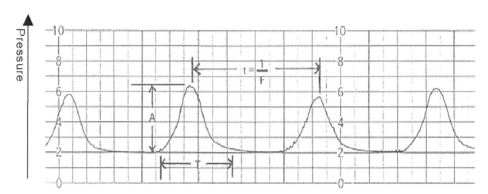

Figure 28-2. Uterine Activities.

lesions or hypoxia
• **Variation**—too much fluctuation indicates stress or hypoxia

UA refers to the frequency and intensity of the contractions of the uterus. During labor, the smooth muscles of the uterus contract rhythmically, thereby increasing the pressure of the amniotic fluid and forcing the fetus against the cervix. UA indicates the progress of labor. Figure 28-2 shows a typical recording of UA.

Some characteristics of UAs are

• **Frequency (F)**—less than once in 3 min is slow progress of labor
• **Duration (T)**—less than 45 sec of contraction is slow progress of labor
• **Amplitude (A)**—more than 75 mmHg usually indicates active labor
• **Shape**—the shape of the contraction pressure is normally bell-shaped. An irregular shape may indicate labor pushing, fetal movement, maternal respiration, or blocked catheter
• **Rhythm**—couplets and triplets indicate abnormal activities
• **Resting tone (pressure between contraction)**—about 5 mmHg for non-labor and rising to 20 mmHg for induced labor

METHODS OF MONITORING FETAL HEART RATE

FHR may be obtained by listening to the heart sound of the fetus, directly connecting electrodes to the fetus, applying electrodes on the abdomen of the mother, or using Doppler ultrasound. The three methods are described in the following sections.

Direct ECG

Direct ECG is an invasive method that connects a spiral electrode to the scalp of the fetus. During application, the electrode is inserted through the vulva. While pushing against the scalp of the fetus, the clinician applies a 360-degree turn to the spiral electrode so that the electrode is screwed and secured into the skin of the scalp (Figure 28-3a). The electrode can be applied only when the head of the fetus is accessible; that is, only after the amniotic sac has ruptured. The other electrode is usually a skin electrode applied to the thigh of the mother. Because the procedure is invasive, it may cause complications (e.g., infection) to the fetus.

Phono Method

The FHR may be derived by listening to the fetal heart sound. Although a microphone can be used, this is usually done manually by the obstetric nurse or physician using a stethoscope placed on the abdomen of the mother. The weak fetal heart sound is usually buried among the louder maternal heart sound and other sounds (such as sound from bowel movement) within the mother's body. The advantage of this method is that it is noninvasive and does not require expensive equipment.

Abdominal ECG

Abdominal ECG is obtained by applying skin electrodes on the abdomen of the mother (on fundus, pubic symphysis, and maternal thigh). The electrodes are attached to a normal ECG machine so that the waveform and heart rate are displayed. Because the electrodes will inevitably pick up the maternal ECG, careful electrode positioning to capture the fetal ECG and differentiate it from the maternal signal is required.

Ultrasound Method

Another noninvasive method to monitor FHR employs a Doppler ultrasound detector. A beam of continuous wave ultrasound (e.g., 2 MHz) from an ultrasound transmitter/receiver pair is applied to the abdomen of the mother (Figure 28-3b). If the ultrasound beam crosses the fetal heart, the Doppler shift detected from the reflected sound will record the motion of the fetal heart wall and thus can be processed to obtain the FHR (*see* Chapter 27). This method provides an accurate beat-to-beat measurement of the heart rate provided that the ultrasound beam covers the fetal heart. To avoid picking up movement artifacts from other organs, a narrow sound beam is

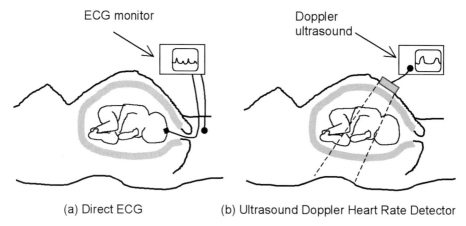

Figure 28-3. Fetal Heart Rate Monitors.

preferred. However, with a narrower sound beam, the transducer position must be checked from time to time to ensure that the sound beam is focused on the fetal heart. In addition, it requires good skin–transducer contact (achieved by application of ultrasound gel) to obtain good signal. Although more complicated and expensive, a pulsed Doppler with time gating can provide better quality signal than a continuous wave Doppler unit.

METHODS OF MONITORING UTERINE ACTIVITIES

Intrapartum UAs may be obtained by using an external pressure transducer applied on the abdomen of the mother or by inserting a fluid-filled catheter into the uterus. The former is an indirect and noninvasive method; the latter is direct and invasive.

External Pressure Transducer Method

Uterine contraction can be monitored by placing a pressure transducer on the abdomen close to the fundus. A pressure-sensitive flat-surfaced contraction transducer, called a tocodynamometer, is affixed to the skin of the abdomen by a band around the belly of the mother (Figure 28-4a). The transducer is often referred to as a toco transducer. Because the pressure required to flatten the wall correlates with the pressure on the other side of the wall, the pressure in the uterus during contractions can be monitored by the externally placed toco transducer. The advantage of this method is its noninvasiveness. However, it has low accuracy (about 20% error), and it requires frequent repositioning and retightening of the belt.

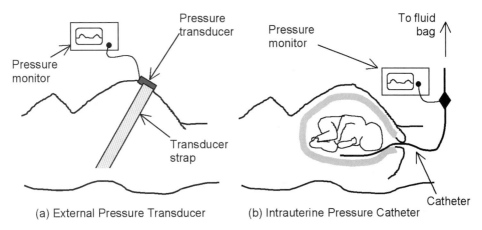

Figure 28-4. Uterine Activity Monitoring.

Intrauterine Pressure Method

The pressure obtained in this method is more accurate than using the toco transducer. It is a direct pressure measurement method using a setup similar to direct blood pressure monitoring. A fluid-filled catheter is inserted into the uterus after the amniotic sac is ruptured. The catheter is connected to a pressure transducer (Figure 28-4b). The pressure inside the uterus is displayed on a blood pressure monitor. Although this method is more accurate, it is invasive. Care should be taken to ensure that there is no obstruction or occlusion of the catheter during labor and that the transducer and setup are properly zeroed before use (*see* Chapter 18 on blood pressure monitors for reasons and methods of pressure transducer zeroing). Disposable pressure transducers are often used.

To enhance patient mobility, telemetry is used to remove the electrical wires and cables connecting the electrodes and transducers on the patient to the monitor.

COMMON PROBLEMS AND HAZARDS

False counting of maternal heart beat as FHR has been reported, which has led to inaccurate diagnoses and inappropriate treatment. Problems associated with telemetry in electronic fetal monitoring are similar to other telemetry devices. Signal fading from attenuation, EMI and transmitter channel conflicts will result in false alarms and momentary loss of monitoring data.

Invasive scalp electrodes may cause complications such as injury to the fetal eye, hemorrhage, and infection. Maternal infection, tissue injury, umbilical cord damage, and compression are some complications that may arise from intrauterine catheter insertion. Although the ultrasound intensity from the ultrasound probe of the electronic fetal monitoring is much lower than that from diagnostic imaging procedures, the potential risk associated with fetal exposure to ultrasound has remained a concern for some investigators

Studies have shown that the false-positive rate for predicting adverse outcomes from fetal monitoring is high and that the increased use of electronic fetal monitoring correlates with the increased rate of instrumental deliveries (cesarean sections, forceps or vacuum extraction, etc.). The American College of Obstetricians and Gynecologists recommends continuous use of electronic fetal monitoring only on high-risk patients.

BIBLIOGRAPHY

Alfirevic, Z., Devane, D., & Gyte, G. M. L. (2006). Continuous cardiotocography (CTG) as a form of electronic fetal monitoring (EFM) for fetal assessment during labour [Online]. Cochrane Database of Systematic Reviews. Available: http://onlinelibrary.wiley.com/doi/10.1002/14651858.CD006066.pub2/abstract

Bailey, R. E. (2009). Intrapartum fetal monitoring. *American Family Physician, 80*(12), 1388–1396.

Dildy, G. A. (1999). The physiologic and medical rationale for intrapartum fetal monitoring. *Biomedical Instrumentation & Technology, 33*(2), 143–151.

Freeman, R. K., Garite, T. J., & Nageotte, M. P. (2003). *Fetal Heart Rate Monitoring* (3rd ed.). Baltimore, MD: Lippincott Williams & Wilkins.

Galazios, G., Tripsianis, G., Tsikouras, P., Koutlaki, N., & Liberis, V. (2010). Fetal distress evaluation using and analyzing the variables of antepartum computerized cardiotocography. *Archives of Gynecology and Obstetrics, 281*(2), 229–233.

Goddard, R. (2001). Electronic fetal monitoring. *British Medical Journal, 322*(7300), 1436–1437.

Liston, R., Crane, J., Hamilton, E., Hughes, O., Kuling, S., MacKinnon, C., . . . , & Trepanie, M. J. (2002). Fetal health surveillance in labour. *Journal of Obstetrics and Gynaecology Canada, 24*(4), 342–355.

Macones, G. A., Hankins, G. D., Spong, C. Y., Hauth, J., & Moore, T. (2008). The 2008 National Institute of Child Health and Human Development workshop report on electronic fetal monitoring: Update on definitions, interpretation, and research guidelines. *Obstetrics and Gynecology, 112*(3), 661–666.

Miesnik, S. R., & Stringer, M. (2002). Technology in the birthing room. *Nursing Clinics of North America, 37*(4), 781–793.

Simpson, K. R. (2004). Monitoring the preterm fetus during labor. *MCN, The American Journal of Maternal/Child Nursing, 29*(6), 380–388.

Zottoli, E. K., & Wood, C. (2003). The fundamentals of electronic fetal monitoring. *Biomedical Instrumentation & Technology, 37*(5), 353–358.

Chapter 29

INFANT INCUBATORS, PHOTOTHERAPY LIGHTS, WARMERS AND RESUSCITATORS

OBJECTIVES

- Describe the clinical functions of infant incubators.
- Explain typical features of an infant incubator.
- Sketch a functional block diagram of a typical infant incubator.
- Explain the construction and major components of an infant incubator.
- Describe the mechanism of phototherapy and its clinical functions.
- Identify the spectral characteristics of a phototherapy light.
- Analyze factors affecting the output intensity of phototherapy lights.
- Describe the applications of infant radiant warmers and resuscitators in delivery rooms and the nursery.
- Explain functional features and parameters of infant radiant warmers and resuscitators.
- Discuss hazards associated with infant incubators, phototherapy lights, infant radiant warmers and resuscitators.

CHAPTER CONTENTS

1. Introduction
2. Purpose
3. Infant Incubators
4. Phototherapy Lights
5. Infant Radiant Warmers
6. Common Problems and Hazards

INTRODUCTION

An infant incubator provides a controlled environment to the infant by regulating the temperature, humidity and oxygen level within the incubator chamber. A phototherapy light is used to break down excessive concentration of bilirubin in the newborn. Incubators and phototherapy lights are found in neonatal care areas to treat preterm or sick infants. Infant warmers and resuscitation units are commonly found in labor and delivery areas, and in neonatal intensive care units to hold infants for emergency intervention and to maintain the body temperature of the infants during intervention or observation.

PURPOSE

At birth, the body temperature of an infant tends to drop significantly due to heat loss from the body. Heat loss can be through conduction (contact with other objects), convection (heat carried away by air circulation), radiation (heat lost to a cooler environment due to infrared radiation from the warm body), and evaporation (latent heat loss from the lungs and skin surface). Most term neonates regulate their body temperature naturally to some extent. Preterm neonates, with thinner skin and higher surface to volume ratio, however, tend to lose more heat and can easily become hypothermic. Infant incubators and radiant warmers are used to provide thermal support for critically ill infants who require constant nursing intervention. An infant radiant warmer radiates heat energy to the infant by using an external heat lamp directed to the infant. Although their objectives are similar, incubators usually provide better temperature regulation than infant warmers do. In addition, incubators provide an enclosed and controlled environment for infants to receive their therapies. When compared to the open design of an infant warmer, however, the enclosed chamber of an incubator is less convenient for clinician access to the infant. Most of the radiant infant warmers are equipped with resuscitation equipment to treat infants when needed.

INFANT INCUBATORS

Principles of Operation

An infant incubator provides an enclosed and controlled environment for the infant. The temperature, oxygen level, and humidity within the enclosed incubator chamber can be precisely controlled. A slightly positive pressure can be maintained inside the chamber relative to the atmosphere to decrease the chance of infection. The clear transparent enclosure with access ports allows observation of and intervention for the infant.

Functional Components and Common Features

An incubator consists of a chamber enclosed by a transparent plastic hood made of Plexiglas™ (a kind of clear acrylic). The infant lies on the mattress inside the enclosed chamber. Access doors and ports through the hood allow relatively easy access to the infant for feeding, examination, and treatment. A blower and heater underneath the mattress provide forced circulation of warm air inside the chamber. An infant incubator usually has two modes of temperature control: skin and air temperature controls. In the skin temperature control mode, a temperature senor is attached to the skin of the infant. The skin temperature signal is compared to the value set by the user to cycle the heater on and off. In the air temperature mode, a temperature sensor is located inside the hood of the incubator to measure the air temperature. This measured value is compared to the set value to turn the heater on or off. To provide better temperature regulation, proportional heating control instead of simple on-off control is used.

In a proportional heating control circuit, instead of being fully switched on or off, the heater can be partially turned on at incremental percentage of power. When there is a large difference between the measured and the set temperature (e.g., at initial startup), the heater is switched on at its full power. When the difference becomes smaller, the heater will run at a lower power setting. This control approach minimizes the fluctuation of temperature within the incubator compartment to provide better infant body temperature regulation. Figure 29-1 shows an example of the power and temperature relationships of a four-level proportional heater controller of an infant incubator. When the temperature difference ΔT is larger than 6°C, the heater is running at 100% power; as the air inside the incubator becomes warmer, the power of the heater is reduced. When the temperature inside the incubator is less than the preset temperature by less than 2°C, the heater is running at only 25%.

Proportional heating control can be implemented by using several banks of heaters (e.g., using four 250 W heaters instead of one 1000 W heater) or, if a single heater is used, it can be achieved by adjusting the duty cycle of the heater supply power. The latter can be designed to produce a continuous variation of heater power according to the measured temperature difference. To further reduce temperature fluctuation, some manufacturers add an additional piece of Plexiglas™ close to the hood inside the infant chamber. A portion of the warm air from the heater compartment is directed to flow inside the gap between the hood and this additional inner wall. This double-

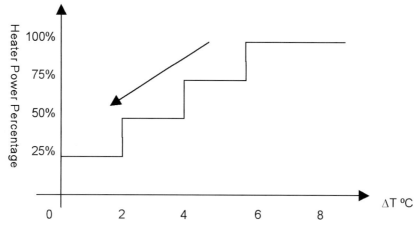

Figure 29-1. Proportional Heater Control Characteristics.

wall design improves temperature isolation of the infant chamber from the external environment.

To increase comfort and prevent dehydration, most incubators allow users to vary the relative humidity inside the infant chamber. A water reservoir is located underneath the infant chamber. Air is blown over the reservoir to add moisture to the air inside the chamber. Humidity is controlled by adjusting the amount of airflow through the reservoir or controlling the temperature of water using a heater.

Most incubators have an oxygen inlet to create an elevated oxygen level within the incubator. Some have a built-in oxygen sensor and controller to maintain a preset elevated level of oxygen inside the chamber. In addition, accessories such as X-ray cassette trays and weighing scales are available. Many infant incubators are connected to an intensive care workstation networked to the hospital information system to allow clinicians to access electronic patient information.

Figure 29-2 shows the functional component diagram of an infant incubator. Common features of infant incubators include

- Easy access to infant with front, side, and rear access ports;
- Access ports are cuffed to minimize heat loss and temperature fluctuation. Instead of a cuff, some manufacturers create an air curtain at the opening ports to reduce heat loss.
- Height-adjustable infant bed (table) with tilt mechanism
- Skin or air temperature sensor options for temperature control
- Adjustable temperature control, with maximum setting of 39°C
- Proportional heater control to minimize temperature fluctuation

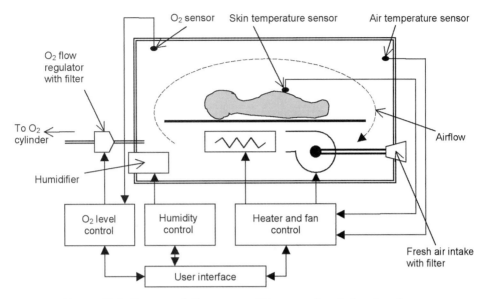

Figure 29-2. Functional Component Diagram of an Infant Incubator.

- Oxygen sensor, regulator, and supply manifold to control oxygen level inside incubator
- Humidity sensor and water reservoir to maintain relative humidity inside incubator
- Low airflow across infant to reduce heat loss and dehydration
- Low internal audible noise (less than 50 dBA) to prevent hearing damage to infant
- Adjustable alarm settings including temperature, oxygen level, and humidity
- Heater over temperature and loss of airflow safety cut off
- Independent maximum air temperature (>41°C) sensor and alarm
- Numerical display including temperature, oxygen level, humidity level, and heater power
- Data trending, alarm log, and networking capability
- Air and oxygen inlet filters
- Built-in baby weight scale
- Construction to allow easy disassembling, cleaning, and disinfection

Transport incubators are used to transport sick infants between healthcare facilities. In addition to providing the same functionalities as a conventional incubator, the weight, size, and portability are design considerations. The longevity of the power supply is a technical challenge of a transport incubator because it can be disconnected from regular line power for an extended period of time.

PHOTOTHERAPY LIGHTS

Principles of Operation

A phototherapy light is used to break down bilirubin in the newborn. Jaundice occurs when the liver of the infant has not reached full detoxification capability, especially in premature infants. During the first week of life, infants have poor liver function to remove bilirubin. A bilirubin level of 1 to 5 mg per 100 ml of blood within the first 3 days of birth is considered normal. This level should decrease as the liver begins to mature. A visible light spectrum of wavelength from 400 to 500 nm (blue) has been shown to be effective in transforming bilirubin into a water-soluble substance that can then be removed by the gallbladder and kidneys. A spectral irradiance of 4 mW/cm^2/nm at the skin surface is considered to be the minimum level to produce effective phototherapy.

Functional Components and Common Features

A phototherapy light can be placed directly over an infant in a bassinet or placed over the hood of an incubator. Instead of the full white light spectrum, blue light sources are used to increase the efficacy of phototherapy. However, blue light can mask the skin tone of the infant and is hard on the eyes of the caregivers. As a compromise, some manufacturers use a combination of white and blue light sources and have built-in features to switch off the blue lights during observation.

Ultraviolet radiation (< 400 nm) emitted from most blue light sources is harmful to the infant. Infants receiving phototherapy are required to wear eye protectors to prevent damage to their retinas. A sheet of Plexiglas placed between the light source and the infant can cut out most of the wavelength below 380 nm. Far-infrared radiation (heat) can create hyperthermia and cause dehydration to the infant. As a precaution, monitoring or periodically checking the skin temperature of an infant undergoing phototherapy treatment is recommended. Some common features of a phototherapy light are as follows:

- A blue light source (e.g., special blue fluorescent tubes, tungsten halogen, etc.) with a high-intensity blue spectrum (e.g., 400–500 nm) or a combination of blue and white light sources is used.
- The output of phototherapy lights measured at skin level should be greater than 4 mW/cm^2/nm within the range of wavelength from 400 to

500 nm. Most devices on the market have output much greater than this minimum level.

- Filters (Plexiglas) to remove ultraviolet (280–400 nm) radiation to avoid damaging the infant's eyes and skin.
- Equipped with white light for observation. Blue light is switched off during observation; a timer automatically switches the light back to phototherapy after observation.
- Light bulb operation timer to signal light bulb end-of-life replacement
- Light source housing on height adjustable stand for light intensity adjustment.

Manufacturers of conventional phototherapy lights use a number of fluorescent tubes (e.g., eight 50-cm tubes) in a metal housing for their phototherapy lights. The lower surface of the light compartment is a piece of Plexiglas for mechanical protection (to prevent tubes from shattering) as well as serving as an ultraviolet (UV) filter. A typical unit is shown in Figure 29-3. The light source housing is either mounted on a height adjustable stand or placed directly on top of the hood of an infant incubator.

Experiments have shown that the spectral output of a fluorescent tube changes with the tube surface temperature as well as with time. A fully enclosed tube housing as shown in Figure 29-3 can reach a temperature of 70°C after 2 to 3 hours of operation. Figure 29-4 shows the typical characteristics of a phototherapy light output with respect to temperature inside the light source housing. It shows that the light output will reach a peak value shortly after it is turned on and will drop to about 70% of its peak value after a few hours when it has risen to a steady temperature.

In addition, from manufacturers' specifications, fluorescent tubes have a limited life span ranging from a few hundred hours (for special blue light

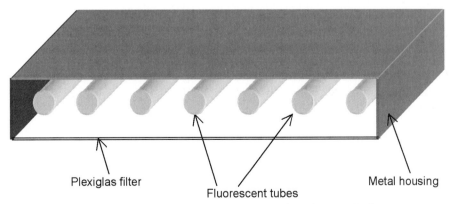

Figure 29-3. Cross-Sectional View of a Phototherapy Light.

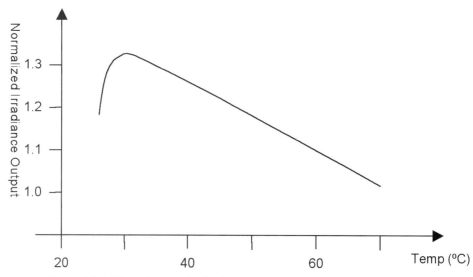

Figure 29-4. Phototherapy Light Output vs. Temperature Characteristics.

tubes) to about 2000 hours (for white light tubes). The tube output also decreases as it ages (e.g., 10% drop after 300 hours). Blue fluorescent tubes generally have a shorter life span than ordinary white tubes have. In order to ensure sufficient light output for phototherapy, some hospitals are measuring the output using a special light meter and replacing the tubes when they are below a certain limit. Instead of performing periodic output measurements, some users implement a fixed schedule (e.g., by monitoring the hours of operation) to replace these light sources. Other than fluorescent tubes, tungsten-halogen and quartz bulbs have been used as light sources for phototherapy. Blue light-emitting diodes (LEDs), which was invented in the early 1990s, enables the creation of high energy efficient, bright, broad spectrum light sources with wide applications in medicine. In recent years, blue LED that emit light with wavelengths from 450 to 475 nm (within the effective phototherapy spectrum of 400 to 500 nm) have gained popularity in phototherapy. LED light sources have a much longer lifespan than conventional phototherapy light sources, produce less heat and require less maintenance. It is now the preferred light source in phototherapy.

Instead of producing therapeutic light irradiating from above, a phototherapy blanket is a special phototherapy device with light coming out from a blanket placed underneath or wrapped around the infant. Thousands of flexible optical fibers transmit light from the light source to the blanket. These fibers terminate at the surface of the blanket to allow therapeutic light to emerge from the ends of the optical fibers on to the bare skin of the infant.

INFANT RADIANT WARMERS AND RESUSCITATORS

Principles of Operation

An infant radiant warmer is a body warming device to provide radiant heat to the infant. Infrared (IR) energy irradiates the surface of the infant and is readily absorbed; the elevated temperature increases blood flow and promotes heat transfer to the rest of the body. The heat loss in some newborn babies is rapid; infant radiant warmers help to maintain the body temperature of the infants.

A typical infant radiant warmer has an overhead radiant heating element located above a bassinet (a basket-shaped bed for the infant). It consists of a heat source, a skin-temperature sensor attached to the skin of the infant, a control unit, a visual display, and audible alarms. Infant radiant warmers usually operate in automatic mode but manual mode may be selected. A built-in timer limits the heating time and prompts the user to assess the infant's status. A resuscitation unit is a radiant warmer equipped with suction and oxygen for resuscitation. Infant resuscitators are equipped with IV poles for fluid therapies and platforms to hold patient monitors and equipment.

Functional Components and Common Features

Instead of using a heater to warm up air surrounding the infant, a radiant warmer uses a heating element (such as a quartz tube) to generate and deliver radiant heat energy directly to the body of the infant. Radiant warmers are designed to produce radiant energy in the far infrared (IR) wavelength region to avoid damaging the retinas and corneas of infants. The radiant output of the heating unit is also limited to prevent thermal injury to the infant.

A temperature sensor (e.g., a thermistor) applied to the infant's abdomen is used to measure the body temperature of the infant. In automatic (or servo) mode, the heating element is turned on and off in response to changes in the infant's skin temperature. Heater power output is controlled by a proportional controller. In manual mode, the user sets an automatic timer to operate the heater at a constant power level for a set period of time. The warmer's alarm sounds after the timer preset period has elapsed, prompting the operator to reexamine the infant's condition and temperature. If the timer is not reset by the user, the heater will remain on for an additional duration (e.g., another 5 min) but will eventually shut down and trigger an alarm until it is manually reset.

Infant resuscitators used in the delivery rooms are fitted with Apgar timers. An Apgar timer reminds the clinician to assess the condition the new born at times of 1 minute and 5 minutes after birth with expanded recording

Table 29-1 Apgar Criteria

Criterion	Score = 0	Score = 1	Score = 2
Skin Color	Blue all over	Blue at extremities	Normal
Heart Rate	Absent	Less than 100	Greater the 100
Reflex	No response	Grimace/feeble cry	Sneeze/cough/pull away
Muscle Tone	None	Some flexion	Active movement
Respiration	Absent	Week or irregular	Strong

at 5-minute intervals if the infant scores seven or less at 5 minutes. Apgar scores are determined by assessing the infant according to the criteria in Table 29-1. The total score is computed by adding the scores from each of the five criteria. In general, a total Apgar score of seven or higher is normal, while a score below three is considered critically low. A low score at the 1-minute mark alert extra attention but not necessary indicate significant problems. A low score at 5-minute mark or beyond indicates that the infant is at significant risk.

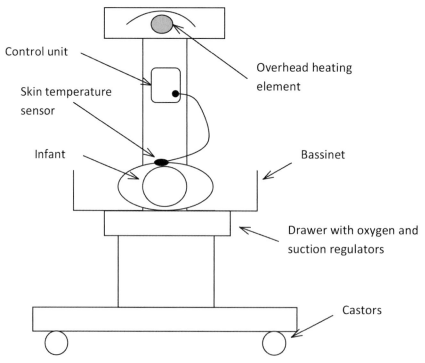

Figure 29-5. Infant Radiant Warmer.

Figure 29-5 shows a freestanding infant radiant warmer. Freestanding units are designed for mobility and provide thermal support during diagnostic or therapeutic treatment. The integrated warmer and bassinet unit allows continuous thermal support of the sick infant and also acts as a short-term resuscitation platform in the delivery suite or operating room.

COMMON PROBLEMS AND HAZARDS

In most cases, infants inside incubators or under radiant warmers are premature with poor body temperature regulation. Temperature control failures leading to extreme hyperthermia can result in skin burns, permanent brain damage, or even death. A very common cause of overheating is due to detached skin temperature sensors when the unit is under the automatic skin temperature control mode. The detached sensor detects air instead of skin temperature. The lower air temperature keeps the heater turned on, which leads to overheating. Periodic checking of sensor condition, mode of temperature control, and temperature setting can prevent such incidents. Sensors and thermostat failures can also create similar problems.

Improper oxygen control may cause hyperoxia or hypoxia. Prolonged exposure of a premature infant to excessively high oxygen concentration in an incubator will cause retrolental fibroplasia (formation of fibrous tissue behind the lens), which can lead to blindness.

Because the infant is always inside the incubator, it is important to reduce the noise level inside the hood to protect the hearing of the infant. The noise level measured inside the hood of the incubator should be below 50 dBA. With proper design, it is not difficult for incubator manufacturers to achieve such low noise level. However, worn-out fan motor bearings and imbalanced fan rotors can produce continuously high noise level inside the incubator chamber. In addition, using a ventilator or a nebulizer, opening and closing access doors, or tapping on the incubator hood can produce temporary noise at a level as high as 100 dBA inside the incubator.

The warm and moist air inside an incubator and the water reservoir used for humidification provide a favorable growth environment for bacteria. Care must be taken to clean and disinfect incubators after every use. Complete disinfection or sterilization should be done periodically.

There have been reported cases of infants falling out of incubators and from radiant warmers due to improper closing or broken latches on doors, access ports, or side panels.

Hazards of phototherapy include hyperthermia and dehydration. Infants with unprotected eyes may suffer eye injuries from the intense light source. Eye protectors (e.g., eye patches or masks) must be worn by infants during phototherapy. Care must be taken to secure the eye protectors to prevent them from being dislodged and causing suffocation of the infants. The blue light will mask the skin tone of the infant. The observation light (white light) built in to the phototherapy unit allows clinician to see the true skin color of the infant. The observation light can be programmed to turn off automatically after a preset period of time. Ultraviolet (UV) radiation will cause skin damage, and erythema has been documented on infants who underwent phototherapy using fluorescent overhead lamps without Plexiglas filters. Radiant heat from the phototherapy light source can induce hyperthermia. UV and IR filters can be used in the light source to prevent such injuries by removing harmful radiation.

BIBLIOGRAPHY

Association for the Advancement of Medical Instrumentation. (2009). Medical electrical equipment—Part 2-19: Particular requirements for the basic safety and essential performance of infant incubators. ANSI/AAMI/IEC 60601-2-19:2009.

Association for the Advancement of Medical Instrumentation. (2009). Medical electrical equipment—Part 2-20: Particular requirements for the basic safety and essential performance of transport incubators. ANSI/AAMI/IEC 60601-2-20:2009.

Association for the Advancement of Medical Instrumentation. (2009). Medical Electrical Equipment—Part 2-21: Particular requirements for the basic safety and essential performance of infant radiant warmers. ANSI/AAMI/IEC 60601-2- 21:2009.

Association for the Advancement of Medical Instrumentation. (2009). Medical electrical equipment—Part 2-50: Particular requirements for the basic safety and essential performance of infant phototherapy equipment. ANSI/AAMI/IEC 60601-2-50:2009.

Bratlid, D., Nakstad, B., & Hansen, T. W. (2011). National guidelines for treatment of jaundice in the newborn. *Acta Paediatrics,* 100(4), 499–505.

De Araujo, M. C., Vaz, F. A., & Ramos, J. L. (1996). Progress in phototherapy. *Sao Paulo Medical Journal, 114*(2), 1134–1140.

De Carvalho, M., Torrao, C. T., & Moreira, M. E. (2011). Mist and water condensation inside incubators reduce the efficacy of phototherapy. *Archives of Disease in Childhood. Fetal and Neonatal Edition, 96*(2), F138–F140.

Ebbesen, F., Madsen, P., Støvring, S., Hundborg, H., & Agati, G. (2007). Therapeutic effect of turquoise versus blue light with equal irradiance in preterm infants with jaundice. *Acta Paediatrics, 96*(6), 837–841.

Fic, A. M., Ingham, D. B., Ginalski, M. K., Nowak, A. J., & Wrobel, L. (2010). Heat and mass transfer under an infant radiant warmer—development of a numerical model. *Medical Engineering & Physics, 32*(5), 497–504.

Finster M, Wood M. (2005). The Apgar score has survived the test of time.

Anesthesiology, 102(4), 855–857

Hill, J. (2009). A primer on infant incubators. *Biomedical Instrumentation & Technology, 43*(4), 295–296.

Kim, S. M., Lee, E. Y., Chen, J., & Ringer, S. A. (2010). Improved care and growth outcomes by using hybrid humidified incubators in very preterm infants. *Pediatrics, 125*(1), e137–145.

Kuboi, T., Kusaka, T., Yasuda, S., Okubo, K., Isobe, K., & Itoh, S. (2011). Management of phototherapy for neonatal hyperbilirubinemia: Is a new radiometer applicable for all wavelengths and light source types? *Pediatrics International, 53*(5), 689–693.

Laroia, N., Phelps, D. L., & Roy, J. (2007). Double wall versus single wall incubator for reducing heat loss in very low birthweight infants in incubators. *Cochrane Database of Systematic Reviews, 18*(2), CD004215.

Lyon, A. J., & Freer, Y. (2011). Goals and options in keeping preterm babies warm. Archives of Disease in Childhood. *Fetal and Neonatal Edition, 96*(1), F71–74.

Maisels, M. J., & McDonagh, A. D. (2008). Phototherapy for neonatal jaundice. *New England Journal of Medicine, 358*, 920–928.

McClelland, P. B., Morgan, P., Leach, E. E., & Shelk, J. (1996). Phototherapy equipment. *Dermatology Nursing, 8*(5), 321–328.

Meyer, M. P., Payton, M. J., Salmon, A., Hutchinson, C., & de Klerk, A. (2001). A clinical comparison of radiant warmer and incubator care for preterm infants from birth to 1800 grams. *Pediatrics, 108*(2), 395–401.

Onishi, S., Isobe, K., Itoh, S., Manabe, M., Sasaki, K., Fukuzaki, R., & Yamakawa, T. (1986). Metabolism of bilirubin and its photoisomers in newborn infants during phototherapy. *Journal of Biochemistry, 100*(3), 789–795.

Plangsangmas, V., Leeudomwong, S., & Kongthaworn, P. (2012). Sound pressure level in an infant incubator. *Mapan–Journal of Metrology Society of India, 27*(4), 199–203.

Poole, D. R. (1997). Challenges in the design of transport incubators. *Biomedical Instrumentation & Technology, 31*(2), 137–142.

Sarici, S. U., Alpay, F., Dündaroz, M. R., Ozcan, O., & Gokçay, E. (2001). Fiberoptic phototherapy versus conventional daylight phototherapy for hyperbilirubinemia of term newborns. *Turkish Journal of Pediatrics, 43*(4), 280–285.

Sherman, T. I., Greenspan, J. S., St. Clair, N., Touch, S. M., & Shaffer, T. H. (2006). Optimizing the neonatal thermal environment. *Neonatal Network, 25*(4), 251–260.

Sinclair, J. C. (2002). Servo-control for maintaining abdominal skin temperature at 36°C in low birth weight infants. *Cochrane Database of Systematic Reviews, 1*, CD0010740.

Sittig, S. E., Nesbitt, J. C., Krageschmidt, D. A., Sobczak, S. C., & Johnson, R. V. (2011). Noise levels in a neonatal transport incubator in medically configured aircraft. *International Journal of Pediatric Otorhinolaryngology, 75*(1), 74–76.

Tan, K. L. (1982). The pattern of bilirubin response to phototherapy for neonatal hyperbilirubinaemia. *Pediatric Research, 16*, 670–674.

Van Imhoff, D. E., Dijk, P. H., & Hulzebos, C. V. (2011). Uniform treatment thresholds for hyperbilirubinemia in preterm infants: Background and synopsis of a national guideline. *Early Human Development, 87*(8), 521–525.

Wubben, S. M., Brueggerman, P. M., Stevens, D. C., Helseth, C. C., & Blaschke, K. (2011). The sound of operation and the acoustic attenuation of the Ohmeda Medical Giraffe OmniBedTM. *Noise Health, 13*(50), 37–44.

Xiong, T., Qu, Y., Cambier, S., & Mu, D. (2011). The side effects of phototherapy for neonatal jaundice: What do we know? What should we do? *European Journal of Pediatrics, 170*(10), 1247–1255.

Chapter 30

BODY TEMPERATURE MONITORS

OBJECTIVES

- Understand the differences between core and peripheral temperatures and list the sites of body temperature measurement.
- Differentiate between continuous and intermittent temperature monitoring.
- Describe the principles of operation of a typical bedside continuous body temperature monitor.
- Analyze the transducer circuit diagram and the functional block diagram of a typical continuous body temperature monitor.
- Describe the principles of operation of IR thermometry.
- Define emissivity and field of view and explain their significance in IR thermometry.
- Analyze the functional building blocks of a typical tympanic (ear) thermometer.
- Describe the principles of temporal artery thermometers
- State the sources of error in body temperature measurement using IR thermometry.

CHAPTER CONTENTS

1. Introduction
2. Sites of Body Temperature Measurement
3. Bedside Continuous Temperature Monitors
4. Infrared Thermometry
5. Tympanic (Ear) Thermometers
6. Temporal Artery Thermometers
7. Common Problems and Hazards

INTRODUCTION

The body temperature of a healthy person is regulated within a narrow range despite variation in environmental conditions and physical activity. Illness is often associated with disturbance of body temperature regulation leading to abnormal elevation of body temperature, or fever. Fever is such a sensitive and reliable indicator of the presence of disease that thermometry is probably the most common clinical procedure in use. Body temperature is also measured during many clinical procedures such as surgery, postanesthesia recovery, treatment of hyperthermic and hypothermic conditions, and so on.

A body temperature monitor allows measurement and display of a patient's body temperature. It may sound an alarm if it is above or below some preset limits and may track temperature variation over a period of time. Most body temperature monitors accept different sensors or probes to measure temperature at different body sites

Body temperature measurement can be continuous or intermittent. A continuous body temperature monitor uses a sensor to acquire the temperature at the measurement site continuously. An example is skin temperature measurement of an infant in an incubator. A temperature sensor is placed on the infant's skin surface to continuously measure and display the infant's body temperature and, using the measurand to control the heater inside the incubator, to achieve temperature regulation. An oral liquid-in-glass thermometer, which takes about 1 or 2 min to obtain a temperature reading between measurements, is an example of an intermittent temperature measurement device. Another probe-type thermometer is a chemical thermometer. This is a paper device with heat-sensitive chemical dots superimposed on the surface; the chemical is designed to change color in accordance with the temperature sensed. It is used as a single-use device with primary intent to prevent cross contamination. Plastic forehead strips containing heat-sensitive liquid crystals in which the indicator changes color according to body temperature are available for fever detection. This chapter describes two types of temperature measurement devices: a continuous temperature monitor that uses a contact temperature sensor and an IR thermometer, which is an intermittent temperature detector that senses the heat radiated from the patient.

SITES OF BODY TEMPERATURE MEASUREMENT

Heat transfer within the body depends mainly on conduction (heat transfer between adjacent tissues) and convection (heat transfer through movement of body fluid). One can simply visualize the body as a central core at

Table 30-1. Normal Temperature Comparisons

	Core	*Oral*	*Rectal*
Adult	36.5–37.6	36.0–37.2	36.3–37.6
Age 7–14 years	36.8–37.3	36.4–36.9	36.4–36.9
Age 3–6 years	37.3–37.6	36.9–37.2	36.9–37.2

uniform temperature surrounded by an insulating shell. In body temperature measurement, it is often desirable to measure the core temperature because it reflects the true temperature of the internal parts of the body. The hypothalamus is considered the site of core body temperature; temperatures measured within the skull, thorax, and abdomen are clinically accepted as the core temperature. Invasive thermometry is considered the gold standard for body temperature measurements. The sensor locations of invasive thermometry include the pulmonary artery, esophagus, bladder, and rectum. The true gold standard for thermal assessment is the pulmonary artery. Using a special catheter, it allows continuous measurement of the blood temperature of the pulmonary outflow of the heart. Peripheral or shell temperature is measured at or near the surface of the body. Temperatures measured at the skin surface or in subcutaneous fat are examples of peripheral temperature. Due to its measurement location, peripheral temperature is often affected by the ambient environment.

It is not practical to define an exact upper level of normal body temperature because there are variations among normal persons as well as considerable fluctuations in a given individual. However, an oral temperature above 37.2°C in a person at rest is a reasonable indication of disease. Table 30-1 lists the range of temperature (in degrees Celsius) measured from different sites on normal individuals of different age groups.

Body sites such as the tympanic membrane, rectum, esophagus, nasopharynx, bladder, and pulmonary artery are used for core temperature measurement. In a normothermic patient, these sites yield very similar temperature values. However, under hypothermic conditions, the rectal site can be cooler than the others by 1°C to 2°C.

BEDSIDE CONTINUOUS TEMPERATURE MONITORS

Continuous temperature monitors are used in situations in which a patient's body temperature is required to be measured continuously.

Table 30-2. YSI 400 Resistance and Temperature Characteristics

Temp °C	Res. Ω	Temp °C	Res. Ω	Temp °C	Res. Ω	Temp °C	Res. Ω
−40	75.79 k	24	2354	40	1200	80	283.1
−35	54.66 k	25	2253	41	1153	85	241.3
−30	39.86 k	26	2156	42	1108	90	206.5
−25	29.38 k	27	2065	43	1065	95	177.5
−20	21.87 k	28	1977	44	1024	100	153.2
−15	16.43 k	29	1894	45	984.2	105	132.7
−10	12.46 k	30	1815	46	946.6	110	115.4
−5	9534	31	1740	47	910.6	115	100.6
0	7355	32	1668	48	876.2	120	88.1
5	5720	33	1599	49	843.2	125	77.4
10	4483	34	1534	50	811.7	130	68.2
15	3539	35	1471	55	672.9	135	60.2
20	2814	36	1412	60	560.7	140	53.4
21	2690	37	1355	65	469.4	145	47.4
22	2572	38	1301	70	394.9	150	42.3
23	2460	39	1249	75	333.5		

Examples of such situations are patients under general anesthesia, or receiving hypothermic or hyperthermic treatment. Many temperature transducers can be used in continuous temperature monitoring. However, the most common type of transducer for clinical application is a thermistor. The YSI series probes (patented by Yellow Spring Instruments) offer true interchangeability between probes and monitors from different manufacturers without the need for recalibration or adjustment. The temperature transducer element in a YSI 400 probe is a precision thermistor manufactured to achieve the resistance and temperature characteristics according to the YSI 400 specifications. The characteristics are shown in Table 30-2 with less than 0.1% resistance tolerance between 0 and 80°C.

The normal transfer characteristics of a YSI 400 temperature probe is illustrated in Figure 30-1. It has a negative temperature coefficient and a highly nonlinear resistance-temperature relationship. It is often desirable to have a temperature transducer with linear output characteristics. A YSI 700 series temperature probe is manufactured by combining two thermistor elements (Figure 30-2a) of different temperature characteristics to form a temperature transducer that can be wired to produce linear output characteristics.

The construction of a YSI 700 series thermistor and its application circuit is shown in Figure 30-2a and b, respectively. Figure 30-3 shows the resistance-temperature relationship of the circuit.

Many continuous temperature monitors on the market can accept either

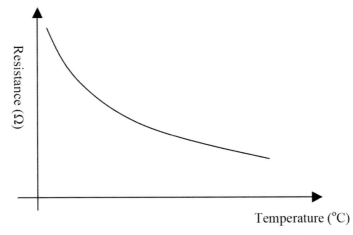

Figure 30-1. YSI 400 Series Temperature Probe Output Characteristics.

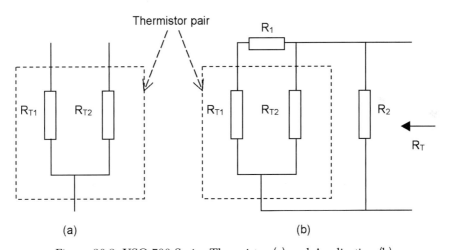

Figure 30-2. YSO 700 Series Thermistor (a) and Application (b).

YSI 400 or YSI 700 temperature probes. Since there are two lead wires for a YSI 400 series transducer, 1/4-inch mono phono jacks are often used as connectors with YSI 400 probes, whereas 1/4-inch stereo phono jacks (three wires) are used with YSI 700 probes.

A YSI 400 temperature probe can be connected to one arm of a Wheatstone bridge as shown in Figure 30-4. The output voltage V_0 will vary according to the change in resistance R_T of the probe. As the bridge output has a nonlinear relationship with the resistance R_T, the resistance values of R_a and R_b can be chosen such that the output voltage follows a piecewise-linear relationship with the temperature.

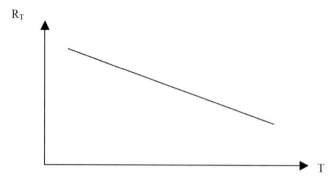

Figure 30-3. Linearized Characteristic of YSI 700 Series Thermistor.

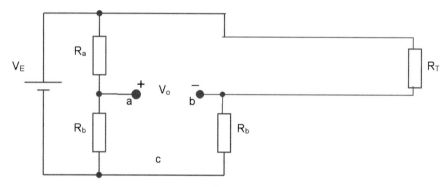

Figure 30-4. Wheatstone Bridge Temperature Monitor.

In a digital thermometer, the resistance-temperature relationship is stored in digital memory in a lookup table. Once the resistance of the probe is measured, the corresponding temperature is determined from the lookup table. No linearization circuit is required.

Block Diagram of a Continuous Temperature Monitor

Figure 30-5 shows a simple functional block diagram of a continuous temperature monitor. It consists of an excitation circuit to provide either a voltage or a current source to convert the change of resistance (due to change in temperature) to a voltage output. This output voltage is amplified, filtered, and sent to the processor and display modules. The purpose of signal isolation is for electrical safety.

Errors in Bedside Continuous Body Temperature Monitors

Continuous temperature monitors are very reliable. Errors are primarily

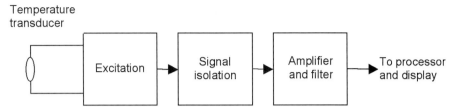

Figure 30-5. Front-end Block Diagram of Temperature Monitor.

associated with the temperature sensor. For skin sensors, failure to make good contact (e.g., with the skin) will cause a temperature discrepancy between the sensing element and the actual temperature of the measurement site. For peripheral temperature measurement, the environmental condition (such as when the temperature sensor is directly exposed to a radiant warmer) will affect the accuracy of the measurement temperature.

INFRARED THERMOMETRY

Liquid-in-glass and electronic thermometers measure temperature by placing the probes under the tongue, in the rectum, or under the armpit. The temperature is read when the probe temperature is in equilibrium with the measurement site by heat conduction. Temperature obtained from these sites may not reflect the true core temperature because they are subject to thermal artifacts. IR thermometers detect the IR spectrum emitting from the patient to determine the body temperature. These thermometers are quick and noninvasive. An example is the IR ear thermometer or tympanic thermometer that allows users to measure temperature by inserting a probe into the patient's ear canal. Tympanic thermometers can measure temperature without touching the mucous membrane and can be used on both conscious and unconscious patients. Temporal thermometers are another example of intermittent body temperature measurement devices using the principle of IR thermometry. Due to their sensing locations, both tympanic and temporal thermometers provide near core temperature values. They provide a convenient and fast alternative to other intermittent temperature measurement methods.

Theory of Infrared Thermometry

IR thermometry (also known as pyrometry) has long been used as a noncontact method to measure temperature in the industry. For example, in foundries, noncontact IR thermometers are used to measure the temperature of molten metals at a distance from the source.

IR thermometry relies on the principle that radiation is emitted by all objects having a temperature greater than absolute zero (0 Kelvin) and that the emission increases as the object becomes hotter. The temperature of the object can be determined from the emission spectrum of the radiation in the IR region. Devices based on IR thermometry are based on Planck's law and Wien's displacement law.

Planck's law states that if the intensity of the radiated energy of a substance is plotted as a function of wavelength (Figure 30-6), the area under the curve represents the total energy radiated from the substance at the associated temperature. Wien's displacement law states that the wavelength λ_{max}, corresponding to the maximum energy intensity in the radiated energy spectrum, is given by the equation

$$\lambda_{max} = \frac{2.89 \times 10^3}{T} \ \mu m,$$

where T is the object temperature in Kelvin.

For example, if the object is at temperature 37°C, T = 37 + 273 = 310 K, substituting this into the Wien's equation gives $\lambda_{max} = 9.32 \ \mu m$.

Another parameter important to IR thermometry is emissivity. Emissivity (ε) is defined as the ratio of the energy radiated by an object at a given temperature to the energy emitted by a blackbody (or perfect radiator) at the same temperature. Therefore, the emissivity of a blackbody is 1.0, whereas the emissivity values for all "nonblack" objects lie between 0.0 and 1.0. The

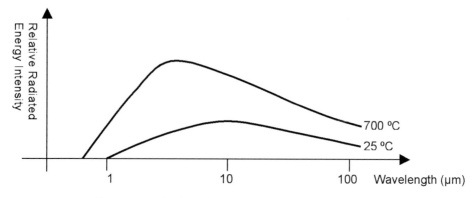

Figure 30-6. Radiated Energy Spectrum of an Object.

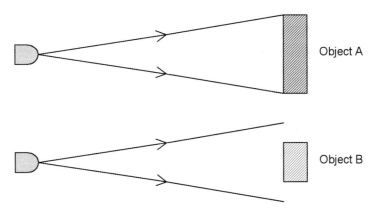

Figure 30-7. Field of View of Optical Sensor.

emissivity of body tissue is about 0.95. For most substances, the emissivity is wavelength dependent except for a graybody. A graybody is defined as an object whose emissivity is the same for all wavelengths in its energy spectrum. Because most objects are nonblackbody, variation of emissivity without proper compensation will lead to error in IR temperature measurement. Most IR thermometers restrict the sensing wavelength to a chosen narrow band so that near-graybody characteristics can be achieved.

In IR thermometry, a sensor is used to collect the radiant energy coming from the object. Therefore, it is important that the sensor's field of view (FOV) be aligned properly with the object to be measured. The FOV is the angle of vision at which the instrument operates. It is determined by the optics of the unit. The IR thermometer in Figure 30-7 will correctly read the temperature of Object A but will read a temperature lower than the actual temperature of Object B if the background is of a lower temperature than that of the object.

TYMPANIC (EAR) THERMOMETERS

A tympanic thermometer measures body temperature by measuring the IR energy emitting from the tympanic membrane in the ear. It is an intermittent temperature measurement device. The tympanic membrane is a good site to measure core temperature because it is sheltered from the external environment and vascularized by the carotid arteries, which also perfuse the hypothalamus. Unlike contact-type electronic thermometers, tympanic thermometers do not rely on thermal conduction; the sensors therefore take less time to reach temperature equilibrium. A typical tympanic thermometer takes less than 5 seconds to take a reading, and it is not subject to external

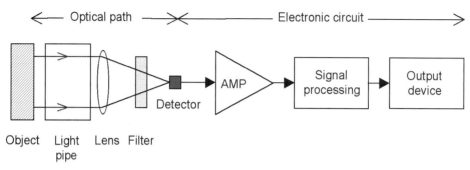

Figure 30-8. Functional Block Diagram of an Infrared Thermometer.

thermal disturbances (such as air temperature fluctuation). Tympanic thermometers are recommended for temperature measurements on unsettled patients.

Block Diagram of a Tympanic Thermometer

A functional block diagram of a tympanic thermometer is shown in Figure 30-8. In the diagram, the object is the tympanic membrane. The IR energy radiated from the membrane is collected by the optical lens and filter. The detector converts the radiated energy to an electrical signal. This electrical signal is processed to obtain the temperature. The following paragraphs provide a more detailed functional description of the building blocks.

The optical path consists of a lens, a filter, and an optical light pipe. The function of the lens and light pipe is to ensure that the FOV covers only the tympanic membrane instead of other tissue, such as the wall of the ear canal. It also focuses the IR radiation to the detector. The optical filter allows only a selected bandwidth of wavelength to reach the detector.

The wavelength of the IR spectrum lies between 0.7 and 20 μm. The bandwidth of a wideband IR thermometer is several microns wide (e.g., 8–14 μm), whereas a narrowband device allows only a single wavelength to reach the detector (e.g., 2.2 ± 0.5 μm). Instead of using a single-band wavelength, a "two-color" thermometer uses two narrowband signals to reduce errors due to emissivity (non-blackbody) and IR absorption in the optical path (e.g., due to water moisture in the atmosphere). Figure 30-9 shows the pass band(s) of a wideband, narrowband, and two-color IR thermometer.

Several types of detectors can be used in IR thermometry. Common detectors are pyroelectric sensors and thermopiles. In a pyroelectric sensor, conductive material is deposited on the opposite surfaces of a slice of a ferroelectric material. The ferroelectric material absorbs radiation and converts it to heat. The resulting rise in temperature changes the polarization of the

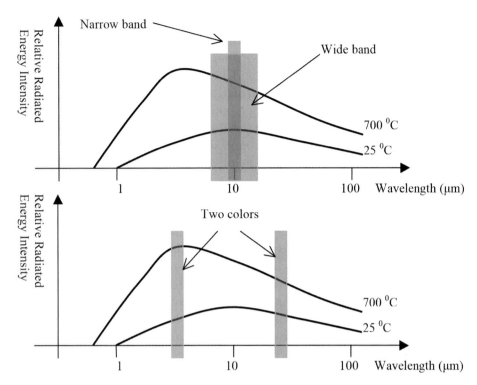

Figure 30-9. Detector Bandwidth of Infrared Thermometers.

material. The current flowing through the external resistor connected across the two conductive surfaces is proportional to the rate of change of temperature of the sensor. A shutter mechanism is often installed to provide a controlled period of exposure to the radiation.

A thermopile is made up of a number of thermocouples connected in series. Radiant energy landing on the hot junction area is first converted to heat, creating a differential temperature between the hot and the cold junctions of the thermocouples. Each thermocouple generates a small voltage (on the order of μV) according to this temperature difference. The output of the thermopile is the summation of all the voltages from the thermocouples in the sensor. An IR thermometer using a thermopile sensor requires another temperature sensor, such as a thermistor, to measure the cold junction temperature. The construction of the thermopile in an IR sensor is shown in Figure 30-10.

To prevent sensing the ear canal temperature, some units use narrow FOV optics with multiple scanning to detect the highest temperature in the ear canal; within the ear canal, the hottest object is the tympanic membrane. To further increase accuracy, some units measure the ambient temperature

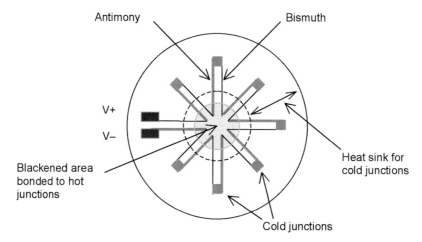

Figure 30-10. Construction of a Thermopile with Eight Thermocouples.

to estimate the heat loss of the blood flowing through the tympanic membrane and the surrounding tissues.

In most IR thermometers, the analog signal from the sensor is first digitized and sent to the signal processor. The main function of the processor is to determine the object temperature based on the signal from the sensor. Calibration curves or lookup tables are stored in the unit's memory for this purpose. Another function of the signal processor is to compensate for nonideal conditions (such as ε <1.0), provide offset to compensate for heat loss; and estimate the oral, rectal, or core temperature from the tympanic reading.

Errors in Tympanic Thermometers

Factors that affect accuracy unique to tympanic thermometers are as follows:

- If the FOV of the sensor includes other tissue inside the ear canal, it will create a lower than normal temperature reading.
- A nonstraight ear canal can prevent a direct line of view to the tympanic membrane with the detector.
- Too much hair or ear wax in the ear canal will block the tympanic membrane from the detector.

In order to provide an accurate temperature reading, it is necessary for the IR sensor to actually "see" the tympanic membrane. To accomplish this, the ear canal must be straightened, with the instrument inserted far enough into the ear canal to see around the bend. The instrument must also scan the ear canal (automatically or maneuver by the user) to ensure seeing the tympanic membrane.

TEMPORAL ARTERY THERMOMETER

The operation principle of a temporal artery thermometer is very similar to the tympanic thermometer. It senses the IR emitted from the temporal artery to determine the temperature of the blood flowing in the artery. The temporal artery branches out from the external carotid, which is about 1 mm below the skin of the lateral forehead. The blood in the temporal artery flows directly from the heart through the aorta and carotid artery. The temporal artery is ideal for temperature measurement due to its proximity to the surface of the skin and provides no risk of injury from being touched. However, the temperature of blood in the temporal artery is lower than that in the heart due to heat lost through the skin to the environment. A characteristic of the temporal artery is that it contains very few arteriovenous anastomoses (channels between the artery and veins) and therefore its blood flow is fairly constant when the patient is stationary. A constant blood flow rate allows calculation of blood temperature reduction due to heat loss to the environment when the ambient temperature is known (referred to by some as heat balance calculation). Temporal artery thermometers have been shown to provide rapid, noninvasive, temperature measurements that are accurate and close to the core body temperature.

The IR transducer in a temporal artery thermometer is usually a pyro-electric sensor; a thermistor is used to measure the ambient temperature. In a measurement process, while maintaining skin contact, the sensor is swept across the forehead and moved along the temporary artery, stopping behind the ear. The IR energy is measured about 1000 times per second to determine the maximum temperature along the scanned path (corresponding to the temporary artery). Together with the ambient temperature and other factors (e.g., emissivity of human skin), the core temperature is computed and displayed at the end of the scan.

A number of factors affect the accuracy of measurements. It is important that the sensor is in good contact with the skin and that it is swept along the temporal artery. Because the skin emissivity is used to calculate the temperature, dirty skin or skin covered with sweat will give incorrect readings. Patient movement affects the blood flow in the temporal artery and will therefore create errors in the heat balance calculation. Since the sensor absorbs heat from the skin and cools down the skin, about 30 sec should be waited before another measurement. Unstable ambient temperature will also introduce errors in measurements.

COMMON PROBLEMS AND HAZARDS

Except when using invasive sensors, body temperature measurement procedures are very safe. The temperature probes (or sensors) in continuous temperature monitors are often single-use disposables to prevent cross contamination. Modern bedside temperature monitors are very accurate and reliable. As with other line-powered medical devices, electrical isolation is used to prevent electric shock to patients.

IR thermometers are noninvasive; they do not emit radiation or produce heat. They are very safe. Because IR probes are in contact with patients, to reduce risk of cross-contamination and nosocomial infections, gas sterilization of the base unit and probes with ethylene oxide is recommended by ASTM International. Single-use disposable probe covers are recommended for nondisposable units.

Some units provide software compensation for the measured temperature reading to estimate the temperature at different body sites (e.g., rectal, oral, etc.). In these units, errors can be caused by inadvertent selection of these compensation factors.

The most common problem reported by users of IR ear thermometers is inaccurate and inconsistent measurement. These are mainly the result of inconsistent technique by health-care providers and the failure of health-care facilities to standardize the brand and model of thermometers being used. Failure to examine the condition of the ear canal of the patient (obstructions, nonstraight ear canal, etc.) can lead to inaccurate temperature readings. Because the FOV of the IR sensor covers tissues other than just the tympanic membrane, some manufactures compensate this by adding a positive offset to the measured value. This offset may produce an inaccurately high reading in some applications such as in neonatal measurement, however. Temporal artery thermometers have proved to have better reliability than tympanic thermometers have.

BIBLIOGRAPHY

Allbutt, T. C. (1870). Medical thermometry. *British and Foreign Medico-Chirurgical Review, 45*(90), 429–441.

Al-Mukhaizeem, F., Allen, U., Komar, L., Maser, B., Roy, L., Stephens, D., . . . , & Schuh, S. (University of Toronto/Hospital for Sick Children). (2004). Comparison of temporal artery, rectal and esophageal core temperatures in children: Results of a pilot study. *Paediatric & Child Health, 9*(7), 461–465.

ASTM E1965-98. (2009). *Standard Specification for Infrared Thermometers for Intermittent Determination of Patient Temperature.* West Conshohocken, PA: ASTM International.

Brinnel, H., & Cabanac, M. (1989). Tympanic temperature is a core temperature in humans. *Journal of Thermal Biology, 14*(1), 47–53.

Carroll, D., Finn, C., Judge, B., Gill, S., & Sawyer, J. (Massachusetts General Hospital). (2004). A comparison of measurements from a temporal artery thermometer and a pulmonary artery catheter thermistor. *American Journal of Critical Care, 13*(3), 258.

Davie, A., & Amoore, J. (2010). Best practice in the measurement of body temperature. *Nursing Standard, 24*(42), 42–49.

Dybwik, K., & Nielsen, E. W. (2003). Infrared temporal artery temperature measurement. *Tidsskrift for Den Norske legeforening (Journal of the Norwegian Medical Association), 123*, 3025–3026.

Farnell, S., Maxwell, L., Tan, S., Rhodes, A., & Philips, B. (2005). Temperature measurement: Comparison of non-invasive methods used in adult critical care. *Journal of Clinical Nursing, 14*(5), 632–639.

Hebbar, K., Fortenberry, J. D., Rogers, K., Merritt, R., & Easley, K. (2005). Comparison of temporal artery thermometer to standard temperature measurements in pediatric intensive care unit patients. *Pediatric Critical Care Medicine, 6*(5), 557–561.

Hooker, E. A., & Houston, H. (1996). Screening for fever in an adult emergency department: Oral vs tympanic thermometry. *Southern Medical Journal, 89*(2), 230–234.

Kistemaker, J. A., Den Hartog, E. A., & Daanen, H. A. M. (2006). Reliability of an infrared forehead skin thermometer for core temperature measurements. *Journal of Medical Engineering and Technology, 30*(4), 252–261.

Kocoglu, H., Goksu, S., Isik, M., Akturk, Z., & Bayazit, Y. A. (2002). Infrared tympanic thermometer can accurately measure the body temperature in children in an emergency room setting. *International Journal of Pediatric Otorhinolaryngology, 65*(1), 39–43.

Latman, N. S., Hans, P., Nicholson, L., Delee Zint, S., Lewis, K., & Shirey, A. (2001). Evaluation of clinical thermometers for accuracy and reliability. *Biomedical Instrumentation & Technology, 35*(4), 259–265.

Lawson, L., Bridges, E., Ballou, I., Eraker, R., Greco, S., Shively, J., & Sochulak, V. (2006). 2006 National Teaching Institute Research Abstracts: Temperature measurement in critically ill adults. *American Journal of Critical Care, 15*(3), 324–346.

Lee, G., Flannery-Bergey, D., Randall-Rollins, K., Curry, D., Rowe, S., Teague, M., . . ., & Schroeder, S. (2011). Accuracy of temporal artery thermometry in neonatal intensive care infants. *Advances in Neonatal Care, 11*(1), 62–70.

Livornese, L.L., Jr., Dias, S., Samuel, C., Romanowski, B., Taylor, S., May, P., . . . , & Levison, M.E. (1992). Hospital-acquired infection with vancomycin-resistant Enterococcus faecium transmitted by electronic thermometers. *Annals of Internal Medicine, 117*(2), 112–116.

Mangat, J., Standley, T., Prevost, A., Vasconcelos, J., & White, P. (2010). A comparison of technologies used for estimation of body temperature. *Physiological Measurement, 31*(9), 1105–1118.

Markin, D. A., Henry, D. A., Baxter, S. S., & Slye, D. A. (1990). Comparison between two types of body surface temperature devices: Efficiency, accuracy, and cost.

Journal of Post Anesthesia Nursing, 5(1), 33–37.

McCarthy, P. W., Heusch, A. I., Kenkre, J. E., Machin, G., & Suresh, J. (2002). Infrared ear thermometers versus rectal thermometers. *Lancet, 360*(9348), 1882–1883.

Modell, J., Katholi, C. R., Kumaramangalam, S. M., Hudson, E. C., & Graham, D. (1998). Unreliability of the infrared tympanic thermometer in clinical practice: A comparative study with oral mercury and oral electronic thermometers. *Southern Medical Journal, 91*(7), 649–654.

Pompei, F., & Pompei, M. (1996). Arterial thermometry via heat balance at the ear. *Medical Electronics,* October.

Pompei, F., & Pompei, M. (2004). *Non-Invasive Temporal Artery Thermometry: Physics, Physiology, and Clinical Accuracy.* Presented at Medical Thermometry for SARS Detection, SPIE Defense and Security Symposium. Proceedings SPIE 5405, Thermosense XXVI, 61(April 12).

Rotello, L. C., Crawford, L., & Terndrup, T. E. (1996). Comparison of infrared ear thermometer derived and equilibrated rectal temperatures in estimating pulmonary artery temperatures. *Critical Care Medicine, 24*(9), 1501–1506.

Rubia-Rubia, J., Arias, A., Sierra, A., & Aguirre-Jaime, A. (2011). Measurement of body temperature in adult patients: Comparative study of accuracy, reliability and validity of different devices. *International Journal of Nursing Studies, 48*(7), 872–880.

Samaras, T. (2006). Thermometry. In J. G. Webster (Ed.), *Encyclopedia of Medical Devices and Instrumentation.* New York: John Wiley & Sons.

Smith, L. S. (2004). Temperature measurement in critical care adults: A comparison of thermometry and measurement routes. *Biological Research for Nursing, 6*(2), 117–125.

Suelman, M., Doufas, A., Akca, O., Ducharme, M., & Sessler, D. (2002). Insufficiency in a new temporal-artery thermometer for adult and pediatric patients. *Anesthesia and Analgesia, 95,* 67–71.

Watmough, D. J., & Oliver, R. (1969). The emission of infrared radiation from human skin—implication for clinical thermography. *British Journal of Radiology, 42,* 411–115.

Chapter 31

PULSE OXIMETERS, OXYGEN ANALYZERS & TRANSCUTANEOUS OXYGEN MONITORS

OBJECTIVES

- Define FiO_2, PaO_2, SaO_2, SpO_2, SvO_2, fractional oxygen saturation, and functional oxygen saturation.
- Discuss the clinical applications of pulse oximetry and blood oxygen measurements.
- Explain Beer-Lambert law.
- Explain the principles of operation of pulse oximeters.
- Describe the construction of a pulse oximeter sensor.
- Sketch a typical block diagram of a pulse oximeter and explain the functions of each block.
- Discuss factors affecting signal quality and accuracy of pulse oximetry.
- Describe the principles of clinical oxygen analyzers.
- Describe the principles of transcutaneous oxygen monitoring.
- Differentiate between the clinical application of oxygen analyzers, transcutaneous oxygen monitors and pulse oximeters.

CHAPTER CONTENTS

1. Introduction
2. Definition of Oxygen Level and Percentage Oxygen Saturation in Blood
3. Principles of Operation of Pulse oximetry
4. Pulse Oximeter Sensor Probes
5. Pulse Oximeter Functional Block Diagram
6. Oxygen Analyzers and Transcutaneous Oxygen Monitors
7. Common Problems and Hazards

INTRODUCTION

The primary goal of the cardio-respiratory system is to deliver adequate oxygen to the tissues to meet their metabolic needs. Hypoxia is a general term describing the condition of lack of oxygen in the body or a part of the body. Acute hypoxia produces impaired judgment and motor incoordination. When hypoxia is long standing, the symptoms consist of fatigue, drowsiness, inattentiveness, and delayed reaction time. More severe hypoxia can affect brain function; in the absence of oxygenation, brain damage will happen within 5 minutes and brain death in another 10 to 15 minutes.

Blood gas analyzers, a medical laboratory instrument, have been used to determine oxygen level in blood (PaO_2) from blood samples drawn from patients. Bedside blood gas analyzers can be deployed to decrease turnaround time. Transcutaneous oxygen monitoring using a sensor placed at skin surface is an alternative method to non-invasively estimate the oxygen level in arterial blood. Until the early 1980s, blood oxygen saturation levels (SaO_2) were measured by drawing arterial blood samples from patients and performing in vitro analysis using laboratory co-oximeters (a multi-wavelength spectrophotometer). Pulse oximeters were developed in the early 1980s, providing a real-time, continuous, and noninvasive mean to monitor the changing level of arterial blood oxygenation in patients therefore allowing earlier clinical intervention to minimize the occurrence of significant hypoxia. In 1986, the American Society of Anesthesiologists (ASA) endorsed the use of pulse oximetry, and in 1990 modified the ASA basic monitoring standards to mandate pulse oximetry during all anesthetic procedures. Since then, the use of pulse oximetry have become a de facto standard of care in anesthesia.

DEFINITION OF PERCENTAGE OXYGEN SATURATION IN BLOOD

Gaseous exchange from pulmonary ventilation allows atmospheric oxygen to diffuse into the blood stream. Oxygen level in inspired air and dissolve oxygen level in blood are measured and reported as fraction of inspired oxygen (FiO_2) or partial pressure of oxygen (PaO_2) respectively. The primary function of red blood cells is to transport oxygen to tissue. This function is carried out by hemoglobin. When blood is circulated into the lungs, oxygen is attached to hemoglobin, forming oxygenated hemoglobin (or oxyhemoglobin). Under normal conditions, hemoglobin in blood becomes almost fully saturated with oxygen before leaving the lungs. When blood is in the capillaries, oxygen is released from the oxyhemoglobin and

Table 31-1. Typical Values of Blood Oxygen Level

Adult	$\%S_aO_2$ 96–98%	P_aO_2 85–100 mmHg	P_aCO_2 38–42 mmHg
	$\%S_vO_2$ 70–75%	P_vO_2 35–40 mmHg	P_vCO_2 41–51 mmHg
Neonates	$\%S_aO_2$ ~94%	P_aO_2 63–87 mmHg	P_aCO_2 31–35 mmHg

delivered to the cells, the hemoglobin becomes deoxygenated hemoglobin (or deoxyhemoglobin). In studying oxygen transport in blood, the terms $\%S_aO_2$, $\%S_pO_2$, and $\%S_vO_2$ are commonly used. Here are their definitions:

- Percentage oxygen saturation of hemoglobin in arterial blood ($\%S_aO_2$) is the percentage of hemoglobin in arterial blood that is bound with oxygen; it is determined by analyzing an arterial blood sample with a co-oximeter (a laboratory instrument).
- Percentage oxygen saturation of hemoglobin in venous blood ($\%S_vO_2$) is the percentage of hemoglobin in venous blood that is bound with oxygen; it is usually determined by analyzing a blood sample taken from the pulmonary artery with a co-oximeter.
- Percentage oxygen saturation of hemoglobin in arterial blood when determined by a pulse oximeter (instead of from a blood sample by a co-oximeter) is denoted as $\%S_pO_2$.

In addition to oxyhemoglobin and deoxyhemoglobin, there are two other forms of hemoglobin. Carboxyhemoglobin is hemoglobin bound with carbon monoxide. Carbon monoxide has higher affinity to hemoglobin than oxygen. Methemoglobin is the oxidized form of hemoglobin. Methemoglobin is incapable of binding with oxygen. A high percentage of carboxyhemoglobin or methemoglobin compromises the oxygen-carrying capacity of blood because there is less hemoglobin available for oxygen transport.

Table 31-1 shows the normal values of $\%S_aO_2$ or $\%S_pO_2$ with the corresponding blood gas values of normal adults and neonates. P_aO_2 and P_aCO_2 are partial pressures of oxygen and carbon dioxide respectively in blood.

PRINCIPLES OF OPERATION OF PULSE OXIMETRY

The principle of pulse oximetry is based on the Beer-Lambert law with differential light absorption of two wavelengths. The wavelengths of the most commonly used light sources are red (RED = 660 nm)

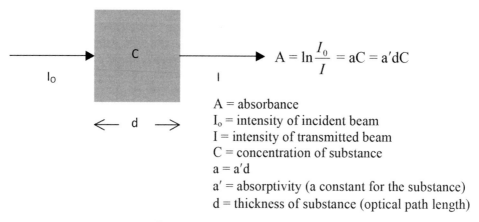

$$A = \ln\frac{I_0}{I} = aC = a'dC$$

A = absorbance
I_o = intensity of incident beam
I = intensity of transmitted beam
C = concentration of substance
a = a'd
a' = absorptivity (a constant for the substance)
d = thickness of substance (optical path length)

Figure 31-1. Beer-Lambert Law.

and infrared (IR = 940 nm). The Beer-Lambert law states that for a substance of concentration C in a fluid, the absorbance A of light, defined as the natural logarithm of the ratio of incident light intensity (I_o) to the transmitted light intensity (I), due to the substance in the fluid, is equal to the product of the absorptivity (a'), the substance's concentration (C) in the fluid, and the distance of the optical path length (d). Figure 31-1 illustrates the concept.

For a mixture of two substances X and Y in the fluid, the total absorbance A is given by the sum of the absorbance due the substance X and the substance Y alone, or $A = A_X + A_Y$

Two equations are used by device manufacturers to calculate oxygen saturation in blood. They are the fractional oxygen saturation and the functional oxygen saturation equations.

i. **Fractional oxygen saturation** ($\%O_2Hb$) is equal to the ratio of the concentration of oxyhemoglobin to the sum of concentrations of all types of hemoglobin in the blood.

$$\%O_2Hb = \frac{C_{O_2Hb}}{C_{HHb} + C_{O_2Hb} + C_{COHb} + C_{metHb}} \times 100\%, \quad (31.1)$$

where C_{O_2Hb} is the concentration of oxygenated hemoglobin in arterial blood,
C_{HHb} is the concentration of deoxygenated hemoglobin in arterial blood,
C_{COHb} is the concentration of carboxyhemoglobin in arterial

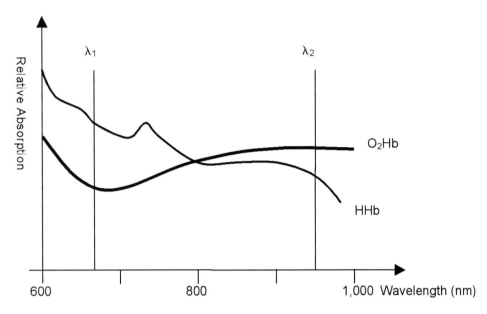

Figure 31-2. Absorption Characteristics of Oxyhemoglobin and Deoxyhemoglobin.

blood, and

C_{metHb} is the concentration of methemoglobin in arterial blood.

ii. **Functional oxygen saturation** ($\%S_aO_2$) is the ratio of the concentration of oxyhemoglobin to the concentration of all functional hemoglobin in the blood. Functional hemoglobin consists of oxyhemoglobin and deoxyhemoglobin which perform the function to transport oxygen.

$$\%SaO_2 = \frac{C_{O2Hb}}{C_{HHb} + C_{O2Hb}} \times 100\%, \qquad (31.2)$$

It is obvious from the two equations that the functional value is higher than the fractional value for the same blood sample. In most cases, the concentration of carboxyhemoglobin as well as that of the methemoglobin in blood is negligible. For a healthy individual, the difference between $\%SaO_2$ and $\%O_2Hb$ is less than 3%.

Figure 31-2 shows the absorption spectrum of oxyhemoglobin (O_2Hb) and deoxyhemoglobin (HHb). To measure the $\%SaO_2$ (or $\%O_2Hb$) in a blood sample, two light sources of wavelengths λ_1 and λ_2 are used.

$$A_1 = A_{10} + A_{1d} = a_{10}C_o + a_{1d}C_d, \qquad (31.3)$$

where A1 is the total absorption due to wavelength λ_1,
>   A10 is the absorption of oxyhemoglobin due to wavelength λ_1,
>   A1d is the absorption of deoxyhemoglobin due to wavelength λ_1,
>   a10 is the product of the optical path length and the absorptivity of oxyhemoglobin due to wavelength λ_1, and
>   a1d is the product of the optical path length and the absorptivity of deoxyhemoglobin due to wavelength λ_1.

At wavelength λ_2, using Beer-Lambert law,

$$A_2 = A_{2o} + A_{2d} = a_{2o}C_o + a_{2d}C_d. \tag{31.4}$$

A1 and A2 are measured from the test set up, and if a10, a1d, a20, a2d are known quantities, one can solve Equations 31.3 and 31.4 for Co and Cd. Once Co and Cd are determined, they are substituted in the functional oxygen saturation equation to determine the oxygen saturation in the blood. Using Equation 31.2,

$$\%SaO_2 = \frac{C_o}{C_o + C_d} \times 100\%$$

$$= \frac{1}{1 + \dfrac{C_d}{C_o}} \times 100\%.$$

This is how $\%SaO_2$ is measured using a laboratory co-oximeter when the blood sample is placed inside a curvette with a fixed optical path length. For a pulse oximeter, the light beams travel through the tissues and are absorbed not only by the hemoglobin in the blood vessels but also by the tissues (skin, muscle, bone, etc.) in the light path. In addition, because the diameters of the capillaries are expanding and contracting according to the blood pressure, the optical path length is not a constant. Therefore, a10, a1d, a20, and a2d, which are the products of the absorptivity and the optical path length, are not constant values. Because of this, Co and Cd cannot be computed analytically from Equations 31.3 and 31.4.

Figure 31-3 shows the absorption waveforms determined by the light sensor in a pulse oximeter finger probe. A RED beam ($\lambda_1 = 660$ nm) and an IR beam ($\lambda_2 = 940$ nm) are commonly used. The solid and dotted waveforms are the results of the absorption characteristics of each of the light beams. Due to the elastic nature of the capillaries, their diameters, and therefore their optical path lengths increase and decrease according to the blood pressure. The rise and fall of the absorption waveform are results of changes in the optical path lengths due to the pulsatile arterial blood pressure. If there is no cardiac pulses, the waveforms will be two horizontal straight lines. Note that the amplitude of fluctuation is only about one percent of the over-all light absorption, most light is absorbed as it travels through other tissues (skin, muscle, bone, etc.) in the finger. As the fluctuation is caused by the pulsatile blood pressure from cardiac activity, the patient's heart rate can be determined from the frequency of fluctuation.

Although the concentrations of oxyhemoglobin and deoxyhemoglobin cannot be obtained analytically to compute %SpO2, %SpO2 is correlated to the differential light absorption by the oxyhemoglobin and deoxyhemoglobin in the capillaries. Instead of measuring light absorption, most pulse oximeter manufacturers derive the %SpO2 values from the optical intensity ratio (r) of the transmitted intensities of the RED (Ird) and IR beam (Iir) measured by the optical sensor in the probe. Note that "p" instead of "a" is used to identify the oxygen saturation value is obtained from a pulse oximeter instead of a co-oximeter.

$$r = \frac{I_{rd}}{I_{ir}} \qquad (31.5)$$

In most cases, an empirical equation or a lookup table between r and the %SpO2 values is established so that during use, the %SpO2 value can be determined from the measured Ird and Iir. This correlation between %SpO2

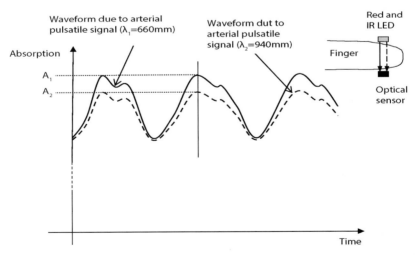

Figure 31-3. Absorption Signal from Pulse Oximeter Finger Probe.

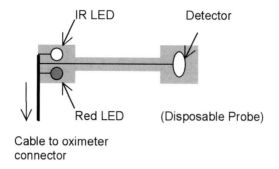

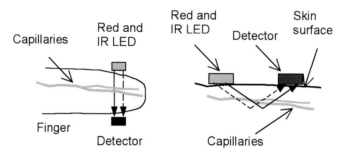

Figure 31-4. Pulse Oximeter Probes.

and the %SaO2 is verified statistically by getting the ratio r from the sensor, and simultaneously taking an arterial blood sample from the patient to obtain %SaO2 from a laboratory co-oximeter. Pulse oximeters are calibrated against oxygen saturation from arterial blood samples using co-oximeter. Manufacturers may use the %O2Hb (fractional) or %SaO2 (functional) equations in the calibration.

When the light source and optical sensor in the pulse oximeter probe are placed on the opposite side of the capillaries, it is called a transmitting probe. When both are placed on the same side, it is called a reflecting probe. In a reflecting probe, the light travels through the skin into the tissue, after being absorbed by the hemoglobin in the capillaries and tissues, it is then scattered back to the skin surface and detected by the optical sensor. Figure 31-4 shows the configurations of a transmitting probe and a reflecting probe.

Using the same principle of pulse oximetry, a cerebral oximeter is a device to detect reduced oxygen supply to the brain during surgery or when the patient is in trauma. With two light sources and a sensor applied to the forehead of the patient, a cerebral oximeter continuously and noninvasively monitors blood oxygen saturation in the area of the brain that lies beneath the forehead. Using similar principle but employing more than 2 wavelength of lights, one manufacturer claimed to be able to determine multiple blood

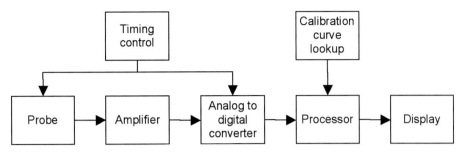

Figure 31-5. Block Diagram of Pulse Oximeter.

constituent parameters. These blood constituent parameters include levels of oxyhemoglobin, deoxyhemoglobin, total hemoglobin, carboxyhemoglobin, methemoglobin, and oxygen content.

PULSE OXIMETER SENSOR PROBES

Many different types of sensor probes are used in pulse oximetry. A typical probe consists of two LEDs; one emits RED light and the other emits IR. These LEDs are pulsed alternately to send a beam of light through the underlying tissues (see the top right part of Figure 31-3). A photo detector in the probe on the other side of the tissue picks up the transmitted light signal and sends it to the processing circuits. Pulse oximeter probes can be classified as reflecting or transmitting, disposable or reusable, or by their sensing locations (finger, earlobe, etc.). A disposable probe is one that will be discarded after being used on a single patient. The LEDs and the photo detector are mounted on each end of a flexible strip. The strip is applied on, and often taped over the tissue (see top diagram in Figure 31-4). Some may reuse probes that are labeled "single patient use" for cost saving.

A reusable probe usually has a more robust and rigid cover to protect the LEDs and the detector. A transmitting probe has the LEDs on one side and the detector on the other side of the capillary bed. The light is transmitted through the capillary bed and tissues. On the other hand, in a reflecting probe, the detector is placed on the same side as the LEDs. The light penetrates the tissue; some is absorbed, and some is scattered and reflected back to the surface. The detector picks up the emerged signal and sends to the processor. In theory, pulse oximeter probes can be placed over any part of the body with capillaries. Common sites for transmitting probes are the index finger and the earlobe. For infants, probes are often taped to the big toe. Transmitting probes are usually placed on the forehead of the patient.

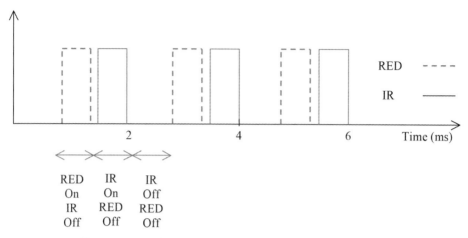

Figure 31-6. Pulsing Sequence of Red and Infrared LEDs.

PULSE OXIMITER FUNCTIONAL BLOCK DIAGRAM

Figure 31-5 shows a typical functional block diagram of a pulse oximeter. It consists of:

- A probe consisting of a RED LED, an IR LED, and a photodetector.
- A timing control circuit to sequence the LEDs and synchronize them with the photodetector. There are three phases in one timing cycle: 1. RED "on" and IR "off", 2. RED "off" and IR "on," and 3. both RED and IR "off". The last phase in the cycle is to measure the "dark signal" to minimize the effect of ambient light. The pulsing sequence of the two LEDs are shown in Figure 31-6.
- Analog and digital electronics to filter, amplify and process the signal.
- A processor to compute the transmitted RED and IR light intensity ratio and match it to the $\%SpO_2$ from the lookup table. It also derives the heart rate from the pulsating waveform and compare the measured values (heart rate, $\%SpO_2$) to the alarm settings
- A display to show the $\%SpO_2$ value, the alarm limits, and the heart rate. A plethysmograph showing the detected signal strength is often displayed to provide the user with an idea of the signal to noise information of the measurement. A strong signal level indicates that the measured value is reliable. A plethysmograph can be a waveform similar to the absorption waveform shown in Figure 31-3 or simply a one-column bar graph proportional to the detected signal strength.

OXYGEN ANALYZERS AND
TRANSCUTANEOUS OXYGEN MONITORS

Oxygen analyzers and pulse oximeters, other than pulse oximeters, are two devices commonly used to monitor a patient's oxygenation in a clinical environment. An oxygen analyzer measures the percentage of oxygen gas in a gas mixture such as the inspired air of a patient. Oxygen analyzers are usually connected to the patient breathing circuit. They are also used to monitor the oxygen level in hospital oxygen supply lines and compressed gas cylinders. Most have alarms to alert clinicians when the oxygen concentration reaches a dangerously low or high level. Polarographic and galvanic oxygen sensors are two sensors commonly used in clinical oxygen analyzers. Within these sensors, molecular oxygen is consumed electrochemically with an accompanying flow of electrical current directly proportional to the oxygen concentration.

Polarographic Oxygen Analyzers

A polarographic oxygen sensor, also known as a Clark cell (see also Chapter 9), consists of an anode (usually silver) and a cathode (usually gold) immersed in an aqueous solution of potassium chloride. The cell is separated from the gas sample by a semipermeable membrane that allows oxygen to diffuse into the sensor. The anode is typically held at a positive potential (e.g., 0.8 V polarizing voltage for silver-gold electrode Clark cells). The half-cell and total reaction equations are shown below:

Anode $2Ag + 2Cl^- \rightarrow 2AgCl + 2e^-$
Cathode $2e^- + \frac{1}{2} O_2 + H_2O \rightarrow 2 OH^-$
Total Reaction $2e^- + \frac{1}{2} O_2 + H_2O + 2Ag + 2Cl^- \rightarrow 2 OH^- + 2AgCl + 2e^-$

From the reaction equations, the oxidation of silver at the anode produces electrons and the reduction of oxygen at the cathode consumes electrons. This electron flow is measured to provide a percent oxygen measurement. Listed below are a few problems with polarographic sensors:

• The AgCl (a low solubility substance) produced at the anode will coat the silver anode over time, slowing down and eventually completely stopping the reaction.
• The hydroxyl ions produced at the cathode will raise the pH level of the electrolyte and cause a zero shift over time.
• The electrolyte will lose its chloride ions over time.

Because of these problems, the sensor will need to be replaced. The time between replacements depends on the condition of use, especially with the concentration of oxygen in the gas sample. In clinical use, polarographic sensors have a life span of about one year. Another shortcoming of polarographic cells is the need for external power (0.8 V) to polarize the electrodes and the rather long initial warm-up time. When power is disconnected and reconnected, a long warm-up time (e.g., 10 min) is needed to repolarize the sensor before it can accurately detect the oxygen concentration again.

Galvanic Oxygen Analyzers

A galvanic cell is an oxygen-powered battery (electro-galvanic fuel cells) in which the output electrical potential (voltage) changes with the concentration of oxygen. A galvanic cell has an anode (zinc or lead) and cathode (platinum, gold or silver) surrounded by electrolytes. The electrodes are separated from the gas sample by a semipermeable membrane. Oxygen diffuses into the cell through the membrane, where it reacts with the cathode to form hydroxyl ions. These ions diffuse to the anode, where they give up electrons to generate a voltage. The measured voltage is proportional to the concentration of oxygen.

Unlike the polarographic cell, the galvanic cell does not require external power for polarization. This is achieved by using two dissimilar metals. In the presence of an electrolyte, there is an electromotive voltage produced between the two metals. A galvanic cell is really a self-polarizing amperometric cell. The reactions of a zinc galvanic cell are:

Anode	$Zn \rightarrow Zn^{2+} + 2e^-$
Cathode	$2e^- + 1/2\ O_2 + H_2O \rightarrow 2OH^-$
Total Reaction	$Zn + 2e^- + 1/2\ O_2 + H_2O \rightarrow Zn^{2+} + 2e^- + 2OH^-$

The net result of the chemical reaction is the production of zinc oxide (ZnO), which is stable and does not deposit on the anode. Additionally, the electrolyte is not consumed and does not need to be replaced. Another advantage is that it does not require external power, and there is no warm-up time required. However, a galvanic cell continues to consume its anode (Zn to ZnO) even when it is turned off; and therefore, may have a shorter life than a polarographic cell.

Transcutaneous Oxygen Monitors

In 1951, Bamberger and Goodfriend demonstrated that when a finger was immersed in an electrolytic solution at 45 °C, the oxygen partial pressure of the solution eventually became the same as the oxygen partial pressure of blood in the capillaries of the finger. Transcutaneous oxygen monitoring is a noninvasive mean to measure the local oxygen released from the capillaries through the skin. Transcutaneous oxygen monitoring is routinely used on preterm neonates, in whom high blood pO_2 levels must be reliably detected to minimize the risk of retinopathy of prematurity. Skin oxygen permeability of newborn and preterm infants are much higher than that of adults due to differences in skin structure, function, and composition. Despite its lower accuracy in adult applications, transcutaneous oximetry is used as a standard mean in wound healing prediction and amputation level determination. Since it is measuring the local oxygen released from the capillaries through the skin, transcutaneous oxygen partial pressure ($tcpO_2$) level reflects the metabolic state of tissue at the injured site. It is also useful to assess the effect of hyperbaric oxygen therapy.

A transcutaneous probe, with skin contact gel on the membrane, is placed on the skin over a vascularized area of the patient. During measurement, an internal heating element inside the probe heats up the tissue beneath the probe. The elevated temperature dilates blood vessels in the underlying tissue and increases oxygen permeability. The thin oxygen permeable membrane separating the electrolyte in the probe and the skin allows oxygen to diffuse into the electrolyte, a Clark cell inside the probe measures the pO_2 level in the electrolyte. Maintaining an accurate and constant temperature (44°C for adults and 43°C for neonates) is critical to ensure accurate measurement. Over temperature produces over reading, under temperature produces under reading. A heater and a temperature sensor inside the probe maintain a constant temperature at the measurement site. Some monitors integrate $tcpCO_2$ with $tcpO_2$ measurements using a single skin probe. Measurement of dissolved CO_2 is made possible by adding a pH electrode inside the probe and using a membrane permeable to both

Table 31-2. Oxygen Analyzer, Transcutaneous Oxygen Monitor, and Pulse Oximeter Comparison

	Oxygen Analyzer	*Pulse Oximeter*	*Transcutaneous Oxygen Monitor*
Principle of Operation	Electrochemical transducer	Beer-Lambert law	Electrochemical transducer
Parameter sensed	Partial pressure of O2 in airway	Oxygen saturation in blood	Partial pressure of O2 in blood
Hypoxia detection	Detects oxygen deficiency in inhaled air	Detects low concentration of oxyhemoglobin	Detects low level of dissolved oxygen in blood

oxygen and carbon dioxide. In neonates, partial pressure readings obtained non-invasively by transcutaneous oxygen monitors highly agree with that from arterial blood samples by blood gas analyzers. In older children and adults, the skin becomes less permeable to oxygen due to different in composition, $tcpO_2$ therefore becomes less than PaO_2.

Ensuring sufficient oxygen in patient's inspired air is the first level of defense to prevent hypoxia. Measuring the percent oxygen saturation and dissolved oxygen gas in blood can detect hypoxia even before other signs such as cyanosis or hyperventilation are observed. Table 31–2 summarizes the main differences between these three devices.

A pulse oximeter measures the percentage of hemoglobin bonded with oxygen in blood, $tcpO_2$ level provides another piece of information. The metabolism of oxygen by the body needs to be measured by other methods, such as expired CO_2. Pulse oximetry is not a substitute for laboratory blood gas analysis, the latter measures dissolved oxygen in blood and includes other important parameters such as blood pH and carbon dioxide levels. As most of the oxygen in blood is carried by hemoglobin, for patient with severe anemia, despite a high $\%SpO_2$ level detected, the patient may not be getting sufficient oxygen to the tissues. Pulse oximeters, transcutaneous oxygen monitors and oxygen analyzers are often used together to detect insufficient oxygen to the patient. They provide complementary protection against hypoxia. For example, in anesthesia, an oxygen analyzer is connected to the inspiratory limb of the patient breathing circuit to sound an alarm if there is a low oxygen level in the patient's inspired gas. A pulse oximeter is connected to the patient (e.g., using a finger probe) to detect the level of oxygen saturation in the patient's bloodstream.

COMMON PROBLEMS AND HAZARDS

Pulse Oximeters

Although we can empirically establish an accurate relationship under ideal situations between the optical intensity ratio (r) and $\%SpO_2$, in the presence of patient motion or other interference, the optical densities will inevitably include these noise components (N). In the presence of noise, the measured beam intensity is composed of signal and noise components; that is, $I = S + N$ (where S is the desired signal and N is the noise). With the noise component, the optical intensity ratio (Equation 31.5) now becomes

$$r = \frac{S_{rd} + N_{rd}}{S_{ir} + N_{ir}}. \tag{31.6}$$

As movement changes the optical path length of blood vessels and tissues, a well-observed source of noise comes from the change in light absorption caused by patient motion. Low blood pressure and low perfusion decreases the signal level. Decrease in signal level and increase in noise lead to a low signal-to-noise ratio (SNR). In a poor SNR situation, the noise level becomes significant in the optical intensity ratio r, which will increase the error in the derived %SpO2. If the noise (N) component is much larger than the signal (S), in other words, N >> S, the optical intensity ratio in Equation 31.6 then becomes

$$r = \frac{N_{rd}}{N_{ir}}.$$

Under such conditions, if the noise levels in the red and infrared regions are similar (i.e., $N_{rd} \approx N_{ir}$), r will approach unity. For most systems with a good SNR, $r = 1.0$ corresponds to a %SpO2 of about 82%. For that reason, a pulse oximeter working under noisy conditions will tend to report a lower oxygen saturation reading. To maximize the SNR, the transmitted beam intensity should be measured during the systolic portion of the blood pressure cycle, where the absorption has its highest value in the cardiac cycle. Most manufacturers are using this technique to improve reliability. A conventional pulse oximeter requires a pulse pressure above 20 mmHg or 2.7 kPa at the measurement site to operate reliably. There have been some reported successes by manufacturers using special digital signal processing techniques such as adaptive filtering or feature extraction to improve detection under low pulse pressure, low perfusion and low SNR. The following are some causes of reduced SNR:

- Poor perfusion and low blood pressure—A patient suffering from poor perfusion usually has lower than normal blood pressure. A lack of blood in the capillaries will decrease the SNR and therefore increase the error of the measurement.
- Low hemoglobin level (< 5 mg/dl)—smaller number of hemoglobin decreases the total hemoglobin absorption and therefore affects the accuracy.
- Excessive signal attenuation—Patients with dark skin pigment, too thick tissue, or nail polish at the measurement site will decrease signal penetration (decrease the detector signal level) and increase measurement error.
- External interference—EMI, ambient light, and heat (IR) sources can introduce errors in the measurement. There were reported incidents that flashing light and fluorescent light sources were misinterpreted by

pulse oximeters as pulsating RED or IR signals. To avoid external light interference, the pulse oximeter probe is enclosed with side covers to block external light from reaching the sensor through the patient's tissue.
- Motion—Motion will cause variations in the optical path length, which will produce measurement errors.
- Substances in blood—Some substances in the bloodstream may affect the absorption of the light sources. High levels of dyshemoglobin in carbon-monoxide poisoning and the presence of artificial dyes (such as methylene blue or indocyanine green) in a patient's blood are examples that reduce the accuracy of the measurements.

There were reports on tissue burns at pulse oximeter probe locations. These incidents were due to the use of functionally incompatible LEDs and sensors. Even though the probe connectors are compatible with the oximeter, the incompatible oximeter may drive too much current into the LEDs to produce excessive heat. To ensure compatibility, a pulse oximeter user must pay attention to use only sensors specified by the manufacturer.

EMI from other devices may cause inaccurate readings. Some manufacturers install a Faraday cage-type of EMI shielding in the finger probe. Some oximeters will keep displaying the last measured value while waiting for the new reading to be computed, this may lead to misinterpretation of oxygen saturation level. A low plethysmograph signal strength provides an indication to warn the user that the sensor is not detecting sufficient signal from the patient. Use of non-MRI-compatible pulse oximetry during MRI procedures will create a burn hazard for the patient.

Oxygen Analyzers and Transcutaneous Oxygen Monitors

Transcutaneous oxygen monitoring requires appropriate probe placement, the sensing site must be at a vascularized area and free from bony prominences. Skin thickness, contact gel, temperature and perfusion all affect the reliability of transcutaneous oxygen measurement. The following patient condition will cause a low $tcpO_2$ reading: shock, acidosis, hypothermia, cyanosis, anemia, or skin edema. High PaO_2 as well as use of vasodilation drugs may cause under reading. The sensing probe of a $tcpO_2$ monitor needs to be moved to another location and be recalibrated every 3 to 4 hours. After a power down or at start up, a $tcpO_2$ monitor needs a warm up period of 10 to 20 minutes to produce a reliable reading. Heat burn injuries and adhesive allergy reactions at sensing sites had been reported.

BIBLIOGRAPHY

Andrews, D. H., & Kokes, R. J. (1962). *Fundamental Chemistry*. New York, NY: John Wiley & Sons, Inc.

Barker, S. J. (2002). Motion-resistant pulse oximetry: A comparison of new and old models. *Anesthesia and Analgesia, 95*(4), 967–972.

Barker, S. J., & Shah, N. K. (1997). The effects of motion on the performance of pulse oximeters in volunteers (revised publication). *Anesthesiology, 86*(1), 101–108.

Barnett, E., Duck, A., & Barraclough, R. (2012). Effects of recording site on pulse oximetry readings. *Nursing Times, 108*(1-2), 22–23.

Bohnhorst, B., Peter, C. S., & Poets, C. F. (2000). Pulse oximeters' reliability in detecting hypoxemia and bradycardia: Comparison between a conventional and two new generation oximeters. *Critical Care Medicine, 28*(5), 1565–1568.

Cannesson, M., Desebbe, O., Rosamel, P., Delannoy, B., Robin, J., Bastien, O., & Lehot, J -J. (2008). Pleth variability index to monitor the respiratory variations in the pulse oximeter plethysmographic waveform amplitude and predict fluid responsiveness in the operating theatre. *British Journal of Anaesthesia, 101*(2), 200–206.

De Felice, C., Leoni, L., Tommasini, E., Tonni, G., Toti, P., Del Vecchio, A., Ladisa, G., & Latini, G. (2008). Maternal pulse oximetry perfusion index as a predictor of early adverse respiratory neonatal outcome after elective cesarean delivery. *Pediatric Critical Care Medicine, 9*(2), 203–208.

Freund, P. R., Overand, P. T., Cooper, J., Jacobson, L., Bosse, S., Walker, B., Posner, K. L., & Cheney, F. W. (1991). A prospective study of intraoperative pulse oximetry failure. *Journal of Clinical Monitoring, 7*(3), 253–258.

Goldman, J. M., Petterson, M. T., Kopotic, R. J., & Barker, S. J. (2000). Masimo signal extraction pulse oximetry. *Journal of Clinical Monitoring, 16*(7), 475–483.

Huch R, Huch A, Lubbers DW. (1973). Transcutaneous measurement of blood Po2 (tcPo2) – Method and application in perinatal medicine. *Journal of Perinatal Medicine, 1*(3):183-191.

Jopling, M. W., Mannheimer, P. D., & Bebout, D. E. (2002). Issues in the laboratory evaluation of pulse oximeter performance. *Anesthesia and Analgesia, 94*(1 Suppl), S62–68.

Jorgensen, J. S., Schmid, E. R., Konig, V., Faisst, K., Huch, A., & Huch, R. (1995). Limitations of forehead pulse oximetry. *Journal of Clinical Monitoring, 11*(4), 253–256.

Lin, J. C., Strauss, R. G., Kulhavy, J. C., Johnson, K. J., Zimmerman, M. B., Cress, G. A., Connolly, N. W., & Widness, J. A. (2000). Phlebotomy overdraw in the neonatal intensive care nursery. *Pediatrics, 106*(2), E19.

Millikan, G. A. (1942). The oximeter, an instrument for measuring continuously the oxygen saturation of arterial blood in man. *Review of Scientific Instruments, 13*(10), 434–444.

Milner, Q. J., & Mathews, G. R. (2012). An assessment of the accuracy of pulse oximeters. *Anaesthesia, 67*(4), 396–401.

Pologe, J. A. (1987). Pulse oximetry: Technical aspects of machine design. *International Anesthesiology Clinics, 25*, 137–153.

Severinghaus, J. W. & Bradley, A. F. (1958). Electrodes for Blood pO2 and pCO2 Determination. *Journal of Applied Physiology. American Physiological Society.* 13 (3): 515–520.

Severinghaus, J. W., & Honda, Y. (1987). History of blood gas analysis. VII. Pulse oximetry. *Journal of Clinical Monitoring, 3*(2), 135–138.

Shah, N., Ragaswamy, H. B., Govindugari, K., & Estanol, L. (2012). Performance of three new-generation pulse oximeters during motion and low perfusion in volunteers. *Journal of Clinical Anesthesiology, 24*(5), 385–391.

Takahashi, S., Kakiuchi, S., Nanba, Y., Tsukamoto, K., Nakamura, T., & Ito, Y. (2000). The perfusion index derived from a pulse oximeter for predicting low superior vena cava flow in very low birth weight infants. *Journal of Perinatology, 30*, 265–269.

Trivedi, N. S., Ghouri, A. F., Shah, N. K., Lai, E., & Barker, S. J. (1997). Effects of motion, ambient light, and hypoperfusion on pulse oximeter function. *Journal of Clinical Anesthesiology, 9*(3), 179–183.

Wukitsch, M. W., Petterson, M. T., Tobler, D. R., & Pologe, J. A. (1988). Pulse oximetry: Analysis of theory, technology, and practice. *Journal of Clinical Monitoring, 4*(4), 290–301.

Chapter 32

END-TIDAL CARBON DIOXIDE MONITORS

OBJECTIVES

- Describe the clinical applications of end-tidal CO_2 monitors.
- Sketch and explain the CO_2 concentration waveform of expired air.
- Explain the principles of operation of end-tidal CO_2 monitors.
- Differentiate between mainstream and sidestream end-tidal $CO2$ monitoring.
- Identify the functional components and construction of a typical mainstream end-tidal CO_2 sensor.
- Sketch the functional block diagram of a typical sidestream CO_2 monitor.
- Discuss factors affecting the signal quality and accuracy of end-tidal CO_2 monitors.

CHAPTER CONTENTS

1. Introduction
2. Carbon Dioxide Concentration Waveform
3. Principles of Operation
4. Mainstream versus Sidestream Monitoring
5. Common Problems and Hazards

INTRODUCTION

Carbon dioxide (CO_2) is a by-product of cellular metabolism and is removed from the body through the circulatory and respiratory systems. The concentration of CO_2 in the exhaled air reflects the metabolic rate and indicates the status of the pulmonary and circulatory systems. CO_2 is continuously produced in the body and transported by blood to the alveoli. The two factors that determine the alveolar concentration of CO_2 are the rate of transport of CO_2 from the blood to the alveoli and the rate of removal of CO_2 from the alveoli by alveolar ventilation. The alveolar CO_2 concentration is directly proportional to the rate of CO_2 excretion and inversely proportional to the alveolar ventilation.

A CO_2 monitor (or capnograph) can noninvasively measure the concentration of CO_2 in breathing air. In operating rooms, it is used to monitor patients' ventilation under general anesthesia. It can detect breathing circuit disconnection, airway leaks, and improper placement of an endotracheal tube. End-tidal carbon dioxide ($EtCO_2$) monitors can be used as a noninvasive method to estimate the arterial partial pressure of carbon dioxide ($PaCO_2$). In intensive care units, $EtCO_2$ is used together with other physiological monitoring parameters to evaluate a patient's cardiopulmonary function. Capnography provides a rapid and reliable method to detect life-threatening conditions. Together with pulse oximetry, capnography reliably prevents most of the avoidable mishaps during general anesthesia procedures.

CARBON DIOXIDE CONCENTRATION WAVEFORM

The expired air from a person in a breathing cycle combines the dead space air and alveolar air. Figure 32-1 shows the changes of CO_2 concentration in the expired air during the course of breathing. The expired air at the beginning of exhalation is dead space air, which is inspired air saturated with moisture. As expiration goes on, more and more alveolar air becomes mixed with the dead space air until the dead space air has been totally washed out. At the end of expiration, the expired air contains 100% alveolar air. For a normal adult breathing in atmospheric air, the partial pressure (or concentration) of the CO_2 in the inspired air (atmospheric air) is 0.3 mmHg or 0.04 kPa (or 0.04%) and that of the alveolar air is about 40 mmHg or 5.3 kPa (or 5.3%). At the beginning of exhalation, the CO_2 concentration is 0.04%, and it rapidly rises until it reaches a plateau. At the end of the exhalation, the CO_2 concentration is at its maximum (about 5%). A recording of the CO_2 concentration against time is called a capnogram.

A normal capnogram includes an almost zero (actually 0.04%) baseline, a sharp upstroke, and a relatively flat alveolar plateau. Significant deviations from this morphology suggest an abnormality in the patient or the gas delivery equipment.

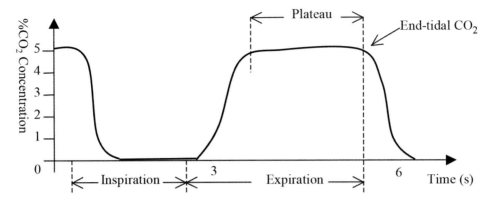

Figure 32-1. CO_2 Concentration in Breathing Air.

PRINCIPLES OF OPERATION

The principle of operation of an end-tidal CO_2 monitor is similar to that of a pulse oximeter. CO_2 monitors use infrared spectroscopy to measure the concentration of CO_2 in the exhaled gas. In normal exhaled gas, only CO_2 and water vapor contribute to the absorption of infrared (IR) radiation. During anesthesia procedures, however, anesthetic gases such as nitrous oxide may affect the IR absorption. Figure 32-2 shows the IR absorption spectrum of water vapor, CO_2, and nitrous oxide. Based on Beer-Lambert law, the concentration of CO_2 in the gas is proportional to its absorption. Selecting the proper wavelength and correcting the effects of other absorptions allow the monitor to accurately determine the CO_2 concentration in the gas sample.

A typical setup to measure CO_2 concentration is shown in Figure 32-3. IR radiation passing through the gas mixture is absorbed by the CO_2 in the exhaled gas sample. The IR beam then passes through two filters; the first filter is within the IR frequency spectrum of CO_2 absorption, and the second is outside. Behind each filter is an IR detector. The detector behind the first filter is for the data channel and the detector behind the second filter is for reference. The reference signal is to compensate for signal fluctuations such as variation of intensity of the light source.

MAINSTREAM VERSUS SIDESTREAM MONITORING

There are two types of capnographs—mainstream and sidestream. They are differentiated by the location of their CO_2 sensors. A mainstream capno-

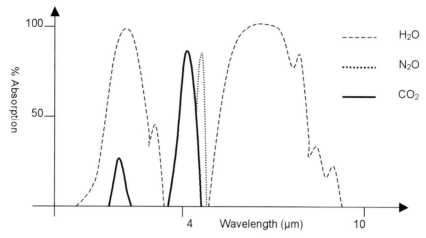

Figure 32-2. IR Absorption Spectra of H_2O, N_2O, and CO_2.

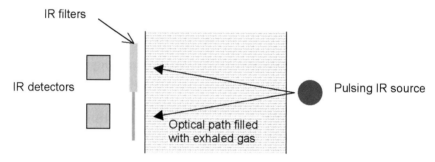

Figure 32-3. Functional Components of CO_2 Sensor.

graph has the sensor located right in the patient breathing circuit while the CO_2 sensor of a sidestream capnograph is inside the machine away from the patient breathing circuit.

Figure 32-4 shows a CO_2 sensor of a mainstream capnograph connected to the patient breathing circuit using a special airway adaptor. The windows on both sides of the airway allow IR from the light source to reach the detectors. As CO_2 gas mixture passes through the airway adaptor, the IR radiation of the pulsing source (e.g., 48 Hz) is partially absorbed by the CO_2 in the gas mixture. It then passes through the two optical filters before reaching the IR detectors. One filter is at the middle of the IR absorption spectrum (e.g., 4.2 µm) for CO_2 and the other is outside (e.g., 3.7 µm) for reference. To accurately measure the level of CO_2 in the gas mixture, the outputs of the detectors are sampled simultaneously and the level of CO_2 is determined from the ratio of the data and reference channels. The ratio is then

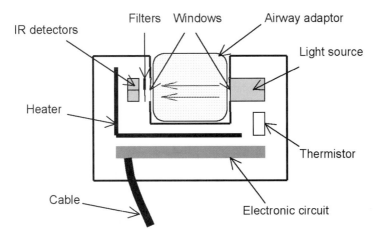

Figure 32-4. Internal View of a Mainstream CO_2 Sensor.

compared to a lookup table in memory to determine the concentration of CO_2.

As the warm expired air is saturated with water vapor, water will condense on the cooler windows of the sensor, causing errors in the IR absorption measurement. To prevent condensation from forming on the window, a heater and a thermistor inside the sensor maintain a sensor temperature above the body temperature (e.g., at 40°C).

The optical bench (light source, filters, detectors, etc.) for a sidestream capnograph is inside the machine. The principle of CO_2 detection is the same as its mainstream counterpart. A small diameter air sampling line connects the patient breathing circuit to the sensor inside the machine. During measurement, the breathing gas sample is drawn by an air pump from the breathing circuit into the sensor. To prevent condensed liquid getting into the optical bench, a water trap is located at the air sampling line near the entrance of the machine. A heater is also used to raise the temperature of the sensor chamber to prevent water condensation. For both mainstream and sidestream capnographs, the derived CO_2 concentration (or partial pressure) must be temperature and pressure corrected because it is measured at a temperature and a pressure different from those inside the patient's body. Figure 32-5 shows a functional block diagram of a sidestream CO_2 monitor.

COMMON PROBLEMS AND HAZARDS

Interference from condensed water and patient secretions that enter the sensing compartment or the sampling tube is the most significant problem

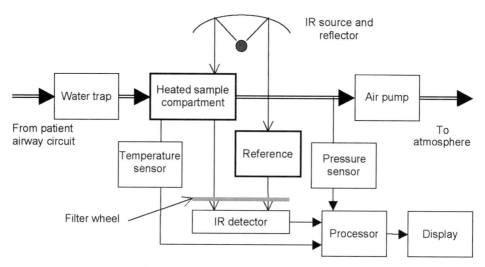

Figure 32-5. Sidestream CO_2 Monitor.

in CO_2 monitors. As discussed previously, moisture as well as its condensation absorb IR radiation. For mainstream monitors, extra attenuation of the IR beam causes falsely high CO_2 readings. In addition, for sidestream monitors, water condensation and patient secretion can block the small-bore air sampling tubing, causing a pressure and flow reduction in the circuit. Anticondensation heaters and water traps are used to prevent such problems. However, the heated mainstream sensors may cause patient burn if they are placed in contact with the patient.

Since the partial pressure of CO_2 is pressure dependent, any pressure fluctuations will introduce errors in the measurements. During anesthesia procedures, halogenated gases and nitrous oxide are present in the patient's exhaled gas. These gases absorb IR radiation and cause measurement errors if they are not properly accounted for. Although the presence of oxygen does not affect IR absorption, the collision of O_2 and CO_2 molecules will broaden the absorption peak and create a falsely low CO_2 concentration reading.

Capnographs may not be able to accurately measure the alveolar CO_2 concentration of neonates or patients with shallow breathing due to the relatively large dead space volume in the breathing circuits and airways. In sidestream monitoring, due to the small tidal volume, gas sampling (removing air from the expiratory breathing circuit) may disrupt the respiratory pattern of nenotes.

BIBLIOGRAPHY

Ahrens, T. (1998). Technology utilization in the cardiac surgical patient: SvO2 and capnography monitoring. *Critical Care Nursing Quarterly, 21*(1), 24–40.

Donald, M. J., & Paterson, B. (2006). End tidal carbon dioxide monitoring in prehospital and retrieval medicine: A review. *Emergency Medical Journal, 23*(9), 728–730.

Eipe, N., & Doherty, D. R. (2010). A review of pediatric capnography. *Journal of Clinical Monitoring and Computing, 24*(4), 261–268.

Gerstenberger, P. D. (2010). Capnography and patient safety for endoscopy. *Clinical Gastroenterology and Hepatology, 8*(5), 423–425.

Jabre, P., Jacob, L., Auger, H., Jaulin, C., Monribot, M., Aurore, A., Margenet, A., Marty, J., & Combes, X. (2009). Capnography monitoring in nonintubated patients with respiratory distress. *American Journal of Emergency Medicine, 27*(9), 1056–1059.

Kodali, B. S. (2013). Capnography outside the operating rooms. *Anesthesiology, 118*(1), 192–201.

Langhan, M. (2009). Continuous end-tidal carbon dioxide monitoring in pediatric intensive care units. *Journal of Critical Care, 24*(2), 227–230.

Lightdale, J. R., Goldmann, D. A., Feldman, H. A., Newburg, A. R., DiNardo, J. A., & Fox, V. L. (2006). Microstream capnography improves patient monitoring during moderate sedation: A randomized, controlled trial. *Pediatrics, 117*(6), 1170–1178.

Nagler, J., & Krauss, B. (2008). Capnography: A valuable tool for airway management. *Emergency Medicine Clinics of North America, 26*(4), 881–897.

Raemer, D. B., & Calalang, I. (1991). Accuracy of end-tidal carbon dioxide tension analyzers. *Journal of Clinical Monitoring, 7*(2), 195–208.

Ratnasabapathy, U., Allam, S., & Souter, M. J. (2002). Evaluation of an expired fraction carbon dioxide monitor. *Anaesthesia, 57*(9), 900–904.

Sakata, D. J., Matsubara, I., Gopalakrishnan, N. A., Westenskow, D. R., White, J. L., Yamamori, S., Egan, T. D. & Pace, N. L. (2009). Flow-through versus sidestream capnometry for detection of end tidal carbon dioxide in the sedated patient. *Journal of Clinical Monitoring and Computing, 23*(2), 115–122.

St. John, R. E. (2003). End-tidal carbon dioxide monitoring. *Critical Care Nurse, 23*(4), 83–88.

Ward, K. R., & Yealy, D. M. (1998). End-tidal carbon dioxide monitoring in emergency medicine. Part 1: Basic Principles. *Academic Emergency Medicine, 5*(6), 628–636,

Westhorpe, R. N., & Ball, C. (2010). The history of capnography. *Anaesthesia and Intensive Care, 38*(4), 611.

Chapter 33

ANESTHESIA MACHINES

OBJECTIVES

- Describe the functions of an anesthesia machine.
- Describe the stages of anesthesia and the effect of inhalation anesthetics.
- Name and explain the functional building blocks of an anesthesia machine.
- Sketch the gas circuit/piping diagram of a continuous-flow rebreathing anesthesia machine.
- Trace the anesthetic gas flow in a typical anesthesia machine and its patient breathing circuit.
- Discuss common problems and potential hazards associated with anesthesia machines and their built-in mitigation features.

CHAPTER CONTENTS

1. Introduction
2. Inhalation Anesthesia
3. Principles of Operation
4. Gas Supply and Control Subsystem
5. Breathing and Ventilation Subsystem
6. Scavenging Subsystem
7. Common Problems and Hazards

INTRODUCTION

Since Dr. Crawford Long of Jefferson, Georgia, administered the first ether anesthetic in 1842 for the painless removal of a neck tumor, anesthesiology has evolved into a branch of medicine that relates to the administration of medication or anesthetic agents to relieve pain and to support physiological functions during surgical procedures. In a surgical procedure, an analgesic or anesthetic agent is administered to the patient for pain relief. There are four main categories of anesthesia: local, sedation, regional and general. For a major or long procedure, the patient is often under general anesthesia. General anesthesia is a reversible state of unconsciousness produced by anesthetic agents in which motor, sensory, mental, and reflex functions are lost. In suppressing nerve signal conduction and cerebral cortex activities to achieve anesthesia, the anesthetic agents will cause respiratory depression and respiratory arrest in a patient under general anesthesia.

An anesthesia machine is a tight integration of medical instrumentations to assist the anesthetist (or anesthesiologist) to induce and maintain anesthesia. It serves three major functions:

1. Dispense a controlled mixture of gases (oxygen and other gases such as nitrous oxide, medical air) and anesthetic agents (such as halothane, sevoflurane) to anesthetize the patient and for patient breathing during surgery.
2. Assist patient's respiration (ventilation function) when normal breathing is compromised due to the anesthetic effect.
3. Monitor the patient's condition, such as vital signs and depth of anesthesia, during the surgical procedure.

Physiological monitoring with alarm functions are integrated into modern anesthesia machines. Automatic record keeping of patient condition (e.g., vital signs), machine parameters, data management, networking and communication functions are essential features in modern anesthesia systems.

This chapter introduces the backbone of a typical anesthesia machine, which includes the gas supply and control, breathing and ventilation, and scavenging subsystems. Physiological monitoring and other functions are discussed in other parts of this book.

INHALATION ANESTHESIA

Anesthesia can be achieved by intravenous administration of agents such as barbiturates or neuromuscular blockers, or by inhalation of anesthetic gases or vapors. The three stages of inhalation anesthesia are: induction, maintenance, and recovery.

1. Induction is the first step in anesthetizing the patient. It is the period between the initial administration of the induction agents and the loss of consciousness. Under the induction stage, the patient progresses from analgesia without amnesia to analgesia with amnesia. During the early time of this stage, the patient is still conscious and can carry on a conversation. The process includes intubation, administration of intravenous drugs, and supply of oxygen and anesthetic gases.
2. When the desired level of anesthesia is established, the patient loses reflexes and rhythmic respiration and attains total muscle relaxation. The maintenance stage keeps the patient at the desired level of anesthesia. Periodic adjustment of the breathing gas composition (such as the concentration of the anesthetic agent) may be necessary.
3. The emergence or recovery stage is for reversing the process of anesthesia. It includes removal of the anesthetic gas from the patient, adjusting the breathing gas composition, and possibly administering drugs to counteract the effect of the anesthetic agents.

The composition of inhalation anesthetic gas dispenses from anesthesia machines consists of oxygen, other medical gasses, and one or more anesthetic agents. At room temperature, anesthetic agents can be in gaseous (e.g., nitrous oxide, cyclopropane) or volatile liquid format (e.g., diethyl ether, halothane, enflurane). The gas and agent combination and their concentrations are determined by the anesthetist based on the nature of the procedure and the condition of the patient. Note that cyclopropane and diethyl ether are no longer used in surgery due to their extremely flammable property.

The minimum alveolar concentration (MAC) of the anesthetic agent represents the potency of the inhaled anesthetic. MAC is defined as the inhalation anesthetic agent concentration that is necessary to produce a lack of response to a standard skin incision in 50% of the test population. The actual concentration of the anesthetic agent administered to the patient is always higher than its MAC level. For example, the MAC level of halothane (a very potent agent) is 0.76%. A concentration of 1 to 2% of halothane in the breathing gas is usually administered to maintain general anesthesia. The MAC level of desflurane is 5 to 10%. This means that to achieve the same anesthetic effect, the concentration of desflurane needs to be about ten times of that used with halothane. In general, a higher level (e.g., 1.4 times) than the MAC value is administered to the patient at the start of the procedure and the level is reduced during the maintenance stage. By monitoring the anesthetic agent concentration level in the patient's exhaled gas and compar-

Table 33-1. Common Inhalation Anesthetics

Name	MAC Level	State at Room Temperature	Color Code
Halothane	0.76%	liquid	red
Enflurane	1.68%	liquid	orange
Isoflurane	1.15%	liquid	purple
Sevoflurane	1.7 to 2%	liquid	yellow
Desflurnae	5 to 10%	gas	blue

ing it to its MAC level, the anesthetist has a guide to maintain the depth of anesthesia during a procedure. The MAC levels of some anesthetic agents commonly used in anesthesia are listed in Table 33-1, note also their physical state at room temperature and color coding.

To prevent hypoxia, a minimum of 25% oxygen must be administered to the patient during anesthesia. Because nitrous oxide is both rapidly absorbed and eliminated by the body, it is often used as a primary gas to accelerate the uptake of a second anesthetic agent. Nitrous oxide has limited potency with a MAC level of 105%, it is therefore not able to provide achieve general anesthesia as a sole agent. The usual practice is to administer a ratio of about 2:1 of nitrous oxide to oxygen along with another agent (such as halothane) to produce a depth of anesthesia sufficient for the surgical procedure. In recent years, use of nitrous oxide in anesthesia has declined due to its adverse health effects on human. In a nitrous oxide free procedure, nitrous oxide is replaced by medical air. To compensate for the loss of anesthetic effect due to elimination of nitrous oxide, the level of volatile anesthetic agent is increased to maintain the same anesthetic effect. In procedures when maintaining low airway flow resistance is critical, Heliox, a mixture of helium and oxygen gas (e.g., 79% helium, 21% oxygen) is used to replace medical air or nitrous oxide. As the density of helium is lower than nitrogen, Heliox has much lower flow resistance than medical air or nitrous oxide in the airway of the patient. Heliox is increasingly being used on small babies or patients with obstructive airway conditions such as COPD.

PRINCIPLES OF OPERATION

The primary function of an anesthesia machine is to facilitate administration of inhalation anesthetics. A basic anesthesia machine consists of three

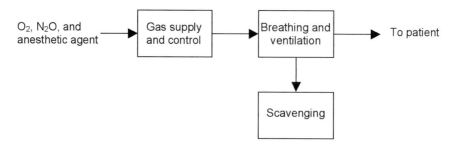

Figure 33-1. Functional Block Diagram of a Basic Anesthesia Machine.

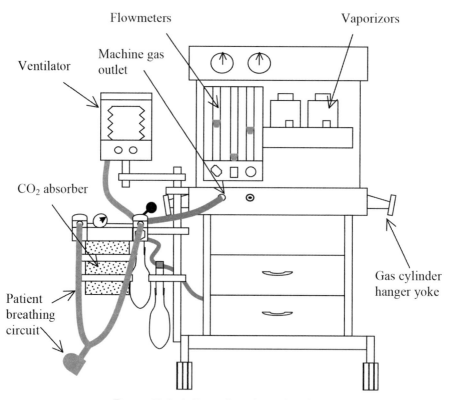

Figure 33-2. A Basic Anesthesia Machine.

main subsystems:

1. Gas supply and control
2. Breathing and ventilation
3. Scavenging

Patient monitoring equipment may be considered as the fourth subsystem (which is not discussed in this chapter). The most commonly used anesthesia machine is the continuous-flow rebreathing anesthesia machine. In this type of machine, the exhaled gas from the patient is channeled back to the patient after it has been processed and mixed with a proportion of fresh anesthetic gas. Figure 33-1 shows a simple functional block diagram of an anesthesia machine. The gas supply and control block takes oxygen and medical gases (air, nitrous oxide, etc.) from the wall outlet and combines them with an anesthetic agent to produce a mixture of anesthetic gas. This gas mixture is supplied to the breathing and ventilation circuit. The function of the breathing and ventilation circuit is to regulate the anesthetic gas mixture to an appropriate flow and pressure level, and then deliver the gas mixture to the patient. It also processes the exhaled gas from the patient and returns it to the patient for rebreathing. The scavenging block is to prevent escaped anesthetic gas from polluting the operating room environment. The scavenging system captures the waste anesthetic gas and discharges it safely outside the operating room. Figure 33-2 shows a basic anesthesia machine. The following sections describe the principles of operation of the three subsystems.

GAS SUPPLY AND CONTROL SUBSYSTEM

Figure 33-3 is a gas flow piping diagram showing the major components of the gas supply and control subsystem of a conventional anesthesia machine. The anesthesia machine shown in the figure is a two-gas machine with oxygen and nitrous oxide as the main gases plus one volatile anesthetic agent forming the inhalation anesthetic gas.

Most anesthesia machines today are three-gas machines (oxygen, nitrous oxide and medical air) and may have Heliox as another gas supply. To simplify the diagram for explanation, a two-gas machine (only oxygen and nitrous oxide) is discussed. The medical air circuit has similar controls and components as the nitrous oxide circuit and connects to the common gas header leading to the anesthetic agent vaporizer.

Under normal operation, oxygen and nitrous oxide are supplied from the piped-wall gas outlets to the respective gas inlets of the machine. The pressure of the gases at the wall outlets is about 50 psi or 350 kPa. Oxygen flow through a check valve (one-way valve) is reduced (e.g., to about 16 psi or 112 kPa) by the second-stage oxygen regulator before it reaches the flow control valve of the oxygen flowmeter. Nitrous oxide gas from the wall outlet passes through the regulators and pressure-sensing shutoff valve before reaching the nitrous oxide flow control valve of the nitrous oxide flowmeter.

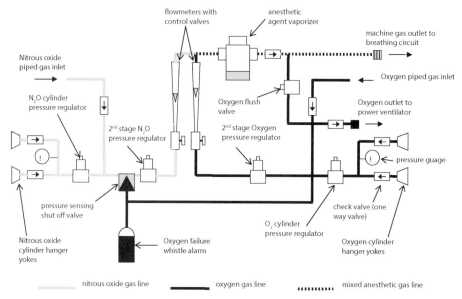

Figure 33-3. Major Components of the Gas Supply and Control Subsystem.

The shutoff valve in the machine shown is held open by the oxygen pressure in the oxygen supply line. If the oxygen pressure drops (e.g., to below 25 psi or 175 kPa), the valve will shut off the nitrous oxide supply to the machine and sound an alarm to alert the anesthetist to the failing oxygen supply. The alarm is sounded by the action of the shutoff valve discharging the oxygen stored in the small gas cylinder through a whistle. This "whistle alarm" design has the advantage that it will function even under total power loss condition. The shutoff and alarm mechanism form the "oxygen failure protection detector" that protects the patient from unknowingly breathing in a low oxygen-level gas mixture in case of oxygen supply failure.

By adjusting the flow control valves, the anesthetist can achieve a suitable mix and flow of the oxygen and nitrous oxide gas mixture. The gas mixture then flows through a calibrated vaporizer, where it picks up a selected amount of an anesthetic agent before it is delivered to the patient via the breathing and ventilation circuit. An oxygen flush valve is available to flush the patient circuit with pure oxygen during setup.

Under rare circumstances in which piped-wall gases are not available, oxygen and nitrous oxide supplies are automatically switched to gas cylinders mounted on the hanger yokes on the sides or the rear of the machine. To provide further backup, some machines have a second oxygen cylinder and a second nitrous oxide cylinder mounted on the anesthesia machine. One cylinder of each gas is on standby (turned on) and the other one serves

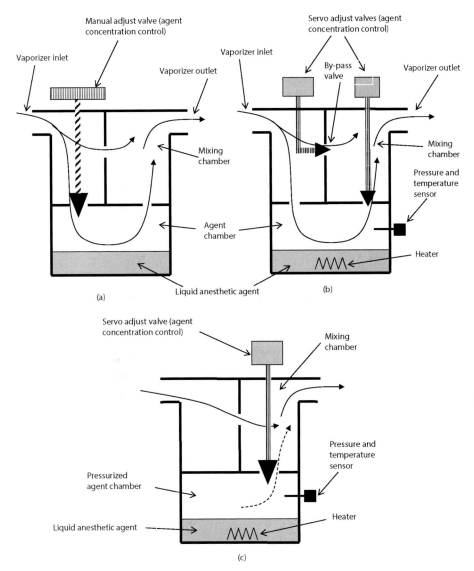

Figure 33-4. Vaporizers: a) Conventional Variable-bypass,
b) Electronic Electronic-Controlled, c) Measured-flow.

as backup in case the first cylinder is depleted. Because the maximum pressure of oxygen and nitrous oxide from the cylinders is 2200 psi or 15 MPa and 750 psi or 5.3 MPa, respectively, cylinder pressure regulators are set to reduce the pressure of these gases to slightly below 50 psi or 350 kPa. Pressure gauges at hanger yokes provide an indication of whether the cylinders are full or empty. Because oxygen in the gas cylinder is in a gaseous state, the level of gas in the cylinder is proportional to the pressure inside the cylinder. However, it is not possible to tell the amount of nitrous oxide left in the cylinder by reading the cylinder pressure. Since compressed nitrous oxide at room temperature is in liquid form inside the cylinder, the cylinder pressure will start to drop only when it is almost empty (no more liquid nitrous oxide in the cylinder). The amount of nitrous oxide left in the cylinder can be estimated by its weight.

A vaporizer is a component in the gas supply and control circuit to introduce a selected concentration of the anesthetic agent into the oxygen and nitrous oxide mixture to form the anesthetic gas. There are usually more than one vaporizers (with a different agent in each) mounted on the output manifold of the anesthesia machine. To prevent anesthetic overdose, an interlock mechanism ensures that only one vaporizer is active at one time. The most commonly used vaporizers are the plenum type variable-bypass vaporizers. Two types of plenum variable-bypass vaporizers are available on anesthesia machines: the conventional variable-bypass vaporizers and the electronic controlled vaporizers. Both conventional variable-bypass vaporizers and electronic controlled vaporizers can be used to deliver liquid anesthetic agents such as halothane, enflurane, isoflurane, and servoflurane. However, desflurane, a more volatile anesthetic agent with lower boiling point (see Table 33-1), requires using the electronic controlled vaporizer. Measured-flow vaporizers, which use a different design, is used by some manufacturers to administer desflurane.

Figure 33-4a shows the simplified construction of a conventional variable-bypass vaporizer. The gas mixture (oxygen, medical air, nitrous oxide) enters the vaporizer from the inlet. It is then branched into two flow paths, one into the vaporizing chamber and the other through a bypass into the mixing chamber. The percentage of the total flow into the vaporizing chamber is determined by the position of the agent concentration control valve. The gas mixture flowing into the vaporizing chamber flows over a reservoir of liquid anesthetic agent, picks up the agent vapor, and exits the vaporizing chamber. The gas then meets and mixes with the bypassed gas and flows to the vaporizer outlet. The concentration of anesthetic agent in the final gas mixture is higher when a larger volume of gas is allowed to flow into the vaporizing chamber. A calibrated control dial is connect to the bypass valve to control the size of the valve opening. The anesthetist uses the control dial to set the agent concentration in the inhalation gas mixture. Since the vapor pressure inside the agent chamber increases and decreases with the rise and fall of temperature, a temperature-sensitive-flow control mechanism is necessary to compensate for temperature variations during the surgical procedure. In a conventional variable-bypass vaporizer, this is achieved by using thermal expansion property to negatively adjust the opening of the gas bypass control valve. In the vaporizer shown in Figure 33-4a, the length of the metal rod connecting to the valve seat will expand and therefore lengthened with temperature increase; this action will decrease the opening of the bypass valve, thus reducing the agent concentration in the gas mixture inside the mixing chamber. Wicks dipped into the liquid anesthetic agent or baffles inside the agent chamber are often added to increase the surface area of agent evaporation.

The design of an electronic-controlled vaporizer (Figure 33-4b) is functionally very similar to a conventional variable-bypass vaporizer. The difference is in the temperature compensation mechanism in which sensors and a microprocessor-based controller are employed. In an electronic-controlled vaporizer, the anesthetic agent inside the agent chamber is heated and pressurized to keep it in its liquid state. The incoming gas mixture is separated into two paths by the by-pass valve. The pressurized vapor of the anesthetic agent is released from the agent chamber to the mixing chamber through a regulating valve. The concentration of the agent in the gas mixture is adjusted by the two valves, one controls the flow of by-passed gas and the other controls the flow of anesthetic agent going into the mixing chamber. To maintain a constant agent concentration at the vaporizer outlet, the electronic controller analyzes the pressure and temperature inside the agent chamber to regulate the positions of valves. To increase accuracy, additional parameters such as gas flow rates, selected and actual (measured) agent concentrations, as well as the pressure and temperature of the inlet gas mixture are measured.

The measured-flow vaporizer is specifically designed to dispense desflurane. Desflurane has high volatility and low boiling point which precludes its use with conventional variable by-pass vaporizers. A measured-flow vaporizer is essentially a precise gas mixer. Its concept is illustrated in Figure 33-4c. The agent chamber is pressurized and the agent is heated and maintained at a constant temperature. The electronic controller, based on the temperature, pressure and the flow rate of the gas mixture, adjust the valve to control the amount of agent in the gas mixture.

BREATHING AND VENTILATION SUBSYSTEM

The function of the breathing and ventilation subsystem of an anesthesia machine is to deliver the anesthetic gas mixture (which is a mixture of oxygen, an anesthetic agent, medical air and/or nitrous oxide) to the patient. During a procedure when the patient's respiration function is suppressed, a mechanical ventilator is used to ventilate the patient. Most anesthesia machines deliver a continuous flow of anesthetic gas and oxygen mixture to

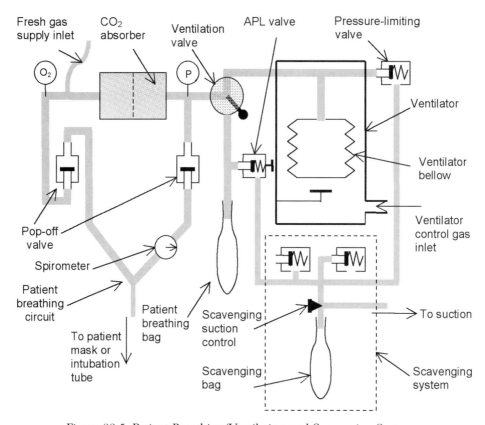

Figure 33-5. Patient Breathing/Ventilation and Scavenging Systems.

the patient. Two types of breathing circuits are used: the circle system and the T-piece system. Both types allow a certain level of rebreathing to conserve moisture and heat as well as anesthetic agents.

Figure 33-5 shows the circle breathing/ventilation subsystem of an anesthesia machine under ventilator mode. The ventilator valve is used to select between manual breathing mode using the breathing (or reservoir) bag and ventilation mode using the built-in mechanical ventilator. The scavenging subsystem for waste anesthetic gas removal is also shown in the diagram.

Figure 33-6 shows the machine in manual breathing mode. The Y-connection of the patient breathing circuit is connected to a face mask covering the patient's mouth and nose or to an endotracheal tube inserted into the trachea of the patient. During patient inhalation, fresh gas from the anesthesia machine enters the inspiratory limb of the breathing circuit into the lungs of the patient (flow direction in solid arrows). During exhalation, expired gas from the lungs goes through the expiratory limb of the breathing circuit into the breathing bag (flow direction in dotted arrows). A pair of check valves

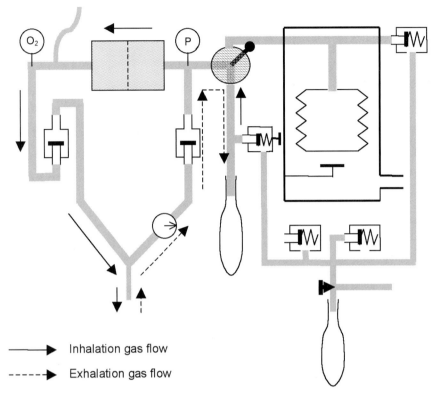

Inhalation gas flow

Exhalation gas flow

Figure 33-6. Patient Circuit Under Manual Breathing Mode.

(one-way valve) are used to prevent reverse gas flow in the inspiratory and expiratory limbs of the breathing circuit. The check valves are also referred to as a pop-off valves due to their construction. The valve seat, in form of a circular disk, sits on top of the valve opening connected to the breathing circuit. The disk will rise (pop-up) when the gas flow pushes the valve to open; reverse gas flow will force shut the disk against the opening.

During manual bagging, the anesthetist squeezes the reservoir or patient breathing bag to create positive pressure in the breathing circuit, the gas collected in the bag is forced into the CO_2 absorption canister, through the inspiratory limb of the breathing circuit, and back to the patient. The canister contains a CO_2 absorbing agent (usually soda lime plus a pH sensitive indicator) to remove CO_2 from the rebreathed gas. The maximum pressure in the breathing circuit is limited by the adjustable pressure-limiting (APL) valve located near the patient breathing bag. The APL valve is adjusted by the anesthetist during the procedure to maintain a slightly inflated breathing bag. Flow and pressure meters measure the volume flow and pressure in the patient breathing circuit. An oxygen sensor monitors the oxygen concentra-

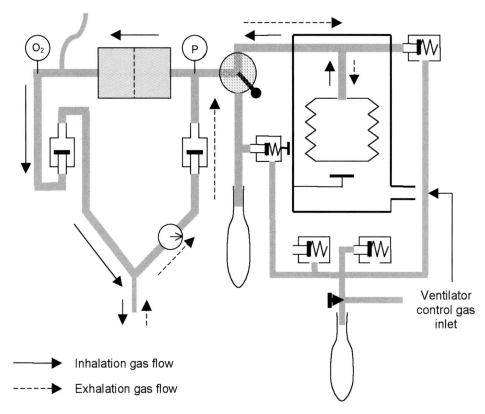

Inhalation gas flow

- - - - ▶ Exhalation gas flow

Figure 33-7. Patient Circuit under Ventilator Mode.

tion in the inhalation gas in the patient circuit.

When the machine is in ventilator mode (Figure 33-7), the patient breathing bag is disconnected. Expired gas from the patient flows into the bellows of the ventilator instead. During the inhalation phase, the positive pressure of the control gas compresses the bellows in the ventilator, forcing the accumulated gas in the bellows to flow into the CO_2 absorber and then through the inspiratory limb of the breathing circuit to the patient. The maximum pressure in the breathing circuit is limited by the pressure-limiting valve of the ventilator. In the controlled mode, the ventilator is cycled to deliver a fixed volume of gas to the patient at a fixed time interval. The tidal volume, respiration rate, and I:E ratio of ventilation are set by the anesthetist.

No CO_2 absorber is used in the T-piece design (Figure 33-8). Fresh gas is mixed with rebreathed gas before entering the patient's lungs. The percentage of rebreathing is controlled by the flow rate of fresh gas from the gas supply and control system. The exhaled anesthetic mixture leaves the circuit through an APL valve. The rate of elimination of exhaled CO_2 in the

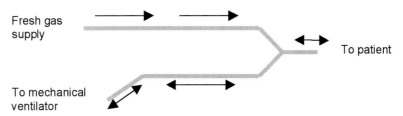

Figure 33-8. Rebreathing Circuit Using T-Piece Design.

rebreathed air mixture is proportional to the flow rate of the fresh gas. This design is often used in pediatric anesthesia.

SCAVENGING SUBSYSTEM

Prolonged personal exposure to anesthetic agents even at low levels, including nitrous oxide, has been shown to be related to some illnesses, such as liver disease and premature infant birth. Exposure to anesthetic agents is classified as an occupational hazard for operating room personnel (e.g., personal exposure of nitrous oxide must be less than 25 ppm time-weighted average over an 8 hours duration). To reduce such health hazards, the scavenging subsystem is designed to capture and remove waste anesthetic gases. Instead of allowing these gases to be released into the operating room environment, waste anesthetic gases are collected from the exhausts of the APL valve and pressure-limiting valve of the breathing/ventilation circuit (Figure 33-9). Scavenging systems remove gas by connecting to either a vacuum or a passive exhaust. A vacuum scavenging (or active) system use the wall suction in the operating room, whereas the passive exhaust scavenging system connects to the exhaust of the room air ventilation system. System pressure adjustment is critical to ensure that the scavenging system is of just enough negative pressure to prevent waste anesthetic gases from being released into the operating room but not so much as to remove too much of the patient breathing gases. A scavenging bag is attached to act as a reservoir to absorb fluctuations. As a safety measure, a pair of pressure-limiting valves is placed near the scavenging bag to limit the pressure to within ±0.5 cmH2O or ±0.05 kPa of the atmospheric pressure. When the gas flow and APL valve are correctly adjusted, with proper functioning of the scavenging system, no anesthetic gas should be released to the atmosphere from these valves. However, anesthetic gas can still escape from a poorly fitted face mask or improperly placed endotracheal tube into the operating room environment.

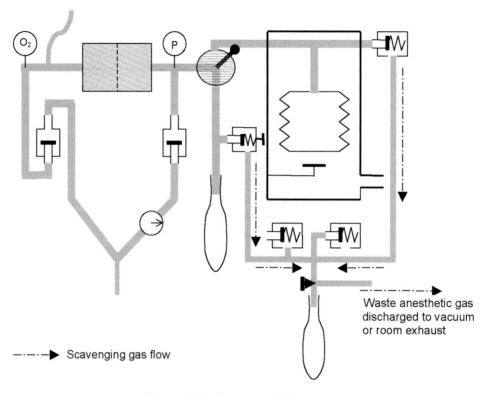

Figure 33-9. Scavenging Subsystem.

$-\cdot-\cdot-\blacktriangleright$ Scavenging gas flow

Waste anesthetic gas
discharged to vacuum
or room exhaust

COMMON PROBLEMS AND HAZARDS

A patient under general anesthesia is unconscious and cannot react or signal for help. Patients under general anesthesia are vulnerable to operating errors or malfunctions of instrumentations. A list of potential safety hazards to patients under general anesthesia follows.

- Insufficient oxygen supply to patient
- Insufficient CO_2 removal from the rebreathing gas
- Excessive or wrong anesthetic agents being delivered to the patient
- Excessive breathing circuit pressure leading to trauma to the lung
- Introduction of foreign particles into the airway

Insufficient oxygen supply to the patient will lead to hypoxia. An oxygen analyzer is used to monitor the oxygen level of the inspired gas. To detect hypoxia in the patient, a pulse oximeter probe is connected to the patient to measure the oxygen saturation level in the arterial blood. An oxygen ratio monitor built into the flow regulators of the anesthesia machine prevents the oxygen level in the gas mixture from accidentally set or dropped below 30%. Touch-coded control knobs to differentiate oxygen, medical air, and nitrous oxide prevent the anesthetist from mistakenly adjusting the wrong gas. Backup cylinders of oxygen are in place to ensure an uninterrupted supply of oxygen during a procedure. Color-coded gas cylinders and hoses (e.g., oxygen, white in Canada or green in the U.S.) are used to prevent misconnection of gases. Different sizes of low pressure hoses (breathing circuit, 22 mm; fresh gas supply and patient Y, 15 mm; scaveng-

Table 33-2. Summary of Hazards and Mitigation Related to Anesthesia Machines

Potential Hazards	Methods to Minimize Hazards
Insufficient oxygen supply to the patient	• Oxygen analyzer • Pulse oximeter • Oxygen ratio monitor • Backup oxygen supply • Color-coded gas cylinders and hose • Touch-coded oxygen and nitrous oxide control knobs • DISS and pin-indexed safety system • Different diameters of hoses used • Vital sign monitoring
Insufficient carbon dioxide removal from the patient	• End-tidal carbon dioxide monitor • Pulse oximeter • Carbon dioxide absorber with color indicator
Excessive or wrong anesthetic agent delivered to the patient	• Agent specific keyed filling spouts on the vaporizers with color coding • Interlock to prevent turning on more than one anesthetic agent • Agent concentration monitors • BIS index monitor
Trauma to the lung caused by excessive pressure	• Pressure monitors • APL (adjustable pressure-limiting) valve Pressure-limiting valve • Particle filters and traps
Foreign matter injuring the airway	• Dust-free carbon dioxide absorber
General	• Regular service by qualified personnel (daily functional check and regular quality assurance inspection and maintenance)

ing hose, 19 mm) are used to prevent misconnection of gas circuits. A diameter-indexed safety system (DISS) employing different diameters of connectors on gas supply hoses and wall outlets prevents machine oxygen gas inlet to be connected to a wrong piped-wall gas outlet. A pin-indexed safety system (using two pins on the cylinder yoke and corresponding holes on the stem) is used to prevent connecting an incorrect gas cylinder to a cylinder yoke of another gas (see Appendix A-3). Vital sign monitors are used to assess the patient's physiological status to detect hypoxic condition.

Failure to remove CO_2 from the patient can lead to hypercapnia. A low oxygen saturation level in arterial blood (from a pulse oximeter) may indicate excessive CO_2 in the patient's system. End-tidal CO_2 monitors are used to detect CO_2 level in patient's exhaled gas. For example, an elevated baseline in the capnogram may indicate problem in the CO_2 absorber. The CO_2 absorber is the main component of the breathing circuit to remove CO_2 from exhaled gas. Color change (typically from white to pink) is an indication of the end-of-life of the CO_2 absorbers. When CO_2 reacts with soda lime, heat and water are produced with change in pH. The change in pH causes the indicator in soda lime to change color, indicating that the absorbent is near the point of exhaustion.

Delivery of a wrong anesthetic agent or elevated concentration can be fatal. The filling spout of a vaporizer is keyed to accept only the bottle of the correct agent. Vaporizers on an anesthesia machine are interlocked to allow only one vaporizer to be turned on at one time. To ensure a correct mixture of anesthetic gases being supplied to the patient, agent monitors are built into modern anesthesia machines to monitor the concentration of the anesthetic agent during the procedure. Bi-spectral (BIS) index monitoring, a special EEG measurement, may also be used to assess the depth of anesthesia (level of consciousness) of the patient.

Overpressure in the airway will create injury to the patient. Pressure monitors and overpressure alarms are installed on all machines. The APL valve and pressure-limiting valves limit the maximum pressure in the patient air circuit. Filters and traps are used to prevent patient injury from foreign particles.

Patients under general anesthesia depend entirely on the anesthesia machine and related equipment for life support. Errors caused by machine failure and faulty adjustments can result in life-threatening situations. Pre-use checklists, regular inspections, and preventive maintenance are critical to minimizing anesthesia related hazards.

Table 33-2 summarizes the preceding discussion. Daily functional verifications of anesthesia machines as well as their accessories are usually performed by the operating room staff, whereas performance inspection is done by biomedical engineering personnel during scheduled inspections.

For many years, using a mixture of oxygen and nitrous oxide as the carrier gas to deliver inhalational agents has been the practice. Recent reviews of the effect of nitrous oxide in patients (such as proven factor of postoperative nausea and vomiting) have raised questions in this practice. Together with its environmental effects (e.g., ozone depleting potential), and adverse health effects in operating room personnel (such as increased incidences of spontaneous abortions and congenital anomalies in offspring), there is an increasing trend in using nitrous oxide-free anesthesia procedures.

Recent studies showed that with the use of newer low-soluble volatile anesthetic agents (such as sevoflurane and desflurane), there is no need to use nitrous oxide in inhalational anesthesia. In many hospitals, nitrous oxide is now used only occasionally and in indicated cases only. The missing analgesic and hypnotic effects can be compensated for by moderately increasing the amount of volatile anesthetic agent and raising its expired concentration. In addition, removing nitrous oxide facilitates the performance of low-flow anesthetic technique. Since the patient only inhales oxygen and the volatile anesthetic, the total gas uptake is reduced significantly.

Faulty or out of action scavenging systems are responsible for most anesthetic gas pollution in the operating room. Improper anesthesia administration technique and leaks in anesthesia equipment are other causes. Common sources of leaks include loose hose connectors, leaky CO_2 absorber connections, disconnected scavenging hoses at APL and ventilator valve, and leaks from poorly fitted endotracheal tube and face mask.

An anesthesia workstation with automated record keeping is designed to centralize system control and to integrate the display and storage of information. This involves continuous acquisition, recording, and presentation of selected physiological parameters and equipment variables on a central display along with limit settings and alarm status. Effective integration of information and alarm management has become a critical requirement of the anesthesia system. An integrated display gives the anesthetist a single point of reference for a wide variety of equipment and physiological information. Anesthesia machines that lack integrated alarms can sometimes cause confusion by sounding numerous alarms simultaneously. In an integrated system of information and alarms, visual alarm messages appear on a central display; audible and visual alarms are prioritized to differentiate the critical alarms from the advisories.

There are reported cases that patients have regained consciousness during surgeries. Instead of directly monitoring brain activity (using EEG) during surgery, anesthetists usually rely on indirect means (such as blood pressure and vital signs) to assess consciousness. Modern anesthesia units often incorporate BIS index monitoring as an additional tool to monitor the patient's level of sedation.

Contamination of any part of the patient breathing circuit may lead to nosocomial infections. Infection control practice recommends single use of disposable or high-level disinfection of reusable between patients to prevent cross-contamination. There were debates on the pros and cons of using disposable bacteria filters to prevent contamination of the machine upstream of the patient breathing circuit. Possible hazards, such as the increased impedance to gas flow and obstruction of the breathing circuit, are associated with installation of these filters. When filters are used, it is important to frequently replace them to prevent inadequate gas delivery due to clogging.

BIBLIOGRAPHY

Association of Anaesthetists of Great Britain & Ireland. (2007). *Recommendations for Standards of Monitoring During Anaesthesia and Recovery* (4th ed.). London, UK: author.

Baillie, J. K., Sultan, P., Graveling, E., Forrest, C., & Lafong, C. (2007). Contamination of anaesthetic machines with pathogenic organisms. *Anaesthesia, 62*(12), 1257–1261.

Braz, L. G., Braz, D. G., Santos da Cruz, D., Fernandes, L., Pinheiro Módolo, N. S., & Braz, J. R. (2009). Mortality in anesthesia: A systematic review. *Clinics (São Paulo, Brazil), 64*(10), 999–1006.

Brodsky, J. B., & Cohen, E. N. (1986). Adverse effects of nitrous oxide. *Medical Toxicology, 1*(5), 362–374.

Centers for Disease Control and Prevention. (1997). Guidelines for prevention of nosocomial pneumonia. *MMWR Recommendations and Reports, 46*(RR- 1), 1–79.

Chakravarti, S. and Basu, S. (2013). Modern Anaesthesia Vapourisers. Indian J Anaesth, Sep-Oct; 57(5): 464–471.

Dorsch, J. A., & Dorsch, S. E. (2007). *Understanding Anesthesia Equipment* (5th ed.). Baltimore, MD: Lippincott Williams & Wilkins.

Goneppanavar, U., & Prabhu, M. (2013). Anaesthesia machine: Checklist, hazards, scavenging. *Indian Journal of Anaesthesia, 57*(5), 533–540.

Hogarth, I. (1996). Anaesthetic machine and breathing system contamination and the efficacy of bacterial/viral filters. *Anaesthesia and Intensive Care, 24*(2), 154–163.

Lagasse, R. S. (2002). Anesthesia safety: *Model or myth? Anesthesiology, 97*(6), 1609–1617.

Myles, P. S., Leslie, K., Silbert, B., Paech, M. J., & Peyton, P. (2004). A review of the risks and benefits of nitrous oxide in current anaesthetic practice. *Anaesthesia and Intensive Care, 32*(2), 165–172.

Quasha A.L., Eger E.I. 2nd, Tinker J.H. (1980). Determination and applications of MAC. *Anesthesiology*, 53:315–334

Springer, R. (2010). Cleaning and disinfection of anesthesia equipment. *Plastic Surgical Nursing, 30*(4), 254–255.

Subrahmanyam, M., & Mohan, S. (2003). Safety features in anaesthesia machine. *Indian Journal of Anaesthesia, 57*(5), 472–480.

Terrell, R. C. (1986). Future development of volatile anesthetics. *Anaesthesiology and Intensive Care Medicine, 188*, 87–92.

Toski, J. A., Bacon, D. R., & Calverley, R. K. (2001). The history of anesthesiology. In P. G. Barash, B. F. Cullen & R. K. Stoelting (Eds.), *Clinical Anesthesia* (4th ed., p. 3). Baltimore, MD: Lippincott Williams & Wilkins.

Whitaker, D. K., & Booth, H. (2013). Immediate post-anaesthesia recovery. *Anaesthesia, 68*(3), 288–297.

Chapter 34

DIALYSIS EQUIPMENT

OBJECTIVES

- List the basic kidney functions and the functions of a hemodialyzer.
- Describe the principles of diffusion, osmosis, and ultrafiltration.
- Define ion/molecular clearance in a hemodialyzer.
- Compare different vascular access.
- Analyze the construction of the artificial kidney (AK).
- State the properties of the membrane in an AK and evaluate the mechanisms of molecular and fluid transport across the membrane.
- Explain dialysate preparation and delivery in hemodialysis.
- Sketch the fluid line diagram and identify basic functional components in the extracorporeal blood and dialysate delivery circuits.
- Differentiate between peritoneal dialysis and hemodialysis.
- Explain the needs and describe methods of water treatment in dialysis.
- Compare different cleaning and disinfection methods for dialysis equipment.

CHAPTER CONTENTS

1. Introduction
2. Basic Physical Principles
3. Kidney Functions Review
4. Mechanism of Dialysis
5. Hemodialysis System
6. Dialyzer (or Artificial Kidney)
7. Patient Interface
8. Dialysate

 9. Basic Components of a Hemodialysis Machine
10. Peritoneal Dialysis
11. Other Medical Uses of Dialysis Treatment
12. Water Treatment
13. Equipment Cleaning and Disinfection
14. Common Problems and Hazards

INTRODUCTION

A healthy kidney maintains the level of body fluid, electrolytes, and acid/base balance and removes some of the metabolic wastes. For patients with impaired renal function, renal dialysis supplements or replaces some of these functions to restore a reasonable state of health to the patient and minimize damage to other organs and physiological systems. The first dialyzer was constructed in 1943 by Dr. Willem Kolff, a Dutch physician. The first successful dialysis treatment was in 1945. Today, hemodialysis and peritoneal dialysis are the two most commonly used treatment procedures for patients with acute renal failure and end-stage renal disease.

The AK, or dialyzer, is a device that supplements or replaces some of the many functions of the human kidneys (e.g., water and electrolyte balance, elimination of waste products, etc.). A very simplified representation of the AK kinetics is shown in Figure 34-1. It consists of a blood compartment and an electrolyte (or dialysate) compartment separated by a semipermeable membrane.

Movement of substances, including water molecules across the membrane, occurs in such an arrangement due to

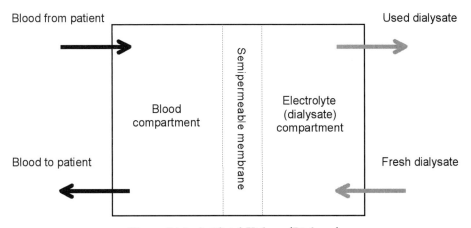

Figure 34-1. Artificial Kidney (Dialyzer).

- Differences in concentrations of various substances in the two compartments
- Particle sizes of these substances with respect to the membrane pores
- Pressure difference between the two compartments.

These phenomena are the result of diffusion and osmosis (or ultrafiltration). The procedure to apply these physical principles in a controlled fashion for medical purposes is known as dialysis.

BASIC PHYSICAL PRINCIPLES

Diffusion

Diffusion is the movement of molecules and ions in a solution as a result of repeated intermolecular collisions. When two regions of a system have different concentrations, a concentration gradient is said to exist between the regions. The rate of diffusion is proportional to the product of the concentration gradient and the cross-sectional area separating the two regions.

In hemodialysis, the factors governing diffusion are complex because of the presence of the membrane. The diffusion of substances (molecules) in a solution through the membrane is substance dependent as well as membrane dependent. The membrane may be characterized for a particular substance i by the permeability P_i of the membrane on the substance. The mass diffusion rate J_i of the substance i of the membrane is given by:

$$J_i = P_i \times \frac{C}{l},$$

where C = the difference in concentration of the substance across the membrane, and
l = the membrane thickness.

Note that not all molecules can go through the membrane.

Osmosis

Osmosis takes place when a concentration gradient exists across a semipermeable membrane. A true semipermeable membrane is one through which only water molecules can pass. In dialysis, the membrane has pores that allow certain sizes of molecules to go through. It is therefore not a true semipermeable membrane.

The net movement of water across a membrane against a solute concentration gradient is called osmosis. By means of osmosis, the system tends to achieve a situation of uniform concentration. This tendency for water to move in response to a solute concentration gradient develops a force called osmotic pressure. Osmotic pressure may also be defined as the hydrostatic pressure that must be exerted on a solution to prevent the movement of water through the semipermeable membrane. Osmotic pressure is a function of the absolute temperature and the concentration gradient of substances in the solution.

Ultrafiltration

If the opposing pressure to prevent movement of water through the membrane is increased to a level above the osmotic pressure of the solution, water is forced to flow from the solution against the osmotic pressure. This event is called ultrafiltration. In hemodialysis, this can be used to remove excess water from the patient. The mass of water transfer per unit time is proportional to the pressure across the membrane.

KIDNEY FUNCTIONS REVIEW

Figure 34-2 shows the daily water transport of an average adult. About 2 L of water is ingested per day; 200 mL of that are excreted from the bowel and the rest is absorbed into the body. About 350 mL are lost to the atmosphere during respiration in the form of water vapor in the expired air. About 1000 mL are excreted as urine through the urinary tract, and 450 mL are evaporated from the surface of the body. There are three water compartments in the body: blood, intracellular (within the cells) fluid, and interstitial (outside the cells) fluid.

An average person (70 kg) has about 40 L of water (or about 40% by weight) in the body. The percentage of water in a newborn is about 75%, much higher than that of an adult. Of the 40 L of water, 25 L are within the cells and 15 L are in the interstitial fluid. There is about 5 L of blood in the body, of which 3 L are plasma and 2 L are blood cells. With a cardiac output of 5 L/min, about 1.2 L/min flow into the renal arteries. The capillaries in the kidneys create about 2.2 m^2 of membrane contact surface area to process and filter the blood. On an average day, 180 L of fluid pass through the membrane inside the kidney, but almost all of it is reabsorbed, leaving only about 1 L of fluid excreted as urine.

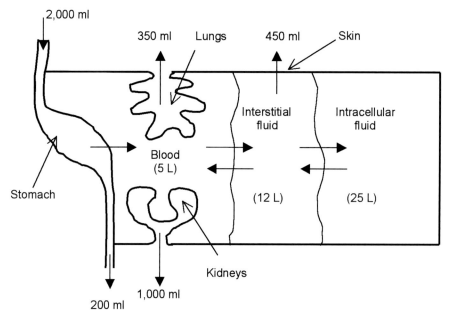

Figure 34-2. Body Fluid Transport.

Kidney Functions

The functions of a healthy kidney include

1. Removal of waste products (urea, uric acid, creatinine) from body fluid (blood and body water)
2. Regulation of blood volume and pressure
3. Regulation of extracellular fluid volume and composition
4. Maintenance of acid-base balance (pH)
5. Control of specific concentration of ions (e.g., Na, K balance)
6. Regulation of externally overtaken products such as glucose
7. Control of volume and composition of urine
8. Regulation of endocrine and metabolic functions

An AK replaces or supplements, to varying degrees, many of these processes (except the last two functions).

One of the main functions of the kidney is the removal of waste products from the blood. The parameter to measure the performance of the kidney in terms of product removal is called plasma clearance. Similar parameters called clearances are also used to evaluate a dialyzer's performance to remove substances from the blood or bodily fluid.

Plasma clearance is substance dependent. CLx (mL/min) of a substance x is given by

$$CL_x = Q_u \times \frac{C_{ux}}{C_{px}},$$

where Q_u = rate of urine production (mL/min),
C_{ux} = concentration of substance x in urine, and
C_{px} = concentration of substance x in plasma.

Renal deficiency refers to underperformance of the renal system. Some of the possible causes of renal deficiency are

• Kidney overload
• Kidney disease or damaged kidneys
• Inadequate renal blood flow

Some of the effects of renal deficiency include

• Uremia (increase in urea and other nonprotein nitrogen)
• Water retention and edema
• Acidosis in renal failure
• Elevated potassium concentration in uremia
• Uremic coma
• Reduced availability of calcium to bones (osteomalacia)

Symptoms of renal failure include

• Loss of appetite, increased blood pressure, nausea, decreased urine output, edema, itching, fatigue, and neurological disturbances. If left untreated, will lead to convulsion, coma, and death.
• One of the best means to assess renal failure is to measure concentrations of the nitrogenous nonprotein substances (e.g., urea).

Treatment of renal problems includes

• Vigorous control of the intake of different materials into the body (diet control)
• Avoidance of excessive supply of water, cations, and nonessential nitrogenous material
• Hemodialysis treatment using an artificial kidney

- Peritoneal dialysis: The use of the peritoneal membrane in the peritoneal cavity as the interface between blood and dialysis
- Kidney transplantation

MECHANISM OF DIALYSIS

The dialyzer, also known as an AK, is the main component of the hemodialysis system in which blood solutes (metabolic wastes) are removed from the blood and dialysate solutes (electrolytes) are added to the blood. During the process of dialysis, blood and dialysate are simultaneously being circulated through the dialyzer, separated only by the semipermeable membrane. Substances (including water) to be added to and removed from the blood are exchanged across the membrane by the principles of diffusion and ultrafiltration. The process usually continues for 4 to 5 hours per treatment and consists of three treatments a week. The principles applicable to dialysis are explained next.

Diffusion

Because the semipermeable membrane separating the blood and dialysate within a dialyzer is not an ideal membrane, ions (solutes) are selectively exchanged across the membrane according to their molecular weight. Water and waste products in the blood (e.g., urea, creatinine, uric acid), which have relatively low molecular weights, can diffuse easily and rapidly through the membrane into the dialysate; higher molecular weight substances, such as glucose and proteins, cannot easily pass through the membrane and are not significantly exchanged. Since the rate of diffusion depends on the concentration gradients of the solutes, it is necessary to maintain a fresh supply of blood and dialysate to facilitate these two-way transports.

Ultrafiltration

Ultrafiltration is the primary method to remove water from the blood using the semipermeable membrane. In ultrafiltration, water is forced through the membrane by applying a hydrostatic pressure across the membrane. The physician may require intermittent adjustment of the ultrafiltration rate based on the observation and the long-term trend of the patient.

HEMODIALYSIS SYSTEM

Figure 34-3 shows a simplified functional block diagram of a hemodial-

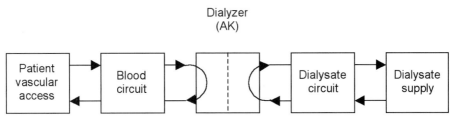

Figure 34-3. Hemodialysis System.

ysis system. It consists of a vascular access serving as the interface between the machine and the blood vessels of the patient. The blood circuit takes the blood from the patient to the dialyzer and returns it to the patient. The blood and dialysate are separated by the semipermeable membrane in the dialyzer where the substance and water exchange between the blood and dialysate take place. The dialysate circuit takes the dialysate from the supply, feeds it to the dialyzer and removes it from the dialyzer, and disposes it to the drain. Table 34-1 lists the functions described above plus other monitoring and control tasks of each of the functional blocks.

DIALYZER (OR ARTIFICIAL KIDNEY)

The dialyzer (or AK) is the heart of the hemodialysis system, where the exchange of substances and removal of water between blood and dialysate take place. The membrane within the AK allows such exchange to occur.

A membrane that permits substances to pass through is said to be permeable to those substances. A true or ideal semipermeable membrane is permeable to water but impermeable to all other substances. Most membranes pass only molecules of certain sizes. Hence they are called selective membranes. Membrane permeability may be passive or active. Active permeability transports molecules against the concentration gradient. Passive permeability depends on the concentration gradient as the driving force. The ability of a particle to pass through a membrane passively is dependent on the molecular size, ionic charge, and the degree of ionic hydration (e.g., OH^- is smaller and passes through the membrane more easily than does Ca^{++}).

The basic properties of a dialyzer depend on

- Type of membrane used (porosity, size of pores, clearances, etc.)
- Effective membrane surface area
- Membrane's ability to withstand hydrostatic pressure

Table 34-1. Functions of a Hemodialysis System

Vascular Access	Blood Circuit	Dialyzer	Dialysate Circuit	Dialysate Supply
Remove blood from patient	Introduce and remove blood to dialyzer	Provide blood/dialysate interface	Introduce and remove dialysate to dialyzer	Prepare and control dialysate composition
Reintroduce blood to patient	Control/monitor blood flow rate	Remove waste from blood	Control/monitor dialysate flow rate	Remove air from dialysate
	Control/monitor blood output pressure	Remove water from blood	Control/monitor dialysate pressure	Control/monitor dialysate temperature
	Control/monitor blood input pressure	Introduce solute to blood	Detect blood leak into dialysate	
	Trap air bubbles and produce alarm		Monitor dialysate pH	
	Prevent blood clot		Monitor dialysate conductivity	

- Transmembrane pressure
- Blood flow rate
- Dialysate flow rate

Performance Parameters

As discussed in the previous section, the parameter to measure the performance of the kidney in terms of product removal is called plasma clearance. A similar parameter known as clearance Clx is also used to evaluate an AK's performance to remove a substance x from the blood or bodily fluid.

In hemodialysis, CLx is redefined as

$$CL_x = Q_B \times \frac{(C_{ax} - C_{vx})}{C_{ax}},$$

where Q_B = blood flow (mL/min),
C_{ax} = arterial concentration of substance x, and
C_{vx} = venous concentration of substance x.

Another performance parameter of a hemodialysis system is the ultrafiltration coefficient (K_{uf}). K_{uf} is a concept used to evaluate an AK's performance to remove water. K_{Uf} is defined as the volume (mL) of fluid that will be transferred per unit time (hour) per unit pressure (mmHg) difference across the membrane of the dialyzer. Typical value of K_{Uf} is 2 to 6 mL/hr/mmHg. However, it can be as high as 50 mL/hr/mmHg for a high permeability dialyzer. Dialyzers with K_{Uf} greater than 20 mL/hr/mmHg are generally referred to as high flux dialyzers.

Example 34.1

In a 4-hour dialysis treatment, it is necessary to remove 2 L of water from the patient. A dialyzer with K_{Uf} = 2.0 mL/hr/mmHg is used. It is estimated that during the treatment, 100 mL of fluid will be ingested by the patient. At the end of the treatment, 300 mL of water will be used to rinse the dialyzer free of blood (back into the patient). What should be the pressure setting in the dialysate compartment if the blood compartment has a positive pressure of 50 mmHg?

Solution:

The total water removal taking into account fluid ingestion and dialyzer rinsing is

$$(2000 + 100 + 300) \text{ mL} = 2400 \text{ mL}.$$

Let x be the pressure setting in the dialysate compartment. The transmembrane pressure P is therefore equal to

$$P = (50 - x) \text{ mmHg}.$$

Using the definition of K_{Uf}: Volume of water removal = K_{Uf} x time x P, and substituting the values into the equation,

$$2400 = 2 \times 4 \times (50 - x) => x = -250 \text{ mmHg}$$

to obtain a total transmembrane pressure P = 300 mmHg.

Types of Dialyzers

Several types of AK with different physical constructions have been used. Coiled tube and parallel plate were used. Hollow fiber AKs are the current choice. These AKs are named according to the construction of the semipermeable membrane. A coiled construction AK consists of a circular cross section tube made of semipermeable membrane material wound into a coil. During dialysis, blood flows inside the tube and the coil is immersed in a container filled with dialysate. A parallel plate AK consists of multiple layers of semipermeable membrane in parallel. Blood is circulated between alternate pairs of plates and dialysate is circulated between the other plates. A hollow fiber AK consists of a large number (10,000 to 15,000) of hollow fibers connected in parallel inside a container (Figure 34-4). Each fiber has an internal diameter of about 0.2 mm and a length of about 150 mm. Blood flows inside the lumens of the fibers with dialysate surrounding them. Although the fiber lumen is small, blood particles can readily pass through (the diameter of an erythrocyte is 8 µm, a monocyte is 14 to 19 µm, and a thrombocyte is 2 to 4 µm). Hollow fiber AKs are the most popular type used today. Common membrane materials are cellulose acetate, cuprophane, nephrophane, and Visking. The total surface area of the membrane ranges from 0.6 to 2 m² and supports a blood flow rate from 100 to 300 mL/min. A typical dialysate flow rate is between 400 and 600 mL/min. Ultrafiltration can be achieved by creating a positive pressure on the blood side or a negative pressure on the dialysate side.

To reduce dialysis time, some AKs are designed to have higher water and substance removal rates. These are referred to as high-efficiency and high-flux hemodialysis. High-efficiency dialysis is defined by a high clearance rate of urea (e.g., >600 mL/min). The membranes of such AKs can be made from cellulosic or synthetic materials. High-flux dialysis removes water at a faster rate (e.g., ultrafiltration coefficient or K_{Uf} greater than 20 mL/hr/mmHg). High blood flow rate (e.g., >350 mL/min) and high dialysate flow rate (e.g., >500 mL/min) are needed to support high-efficient and high-flux hemodialysis.

Tables 34-2 and 34-3 list the construction and performance specifications of a hollow fiber AK for hemodialysis. Dialyzers are evaluated by comparing

- Clearances for different substances (e.g., urea, creatinine, phosphate, vitamin B12, uric acid, blood serum phosphate, glucose, sodium chloride)
- Ultrafiltration rate (K_{Uf})
- Priming volumes
- Cost (taking into account single or multiple use)
- Clotting properties

Note that the clearance values of an AK are substance dependent and vary with the blood and dialysate flow rates. As blood and dialysate flow rates increase, the rate of increase of the clearance decreases until the clearance reaches a maximum (zero rate of increase). This maximum clearance at infinite blood and dialysate flow for urea is defined as the mass transfer

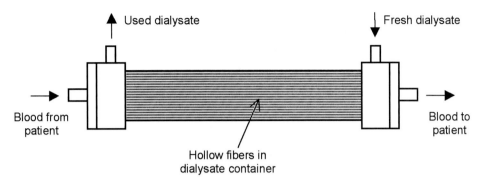

Figure 34-4. Hollow Fiber Artificial Kidney.

area coefficient (K₀A) or intrinsic clearance of the dialyzer. The dialyzer mass transfer area coefficient is a useful parameter for comparing dialyzer performance.

PATIENT INTERFACE

Because hemodialysis is often done three times a week, several hours per session, a semipermanent interface between the dialysis unit and the patient's circulatory system to allow repeated dialysis is required. This interface is called the vascular access. Under hemodialysis, it is necessary to draw blood from a vessel that has a high volume of blood flow. The radial artery

Table 34-2. Physical Specifications of AK

Housing Construction	Rigid transparent plastic
Tube Sheets Material	Medical-grade silicon rubber
Dimensions	21 cm long x 7.0 cm diameter
Weight	650 g (filled)
Blood Volume	135 mL
Dialysate Volume	100 mL
Fiber Material	Regenerated cellulose
Number of Fibers	11,000
Effective Length per Fiber	13.5 cm
Fiber Lumen	225 μm
Fiber Wall Thickness	30 μm
Effective Membrane Area	1.0 m²

Table 34-3. Performance Specifications of AK

Blood Compartment Flow Resistance	At blood flow rate of 200 mL/min	15 to 55 mmHg
Dialysate Compartment Flow Resistance	At dialysate flow rate of 500 mL/min with negative pressure of 400 mmHg	<50 mmHg
Average Ultrafiltration Rate	At 500 mL/min dialysate flow and 300 mmHg negative pressure	300 mL/hour
Urea Clearance Creatinine Clearance Phosphate Clearance	At 200 mL/min blood flow and 500 mL/min dialysate flow	135 mL/min 105 mL/min 85 mL/min

in the arm is a good candidate. However, it is not convenient or safe to have repeated needle punctures to the artery. A large vein that is closer to the skin surface can be easier to access and stick with a needle. Unfortunately, blood in a vein does not flow fast enough for dialysis purpose. In addition, repeated needlesticks into the vein will cause a blood clot. When an artery is surgically connected directly to a vein, both the artery and the vein dilate and elongate in response to the greater blood flow and pressure, but the vein dilates more and becomes "arterialized." As a result, the blood flow will increase. For example, normal blood flow in the brachial artery is 85 to 110 mL/min. After the creation of a brachiocephalic fistula (shunting the brachial artery to the cephalic view right above the elbow), the blood flow increases to 400 to 500 mL/min immediately and to 700 to 1000 mL/min within 1 month. It is a common practice to connect an artery directly to a vein to form the vascular access for hemodialysis. Several common methods to access the bloodstream are described in the following sections.

Arteriovenous Shunt

An arteriovenous shunt (A-V) is a pair of cannulae of polytetrafluoroethylene (PTFE) inserted through the skin into an artery and a vein near the inner surface of the forearm or the lower leg. Between dialysis treatments, the two cannulae, which are permanently implanted, are joined by a short length of Silastic® tubing to allow blood circulation. During dialysis, the Silastic tubing is removed and replaced by two lengths of plastic tubing that direct the blood to and from the dialyzer. These arteriovenous connections can provide a natural blood pressure differential to circulate blood through the dialyzer. Due to the low differential pressure, however, it requires a low flow resistance dialyzer; otherwise a pumping mechanism is necessary to provide enough blood flow. Arteriovenous shunts have been used in acute as well as chronic therapies. Due to the external cannula connection through the skin, this method is no longer used.

Arteriovenous Fistula

In this method, an internal shunt is developed by joining an artery and a vein within the limb directly or by a short length of fibrin tubing. Due to the arterial pressure, the vein will increase in size and its wall will thicken. (In one study, the cephalic vein increased from 2.3 mm to 6.3 mm diameter after 2 months.) It usually takes about 3 to 7 weeks for the fistula vein to mature before it is ready to be used for dialysis blood access. Blood is obtained by venous puncture using either one or two large-bore needles. A blood pump is necessary to create enough blood flow.

In the double-needle technique, blood is continuously and simultaneously withdrawn from one needle and returned through the other. The single-needle technique requires a Y-connection and a controller to alternately infuse and withdraw blood to and from the patient. A special pump, or a pair of synchronized pumps, is required for this technique. To provide continuous blood transfer, a special double-lumen needle/catheter is used.

Arteriovenous fistula is the most commonly used method for vascular access because it requires no permanent open site, which minimizes infection problems encountered with an arteriovenous shunt. Other advantages of the arteriovenous fistula are higher blood flow rates and a lower incidence of thrombosis. It has a 3-year, 70% average site survival rate.

Arteriovenous Graft

If an arteriovenous fistula cannot be created, a graft is used to connect the artery to the vein. Similar to an arteriovenous fistula, this vascular access method surgically puts in a graft (e.g., a section of autogenous saphenous vein or PTFE [Teflon®]) to connect an artery to a vein (e.g., graft between the radial artery and basilic vein). Needles are inserted into the graft for vascular access. An arteriovenous graft has a 3-year, 30% average site survival rate.

Percutaneous Venous Cannula

This method is only for acute cases. To create the vascular access, a soft cannula can be inserted into the subclavian, femoral, or internal jugular vein. A percutaneous venous cannula can be single or double lumen. However, blood flow is almost always less than that of a well-established fistula or graft.

DIALYSATE

Dialysate is a solution used to effect diffusion and ultrafiltration. It is also used to carry away waste products and rectifies metabolic acidosis. Excess body acid is neutralized (pH balance) through the buffer effect of either acetate or bicarbonate in the dialysis solution. Studies showed that adjusted survival of hemodialysis patients is reduced when predialysis bicarbonate concentration is less than 18 mmol/L or greater than 24 mmol/L; maintaining a predialysis plasma bicarbonate level of 22 mmol/L is recommended. There are two basic types of dialysate: acetate base and bicarbonate based.

Acetate based dialysate was used in the 1980s because batch prepared bicarbonate dialysate was not stable (tends to lose CO_2 gas from the solution) and was easily contaminated by bacteria. However, acetate may induce adverse side effects, such as hypoxemia, vasodilatation, and depressed left ventricular function during high-efficiency dialysis applications. Without these side effects and with the development of better mixing mechanisms that prevent loss of CO_2, bicarbonate-based dialysate has regained popularity. Most dialysis centers today are using bicarbonate-based dialysate. Dialysate is prepared in batch before dialysis or on demand during dialysis. Batch processing is done only in large dialysis centers where the dialysate is prepared and stored in large holding tanks. In a smaller center or when prescription dialysis is required, the dialysate is prepared by blending dialysate concentrate in proportion with treated water during the dialysis process. The pH and concentration of the dialysate must be continuously monitored. The mechanism for dialysate blending, monitoring, and control is often an integral part of the dialysis machine. Examples of dialysate compositions compared to the same in patient's blood are listed in Table 34-4. The unit of concentration is mEq/L.

The following processes are necessary before the dialysate is introduced into the AK or during dialysis:

- Treat water before it is used to prepare the dialysate (water treatment is discussed later in the chapter)
- Warm the dialysate to body temperature (37°C) before entry into the dialyzer
- Deaerate the dialysate to prevent gas evolution at body temperature and subatmospheric pressure (for AK using negative pressure ultrafiltration) and to prevent supersaturating blood with nitrogen at body temperature. Air bubbles will also decrease the efficiency of the AK.
- Monitor dialysate pressure at entry to and exit from the dialyzer to ensure that the blood pressure is always greater than that of the dialysate
- Detect blood leaks across the membrane using a photoelectric detector

Table 34-4. Dialysate versus Blood Compositions

	Na	*K*	*Ca*	*Mg*	*Cl*	*Urea*	*Creatinine*	*Bicarbonate*
Blood	140	6.3	1.1	0.54	109	65	896	18
Dialysate	135	2.0	1.5	0.75	106	0	0	35

(Unit in mEg/L)

In modern-day dialysis, the dialysate is continuously fed into and removed from the AK. The used dialysate is disposed of during dialysis. This is referred to as a single-pass dialysate system. In the past, in order to conserve chemicals (and cost), dialysate was reused on the same patient. Many methods to reuse dialysate have been employed. For example, sorbent materials are used to remove some chemicals from the dialysate so that it can be regenerated and reintroduced into the AK (sorbent regenerative system). In another method, the used dialysate may be mixed with a certain proportion of fresh dialysate and reintroduced into the AK (single-pass recirculation system). In the extreme case, the used dialysate is recirculated back into the AK until the performance of dialysis has decreased to such a level that the old dialysate must be replaced with a fresh batch (total recirculating system).

BASIC COMPONENTS OF A HEMODIALYSIS MACHINE

Figure 34-5 shows the basic components and fluid flow diagram of a typical hemodialysis machine. Blood enters the machine from the vascular access via the arterial blood line, goes through the AK (dialyzer), and returns to the patient via the venous blood line. Dialysate is prepared and fed into the AK, where water and substance exchange take place. After passing through the AK, the dialysate is dumped into the drain. The blood circuit is separated from the dialysate circuit. Blood is separated from the dialysate as long as the membrane in the AK remains intact. The basic components in the extracorporeal blood and dialysate delivery circuits are described in the following sections.

Extracorporeal Blood Circuit

Heparin is administered intermittently or continuously (e.g., at a rate of 1.5 mL/hr) to the incoming blood from the venous access. Blood clots decrease the efficiency of the AK and are hazardous if they are reintroduced into the patient. Regional heparinization is used in some clinical conditions

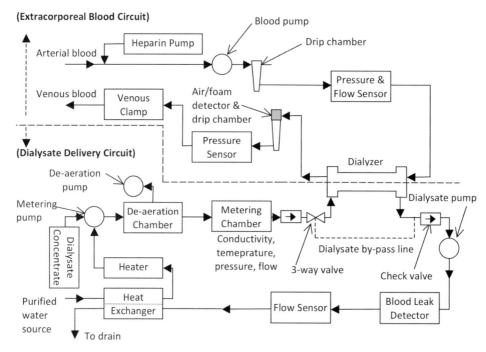

Figure 34-5. Hemodialysis Machine Basic Components and Flow Diagram.

(e.g., to avoid worsening of existing bleeding sites). During regional heparinization, careful neutralization of heparin is achieved by injecting coagulants (such as protamine) into blood returning to the patient.

A roller pump in the dialysis machine controls the blood flow rate and creates a negative pressure in the arterial blood line to draw blood from the vascular access and a positive pressure in the blood compartment of the AK. The blood flow rate is controlled by varying the speed of the pump. The pressure in the blood line is regulated by the roller pump and the venous clamp located at the end of the blood circuit in the machine. The positive pressure in the blood circuit also prevents ingress of air and dialysate fluid into the blood circuit.

An arterial drip chamber provides a visual indication of the blood flow in the circuit. The pressure and flow rate of blood before entering the AK are monitored. As the blood exits the AK, it passes through another drip chamber. As the pressure becomes less, foaming occurs as air exits the blood. To prevent excessive air from getting into the bloodstream of the patient via the returning blood, an air/foam detector is located at the drip chamber. A pressure sensor monitors the blood pressure at the exit of the AK. Together with the upstream pressure sensor, the sensors monitor the blood average compartment pressure in the AK for ultrafiltration control as well as measuring the resistance of the AK. If the hollow fibers are blocked by blood clots, a large pressure drop will develop across the AK in the blood circuit.

Dialysate Delivery Circuit

Water from the purified water source enters the dialysate circuit of the dialysis machine. It is first heated when passing through the heat exchanger and the heater compartment. One or more metering pumps accurately introduce concentrated chemicals into the water path to produce the right concentration of dialysate to the patient. Three-stream (water, bicarbonate, and acid), dual concentrate (bicarbonate concentrate and acid concentrate) proportioning methods are often used to produce the final dialysate. Gas in the dialysate is removed by passing the dialysate through the deaeration chamber, where a reduced pressure is created by the deaeration pump. Excessive gas in the dialysate can diffuse across the membrane into the patient's blood.

Temperature, pressure, flow rate, pH, and conductivity of the dialysate are measured in the metering chamber. To prevent cooling or heating the patient's blood, the temperature sensor ensures that the dialysate is at body temperature before entering the AK. Correct flow rate and pressure are required to ensure removal of substances and water from the patient. The dialysate pH and conductivity are indications of the dialysate composition and concentration.

After passing through the metering chamber, the dialysate is introduced via a check valve (one-way valve) into the AK. A dialysate pump maintains the flow rate and produces a negative pressure in the dialysate chamber of the AK. The positive pressure in the blood circuit and the negative pressure in the dialysate circuit create the transmembrane pressure that controls the ultrafiltration rate, whereas the blood flow and dialysate flow rates control the substance removal rate (clearances) of dialysis. Before going through the heat exchanger and being dumped into the drain, the dialysate passes through a blood leak detector. If blood is detected in the dialysate, which indicates rupture of the membrane in the AK, the machine will sound an alarm and have to be shut down. The dialysate bypass line can be used to facilitate replacement of the AK. It can also be used to temporarily suspend the flow of dialysate into the AK when there is a problem with the dialysate preparation and delivery system.

Table 34-5 shows the typical range of control and monitoring parameters of a hemodialysis machine.

Table 34-5. Range of Control and Monitoring Parameters

Parameters	Typical Range
Dialysate flow rate	200–1000 mL/min
Dialysate temperature	35–39°C
Conductivity	7–17 ms/cm
Blood flow rate	50–650 mL/min
Heparin flow rate	0.0–5.5 mL/hr
Venous/arterial pressure display	–300 to +600 mmHg
Transmembrane pressure	–100 to +500 mmHg
Blood leak detector sensitivity	0.35–0.45 mL/min
Air detector sensitivity	Air bubble size >5–25 µl in blood line
Ultrafiltration rate	0–4 L/hr

PERITONEAL DIALYSIS

Hemodialysis requires removing blood from the patient and processing the blood in the AK external to the patient's body. Another form of dialysis is performed using a natural membrane inside the human body to achieve substance and water exchange. Because the peritoneal cavity is lined with blood vessels and capillaries, peritoneal dialysis uses the peritoneal membrane as the blood-dialysate interface instead of an artificial membrane in an external dialyzer. A peritoneal dialysis setup is much simpler than a hemodialysis system is. It uses a gravity feed and drain instead of blood and dialysate pumps. The dialysate stays in the peritoneal cavity instead of being circulated through the external AK. Figure 34-6 shows a setup for continuous cycler-assisted peritoneal dialysis (CCPD).

At the start of a dialysis cycle, valve number 4 opens (all others remain shut) to allow a selected volume of dialysate to flow from the supply reservoir to the volume control and heater compartment. The dialysate stays in the compartment until it is warmed to body temperature. Valve number 2 is then opened so that the dialysate flows by gravity to the patient's peritoneal cavity through an indwelling catheter. The dialysate stays inside the peritoneal cavity for a period of time (e.g., 45 min) to allow substances and water exchange between the blood in the capillaries and the dialysate. After the preset time, valve number 3 opens to drain the used dialysate (together with the additional water from osmosis) from the peritoneal cavity to the first drain bag. The scale measures the weight of the dialysate to monitor the fluid removed from the patient. After the measurement, valve number 1 is opened to allow the dialysate to flow into the disposal bag to complete the cycle.

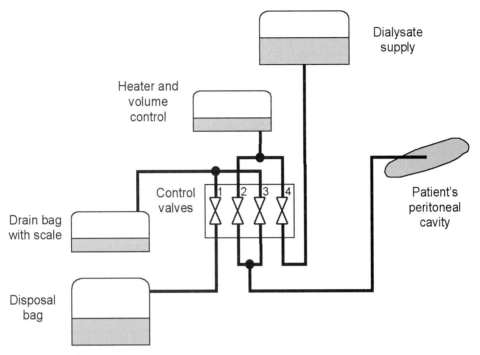

Figure 34-6. Continuous Cycler-Assisted Peritoneal Dialysis Setup.

Although peritoneal dialysis takes more time due to slower fluid and substance transports, the rate and process have more resemblance to those of the natural kidneys and therefore reduce the likelihood of shock to the patient. Peritoneal dialysis is often performed at home due to its relatively simple operation and less sophisticated equipment setup. Because of the high risk of developing peritonitis (infection of the peritoneum due to careless handling of indwelling catheters by patients or home caregivers), however, patients are often forced to switch to hemodialysis due to reduction in dialysis efficiency after repeated occurrences of peritonitis. Continuous ambulatory peritoneal dialysis (CAPD) and CCPD are the two commonly performed types of peritoneal dialysis.

Continuous Ambulatory Peritoneal Dialysis

In CAPD, the dialysate is constantly present in the abdomen but is changed three to five times daily with a per-fill-volume from 1.5 to 3.0 L (typically 2 L). Dextrose (1.5, 2.5, or 4.25%) in the dialysate is used to create osmosis for water removal. Drainage and replenishment of dialysate are performed using gravity.

Continuous Cycler-Assisted Peritoneal Dialysis

CCPD is performed at bedtime using an automated cycler to change dialysate four to five times during the night. The dialysate used is similar to that used in CAPD.

OTHER MEDICAL USES OF DIALYSIS TREATMENT

Other than treating patients with renal problems, dialysis may be used to eliminate toxic materials in the blood, to perfuse isolated organs, to reduce abnormally high ammonia concentration found in the blood following liver malfunction, and to supplement renal function during and after major surgery.

Similar to using an intermittent hemodialysis machine to treat a patient with chronic renal diseases, continuous renal replacement therapy (CRRT) is used to treat patients suffering from acute kidney injury (AKI). CRRT offers extracorporeal blood purification therapies to continuously replace impaired renal function over an extended period of time (e.g., 24 hours a day for several days) until the kidneys can resume their usual function.

CRRT is slower and better mimics the physiological processes of the kidney than intermittent hemodialysis. They are usually found in intensive care settings treating AKI patients who are often hemodynamically unstable. Patients with cerebral edema who cannot handle rapid fluid shift or solute removal are good candidates for CRRT. CRRT also allows clinicians to administer drugs, antibiotics, and nutrition without the need to increase fluid intake.

WATER TREATMENT

Normal tap water contains traces of metal ions (e.g., copper, lead) and chemicals (e.g., chlorine or chloramines). There are five categories of contaminants in water for dialysis. They are

1. Particulates—dirt, debris, etc.
2. Gases—carbon dioxide, methane, and so on that are soluble in water
3. Organics—carbon-containing compounds, including pesticides, herbicides, and chloramines
4. Inorganics—salts and heavy metals
5. Bacteria and pyrogens—various living microorganisms

During a 4-hour dialysis treatment using a dialysate flow of 500 ml/min, 120 liters of dialysate interfaces with the patient's blood, whereas a healthy individual consumes only about 2 liters of water per day. Under repeated dialysis, if untreated water is used to prepare the dialysate, contaminants that normally are not harmful under usual consumption can quickly accumulate in a patient's body. Therefore, it is necessary to remove ions (e.g., Cu^{++}) and other molecules (suspended particles) in the water used for dialysate preparation. Raw water for dialysis usually goes through a pretreatment process that consists of

- Cartridge filter to remove particles greater than 0.5 μm (but cannot remove dissolved toxin or endotoxin)
- Water softener to remove ions such as Ca^{++} or Mg^{++}
- Activated carbon to remove chlorine or chloramines

A second-stage water treatment process is carried out to remove smaller particles and remaining chemicals in the water. Reverse osmosis (RO) is the most commonly used method. Distillation can also be used. Reverse osmosis (using the same principle as ultrafiltration but with smaller pore size) is achieved by applying pressure to the supply water compartment, forcing water to pass through a near ideal semipermeable membrane while leaving impurities behind. Reverse osmosis removes over 90% of impurities (including dissolved minerals, organic compounds, bacteria, and endotoxins) and is acceptable for dialysis in most cases. Heat or ultraviolet radiation can be used for water sterilization. However, there is currently no requirement for water sterilization because reverse osmosis or distillation can remove most bacteria and endotoxins.

Figure 34-7 shows the usual stages of water treatment in hemodialysis. Each stage of the treatment process is described below.

1. Cartridge filters for particulate removal. Multiple cartridges from coarse to fine (sub-micron) are employed.
2. Activated carbon to remove dissolved chlorine and chloramines
3. Cation exchange resin (water softener) to exchange Ca^{++}, Mg^{++}, Fe^{++}, etc. for Na^+. A brine tank is connected to supply Na ions. Removal of these cations also protects the RO membrane.
4. Reverse osmosis to remove organic matter, bacteria, and endotoxins.
5. Deionizer resins to remove cations and anions.
6. Ultraviolet (UV) irradiation to kill any remaining pathogens.

Water is continuously circulated to avoid stagnation, and siphoned off to different dialysis stations during treatment.

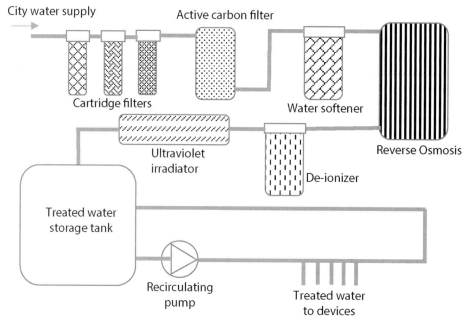

Figure 34-7. Hemodialysis Water Treatment System.

The Association for the Advancement of Medical Instrumentation (AAMI) has published a set of quality standards for dialysate purity. The most recent recommendations require dialysates to have less than 200 CFU/ml of bacteria and less than 2 EU/ml of endotoxins (CFU stands for colony forming unit, which is a measure of viable colonogenic cell numbers; EU stands for endotoxin unit, one EU equals approximately 0.1 to 0.2 ng endotoxin/ml of solution). Studies in recent years suggested that the presence of high levels of bacteria and endotoxins in the dialysate can pave the way for substantial degree of inflammation in patients. The European Renal Association now strongly recommends the use of ultrapure dialysate, which has a maximum bacteria level of 0.1 CFU/ml and a maximum endotoxin level of 0.03 EU/ml.

EQUIPMENT CLEANING AND DISINFECTION

Because dialysis is an invasive procedure, care must be taken to reduce the chance of infection. Dialysis machines are required to be flushed and disinfected between patients. Heat disinfection (e.g., heat fluid to above 85 °C for at least 15 min.) can eliminate waterborne bacteria such as *Pseudomonas cepacia* (a gram-negative bacterium). Although heat disinfection is convenient, it is ineffective to kill spore-forming bacteria such as Bacillus varieties. To eliminate spore-forming bacteria, chemical disinfection (e.g., using formaldehyde) is used instead of heat treatment. Disinfection treatment using heat or chemicals is usually done daily on every machine and after each patient's treatment. In addition to daily disinfection, sodium hypochlorite (bleach) is used to disinfect the machine weekly. To prevent injury from residual chemicals, it is important to rinse the machine thoroughly after chemical or bleach disinfection to remove all chemical residuals before the machine is used on patients.

A dialysis center must set up standard operation procedures to take regular bacteria cultures in order to monitor the effectiveness of disinfection and detect possible contamination. The blood and dialysate lines are single-use disposable items. Although all dialyzers are labeled for single use, some centers reuse a dialyzer on the same patient. Studies have shown that if proper cleaning and disinfection procedures are followed, reusing a dialyzer on the same patient is easier on the patient (i.e., it has less chance to cause adverse reactions). Others have reported that after taking into account the time and materials for processing, reusing dialyzers can achieve cost saving. However, health agencies in general do not recommend reuse of single-use products, including dialysis supplies.

COMMON PROBLEMS AND HAZARDS

Infections are a leading cause of morbidity and mortality in chronic hemodialysis patients. Specific policies and procedures designed to reduce infection risks should be implemented and strictly followed. These policies should address issues such as cleaning, sterilization and disinfection, maintenance, waste disposal, and infection precautions.

Although arteriovenous fistula is the vascular access of choice for chronic hemodialysis, too much blood may be drawn into the fistula and returned to the general circulation without entering the limb's capillaries. This may cause cold extremities, develop cramps, and eventually lead to tissue damage. Repeated needle access may weaken the wall of the vein, leading to aneurysm. Aneurysms will shorten the useful life of the fistula and require corrective surgery.

Adequate water purification is essential in hemodialysis because a long term dialysis patient is exposed to a much higher volume of water than a healthy individual is. Excessive accumulation of water contaminants in the body may cause hemolysis, bone disease, neurological damage, metabolic acidosis and anemia. Dialysis removes substances such as vitamin B1 and amino acids from the blood. These useful substances are normally retained by a healthy kidney. Replenishment of substances (such as iron and zinc) and vitamin therapy are often prescribed for dialysis patients. The following symptoms in patients may be due to water treatment problems:

- Anemia—cause by aluminum, chloramines, copper, zinc
- Hemolysis—cause by chloramines, copper, nitrates
- Metabolic acidosis—cause by low pH
- Bone disease—cause by aluminum, fluoride
- Hypertension—cause by calcium, sodium
- Hypotension—cause by bacteria, endotoxins, nitrates
- Encephalopathy—aluminum
- Nausea/vomiting—bacteria, endotoxins, low pH, nitrates, sulfates, zinc

Improper preparation of dialysate is possible due to either machine failures or human errors. Because solution conductivity reflects only the total ionic content of the dialysate rather than measure its actual composition, both pH and conductivity of the dialysate should be checked before each dialysis treatment.

Although reuse of dialyzers is discouraged, financial considerations have enticed some hospitals and clinics to reuse AKs. In addition to the financial incentive, reprocessed dialyzers may benefit some patients in reducing first-use allergic reaction. For centers practicing dialyzer reuse, safe and effective methods for reprocessing dialyzers are critical.

Proper cleaning, disinfection, and sterilization of dialysis equipment are crucial. Most machines have built-in automatic cycles to ensure proper cleaning and disinfection of the internal components and lumens. However, there have been many incidents in which disinfection was not properly completed (e.g., disinfectant was not drawn into the machine and the alarm failed to indicate this problem). Incidents of inadequate flushing to remove all residual chemicals after machine disinfection or sterilization have been reported.

Many patients undergoing dialysis treatment, especially under high-efficiency dialysis, have experienced symptoms, including drowsiness, convulsions, and, on some occasions, coma and death. These adverse reactions were suspected to be caused by the inability of the vascular system to adjust to the change in fluid volume during dialysis. Proper clinical assessment of a patient's condition before and during dialysis is essential.

Patient cross-contamination due to machine failures and component breakdown is possible. A hazard report described failure of a transducer protector leading to potential patient cross-contamination. These protectors act as a barrier to prevent blood from contacting the pressure transducer element within dialysis machines. Such failure may allow blood from one patient to come into contact with blood from subsequent patients.

Peritonitis (inflammation of the peritoneum) is the most serious complication of peritoneal dialysis. Poor aseptic technique may introduce bacteria through the catheter insertion site into the peritoneal cavity, resulting in peritonitis and catheter-site infections. Peritoneal scarring from peritonitis will decrease the efficiency of dialysis and is one of the most common reasons for interrupting the therapy. User errors may introduce other solutions (e.g., disinfectants) into or overfill the peritoneal cavity. Proper training and using caution prevent bacterial infection and other complications.

BIBLIOGRAPHY

Abdeen, O., & Mehta, R. L. (2002). Dialysis modalities in the intensive care unit. *Critical Care Clinics, 18*(2), 223–247.

Association for the Advancement of Medical Instrumentation. (2004). *Dialysate for Hemodialysis.* ANSI/AAMI RD52:2004. Arlington, VA: author.

Bagdasarian, N., Heung, M., & Malani, P. N. (2012). Infectious complications of dialysis access devices. *Infectious Disease Clinics of North America, 26*(1), 127–141.

Cheung, A. K., Levin, N. W., Greene, T., Agodoa, L., Bailey, J., Beck, G., ..., & Eknoyan, G. (2003). Effects of high-flux hemodialysis on clinical outcomes: Results of the HEMO study. *Journal of the American Society of Nephrology, 14*(12), 3251–3263.

Daugirdas, J. T., Black, P. G., & Ing, T. S. (Eds.). (2007). *Handbook of Dialysis* (4th ed.). Philadelphia, PA: Lippincott Williams & Wilkins.

Feldman, H. I., Kinosian, M., Bilker, W. B., Simmons, C., Holmes, J. H., Pauly, M. V., & Escarc, J. J. (1996). Effect of dialyzer reuse on survival of patients treated with hemodialysis. *JAMA, 276*(8), 620–625.

Fissell, W. H., Shuvo, R., & Davenport, A. (2013). Achieving more frequent and longer dialysis for the majority: Wearable dialysis and implantable artificial kidneys. *Kidney International, 84,*256–264.

Gibney, R. T., Kimmel, P. L., & Lazarus, M. (2002). The Acute Dialysis Quality Initiative—Part I: Definitions and reporting of CRRT techniques. *Advances in Renal Replacement Therapy, 9*(4), 252–254.

Kanno, Y., & Miki, N. (2012). Development of a nanotechnology-based dialysis device. *Contributions to Nephrology, 177,* 178–183.

Khan, A., Rigatto, C., Verrelli, M., Komenda, P., Mojica, J., Roberts, D., & Sood, M.M. (2012). High rates of mortality and technique failure in peritoneal dialysis patients after critical illness. *Peritoneal Dialysis International, 32*(1), 29–36.

Layman-Amato, R., Curtis, J., & Payne, G. M. (2013). Water treatment for hemodialysis: An update. *Nephrology Nursing Journal, 40*(5), 383–404.

Ligtenberg, G. (1999). Regulation of blood pressure in chronic renal failure: Determinants of hypertension and dialysis-related hypotension. *The Netherlands Journal of Medicine, 55*(1), 13–18.

Locatelli, F., Martin-Malo, A., Hannedouche, T., Loureiro, A., Papadimitriou, M., Wizemann, V., . . . , & Vanholder, R. (2009). Effect of membrane permeability on survival of hemodialysis patients. *Journal of the American Society of Nephrology, 20*(3), 645–654.

Locatelli, F., Mastrangelo, F., Redaelli, B., Ronco, C., Marcelli, D., La Greca, G., & Orlandini, G. (1996). Effects of different membranes and dialysis technologies on patient treatment tolerance and nutritional parameters. The Italian Cooperative Dialysis Study Group. *Kidney International, 50*(4), 1293–1302.

Luehmann, D. A., Keshaviah, P. R., Ward, R. A., & Klein, E. (1989). *A Manual on Water Treatment for Hemodialysis.* U.S. Department of Health and Human Services (HHS publication FDA 89-4234). Rockville, MD.

Medical Devices Agency. (1996). Peritoneal Dialysis Equipment (Kimal Proteus). London, UK: Department of Health.

Misra, M. (2005). The basics of hemodialysis equipment. *Hemodialysis International, 9*(1), 30–36.

Moran, J. (2007). The resurgence of home dialysis therapies. *Advances in Chronic Kidney Disease, 14*(3), 284–289.

Parker, T. F., 3rd. (2000). Technical advances in hemodialysis therapy. *Seminars in Dialysis, 13*(6), 372–377.

Ronco, C. (2006). Recent evolution of renal replacement therapy in the critically ill patient. *Critical Care, 10*(1), 123.

Sam, R., Vaseemuddin, M., Leong, W. H., Rogers, B. E., Kjellstrand, C. M., & Ing, T. S. (2006). Composition and clinical use of hemodialysates. *Hemodialysis International, 10*(1), 15–28.

Sritippayawan, S., Nilwarangkur, S., Aiyasanon, N., Jattanawanich, P., & Vasuvattakul, S. (2011). Practical guidelines for automated peritoneal dialysis. *Journal of the Medical Association of Thailand, 94*(Suppl 4), S167–174.

Szeto, C. C., Kwan, B. C-H., Chow, K-M., Law, M. C., Pang, W-F., & Leung, C-B. (2011). Repeat peritonitis in peritoneal dialysis: Retrospective review of 181 consecutive cases. *Clinical Journal of the American Society of Nephrology, 6*(4), 827–833.

Tomazic, P. V., Sommer, F., Treccosti, A., Briner, H. R., Leunig, A. (2021). 3D endoscopy shows enhanced anatomical details and depth perception vs 2D: a multicentre study. *Eur Arch Otorhinolaryngol.* Jul;278(7):2321-2326.

Ward, R. A. (2004). Ultrapure dialysate. *Seminars in Dialysis, Wiley Online Library 17*(6), 489–497.

Chapter 35

SURGICAL LASERS

OBJECTIVES

- Describe the characteristics of lasers and their applications.
- Explain the effect of lasers on tissues.
- Describe the physics of laser action.
- List different surgical lasers and their applications.
- Explain the two common methods of laser delivery
- Identify the benefits and limitations of laser surgery over conventional surgical methods.
- List the hazards associated with the use of lasers and the methods to mitigate the risks.
- Review the maintenance requirements and handling precautions of laser systems

CHAPTER CONTENTS

1. Introduction
2. Characteristics of Lasers
3. Laser Action
4. Applications of Lasers
5. Laser Tissue Effects
6. Characteristics and Use of Surgical Lasers
7. Functional Components of a Surgical Laser
8. Laser Beam Delivery System
9. Advantages and Disadvantages of Laser Surgery
10. Laser Safety
11. Maintenance Requirements and Handling Precautions
12. Common Problems and Hazards

INTRODUCTION

Laser is the acronym for Light Amplification by Stimulated Emission of Radiation. In 1917, Albert Einstein described the absorption, spontaneous emission, and stimulated emission of light, which eventually led to the development of the first optical laser, a ruby laser, in 1960 by Theodore Maiman, an American engineer and physicist. The first CO_2 laser was invented by Kumar Patel of Bell Labs in 1963. Lasers have found applications in all walks of life, including medicine.

When struck by a photon, an electron at its resting or ground state can absorb the energy of the photon, become excited, and move to a higher energy level. On spontaneous return to its ground energy level, the electron emits a photon of energy equal to the difference between the two energy levels. This emitted photon can interact with an atom with an excited electron to produce another photon with the same frequency and phase traveling in the same direction. When there are many excited atoms in the medium (known as having a high degree of population inversion), this mechanism will set up a chain reaction of stimulated emission. Stimulated emission under the right conditions creates light amplification.

The first laser used in surgery was a ruby laser for treatment of retinal hemorrhages in the United States in the 1960s. It was not until 1972, when Gezo Jako adapted a CO_2 laser to an operating microscope, that the widespread use of lasers in operating rooms started. This chapter discusses some of the applications of lasers in medicine in particular in surgery.

CHARACTERISTICS OF LASERS

Although both are electromagnetic waves, a laser is quite different from a common light source. Lasers have the following characteristics:

- **Monochromatic**—Lasers have one or a few discrete wavelengths due to the fixed energy band gaps of the atoms, whereas normal light consists of a relatively wide spectrum of wavelengths. In practice, however, a laser has a finite (but narrow) width of wavelength (Figure 35-1a).
- **Coherent**—Due to stimulated emission, the waves or photons coming from the laser are all in phase, whereas those from normal lights have different phase angles (Figure 35-1b).
- **Collimated**—Due to the repeated reflection between the parallel mirrors (generation of laser is discussed later in this chapter), all the waves of the laser beam are parallel along the longitudinal axis of the mirrors. Compared to normal light, the trajectory of the laser beam coming from

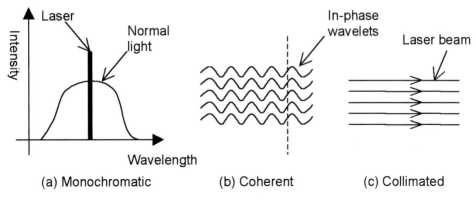

Figure 35-1. Laser Characteristics.

the lasing medium has minimal divergence or convergence (Figure 35-1c).

LASER ACTION

Although there are many types of lasers, they all have three basic functional components:

1. A lasing medium, which can be gas, liquid, or solid
2. An external excitation source that pumps energy into the lasing medium
3. A resonator or optical cavity with two parallel mirrors housing the lasing medium. One mirror is totally reflective and the other is partially reflective.

A ruby laser, for example, consists of a flash lamp (excitation source), a ruby crystal (lasing medium), and two mirrors (resonator) as shown in Figure 35-2. The flash lamp ignites and pumps energy into the ruby atoms. Light energy is absorbed by the atoms in the ruby crystal to excite electrons to higher energy levels. Some of the excited electrons return to their ground state and emit photons. The photons traveling in the direction perpendicular to the mirrors are bounced back and forth between the two mirrors. As they travel inside the crystals, they stimulate more photon emissions from the excited atoms. The beam intensity therefore increases as it undergoes multiple reflections and travels along the longitudinal axis between the mirrors. A portion of the beam is allowed to leave the laser through the partially reflective mirror. A ruby laser is a solid laser; a CO_2 laser is a gas laser; a dye laser is an example of a liquid laser. Gas lasers are the least efficient because they require a large amount of energy to excite the ionic transitions.

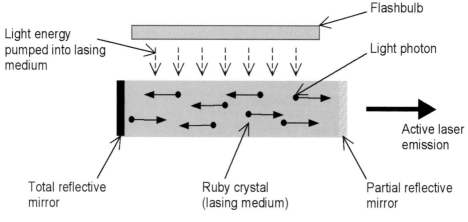

Figure 35-2. Laser Action.

APPLICATIONS OF LASERS

Lasers are used in a wide range of products and technologies. Applications of lasers can be classified according to laser properties.

Thermal Effect

When a laser is absorbed by a target, it converts to heat energy. A lens system or a light pipe can be used to focus and redirect a laser beam. This property is used in the industry in cutting materials such as metal or in burning a compact disc. In the military, its heating effect is used in laser guns to destroy military targets. In medicine, the heat energy of lasers is used in surgery and physiotherapy.

Straight Collimated Beam

The collimated property of a laser beam produces a parallel beam of light that has little convergence and divergence. That is, the beam diameter of an ideal laser will stay constant irrespective of the distance. Due to this property, a laser beam can travel a long distance without losing its intensity (except from absorption in the optical path). Lasers are widely used in industry such as for land survey's and in precision alignment. In the military, laser beams are used in weapon guidance systems such as laser-guided missiles. In medicine, this property is used in position alignment such as beam alignment of linear accelerators in cancer treatment.

Photostimulation Effect

Because a laser produces a monochromatic beam of high-intensity light, it can be used as a stimulant. In medical applications, a laser beam can be used to stimulate blood circulation and to promote cell healing in physiotherapy. In addition, it can be used in conjunction with a photodynamic drug to selectively activate the drug by a laser beam.

Photomechanical Effect

Some materials will change in shape when exposed to light. This photomechanical effect was first documented by Alexander Graham Bell in 1880. An example of the application of photomechanical effect is in intraluminal pneumatic lithotripsy to removed encrusted urinary catheters. Another example of photomechanical effect is light-induced heating; for example, heating of a conductor to thermionic electron emission temperature by a laser.

LASER TISSUE EFFECTS

The surgical effect of a laser is primarily due to its thermal effect on tissues. Laser tissue effect depends on

1. Type of tissue
2. Type of laser
3. Power density at the lasing site
4. Exposure time (including pulsed characteristics)

The general tissue effect inflicted by a surgical laser is shown in Figure 35-3. In essence, when the laser beam hits the tissue, the laser energy is absorbed by the tissue to create three zones of injury. Due to the intense heat, the cell membranes rupture and vaporize at the center of the laser beam (zone 1). Next to the vaporized zone is a zone of cell necrosis where the tissues undergo irreversible heat damage (zone 2). Beyond the necrosis zone is a layer of cells that were injured due to the elevated temperature (zone 3). Tissues in zone 3 are able to repair and recover. This is often referred to as the three zones of laser tissue injury.

The rise in temperature and temperature distribution in the tissue during laser irradiation depends on the energy absorbed and the thermal characteristics of the tissue. Tissue heated to less than 60°C undergoes little or no permanent damage. Denaturation of protein will result when the tissue temper-

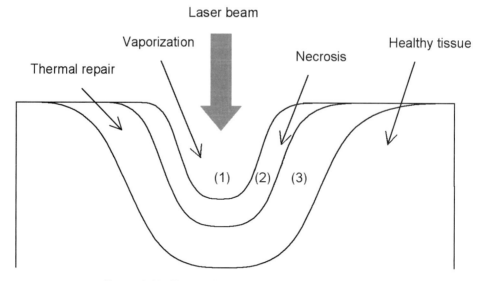

Figure 35-3. Zones of Injury from a Surgical Laser.

ature is above 60°C. Coagulation of blood occurs when the blood temperature is above 82°C. Tissue charring and vaporization starts to occur above 90°C. The heat absorption effect of tissue when hit by a laser depends on the characteristics of the tissue and the type of laser. For example, an argon laser is highly absorbed by hemoglobin but not by water. Using this property, an argon laser can be used as a photocoagulator to stop bleeding at the back of the eye. In the procedure, the laser beam passes through the cornea with very little or no absorption and delivers its energy to the blood vessels on the retina. On the other hand, a CO_2 laser is highly absorbed by water, which makes it a general surgical laser because all soft tissues contain a high percentage of water. The choice of laser depends on whether the tissue absorbs the laser energy and is heated by the beam or, alternatively, is transparent to the laser beam.

The power density or intensity of a laser beam striking an object is equal to the power divided by the beam area on the object. A laser, like light, can be reflected by a mirror or focused by a lens. A laser beam can be focused to a tiny spot to produce a very high intensity beam or defocused to cover a larger area with lower intensity. Figure 35-4 shows the tissue effects of different focal spot sizes of the same laser. In general, the higher the beam intensity is, the deeper the vaporization zone is.

For a continuous laser, the longer the exposure time, the more energy the tissue will absorb. In terms of a surgical laser, long exposure time will produce a deeper and wider zone of injury. A laser can be pulsed to increase its peak power while maintaining the same total output power. In pulsed mode, the laser fires repetitive short pulses at a selected exposure duration. Users can adjust the average power output, total irradiation time, pulse intensity and duration, and the pulse repetition frequency. A pulsed laser produces intense heat during the short duration of the high powered pulse but allows periods of cooling between pulses. Such cooling periods slow heat conduction to adjacent tissues. A pulsed laser will provide a deeper vaporization with less surrounding tissue damage than with a continuous laser at the same power output. Increasing the average power enables faster tissue cutting or removal. The higher the energy per pulse, the deeper the cut or the more tissue ablated. The faster the pulse rate, the more precise and smoother the cut or the surface ablated. By manipulating these parameters, different surgical effects can be created.

When an intense beam of laser slides across the surface of a soft tissue, it produces a zone of vaporization along the path of the laser. This action produces a sharp, clean cut. Abrasion effect (removal of a thin layer of surface tissue) is created by moving a defocused beam of laser with sufficient intensity over the tissue. Laser energy absorbed by the tissue may create destruction and charring effects on the tissue. Absorption of laser energy by blood produces a coagulation effect. Retina reattachment, vision correction, vascular surgery, and microsurgery are some of the many examples of laser surgical applications.

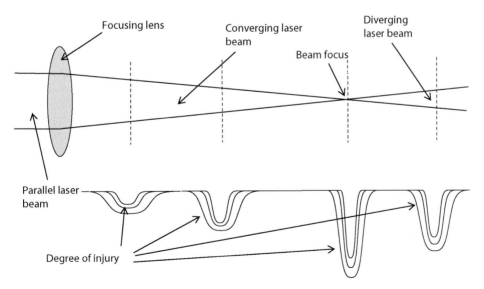

Figure 35-4. Zones of Injury from a Surgical Laser.

CHARACTERISTICS AND USE OF SURGICAL LASERS

There are many types of lasers. There are solid, fluid, and gas lasers and each one has different lasing media. Lasers are often named after their lasing media. Examples of gas lasers are helium-neon (HeNe) and CO_2 lasers. The name *Excimer laser* is derived from "excited and dimmers." A reactive gas mixture is electrically stimulated to form a pseudomolecule (dimer) and when excited produces a cool laser. A dye laser uses a fluorescent liquid dye as the lasing medium. When exposed to an intense laser such as an argon beam, it absorbs the laser energy and fluoresces over a broad spectrum. A tunable prism can be used to adjust the wavelength. A semiconductor diode can be manufactured to emit laser. The characteristics and applications of some surgical lasers are discussed in the following. Table 35-1 summarizes the characteristics and applications of these surgical lasers.

CO_2 Lasers

Surgical CO_2 lasers typically consist of a sealed laser tube, a laser pump, a cooling system, an aiming laser, and a delivery system. High-voltage discharge acts as the laser pump, supplying energy to the gases in the laser tube. A small amount of laser gas is supplied to replenish the CO_2 molecules that break down during use. For cooling purposes, the laser tube is surrounded by water circulated through a fan-cooled radiator. High power CO_2 lasers can produce over 100 W of continuous output power. Since CO_2 laser is invisible to the human eyes, an aiming beam (e.g., low power HeNe laser) is required.

CO_2 lasers emit infrared energy at a wavelength of 10,600 nanometers. Since this wavelength is readily absorbed by water and soft tissues are composed mainly of water, the energy of a CO_2 laser is absorbed superficially at the tissue with little penetration. CO_2 lasers are used to cut, ablate, and char tissue depending on the power density.

Nd:YAG Lasers

The laser medium is a rod-shaped yttrium, aluminum, and garnet (YAG) crystal doped with neodymium (Nd); the laser pump is a flash lamp. This laser emits a near-infrared wavelength of 1064 nm in the invisible portion of the spectrum. A visible aiming beam is needed.

An Nd:YAG laser is poorly absorbed by water but readily absorbed by protein. It passes through water with less absorption and penetrates more deeply into tissue than a CO_2 laser does. It is used in gastroenterology, urology, gynecology, and dermatology to cut and coagulate tissues. Examples of

Nd:YAG laser surgical applications are to control excessive uterine and gastrointestinal ulcer bleeding; to destroy prostate, rectal, and bladder tumors; to remove hair, rejuvenate skin, remove tattoos in dermatologic procedures.

Ho:YAG Lasers

Like most lasers, Holmium:YAG laser systems consist of a laser cavity or tube, a pumping system, an aiming laser beam, and a cooling system. The laser cavity contains the solid rod (laser medium) and mirrors. The Ho:YAG laser rod comprises a YAG crystal doped with holmium (Ho). When the laser pump supplies energy (e.g., a krypton arc lamp) to the rod, it emits a monochromatic beam of high-energy radiation in the near-infrared spectrum (2100 nm). The average power of Ho:YAG lasers ranges from 3 to 80 W, but some can reach up to 100 W. It can be operated in continuous or pulsed mode.

Ho:YAG lasers are used in hospitals and outpatient surgical facilities for a wide range of surgical applications, including orthopedics, ophthalmology, otolaryngology, cardiology, urology, oral/maxillofacial surgery, and pulmonary medicine. Because Ho:YAG lasers emit energy near the absorption peak of water (approximately 2100 nm), it is absorbed superficially by tissue. Ho:YAG lasers are used in superficial cutting or ablation of tissue. They can cut or ablate tissue with moderate hemostasis, little charring, and with a thin zone of necrosis. Ho:YAG laser energy can be delivered through a small-diameter (e.g., 550 µm) silica quartz fiber and can be used in contact procedures or operate away from the tissue in air and at a short distance in liquid. The flexible small fiber allows access to narrow spaces such as the wrist and posterior knee, enabling removal of torn ligaments and smoothing of rough cartilage without injuring nearby tissue. In the head and neck region, Ho:YAG lasers are used to cut bone, open tear ducts, and treat temporomandibular joint problems. It is also used in fragmentation of stones in the urinary tract (lithotripsy) and in treating herniated intervertebral disks. Laparoscopically, an Ho:YAG laser fiber can be more readily maneuvered than a CO_2 laser delivery device can to perform procedures in the gastrointestinal tract, such as removal of sessile polyps.

Comparing the three general surgical lasers (CO_2, Nd:YAG, and Ho:YAG) at similar energy level, the depth of the three zones of damage created by a CO_2 laser is about 0.05 mm, 0.5 mm by a Ho:YAG laser, and 4.0 to 6.0 mm by a Nd:YAG laser. Although CO_2 lasers can produce precise cuts with very narrow zones of thermal damage, their residual thermal energy is not enough to provide for hemostasis in vascularized tissue. In contrast, Ho:YAG lasers produce adequate hemostasis during ablation, and their wavelength can be transmitted in optical fibers, thereby allowing their use to be extended to endoscopic applications.

Ruby Lasers

A ruby laser is a solid laser that uses a synthetic ruby crystal as its laser medium. It must be pumped with very high energy (usually from a flashtube) to achieve a population inversion. The rod is placed between two mirrors, forming an optical cavity, which oscillate the light produced by the ruby's fluorescence, causing stimulated emission. Ruby lasers produce pulses of visible light at a wavelength of 694 nm, which is a deep red color. Typical ruby laser pulse lengths are on the order of a millisecond.

Ruby lasers were used extensively in dermatology procedures, such as in port wine stain removal, and in tattoo and hair removal. Today, such procedures are often done with other more efficient lasers such as tunable dye and Nd:YAG lasers.

KTP/532 Lasers

KTP/532 or frequency-doubled Nd:YAG laser is produced by passing the Nd:YAG laser through a potassium titanyl phosphate (KTP) crystal. It doubles the Nd:YAG frequency to produce a visible green laser with wavelength of 532 nm. Similar to Nd:YAG lasers, the laser production mechanism of KTP/532 is stable, has a long lifetime, and has low operating costs. They are smaller and more maneuverable than argon, krypton, and dye lasers.

KTP/532 passes through clear fluids, unpigmented tissues, and the top layer of the skin. It is excellent for hemostasis to a depth of 1 to 2 mm. It can photocoagulate blood vessels at low power densities (e.g., for treatment of vascular anomalies) and vaporize tumors at high power densities. KTP/532 lasers are widely used in dermatology, otolaryngology, and gynecology; in endoscopic procedures such as laparoscopic cholecystectomy; and in treating benign prostate hyperplasia.

Argon and Krypton Lasers

Both argon and krypton lasers are noble gas ion lasers. A typical noble gas ion laser is produced from high current density glow discharge gas plasma containing the noble gas in the presence of a strong magnetic field. Both lasers emit discrete multiple wavelengths of light.

Argon lasers used in ophthalmology applications emit light in the blue-green (488 nm) and green (514 nm) regions. Krypton medical lasers emit light in the yellow-red region (647 nm). Because they are highly absorbed by retina pigments, argon and krypton lasers are used in retinal vascular and neovascular disease procedures such as retinal phototherapy for treating diabetic macular edema. Since the production of gas ion lasers is highly ineffi-

Table 35-1. Laser Characteristics and Applications

Laser	Wavelength	Color	Lasing Medium	Applications
CO2	10,600 nm	Far infrared	Mixture of carbon dioxide, nitrogen, and helium gases	Readily absorbed by water. For vaporization and cutting tissue.
Ho:YAG	2100 nm	Mid infrared	Crystal of holmium, thulium, and chromium	Absorbed by tissue containing water. Precise cutting and less generalized heating of tissue.
Nd:YAG	1064 nm	Near infrared	Crystal of neodymium, yittrium, aluminum, and garnet	Poorly absorbed by hemoglobin and water, but absorbed by protein. For denaturing protein and shrinking tissue and coagulation.
Ruby	694 nm	Red	Ruby crystal	Not absorbed by transparent tissues and blood vessel wall. High-energy pulses selectively vaporize tissue; use in dermatology and plastic surgery such as port-wine stain removal.
HeNe	630 nm	Red	Helium-neon gas	Use as aiming beam for invisible medical lasers.
KTP/532	532 nm	green	Crystal of KTP	Highly absorbed by red or dark tissue. Use for coagulation and precision work.
Argon	514 nm 488 nm	Green Blue-green	Argon gas	Passes through water and clear fluid but highly absorbed by red-brown pigments. Use in coagulation, ophthalmology, dermatology, and plastic surgery.
Dye tunable	400 to 900 nm	Entire visible spectrum	Fluorescent liquid dyes	Use in photodynamic therapy, dermatology, and plastic surgery.
Excimer	193 nm 308 nm 351 nm	Ultraviolet	Argon fluoride gas Xenon chloride gas Xenon fluoride gas	Precision cutting and coagulation with little thermal damage to surrounding tissue. Use in ophthalmology, angioplasty, orthopedics, neurosurgery.

cient, they require extensive cooling and are large and heavy. For ophthalmic applications, argon and krypton lasers are being replaced by other lasers (such as diode lasers).

Dye Lasers

Dye laser tubes contain organic dye solutions that are optically excited (pumped) by an argon laser or flash lamp. Other than the usual liquid state, solid state dye-doped organic matrices can be used as laser media in solid state dye lasers. Depending on the dye used, dye lasers emit radiation adjustable from 340 to 1000 nm. A prism or diffraction grating is usually mounted in the beam path to allow tuning (wavelength selection) of the beam. Their ability to be tunable, with a narrow bandwidth and high intensity, as well as being able to produce ultrashort pulses to continuous wave, makes them suitable for a wide range of applications.

Dye lasers are used extensively in dermatology. The wide range of wavelengths allows close matching to the absorption of specific tissues, such as melanin or hemoglobin. The narrow bandwidth at specific wavelengths produces the desired tissue effect while reducing damage to the surrounding tissue. Dye lasers are used to treat port wine stains, scars, and other blood vessel disorders, as well as being used in cosmetic treatments to improve skin tone and remove skin pigments. It is also used for tattoo removal.

Excimer Lasers

Excimer lasers are produced by exciting noble gas halides with an electron beam. Different noble halide gases are used in the laser medium to produce different wavelengths of lasers. An excimer laser emits a wavelength of 193 nm when argon fluoride gas is used, 308 nm when xenon chloride gas is used, and 351 nm when xenon fluoride gas is used. Excimer lasers emit relatively low power (0 to 3 W) at tissue; they can be operated in pulse or continuous modes.

With high-precision focus and delivery mechanism, excimer lasers are used in phototherapeutic keratectomy to remove calcification and smooth out scarring over the corneal. In photorefractive keratectomy and in laser-assisted in-situ keratomileusis (LASIK), excimer lasers are used to shape the cornea to correct myopia (nearsightedness), hyperopia (farsightedness), and astigmatism.

Diode Lasers

Similar to LEDs, semiconductor diode lasers convert energy to light. The specific wavelength of the emitted light is determined by the semiconductor material used in the active medium. Current high-power surgical diode lasers use either AlGaAs, which emits light at a nominal wavelength of 810 nm, or InGaAs, which emits light at a nominal wavelength of 980 nm. Common wavelengths for medical applications are around 532 and 810 nm. The maximum power produced by a laser diode is limited to about 5 W. To produce high power for surgical applications, an array of laser diodes are used; the laser emissions from each diode are focused onto the laser fiber by a lens system. Using this cumulative mechanism, diode lasers of up to 60 W are available. Because diode lasers convert electrical energy to optical energy at an efficiency of 30% to 50% (compared to less than 10% for conventional lasers), they do not require water-cooling or special gas-cooling systems. Air cooling with a fan is sufficient to dissipate the heat generated. Surgical diode lasers are much more reliable than other conventional lasers are. A laser diode has a life span of 10,000 to 25,000 hours of use. They are nearly maintenance free because they have no mirrors or moving parts. Use of diode lasers may contribute to significant cost savings over time.

Laser wavelengths between 800 to 900 nm are highly absorbed by hemoglobin. High-power surgical diode lasers are used to cut soft tissue with hemostasis and to photocoagulate soft tissue in surgical specialties such as general surgery, gastroenterology, gynecology, neurology, otorhinolaryngology, plastic surgery, and urology. Other wavelengths and lower power applications include ophthalmic phototherapy, surgery for interstitial laser photocoagulation (a minimally invasive technique for tumor destruction), and various tissue-welding applications.

Transverse electromagnetic modes (TEMs) of the laser beam are due to the oscillatory behavior of the electric and magnetic fields at the boundary of the laser resonator. The transverse mode is defined by the shape of the output beam. Figure 35-5a shows the patterns of selected rectangular transverse laser modes. These modes can be visualized by the burn mark on a wooden tongue blade after irradiating it vertically with the laser beam. The fundamental mode of TEM_{00} with a single Gaussian beam intensity distribution is the best profile in surgical applications because it maximizes the energy density at the center of the beam. Some TEMs produced by lasers with cylindrical symmetry are shown in Figure 35-5b Misalignment of the laser delivery system also affects the beam output geometric profile.

FUNCTIONAL COMPONENTS OF A SURGICAL LASER

Figure 35-6 shows the functional components of a surgical laser. A low power HeNe laser producing red light is often used as the "aiming beam." An alignment optical system is used to align the HeNe beam with the main laser beam. Conventional methods to produce lasers are highly inefficient (some are below 10% efficiency). Most surgical lasers require high power

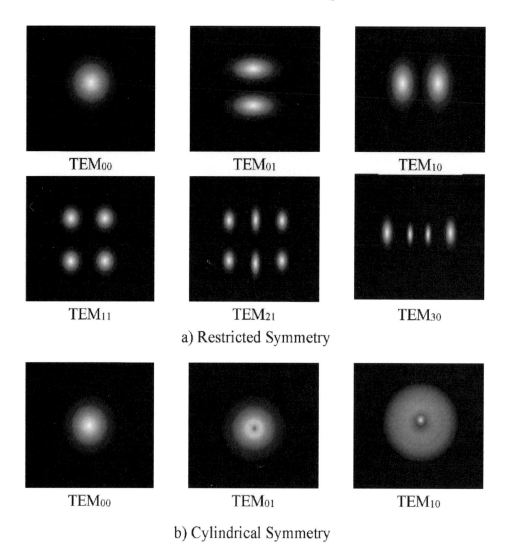

TEM00   TEM01   TEM10

TEM11   TEM21   TEM30

a) Restricted Symmetry

TEM00   TEM01   TEM10

b) Cylindrical Symmetry

Figure 35-5. Transverse Electromagnetic Modes.

output. A cooling system therefore must be installed to remove the heat generated from laser production. High capacity cooling such as water circulation with a forced air-cooled radiator is often used to remove the heat from the laser medium. The laser generated will be coupled to the surgical site via a system of delivery devices (or transport media).

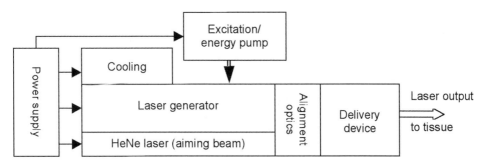

Figure 35-6. Functional Components of a Surgical Laser.

LASER BEAM DELIVERY SYSTEMS

In laser procedures, the delivery of the laser to the surgical site can be via a system of mirrors or through an optical fiber. Some shorter wavelength lasers (such as Nd:YAG, argon) can be delivered using optical fibers, whereas others (far-infrared lasers such as CO_2) are delivered through a system of mirrors. Figure 35-7a shows an articulation arm system for a conventional mirror laser delivery system. The laser travels inside a hollow rigid tube and is reflected to another tube segment by a first surface mirror located at the junction of the two tube segments. After several reflections, the laser will reach the handpiece and can be directed at the surgical site.

Instead of using mirrors, ultraviolet, visible, and near infrared lasers may travel inside a flexible optical fiber by total internal reflection (Figure 35-7b). A fiberoptic delivery system consists of a laser machine connector, a flexible fiber, and a handpiece. The laser from the machine travels inside the optical fiber until it reaches its other end and exits from the handpiece. Due to its small diameter and flexibility, optical fiber has the advantage of being easy to move around the surgical site. Fibers used in laser surgery usually have a silica core with an outer cladding and protective sheath. Typical core diameters of the fibers are 400, 600, and 1000 µm. The larger the core size, the higher the laser power it can deliver. Optical fibers can be disposable or reusable. Reusable fibers need to be inspected, cleansed, and sterilized between uses.

In recent years, flexible hollow fibers have emerged to replace rigid (mirrors) delivery systems for some CO_2 lasers. Such technological innovation was pioneered by an MIT group and published in 2002. The technology, known as "photonic bandgap reflectors," uses a dielectric mirror to create a photonic bandgap through which photons cannot propagate but can be reflected. The result is a highly reflective surface for any angle of incidence and can be fabricated to work in a large band of wavelengths. Manufacturers

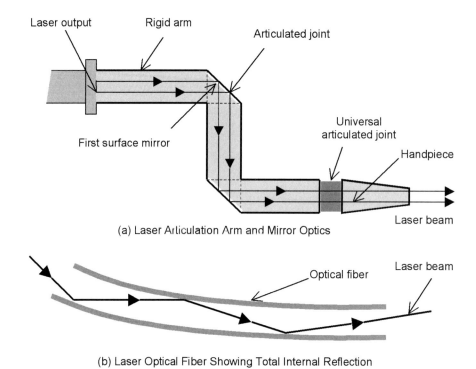

(a) Laser Articulation Arm and Mirror Optics

(b) Laser Optical Fiber Showing Total Internal Reflection

Figure 35-7. Laser Delivery Systems.

have produced some highly flexible polymer hollow fibers to transmit lasers by internal reflection. Some of these fibers have been adapted in conventional CO_2 lasers that are capable to deliver up to about 20 W of laser energy with relatively little loss (10% loss per meter). CO_2 laser can now be used with flexible fiber-connected handpieces and delivered through flexible endoscopes.

Laser procedures can be contact or noncontact. For noncontact procedures, bare fibers with polished tips are used to vaporize and coagulate soft tissues. A reflector placed at the fiber tip can redirect the laser beam to exit at an angle to the axis of the fiber (e.g., 90° for side-firing fibers). A lens can also be included to reshape the beam dimension.

Contact laser tips offer a completely different method of delivering laser energy to tissue. Instead of the laser directly transferring its energy to the tissue, the laser first heats the contact tip, and the heat of the tip in contact with the tissue is used to create the surgical effect. A contact laser tip can be heated to 2000°C. Similar to noncontact lasers, contact lasers will cut, coagulate, vaporize, and ablate. There are different types and sizes of tips. Conical tips are typically used for precise cutting and vaporization with hemostasis. Ball tips are used for wider incisions or hemostasis on large tissue surfaces. These tips can be attached to a variety of handles for use in open surgical procedures or can be affixed to a standard optical fiber and passed through any rigid or flexible endoscope. Contact laser tips are often made of synthetic sapphire crystals with great mechanical strength, low thermal conductivity, and high melting temperature.

ADVANTAGES AND DISADVANTAGES OF LASER SURGERY

Lasers are selectively absorbed by different tissues to produce different surgical effects. Lasers offer many advantages over other surgical techniques.

- The laser beam can be precisely focused for localized destruction of tissue. The depth of penetration can also be regulated by the power density, pulse duty cycle, duration, and focal size.
- Using flexible optical fibers, lasers can access areas that are inaccessible to other surgical instruments.
- The laser beam simultaneously cuts and coagulates blood vessels and seals lymphatic vessels, resulting in less bleeding and less swelling.
- The laser seals nerve endings as it cuts. So the patient will have less pain.
- Noncontact procedure reduces risk of contamination and infection.
- Less tissue trauma due to no pressure and no traction applied on tissue.
- The laser sterilizes the surgical site as it cuts. Bacteria and viruses are vaporized by the laser during laser surgery.
- All the above will lead to faster patient recovery and reduce the patient's hospital length of stay.

Some of the disadvantages of laser surgery are as follows:

- Safety risks to patient and staff in terms of potential eye injuries, burns, and fire hazards.
- Higher cost per procedure due to expensive equipment, accessories, and disposables.
- Need for special facility support, supplies, and special staff training.

LASER SAFETY

Surgical lasers present hazards to patients and to clinicians. Their use must comply with regulations, standards, manufacturers' recommendations, and professional practices. Lasers must be operated only by trained users, in designated areas, and adhering to institutional policies and procedures. To ensure laser safety in a surgical procedure, all users must be familiar with the specific laser to be used, its accessories, modes of operations, tissue effects, and potential risks.

Eye Protection

In addition to hazards common to all electromedical devices, a high-energy laser beam (with its collimated property) can cause damage at a distance far from its source. Inadvertent firing of a laser may cause burns on patient or staff, start a fire, or even cause an explosion in an oxygen-enriched environment. Many lasers are in the infrared or ultraviolet range in the electromagnetic spectrum and are invisible to the human eye. An operator may not be aware of the laser path until damage has been done. When a laser beam is directed to the eye, the collimated beam of a laser will be focused by the lens to a small area with high-power density on the back of the eye. This high-intensity beam will create irreversible damage to the eye. Even a low-energy laser beam, which normally will not create tissue burns, will have enough power density to inflict ocular injuries after being focused by the lens of the eyes. An acute exposure to laser can cause a scotoma (permanent damage to a small area of the retina), resulting in a blind spot in the field of vision. Long-term exposure to low-energy laser may lead to slow degenerative changes due to thermal or photochemical injuries. Examples of such injures are slow cataract formation in damaged lens and chronic reduction of color-contrast sensitivity from a damaged retina.

Laser Classifications

Based on these potential hazards, lasers are classified according to their risks, especially in ocular exposure. According to these classifications, safety measures and special precautions are required during laser procedures. The following is a summary of the laser classification system based on the Canadian Standard for Safe Use of Lasers in Health Care (Z386-14) and the American National Standard (ANSI Z136.3–2011). The standards also stipulate responsibilities of health care facilities; composition and responsibilities of the laser safety committee and the laser safety officer; safety control measures; risk management and quality assurance guidelines; and training, education, and credentialing of laser users.

Class 1.   Lasers consider to be incapable of producing damaging radiation levels under conditions of normal use.

Class 1M. Lasers consider to be incapable of producing hazardous exposure conditions during normal operation unless the beam is viewed with an optical instrument such as an eye-loupe (diverging beam) or a telescope (collimated beam).

Class 2. Lasers emit in the visible portion of the spectrum (0.4 to 0.7 μm); eye protection is normally afforded by the blink reflex or aversion response.

Class 2M. Lasers emit in the visible portion of the spectrum (0.4 to 0.7 μm); eye protection is normally afforded by the blink reflex for unaided viewing (i.e., they are potentially hazardous if viewed with certain optical aids).

Class 3R. Lasers potentially hazardous under direct and specular reflection viewing condition when the eye is appropriately focused and stable; but will not pose a fire hazard or diffuse-reflection hazard.

Class 3B. Lasers may be hazardous under direct and specular reflection viewing conditions; but is normally not a diffuse reflection or fire hazard.

Class 4 High-power lasers hazardous to the eyes or skin from the direct beam, and may a pose diffuse reflection or fire hazard.

The blink reflex or aversion response mentioned in Class 2 laser is the average human reflex time for eye closure (about 0.25 sec.) when a visible beam of light hits the eye. For high-power lasers, even a beam reflected from a shiny surface or scattered from a dull surface can cause injuries. In medicine, most lasers are Class 3 and Class 4. The HeNe lasers used in the aiming beam for invisible lasers are usually Class 2 lasers.

To prevent eye damage, all personnel inside the operating room during laser surgery must wear appropriate protective eyewear (eyeglasses or goggles). The lens of the protective eyewear must attenuate the laser beam to an acceptably safe level while allowing enough visible light to pass through. Protective eyewear must be certified for the type of laser and specify its optical density and visible light transmission characteristics. Optical density defines the attenuation of the laser beam by the lens of the protective eyewear; visible light transmission provides an indication of the amount of visible light that can be transmitted through the lens. Protective eyewear must shield the wearer's eyes from all directions of the visual field and be free from scratch.

Skin Protection

Skin burns (patient or operating room personnel) can occur from exposure to direct or reflected laser energy. Overexposure to ultraviolet lasers may create skin sensitivity. To reduce the power density of the reflected laser beam, metallic instruments with a polished surface should not be used during laser procedures. The area surrounding the surgical site should be covered with fire-retardant materials such as wet cotton drapes.

Laser Plume Hazards

The smoke or laser plume arising from vaporization and charring of tissues may become airborne from the surgical site into the surrounding atmosphere. The plume has a distinct odor and may pose health hazards to the patient and staff in the operating room. Analysis of laser plume samples revealed that they contain water, carbonized particles, DNA, and even intact cells, and therefore inhalation should be avoided. Furthermore, the plume can scatter and attenuate the laser beam and obscure the surgical site (especially in endoscopic procedures). Removal of the laser plume enhances the visibility of the target site for the surgeon. Removing the laser plume from the surgical site and wearing face masks can prevent personnel from inhaling the laser plume. Laser smoke evacuators are highly efficient vacuum machines specially designed to fit onto laser handpieces to capture laser plume before it is released into the surrounding air. It is fitted with submicron filters to remove bacteria and viruses and active carbon filters to remove odor and some chemicals.

Fire Hazards

Since a high-energy laser beam is used in laser surgery, operating room personnel should be aware of and prepared for fire hazards. Flammable prep solution should not be used. Fire-resistant drapes and gowns should be used. A basin of sterile water should be available at the sterile site to put out fire on the patient if needed. A halon fire extinguisher must be available in the operating room. The oxygen concentration in the room should be as low as possible. Instruments and accessories (such as an endotracheal tube) used near the surgical site must be nonreflective and nonflammable.

Access and Environmental Control

During a laser procedure, only properly trained personnel with protective eyewear should be allowed to enter and stay in the operating room. Others must be aware of the hazards. To maintain a safety zone, the laser operating location must be enclosed with access control. See-through windows should be covered to prevent the laser beam from passing outside the operating room (except for CO_2 laser, which is absorbed by glass). Walls and ceilings should have nonreflective surfaces. Reflective surfaces (glass on win-

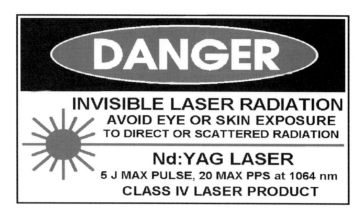

Figure 35-8. Laser Warning Sign.

dows, mirrors, X-ray view boxes, etc.) should be covered with nonreflective materials to prevent reflection of the laser beam.

Warning signs should be posted on the doors outside the operating room when lasers are being used. The wording and symbols on these signs should be specific for the type of laser in use. An example of a laser warning sign is shown in Figure 35-8. Indicator lights mounted above the main operating room entrance to signal laser procedures in progress are advised. Some facilities install an automatic door locking mechanism to prevent inadvertent personnel entry to the operating room when the laser machine is turned on.

Laser Safety Program

The CAN/CSA Z386 and ANSI Z136.3 standards recommend laser safety programs to be set up in workplaces using Class IIIB or Class IV lasers. The program should include the following components: administrative (develop laser policy, establish laser safety committee, etc.), engineering (install and maintain exhaust ventilation, window covers, etc.), and personal protection (provide eye protections, appropriate training, etc.).

The standards also recommend the appointment of a laser safety officer (LSO) whose duty is to ensure the safe use of lasers in the workplace. Duties of the laser safety officer include

- Determine laser classifications
- Ensure that laser equipment is properly installed and maintained
- Limit access to laser areas
- Arrange training for workers in safe use of lasers
- Recommend and ensure appropriate personal protection such as eyewear and protective accessories

MAINTENANCE REQUIREMENTS
AND HANDLING PRECAUTIONS

In addition to knowledgeable and trained staff, the performance and safety of lasers rely on an effective preventive maintenance program. Performance inspection of medical lasers and their accessories should be carried out periodically to ensure that they conform to the manufacturers' specifications as well as current performance and safety standards. Output characteristics, including laser power output, pulsing sequence, and timing accuracy are measured by calibrated laser power meters. For noncontact lasers, the beam geometry and energy distribution are to be measured. For some lasers, the laser gas must be replaced or recharged after being used for a period of time. Failure to replace the gas will result in reduced laser power output. The lenses and mirrors in the laser delivery system are fragile; they are easily scratched and damaged. Shock and motion from rough handling and even from normal use will cause misalignment of the optical path in the handpieces and laser arms. A misalignment in the delivery system of a laser will result in little or no laser output. Optical alignment of the system should be performed according to manufacturers' procedures. The TEM of the laser beam should be verified after every optical alignment. In most lasers, it should be as close as possible to TEM_{00}. Dirt on the first surface mirror will eventually lead to heat damage of the mirror from absorbing the laser energy. Glass or silica optical fibers are brittle and therefore cannot be bent too much. Care must be taken to inspect the tips of bare fibers or the tips of contact laser probes for signs of cracks and heat damage.

COMMON PROBLEMS AND HAZARDS

Staff and patient safety in terms of ocular injury and skin burn have been discussed earlier in this chapter. Use of protective eyewear, nonreflective and nonflammable instruments and accessories, installation of proper signage, room access control, establishment of laser safety policies and procedures, plus staff training can reduce such risks. Smoke evacuators to remove laser plume are now standard equipment in laser procedures.

A laser fiber is not a perfect transmitter; it will absorb some of the laser energy. With repeated use, the transmission efficiency will deteriorate. Some lasers have a built-in power meter that measures the power output at the laser fiber tip. The process automatically adjusts the laser's output so that the power delivered from the fiber matches the set value. Because laser fibers are very small and delicate, they are easily broken or damaged from mishandling or overheating. Excessive heat generated in the laser fiber may damage or even ignite the cover sheath of the fiber. Damaged fibers suffer from significant power transmission loss (e.g., greater than 30%) resulting in reduced surgical effect. Most lasers include a calibration mode, which assesses the transmission loss of the fiber.

Below are some of the common safety features to alert users and other personnel in the operating room that the machine is emitting a laser, or to prevent inadvertent emission of the hazardous radiation:

- Emission of audible tones during activation
- Visual laser activation indicators
- Interlocks that turn the laser off or shutters that block the beam when a laser fiber is not connected
- Removable lockout key to prevent unauthorized operation of the laser
- Activation of an alarm when malfunctions of critical components (such as cooling system) are detected

Single-use (disposable) fibers can cost up to $200 each and may become unusable before the end of the procedure. A contact laser tip alone can cost a few hundred dollars. Some fibers are reusable, and some damages are repairable. To decide if disposable or reusable fibers are to be used, in addition to considering its applications, it is critical to determine the number of times that a product can be reprocessed before direct and associated costs can be computed for consideration.

BIBLIOGRAPHY

Absten, G. T., & Joffe, S. N. (1993). *Lasers in Medicine and Surgery: An Introductory Guide* (3rd ed.). London, UK: Chapman and Hall.

Ahmed, F., Kinshuck, A. J., Harrison, M., O'Brien, D., Lancaster, J., Roland, N. J., ..., & Jones, T. M. (2010). Laser safety in head and neck cancer surgery. *European Archives of Oto-Rhino-Laryngology, 267*(11), 1779–1784.

Allen, K. B. (2006). Holmium: YAG laser system for transmyocardial revascularization. *Expert Review of Medical Devices, 3*(2), 137–146.

Alster, T. S., & Hirsch, R. (2003). Single-pass CO_2 laser skin resurfacing of light and dark skin: Extended experience with 52 patients. *Journal of Cosmetic and Laser Therapy, 5*(1), 39–42.

American National Standards Institute (ANSI). (2007). *American National Standard for Safe Use of Lasers*. ANSI Z136.1-2007. Orlando, FL: Laser Institute of America.

American National Standards Institute (ANSI). (2011). *American National Standard for Safe Use of Lasers in Health Care*. ANSI Z136.3-2011. Orlando, FL: Laser Institute of America.

Bach, T., Herrmann, T. R. W., & Gross, A. J. (2012). Radiopaque laser fiber for holmium:yttrium-aluminum-garnet laser lithotripsy: Critical evaluation. *Journal of Endourology, 26*(6), 722–725.

Bridges, W. B. (1964). Laser oscillation in singly ionized argon in the visible spectrum. *Applied Physics Letters, 4*(7), 128–130.

Canadian Standards Group (2014) *Safe use of lasers in health care.* CSA Z386-14. Mississauga, Ontario, Canada.

Canby-Hagino, E. D., Caballero, R. D., & Harmon, W. J. (1999). Intraluminal, pneumatic lithotripsy for the removal of encrusted urinary catheters. *Journal of Urology, 162*(6), 2058–2060.

Costela, A., García-Moreno, I., & Sastre, R. (2008). Medical applications of dye lasers. In F. J. Duarte (Ed.), *Tunable Laser Applications* (2nd ed., pp. 227–244). Boca Raton, FL: CRC Press.

Fournier, G. R., Jr., & Narayan, P. (1994). Factors affecting size and configuration of neodymium:YAG (Nd:YAG) laser lesions in the prostate. *Lasers in Surgery and Medicine, 14*(4), 314–322.

Hallock, G. G. (2001). Expanding the scope of the UltraPulse carbon dioxide laser for skin deepithelialization. *Plastic and Reconstructive Surgery, 108*(6), 1707–1712.

Jacobson, A. S., Woo, P., & Shapshay, S. M. (2006). Emerging technology: Flexible CO_2 laser WaveGuide. *Otolaryngology–Head and Neck Surgery, 135*(3), 469–470.

Jayarao, M., Devaiah, A. K., & Chin, L. S. (2011). Utility and safety of the flexible-fiber CO_2 laser in endoscopic endonasal transsphenoidal surgery. *World Neurosurgery, 76*(1-2), 149–155.

Jelinková, H. (Ed.). (2013). *Lasers for Medical Applications: Diagnostics, Therapy, and Surgery.* Oxford, UK: Woodhead.

Koo, V., Young, M., Thompson, T., & Duggan, B. (2011). Cost-effectiveness and efficiency of shockwave lithotripsy vs flexible ureteroscopic holmium:yttrium-aluminum-garnet laser lithotripsy in the treatment of lower pole renal calculi. *BJU International, 108*(11), 1913–1916.

Luttrull, J. K., & Dorin, G. (2012). Subthreshold diode micropulse laser photocoagulation (SDM) as invisible retinal phototherapy for diabetic macular edema: A review. *Current Diabetes Reviews, 8*(4), 274–284.

Maiman, T. H. (1960). Stimulated optical emission in ruby. *Journal of the Optical Society of America, 50*(11), 1134.

Mainster, M. A., & Turner, P. L. (2004). Retinal injuries from light: Mechanisms, hazards and prevention. In S. J. Ryan, T. E. Ogden, D. R. Hinton, & A. P. Schachat (Eds.), *Retina* (4th ed.). London, UK: Elsevier Publishers.

Marks, A. J., & Teichman, J. M. (2007). Lasers in clinical urology: State of the art and new horizons. *World Journal of Urology, 25*(3), 227–233.

McNab, D. C., & Schofield, P. M. (2002). Transmyocardial and percutaneous myocardial laser revascularization. *Circulation, 105*(19), e171–172.

Olk, R. J. (1990). Argon green (514 nm) versus krypton red (647 nm) modified grid laser photocoagulation for diffuse diabetic macular edema. *Ophthalmology, 97*(9), 1101–1112.

Polanyi, T. G., Bredemeier, C., & Davis, T. W. (1970). A CO_2 laser for surgical research. *Medical & Biological Engineering, 8*(6), 541–548.

Sivaprasad, S., Elagouz, M., McHugh, D., Shona, O., & Dorin, G. (2010). Micropulsed diode laser therapy: Evolution and clinical applications. *Survey of Ophthalmology, 55*(6), 516–530.

Sliney, D. H., Mellerio, J., Gabel, V-P., & Schulmeister, K. (2002). What is the meaning of threshold in laser injury experiments? Implications for human exposure limits. *Health Physics, 82*(3), 335–347.

Spanier, T. B., Burkhoff, D., & Smith, R. (1997). Role for holmium:YAG lasers in transmyocardial laser revascularization. *Journal of Clinical Laser Medicine & Surgery, 15*(6), 287–291.

Tanzi, E. L., Lupton, J. R., & Alster, T. S. (2003). Lasers in dermatology: Four decades of progress. *Journal of the American Academy of Dermatology, 49*, 1–22.

Temelkuran, B., Hart, S. D., Benoit, G., Joannopoulos, J. D., & Fink, Y. (2002). Wavelength-scalable hollow optical fibres with large photonic bandgaps for CO_2 laser transmission. *Nature, 420,* 650–653.

Yaghoobi, P., Moghaddam, M. V., & Nojeh, A. (2011). "Heat trap": Light-induced localized heating and thermionic electron emission from carbon nanotube arrays. *Solid State Communications, 151*(17), 1105–1108.

Chapter 36

ENDOSCOPIC VIDEO SYSTEMS

OBJECTIVES

- Describe clinical applications of endoscopic video systems.
- Analyze the construction and function of rigid and flexible endoscopes.
- Differentiate between fiberscopes and videoscopes.
- Describe and evaluate features and functional characteristics of endoscopic video components.
- Contrast common light sources for surgical video illumination.
- Describe the function and characteristics of laparoscopic insufflators.
- Discuss common problems and hazards.

CHAPTER CONTENTS

1. Introduction
2. Applications
3. System Components
4. Endoscopes
5. Light Sources
6. Video Cameras, Image Processors, and Displays
7. Image Management Systems
8. Insufflators
9. Related Technological Advancement
10. Common Problems and Hazards

INTRODUCTION

An endoscopic video system allows the physician to look inside the patient's body by inserting a viewing scope and light pipe into the body through a natural lumen or a small surgical incision. The first endoscopic instrument was developed in 1876 by Maximilian Nitze in Austria. Endoscopic inspection of the abdominal cavity was introduced in 1902 and has since been refined and become widely used in many diagnostic and therapeutic procedures. Endoscopy such as laparoscopic cholecystectomy (removal of gallbladder using a surgical video system) or arthroscopy has replaced many open surgical procedures. Endoscopic procedures are less traumatic to patients, cause less discomfort, and reduce recovery time. Surgical procedures using endoscopy instead of open surgery are often called minimally invasive surgeries or keyhole surgeries.

APPLICATIONS

Laparoscopy refers to the minimally invasive treatment and examination of organs and tissues in the peritoneal cavity using an endoscope and other special instruments. In a multipuncture laparoscopic procedure, a small incision is made to allow the insertion of a cannula with the aid of a trocar. A trocar is a pointed and solid metal (or plastic) rod inserted in the lumen of the cannula with the tip exposed to aid insertion. A cannula is a hollow tube with a sealing cap to close off the lumen when the trocar is removed or when no instrument is inserted. After the trocar is removed, a viewing laparoscope is put through the lumen of the cannula. A second incision is made for the insertion of another cannula for introducing surgical instruments. Alternatively, a surgical instrument may be inserted directly through the incision. Procedures such as cholecystectomy and appendectomy can be performed by viewing the surgical site through the laparoscope and inserting the surgical instrument through the second incision without opening the abdominal cavity. An external light source connected to the laparoscope is needed to illuminate the surgical site. An insufflator helps to maintain a pneumoperitoneum. The purpose of pressurizing the peritoneal cavity is to enlarge the working space of the surgical instruments and increase the surgeon's field of view within the peritoneal cavity. The insufflator gas may be supplied via a port in the cannula or through a Veress needle. Similar to laparoscopy, arthroscopy allows the diagnosis and treatment of some joint injuries and diseases without open arthrotomy. Laparoscopic procedures using more than one abdominal wall puncture are called multiple-puncture laparoscopy. Single-incision laparoscopy, with the surgeon manipulating through only one abdominal wall puncture, is becoming more popular.

Although single-incision laparoscopy allows quicker recovery time and better cosmetic outcomes, multiple-puncture laparoscopy allows a better view of the operating field, offers greater flexibility in manipulating tissue and instruments, and permits independent movement of the laparoscope and surgical instruments.

Compared to open procedures, laparoscopic surgeries have the following advantages:

- Smaller incision—less postoperative scarring and a faster recovery, leading to shorter hospital stay
- Less bleeding—minimizes blood loss and reduces need for blood transfusion
- Less pain—requires less pain medication
- Lower risk of infection—due to reduced exposure of internal organs to external environment

Disadvantages of laparoscopic surgeries are mainly due to surgeons using long, narrow instruments to interact with tissues rather than their own hands and the limited visibility of the surgical sites. They include

- Limited dexterity and range of motion at the surgical site
- Loss of touch sensation and poor tactile feedback
- Procedures are not as intuitive due to the unconventional maneuvering mechanism of specialized instruments
- Poor depth perception from viewing anatomy on two-dimensional display monitor
- Cannot see surrounding and behind camera anatomy, may miss lesions or injuries outside the field of view (e.g., secondary ESU burn)
- Longer procedure time

In addition to rigid endoscopes, some endoscopes have insertion tubes that are flexible and can be bent to facilitate insertion into non-straight body lumens. Flexible endoscopes are predominantly inserted through natural openings of the body. In a gastrointestinal endoscopic procedure, a flexible endoscope is inserted through the esophagus into the stomach. In bronchial endoscopy, flexible endoscopes are inserted through the trachea into the lungs. These procedures allow the diagnosis and treatment of diseases in the gastrointestinal and respiratory tracts. Below is a list of some flexible endoscopes named after the anatomy of applications.

- Bronchoscopes
- Gastroscopes

- Choledochoscopes
- Duodenoscopes
- Colonoscopes
- Sigmoidoscopes

Although endoscopes are inserted into the patient's body, most explorative flexible endoscopic procedures do not puncture the skin or injure tissue. They can be considered semi-invasive procedures. However, some flexible endoscopic procedures are invasive because they are intended for treatment (e.g., removal of cysts in the colon). In some cases, intervention procedures (such as taking tissue for biopsy) are found necessary during explorative procedures.

SYSTEM COMPONENTS

A typical endoscopic video system consists of the following functional components:

- An endoscope
- A light source
- A video camera
- An image processor
- One or more video display monitors
- An image management system

Depending on the procedure, some of the following instruments and devices may be used in endoscopic procedures:

- Trocars and cannulae
- Gas insufflators
- Air, water, and suction pumps
- Laser, electrosurgical instruments, ultrasound ablators, cutters, forceps, scissors, biopsy snares, and so on

In addition, a special flexible endoscope washer to clean and disinfect or sterilize the scopes is required. The following sections describe the basic components and characteristics of a typical video endoscopic system.

ENDOSCOPES

An endoscope is used by the surgeon to view anatomical structures and to perform therapy in the interior of the body. The diameter of an endoscope varies from the 2 mm needle fetoscope, to the 5 mm arthroscope, to the 20 mm colonoscope. The length of the endoscope must be appropriate to reach the desired structure. Depending on the procedure, the insertion tube of an endoscope can be rigid or flexible.

Rigid Endoscope

A rigid scope (Figure 36-1) either has a straight hollow shaft that allows straight viewing (such as laryngoscopes) or has an eyepiece and lens system that allows viewing in a variety of directions (such as cystoscopes). The sheaths of most rigid scopes are made of stainless steel, although plastic-sheathed scopes (mostly disposable) are available.

A laparoscope is an example of a rigid endoscope. A viewing laparoscope employs a series of rod lenses to convey high-resolution, wide field of view (FOV) images to the eyepiece. Objects seen through a laparoscope may be magnified or reduced depending on the distance between the object and the tip of the scope. Optical fibers surrounding the rod lenses transmit illumination to the object from an external light source connected to the laparoscope via a fiberoptic light cable (or light guide). The eyepiece of an operating laparoscope can be offset from the shaft so that a surgical instrument is inserted through a separate instrument channel. Operating laparoscopes use prisms or mirrors to reflect light from the object to the eyepiece. They usually have a larger diameter (8 to 12 mm) than viewing laparoscopes (5 to 10 mm) have. During a procedure, the object can be viewed directly through the eyepiece. In practice, the eyepiece is often coupled to a video camera, and the images are displayed on a video monitor.

Flexible Endoscope

Instead of a rigid shaft, a flexible fiberscope has a long flexible insertion tube connected to a proximal housing (Figure 36-2). Flexible endoscopes can be inserted into curved orifices of organs such as colon, lung, and stomach. To facilitate scope insertion and viewing, wires running from the control head to the distal tip enable the user to angulate the distal end of the endoscope. A flexible endoscope consists of the following main components:

- Insertion tube
- Control head
- Light guide connector
- Universal cord (or light guide tube)

In a typical fiberscope, the insertion tube contains two bundles of optical

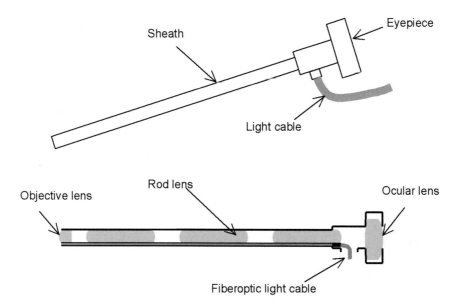

Figure 36-1. External and Cross-Sectional View of a Rigid Viewing Endoscope.

fibers, one for illumination and the other for transmitting the image. A water channel, an air channel, and an instrument channel are also included in the insertion tube. Figure 36-3 shows the viewing and illumination optical pathway.

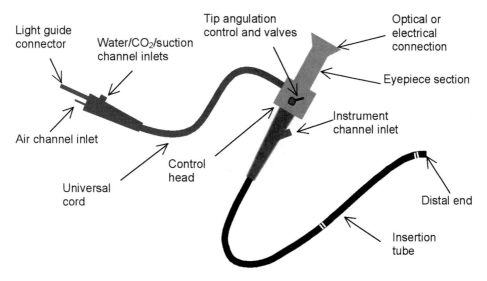

Figure 36-2. Flexible Endoscope.

During an endoscopic procedure, the physician holds the control head to manipulate the insertion tube, introducing water or air to flush the site. The control head houses the up/down and left/right angulation control knobs to move the distal tip of the insertion tube as well as the air/water and suction control valves. The opening of the instrument channel is also located on the control head.

The light guide tube is a flexible tube containing the fiber optic bundle for the light source. It also has separate air, water, suction, and CO_2 channels connected to those in the insertion tube via valves on the control head. The

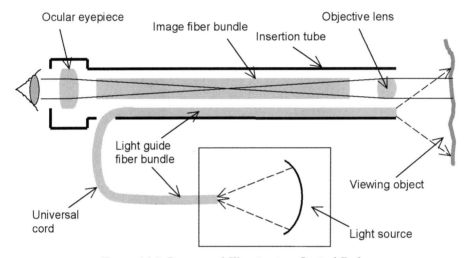

Figure 36-3. Image and Illumination Optical Path.

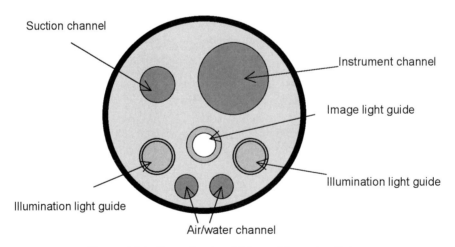

Figure 36-4. Cross-Sectional View of Insertion Tube.

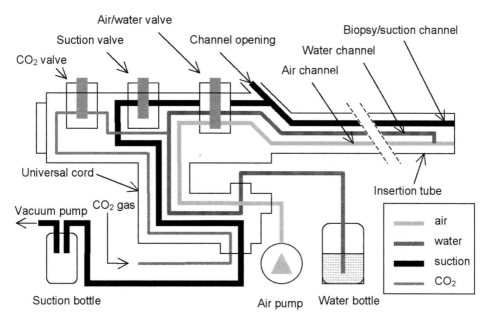

Figure 36-5. Air, Water, Suction, and CO₂ Channels of an Endoscope.

light guide connector houses the adaptor for the fiber optic bundle to the light source. The connectors for air, water, suction, and CO2 (as well as the electrical connector for videoscope) are also located on the light guide connector. Figure 36-5 shows the water, air, suction, and CO2 channels of a typical gastrointestinal endoscope.

LIGHT SOURCES

A light source is connected to the illumination light guide of the rigid or flexible endoscope to provide illumination for viewing the surgical fields or body cavities. Light sources are intended to provide the physician with a sufficient level of visible light for diagnostic observations and surgical procedures. A light source usually emits a wide spectrum covering the visible, infrared, and sometimes ultraviolet radiation. Infrared filters are installed in the light source to prevent infrared radiation from entering the body, which otherwise can cause thermal burn or even fire. A surgical light source can use a variety of lamps, including xenon, quartz halogen, metal halide, mercury vapor, and, recently, LED. Xenon (color temperature from 5600 to 6600 K) and quartz halogen (from 3200 to 5500 K) are popular lamps for endoscopic procedures due to their high intensity and near-daylight spectrum (5000 to 6000 K). LED light sources are gaining popularity because they are more energy efficient, generate less heat, and last much longer than conventional light sources do. In addition, LEDs can be fabricated to emit high intensity white light (e.g., 6000 K color temperature), or blended (e.g. with a red, a green, and a blue LED) to produce light of different color and intensity.

The output intensity of a light source can be adjusted either by an adjustable aperture or by changing the brightness of the lamp. Changing the brightness of the source by changing its supply voltage or current may be more energy efficient, but doing so may alter the color temperature of the light. In systems that have automatic brightness control, the light source is connected to the video processor to automatically maintain the level of illumination throughout the procedure.

The output intensity and color temperature of light sources usually decrease with time. A typical xenon lamp has an approximate useful life span of 500 operating hours and costs about $1000 (USD). A typical quartz halogen lamp has an approximate useful life span of 100 operating hours and costs about $50. A typical LED lamp has an approximate useful life span of over 2000 operating hours and costs about $2000. Most lamps (except LED) require forced cooling to maintain a safe operating temperature. Some light sources have a built-in light meter to monitor the output intensity and a timer to track the operating time. The light source should come with a backup lamp to avoid interruption. Light from the light source is often transmitted to the tissue through a flexible fiber optic light guide.

VIDEO CAMERAS, IMAGE PROCESSORS, AND DISPLAYS

The video camera, processor, and display serve as the eye of the physician located inside the body of the patients. The performance and quality of these components are critical for the accurate diagnosis and treatment of the patients.

Video Cameras

With traditional rigid endoscopes and fiberscopes, an endoscopic camera head is attached to the eyepiece (through an adapter) of the rigid scope or the proximal end of the flexible endoscope. A single-chip mosaic color filter CCD camera consists of a single CCD chip with red-, green-, and blue-colored filters overlaying each CCD pixel. The light reflected from the object is filtered by each of the color filters and incident on the underlying CCD elements. Each group of red, green, and blue filters and CCD elements forms one color image pixel. The intensity of light reaching the CCD is measured and converted into an electrical signal. After each exposure, the mosaic signal from the CCD pixels is sent to the image processor. The three signals (RGB) from each group of pixels are combined to reconstruct the color and intensity of the incident light source. Another single-chip design uses a rotating color wheel containing segments of red, blue, and green color filters. In essence, each CCD element is time shared by the filters to measure the intensity of the color components of the incident light. In a three CCD-chip system, the incoming light is split into red, green, and blue beams by a prism and each beam is aligned with one of the three dedicated CCDs. A three CCD-chip system provides a higher resolution image than the mosaic filter system does and a higher refreshing rate than the rotating color wheel system does. For the camera to record smooth motion pictures, discrete pictures are captured at a rate fast enough to appear continuous to the human eye. A frame rate greater than 30 frames per second (fps) is required.

In a videoscope, the CCD-chips are integrated into the tip of the scope to provide a high-quality picture free from image distortion and degradation from optical misalignment and deterioration of the optical fibers and lens system. Placing the CCDs at the tip of the insertion tube, however, increases the size and diameter of the endoscope.

Video Image Processors

Image or video processors take the electrical signal from the camera head connected to a rigid endoscope or flexible fiberscope, or from the cable output of a videoscope. Users can select one or more images from connected video sources. The processor is responsible for white balance, brightness, contrast, and color control. It may also adjust the focus, zoom, shutter, and aperture of the camera. Some video processors can support automatic gain control, multi-image formatting, and character generation. In some units, special image processing functions such as filtering, enhancement, color mapping, edge detection, and segmentation are available. Diagnostic algorithms (such as cancer detection in colonoscopy) may be built into some special applications.

The processor compiles the electric signal from the camera to produce a full-color image to be displayed on one or more display monitors. The image may also be exported to a storage device or routed to remote sites such as a physician's office. Video processors may support a number of video formats to interface with other system components. Today, analog video (composite, S-video, RGB, YPBPR) have been replaced by digital video (DVI, HDMI, DisplayPort). Digital signals can be compressed and organized into packets and therefore allow signal transmission at a higher data rate. Modern cameras and displays support high definition video, high frame rate, and three-dimensional images.

Video Displays

During a minimally invasive surgery, video or still images are displayed on one or more color monitors. Typical medical-grade video monitors have high resolution, brightness, contrast, and support gamma curve calibration. They also produce high frame rate and have low leakage current for medical applications. Both CRT monitors and LCDs have been used in endoscopic systems. High definition flat panel displays are currently the display of choice for video endoscopic procedures. High definition video systems enhance detail and visibility, as well as im-prove depth of perception, and may cause less visual fatigue for clinicians.

Three Dimensional Video Systems

Conventional endoscopy projects 3D anatomy onto a 2D display. A surgeon performs the procedure by watching a 2D image instead of a normal life 3D will lose the depth perception. 3D endoscopic systems, which restore the stereoscopic depth, are now available in the market. Some suggested that 3D will eventually replace all 2D systems in minimal invasive laparoscopic procedures. Studies have demonstrated the following advantages and disadvantages of 3D over 2D procedures:

Advantages

- Easier to identify plane
- More precise
- Feel more confident
- Less mistakes
- Reduce procedure time
- Shorten learning curve for new surgeons

Disadvantages

- Users complained about tiredness, headaches, ocular fatigue and nausea
- More expensive
- Require more equipment, such as googles
- Technical issues, such as misalignment between the eye plane and the original camera plane from scope rotation.

Figure 36-6 is a 3D endoscopic video system showing the image capture and display subsystems. The convergence is the angle formed by the eyes and the observed object. The higher the angle value is, the nearer is the observed object, and vice versa. While watching an object, the left and right

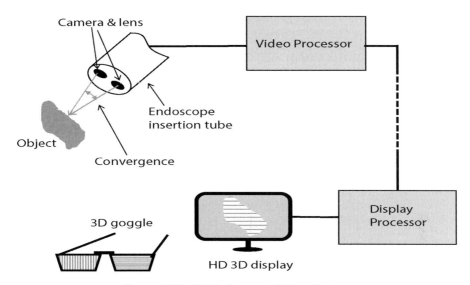

Figure 36-6. 3D Endoscopic Video System.

human eyes see the object at a slightly different angle. These two images, after processed by the brain, create the depth perception of the object. In a 3D video system, two cameras (instead of one camera in a 2D system) separated by a small distance are used to capture two parallax images of the same object. When the right camera image is projected to the right eye and left camera image to the left eye of the observer, a 3D object is seen. In a 3D endoscopic video system, both images are displayed on the same display screen. 3D image is created using either polarizing technology or active switching. In a system using polarizing technology, the parallax images are line interlaced on the display screen, the surgeon is wearing a pair of goggle with different circular polarizing filters on the left and right eyes, so that only the left image is seen by the left eye and the right image by the right eye. In a system using active switching, the two parallax images are displayed alternatively at a high frequency, the left and right filters of the goggle are switched on and off in synchronous with the display so that the left parallax image is only seen by the left eye and the right image by the right eye.

IMAGE MANAGEMENT SYSTEMS

Some surgical video systems are integrated with a computerized information management system. The basic functions of such a system include organization of patient data; image storage, retrieval, and transfer; and production of hard copies. More sophisticated systems comply with Digital Imaging and Communications in Medicine (DICOM) standards and Health

Level Seven (HL7) standards which allow them to network with Picture Archiving and Communication Systems (PACS) and Hospital Information Systems (HIS). Such systems can retrieve patient information, download work lists and upload test results to electronic patient records. Digital capture devices are now standard features, allowing instantaneous capture of still and video images. In some systems, the display screen can be split in multiple partitions to display data, previously stored images, and so on, alongside real-time images. Many systems offer remote controls for operating the processor and video systems. Data entry keyboards are available for staff to edit images and to enter annotations, notes, and comments into the video image files.

INSUFFLATORS

An insufflator helps to maintain a pneumoperitoneum to provide more working space to maneuver surgical instruments and increase the field of view of the surgeon within the peritoneal cavity. A gas is introduced into the peritoneal cavity to distend the abdomen during the procedure. CO_2 is the most commonly used insufflation gas; others include nitrous oxide, helium, and argon. The device includes a pressure-controlled flow regulator converting the high-pressure gas source (either from a cylinder or from a gas wall outlet) to about 10 to 15 mmHg (or 1.3 to 2.7 kPa) before delivering it to the patient. An insufflator automatically regulates the flow to maintain a user-selected pressure throughout the procedure. Pressure regulators and flow-restricting orifices in the device control pressure and gas flow during insufflation. Most insufflators offer both low pressure (10 to 20 mm Hg or 1.3 to 2.0 kPa) and high pressure (30 to 40 mm Hg or 4.0 to 5.3 kPa) settings. Most units allow the user to manually adjust the flow to a specific rate. For many types of surgery, low flow settings are typically from 1 to 3 L/min and high flow settings are typically from 4 to 6 L/min. Some units can provide a gas output flow rate up to 45 L/min to maintain pneumoperitoneum under aspiration conditions. High and low flow as well as high and low pressure alarms are built in to ensure patient safety. Most insufflators have sensors and displays to indicate the pressure in the peritoneal cavity. For patient safety, many electronic insufflators have an automatic pressure-relief mechanism. When the abdominal pressure exceeds the set pressure, the mechanism will release the insufflated gas to the atmosphere.

Gases used for the pneumoperitoneum include CO_2, air, oxygen, nitrous oxide, argon, helium, and mixtures of these gases. CO_2 is the preferred insufflation gas because it is colorless and nonflammable, has a high diffusion rate, and is a normal metabolic end product that can be rapidly excreted from the body. Among other gases used for insufflation, the risk of gas embolism is lowest with CO_2. Because of possible CO_2-induced hypercarbia, which can lead to tachycardia and acidosis, N_2O is preferred over CO_2 in patients with cardiac disease.

To avoid causing patient hypothermia and dehydration during a long procedure, some units are equipped with a heater and humidifier to heat up and moisturize the output gas. An in-line hydrophobic bacterial filter is used between the insufflator and the patient to prevent the passage of abdominal fluids and airborne bacteria from the patient into the insufflator (and vice versa) during the procedure.

RELATED TECHNOLOGICAL ADVANCEMENT

The field of endoscopy is evolving very rapidly. New applications and procedures are being developed along with new instruments and devices. A few new developments of endoscopic video systems are as follows:

- Three-dimensional endoscopy — the major advantage of 3D over 2D endoscopy is the presence of depth perception. However, there is still much room for improvement, such as issue with visual horizon inconsistency from scope rotation.
- Self-propelling endoscope — navigating the turns by pushing the endoscopy can be difficult and may lead to complications such as bowel perforation. Researchers have been working on endoscopic insertion tubes that can propel themselves rather than just relying on physical manipulation by the clinician.
- Cancer detection — special light spectrum or dyes to create fluorescence are used to detect malignant tissues in patients during endoscopic procedures.
- Wireless endoscopic camera capsules — the pill size capsule combines a miniature camera and wireless signal transmitter in one unit. After swallowed by the patient and traveling through the digestive tract, still and video images are captured and transmitted to an external receiver. Although noninvasive capsule endoscopy is considered a less stressful imaging method, it cannot totally replace conventional endoscopic procedures as it does not support interventional procedures such as tissue biopsy.
- Computer tomography (CT) colonoscopy — virtual CT colonoscopy is a special x-ray examination of the colon using low dose CT. Special software algorithm creates images (video of still) inside the colon. A radiologist reviews the images to look for polyps inside the colon that may turn

into colon cancer. Although it is a less invasive procedure than a conventional colonoscopy, it is not able to capture tissue color to support diagnoses nor perform interventional procedures. In addition, x-ray exposes patients to potentially harmful ionization radiation.

- Robotic Laparoscopic Surgery - using robots in laparoscopic procedures has gain much attention in recent years. The guided surgical robot system is the most commonly used type in laparoscopic surgeries. In a guided system, the surgeon operates at a remote control console with real time image feed from the surgical site. Following the control signal from the surgeon, the robotic arms performs the operation on the patient. Below listed some benefits of robotic surgery:

  o Improve hand-eye coordination of the surgeon
  o Motion can be diminished and filtered for more precise and accurate micro-manipulations
  o Effectively adding more hands to the surgeon
  o Provide more degree of freedom than human hand
  o Reduce surgeon fatigue in long procedures
  o Less trauma to neighboring tissues and less blood loss, leading to quicker patient recovery.

High capital and operating costs is the major barrier for using surgical robots. Steep learning curve is a disincentive for users. Suboptimal tactile feedback is a challenged yet to be improved. Benefit in clinical outcomes has not yet been demonstrated to justify wide spread adoption of the technology in healthcare facilities.

COMMON PROBLEMS AND HAZARDS

Among all problems, the two major hazards that may lead to serious complications during endoscopy procedures are perforation and internal bleeding.

1. Perforation is a major cause of concern when rigid scopes are used. Trocar injuries during insertion into the abdominal cavity are not uncommon. The risk of perforation for flexible scopes is lower but it remains a potential complication. A perforated bowel can occur during colonoscopy when the insertion tube accidentally punctures the wall of the colon. Bowel perforation is a medical emergency because the leakage of the bowel contents into the abdominal cavity will cause sepsis or blood infection, which if not treated properly can cause almost immediate death.

2. Internal bleeding can occur from areas where tissue has been cut, for example, from a biopsy site or from the removal of a polyp. The patient is put into danger if excessive bleeding cannot be controlled. Sometimes, bleeding may recur after the procedure or may not be noticed during the procedure due to the limited FOV of the camera.

In addition to perforation and bleeding, problems associated with endoscopic system can be grouped into the following categories:

1. Heat-related injuries
2. Risk of infection
3. Optics and video quality
4. Equipment breakdown
5. Other problems

Heat-Related Injuries

Despite the remote location of the light source, many light sources can produce visible irradiances of up to twenty times that of sunlight at earth surfaces (2 W/cm2) at the tip of the fiber optic light guide. The heat from the intense light source may cause second-degree burns, retinal damage, and fires at the site of illumination. Although infrared filters in the light source are in place to remove most IR radiation, care must be taken not to shine the light onto the same position for an extended period of time and to prevent the tip of the endoscope from prolonged contact with tissue.

Fires have been reported in association with fiber optic light sources. Users should use caution when operating high intensity light sources, especially in an oxygen-enriched environment. The tip of the light guide can still be very hot when it is first turned off. Users should be cautioned not to place the light guide in contact with the patient or flammable materials during or after the procedure. Users should ensure that the correct light guide is used and is properly connected before being activated. It must not be allowed to irradiate drapes covering the patient. The light source must be turned off or be placed in standby mode before disconnecting the fiber optic light cable. When not in use, it should be switched to the standby mode to reduce the risk of skin or eye injury and fires. The light source should be set to the lowest level at the start and its intensity adjusted upward until it is adequate for viewing.

In endoscopic procedures using electrosurgery, RF leakage current may cause secondary site burns on a patient. High frequency electrosurgical current is delivered to the tissue using a special endoscopic ESU handpiece. A typical handpiece has a long insulated conductor with the ESU active electrode exposed at the tip. Electrosurgical current passes from the tip to the tissue when the ESU is being activated. ESU handpieces for endoscopic procedures can be rigid or flexible. The long shaft of the ESU handpiece is inserted through the instrument channel of the endoscopy or through the opening of a cannula during the procedure.

Electrical leakage and insulation tests are performed periodically on endoscopes to detect potential current leakage problems. Tissue in contact with the shaft of the ESU handpiece may receive burns caused by conductive, capacitive, or inductive leakage current from the ESU. The insulation over the conductor can be damaged from colliding with other sharp instruments or from poor handling during cleaning and sterilization. ESU current will conduct from the instrument to the patient when tissue or body fluid is in contact with the exposed conductor. When the tissue is in close contact with the insulated shaft of the handpiece, because of capacitive coupling, a high frequency current will flow from the conductor to the tissue when the ESU is energized. This high frequency current has the potential to create a secondary burn, especially at a high power setting and prolonged activation time. Secondary burns often happen outside the FOV of the surgeon and therefore may not be noticed until complications (such as internal bleeding) appear after the surgery. In addition, an ESU burn may occur if the ESU is accidentally activated with the handpiece inside the patient and touching other tissue or organs.

Risk of Infection

All parts of endoscopes and accessories must be thoroughly cleaned, disinfected, or sterilized after every use. Endoscopic instruments, especially flexible endoscopes that have multiple, long, narrow channels, are difficult to get thoroughly clean and properly sterilized. Endoscopic procedures can cause nosocomial infection if the instruments are not disinfected or sterilized properly. Whenever possible, endoscopic equipment should be autoclaved. Flexible endoscopes, however, cannot withstand autoclaving nor can they withstand frequent steam sterilization. Tissue, mucus, blood, feces, and protein residue can become trapped in the channels and are difficult to remove when dried. It is important to clean all parts of the endoscope as soon as possible after use while organic debris is still moist. Liquid detergent that leaves no residue must be applied by small brushes to clean the inside lumen of all channels. Special ultrasonic cleaners may be used to break loose the debris. Endoscopes and instruments must be thoroughly cleaned and dried before ethylene oxide sterilization. Instruments that cannot be sterilized are high-level disinfected by soaking in activated glutaraldehyde or Cidex solution. Care must be taken to fill and soak all the lumens with the disinfectant.

Automatic endoscope reprocessors are available to wash and disinfect flexible scopes. Once the scope is set up properly in the reprocessor, it will automatically cycle through wash, disinfect, rinse, and dry phases. Automatic endoscope reprocessor manufacturers, using specially formulated chemical solutions (e.g., orthophthalaldehyde, peracetic acid), claim to achieve high-level disinfection or even sterilization. There has been debate over the need for sterilization of flexible endoscopes. Generally, the infection/risk control departments make decisions and provide clinical guidelines on use and reprocessing of endoscopes and associated instruments.

Optics and Video Quality

Clinicians often have preferred display settings, such as color, hue, and contrast. Unless multiple user settings can be stored, the settings on the systems should be locked to avoid unauthorized and improper adjustment that may impair accurate color display. The system should accurately reproduce colors, especially in the expected blood and tissue range. Dynamic response should be wide enough to pick up small differences in object brightness while avoiding saturation (or blooming) when exposed to bright light. Geometric distortion can cause an object near the tip of the endoscope to appear closer than it actually is and objects away from the tip to appear farther away than they actually are. For inexperienced users, this can lead to a miscalculation of the size and distance of objects. The characteristics of video components (light source, fiber optic light guide, video camera, etc.) may drift over time. It is important that the system performance can be measured and adjusted to return to acceptable condition.

Optical fibers in the light guide may break, causing lower light illumination and darker image. Damaged image fibers can create a hazy or spotty image with dark pixels at fixed locations. Moisture inside the sheath of the fiber optic cable may decrease light transmission and damage the optical components. Poor handling of the scopes may damage the lens or cause misalignment. Autoclave and chemical disinfection may cause deterioration in the camera heads and optical components. Rough handling can knock out lens and prism alignments inside cameras.

Equipment Failures

Although electronic components in endoscopic systems have become very reliable, EMI (e.g., radiated high frequency signals from electrosurgical generators) affecting video performance is common. Proper grounding of system components reduces the effect of EMI.

High intensity lamps must not be handled by hand. Dirt or fingerprints on light source lamps can cause premature lamp failures. Fingerprints, grease, or dirt must be wiped off according to manufacturers' instructions.

An insertion tube of a flexible endoscope is covered with a waterproof sheath. If this waterproof sheath is compromised, water or bodily fluid will enter the internal part of the scope. Moisture will fog up and damage optical components. Water leaks will cause deterioration of internal components (e.g., rusting of the angulation cables) and prevent effective disinfection or sterilization. Visual inspection for nicks and punctures should be performed during cleaning after every procedure. Leak tests on flexible endoscopes should be done on a regular basis and preferably during each reprocessing. Damage from fluid can be avoided if leaks are detected early.

Proper use, handling, cleaning, and storage of flexible endoscopes will prevent unexpected failures and minimize costly repairs. Bending or twisting the endoscope with excessive force and hitting the distal tip against a hard surface can damage the control wires, rendering the endoscope unusable. Broken control wires may cause the insertion tube to freeze, making withdrawal difficult.

Other Problems

Debris may build up inside the channel lumens of flexible endoscopes if they are not thoroughly cleaned. There have been reports of difficulty in inserting forceps through the instrument channel of endoscopes during procedures. It is important to follow proper cleaning and reprocessing procedures. In addition, periodic quality assurance inspections must be performed; a backup scope should be available.

Running out of insufflation gas during a procedure should be avoided because cylinder replacement will delay the procedure and unnecessarily prolong the amount of time that the patient is under anesthesia; loss of flow may allow the pneumoperitoneum to collapse, obstructing visualization of the operating field and limiting the surgeon's ability to react quickly in the event of complications. Most units have an alarm that indicates when the external cylinder's pressure is low.

To protect against over pressurization of the peritoneum, most insufflators are equipped with gauges or displays that indicate the pressure in the pneumoperitoneum. Many insufflators have an automatic pressure-relief mechanism, usually a solenoid valve that is activated when the abdominal pressure exceeds the set pressure, causing the insufflated gas to flow back into the insufflator and to vent into the room.

Residue gas in the abdominal cavity can cause temporary postsurgical pain and discomfort to the patient. Some patients cannot tolerate pneumoperitoneum, resulting in a need for conversion to open surgery. There is an increased risk of hypothermia and peritoneal trauma due to increased exposure to cold, dry gases during insufflation. Heaters and moisturizers are built into modern insufflators to alleviate such conditions.

Residue chemicals used in disinfection (such as glutaraldehyde), if not rinsed off thoroughly, are toxic and can have deleterious effects on patient's mucous membranes. Exposure to glutaraldehyde and other disinfection chemicals from endoscope cleaning and processing poses human health hazards. Patients and staff should avoid prolonged exposure to these chemicals. Endoscope clinics should have adequate ventilation and sufficient air change, especially in scope cleaning and disinfection areas.

BIBLIOGRAPHY

American Medical Association. (1990). Diagnostic and Therapeutic Technology Assessment: Rigid and flexible sigmoidoscopies [technology assessment report]. *JAMA, 264*(1), 89–92.

American Society for Gastrointestinal Endoscopy/Society for Healthcare Epidemiology of America. (2003). Multi-society guideline for reprocessing flexible gastrointestinal endoscopes. *Infection Control and Hospital Epidemiology, 24*(7), 532–537.

Becker, H. D., Melzer, A., Schurr, M. O., & Buess, G. (1993). 3-D video techniques in endoscopic surgery. *Endoscopic Surgery and Allied Technologies, 1*(1), 40–46.

Benson, K. B., & Whitaker, J. C. (Eds.). (2003). *Standard Handbook of Video and Television Engineering.* New York, NY: McGraw-Hill.

Borten, M., Walsh, A. K., & Friedman, E. A. (1986). Variations in gas flow of laparoscopic insufflators. *Obstetrics and Gynecology, 68*(4), 522–526.

Cotton, P. B., & Williams, C. B. (2008). *Practical Gastrointestinal Endoscopy: The Fundamentals* (6th ed.). Boston, MA: Blackwell Scientific Publications.

Cuschieri, A. (2005). Laparoscopic surgery: Current status, issues and future developments. *Surgeon, 3*(3), 125–130, 132–133, 135–138.

El-Minawi, M. F., Wahbi, O., El-Bagouri, I. S., Sharawi, M., & El-Mallah, S. Y. (1981). Physiologic changes during CO2 and N2O pneumoperitoneum in diagnostic laparoscopy. A comparative study. *Journal of Reproductive Medicine, 26*(7), 338–346.

Fraser, V. J., Zuckerman, G., Clouse, R. E., O'Rourke, S., Jones, M., Klasner, J., & Murray, P. (1993). A prospective randomized trial comparing manual and automated endoscope disinfection methods. *Infection Control and Hospital Epidemiology, 14*(7), 383–389.

Fritscher-Ravens, A., & Swain, C. P. (2002). The wireless capsule: New light in the darkness. *Digestive Diseases, 20*(2), 127–133.

Gordon, A. G., & Magos, A. L. (1989). The development of laparoscopic surgery. *Baillière's Clinical Obstetrics and Gynaecology, 3*(3), 429–448.

Griffin, W. P. (1995). Three-dimensional imaging in endoscopic surgery: A look at the benefits and limitations of the various techniques. *Biomedical Instrumentation & Technology, 29*(3), 183–189.

Jacobs, V. R., Morrison, J. E., Jr., & Kiechle, M. (2004). Twenty-five simple ways to increase insufflation performance and patient safety in laparoscopy. *Journal of the American Association of Gynecologic Laparoscopists, 11*(3), 410–423.

Kaban, G. K., Czerniach, D. R., & Litwin, D. E. M. (2003). Hand-assisted laparoscopic surgery. *Surgical Technology International, 11*, 63–70.

Kozarek, R. A., Raltz, S. L., Brandabur, J. J., Bredfeldt, J. E., Patterson, D. J., & Wolfsen, H. W. (1997). Virtual Vision for diagnostic and therapeutic esophagogastroduodenoscopy and colonoscopy. *Gastrointestinal Endoscopy, 46*(1), 58–60.

Lynch, D. A., Parnell, P., Porter, C., & Axon, A. T. (1994). Patient and staff exposure to glutaraldehyde from Keymed Auto-Disinfector endoscope washing machine. *Endoscopy, 26*(4), 359–361.

Marshall, R. L., Jebson, P. J. R., Davie, I. T., & Scott, D. B. (1972). Circulatory effects of carbon dioxide insufflation of the peritoneal cavity for laparoscopy. *British Journal of Anaesthesia, 44*(7), 680–684.

Martin, M. A., & Reichelderfer, M. (1994). APIC guideline for infection prevention and control in flexible endoscopy. *American Journal of Infection Control, 22*(1), 19–38.

Muscarella, L. F. (1996). Advantages and limitations of automatic flexible endoscope reprocessors. *American Journal of Infection Control, 24*(4), 304–309.

Neuhaus, S. J., Gupta, A., & Watson, D. I. (2001). Helium and other alternative insufflation gases for laparoscopy. *Surgical Endoscopy, 15*(6), 553–560.

Ofstead, C. L., Wetzler, H. P., Snyder, A. K., & Horton, R. A. (2010). Endoscope reprocessing methods: A prospective study on the impact of human factors and automation. *Gastroenterology Nursing, 33*(4), 304–311.

Phillips, E., Daykhovsky, L., Carroll, B., Gershman, A., & Grundfest, W. S. (1990). Laparoscopic cholecystectomy: Instrumentation and technique. *Journal of Laparoendoscopic Surgery, 1*(1), 3–15.

Rutala, W. A., & Weber, D. J. (1999). Disinfection of endoscopes: Review of new chemical sterilants used for high-level disinfection. *Infection Control and Hospital Epidemiology, 20*(1), 69–76.

Salky, B. A., Bauer, J., Gelernt, I. M., & Kreel, I. (1988). The use of laparoscopy in retroperitoneal pathology. *Gastrointestinal Endoscopy, 34*(3), 227–230.

Seidlitz, H. K., & Classen, M. (1992). Optical resolution and color performance of electronic endoscopes. *Endoscopy, 24*(3), 225–228.

Sinha, R. Y., Raje, S. R., Rao, G. A. (2017). Three-dimensional laparoscopy: Principles and practice. *J Minim Access Surg. Jul-Sep; 13*(3):165-169

Skreenock, J. J., Mead, D. S., & Stalker, J. H., Jr. (1981). A study of the common characteristics, hazards, and risks of endoscopes and endoscopic accessories. *U.S. Department of Health and Human Services,* April 1.

Society of Gastroenterology Nurses and Associates. (1997). Standards for infection control and reprocessing of flexible gastrointestinal endoscopes. *Gastroenterology Nursing, 20*(2), 1–13.

Voorhorst, F. A., Overbeeke, K. J., & Smets, G. J. (1996-97). Using movement parallax for 3D laparoscopy. *Medical Progress Through Technology, 21*(4), 211–218.

Yu, J., Wang, Y., Li, Y., Li, X., Li, C., Shen, J. (2014). The Safety and Effectiveness of Da Vinci Surgical System Compared with Open Surgery and Laparoscopic Surgery: a Rapid Assessment. *Journal of Evidence-Based Medicine,* May;7(2):121-34.

Chapter 37

CARDIOPULMONARY BYPASS UNITS

OBJECTIVES

- Describe the applications of cardiopulmonary bypass (CPB).
- Explain the principle of extracorporeal membrane oxygenation (ECMO).
- Describe the clinical setup of CPB and ECMO.
- Draw and explain the functional building blocks of CPB and ECMO units.
- Discuss the construction and characteristics of key system components of a heart-lung machine (HLM).
- Discuss problems and hazards associated with CPB and ECMO.

CHAPTER CONTENTS

1. Introduction
2. Functions of Cardiopulmonary Systems
3. Principle of Extracorporeal Oxygenation
4. System Setup and Operation
5. Monitoring and Peripheral Components
6. Common Problems and Hazards

INTRODUCTION

Cardiopulmonary bypass (CPB) is a method to replace the function of the heart and lungs. During open heart surgical procedures, such as heart valve replacement or coronary artery bypass, CPB maintains the circulation of blood and the oxygen content of the body by an extracorporeal system including a blood pump and an oxygenator. A heart-lung machine (HLM) with ancillary equipment provides the CPB function to support open heart surgery. In patients with compromised respiratory functions such as respiratory distress syndrome or severe respiratory deficiency, modified CPB procedures are used to complement the oxygenation of blood and removal of metabolic CO_2. Extracorporeal oxygenation using membrane oxygenators to supplement oxygen uptake are called extracorporeal membrane oxygenation (ECMO). Procedures using CPB to remove CO_2 are referred to as extracorporeal CO_2 removal ($ECCO_2R$) procedures. Standard CPB procedures using HLMs provide short-term support during various types of cardiac surgical procedures. ECMO units are used for longer-term support ranging from 3 to 10 days to allow time for intrinsic recovery of the lungs and heart. Both HLMs and ECMO units provide cooling and heating functions to regulate blood temperature. Blood cooling to less than 28°C (deep cooling) is selected to reduce oxygen demand of the patient during open heart surgical procedures.

Sergei Brukhonenko, a Soviet scientist, developed the first heart-lung machine for total body perfusion in 1926. The first known human open heart operation with temporary mechanical takeover of both heart and lung functions was performed by Clarence Dennis and his team in 1951 at University of Minnesota Hospital; the patient did not survive due to an unexpected complex congenital heart defect. The first successful open heart procedure on a human utilizing the heart-lung machine was performed by John Gibbon in 1953 at Thomas Jefferson University Hospital in Philadelphia. In 1954, Walton Lillehei developed the cross-circulation technique by using slightly anesthetized adult volunteers as living CPB machines during the repair of cardiac disorders. Rashkind and coworkers were the first in 1965 to use a bubble oxygenator as life support in a neonate dying of respiratory failure. In 1970, Baffes and colleagues reported the successful use of ECMO as support in a neonatal cardiac surgery. In 1975, Bartlett and associates were the first to successfully apply ECMO in neonates with severe respiratory distress.

The main differences between ECMO and CPB are as follows:

- The purpose of ECMO is to allow time for intrinsic recovery of the lungs and heart; a standard CPB provides support during cardiac surgeries.
- Cervical cannulation, which can be performed under local anesthesia, is often used in ECMO; transthoracic cannulation under general anesthesia is standard in CPB
- Unlike standard CPB, which is used for short-term support measured in hours, ECMO is used for longer-term support ranging from 3 to 10 days.

FUNCTIONS OF CARDIOPULMONARY SYSTEM

The heart circulates blood around the body, including the lungs; the function of the lungs is to remove CO_2 from the venous blood and provide oxygen to the pulmonary blood. Perfusion, the passage of blood through blood vessels to tissues, delivers oxygen and nutrients to the tissues. Venous blood carries away metabolic wastes from tissues including CO_2. Respiration allows oxygen to diffuse from the alveoli into the pulmonary blood. Oxygen is primarily transported by hemoglobin in the blood. The presence of hemoglobin allows the blood to transport 30 to 100 times more oxygen as could be transported by dissolved oxygen in the blood plasma. In arterial blood, oxygen is bound to hemoglobin to form oxygenated hemoglobin. The hemoglobin in 100 ml of blood can carry about 20 ml of oxygen. The cells in the tissue consume oxygen and produce CO_2 as a metabolic waste. The CO_2 produced by tissues enters the blood in capillaries and eventually reaches the lungs. Although some CO_2 combines with hemoglobin to form carbamino-hemoglobin, unlike O_2 transport, association and dissociation of CO_2 with hemoglobin only accounts to about 20% of the total CO_2 transport. Removal of CO_2 is primarily achieved through diffusion from the pulmonary blood to the alveoli gas. The following paragraphs explain oxygen and CO_2 transport in the cardiopulmonary system.

About 97% of the oxygen transported from the lungs to tissues is carried by hemoglobin in the red blood cells. The remaining 3% is carried as dissolved oxygen in the blood plasma. Oxygen molecules combine loosely and reversibly with the heme of the hemoglobin. When the oxygen partial pressure in blood (PO_2) is high, oxygen binds with the hemoglobin; when the PO_2 is low, oxygen is released from the hemoglobin. Figure 37-1 shows the oxygen-hemoglobin dissociation curve.

The oxygen partial pressure in air is about 159 mmHg (760 mmHg × 20.8%) or 21 kPa. When it reaches the alveoli, with elevated CO_2 level and saturated water pressure, it is about 104 mmHg or 14 kPa (Table 37-1). Oxygen in the alveoli enters into the capillary blood by diffusion. The difference in oxygen pressure in the alveolar gas and in the pulmonary capillaries is an important factor governing the rate of diffusion. A higher differential pressure will create higher diffusion rate. For a heathy individual, under normal activities, blood inside alveolar capillaries will eventually reach the same PO_2 as the alveolar gas (i.e., 104 mmHg or 14 kPa). These oxygen molecules will combine with heme to form oxygenated hemoglobin. When this oxygenated blood combines with the blood returning from the bronchial ventilation (which contains blood at PO_2 = 40 mmHg or 5.3 kPa) and flows into the left heart chambers, the PO_2 of the blood is reduced to about 95 mmHg or 13 kPa. In the arterial blood, according to Figure 37-1, a PO_2 of 95 mmHg or 13 kPa produces 97% oxygen saturation. In venous

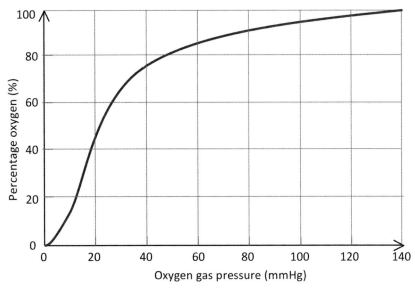

Figure 37-1. Oxygen-Hemoglobin Dissociation Curve.

blood for which the PO_2 is 40 mmHg or 5.3 kPa, this drops to 75% oxygen saturation.

When the arterial blood reaches the peripheral tissues, its PO_2 is still 95 mmHg or 13 kPa. The PO_2 in the interstitial fluid is about 40 mmHg or 5.3 kPa and inside a cell is about 23 mmHg or 3 kPa. This pressure difference causes oxygen to diffuse rapidly from the capillary blood into the interstitial fluid. As the PO_2 decreases in the blood, the oxygen is dissociated from the oxygenated hemoglobin until it is in equilibrium with the partial pressure in the interstitial fluid (i.e., PO_2 = 40 mmHg or 5.3 kPa). Figure 37-2 illustrates the diffusion of oxygen from tissue capillary to the interstitial fluid. Note that the rate of diffusion (indicated by the magnitude of the arrow) decreases as the PO_2 of the blood drops when it flows along the capillary. The removal of CO_2 from the tissues to capillaries is similar to the oxygen diffusion but in the opposite direction. The PCO_2, partial pressure of carbon dioxide, is about 45 and 46 mmHg (or 6.0 to 6.1 kPa) respectively, in interstitial fluid and inside the cells. The PCO_2 in arterial blood is 40 mmHg or 5.3 kPa and in venous blood is 45 mmHg or 6 kPa. Diffusion of CO_2 from interstitial fluid to capillary blood is shown in Figure 37-2.

PRINCIPLE OF EXTRACORPOREAL OXYGENATION

In CPB procedures, the oxygenator replaces the functions of the lungs.

Table 37-1. Partial Pressures of Respiratory Gases

	Atmospheric Air		Alveolar Air	
	Partial Pressure (mmHg/kPa)	Percentage	Partial Pressure (mmHg/kPa)	Percentage
O_2	159.0/21.2	20.84	104/13.9	13.6
CO_2	0.3/0.04	0.94	40.0/5.3	5.3
H_2O	3.7/0.49	0.50	47/6.3	6.2
N_2	597.0/79.6	78.63	569/75.9	74.9
Total	760/101	100	760/101	100

Thin film oxygenators were used in early days. In a thin film oxygenator, blood flows over a solid surface in an oxygen-enriched compartment to allow oxygen to diffuse into the blood. Thin film oxygenators were replaced by bubble oxygenators in the late 1960s. In a bubble oxygenator, ventilated gas is bubbled into the venous blood to allow oxygen and CO_2 exchange across the bubble-blood interface. To prevent air embolism, the oxygenated blood had to be defoamed before returning to the patient. In a membrane oxygenator, gas exchange occurs across a hydrophobic, gas-permeable membrane that separates the blood from the ventilating gas. Although the technology of membrane oxygenators was available in the 1960s, it was not used in CPB until the 1980s due to its low gas permeability and low effective surface area. Technological improvement of the membrane has led to its current widespread use in CPB and ECMO.

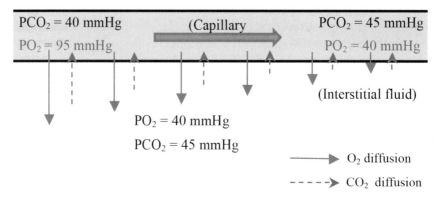

Figure 37-2. Diffusion of Oxygen (solid line) and Carbon Dioxide (dotted line) Between Capillary and Interstitial Fluid.

Thin film and bubble oxygenators have no physical barrier between blood and oxygen; these are called direct contact oxygenators. Membrane oxygenators use a gas-permeable membrane to separate the blood and oxygen compartments. Such a design decreases the blood trauma of direct-contact oxygenators as well as lower the chance of air embolism. Much work since the 1960s had been focused on overcoming the low rate of gas exchange across the membrane barrier, leading to the development of high-performance, microporous hollow-fiber, plasma-tight oxygenators that eventually replaced direct-contact oxygenators in CPB surgeries. The membrane of modern oxygenators is constructed from thermoplastic polymer (e.g., polymethylpentene) hollow fibers. The fiber diameter is about 100 to 200 μm with gas flow inside the fibers and blood flow over and across the surface of the fibers (Figure 37- 3). Compared to the ones with blood flow inside the fibers and gas flow outside, this configuration reduces the pressure drop of blood across the oxygenator. Figure 37-4 and Figure 37-5 show the gas transport characteristics of a membrane oxygenator. Notice that the gas transfer rates increase with the blood flow. However, a higher pump pressure is needed to maintain a higher blood flow as the flow resistance increases with the flow velocity. Table 37-2 lists the specifications of a typical membrane oxygenator.

The preferred characteristics of an oxygenator are as follows. The specifications of membrane oxygenators fit well into these characteristics:

- Good O_2 and CO_2 transfer performance
- Low blood pressure drop

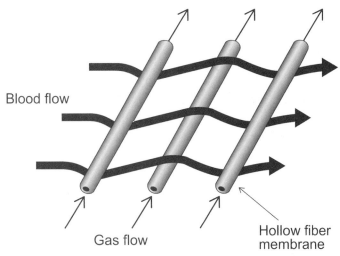

Figure 37-3. Blood Flow Across a Hollow Fiber Membrane.

Figure 37-4. Oxygen Transfer of Membrane.

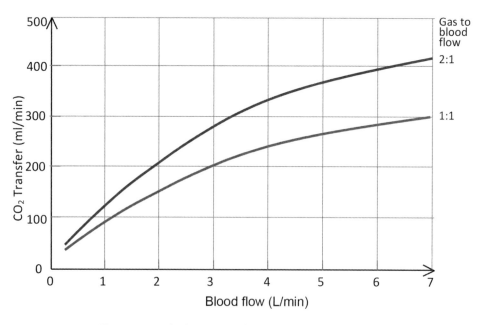

Figure 37-5. Carbon Dioxide Transfer of Membrane.

Table 37-2. Specifications of a Typical Membrane Oxygenator

Blood flow rate	0.5 to 7 l/min
Pressure drop at 5 l/min of blood flow	45 mmHg (6.0 kPa)
Priming volume	200 ml
Material of microporous membrane	Polypropylene
Surface area of gas exchanger	2 m^2
Material of heat exchange capillary	Polyurethane
Surface area of heat exchanger	0.6 m^2
Recommended air and blood ratio	1:1

- Low priming volume
- Hemo-compatible materials
- Optimum surface refinement
- High efficiency heat exchange

SYSTEM SETUP AND OPERATION

Heart-lung machines (HLMs) for bypass surgeries are operated by perfusionists. They provide short term replacement of the heart and lung functions and a blood-free operating zone for the surgeon. An open heart surgical procedure usually runs for a few hours. ECMO and ECCO$_2$R are used in acute care areas (such as intensive care units) to complement the compromised cardiopulmonary functions of the patient. To maintain the blood at body temperature, a heat exchanger is used. ECMO and ECCO$_2$R usually run for a few days and are operated by nurses.

Heart-Lung Machines

Under normal cardio-pulmonary condition, the venous blood flows into the right heart chambers is pushed through the pulmonary arteries into the lungs. The oxygenated blood from the lungs flows into the left atrium is pumped out from the left ventricle through the aorta to the rest of the body. In a CPB procedure, the functions of the lungs (blood oxygenation and CO$_2$ removal) and heart (pumping blood) are replaced by the CPB unit. Venous blood is draw from a large vein into an extracorporeal blood circuit. A blood pump pushes the venous blood through an oxygenator before it is re-introduced back to the patient via a major artery (Figure 37-6).

Figure 37-7 is a typical HLM showing the blood flow paths and functional components. A crystalloid solution (aqueous solutions of mineral salts) is usually used to prime the silicon pre-sterilized bypass tubing. During a total bypass procedure, the patient's blood is heparinized to prevent blood clots

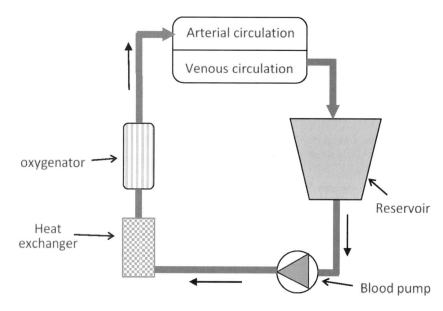

Figure 37-6. Blood Flow Paths of Pulmonary Bypass Unit.

within the bypass circuits especially in the membrane oxygenator. Adequacy of anticoagulation is measured by laboratory analysis of a blood sample or point-of-care testing every 30 minutes to maintain an activated clotting time (ACT) of 480 seconds or higher (normal range of ACT is from 70 to 120 sec.). In addition, to reduce the chance of inflammation and blood clot formation, the inner surface of the blood tubing is heparin bonded. After the bypass procedure, the effect of the anticoagulant is reversed by administration of an antigen such as protamine sulfate.

A number of cannulae are used to connect the patient to the machine circuit. They are made of plasticized polyvinylchloride (PVC) and are wire enforced to prevent occlusion due to compression or kinking. A pair of venous cannulae are surgically inserted into the venae cavae to redirect blood that normally returns to the right atrium to the venous reservoir. An aortic cannula to return blood to the patient is inserted in the ascending aorta. A cardioplegia cannula is inserted at the root of the aorta. A normally closed aortic root vent branched out from the cardioplegia cannula is used to aspirate the aorta and left ventricle when needed. An aorta cross-clamp placed between the aortic cannula and the cardioplegia cannula is then applied to isolate the heart from the systemic circulation. Once the heart is isolated from the rest of the systemic circulation and the left ventricle is sufficiently unloaded, about 1 liter of cold (usually 4°C) crystalloid cardioplegia solution (e.g., a 20–30 mmol/L KCl-based solution) is infused via the cardioplegia cannula into the heart and coronary arteries. This induced hypothermia lowers the metabolic rate of the heart muscle, reduces myocardial oxygen consumption, and causes cardioplegia arrest. A fibrillator applying 50 or 60 Hz AC voltage to the myocardium is another way of inducing cardiac arrest. Blood is commonly added to the cardioplegia solution (at a ratio of 1:1 to 1:8) to prevent cell death during the ischemic period of time. Heart-lung bypass machines are equipped with a cardioplegia pump to assist in the intermittent administration of cardioplegia solutions during long CPB procedures. However, the additional volume of fluid from the cardioplegia solution contributes substantially to hemodilution during CPB. A hemoconcentrator can be added to the extracorporeal blood circuit to remove this extra fluid from the circulation.

Venous blood flows by gravity via the venous cannula into the venous reservoir. The venous clamp occludes the line before initialization of the CPB procedure. A level sensor monitors the blood level in the reservoir to prevent air from entering into the blood circuit. An arterial blood pump moves the blood from the venous reservoir to the membrane oxygenator for gaseous exchange before it is returned to the patient's circulation via the aortic cannula distal to the aortic cross-clamp. A heat exchanger is used to regulate the blood temperature before it is administered back to the patient. A temperature control unit supplies warm or cold water to the heat exchanger. Moderate hypothermia (28°C to 32°C) is employed to slow down the patient's basal metabolic rate and to reduce the oxygen demands of other organs such as the brain, kidneys, and liver during the procedure. Deep hypothermia (< 28°C) is selected for procedures requiring a temporary period of circulatory arrest (e.g., aortic arch repair). To prevent air embolism and remove particulates, blood from the oxygenator passes through a defoamer (e.g., made of polyurethane foam) and blood filter (20 to 40 µm) before returning to the patient.

A mixture of medical air and oxygen is supplied through a sub-micron bacterial filter to the membrane oxygenator. The oxygen concentration and flow of the gas mixture is used to control the PO_2 in arterial blood (usually kept at 150 to 250 mmHg or 20 to 33 kPa) during CPB. Point-of-care arterial blood gas is measured at intervals of approximately 30 minutes. A continuous venous oximeter is located at the venous return line. Mixed venous oxygen saturation (%SvO_2) should be maintained at 75% or higher as an indication of sufficient peripheral perfusion. Low %SvO_2 level may be caused by low PO_2 level or low hemoglobin level in the arterial blood. Low %SvO_2 level can be treated by increasing the CPB blood flow rate. During CPB, general anesthesia is maintained by introducing a controlled level of volatile anesthetic agent to the gas mixture to the oxygenator. In some cases, intravenous anesthetic such as propofol may be used instead.

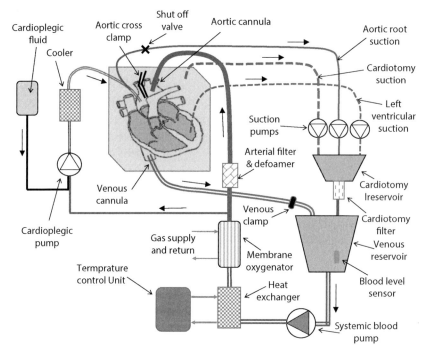

Figure 37-7. Blood Flow Paths and Main Components of a Heart-Lung Machine.

Roller pumps or centrifugal pumps are used as arterial blood pumps. A centrifugal pump may cause less hemolysis to blood cells but requires the addition of a one-way valve to prevent reverse blood flow when the pump is de-energized. The impellers of modern centrifugal pumps are magnetically coupled to the pump's driving mechanism to allow the pump cartridge to be totally sealed in order to maintain sterility during operation. Both roller and centrifugal pumps mays be operated in continuous or pulsatile mode. The target flow rate during CPB is 2.2 to 2.4 L/min/m^2 where the last quotient is the patient's body surface area. Mean arterial pressure (MAP) is usually targeted at 50 to 80 mmHg or 6.7 to 11 kPa during CPB, but it should not exceed 100 mmHg or 13 kPa. Increasing arterial blood flow may increase MAP. Low MAP can be treated with vasopressors.

To prevent excessive blood loss and minimize transfusions, blood that pools in the surgical site is suctioned and collected in the cardiotomy reservoir by a suction pump. A cardiotomy filter (typically 150 to 200 μm) in the reservoir removes debris from the recovered blood before it is mixed with the venous blood and moved to the systemic circuit. A hemoconcentrator may be inserted into the extracorporeal blood circuit during CPB to circumvent hemodilution and to maintain hematocrit levels. Hemoconcentrators use ultrafiltration to removes excess plasma water while retaining blood cells and plasma protein. Use of a hemoconcentrator can reduce the needs of blood products administration during and after the bypass.

In addition, a ventricular venting circuit is employed in long CPB procedures to prevent stasis of the blood which may result in thrombosis. A suction pump intermittently removes blood pooled inside the left ventricle from retrograde blood flow and drains the blood into the cardiotomy reservoir.

Extracorporeal Membrane Oxygenation Machines

Similar to the heart-lung bypass machine, an ECMO machine draws venous blood from a large vein, pushes it through the membrane oxygenator for gaseous exchange, into the defoamer to remove air, and through the heat exchanger for temperature regulation (Figure 37-8). After going through gaseous exchange, the extracorporeal blood is returned to the patient's circulatory system through a vascular insertion. During the procedure, blood is still circulated by the patient's heart through the lungs and around the body. The quantities of oxygen introduced and CO_2 removed by the ECMO machine depend on the extracorporeal blood flow, the gas composition from the oxygen blender, and the gas flow rate into the membrane oxygenator.

There are different methods of vascular access for ECMO; the two most common ones are venoarterial (VA) and venovenous (VV). In both methods, blood is drained from the venous system and pumped through the extracorporeal membrane oxygenator. The blood in VA ECMO systems is returned to the arterial system. The blood in VV ECMO systems is returned to the venous system. In a VA ECMO, the venous cannula for blood removal is often placed in the right common femoral vein near the junction of the inferior vena cava and right atrium, and the arterial cannula for blood return is positioned in the right femoral artery with the tip in the iliac artery. In VV ECMO, the venous cannula for blood removal is usually placed in the right common femoral vein, and the cannula for blood return is placed in the right internal jugular vein. VV ECMO is typically used for respiratory failure; VA ECMO is used for cardiac failure.

MONITORING AND PERIPHERAL COMPONENTS

A HLM has an arterial pump, a cardioplegia pump, and two to three suction pumps. Most have an additional arterial pump as back up. The pumps may be a roller or centrifugal type. Arterial pumps in current HLMs can operate in either continuous or pulsatile mode. The pulsatile mode is to simulate the natural action of the ventricles.

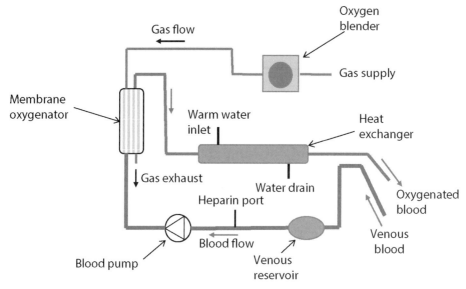

Figure 37-8. Extracorporeal Membrane Oxygenation Machine.

In order for the CPB and ECMO systems to operate effectively and safely, a number of monitoring and control components are in place. Heat exchangers are used to maintain the desired temperature of blood and the cardioplegia solution. Temperature-regulated water from an external temperature control unit circulates through the heat exchanger to maintain the blood and solution temperature. Temperature sensors (e.g., thermistors) are placed at various locations along the patient circuit to monitor temperature for system control and safety. Pressure sensors to record venous and arterial pressures are incorporated into blood circuit and are displayed on monitoring equipment. Flow sensors are installed in fluid circuits to monitor and control flow velocities. Ultrasonic bubble detectors are used to monitor the arterial blood circuit before the blood is reintroduced into the patient's body. When excessive blood foaming or air is detected, an alarm will sound and may cause the pumps to shut down. Point-of-care blood gas, oxygen saturation and hematocrit sensors are inserted in the blood return line. One or more level detectors monitor the blood level in the venous reservoir. Blood gas, electrolytes, hematocrit, coagulation factor, and other blood chemistries are performed periodically (e.g., every 30 min) during the procedure.

COMMON PROBLEMS AND HAZARDS

During a CPB or ECMO procedure, vital functions of the body are replaced by an external device. Blood is continuously removed from the patient, processed, and returned to the patient during the procedure. This complicated process poses severe risks on the patient and may lead to serious injury or death. This section describes some of the hazards and methods of mitigation.

Embolism, created by either air or particulates in the patient's circulation, creates one of the most severe hazards during a CPB procedure. Gross embolism (e.g., caused by an air bubble greater than 1 mL) in arterial circulation can result in death. Smaller emboli and debris in the blood may cause neurological disorders or pathological damage in organs and may contribute to immediate cognitive decline. Despite adding filters, bubble traps, and defoamers to remove air and particulates in the blood, embolism still occurs. Improper or defective connections in tubing allow air to be drawn into the negative pressure circuits. Rapid inadvertent emptying of a venous reservoir while it is receiving a low flow of venous blood can cause a massive infusion of air. In addition to surveillance by the perfusionist, bubble and blood-level detectors are essential safeguards against gross air embolism. Systemic heparinization and the use of heparin-coated extracorporeal circuits have substantially inhibited blood clotting and systemic inflammation. Newer biocompatible coating materials are continually being researched. Phosphorylcholine-coated tubing and circuit components, intended to mimic the lining of the body's blood vessels, are available in some models.

Blood damage is inevitable in all perfusion procedures. Hemolysis, platelet damage, and leukocyte damage can result from blood being in contact with foreign materials. Agitating and mixing blood with air within the extracorporeal circuit can also create blood damage. Excessive pressurization and suction are traumatic to blood. These damages to blood will accumulate during a long procedure. Centrifugal blood pumps are considered to impose less damage to blood cells than roller pumps. A drop in platelet count also occurs because of platelet aggregation and destruction inside the membrane oxygenator. To limit blood loss, blood from the surgical field is returned to the venous reservoir via the cardiotomy suction.

Technical advances in electronics have made current heart-lung bypass systems easier to use than their predecessors were. Patient adverse outcomes due to hardware failures are no longer common in today's systems. Backup pumps are available and can be easily swapped in case of problems. To circumvent power failure, the primary pump often has backup power from a battery source or from an uninterruptible power supply. A hand-cranked pump to circulate blood in the circuit is often incorporated to overcome catastrophic failures.

There are many sensors and alarms in a CPB system. Computerized perfusion controllers can monitor basal temperature and blood pressure; the cardioplegia delivery system can alert the perfusionist to imminent danger by monitoring the temperature, pressure, and flow of the solution to the patient. In some models, the air embolism protection system will stop the arterial pump. CPB units have in-line arterial filters to trap particulate and gaseous emboli.

A heart-lung bypass system should be properly maintained by qualified service professionals and must be inspected before the bypass is initiated. It is important for the perfusionist to ensure all connections are securely tightened and that the tubing is neither twisted nor kinked. The perfusion, oxygenation, and suction settings should be properly adjusted before the procedure begins. The perfusionist must remain vigilant during the entire procedure.

Bleeding occurs in 30% to 40% of patients receiving ECMO and can be life threatening. It is due to both the necessary continuous heparin infusion and the platelet dysfunction. Bleeding tendency can be identified by periodically accessing coagulation and platelet functions in blood. A patient can rapidly exsanguinate (lose blood) if a line becomes disconnected. A variety of complications can occur during cannulation, including vessel perforation with hemorrhage, arterial dissection, distal ischemia, and incorrect placement.

Despite the high cost of disposable components (about US$1000 per procedure) used in conjunction with CPB procedures, their use is one of the main contributing factors in reducing the danger of cross-contamination by blood-borne pathogens, such as hepatitis B and HIV, through exposure to contaminated equipment. The disposables used with current day systems, which include Luer-lock connectors, color-coded ports, and rotatable lids or bases, facilitate tubing connections and reduce accidental misconnection. In addition, the use of disposable components reduces the time and labor involved in equipment cleaning and preparation before surgery.

BIBLIOGRAPHY

Ayad, O., Dietrich, A., & Mihalov, L. (2008). Extracorporeal membrane oxygenation. *Emergency Medicine Clinics of North America, 26*(4), 953–959.

Castiglioni, A., Verzini, A., Pappalardo, F., Colangelo, N., Torracca, L., Zangrillo, A., & Alfieri, O. (2007). Minimally invasive closed circuit versus standard extracorporeal circulation for aortic valve replacement. *Annals of Thoracic Surgery, 83*(2), 586–591.

Cheng, R., Hachamovitch, R., Kittleson, M., Patel, J., Arabia, F., Moriguchi, J., . . ., & Azarbal, B. (2014). Complications of extracorporeal membrane oxygenation for treatment of cardiogenic shock and cardiac arrest: A meta-analysis of 1,866 adult patients. *Annals of Thoracic Surgery, 97*(2), 610–616.

De Vroege, R., Wagemakers, M., te Velthius, H., Bulder, E., Paulus, R., Huybregts, R., . . . , & Wildevuur, C. (2001). Comparison of three commercially available hollow fiber oxygenators: Gas transfer performance and biocompatibility. *ASAIO Journal, 47*(1), 37–44.

Dennis, C., Spreng, D. S., Nelson, G. E., Karlson, K. E., Nelson, R. M., Thomas, J. V., . . . , & Varco, R. L. (1951). Development of a pump-oxygenator to replace the heart and lungs; an apparatus applicable to human patients, and application to one case. *Annals of Surgery, 134*(4), 709–721.

Gravlee, G. P. (2008). *Cardiopulmonary Bypass: Principles and Practice* (3rd ed.). Philadelphia, PA: Lippincott.

Gay, W. A. (1994). Crystalloid potassium cardioplegia: Concepts and early studies. *Annals of Thoracic Surgery, 58*(4), 1285–1286.

Hamada, Y., Kawachi, K., Nakata, T., Kohtani, T., Takano, S., & Tsunooka, N. (2001). Anti-inflammatory effect of heparin-coated circuits with leukocyte-depleting filters in coronary bypass surgery. *Artificial Organs, 25*(12), 1004–1008.

Hemmila, M. R., Rowe, S. A., Boules, T. N., Miskulin, J., McGillicuddy, J. W., Schuerer, D. J., . . . , Bartlett, R. H. (2004). Extracorporeal life support for severe acute respiratory distress syndrome in adults. *Annals of Surgery, 240*(4), 595–607.

Just, S. S., Müller, T., Hartrumpf, M., & Albes, J. M. (2006). First experience with closed circuit/centrifugal pump extracorporeal circulation: Cellular trauma, coagulatory, and inflammatory response. *Interactive Cardiovascular and Thoracic Surgery, 5*(5), 646–648.

Lim, M. (2006). The history of extracorporeal oxygenators. *Anaesthesia, 61*(10), 984–995.

Mateen, F. J., Muralidharan, R., Shinohara, R. T., Parisi, J. E., Schears, G. J., & Wijdicks, E. F. (2011). Neurological injury in adults treated with extracorporeal membrane oxygenation. *Archives of Neurology, 68*(12), 1543–1549.

Murphy, G. S., Hessel, E. A., & Groom, R. C. (2009). Optimal perfusion during cardiopulmonary bypass: An evidence-based approach. *Anesthesia and Analgesia, 108*(5), 1394–1417.

Øvrum, E., Tangen, G., Tølløfsrud, S., Øystese, R., Ringdal, M. A. L., & Istad, R. (2004). Cold blood cardioplegia versus cold crystalloid cardioplegia: A prospective randomized study of 1440 patients undergoing coronary artery bypass grafting. *Journal of Thoracic and Cardiovascular Surgery, 128*(6), 860–865.

Pearson, D. T., Holden, M. P., Poslad, S. J., Murray, A., & Waterhouse, P. S. (1986). A clinical evaluation of the performance characteristics of one membrane and five bubble oxygenators: Gas transfer and gaseous microemboli production. Perfusion, 1(1), 15–27.

Pearson, D. T., McArdle, B., Poslad, S. J., & Murray, A. (1986). A clinical evaluation of the performance characteristics of one membrane and five bubble oxygenators: Haemocompatibility studies. *Perfusion, 1*(2), 81–98.

Peek, G. J., Moore, H. M., Moore, N., Sosnowski, A. W., & Firmin, R. K. (1997). Extracorporeal membrane oxygenation for adult respiratory failure. *Chest, 112*(3), 759–764.

Ranucci, M., Balduini, A., Ditta, A., Boncilli, A., & Brozzi, S. (2009). A systematic review of biocompatible cardiopulmonary bypass circuits and clinical outcome. *Annals of Thoracic Surgery, 87*(4), 1311–1319.

Shaw, C. I. (2008). Heart lung machines. *Biomedical Instrumentation and Technology, 42*(3), 215–218.

Ullrich, R., Lorber, C., Röder, G., Urak, G., Faryniak, B., Sladen, R. N., & Germann, P. (1999). Controlled airway pressure therapy, nitric oxide inhalation, prone position, and extracorporeal membrane oxygenation (ECMO) as components of an integrated approach to ARDS. *Anesthesiology, 91*(6), 1577–1586.

Chapter 38

AUDIOLOGY EQUIPMENT

OBJECTIVES

- Explain the field of audiology.
- Discuss sound physics and measurements related to human hearing.
- Explain the loudness scale.
- Analyze the structure and functions of the human ear.
- Describe hearing problems and disorders.
- Discuss the principles and constructions of diagnostic and measurement instrumentations in audiology.
- Review the principles and construction of hearing aids and cochlear implants.
- Discuss problems and hazards associated with devices used in audiology.

CHAPTER CONTENTS

1. Introduction
2. Physics of Sound
3. Mechanism of Hearing
4. Instrumentations in Audiology
5. Audiometers
6. Middle Ear Analyzers
7. Otoacoustic Emission Detectors
8. Auditory Brainstem Response Units
9. Hearing Aids, Cochlear Implants, and Sound Booths
10. Common Problems and Hazards

INTRODUCTION

Hearing loss affects about 5% of the world's population. About 15% (32.5 million) of American adults aged 18 or above reported some trouble in hearing. The causes for hearing impairment are varied within the population. Older adults who develop hearing impairments are usually due to life-related factors, such as aging, noise, and ototoxic exposure. In children and young adults, hearing loss is usually syndrome related, due to acoustic trauma and vestibular disorders.

Audiology is a branch of science that studies hearing disorders, including balance and other related impairments. Through tests and measurements, audiology aims to determine if someone can hear within the normal range and, if not, which portions of hearing (high, middle, or low frequencies) are affected and to what degree. If it is determined that a hearing loss or vestibular abnormality is present, rehabilitation options (e.g., hearing aid, cochlear implants, or further medical referrals) will be recommended and carried out.

Audiometry measures a subject's hearing levels with the help of specialized instruments (such as an audiometer or tympanometer) but may also measure the ability to discriminate between different sound intensities, recognize pitch, and distinguish speech from background noise. Otoacoustic emission measurement and auditory brainstem response may also be performed. Results of audiometric tests are used to diagnose hearing loss or diseases of the ear.

This chapter studies audiometric equipment, including audiometers, tympanometers, otoacoustic emission (OAE) detectors, and auditory brainstem response (ABR) devices. Hearing aids are also briefly described.

PHYSICS OF SOUND

Sound is a longitudinal mechanical wave in which particles are moving back and forth parallel to the direction of wave travel. Sound does not travel in a vacuum because it requires a medium to transmit. The human ear can detect sound in the range of about 20 to 20,000 Hz. Although human hearing is limited to an upper frequency of about 20 kHz, many animal species can detect sound frequencies beyond this upper limit. The human ear is most sensitive between 2 to 5 kHz and less sensitive to low frequencies and impulse sound (less than 1 sec in duration). The wavelength λ of a sound wave is equal to the propagation speed of the sound c divided by its frequency f or

$$\lambda = \frac{c}{f}. \tag{38.1}$$

The speed of sound in dry air at 0°C is 331 m/sec. Sound travels slightly faster at higher temperature (e.g., 343 m/sec at 20°C) and faster in materials with molecules closer together (e.g., at 1480 m/sec in water). The propagation speeds of sound in different media are shown in Table 38-1.

The acoustic impedance Z is the interference of sound propagation by objects in the path of the sound waves. It is defined as the ratio of acoustic pressure P to the volume flow Q or

$$Z = \frac{P}{Q}. \tag{38.2}$$

The unit of Z is $Pa.s.m^{-3}$. Its value depends on the medium and the sound frequency. Acoustic admittance is the inverse of acoustic impedance.

Specific acoustic impedance is the ratio of acoustic pressure to the propagation speed of sound c, or

$$z = \frac{P}{c}. \tag{38.3}$$

Its unit is $Pa.s.m^{-1}$ (or rayl). It can be shown that the specific acoustic impedance is also equal to the density of the medium multiplied by the velocity of sound in the medium or $z = \rho c$. For air, the density is 1.2 $kg.m^{-3}$ and c is 343 $m.s^{-1}$, so the specific acoustic impedance for air is 420 $kg.s^{-1}.m^{-2}$ = 420 $Pa.s.m^{-1}$ (or 420 rayl). For water, the density ρ is 1000 $kg.m^{-3}$ and c is 1480 $m.s^{-1}$, so the specific acoustic impedance for water is 1.48 $MPa.s.m^{-1}$ (or 1.48 Mrayl). For human soft tissues, the values are comparable with those of water. Examples of specific acoustic impedances are presented in Table 38-1.

Table 38-1. Acoustic Properties of Biological Tissues and Materials (at 20°C)

Medium	Propagation Speed (m/sec)	Specific Acoustic Impedance (Mrayl)
Air	343	0.0004
Water	1480	1.48
Soft Tissue	1440 to 1640	1.3 to 1.7
Bone	4080	6.00
Aluminum	6400	17.00

The Level of Sound

Sound waves exert pressure on objects in their path. The unit of sound pressure level (SPL) is in Pascal (Pa). The intensity of the sound I is the amount of energy flowing through per unit time per unit area. Therefore, it is the power of sound divided by the area over which the power is spread (assuming it is perpendicular and uniformly applied over the area):

$$Intensity\ (W/cm^2) = \frac{power\ (W)}{area\ (cm^2)} = \frac{PQ}{A} = Pc,$$

where P is the sound pressure, Q is the volume flow, A is the cross-sectional area, and c is the velocity of sound.

Substituting $c = P/z$ (from equation 2 gives):

$$I = P \times \frac{P}{z} = \frac{P^2}{z}. \tag{38.4}$$

Therefore, the sound intensity (and the power) is proportional to the square of the sound pressure. The level of sound is usually measured using a microphone, which is a transducer sensitive to sound pressure. The minimum audible sound intensity at 1 kHz for a young healthy adult is considered to be 10^{-16} W/cm^2. The corresponding SPL is 20 µPa, which is defined as the threshold of hearing. The human ear can tolerate sound pressure more than one million times than this minimum level. Because of this big difference and the fact that human hearing reacts to logarithmic changes to sound, sound level is often expressed in decibels (dB), and it is referenced to the threshold of hearing.

$$Sound\ level\ in\ dB = 10log\frac{I}{I_R} = 10log\frac{P^2}{P_R^2} = 20log\frac{P}{P_R}.$$

At the threshold of hearing (20 µPa), the sound level in dB is

$$20log\frac{20\ \mu Pa}{20\ \mu Pa} = 20log1 = 0\ dB.$$

A sound level of 1 million times the threshold of hearing (1,000,000 × 20 µPa = 20,000,000 µPa) is

$$20\log \frac{20,000,000 \; \mu Pa}{20 \; \mu Pa} = 20\log 1000000 = 20 \times 6 = 120 dB.$$

Doubling the sound level (= 20log2) is equivalent to adding 6 dB to the existing sound.

Attenuation and Transmission of Sound

When sound travels in air, energy is lost due to a number of mechanisms including air viscosity and temperature. Sound attenuation or absorption is approximately proportional to the square of sound frequency. In addition, the attenuation of sound intensity in air varies significantly with temperature and humidity. Table 38-2 gives values of attenuation of sound in air in dB km^{-1} for a temperature of 20°C and a pressure of 101 kPa (at room temperature and atmospheric pressure) at different levels of humidity.

When a sound wave travels from one medium to another, a portion of it will be reflected at the boundary and the rest will be transmitted across the boundary. The intensities of reflected and transmitted sound waves are dependent on the incident intensity and the specific acoustic impedance of the two media.

The incident reflection coefficient (IRC) is equal to the ratio of the reflected intensity to the incident intensity. The incident transmission coefficient (ITC) is equal to the ratio of the transmitted intensity to the incident intensity. Assuming no loss at the boundary,

$$\text{IRC} + \text{ITC} = 1. \tag{38.5}$$

For normal or perpendicular incident,

$$\text{IRC} = \frac{I_r}{I_i} = \left[\frac{z_2 - z_1}{z_2 + z_1} \right]^2. \tag{38.6}$$

If $z_2 \neq z_1$, reflection will occur, if $z_2 = z_1$, all incident sound wave will be transmitted. Less sound will transmit through the boundary if the media have a large difference between their acoustic impedances. This is why ultrasound gel is used to reduce the acoustic impedance difference at the transducer-skin interface in medical ultrasonography.

With oblique incident, the transmitted sound beam will not travel in the same direction as the incident sound beam. The angle difference depends on the propagation speed c of the sound in the media. This change in direction of the sound beam at the boundary is called refraction.

Table 38-2. Attenuation of Sound in Air (dB km[-1])

Freq (Hz)	Relative Humidity (%)			
	20	40	60	80
1	6.5	4.7	4.8	5.1
2	22	11	9.3	9
5	110	55	38	31
10	280	190	130	100
20	510	580	470	380

$$\frac{\sin \angle i}{\sin \angle R} = \frac{c_1}{c_2}, \tag{38.7}$$

where $\angle i$ = angle of incident and $\angle R$ = angle of refraction.

MECHANISM OF HEARING

The Human Auditory System

The ear houses the receptors for hearing and for body equilibrium. The auditory impulses travel along the cochlear branch of the vestibulocochlear nerve to the cochlear centers in the brain's medulla and terminate in the hearing area of the temporal lobe cortex, where sound is recognized and interpreted. The vestibular impulses from the ear travel along the vestibular branch of the vestibulocochlear nerve to the brain, setting up reflexes to skeletal muscles for necessary adjustments to maintain dynamic equilibrium. The ear consists of three parts: the external ear, the middle ear, and the inner ear (Figure 38-1).

The External Ear

The external (or outer) ear consists of the pinna (or auricle) and the meatus (or external ear canal). The pinna is a flap of elastic cartilage covered by thick skin. The meatus is a tube about 2.5 cm long leading to the tympanic membrane (or the eardrum). The canal contains hairs and specialized glands that secret cerumen (earwax). The combination of hair and earwax helps to prevent foreign objects from entering the ear. The tympanic membrane is a thin fibrous tissue with a concave external surface covered with skin; its inner surface is convex and covered with mucous membrane.

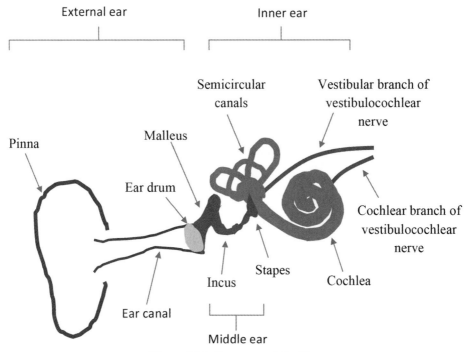

Figure 38-1. Structure of the Ear.

Ambient sounds are collected by the pinna and directed to the middle ear by the external ear canal. The ear canal serves as a waveguide, with one end open to the source of the sound and the other end closed by the tympanic membrane. The acoustic properties of the external ear depend on the dimension of the ear canal and the acoustic impedance of the eardrum. The ear canal can resonate or cause attenuation at certain sound frequencies; it is therefore affecting the frequency response of the overall auditory system.

The Middle Ear

The middle ear, also called the tympanic cavity, is a small air-filled cavity inside the temporal bone. It is separated from the external ear by the eardrum and from the internal ear by a thin bony partition. The middle ear is lined with epithelium and connects to the nasopharynx of the throat by the eustachian (or auditory) tube. The auditory tube equalizes the pressure on both sides of the tympanic membrane to protect it from damage by abrupt changes in external or internal pressure. During swallowing or yawning, the tube opens to the atmosphere to allow air to enter or leave the middle ear.

Immediately behind the tympanic membrane are three small bones called auditory ossicles: the malleus, incus, and stapes (or hammer, anvil, and stirrup). The ossicles are connected by synovial joints. One side of the malleus is connected to the inner surface of the tympanic membrane and the other side is connected to the incus. The distal end of the incus articulates with the stapes. The base of the stapes fits into a small opening called the oval window in the thin bony partition between the middle and the inner ear. Another thin window, called the round window, is located below the oval window. Both windows are connected to the base of the cochlea in the inner ear.

The ossicles are attached to the middle ear cavity by ligaments. There are two small muscles attached to the ossicles to prevent excessive movement of the tympanic membrane. The tensor tympani muscle connects the malleus and the stapedius muscle to the stapes. These muscles protect the tympanic membrane by dampening excessive vibrations that result from loud external noise.

The malleus picks up the sound vibrations from the eardrum and passes them through the incus and stapes on to the inner ear. The relatively large size of the tympanic membrane and the lever mechanism of the ossicles amplify the vibration before it is transmitted to the inner ear. This mechanical combination amplifies the sound level by 15 to 20 dB.

The Inner Ear

The inner ear is also called the labyrinth because it consists of a complicated collection of canals and chambers inside the temporal bone. The bony labyrinth is divided into three areas: the semicircular canals, vestibule, and cochlea. Inside the bony labyrinth runs the membranous labyrinth (can be viewed as a membranous tube inside a hollow tubular cavity). A fluid called perilymph fills the space between the bony labyrinth and the membranous labyrinth. The interior of the membranous labyrinth is filled with a fluid called endolymph.

In response to external sound, the vibration of the stapes is transmitted across the oval window into the entire fluid system of the inner ear. The cochlea is responsible for hearing. It is shaped like a snail with $2\frac{3}{4}$ turns. The coil of the cochlea is broad at the base, tapers toward the apex, and is lined with hair cells. Neuron dendrites extending from the cochlea branch of the vestibulocochlear nerve are positioned in close proximity to the hair cells. The vibration causes some of the hair cells to bend and stimulate the neuron endings creating biopotential impulses. These nerve impulses travel to the cochlea centers in the brain's medulla and eventually reach the hearing area of the temporal lobe of the brain cortex, where sounds are recognized and

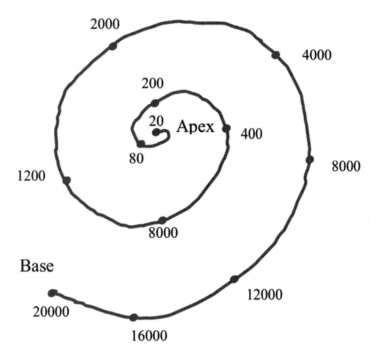

Figure 38-2. Frequency-Sensitive Locations of the Cochlea.

interpreted. Due to the construction of the cochlea, sound waves of different frequencies cause stimulation to hair cells in specific region of the cochlea. High-frequency sounds stimulate the hair cells near the base of the cochlea, and low-frequency sounds stimulate those near the apex. Higher intensity sounds produce higher nerve impulses due to greater vibration intensity. The frequency-sensitive locations of the cochlea are shown in Figure 38-2. The numbers marked are frequencies of the sound waves in hertz.

The vestibule and semicircular canals are responsible for body equilibrium: the vestibule for static equilibrium (orientation of the body) and the semicircular canals for dynamic equilibrium (maintenance of body position). Similar to the cochlea, hair cells are present in different parts of the vestibule and semicircular canals to detect position and orientation of the body. The three ducts of the semicircular canals are positioned at right angles in each of the three dimensional planes. Leaning of the head creates and imbalance in the flow of the endolymph, causing hair cells to bend and stimulate sensory neurons inside the ducts. These nerve impulses pass over the vestibular branch of the vestibulocochlear nerve to the cerebellum and the brain. Using this information, equilibrium is maintained by regulating the stimulation to the corresponding skeletal muscles.

Table 38-3. Sound Pressure Level in dB in Different Environment

Environment	SPL (dB)
Threshold of hearing	0
Library	35
Business office	65
Jet plane takeoff	125
Threshold of pain	130

Characteristics of Human Hearing

The human ear reacts to a logarithmic change in sound level that corresponds to the decibel scale of change. The lowest SPL the human ear can detect is about 20 μPa (threshold of hearing).

Although a 6-dB increase represents a doubling of the sound pressure, an increase of about 10 dB is required before sound appears to be twice as loud to the ear. The smallest change in sound level detectable by the human ear is 3 dB. The threshold of pain is about 130 dB. Table 38-3 shows typical human response to sound and levels in different environment.

We have mentioned that human ears of young adults can respond to the range of sound frequencies from 20 Hz to 20 kHz. However, the sound reception is not equally sensitive at all frequencies. Hearing is most sensitive

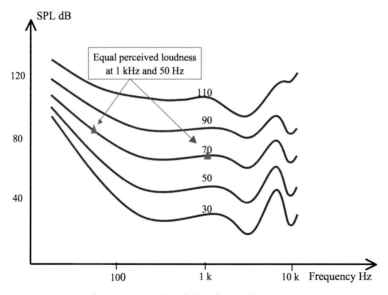

Figure 38-3. Equal Loudness Contours.

between 2 kHz to 5 kHz and less sensitive at higher and lower frequencies. The differences in sensitivity to different sound frequencies are more pronounced at low SPLs than at high SPLs. Figure 38-3 shows the equal loudness contours at different SPLs. An equal sound loudness contour represents the SPL required at any frequency in order to give the same apparent loudness as a 1 kHz tone. For example, a tone at 50 Hz at 85 dB will appear to be as loud as a 1 kHz tone at a level of 70 dB.

Another factor affecting hearing is the duration of sound. Impulse sound with duration less than 1 second is less sensitive to the human ear. An example of impulse sound is noise from a jack hammer or the sound from a typewriter.

A sound pressure meter is a device that measures sound pressure level over a wide frequency range. As human's hearing perception of loudness is different at different sound frequencies, a weighting curve is used to approximate human's hearing response to sound in human sound exposure measurements. The dBA and dBC weighting scales are used in sound measurements in human hearing applications. The "A" weighting scale covers the full frequency range of 20 Hz to 20 kHz and is adjusted to the sensitivity of human hearing at different frequencies. The "C" weighting resembles the effect of low frequency sounds on the human ear and is flat between 31.5 Hz and 8 kHz. dBA is commonly used to approximate human hearing at normal sound level while dBC is used for peak or very loud sound level measurements. In occupational health hearing protection, dBA is used to assess workplace noise exposure to workers.

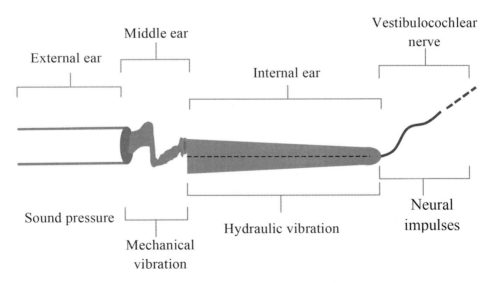

Figure 38-4. System Components of the Auditory Tract.

Disorder of the Ear

The ear is a complex system. Sound pressure is collected by the external ear, then converted into mechanical vibration by the eardrum. The ossicular chain amplifies and transmits the vibration to the oval window of the middle ear and causes fluid in the cochlea to vibrate. The vibration pressure of the fluid in the internal ear is transformed into biopotential impulses by the hair cells. These neural impulses are then conducted to the brain via the vestibu-locochlear nerve. Figure 38-4 shows the system components of the auditory tract.

Hearing loss can result from problems at any location along the auditory tract. Disorders in the external ear are often caused by blockage of sound in the ear canal by an accumulation of earwax or perforation of the eardrum. Disorders in the middle ear are often caused by infection, which may give rise to accumulation of pus pushing on the eardrum. Disconnection of the ossicles disrupts the transmission of sound vibration. Freezing of the stapes hinders the amplification action of the ossicles. Problems affecting the outer or middle ear are termed conductive. Disorders in the inner ear are usually due to damage to the hair cells; depending on the location, hair cell damage affects the frequency response of hearing. Hair cell damage may be caused by infection or ingestion of ototoxins. Damage or diseases of the vestibulo-cochlear nerve or the auditory center of the brain will cause hearing problems.

Problems affecting the inner ear and auditory nerves are termed sensorineural. In addition to hearing problems, disorders affecting the semicircular canals and the nerve pathways or brain centers will result in nausea, dizziness, and loss of balance.

Simple ear disorders can often be satisfactorily treated; more complex cases will require special medical procedures. Therapeutic medical devices such as hearing aids and cochlea implants can be used to correct some hearing problems. Although most people are born with normal hearing, aging and exposure to high intensity noise can develop hearing loss. Accurate diagnosis is the first critical step in mitigating hearing problems. The following sections cover audiology diagnostic and therapeutic devices.

INSTRUMENTATIONS IN AUDIOLOGY

A number of devices and procedures are used to assess human hearing, perform diagnoses, and carry out treatments. Devices used in diagnosis of hearing problems are listed below:

- Audiometer—assess the entire auditory tract

- Middle ear analyzer (tympanometer)—assess the middle ear
- OAE detector—assess the inner ear
- Auditory brain stem response (ABR) units—assess the entire auditory tract

  Devices to correct hearing problems:

- Hearing aids
- Cochlear implants

  Devices to support hearing measurements:

- Hearing aids analyzers
- Sound booths
- Other auditory calibration equipment

AUDIOMETERS

An audiometer is a device used for evaluating and quantifying hearing loss of an individual. It measures the hearing sensitivity by determining the individual's hearing threshold for pure tones and speech. These thresholds are then compared with standard threshold values (references derived from a group of young adults). For patients who are losing their hearing, the progress of hearing loss can be tracked over the years to assess the rate at which the hearing is lost. Pure tone audiometry measures hearing sensitivity for a series of single frequency sounds within the range of normal hearing. Speech audiometry measures hearing sensitivity and speech discrimination in conversation.

In pure tone audiometry, a low frequency pure tone (e.g., 125 Hz pure sine wave), is first selected, and the sound intensity is slowly increased until the subject signals the operator that he or she can hear the tone. The sound intensity is recorded as the threshold of hearing at the set frequency. Another frequency higher than the previous one is then selected and the measurement is repeated. The process continues until the highest testing frequency is reached (e.g., 8 kHz). An audiogram (see Figure 38-4) is created from the measurements. Conventional audiometry examines hearing frequencies between 250 Hz and 8 kHz; high-frequency audiometry tests from 8 kHz to 20 kHz. High-frequency audiometry is used to assess hearing loss associated with environmental factors such as ototoxic medication and noise exposure, which appear to be more detrimental to high-frequency sensitivity than to that of middle or low frequencies. It is also used in detecting the auditory sensitivity changes that occur with aging. Because pure tone audiometry relies on the patient's response to pure tone stimuli, it is a subjective measurement of the hearing threshold and is limited to use on adults and children old enough to cooperate with the test procedure. To avoid ambient noise affecting the measurements, testing is often conducted inside a sound booth.

When sound is applied to one ear, its vibration is conducted through the bone of the skull to the contralateral cochlea; this effect is known as cross hearing. Masking is a technique to remove the effect of cross hearing during testing by temporarily presenting noise at a predetermined level. The masking noise temporarily elevates the threshold of the non-test ear, thereby preventing the non-test ear from detecting the test signal presented to the test ear. Therefore, hearing thresholds obtained with masking provide an accurate representation of the true hearing threshold level of the test ear. There are three types of masking noise: white noise, pink noise, or narrow-band noise. White noise consists of a wide range of frequencies with uniform loudness and is used to mask tones or speeches. Pink noise is similar to white noise except that it consists of a higher proportion of lower frequencies and is used to mask speech alone. Narrow-band noise consists only of frequencies close to the ones being tested and is used to mask pure tones.

The tone generator in the audiometer supplies sound of specific frequency and level to the subject by using a pair of earphones, insert earphones, bone conductors, or loudspeakers. Bone conduction is performed by placing a vibrator on the mastoid bone behind the ear. Since the use of bone conductors bypasses the air channel of the external ear, when the thresholds

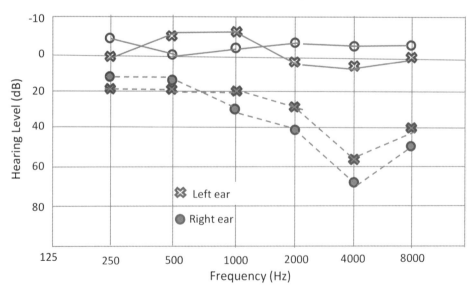

Figure 38-5. Audiogram (solid line: normal hearing; dotted: hearing loss at high frequency).

obtained from air conduction are examined alongside thosefrom bone conduction, the possible cause of hearing loss can be revealed. In conductive hearing loss, air-conduction thresholds are higher than bone-conduction thresholds. In sensorineural hcaring loss, both air- and bone-conduction thresholds are equal and elevated. When earphones are not practical (e.g., when young children are being tested), free-field testing is used. The patient is placed equidistant between two loudspeakers in a sound isolation room.

In an audiogram, hearing thresholds at their corresponding sound frequencies are expressed in dB and normalized to the standard hearing thresholds (established by averaging the thresholds from a group of young adults). Figure 38-5 shows the pure tone test results for both ears from two individuals, one normal (solid lines) and the other (dotted lines) with hearing loss at high frequencies.

Speech recognition threshold (SRT) is defined as the sound pressure level at which 50% of the speech is correctly identified. For a person with a conductive hearing loss or a sensorineural hearing loss in quiet environment, the SRT is higher than for a person with normal hearing. The increase in SRT depends on the degree of hearing loss. In noise, the person with a sensorineural hearing loss requires a better SNR to achieve the same performance level as does the person with normal hearing and the person with a conductive hearing loss.

The following four types of audiometer are specified by ANSI S3.6, 2004: American National Standard Specification for Audiometers, with Type 1 being the highest precision and having the most features.

1. Type 1 (advance diagnostic) audiometer is for precision clinical testing and more advanced diagnostic procedures, including pure tone and speech. It is able to carry out air and bone conduction and sound field examinations. Two sound channels are used; one is for the hearing test and the other is used for masking.
2. Type 2 (diagnostic) audiometer is able to carry out air and bone conduction, as well as sound field examinations, including pure tone and speech; masking is available.
3. Type 3 (portable screening) audiometer is able to carry out air and bone conduction examinations without speech; masking is available.
4. Type 4 (industrial screening) audiometer is able to carry out air conduction examinations only with no masking.

The basic components of an audiometer include a precision function generator with frequency control, an audio amplifier with output sound level control, as well as output devices such as earphone, bone conductor, and audio speakers. Figure 38-6 is the functional block diagram of a typical

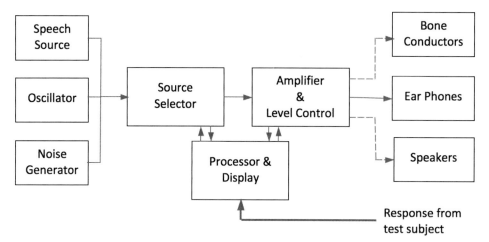

Figure 38-6. Audiometer Functional Block Diagram.

audiometer.

A tunable oscillator generates the required pure tone frequencies. A source selector selects the signal source to be introduced to the test subject. An attenuator adjusts the signal level before it is fed to a fixed gain amplifier. A number of output devices such as earphones or bone conductors can be selected. Moving coil earphones are commonly used because they provide reasonably flat frequency response up to about 6 kHz. In addition to the pure tone signals from the oscillator, noise (for masking) from the noise generator and spoken voices (for speech audiometry) recorded in the speech source module can be selected. The processor receives the responses (e.g., a push-button switch) from test subject, and correlates those to the sound frequencies and levels to generate the audiogram.

Transducers in Audiometry

There are a number of fundamental transducers in an audiometer. Described below are the characteristics of such transducers.

Microphones

A microphone is a transducer that converts sound in air into an electrical signal. Common microphones are condenser (capacitor) microphones, dynamic (induction coil) microphones, and piezoelectric microphones. Due to its wider frequency response and faster transient response, a condenser microphone is commonly used in audiology applications.

Earphones

An earphone is a transducer that converts electrical signals into sound in air. Moving coil earphones are commonly used because they provide reasonably flat frequency response up to about 6 kHz. Earphones, in general, are not interchangeable because they are calibrated together with the other audiology system components.

Ear Cups

Specially designed audio cups are used to enclose fully the external ears and the unshielded earphones to exclude ambient noise.

Bone Vibrators

Bone vibrators convert electrical signals into mechanical vibrations. Diaphragm-type bone vibrators are commonly used in hearing applications. They are becoming less frequently used today due to their limited and non-flat frequency response.

Loud Speakers

When coupling of the transducer to the ear is not feasible, loudspeakers are used to deliver auditory stimuli. Acoustic energy loss into the surrounding area will be much greater than when the stimulation is applied directly using an earphone. Room acoustics is an important factor when using loud speakers; masking of the ear not-under-test is often required.

Typical Specifications

The following list shows the typical specifications for an audiometer.

- Test frequencies: 256 to 8000 Hz in step increment of one octave
- Test modes: Automatic, semiautomatic and manual
- Frequency accuracy: Better than 1%
- Distortion: Total harmonic distortion below −40dB (1%)
- Hearing loss attenuator: 0 dB to 100 dB in 5-dB steps, accuracy ±1 dB
- Rise/fall time: Meets ANSI specifications
- Earphones: Matching 10 ohm cushioned earphones
- Physical: Width 32 cm, depth 28 cm, height 12 cm; weight 3.2 kg
- Power: Auto selectable 90 to 240 V with transient suppression for power line spikes

- Computer/printer interface: RS232 port
- Data output: 600 to 19,200 baud, selectable
- Real-time clock: With battery backup; time of day printout

Problems and Precautions

Hearing threshold determination relies on the subjective response from the test subject. After the examination, a second threshold check should be carried out to ensure they are within an acceptable difference (e.g., < 10 dB); otherwise, a retest must be performed. The accuracy can be affected by the following factors:

- Machine limitations—accuracy of frequency settings and sound output levels.
- Headphone fit—the location and fit of the headphones affect the detection of the threshold.
- Ambient noise—audiometric exams should be carried out in a sound booth to eliminate external sounds from influencing the test.
- Learning curve—the subject becomes more proficient in detecting the threshold in subsequent trials.
- Cooperation of subject—the test results will not be reliable if the subject is unable or unwilling to cooperate with the test.

MIDDLE EAR ANALYZERS

Tympanometry is an examination to test the condition of the middle ear, including the mobility of the eardrum (tympanic membrane) and the ossicular chain. Tympanometry is an objective test of middle-ear function. It is not a test to assess the sensitivity of hearing but rather to measure the effectiveness of energy transmission through the middle ear. In conjunction with pure tone audiometry, tympanometry can be used to differentiate sensorineural and conductive hearing loss. Tympanometry is an effective test to reveal the presence of a middle ear effusion and is therefore helpful in making a diagnosis of tympanitis.

To produce a tympanogram, the middle ear analyzer generates a 226-Hz pure tone into the ear canal. When the sound strikes the tympanic membrane, it causes vibration of the middle ear. A portion of the sound is reflected back from the tympanic membrane and is picked up by a microphone. The percentage of sound reflection depends on the stiffness of the tympanic membrane. Most middle ear problems result in stiffening of the middle ear. A stiffer middle ear has higher acoustic impedance, which increases sound

reflection. When all other factors remain unchanged, the acoustic impedance is the lowest when the pressure on both sides of the tympanic membrane is the same. Pressure (positive or negative) is applied to the tympanic
membrane during measurements.

A tympanogram is a plot of sound admittance (reciprocal of impedance)
against the pressure inside the ear canal. Since the eustachian tube connects
the middle ear to the ambient atmosphere, the air pressure in the middle ear
is normally at ambient pressure. In a healthy individual, the maximum
sound is transmitted through the middle ear when the ambient air pressure
in the ear canal is equal to the pressure in the middle ear. When the pressure
is higher on one side of the tympanic membrane than the other (e.g., when
fluid accumulates inside the middle ear), the tympanic membrane becomes
stiffer. The higher the pressure difference, the stiffer the tympanic membrane will be; this results in higher acoustic impedance (lower acoustic
admittance).

To conduct a middle ear examination, an otoscopy (examination of the
ear with an otoscope) is first performed to ensure that the path to the
eardrum is clear and that there is no tympanic membrane perforation. A
probe is then inserted into the ear canal and sealing it off. The analyzer sets
the pressure in the ear canal, generates a pure tone, and measures the
eardrum responses to the sound. The process is repeated at different pressure settings. Figure 38-7 displays three tympanograms with the admittance
(in milli mhos) as the vertical axis and the ear canal pressure (in decapascal

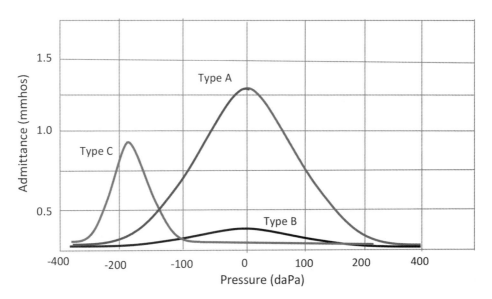

Figure 38-7. Tympanograms.

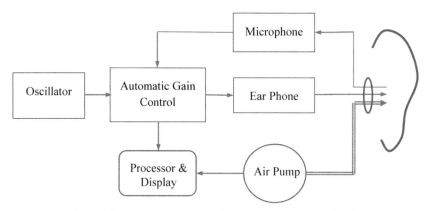

Figure 38-8. Middle Ear Analyzer (Tympanometer) Block Diagram.

or daPa) as the horizontal axis. There are three types of tympanograms, classified according to their shapes. Type A is a normal tympanogram with the peak at zero (atmospheric) pressure. Type B is a flat tympanogram, and type C is a tympanogram with peak at a negative pressure. Type A indicates a normal pressure in the middle ear with normal mobility of the eardrum and the conduction bones. If there is disruption of the ossicles, or if a portion of the tympanic membrane is flaccid, more sound energy will be absorbed and the tracing will display an abnormally high peak. Type B (flat) may be caused by middle ear infection with effusion, immobile or perforated membrane, or impacted cerumen. Perforated tympanic membranes and impacted cerumen can be diagnosed by otoscopy or by measuring the volume of the ear canal. Type C (significant negative pressure in the middle ear) is possibly caused by anomaly of pathology such as blockage of the eustachian tube.

By measuring the rate of pressure rise, a middle ear analyzer also calculates the ear canal volume. If there is cerumen or other material occluding the ear canal, the volume will measure abnormally low. If there is a perforation of the tympanic membrane, due to the additional space of the middle ear and mastoid air cells, the device will measure an abnormally large canal volume. The functional block diagram of a middle ear analyzer is shown in Figure 38-8. The oscillator provides a 226-Hz pure sinusoidal signal to the amplifier circuit. A probe is inserted into the ear canal and creates an air seal for the ear canal from the ambient atmosphere. The probe includes an earphone connected to the output of the amplifier and a microphone to pick up the sound reflected from the tympanic membrane. Through the probe, a tube connects the ear canal cavity to an external air pump. The pump regulates the pressure inside the ear canal according to the processor commands.

The intensity of the echoes measured by the microphone depends on the energy transmission (or reflection) characteristics of the middle ear. The admittance is derived from the measurement. In one configuration, at each ear canal pressure, the intensity of the pure tone emitting from the earphone is automatically adjusted by the amplifier until a constant preset intensity is picked up by the microphone. The amplification factor is used to calculate the admittance of the middle ear.

Problems and Precautions

Due to the fact that the skin and cartilage in the ear canal of children under 6 months of age are quite lax, it may show falsely increased compliance. In such situations, the child may have a middle ear filled with fluid but still register a compliance within the normal range. In ear canal volume calculation, if there is cerumen or other material occluding the ear canal, the volume will measure abnormally low. An accurate reading cannot be obtained until this material is removed. If there is perforation in the ear drum, the volume will measure abnormally high.

OTOACOUSTIC EMISSION DETECTORS

OAE detectors are used in assessing the inner ear. Although its exact mechanism is still not fully understood, OAE is considered to be related to the amplification function of the cochlea. It is suggested that the outer hair cells of the cochlea are responsible for enhancing hearing sensitivity and frequency selectivity of the cochlea. OAEs are not echoes; they are sounds generated spontaneously within the cochlea by the outer hair cells or in response to acoustic stimulation. A sensitive microphone measures the sound excitation from the cochlea back to the ear canal. In the absence of external stimulation, the amplification of the cochlea increases, resulting in the production of sound (spontaneous OAE or SOAE). The absence of SOAEs may be an indication of a hearing problem related to the inner ear. In addition to SOAEs, evoked OAEs are commonly used in screening and hearing assessments. The two major types of evoked OAE testing in clinical diagnostic hearing evaluations are transient-evoked and distortion product OAEs. OAEs are recorded in the external ear canal. For all OAE testing, an optimal probe fit is critical. The recorded OAE is transformed into its frequency domain for interpretation.

In diagnostic hearing evaluations, OAEs are used to determine cochlear status and hair cell functions for

- Screening for hearing loss in infants and young children, as well as uncooperative and unconscious patients
- Testing for functional hearing loss

- Estimating hearing sensitivity in individual cochlear frequency regions
- Differentiating between the sensory and the neural components of sensorineural hearing loss

Figure 38.9 shows a typical functional block diagram of an OAE setup. Because the signal levels of OAEs are very small and easily corrupted by noise, measurements should be performed under low ambient noise environment. The technique of signal averaging is used in some devices to increase the signal-to-noise ratio (SNR).

Spontaneous Otoacoustic Emissions

In spontaneous otoacoustic emission (SOAE) testing, no stimulus is required. SOAE recordings are usually obtained in steps of narrow bands (< 30 Hz bandwidth) spanning from 500 Hz to 7000 Hz. Multiple recordings are made to ensure repeatability and to distinguish the response from the noise floor. In general, SOAEs are not found in individuals with hearing thresholds worse than 30 dB. SOAEs generally occur in only 40% to 50% of individuals with normal hearing; in neonates, the range is approximately 25% to 80%. Therefore, the presence of SOAEs is usually considered a sign of cochlear health, but the absence is not necessarily a sign of abnormality. When present, SOAEs usually occur in the frequency region from 1000 to 2000 Hz, with amplitudes between –5 and +15 dB SPL. Some individuals may have multi-frequency SOAEs over a broader frequency range. Figure 38-10 shows the amplitude-time recording and its corresponding frequency spectrum of an SOAE.

Transient-Evoked Otoacoustic Emissions

Transient-evoked OAEs (TEOAEs) are evoked using a click or tone burst

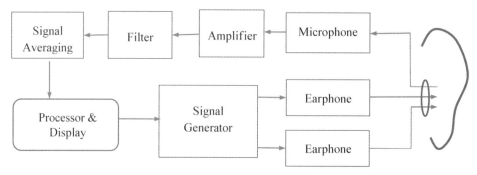

Figure 38-9. Otoacoustic Emission Detector.

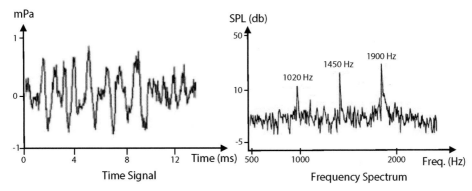

Figure 38-10. Spontaneous Otoacoustic Emission Signal.

stimulus. In clinical TEOAE testing, 80 to 85 dB SPL stimuli at a rate of about 60 stimuli per second are used. The evoked response from a click covers the frequency range from 500 Hz up to around 4 kHz; a tone burst response covers the region that has the same frequency as the pure tone. TEOAEs are recorded only in response to very short or transient stimuli. Signal averaging is employed to extract the response from noise. Recent analysis techniques allow the TEOAE to be separated into various frequency bands for analysis. In general, the presence of a TEOAE in a particular frequency band suggests that cochlear sensitivity in that frequency region is approximately 20 to 40 dB hearing level or better.

Distortion Product Otoacoustic Emissions

Distortion product OAEs (DPOAEs) are evoked using a pair of primary tones f_1 and f_2 ($f_2 > f_1$) either of equal intensity (e.g., L = 65 dB) or slightly different intensities (e.g., L_1 = 65 and L_2 = 55 dB). In a DPOAE examination, the device generates a series of test tones, directs them into the ear canal, and then measures the level of the DPOAE tone generated by the cochlea. By using different test frequencies, the device provides an estimate of outer hair cell function over a wide range of frequencies. The evoked responses from these stimuli occur at frequencies mathematically related to the primary frequencies with the two most prominent ones at $2f_1-f_2$ (also known as the cubic distortion tone) and f_2-f_1 (the quadratic distortion tone). The analysis of the frequency components of the OAE recording are used to estimate cochlear integrity and its transfer functions. Compared to TEOAE, DPOAEs allow greater frequency specificity and can be used to record higher frequencies (> 8 kHz) than TEOAEs. DPOAEs may, therefore, be more useful for early detection of cochlear impairment due to ototoxicity and noise-induced damage. For individuals who have moderate hearing loss and

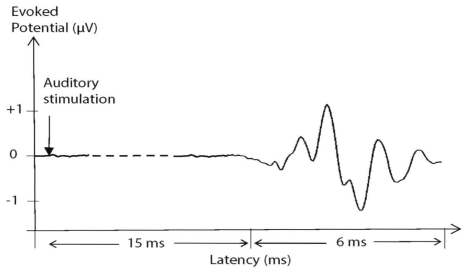

Figure 38-11. Brainstem Auditory Evoked Potential.

whose TEOAEs are absent, DPOAEs can often be recorded. However, the accuracy of DPOAEs in estimating actual hearing sensitivity has not been fully determined.

Problems and Precautions

OAEs are very low-level sounds. Any noise picked up by the earphones during the examination can mask this emission. The most prominent source of noise is usually generated from any patient movement including coughing and talking. The patient must remain calm and refrain from moving or talking. Ambient noise in the testing environment is another major source of noise during the test. Correct fitting of the ear probe is critical. A properly sealed ear probe can block much of the ambient noise but performing the testing in a relatively quiet environment is recommended. Listed below are factors that can affect detection of OAEs even when they are present.

Nonpathological Problems

- Uncooperative patient: recordings need to wait until patient calms down
- Cerumen occluding the canal or blocking a probe port: can be prevented by initial inspection
- Debris (including vernix caseosa in neonates) and foreign objects in the outer ear canal: can be prevented by initial inspection
- Improper probe tip placement or poor seal: most equipment alerts clini-

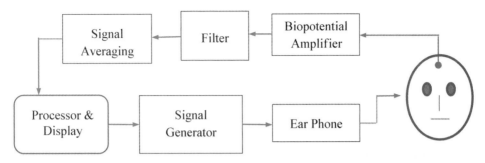

Figure 38-12. Auditory Brainstem Response Detector.

cians to these problems
• Standing waves: most equipment alerts clinicians to standing waves

Pathological Problems

• Outer ear stenosis and cysts
• Middle ear cysts
• Middle ear disarticulation
• Tympanic membrane perforation
• Otosclerosis (abnormal bone growth in middle ear)
• Cholesteatoma (skin growth that occurs in the middle ear behind the eardrum; often due to repeated infection)
• External otitis
• Abnormal middle ear pressure

AUDITORY BRAINSTEM RESPONSE UNITS

ABR is an electrophysiological assessment of the entire auditory system's response to sound. Sound stimulation (a soft click or a short tone burst at about 1 to 5 kHz, 30 to 40 dB) is presented to the ear(s) via earphones or probes. Biopotential electrodes placed on the patient's scalp surface are used to obtain the electrical response from the auditory nervous system and the brain. Within 20 ms after the stimulus is delivered, five to seven identifiable ABR waves appear in the EEG. The activities the EEG in synchronized with the stimuli are referred to as auditory evoked potentials. ABR testing is often considered the "gold standard" in assessing the integrity of the entire auditory system. The EEG, a neurological biopotential device and evoked potential measurements are discussed earlier in Chapters 16 and 17. Figure 38-12 is a functional block diagram of ABR unit.

HEARING AIDS, COCHLEAR IMPLANTS, AND SOUND BOOTHS

Hearing Aids

Hearing aids are prostheses (therapeutic medical devices) to overcome certain deficiencies associated with hearing loss. The most common form of hearing loss is sensorineural related, which involves multiple factors compromising the hearing ability of an individual. Only a relatively small portion of adult hearing problems, such as ear infection and middle ear diseases, are medically or surgically treatable. If the condition cannot be treated, use of hearing aids may be beneficial. Hearing aids are prescribed according to the results from audiometric examination. In certain cases where behavioral thresholds cannot be attained, ABR thresholds can be used for hearing aid fittings.

A hearing aid is basically an audio amplifier that picks up sounds, increases its intensity, and delivers the amplified sounds to the ears of the subject. Modern hearing aids can be programmed to amplify different sound frequency bands with different gains. The gains at different frequency bands are programmed to compensate for the deteriorated regions revealed by the audiogram. The microphone, amplifier, speaker, and power source components are packaged into a small discreet device that can be worn behind the ear or in the ear. More sophisticated hearing aids use digital signal processing to classify the sounds received (e.g., music, speech, noise) and amplify them selectively. They can also determine the environment (e.g., indoor, outdoor, theater room, classroom) from which the sound is being received and apply the optimal amplification patterns to improve performance.

A bone-anchored hearing aid (BAHA) is an option for patients without external ear canals, when conventional hearing aids cannot be used. It is an auditory prosthesis based on bone conduction that can be surgically implanted. The BAHA uses the skull as a pathway for sound to travel to the inner ear. For people with conductive hearing loss, the BAHA bypasses the external auditory canal and middle ear, directly stimulating the functioning cochlea. For people with unilateral hearing loss, the BAHA uses the skull to conduct the sound from the deaf side to the functioning side of the cochlea.

Cochlear Implants

The clinical application of cochlear implant is mainly to improve the hearing of people who have severe sensorineural hearing loss in the cochlea but with healthy auditory nerves. Those with mild or moderate hearing loss are not in the targeted group because they are able to get help through hearing aids. More specifically, a cochlear implant replaces the function of dam-

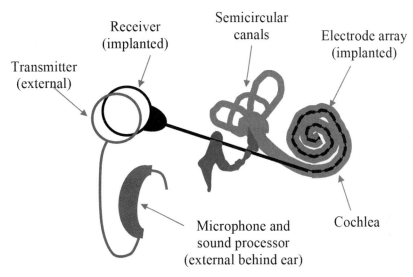

Figure 38-13. Cochlear Implant.

aged hair cells inside the patient's cochlea. Adults losing their hearing due to some diseases (such as meningitis) can benefit from a cochlear implant by regaining the ability of speech comprehension. Another targeted group for cochlear implants is babies who were born deaf. An earlier implant is encouraged in order for them to develop their comprehension and spoken language skills.

A cochlear implant system consists of an implantable module and an external module. The implantable module consists of an array of electrodes, a signal receiver, and a receiving coil (Figure 38-13). ABR is often used to determine the need for a cochlear implant and to assess if the cochlear implant is working post implantation.

The functional block diagram of the system is shown in Figure 38-14. The external module consists of a microphone to pick up the sound signal, a sound processor to separate the sound signal into its frequency components, and a transmitter and transmitting coil to send the processed signal to the implantable module. The transmitting coil is positioned on top of the receiving coil separated by the skin behind the ear. The implantable electrode array is in the shape of a long flexible wire with a number of electrodes (e.g., precurved array with twenty-two platinum electrodes) along the length of the wire. The array is inserted surgically inside the membranous labyrinth of the cochlear. The receiver and receiving coil are surgically implanted into the mastoid bone behind the ear. During operation, the sound signal is converted into its frequency components and grouped into a number of frequency bands according to the number of electrodes in the array. The signal in

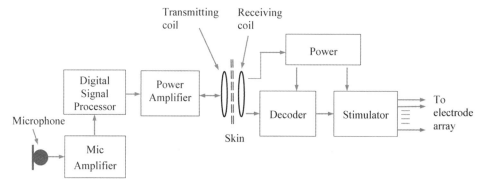

Figure 38-14. Function Block Diagram of Cochlear Implant.

each frequency band is converted by a digital signal processor into electrical impulse. Impulses corresponding to higher frequency bands are directed to electrodes closer to the base of the cochlear while impulses from lower frequency bands are sent to the electrodes closer to the apex. The impulses from the electrodes stimulate the auditory nerve fibers in the cochlea, which carry the signal on to the brain, where it is processed. The energy to power the implantable module is also delivered from the external module across the coils. The stimulation signal amplitude is about 20 µA to 2 mA, with a pulse width of 10 to 500 µs, and at a repetition frequency of up to 30 kHz.

The performance of a cochlear implant depends on many factors, which can be grouped under personal, electrophysiological, and device. Listed below are some of these factors and their implications:

- Personal Factors—patients who have better outcomes from the implants are younger, with post-lingual onset of deafness, suffer shorter duration of deafness, and are able to receive effective and appropriate rehabilitation training.
- Electrophysiological Factors—patients who have more auditory neurons and ganglion cells that have survived will regain better listening ability with the implant.
- Device Factors—The correct electrode placement and the number of electrodes will positively affect the outcomes. Multi-electrode arrays can provide patients with much more speech information by allowing better performance on speech recognition. Too many electrodes may lead to cross talks, however. Modern systems with better speech processing algorithms and using non-simultaneous (or interleaved) electrode activation can reduce cross talks between adjacent channels. Misalignment of the transmitter and receiver coils may degrade signal transfer performance between the external processor and the internal module.

A cochlear implant requires surgical insertion of the electrode array into the cochlea and shaping of the mastoid bone to house the receiver module. The procedure expose the patient to hazards related to general anesthesia and surgical complications such as surgical site infection, meningitis, cerebral-spinal fluid (CSF) leak, and perilymph fluid leak. It is also possible that the patient's residual hearing will be weakened or totally lost after accepting the cochlear implant.

Audiometric Booths

Audiometric booths, also known as sound booths or acoustic chambers, provide a consistent and controlled acoustic environment and keep background noise at an acceptably low level for clinical audiometry or research. The walls of audiometric booths consist of steel panels on the outside, sound-absorbing perforated steel panels on the inside, and an incombustible acoustic insulating material (e.g., fiberglass) in between. Single-wall booths are constructed from a single wall with a layer of sound-absorbing material and are used primarily for pure-tone air-conduction testing. Double-wall booths consist of an examination booth inside a sound booth, with the walls of the inner examination booth separated from the outer panels by an air space. A double-wall booth further reduces outside noise, which is preferred for bone conduction, sound field, and speech testing. In general, for booths with similar dimension and construction, a heavier booth provides better sound proofing.

COMMON PROBLEMS AND HAZARDS

Specific problems and hazards associated with each device have been described earlier. The following lists some common hazards of audiology devices and related procedures:

- Electrical shock hazards similar to any line-powered electrical equipment
- Risk of infection and biocompatibility for implantable devices (e.g., cochlear implant)
- Cross infection when earphones and bone conduction transducers are not properly cleaned and disinfected between uses
- Blockage of pressure tubing and insertion tubes
- Potential rupturing of the eardrum if the probe gets inserted too deep with high force
- Device malfunctions could create a tone intensity high enough to potentially damage the ears

- Device out of calibration will produce over or under measurements that will result in misdiagnosis
- Damage to probes and broken cables due to mishandling and bad connections

BIBLIOGRAPHY

Abdala, C. (1996). Distortion product otoacoustic emission (2f(1) 2f(2)) amplitude as a function of f(2)/f(1) frequency ratio and primary tone level separation in human adults and neonates. *Journal of the Acoustical Society of America, 100*(6), 3726–3740.

American National Standards Institute (ANSI). (Rev. 2013). *American National Standard Maximum Permissible Ambient Noise Levels for Audiometric Test Rooms.* ANSI S3.1-1999. Melville, NY: Acoustical Society of America.

American National Standards Institute (ANSI). (2004). *American National Standard Specification for Audiometers.* ANSI S3.6-2004. Melville, NY: Acoustical Society of America.

American National Standards Institute (ANSI). (1987). *American National Standard Mechanical Coupler for Measurement of Bone Vibrators.* ANSI S3.13-1987. Melville, NY: Acoustical Society of America.

Avan, P., Bonfils, P., Gilain, L., & Mom, T. (2003). Physiopathological significance of distortion-product otoacoustic emissions at 2f1–f2 produced by high- vs. low-level stimuli. *Journal of the Acoustical Society of America, 113*, 430–441.

Beck, D. L., Speidel, D. P., & Petrak, M. (2007). Auditory Steady-State Response (ASSR): A beginner's guide. *The Hearing Review, 14*(12), 34–37.

Bian, L., & Chen, S. (2008). Comparing the optimal signal conditions for recording cubic and quadratic distortion product otoacoustic emissions. *Journal of the Acoustical Society of America, 124*(6), 3739–3750.

Billings, C. J., Tremblay, K., Souza, P. E., & Binns, M. A. (2007). Effects of hearing aid amplification and stimulus intensity on cortical auditory evoked potentials. *Audiology of Neurotology,* 12(4), 234–46.

Bromwich, M. A., Parsa, V., Lanthier, N., Yoo, J., & Parnes, L. S. (2008). Active noise reduction audiometry: A prospective analysis of a new approach to noise management in audiometric testing. *Laryngoscope, 118*(1), 104–109.

Burkard, R. F., & Manuel, D. (2007). *Auditory Evoked Potentials: Basic Principles and Clinical Application.* Hagerstown, MD: Lippincott Williams & Wilkins.

Casselbrant, M. L., & Mandel, E. M. (2010). Acute otitis media and otitis media with effusion. In P. W. Flint, B. H. Haughey, V. J. Lund, J. K. Niparko, M. A. Richardson, K. T. Robbins, & J. R. Thomas (Eds.), *Cummings Otolaryngology: Head f Neck Surgery* (5th ed.; Vol. 3., pp. 2761–2777). Philadelphia, PA: Mosby Elsevier.

Ceponien, R., Cheour, M., & Näätänen, R. (1998). Interstimulus interval and auditory event-related potentials in children: Evidence for multiple generators.

Electroencephalography and Clinical Neurophysiology/Evoked Potentials Section, 108(4), 345–354.

Choi, J. M., Lee, H. B., Park, C. S., Oh, S. H., & Park, K. S. (2007). PC-based teleaudiometry. *Telemedicine Journal and E-Health, 13*(5), 501–508.

Don, M., Kwong, B., Tanaka, C., Brackmann, D., & Nelson, R. (2005). The stacked ABR: A sensitive and specific screening tool for detecting small acoustic tumors. *Audiology & Neurotology, 10*(5), 274–290.

Eggermont, J. J., & Ponton, C. W. (2003). Auditory-evoked potential studies of cortical maturation in normal hearing and implanted children: Correlations with changes in structure and speech perception. *Acta Oto-Laryngologica, 123*(2), 249–252.

Eggermont, J. J., Ponton, C. W., Don, M., Waring, M. D., & Kwong, B. (1997). Maturational delays in cortical evoked potentials in cochlear implant users. *Acta Oto-Laryngologica, 117*(2), 161–163. [AU: please note the article title has been changed per verification with journal.]

Eiserman, W., Hartel, D., Shisler, L., Buhrmann, J., White, K., & Foust, T. (2008). Using otoacoustic emissions to screen for hearing loss in early childhood care settings. *International Journal of Pediatric Otorhinolaryngology, 72*(4), 475–482.

Erdman, S. A., & Demorest, M. E. (1998). Adjustment to hearing impairment I: Description of a heterogeneous clinical population. *Journal of Speech, Language, and Hearing Research, 41*, 107–122.

Everest, F. (2001). *The Master Handbook of Acoustics.* New York, NY: McGraw-Hill. Ho, A. T., Hildreth, A. J., & Lindsey, L. (2009). Computer-assisted audiometry versus manual audiometry. *Otology & Neurotology, 30*(7), 876–883.

House, W. F. (1976). Cochlear implants. *Annals of Otology, Rhinology & Laryngology, 85* (Suppl. 27):1–93.

International Electrotechnical Commission. (2001). International Standard: Electroacoustics—Audiometric equipment. Part 1: Equipment for pure-tone audiometry. IEC 60645-1:2001. Geneva, Switzerland: IEC.

Kemp, D. T. (1978). Stimulated acoustic emissions from within the human auditory system. *The Journal of the Acoustical Society of America, 64*(5), 1386–1391.

Kiessling, J. (1982). Hearing aid selection by brainstem audiometry. *Scandinavian Audiology, 11*(4), 269–275.

Kratz, I. C. (1997). Using equipment in unfamiliar clinical settings: Audiology screening. *Journal of Pediatric Nursing, 12*(5), 307–310.

Kujawa, S. G., Fallon, M., & Bobbin, R. P. (1995). Time-varying alterations in the f2-f1 DPOAE response to continuous primary stimulation I: Response characterization and contribution of the olivocochlear efferents. *Hearing Research, 85*(1-2), 142–154.

Lieberthal, A. S., Carroll, A. E., Chonmaitree, T., Ganiats, T. G., Hoberman, A., Jackson, M. A., . . . , & Tunkel, D. E. (2013). The diagnosis and management of acute otitis media. *Pediatrics, 131*(3), e964–999.

Lilaonitkul, W., & Guinan, J. J. (2009). Reflex control of the human inner ear: A half-octave offset in medial efferent feedback that is consistent with an efferent role in the control of masking. *Journal of Neurophysiology, 101*(3), 1394–1406.

Martin, F. N., & John, J. G. (2011). *Introduction to Audiology* (11th ed.). Boston, MA: Allyn & Bacon.

Montaguti, M., Bergonzoni, C., Zanetti, M. A., & Rinaldi Ceroni, A. (2007). Comparative evaluation of ABR abnormalities in patients with and without neurinoma of VIII cranial nerve. *Acta Otorhinolaryngologica Italica, 27*(2), 68–72.

Musiek, F. E., & Rintelmann, W. F. (1999). *Contemporary Perspectives in Hearing Assessment* (3rd ed.). Boston, MA: Allyn & Bacon.

Picton, T. W., Dimitrijevic, A., Perez-Abalo, M-C., & Van Roon, P. M. (2005). Estimating audiometric thresholds using auditory steady-state responses. *Journal of the American Academy of Audiology, 16*(3), 140–156.

Rahne, T., Ehelebe, T., Rasinski, C., & Gotze, G. (2010). Auditory brainstem and cortical potentials following bone-anchored hearing aid stimulation. *Journal of Neuroscience Methods, 193*(2), 300–306.

Sharma, A., Gilley, P. M., Dorman, M. F., & Baldwin, R. (2007). Deprivation-induced cortical reorganization in children with cochlear implants. *International Journal of Audiology, 46*(9), 494–499.

Swanepoel, D. W., Mngemane, S., Molemong, S., Mkwanazi, H., & Tutshini, S. (2010). Hearing assessment—reliability, accuracy, and efficiency of automated audiometry. *Telemedicine Journal and E-Health, 16*(5), 557–563.

Teas, D. C., Eldredge, D. H., & Davis, H. (1962). Cochlear responses to acoustic transients: An interpretation of whole-nerve action potentials. *Journal of the Acoustical Society of America, 34*(9B), 1438–1489.

APPENDICES

Appendix A-1

A PRIMER ON FOURIER ANALYSIS

Consider a symmetrical square waveform with amplitude of 1 V and a period of 1 sec. (Figure A-1.1)

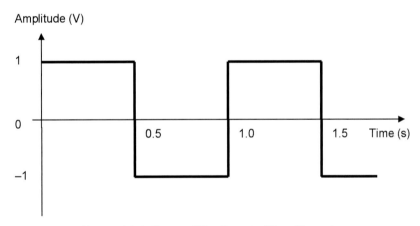

Figure A1-1. Square Waveform in Time Domain.

This signal can be expressed by the following summation of sinusoidal signals, known as the Fourier series:

$$V(t) = 4/\pi \ [\sin(2\pi t) + 0.33 \ \sin(6\pi t) + 0.20 \ \sin(10\pi t) + 0.14 \ \sin(14\pi t) + \ ....] \ V$$

The sinusoidal components in the Fourier series constitute the frequency spectrum of the square wave signal. Such a spectrum can be graphically represented in Figure A-1.2, where the horizontal axis represents the frequency in Hz. The lowest non-zero frequency of the spectrum is called the fundamental frequency f_0, in this case f_0 is equal to 1 Hz (the first sinusoidal term in the equation), and the others, which are in multiples of f_0, are called the harmonics. In the case of this square wave signal, the frequency spectrum contains the fundamental frequency and only the odd harmonics (i.e., third, fifth, seventh, etc.).

Amplitude (V)

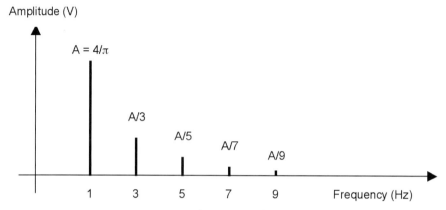

Figure A1-2. Square Wave (Figure A1-1) in Frequency Domain.

Note that the amplitude of the higher harmonics decreases with increasing frequency. In general, harmonics of very low amplitude are insignificant and can be ignored (which means that the signal is considered to have a finite bandwidth).

Figure A-1.3a shows three sinusoidal waveforms of frequencies 1 Hz, 3 Hz, and 5 Hz with amplitudes equal to 1.0 V, 0.33 V, and 0.20 V, respectively. The three waveforms can be represented in mathematical form as:

$$V_1 = 1.0 \sin(2\pi t)$$
$$V_3 = 0.33 \sin(6\pi t)$$
$$V_5 = 0.20 \sin(10\pi t)$$

Figure A-1.3b shows the sum of the waveforms $V_1 + V_2$ and Figure A-1.3c shows $V_1+V_2+V_3$. Note that as more components of the frequency spectrum are added together, the more the waveform will resemble a square wave. As it turns out, if we keep adding the higher harmonics, we will eventually get back a perfect square wave like we have seen in Figure A-1.1. This simple example illustrates that a signal in the time domain can be fully represented by its frequency and phase spectrum in the frequency domain (note that we have not discussed the effect of phase shifts of the harmonics).

We have shown that the frequency spectrum for a periodic square wave in the time domain is composed of discrete frequency components. These frequency components are at multiple frequencies of the fundamental frequency f_0. Other than periodic signals, Fourier transform may be applied to a non-periodic function of time.

We have shown that the frequency spectrum for a periodic square wave in the time domain is composed of discrete frequency components. These frequency components are at multiple frequencies of the fundamental frequency f₀. Other than periodic signals, Fourier transform may be applied to a non-periodic function of time.

A non-periodic signal produces a continuous frequency spectrum (instead of a discrete spectrum). That is, unlike the spectrum of a periodic function of time, the frequency spectrum of a non-periodic waveform spreads over the entire bandwidth,

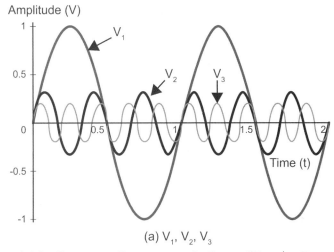

(a) V_1, V_2, V_3

Figure A-1.3a. Frequency Component of a Square Wave (*see* Figure A1-1).

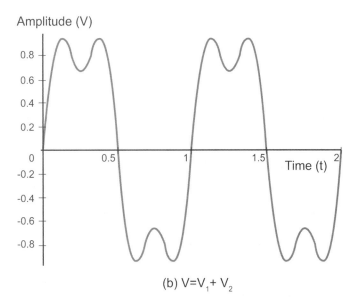

(b) $V=V_1+V_2$

Figure A-1.3b. *Continued.*

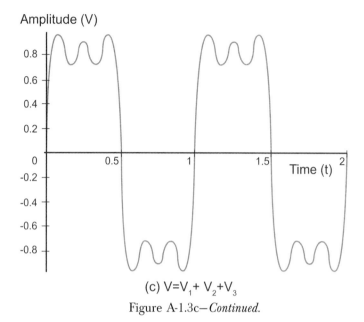

(c) $V = V_1 + V_2 + V_3$

Figure A-1.3c—*Continued.*

as shown in Figure A-1.4.

Figure A-1.5 shows a time domain signal of a blood pressure waveform and its frequency spectrum. (Note that the signal has no frequency components above 10 Hz and it has positive amplitude at zero frequency due to a nonzero average pressure in the time domain). In another words, this pressure-time waveform has a band-

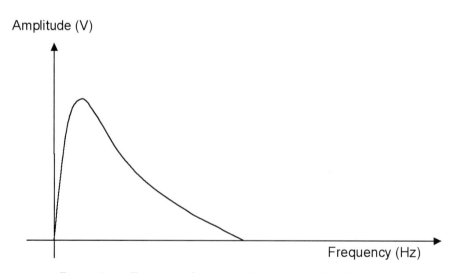

Figure A-1.4. Frequency Spectrum of a Nonperiodic Waveform.

width of 10 Hz (with continuous frequency components from 0 to 10 Hz). All physiological signals have a finite bandwidth.

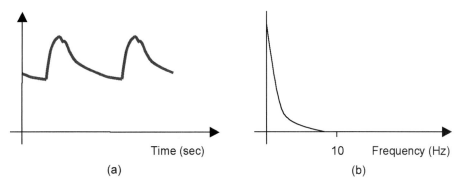

Figure A-1.5. Blood Pressure Waveform in (a) Time and (b) Frequency Domains.

Appendix A-2

OVERVIEW OF MEDICAL
TELEMETRY DEVELOPMENT

INTRODUCTION

Purpose

Medical telemetry is defined by the American Heart Association (AHA)'s Spectrum Selection Workgroup as

> *The wireless transfer of information associated with the measurement, control, and/or recording of physiological parameters and other patient related information between points separated by a distance, usually within the healthcare institution.*

Note that the modulated signal can also be transmitted via hard wires, such as a telephone network. The most common patient vital sign transmitted in patient monitoring systems is the ECG waveform. Telemetry has substantially reduced the risk to patients who may otherwise require besides continuous monitoring.

Typically, a telemetry ECG transmitter would be used on a patient who has been released from ICU and is now in a step-down inpatient ward. In a lot of cases, it is critical for the patient's recovery that they become ambulatory. In the past, without telemetry, this was not possible due to the wired monitoring requirements.

Advantages of Telemetry

- Mobility of subject under study or monitor
- Minimal disturbance of treatment routines
- Centralization of expensive equipment (e.g., automatic arrhythmia monitors)

Examples of Biomedical Applications

- Wireless transmission of physiological signals (e.g., ECG, EEG, temperature, respiration, SPO2, EMG, pH)

- Location detection
- Remote stimulation
- Pacemaker programming and interrogation
- Transmission of clinical information over long distances (e.g., tele-radiography)

PROBLEMS WITH TELEMETRY

Problems with telemetry are primarily related to data rate and reliability. Some factors are:

- Bandwidth requirements
- Channel overcrowding
- EM interference and immunity
- Transmission range
- Power requirement for mobile units
- Industry Canada/FCC licensing
- Primary users (registered users who have the right to use the bandwidth) versus secondary users (users who do not have the exclusive right to use the bandwidth)
- Cybersecurity through wired or wireless networks

DEVELOPMENT AND TREND IN MEDICAL TELEMETRY

Industry Trend

- Started in the VHF band (174–216 MHz) using analog modulation (AM) techniques.
- Migrate to UHF band (460–470 MHz) using digital modulation (e.g., pulse width modulation).
- New wireless medical telemetry system (WMTS) bands: 608 to 614, 1395 to 1400, and 1427 to 1432 MHz.
- The Industrial, Scientific and Medical (ISM) bands, 2.45 GHz and 5.8 GHz (bandwidth ±50 MHz and ±75 MHz respectively), using spread-spectrum technology with limits in bandwidth and transmission power, allow users to share the bandwidth with equal rights.
- Currently, some manufacturers use the ISM bands, others choose to use the WMTS bands.

Technology Development in Medical Telemetry

Medical Telemetry Using General VHF Band (obsolete)

- 174 to 216 MHz

- Non-primary user
- Sharing frequencies with other nonmedical users (e.g., TV Channels 7 to 13)
- Unidirectional data transfer
- Does not support voice or video
- Is now obsolete

Medical Telemetry using General UHF Band (phased out)

- 460 to 470 MHz
- Non-primary user
- Sharing frequencies with other no-medical users
- Unidirectional data transfer
- Does not support voice or video
- Is being phased out

Medical Telemetry using WMTS Band (current practice)

- 608 to 614 MHz, 6 MHz bandwidth, and 1,395 to 1,400 MHz, 1,427 to 1,431.5 MHz (1427 to 1429.5 or 1429 to 1431.5 MHz depending on the jurisdiction)
- Dedicated band for medical telemetry
- Unidirectional or bidirectional data transfer
- Does not support voice or video
- Primary user protected against intentional interference but not from out-of-band interference (e.g., EMI sources from other medical or nonmedical devices such as foot massagers)

Medical Telemetry Using ISM Band (current practice)

- 2.4000 to 2.4835 GHz, 83.5 MHz bandwidth
- Bidirectional data transfer
- Allow voice and video data (e.g., can use VoIP)
- All users must use spread-spectrum technology
- WLAN connects to LAN via an access point (AP)
- Wireless Ethernet: IEEE802.11 (a wireless extension of Ethernet: IEEE802.3)
  - 802.11: 2.4 GHz, 1 and 2 Mb/s using DSSS or FHSS
  - 802.11b: 2.4 GHz, 11 Mb/s using DSSS
  - 802.11a: 5.8 GHz, 54 Mb/s using OFDMSS
  - 802.11g: 2.4 GHz, 54 Mb/s using OFDMSS, OFDMSS
  - 802.11n: 2.4 or 5 GHz, 600 Mb/s using OFDMSS (64 QAM) with multiple antennas to increase data rate.
  - 802.11ac: 5 GHz, 6.7 Gb/s using OFDMSS (256 QAM), detects location of connected devices and increases signal strength specifically in their direction

SPREAD-SPECTRUM TECHNOLOGY CONCEPT

The characteristics of spread-spectrum technology are listed below:

- Use packet (short burst of data) transmission – data is separated into multiple small segments (packets). Each packet is transmitted independently. The receiving end will thread the received packets together back to its original sequence before decoding
- Bandwidth of transmitted signal is spread over a much greater band-width than the original signal
- 79 channels available in North America (79 × 1 MHz wide channels from 2.402 to 2.480 GHz)
- A two-step modulation process - one modulation step to spread the data and the second step to modulate the spread signal (e.g., the second step uses frequency modulation to modulate the signal before transmission)
- Increased processing gain
- Robust—high resistance to noise and interference
- Bidirectional—allow data retransmission in case of interference (receive acknowledge). May reduce throughput but ensure data integrity
- Allow two-way communications
- Low spectral power density: average energy in a specific frequency band is very low; less chance of signals interfering with other systems
- Three methods: direct sequence spread spectrum (DSSS), frequency hopping spread spectrum (FHSS), orthogonal frequency division multiplexing spread spectrum (OFDMSS)
- May still experience overloading if too many users are transmitting at the same time

Appendix A-3

MEDICAL GAS SUPPLY SYSTEMS

Piped-in Gas Supplies

In a typical acute care hospital, medical gases are available through piped-in wall outlets. A typical operating room is equipped with oxygen, nitrous oxide, medical air, nitrogen gas, and suction outlets on the wall or on the ceiling column. At the bedside of a typical patient ward, oxygen, medical air, and suction are available. Cylinder gases are used to supply some less common gases and also used as backup in case the central supply is interrupted. The pressure for piped-in wall gas supply for oxygen, nitrous oxide, medial air, and carbon dioxide is 50 to 55 psig (345 to 380 kPa) in North America, or 4 to 5 bar (400-500 kPa or 58-73 psi) in Europe. For nitrogen, it is greater than 160 psig (1,100 kPa). All piped-in wall gas supplies must not be lower than the minimum pressure values at or below a gas flow of 100 L/min. The absolute pressure of wall suction is usually about 200 mmHg (27 kPa), in gauge pressure it is -560 mmHg (-75 kPa).

Figure A-3.1 shows a typical cryogenic bulk central supply system. During normal operation, gas is drawn from the primary operating supply reservoir. A secondary operating supply reservoir is used when the primary is depleted. The primary reservoir is filled soon after the supply has been switched to the secondary reservoir. To ensure uninterrupted supply, a number of gas cylinders are connected to the central supply line as reserve supply. These cylinders are checked regularly but will not normally be used to supply the system.

Four sizes of gas cylinders are available. Their dimensions (according to NFPA 99) are tabulated in Table A-3.1.

Table A3-2 shows the characteristics of common medical gas cylinders. The pressure at room temperature, capacity (H-cylinder), color coding, and the state of the gas inside the cylinder of the gases are listed.

In order to prevent connecting the wrong medical gas cylinder to the gas line or the inlet of a device, a pin-indexed safety system (PISS) is used. In this safety system, the stem of a gas cylinder can only connect to the cylinder yoke of the same gas. A pin-indexed safety system uses a set of pins on the yoke and a set of holes on the stem to encode the medical gases. The cylinder can connect to the yoke only when the pins are at the same matching location as the holes. For the same gas, the

Table A-3.1. Gas Cylinder Dimensions

Cylinder Size	Height w/Cap (in/mm.)	Outside Diameter (in/mm.)
D	17/432	$4\frac{1}{4}$/108
E	26/660	$4\frac{1}{4}$/108
M	43/1092	7/178
H	51/1295	$9\frac{1}{4}$/235

location of the pins on the yoke aligns with the locations of the holes on the stem.

A diameter-indexed safety system (DISS) is designed to prevent a wrong hose from being connected to the piped-in outlets. In this system, the diameter of the connector on the flexible hose is encoded together with the connector of the wall outlet for the medical gas. Only the hose connector of the gas can be connected to the piped-in wall gas outlet of the same gas. The flexible hoses are color-coded according to the gases to further minimize connection errors.

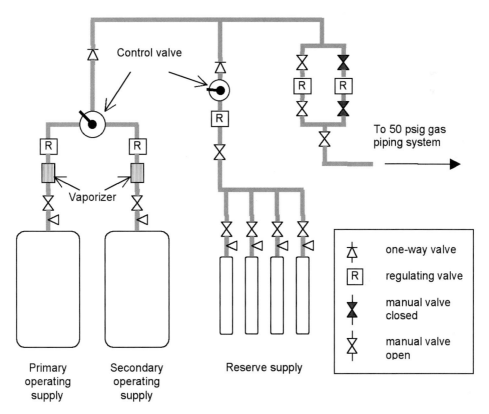

Figure A-3.1. Medical Gas Central Supply System.

Table A-3.2. Cylinder Data of Common Medical Gases

Gas	Pressure (psig/kPa)	Capacity (liters)	State	Color Code
Oxygen	2,217/15,300	7,000	Gas	Green (white-EU)
Nitrous oxide	745/5,200	15,540	Liquid	Blue
Medical air	2,217/15,300	6,500	Gas	Black and white
Helium	2,217/15,300	8,200	Gas	Brown
Carbon dioxide	838/5,780	12,360	Liquid	Grey
Nitrogen	2,217/15,300	6,400	Gas	Black

Oxygen Concentrators

Oxygen is a common gas in healthcare facilities. Patients with respiratory problems such as chronic obstructive pulmonary disease are relying on elevated oxygen level to support ventilation. In smaller facilities such as long-term care or homes of patients where piped-in oxygen supply is not available, oxygen concentrators are cost-effective options to provide oxygen enrich respiration gas for patients in small to medium scale systems. An oxygen concentrator is a standalone electrically powered device using pressure swing absorption (PSA) to extract and purge nitrogen from room air to increase oxygen concentration. The oxygen enriched gas is continuously supplied to the patient.

During operation, room air is drawn through filters to remove particulates and bacteria, it is then pressurized to about 90 psi (620 kPa) and pushed into a cylinder containing a sieve material; nitrogen is absorbed but allowing oxygen and other gases to pass through into an accumulation tank to be supplied to the patient. Nitrogen is released from the sieve material when the pressure is reduced. A reverse flow of gas flushes or purges nitrogen out of the sieve and out of the system, allowing the sieve material to absorb nitrogen in the next cycle. Zeolite is commonly used as the sieve material. To allow continuous production of enriched oxygen gas, two cylinders of sieve material are used in tandem in an oxygen concentrator. In half of the cycle, while one cylinder is absorbing nitrogen, the other cylinder is being purged; in the next half of the cycle, the processes are reversed. A typical single patient use oxygen concentrator produces pressurized oxygen (about 7 psi or 48 kPa) at the outlet with concentration between 82% and 95%. The oxygen flow rate depends on capacity of the unit (e.g., 5 L/min).

Appendix A-4

CONCEPTS OF INFECTION CONTROL IN BIOMEDICAL DEVICE TECHNOLOGY

Infection prevention and control (IPAC) practices are evidence-based procedures and practices that can prevent and reduce disease transmission, and can eliminate sources of potential infections. IPAC guidelines are in place to protect patients, health care personnel, and visitors from the transmission of microorganisms that cause infections. Understand IPAC is important for engineering professionals who work in healthcare facilities as well as those who design and develop medical devices. The chain of infection can be broken by understanding how microorganisms survive, how they can be eliminated and how they travel.

The following procedures will reduce the risk of transmission of infection:

- washing hand frequently
- using Personal Protective Equipment (PPE)
- cleaning and sanitizing
- using proper disposal techniques
- reporting exposures
- modifying the environment via engineering controls (for example, using negative pressure rooms for isolation patients)

A critical element of IPAC is aseptic techniques which include medical asepsis and surgical asepsis. Medical asepsis is concerned with eliminating the spread of microorganisms through facility practices. Surgical asepsis, also called "sterile technique," eliminates microorganisms before they can enter an open wound.

Medical Asepsis

Adequate medical asepsis procedures by healthcare professionals can limit the transmission of microorganisms that can spread disease. This is especially critical for patients with compromised immune systems. A big part of medical asepsis is cleaning and disinfection of the equipment. It is important to understand the effectiveness of disinfectants on microorganisms, their effects on the equipment, and their safety

Table A-4.1. Effectiveness of Disinfectants on Microorganisms

Types of microorganism	Examples of microorganism	Low level disinfectant	Intermediate level disinfectant	High level disinfectant	Chemical Sterilant
Vegetative Bacteria	Salmonella, Coliforms	Effective	Effective	Effective	Effective
Enveloped Viruses	Herpes simplex, Influenza virus	Effective	Effective	Effective	Effective
Fungi	Crytoccocus species, Dermatophytes		Effective	Effective	Effective
Mycobacteria	M. tuberculosis, M. chelonae			Effective	Effective
Non-enveloped Viruses	Norwalk-like virus, Hepatitis A virus			Effective	Effective
Bacteria with Spores	C.botulinum, Protozoa with Cysts				Effective

on human exposure.

Classes of Microorganisms

The microorganisms in Table A-4.1 are ranked in order of resistance to disinfectants with lowest to highest resistance from top to bottom.

Disinfectants and Sterilants

Table A-4.2 listed the properties of some commonly used disinfectants in healthcare. Their intended uses, advantages and disadvantages are included. Care should be taken to follow the instructions of use (concentration, application time, temperature, etc.) to ensure effectiveness and safety.

Other than using disinfectants and chemical sterilants, heat can be used in disinfection or sterilization. The level of microbial inactivation depends on the temperature and time of exposure. For heat sensitive and intermediate risk items such as respiratory therapy accessories, pasteurization (65-77°C, 30 min.) is capable to achieve high level disinfection. Heat tolerant high risk items such as surgical instruments can be sterilize by using steam (e.g., 121–132°C steam with pre-vacuum, 60 min.). Dry heat sterilization, although not as effective as steam sterilization, can achieve the same effect with higher temperature and longer duration (e.g., 160–170 °C, 2-4 hours).

Table A-4.2. Disinfectants Selection Guidelines

Disinfectant	Examples of Intended Use	Advantages	Disadvantages
Alcohol Intermediate level disinfectant	• Disinfect external surfaces of some equipment (e.g., stethoscopes) • Used as a skin antiseptic	• Fast acting • No residue • Non staining	• Volatile • Evaporation may diminish concentration • May harden rubber or cause deterioration of glues
Chlorine Intermediate (diluted) and high (undiluted) level disinfectant	• Disinfect hydrotherapy tanks, dialysis equipment, etc. environmental & surface disinfection • Effective disinfectant following blood spills	• Low cost • Fast acting • Readily available in non-hospital settings	• Corrosive to metals • Inactivated by organic material • Irritant to skin and mucous membranes • Shelf life shortens when diluted
Formaldehyde Chemical sterilant	• Gaseous form used to decontaminate laboratory safety cabinets • Sometimes used to reprocess hemodialyzers	• Active in presence of organic materials	• Carcinogenic • Toxic • Strong irritant • Pungent odour
Glutaraldehydes High level disinfectant (2% formulation)	• For heat sensitive equipment • Most commonly used for endoscopes, respiratory therapy equipment and anaesthesia equipment	• Noncorrosive to metal • Active in presence of organic material • Compatible with lensed instruments • Sterilization may be accomplished in 6-10 hours	• Extremely irritating and toxic to skin and mucous membranes • Shelf life shortens when diluted (effective for 14-30 days depending on formulation) • High cost • Need to monitor concentration in reusable solutions
Hydrogen Peroxide Low (3%) and High (6%) level disinfectant. Chemical sterilant (high concentration)	• Low level disinfection for equipment surface, floors, walls, and furnishings • High level disinfection for flexible endoscopes, soft contact lenses, etc. • Chemical sterilant for heat sensitive medical devices	• Strong oxidant • Fast acting • Breaks down into water and oxygen	• Can be corrosive to aluminum, copper, brass or zinc • Surface active with limited ability to penetrate
Iodophors Intermediate level disinfectant	• Intermediate level disinfectant for some equipment, e.g., hydrotherapy tanks, thermometers. • Low level disinfectant for hard surfaces and equipment that does not touch mucous membranes (e.g., IV poles, wheelchairs, beds, call bells)	• Rapid action • Relatively free of toxicity and irritancy	• Corrosive to metal unless combined with inhibitors • May burn tissue • Inactivated by organic materials • May stain fabrics and synthetic materials
Peracetic Acid High level disinfectant and chemical sterilant (high concentration)	• High level disinfectant or sterilant for heat sensitive equipment • Chemical sterilants in specially designed equipment for decontamination of heat sensitive medical devices	• Innocuous decomposition (water, oxygen, acetic acid, hydrogen peroxide) • Rapid action at low temperature • Active in presence of organic materials	• Can be corrosive • Unstable when diluted
Phenolics Low and Intermediate level disinfectant	• Clean floors, walls and furnishings • Clean hard surfaces and equipment that does not touch mucous membranes (e.g., IV poles, wheelchairs,beds, call bells)	• Leaves residual film on environmental surfaces (prolong disinfection) • Commercially available with added detergents to provide one-step cleaning and disinfecting	• Do not use in nurseries • Not recommended for use on food contact surfaces • May be absorbed through skin or by rubber • Some synthetic flooring may become sticky with repetitive use
Quaternary ammonium compounds Low level disinfectant	• Clean floors, walls and furnishings • Clean blood spills	• Generally non-irritating to hands • Usually have detergent properties	• DO NOT use to disinfect instruments • Non-corrosive • Limited use as disinfectant because of narrow microbiocidal spectrum

An autoclave is a moist heat sterilizer for surgical instruments, laboratory articles and pharmaceutical items. It uses saturated steam under pressure to kill microorganisms including heat-resistant endospores. Autoclaves can be used to sterilize liquid, solid or hollow objects. A typical unit creates a chamber temperature of 121°C at a pressure of 15 psi (103 kPa). Depending on the load, it takes 15 to 20 minutes to achieve sterilization. For effective sterilization, air should be evacuated before the chamber is filled with steam and the articles should be placed so that steam can easily penetrate them.

Ethylene oxide (EtO) sterilization is widely used in hospital settings to sterilize heat- and moisture- sensitive devices that would be damaged by pure steam or liquid chemical sterilization. It is a low-temperature process (typically between 37 and 63°C) that uses EtO gas to eliminate infectious agents. For some medical devices (e.g., with heat sensitive materials), EtO sterilization is the only effective sterilization method that does not damage the device during the sterilization process. Medical devices made from certain materials such as plastic, resin, metals, glass, or devices that have difficulty to access locations (for example, catheters) are candidates to be EtO sterilized. EtO is used in gas form and is usually mixed with other substances, such as CO_2 and steam. It is a colorless gas that is toxic, flammable and explosive.

Factors that affect effectiveness of EtO sterilization are: EtO gas concentration, temperature, relative humidity and exposure time. Typical values used are (450-1200 mg/L, 37-63°C, 40-80% RH, 1-6 hrs.). A typical EtO sterilization cycle takes approximately 2.5 hours excluding aeration time, and consists of preconditioning and humidification phase, gas introduction phase, exposure phase, evacuation phase, and air washing phase. Mechanical aeration for 8 to 12 hours at 50 to 60°C is needed to remove the toxic EtO residual absorbed in materials. Acute exposure to EtO may cause eye pain, sore throat, difficulty breathing, blurred vision, nausea, headache, convulsions, blisters and vomiting. In vitro studies have shown EtO to be carcinogenic. Occupational long term exposure to EtO has also been linked to spontaneous abortion, genetic damage, nerve damage, peripheral paralysis, muscle weakness, and impaired memory. EtO gas level monitors are mandated by Occupational Health and Safety in locations where EtO sterilization is used.

Surgical Asepsis

Surgical asepsis is the absence of microorganisms and their spores in all types of invasive procedures. Sterile technique is practiced to ensure equipment and operating environment are free from microorganisms and their spores (in another word, to maintain sterility). Operating environment includes operating theatres, labour and delivery rooms, diagnostic and treatment areas, or patient bedside where invasive procedures are performed. While intact skin serves as a natural barrier against microorganism invasion, a break in the skin provides a point of entry. Invasive procedures include surgical incisions, dressing of wounds, inserting devices into the body or cavities (e.g., insertion of chest tube, central venous line, or indwelling urinary catheter) where the integrity of the skin is accessed, impaired, or broken.

General Safety Considerations

A patient may contact microorganisms from people, environment, or equipment. The environment may spread pathogens through movement, touch, or proximity. Interventions such as restricting traffic, isolating patient, controlling air flow, filtering air to remove airborne bacteria and contaminants, or using low-particle generating clothing can minimize environmental infection hazards.

Healthcare professionals can spread pathogens inadvertently. In addition to strictly following IPAC practices, below listed a few general precautions before and during surgical procedures.

- Hand hygiene is a priority, proper hand washing is required before any aseptic procedure.
- Review hospital procedures and requirements for sterile technique prior to initiating any invasive procedure.
- Choose appropriate PPE to decrease the transmission of microorganisms from patients to health care worker.
- When performing a procedure, ensure the patient, if conscious, understands how to prevent contamination of equipment and refrain from movements.
- Health care providers who are ill should avoid taking part in invasive procedures.

Principles of Surgical Asepsis

Sterile technique is the method of maintaining an area that is free of microorganisms with the goal to protect the patient from infection. For surgical patients, infection can be acquired even before their hospitalization, such infections are called community-based infection. A patient can acquire infection in preoperative, during the surgery, or post-operative while in the hospital, such infections are called nosocomial, or hospital-acquired infections. A patient under surgery is especially vulnerable to infection due to open incisions. An infection is a health hazard to any surgical patient as it may create complications and significantly affect their recovery. The following sections focus on aseptic techniques in surgery.

A typical surgical team consists of the operating surgeon, assistant to the surgeon, an anesthetist, a scrub nurse, and a circulating nurse. The surgeon, surgical assistant and scrub nurse are the only people allowed to be inside the sterile field. The scrub nurse sets up the operating room for the patient, ensure all the tools are ready and sterile, passes tools to the surgeons during the surgery, performs tasks such as applying suction to the surgical site, or cutting suturing material. The anesthetist is a non-sterile member responsible for the patient's anesthesia during the surgery. The circulating nurse is responsible for managing all nursing care within the operating room, observing the surgical team from a broad perspective, and assisting the team to create and maintain a safe, comfortable environment for the patient. Sterile packages may be retrieved and handed over to the scrub nurse using sterile techniques during the surgical procedure.

Table A-4.3. Principles of Sterile Technique

1. All objects used in a sterile field must be sterile.
2. A sterile object becomes non-sterile when touched by a non-sterile object.
3. Sterile items that are below the waist level, or items held below waist level, are considered to be non-sterile.
4. Sterile fields must always be kept in sight to be considered sterile.
5. When opening sterile equipment and adding supplies to a sterile field, take care to avoid contamination.
6. Any puncture, moisture, or tear that passes through a sterile barrier must be considered contaminated.
7. Once a sterile field is set up, the border of one inch at the edge of the sterile drape is considered non-sterile.
8. If there is any doubt about the sterility of an object, it is considered non-sterile.
9. Sterile persons or sterile objects may only contact sterile areas; non-sterile persons or items contact only non-sterile areas.
10. Movement around and in the sterile field must not compromise or contaminate the sterile field.

Sterile techniques may include the use of sterile equipment, instruments, gowns, and gloves. Proper sterile asepsis minimizes patient exposure to infection-causing agents and thus reduces patient risk of infection. In general, before and during an aseptic procedure, a sterile field must be established, only sterile equipment and supplies can be brought into the sterile field, and all personnel involved in an aseptic procedure are required to follow the principles and practice of sterile technique. These principles (Table A-4.3) must be strictly followed whenever aseptic procedures are performed.

Although there may be variations, healthcare facilities all have approved and documented sterile procedures to ensure strict adhesion to these principles.

Infection Control of Medical Equipment

Medical equipment, accessories and associated supplies can be media for transmission of microorganisms. The previous sections focused on protecting patients from infection, contaminated equipment and accessories also pose risk to users and

Table A-4.4. Medical Device Infection Risk Categories

Risk Category	Description	Example	Decontamination
Low Risk (non-critical)	items that come into contact with intact skin	stethoscope	cleansed with detergent and dried
Intermediate Risk (semi-critical)	Items that do not penetrate the skin or enter sterile areas of the body, but are in close contact with mucous membranes or non-intact skin.	endoscopic equipment, endotracheal tubes and thermometers	require high level disinfection or sterilization
High Risk (critical)	Items that enter the body cavities and vascular system	surgical instruments, intrauterine devices, vascular catheters and implants	require sterilization
Single Use	may be used on the skin, mucous membranes or inside the body cavity	gloves, tongue depressors, and needles	disposed of after use or handling

support staff. While healthcare professionals in general understand the importance of following ICAP procedures when treating patients, many may not be so careful in handling patient care equipment. With particular concern to biomedical engineering personnel, these equipment and accessories may be contaminated before being sent to the department for repair or inspection.

Decontamination is a process or treatment that renders a medical equipment, instrument, or environmental surface safe to handle. Disinfection and sterilization are forms of decontamination. Medical equipment and instruments all require decontamination after use and before being handled or reuse. Depending on the type of medical devices and the kind and level of suspected microorganism, the items are either sterilized or disinfected.

The risk of infection from equipment, instruments and accessories is generally classified into three categories. They are low risk or non-critical, intermediate risk or semi-critical, and high risk or critical. Single Use Items may be considered as the forth category. These items are typically constructed of plastic and may be used on the skin, mucous membranes or inside the body cavity. These are described in Table A-4.4.

Medical equipment can be sterilized by several methods that may include chemical treatment, heat, or radiation. Figure A-4.1 illustrates the process of equipment decontamination.

Cleaning is the first step in equipment (or instrument) decontamination. It is important to remove any debris (tissue fragments, blood stain, dried bodily fluid, etc.) from the equipment surface before moving to the next step of decontamination.

Figure A-4.1. Process of Equipment Decontamination

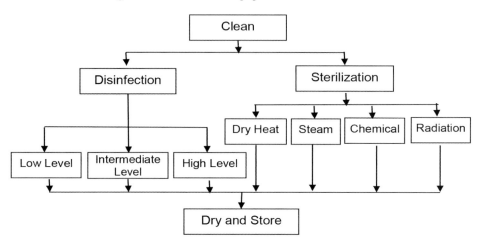

Dried protein forms a barrier which may prevent chemical or heat to reach the microorganism during decontamination. After disinfection or sterilization, the equipment or instrument will be dried, tagged and packed inside a protective package.

To reduce the risk of infection from medical equipment, the following basic infection precautions should be observed by biomedical engineering personnel.

- Follow hospital ICAP procedures when entering and leaving patient care areas, and retrieving and returning medical equipment
- Inspect devices for signs of contamination (blood stain, dried fluid, etc.); if present, send to Central Supply Department for proper decontamination
- Handle all devices with caution, especially those that likely have contact with blood or bodily fluid (such as dialysis equipment, breathing circuits, surgical instruments)
- Practice proper hand washing
- Do not eat or drink while working
- Avoid touching yourself (rub eyes, nose, etc.) while working
- Wear appropriate PPE (gloves, lab coat, mask, eye protector, etc.) to prevent inhalation and splash
- Use vacuum rather than dusting
- Dispose of all cleaning and disinfection materials and disposables
- Report any percutaneous or mucous membrane exposure

When choosing cleaning and disinfection agents on medical equipment, users must also pay attention to their adverse effects in damaging the equipment (e.g., alcohol will harden rubber). In addition, the department should have a designated area to hold incoming equipment and to perform decontamination. Segregating incoming and decontaminated equipment is critical in medical equipment infection control.

REVIEW QUESTIONS

Chapter 1

1. What is the purpose of the signal isolation functional block in an electro-medical device?

2. Which of the following is NOT part of a usability evaluation for an intensive care ventilator?

   a. Calculate the average time it takes for a user to re-program the device for a new patient.
   b. Tabulate user errors made during the re-assembly of the unit after cleaning and sterilization.
   c. Investigate the impact of user error on the recovery of post-operative patients.
   d. Evaluate users' responsiveness to patient alarms.

3. Which of the following is not a diagnostic medical device?

   a. Infusion Pump
   b. Vital Signs monitor
   c. Patient Monitor
   d. Electroencephalogram

4. Which of the following is not a therapeutic medical device?

   a. Electroconvulsive therapy device
   b. Electrosurgical unit
   c. Electrocardiograph
   d. Muscle stimulator

5. The path of a physiological signal, from patient to ECG printout, travels through which of the following functional sequences?

    a. Patient > Electrode > Signal analyzer > Amplifier > Recorder > ECG printout
    b. Patient > Recorder > Amplifier > Signal Analyzer > Electrode > ECG Printout
    c. Patient > Electrode > Amplifier > Signal Analyzer > Recorder > ECG Printout
    d. Patient > Electrode > Recorder > Amplifier > Signal Analyzer > ECG Printout

6. The objective of in vivo functional tests for biocompatibility is:

7. While designing medical devices, human factors are taken into consideration alongside its safety and efficacy levels. What is the main focus when evaluating the usability of the medical device?

    a. Evaluation of the users
    b. Ergonomics of maintenance tasks
    c. How the device fits into their workflow
    d. Potential problems caused by the operating environment

8. Medical devices can be classified according to risk classes. Which of the following is organizing the listed devices in order of risk from Class I (lowest risk) to Class 4 (highest risk)?

    a. Band-aids, indwelling catheters, heart valve implants, defibrillators
    b. Band-aids, Latex Gloves, IV bags, heart valve implants
    c. Contact Lenses, Conductive Electrode Gel, defibrillators, IV bags
    d. Contact Lenses, Conductive Electrode Gel, IV bags, defibrillators

9. Which of the following is a potential problem to be considered which is unique in medical device design and development?

    a. small signal amplitude
    b. signal variation with time
    c. size of transducer
    d. biocompatibility

10. What is refractory period in single cell action potential?

11. When a person failed to carry out a correct intention, the human error is considered as a

a. slip
b. lapse
c. mistake
d. crime

12. What is risk index of a hazardous situation in risk analysis?

13. In performing human factors analysis of an acute care patient monitor in the device acquisition, which of the following is usually not a participating group?

a. acute care nurses
b. central supply (cleaning) staff
c. purchasing staff
d. biomedical engineering staff

Chapter 2

1. Which of the following is not part of the specifications of a medical device?

a. cost
b. dimension
c. function
d. regulatory certification

2. A student designed and built a regulated 9 V dc power source. The output is measured with a digital voltmeter. The meter reads 9.08 V when the range selection is at 0 to 25 V. If the dc accuracy of the voltmeter is stated as ±3% of full scale. Which of the statement can be concluded:

a. the regulator is likely working according to the design
b. the regulator output is too high
c. the regulator output is too low
d. the regulator output cannot be verified

3. In an experiment, the output Z is the difference between two measured quantities given by $Z = 0.5A - 2B$, what is the uncertainty of Z is the uncertainties of A and B are 0.3 and 0.8 respectively?

4. Find $\dfrac{\Delta Z}{Z}$ when $Z = xy^{-1}\, w^2\, (1 + sin\theta)^{1/2}$ in terms of $\Delta x, \Delta y, \Delta w$ and $\Delta \theta$

Questions 5 to 8. According to the specifications, the transfer function of a measuring device (shown below) is a straight line and the input range is from 0 to +B. In questions 5, 6, & 7, express the answers in terms of the absolute quantities A, B, C & D

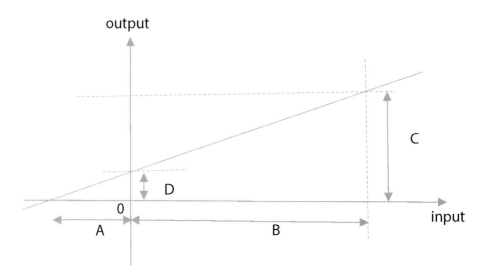

5. Zero offset

6. Sensitivity

7. Output span (the change of output with respect to the input range)

8. What is the output when the input is − (B + 2A)?

    a. 0
    b. C
    c. C − D
    d. cannot be determined

9. A wide bandwidth biopotential amplifier with an amplification factor of 1,000 is connected to a ±15 V dc power supply. The peak of the biopotential signal at the output is clipped when the amplitude of the input is larger than 14 mV. This distortion is due to

    a. too low amplification factor
    b. too low amplifier bandwidth
    c. saturation of the amplifier
    d. non-zero dc component of the signal

10. The non-linear characteristic of the transformer magnetization curve is called

    a. saturation
    b. hysteresis
    c. bang-bang
    d. break down

11. A notch filter (e.g., to remove 60 Hz power frequency noise in electrical circuit) is a

    a. high pass filter
    b. low pass filter
    c. band pass filter
    d. band reject filter

12. Which of the following is/are essential elements in process calibration?

    a. timely performed by trained professionals
    b. traceable
    c. fully documented
    d. all of the above

Chapter 3

1. Which of the following is not a transducer?

    a. a cell culture petri disk
    b. a biopotential electrode
    c. an electric motor
    d. a solar cell

2. Which of the following is an active transducer?

    a. a cell culture petri disk
    b. a biopotential electrode
    c. an electric motor
    d. a solar cell

3. Which of the following is a passive transducer?

    a. a piezoelectric ultrasound detector
    b. a resistive strain gauge
    c. a thermocouple
    d. a solar cell

4. Which of the following is an indirect transduction process?

    a. measure blood pressure by an invasive pressure transducer
    b. measure light intensity by a photo-voltaic cell
    c. measure blood glucose level using an oxygen electrode
    d. measure air flow velocity using a hot wire flow meter

5. Which of the following output (y) versus input (x) relationship is not a linear transfer characteristics?

    a. $y = 0.2x + 5$
    b. $y = x$
    c. $y = - x - 5$
    d. $y = x^2$

6. Which of the following transducer characteristic is not required for faithful reproduction of the input signal?

    a. output amplitude is proportional to the input
    b. adequate frequency response
    c. free from phase distortion
    d. amplitude linearity

7. Which is the following is the effect of a low pass filter (fc < 40 Hz) on an ECG waveform?

    a. remove dc offset
    b. reduce amplitude or the R-wave
    c. impose delay on the R-wave
    d. slow down the ECG repetitive frequency

8. Derive the full bridge equation in Fig 3.8

9. A half bridge circuit with 2 resistive strain gauges and excitation voltage of 10 V is used in a baby scale. If the nominal resistance of the strain gauges is 2.0 kΩ and a load of 1.0 kg will produce a change of 0.5 Ω, what is the sensitivity (in V/kg) of bridge inside the scale?

10. What is the bridge output voltage of the above when a baby of 6 kg is placed on the scale?

Chapter 4

1. Which of the following is not a unit of pressure?

    a. psi
    b. mmHg
    c. Pa
    d. mV

2. What is 1 standard atmospheric pressure (760 mmHg) in mH_2O?

3. What is 1 standard atmospheric pressure (760 mmHg) in Psi?

4. For the water tube manometer in Fig 4-2, what is P_b if P_a = 1 atm and h = 50 cm?

5. A patient's systolic and diastolic blood pressure is 120 mmHg and 80 mmHg respectively, what are the corresponding pressures in kilo Pascal?

6. Which of the following mechanic pressure gauges is more suitable to sense change in atmospheric pressure?

    a. Bellow
    b. Diaphragm
    c. Bourdon tube
    d. Load cell

7. Which of the following strain gauge has the highest gauge factor?

    a. copper wire
    b. platinum wire
    c. quartz crystal
    d. an alloy of aluminum and copper

8. A metal wire resistive strain gauge has a nominal resistance R_0 of 100.0 Ω, with an axial stain of 6000 $\mu\varepsilon$, the change in resistance is 1.50 Ω. What is the gauge factor of the strain gauge?

9. The above strain gauge is used in force measurement. If the 1.0 mm strain wire with nominal resistance of 100 Ω extends by 5.0 μm when a force of 2.00 N is applied? What is applied force when the strain wire resistance is 102.5 Ω?

Chapter 5

1. Which of the following is not an IPTS temperature reference?

    a. Triple point of water
    b. Boiling point of water
    c. Melting point of ice
    d. Boiling point of oxygen

2. The body temperature of a patient under hypothermia is 94 °F, what is the corresponding temperature in degree Celsius?

3. The body temperature of a patient under hypothermia is 94 °F, what is the corresponding temperature in kelvin?

4. Which of the following sensing element is using the property of differences in temperature expansion coefficient to measure temperature?

    a. Bimetallic sensor
    b. Red-dye encapsulated in a thin tube
    c. Thermocouple
    d. Thermistor

5. A RTD is used to measure the temperature inside an infant incubator. Given R_0 = 100 Ω at 0.0°C and α = 0.0038 Ω/Ω/°C, Determine the incubator temperature when the resistance of the RTD is 119 Ω.

6. For the RTD in the above question, calculate the **change** in temperature when the resistance has increased by 5.0 Ω from 119 Ω.

7. The temperature of a 37 °C water bath is monitored using a standard platinum RTD, a pair of long lead wires are used to connect the RTD to an ohmmeter. If α = 0.00385/°C and each lead wire has an overall resistance of 0.5 Ω, find the lead error.

8. If the water bath in the above question is monitored instead by a thermistor with β = 4000 K and R_0 = 8.00 kΩ, what is the resistance of the thermistor?

9. What is the lead error in Question 7 when the a thermistor in Question 8 is used instead of a RTD?

10. Which of the following parameter(s) is not required when using a thermocouple to measure temperature at the "hot" junction?

a. Cold junction temperature
b. The type of metals forming the thermocouple
c. The bonding resistance at the hot junction
d. The voltage measured at the thermocouple terminals

11. Which is the most appropriate transducer to measure the temperature inside a medical sterilizer to an accuracy of ±5°C with a temperature range from room temperature to +350°C. The only instrument available is an analog multi-meter (volt/amp/Ohm).

a. J-type thermocouple with the thermocouple lookup table
b. Platinum RTD with the RTD lookup table
c. Negative temperature coefficient thermistor with known thermistor material constant β
d. IC temperature sensor

12. Which of the following type of temperature sensors in general has the best linearity characteristics?

a. Thermistors
b. RTDs
c. Thermocouples
d. IC temperature sensors

Chapter 6

1. A one-turn linear scale potentiometer is used as a rotational transducer. If the resistance measured is 100 kΩ when it is rotated by 270°, what is the rotational angle when the output resistance is 33 kΩ?

2. Which is an undesirable characteristics of a linear variable differential transformer in displacement sensing?

a. it is using inductance as sensing elements
b. it has a non-linear transfer function
c. it cannot differentiate the direction of displacement from its zero position
d. the object to be measured is connected to the moving magnetic core

3. Which of the following will not affect the capacitance of a parallel plate capacitor?

a. distance between the parallel plates
b. overlapping area of the parallel plates
c. the permeability of the material between the parallel plates
d. the permittivity of the material between the parallel plates

4. The Hall Effect voltage produced by a current flowing through the Hall sensor and a magnetic field perpendicular to the current is?

    a. proportional to the magnetic field
    b. proportional to the thickness of the Hall sensor element
    c. proportional to the product of the magnetic field and thickness of the Hall sensor element
    d. proportional to the product of the current and thickness of the Hall sensor element

5. Choose the transducer combination to implement an accelerometer.

    a. a linear variable differential transducer and a piezoelectric transducer
    b. a linear variable differential transducer and a linear resistive potentiometer transducer
    c. a Hall Effect transcuder and a linear resistive potentiometer transducer
    d. a thermistor and a piezoelectric transducer

6. A setup is used to measure displacement d versus time t. State the mathematical equations to compute velocity and acceleration.

7. The diagram below is the transfer function of an object captured by a motion transducer. What is the velocity of the object?

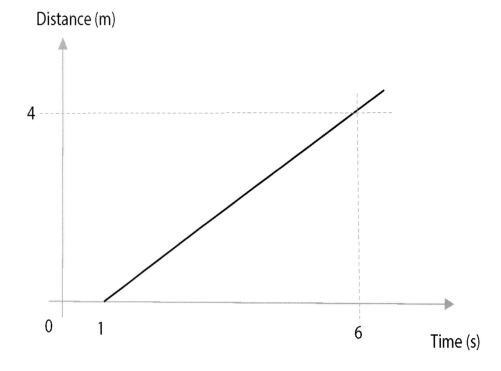

8. What is the acceleration of the object in Question 7?

9. The diagram below is the transfer function of an object captured by a motion transducer.

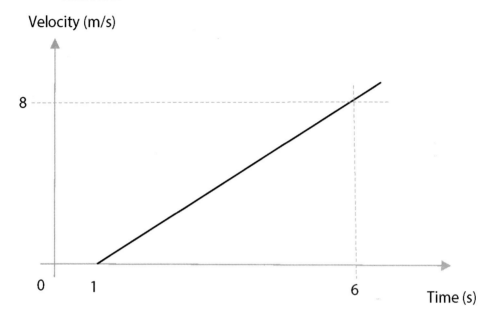

Velocity (m/s)

What is the acceleration the object?

10. If the force acting on the object is 1.2 N, what is the mass of the object?

Chapter 7

1. What is the average blood flow velocity in the common aorta if its cross sectional area is 2.5 cm/s and the cardiac output is 4.5 L/min?

2. What is the average blood velocity when the diameter of the aorta is reduced by 20 percent in Question 1 (assuming the aorta has a circular cross section and all other parameters remains the same)?

3. Which of the following statement regarding laminar and turbulent fluid flow is correct?

   a. in general, velocity of laminar flow is higher than turbulent flow.
   b. turbulent flow is characterized by flow with smooth flow layers
   c. laminar flow is characterized by flow with eddies
   d. fluid with higher viscosity has higher tendency to produce laminar flow than turbulent flow

4. A blood vessel has a short section with serious stenosis (narrowing). Which of
    the following is the most unlikely at the stenotic section when compared to
    the other part of the blood vessel?

    a. blood flow velocity is higher
    b. volume blood flow is the same
    c. blood flow is turbulent
    d. the Reynolds number of blood is less than 2000

5. Which of the following fluid has the highest viscosity?

    a. whole blood
    b. blood plasma
    c. water
    d. ethyl alcohol

6. An orifice plate flow meter measures the volume flow by measuring the:

    a. cross sectional area of the orifice
    b. the flow velocity at the orifice
    c.   the fluid pressure readings upstream and downstream of the orifice
    d. the fluid pressure and temperature upstream of the orifice

7. Which is a critical characteristic specific to a flow transducer used in monitor-
    ing respiration gas flow?

    a. low flow resistance
    b. high accuracy
    c. high upper cutoff frequency
    d. linear transduction characteristic

8. Calculate the volume flow rate of water measured by a Venturi tube (Figure 7.5)
    if D_1, P_1, D_2, and P_2 are 2.0 cm, 500 Pa, 1.0 cm, and 300 Pa respectively.

9. A 10 cm long artery with inner diameter of 5.0 mm is detected to have an aver-
    age blood flow velocity of 150 cm/s. What is the drop of blood pressure in
    this length of artery given the viscosity of blood at body temperature is
    0.0027 N.s/cm^2.

10. Which of the following flow sensor is using a piezoelectric transducer element?

    a. rotameter flowmeter
    b. Pitot tube flowmeter
    c. electromagnetic flowmeter
    d. ultrasound vortex flowmeter

11. Which of the following flowmeter is based on the Bernoulli's equation?

    a. Turbine flowmeter
    b. pitot tube flowmeter
    c. electromagnetic flowmeter
    d. ultrasound vortex flowmeter

12. Which of the following fluid property or flow condition will invalidate the Bernoulli's equation?

    a. at room temperature
    b. under laminar flow
    c. with negligible viscosity
    d. incompressible

Chapter 8

1. What is not an electromagnetic wave?

    a. ultrasound
    b. x-ray
    c. sun light
    d. microwave

2. What is frequency of x-ray with a wavelength of 1.0 nm?

3. What is the photon energy of 1 nm x-ray?

4. Differentiate photometry from radiometry

5. A photo detector with electrical output proportional to the rate of absorption of light photon is

    a. a thermal detector
    b. a radiation detector
    c. a photon detector
    d. a light detector

6. Define black body

7. The ratio of the output power a radiator at wavelength λ to that of a blackbody radiator at the same temperature and wavelength known as

    a. whitebody

b. graybody
c. non-ideal blackbody
d. spectral emissivity

8. A tungsten filament lamp is known to behave almost like a black body. If the tungsten filament is heated up to 3000 K, what is the approximate color temperature of the light emitting from the tungsten filament?

a. much less than 3000 K
b. about 3000 K
c. much higher than 3000 k
d. 6500 k

9. The photo-sensing elements of a thermopile are

a. solar cells
b. pyroelectric sensors
c. thermocouples
d. photodiodes

10. A photodiode in photovoltaic mode is also called a

a. solar cell
b. pyroelectric sensor
c. thermopile
d. photoemissive sensor

11. The emitter current of a phototransistor is approximately proportional to the illumination intensity at

a. the base
b. the emitter
c. the collector
d. the P-N junction

12. Noisy picture produced by CCD due to dark current can be reduced by

a. reading the captured signal slower
b. reading the captured signal more than once
c. cool down the CCD sensor
d. maintain CCD sensor at a constant temperature

13. What type of optical fiber is needed to transmit the image at the distal end of an endoscopy?

a. coherent optical fiber bundle

b. a non-coherent optical fiber bundle
c. a multi-mode optical fiber
d. a single-mode optical fiber

14. Which of the following fiber and light source supports the highest transmission bandwidth?

a. a step-index fiber with a monochromatic laser light source
b. a graded- index fiber with a LED light source
c. a single-mode fiber with a monochromatic laser light source
d. a single-mode fiber with a LED light source

Chapter 9

1. Dissociation of Zn metal into Zn^+ ion and releasing electrons in a solution is

a. an oxidation reaction
b. a reduction reaction
c. an oxidation reaction first and then a reduction reaction
d. a reduction reaction first and then an oxidation reaction

2. The magnitude of the half-cell potential of Ag/AgCl skin electrode is about +0.22 V. Which of the following will not affect this magnitude?

a. the temperature at the electrode site
b. the concentration of chloride ion in the electrolyte at the electrode site
c. the proportion of solid silver to AgCl of the electrode
d. the absence (or dried out) of electrolyte interfacing the electrode and the skin

3. What is a standard cell potential of the cell Zn(s) | Zn2+ || Cu2+ | Cu(s)

4. Which is an important characteristic of a reference electrode used in determining the concentration of an analyte?

a. it has a known constant half-cell potential independent of the analyte
b. it creates a known oxidation reaction with the analyte
c. it provides a known reduction reaction with the analyte
d. it must not react with the analyte

5. Name a common reference electrode and its pratical half-cell potiential.

6. The half-cell potential E of a Ag electrode in a saturated solution of AgCl is +0.74 V. What is the concentration of Cl^- ion in the solution if the solubility product Ksp of AgCl at 25 °C in aqueous solution is 1.6×10^{-10}?

7. An ion selective electrode selects the ion to be analyzed by a special

    a. voltage
    b. concentration
    c. temperature
    d. membrane

8. In a pK electrode to measure concentration of K^+ ion in a solution, valin-
    imycin is used to

    a. react with K^+ ion in the solution to produce a voltage related to the concen-
        tration of K^+ in the solution
    b. bind with K^+ ion and move across the synthetic membrane
    c. push the K^+ ion across the synthetic membrane
    d. geminate bacteria to react with the K^+ ion in the solution

9. A pH electrode measures

    a. the partial pressure of dissolved hydrogen gas in a solution
    b. the concentration of H^+ ion in a solution
    c. the half-cell potential of a hydrogen cell
    d. the half-cell potential of a Ag/AgCl electrode different temperature

10. Which of the following is not a correct for a pCO_2 electrode in blood CO_2
    measurement?

    a. a membrane is used to separate the blood sample and the buffer solution in
        the pCO_2 electrode
    b.   the membrane allows CO_2 to diffuse from the blood sample to the buffer
        solution in the pCO_2 electrode
    c. The CO_2 level in blood is correlated to the change in pH level in the buffer
    d. The CO_2 level in the blood sample is determined by the change in potential
        of the Calomel electrode in respond to change in CO_2 concentration in
        the buffer.

11. A Clark electrode is

    a. a polarographic oxygen electrode
    b. a pCO_2 electrode
    c. a pH electrode
    d. a galvanic oxygen cell

12. Which is not true about a fuel cell?

    a. it is a galvanic cell

b. the by-product of a hydrogen fuel cell is water
c. platinum in the electrodes is not consumed in the cell reaction
d. hydrogen is oxidized at the cathode and oxygen is reduced at the anode

Chapter 10

1. Biopotential is the result of

   a. flow if ions across the biological tissue
   b. current induced to the tissue from external electromagnetic field
   c. electrons produced by oxidation reaction in tissue
   d. half-cell potential from the interface of electrodes and tissue

2. State six ideal characteristics of biopotential electrodes

3. Electrodes in which no ion transfer across the metal-electrolyte interface are said to be

   a. inert electrodes
   b. ideal electrodes
   c. perfectly polarized electrodes
   d. perfectly nonpolarized electrodes

4. Sketch the electrical equivalent circuit of a pair of electrodes interfacing with the patient and a biopotential amplifier.

5. The non-zero dc offset voltage appears across a pair of biopotential electrodes is due to

   a. induced electromagnetic interference
   b. unequal half-cell potentials at each of the electrode-skin interface
   c. ion flow across the electrode-skin interface
   d. charges accumulated at the electrode double layer

6. The electrode double layer is represented by which component(s) in the biopotential electrode equivalent circuit in Figure 10.5?

   a. R_{se} & C_d
   b. R_d & C_d
   c. C_d & V_{hc}
   d. V_{hc} & R_{se}

7. The effect of electrode double layer will

   a. increase the impedance of high frequency biological signal

b. decease the impedance of high frequency biological signal
c. attenuate high frequency biological signal
d. attenuate low frequency biological signal

8.   A common electrolyte applied between a Ag/AgCl electrode and tissue is

a. Potassium Chloride
b. Copper Chloride
c. Copper Sulphate
d. Potassium Sulphate

9.   Which is not a desired characteristic of Ag/AgCl electrodes used in biopotential measurement?

a. low toxicity to tissue
b. formation of electrode double layer
c. low impedance
d. low dc offset potential

10. Floating biopotential electrodes is used to minimize

a. electromagnetic interference
b. dc offset potential
c. formation of electrode double layer
d. noise arising from mechanical disturbances at the skin-electrode interfae

Chapter 11

1.   List the characteristics of a good instrumentation amplifier

2.   For the differential amplifier shown below, when a voltage of 4.0 V is applied to both V_1 and V_2 (i.e., $V_1 = V_2 = 4.0$ V), the amplifier output V_{out} is 1.0 mV. When $V_1 = 3.0$ mV and $V_2 = 5.0$ mV, $V_{out} = 8.0$ V. What is the common mode gain of the amplifier?

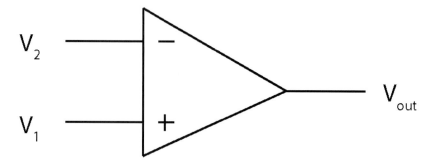

3. For the amplifier in Question 2, what is the differential mode gain of the amplifier?

4. For the amplifier in Question 2, what is the common mode rejection ratio of the amplifier in decibels?

5. Which is the most common source of noise in biopotential signal acquisition?

   a. power frequency electromagnetic interference
   b. movement of patient
   c. other biopotential signal from the patient body
   d. external light stimulation

6. In biopotential signal (such as ECG, EMG) acquisition, which of the follow is the most effective to reduce 50 Hz common mode noise from the signal?

   a. a high gain amplifier
   b. a low pass filter with cutoff frequency of 60 Hz
   c. a differential amplifier
   d. an amplifier with high input impedance

7. A patient has 10 mV of 60 Hz common mode signal induced on the body. An instrumentation amplifier with differential gain of 1,000, CMRR 80,000 and 2 MΩ input to ground impedance is built to acquire the patient's ECG. If he ECG signal amplitude is 0.5 mV, what is the amplitude of the ECG signal at the simplifier output?

8. For the ECG amplifier in Question 7, what is the common mode gain of the amplifier?

9. For the amplifier in Question 7, if the resistances of the two skin/electrode interfaces are 50 and 40 kΩ, what is the magnitude of the 60 Hz signal at the amplifier output?

10. In biopotential signal acquisition with long lead wires connecting the patient to the instrumentation amplifier, interference from external magnetic field can be reduced by

    a. magnetic shielding of lead wires
    b. placing lead wires closer together
    c. twisting the lead wires together
    d. all of the above

11. The defibrillator protection circuit to protect a biopoential signal amplifier is a

    a. high power resistor

  b. a voltage follower
  c. a fast acting fuse
  d. a voltage limiter

Chapter 12

1. Tissue effect due to electrical current is not dependent on

   a. The condition of the patient
   b. The frequency of the electrical current
   c. The source of the electrical current
   d. The duration of current flow though the patient's body

2. The reason for hospital patients to be more susceptible to electrical shock is

   a. Patient's skin resistance are often compromised
   b. Patient's may have conductive pathway for electricity leading to heart
   c. Patients are often connected electrically powered medical devices
   d. all of the above are reasons

3. A medical procedure requires delivery a constant current into the body of the patient. A method to avoid triggering physiological effect to muscle and nerve is to

   a. increase the current path resistance
   b. increase the frequency of the risk current
   c. decrease the voltage of the current source
   d. increase the time duration of the procedure

4. Which is not a hazard due to electricity in healthcare settings?

   a. tissue burn
   b. electrical micro-shock
   c. explosion
   d. all of the above are possible hazards

5. A patient applied part has a resistance of 50 kΩ between the medical device and the patient. What is the maximum risk current to the patient if the medical device end of the patient applied part is exposed to a 120 V 60Hz power source?

6. In Question 5, if the patient applied part is a temperature probe on the skin surface of the patient. What is the worst case electric shock to the patient?

7. In Question 5, if the patient applied part is a heart catheter. What is the worst case electric shock to the patient?

8. What is the maximum allowable patient leakage current of a cardiac catheter under normal condition in a Canadian hospital?

   a. 10 µA
   b. 50 µA
   c. 100 µA
   d. 500 µA

9. Which hazard will an isolated power system prevent?

   a. patient burn
   b. macro shock on the patient
   c. respiratory arrest of the patient
   d. all of the above

10. Which hazard will a line isolation monitor (LIM) not able to prevent?

    a. patient burn
    b. macro shock of the patient
    c. micro shock of the patient
    d. respiratory arrest of the patient

11. Which hazard will the signal isolation circuit in a medical device prevent?

    a. patient burn
    b. macro shock of the patient
    c. micro shock of the patient
    d. respiratory arrest of the patient

12. The function of the capacitor in the passive network of the IEC601-1 leakage current measurement device is to

    a. attenuate the leakage current to a measurable level
    b. protect the milli-voltmeter
    c. match the human body impedance to the milli-voltmeter
    d. simulate the frequency response of human body to leakage current

Chapter 13

1. In a thermal dot array paper chart recorder, the vertical resolution is limited by

    a. the paper speed
    b. the number of thermal elements on the print head
    c. the size of the thermal dots
    d. the heat response of the thermal paper

2.   What is the likely problem if the toner is not sticking to the paper of the print?

    a. Photosensitive drum is scratched
    b. the fuser temperature is not high enough
    c. the power to the writing laser is cut off
    d. the cleaning blade cannot remove all residual toner on the photosensitive drum

3.   In an erase bar display, the newest information appears

    a. at the trailing edge of the erase bar
    b. at the leading edge of the erase bar
    c. at the left side of the display
    d. at the right side of the display

4.   How many pixel elements are required to form an ultra-high definition 4k (3840×2160) color LCD display?

5.   Which type of display is the best under bright external light?

    a. Plasma
    b. OLED
    c. Electroluminescent
    d. LCD

6.   Which of the following video signal interface does not support digital signal?

    a. VGA
    b. DVI-I
    c. HDMI
    d. DisplayPort

7.   The following electrocardiogram was recorded by a single channel electrocardiograph when an ECG simulator was attached to the input.

    If the sensitivity setting of the electrocardiograph is 5 mm/mV, what is the signal amplitude (in mV) from the simulator (note: each small vertical and horizontal division on the chart equals to 1 mm).

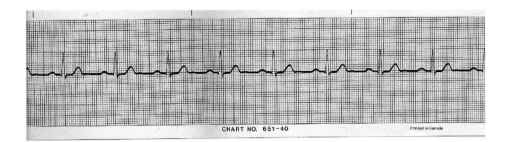

CHART NO. 651-40   Printed in Canada

8. In the previous question, calculate the rate of the ECG signal in bpm (beats per minute) if the chart speed is 25 mm/s.

9. A paper chart recorder set at a sensitivity of 10 mm/mV produces a vertical deflection of 19 mm when the input signal amplitude is 2.0 mV. What is the percentage error of the recorder?

10. A 1.0 mV ECG signal has a bandwidth of 0.01 Hz to 150 Hz and a dc offset of 0.01 mV. If the high frequency cutoff is the display is reduced to 70 Hz, what is the effect on the ECG signal waveform?

    a. dc offset is removed
    b. signal becomes more oscillatory
    c. amplitude of R-wave is reduced
    d. no change to the waveform

11. When a step waveform is applied to a biopotential amplifier, the output shows an exponential decay waveform. If the amplitude of the waveform is 4.0 V at time 1.0 second (from the start of the step) and 2.5 V at time 2.0 s, find the approximate low cut-off frequency of the biopotential amplifier assuming that it behaves as a single-pole high pass filter.

12. When a step input is applied to a paper chart recorder, the chart recorded an exponential decay waveform. If the amplitude is decayed to 50% after 2.0 seconds, what is the low cut off frequency of the recorder assuming that is behaves a s single-pole high pass filter.

Chapter 14

1. Which is not a function of a patient monitor

    a. analyze physiological signal
    b. provide communication between nurses and physicians
    c. generate alarm to alert violation of preset physiological limits

    d. display physiological signal in a visual format

2. Which of the following is a parameter being captured by a patient monitor?

    a. respiration waveform
    b. heart sound
    c. urinary flow
    d. kidney function

3. Which of the following is not a feature of most patient monitors?

    a. Trending of physiological signals
    b. Alarms that can be preset
    c. recording of physiological parameters and waveform
    d. Wireless transmission of physiological signals

4. Which of the following is not an advantage of medical telemetry?

    a. reduce the number of hardwire connections between patients and machines
    b. increase patient mobility
    c. reduce system complexity
    d. increase rate of recovery for cardiac patient

5. The arrhythmia detection technique by characterizing and comparing the height, width and slopes of the electrocardiogram is called a

    a. waveform feature extraction
    b. data compression
    c. waveform template matching
    d. machine facial recognition

6. List 4 desirable characteristics of a patient monitoring network.

7. The network protocol employing carrier sense, multiple access and collision detection is a

    a. Token ring
    b. ARCnet
    c. Windows 2000
    d. Ethernet

8. DICOM standards facilitates

    a. interoperability among medical imaging devices
    b. computation of medical data

c. deep learning in artificial intelligence

d. data integration in healthcare communication

9. Which of the following about HL7 is not correct?

a. it is a network standard

b. it supports management, delivery, and evaluation of clinical patient information

c. it is at layer 7 of the ISO OSI network communication model

d. it is the acronym of High Level 7

10. Which of the following policies and procedures will enhance patient information privacy and confidentiality in patient monitoring systems?

a. user authorization

b. network security

c. removal of patient information before disposal

d. all of the above

11. A fiber optic cable in the network is considered as a

a. transmission link

b. interconnection device

c. network operating system

d. backbone

12. A network device to direct traffic across multiple local or wide area networks is

a. a repeater

b. a hub

c. a NIC

d. a router

Chapter 15

1. How many ECG leads can be obtained from the 3 limb (RA, LA, & LL) electrodes?

a. 3

b. 4

c. 5

d. 6

2. In practice, how many electrodes are used in a resting 12-lead ECG acquisition?

    a. 9
    b. 10
    c. 11
    d. 12

3. In ECG acquisition, the right-leg driven circuit is designed to reduce

    a. patient motion artifact
    b. other bipotential signals such as EMG
    c. common mode noise induced on the patient body
    d. 60 Hz electromagnetic interference

4. The purpose of signal isolation in ECG acquisition is to

    a. prevent micro shock on patient
    b. reduce electromagnetic interference
    c. eliminate power frequency noise
    d. improve signal quality

5. Comparing with monitoring ECG, diagnostic ECG waveform

    a. has a wider bandwidth
    b. has a bigger amplitude
    c. includes dc (zero frequency) component
    d. requires better quality skin electrodes

6. Motion artifact in ECG waveform can be removed by

    a. increasing the high cut off frequency
    b. decreasing the high cut off frequency
    c. increasing the low cut off frequency
    d. decreasing the low cut off frequency

7. Which of the following regarding ECG leads limb are not correct?

    a. I = II + III
    b. II = I + III
    c. III = I + II
    d. II = III – I

8. In surface ECG waveform, the start of ventricular depolarization is marked by the

    a. P wave

  b. T wave
  c. U wave
  d. QRS complex

9. How many electrodes are needed to obtain the 6 chest (V) leads?

  a. 3
  b. 4
  c. 6
  d. 9

10. Defibrillator Protection in an electrocardiograph is implemented by

  a. a high pass filter
  b. a voltage limiter
  c. a low pass filter
  d. a voltage follower

11. In the 12-lead ECG recording in Figure 15-21, the QRS axis is in the

  a. Normal segment
  b. EAD segment
  c. LAD segment
  d. RAD segment

12. The function of signal isolation is ECG is for

  a. power noise reduction
  b. common mode signal elimination
  c. micro shock prevention
  d. high voltage protection

13. List four external sources of ECG waveform artifacts

14. List sources of ECG waveform artifacts caused by the patient

Chapter 16

1. EEG signals are generated by the inhibitory and excitatory postsynaptic potentials of the cortical nerve cells.

  a. True
  b. False

2. Due to the small signal level, it needs low input impedance high gain amplifiers in scalp EEG acquisition.

    a. True
    b. False

3. Polysomnography studies are often done in areas with electromagnetic shielding.

    a. True
    b. False

4. Bispectral Index is used as one of the indications of the depth of anesthesia in surgery.

    a. True
    b. False

5. In EEG studies using scalp electrodes, collodion is used to reduce the skin-electrode impedance.

    a. True
    b. False

6. The right and left auricular points, nasion, and Inion are four electrode placement locations in 10-20 system EEG acquisition.

    a. True
    b. False

7. In scalp EEG acquisition, which of the following value (in Ohms) is not acceptable when measuring impedance between a pair of electrodes?

    a. 50
    b. 200
    c. 500
    d. 1500

8. The normal EEG spectral waveform recorded when the subject is in deep sleep is primarily

    a. Beta rhythm
    b. Alpha rhythm
    c. Theta rhythm
    d. Delta rhythm

9. A montage in EEG acquisition

   a. is the signal captured between a pair of electrodes
   b. is a unipolar connection configuration
   c. is a distinct combination of differential signals from electrode pairs
   d. forms the 10-20 system

10. Which of the following is used to reduce random noise in evoked potential studies?

   a. Signal isolation
   b. Active noise cancellation
   c. 60 Hz Notch filter
   d. Signal averaging

11. The strength of voluntary muscle contraction is independent of

   a. the integrity of the motor nerve innervation with the muscle
   b. the shape of the single motor unit action potential (MUAP)
   c. the number of motor units activated
   d. the firing rate of individual motor unit

12. The EMG waveform produced during the application of a needle electrode is called

   a. End plate noise
   b. End plate spikes
   c. Spontaneous activity
   d. Insertion activity

13. Nerve conduction velocity can be determined by

   a. The time delay between the stimulation and the negative peak of the corresponding sensory nerve action potential
   b. The time delay between the stimulation and the negative peak of the corresponding motor response
   c. The time duration between the onset to return-to-baseline of the sensory corresponding nerve action potential
   d. The time duration between the onset to return-to-baseline of the corresponding motor response

Chapter 17

1. Which of the following is used to reduce random noise in evoked potential studies?

   a. Signal isolation
   b. Active noise cancellation
   c. 60 Hz Notch filter
   d. Signal averaging

2.   The strength of voluntary muscle contraction is independent of

   a. the integrity of the motor nerve innervation with the muscle
   b. the shape of the single motor unit action potential (MUAP)
   c. the number of motor units activated
   d. the firing rate of individual motor unit

3.   The EMG waveform produced during the application of a needle electrode is called?

   a. End plate noise
   b. End plate spikes
   c. Spontaneous activity
   d. Insertion activity

4.   Voluntary muscle effort involves recruitment of motor units, explain the relationship of muscle strength in relation to spatial and temporal recruitment of motor units.

5.   A motor response is

   a. obtained by stimulating a nerve and recording from a muscle that it innervates
   b. obtained by stimulating a nerve and recording from it or one of its branch
   c. obtained by recording the biopotential from a muscle under contraction
   d. obtained by recording the biopotential from a nerve fiber

6.   Nerve conduction velocity can be determined by

   a. The time delay between the stimulation and the negative peak of the corresponding sensory nerve action potential
   b. The time delay between the stimulation and the negative peak of the corresponding motor response
   c. The time duration between the onset to return-to-baseline of the sensory corresponding nerve action potential
   d. The time duration between the onset to return-to-baseline of the corresponding motor response

7.   Which is a suitable filter bandwidth for muscle action potential acquisition?

   a. 0.05 to 150 Hz

b. dc to 800 Hz
c. 20 to 8000 Hz
d. 20 to 800 Hz

8. What is a typical amplifier gain needed to record motor unit action potentials captured by surface electrodes?

a. 50
b. 5,000
c. 50,000
d. 100,000

Chapter 18

1. In invasive blood pressure monitoring, mean blood pressure is obtained by averaging the values of diastolic and systolic blood pressure.

a. True
b. False

2. The dicrotic notch in arterial blood pressure waveform is due to

a. the opening of the mitral valve
b. the closing of the mitral valve
c. the opening of the aortic valve
d. the closing of the aortic valve

3. The purpose of the continuous flush valve in the arterial line in invasive blood pressure monitoring is to

a. prevent blood clot at the tip of the indwelling catheter
b. maintain a pressure in the arterial line to prevent back flow of blood
c. prevent blood adhesion on the surface of the pressure transducer
d. prevent air bubble forming in the line

4. Which of the following will happen to the readings on the monitor when a patient sits up while under invasive blood pressure monitoring?

a. Both the systolic and diastolic pressure will be higher
b. The systolic pressure will be higher and the diastolic pressure will be lower
c. The systolic pressure will be lower and the diastolic pressure will be higher
d. Both the systolic and diastolic pressure will be lower

5. In invasive blood pressure monitoring, the zeroing procedure will

a. compensate the non-linearity of the transducer
b. remove the zero offset of the transducer
c. align the diastolic blood pressure with the atmospheric pressure
d. reduce noise due to external electromagnetic interference

6. A patient under invasive blood pressure monitoring has a systolic and diastolic pressure of 130 and 100 mmHg respectively. The pressure transducer is mounted on a floor standing IV pole. What will be the blood pressure readings if the patient bed is raised by 200 mm? (given that the density of saline in the arterial line is 1020 kg/m^3)

7. A disposable invasive blood pressure transducer with sensitivity of 5.0 μV/V/mHg is used with a blood pressure monitor. If the monitor is to display 100 mmHg when the input is 1.0 mV, what is the excitation on the transducer?

    a. 1 V
    b. 2 V
    c. 5 V
    d. 10 V

8. Which of the following problems in the arterial line will lower the upper cut off frequency of the blood pressure monitoring system?

    a. blood clot at the tip of the indwelling catheter
    b. add an extra length of tubing between the transducer and the arterial access
    c. switch to a narrower tubing
    d. all of the above

9. Which of the following change in the arterial line will not cause distortion to the invasive blood pressure waveform?

    a. increase the length of the line
    b. increase the diameter of the line
    c. introduce air bubbles in the line
    d. pinching the line

10. What is the output voltage of a standard disposable blood pressure transducer when the pressure at the input is 100 mmHg, the transducer excitation is 5.0 Vac.

Chapter 19

1. Which of the following will affect the accuracy of non-invasive blood pressure measurement using the oscillometric method?

a. movement of the patient
b. heavy metal music
c. 60 Hz power noise
d. high frequency electromagnetic noise

2. Which of the following will affect the accuracy of non-invasive blood pressure measurement using the auscultatory method?

a. movement of the patient
b. heavy metal music
c. 60 Hz line power noise
d. high frequency electromagnetic noise

3. Which of the following is not a component of a sphygmomanometer?

a. pressure gauge
b. air pump
c. stethoscope
d. pressure cuff

4. The mean blood pressure can be accurately determined by a skilled clinician using the manual auscultatory method.

a. True
b. False

5. The mean blood pressure is obtained by correlating the cuff pressure to the occurrence of the highest pressure oscillation magnitude inside the blood pressure cuff.

a. True
b. False

6. In a non-invasive blood pressure monitor, the filter circuit used to extract the oscillometric waveform from the cuff pressure sensor is a

a. low pass filter
b. high pass filter
c. band pass filter
d. band reject filter

7. In non-invasive blood pressure monitoring using the oscillometric method, a large adult cuff being used on a small adult will lead to

a. pressure injury
b. falsely-high reading of blood pressure

c. falsely-low reading of blood pressure
d. longer measurement time

8. The American Heart Association recommends that the width of the bladder under the cuff be 80% of the circumference of the patient's arm for non-invasive blood pressure measurements.

a. True
b. False

9. The advantage of blood pressure measurement using oscillometric method over invasive blood method is

a. oscillometric method is non-invasive
b. oscillometric method provides a more accurate reading of the mean blood pressure
c. it oscillometric method is not affected by ambient noise
d. all of the above are advantages of oscillometric method over invasive method

10. Too fast cuff pressure deflation rate in non-invasive blood pressure measurements will lead to

a. pressure injury of tissue under cuff
b. falsely-high reading of systolic blood pressure
c. falsely-low reading of systolic blood pressure
d. prolonged measurement time

Chapter 20

1. Cardiac output is equal to

a. Stroke volume multiplied by the heart rat
b. Heart rate multiplied by the tidal volume
c. Aortic blood flow multiplied by the heart rate
d. Blood flow measured by temperature drop in a pulmonary artery multiplied by the heart rate

2. Which of the following has the greatest potential to increase the cardiac output of a person?

a. Increase in patient resting heart rate
b. Mental status of the patient
c. Intensity of physical activity of the patient
d. Sleep pattern of the patient

3. In cardiac output measurement using the direct Fick method, if the arterial and mixed venous oxygen contents are 180 ml/L and 120 ml/L respectively, and the oxygen consumption rate is 240 ml/min, what is the nearest cardiac output of the patient?

   a. 3 L/min
   b. 4 L/min
   c. 5 L/min
   d. 6 L/min

4. In cardiac output measurement using the thermal dilution method, the indicator is

   a. indocyanine green dye
   b. saline solution
   c. heat
   d. water at 0 degree Celsius

5. A Swanz Ganz catheter is used in

   a. Direct Fick method cardiac output measruement
   b. PAC thermodilution cardiac output measurement
   c. TD thermodilution cardiac output measurement
   d. PC cardiac output measurement

6. Which of the following method is used to confirm proper placement of a Swanz Ganz catheter in a patient?

   a. monitoring blood pressure waveform at the distal lumen of the catheter
   b. monitoring temperature near the tip of the catheter
   c. observing the distance indicator marks along the length of the catheter
   d. inflation of the balloon at the catheter tip

7. In cardiac output measurement using the a Swanz Ganz catheter, injectate warming will lead to

   a. unpredictable measurement error
   b. over-reading of cardiac output
   c. under-reading of cardiac output
   d. patient hyperthermia

8. In cardiac output measurement using the transpulmonary thermodiluton method, injectate warming will lead to

   a. unpredictable measurement error
   b. over-reading of cardiac output

c. under-reading of cardiac output
d. patient hyperthermia

9. In cardiac output measurement using the a Swanz Ganz catheter, recirculation will lead to

a. unpredictable measurement error
b. over-reading of cardiac output
c. under-reading of cardiac output
d. patient hyperthermia

10. Comparing to pulmonary artery catheter method, thermodilution cardiac output measurement using transpulmonary method will need to use

a. Larger volume and higher temperature injectate
b. Larger volume and lower temperature injectate
c. Smaller volume and lower temperature injectate
d. Smaller volume and higher temperature injectate

11. Which of the following can be a temperature detection site to create the thermodilution curve in transpulmonay thermodilution cardiac output measurement?

a. Common aorta
b. Pulmonary artery
c. Brachial artery
d. All of the above

12. The principle assumption behind pulse contour cardiac output measurement is

a. Cardiac output is proportional to arterial pulse pressure
b. Stroke volume is equal to the area under the arterial pressure curve
c. Stroke volume is proportional to the compliance of the arterial blood vessel
d. Heat loss by the injectate is equal to the heat gain of the blood

Chapter 21

1. Muscle stimulation problem in implantable cardiac pacemaker systems can be prevented by

a. increasing the sensitivity level
b. lowering the sensitivity level
c. reducing the pacing energy
d. using bipolar pacemaker leads

2. Muscle sensing problem in implantable cardiac pacemaker systems can be prevented by

   a. increasing the sensitivity level
   b. lowering the sensitivity level
   c. reducing the pacing energy
   d. using bipolar pacemaker leads

3. Cross talk problem in implantable cardiac pacemaker systems can be prevented by

   a. increasing the sensitivity level
   b. lowering the sensitivity level
   c. increasing the refractory period
   d. using bipolar pacemaker leads

4. Which of the following cardiac pacemakers delivers the highest pacing energy per stimulation?

   a. implantable
   b. external invasive
   c. external non-invasive
   d. all pacemakers deliver similar pacing energy

5. An external invasive pacemaker delivers similar amount of energy per stimulation as an implantable pacemaker.

   a. True
   b. False

6. For an implantable pacemaker, if the pacing rate is increased by 20%, the battery life will

   a. increase by 20%
   b. decrease by 20%
   c. increase by 17%
   d. decrease by 17%

7. A programmable implantable pacemaker employs _____ to block electromagnetic noise being picked up by the telemetry sensing coil of the pacemaker when it is under normal pacing mode.

   a. an isolation transformer
   b. a low pass filter
   c. a magnetic reed switch
   d. electromagnetic shielding

8.   In the NBG pacemaker mode, the heart chamber sensed is indicated in

     a. position I
     b. position II
     c. position III
     d. position IV

9.   A pacemaker which senses the atrium and delivers a stimulation to the ventri-
     cle when an atrial contraction is detected is a

     a. AOO pacemaker
     b. VOO pacemaker
     c. VAT pacemaker
     d. VVI pacemaker

10. Which is the following modes is not supported by an SSI implantable pacemak-
     er?

     a. AOO pacemaker
     b. VOO pacemaker
     c. VAT pacemaker
     d. VVI pacemaker

11. A common test load (in Ohms) to test cardiac pacemakers is

     a. 8
     b. 50
     c. 500
     d. 5000

12. Which of the following pacemakers delivers the highest energy per pacing pules?

     a. implantable pacemaker
     b. capsule pacemaker
     c. non-invasive (transthoracic) pacemaker
     d. external invasive pacemaker

Chapter 22

1.   Which of the following cardiac defibrillator waveform has the highest output
     voltage at the same energy setting.

     a. pulsed biphasic
     b. critically damped monophasic sinusoidal
     c. truncated exponential

d. biphasic

2. A defibrillator is designed to deliver a maximum of 400 J of energy to the patient. What is the minimum capacitance (in micro-farad) of the energy storage capacitor if the maximum charging voltage is 6,500 kV?

   a. 10
   b. 20
   c. 30
   d. 40

3. In cardiac defibrillation, which of the following is not a method to reduce the risk of high voltage shock to the operator when the energy is being delivered to the patient?

   a. use defibrillator with built-in signal isolation
   b. use defibrillator with built-in output isolation
   c. do not touch the patient
   d. do not touch the paddle of the defibrillator

4. In synchronous cardioversion, the purpose of delivering the energy to the patient after detecting the R-wave from the patient's ECG is to

   a. prevent damage to electronic components from the defibrillator high voltage
   b. allow the patient to breath before the defibrillation current enters the patient
   c. reduce the chance of triggering ventricular fibrillation
   d. allow maximum energy to be delivered to the patient

5. In a cardiac defibrillator, the energy delivered to the patient is always smaller than the initial energy stored in the energy storage capacitor. Which of the following is not a cause?

   a. leakage in the energy storage capacitor
   b. non-zero electrical resistance in the pathway from capacitor to patient
   c. energy discharge time is limited (not infinite)
   d. not all defibrillation current entering the patient will go through the heart of the patient

6. Which of the following cardiac defibrillator waveform has the shortest discharge time at the same energy setting.

   a. pulsed biphasic
   b. critically damped monophasic sinusoidal
   c. truncated exponential
   d. biphasic

7. In a cardiac defibrillator, energy dumping is to remove the charge stored in the energy storage capacity. Which of the following will not activate the energy dumping function?

    a. when the operator change the defibrillation energy to a higher energy setting
    b. when the operator change the defibrillation energy to a lower energy setting
    c. when the defibrillator is idled for a period of time (e.g., 60 sec.)
    d. when the operator moves the discharge paddles off from the chest of the patient

8. Which of the following is not true about the battery in a cardiac defibrillators?

    a. It supports rapid charging of the energy storage capacitor
    b. it allows the defibrillator to be used without the need to connect to AC power
    c. defibrillators are not recommended to be plugged in (charging the batteries) all the time
    d. it has large capacity to allow multiple consecutive full energy defibrillations

9. A common test load (in Ohms) to test cardiac defibrillators is

    a. 8
    b. 50
    c. 500
    d. 5000

10. The wave shaping circuit of a cardiac defibrillator to produce a truncated exponential waveform comprises of

    a. RLC circuit
    b. LC circuit
    c. RC circuit
    d. RL circuit

11. In the defibrillator circuit in Figure 22.5, the function and purpose of the resistor R_L is

    a. wave shaping circuit component to produce the MTE waveform
    b. limit the initial capacitor charging current to prevent damage to circuit components
    c. reduce the voltage to prevent patient injury
    d. a component of output isolation to reduce macro shock risk to the operator

12. Whish of the following heart problem may be corrected by the synchronous cardioversion mode of a defibrillator?

    a. ventricular fibrillation
    b. atrial flutter
    c. cardiac asystole
    d. all of the above

13. What is the energy stored in a 64 μF capacity when it is charged to a voltage of 3000 V?

14. A 12 V Lithium battery is used in an AED which is designed to deliver 200 J of energy in a single discharge. What is the minimum capacity of the battery in ampere hour (mAh) if the unit can deliver 10 consecutive defibrillation discharges?

Chapter 23

1. In an IV infusion set up, the time between adjacent solution drops observed in the drip chamber is 1.0 seconds. If the size of the drip chamber nozzle is 20 drops per ml, what is the volume flow rate of the infusion?

    a. 180 ml/hr
    b. 60 drops/min
    c. 30 ml/hr
    d. 3600 drops/hr

2. Name the components of a typical IV set.

3. In manual gravity infusion, the component to control the solution flow rate is

    a. drip chamber
    b. regulating clamp
    c. occlusion clamp
    d. y-connection

4. In the piggy-back infusion set up in Figure 23.3, when both the primary and secondary lines are fully open, which of the following statement is correct?

    a. only the primary solution is delivered to the patient
    b. only the secondary solution is delivered to the patient
    c. both primary and secondary solutions are delivered to the patient
    d. none of the solution is delivered to the patient

5. Which of the following pumping mechanism will produce the least flow rate fluctuation?

  a. piston cylinder
  b. piston diaphragm
  c. linear peristaltic
  d. syringe

6. What is the flow rate of the syringe pump in Figure 23-12 if the screw pitch is 0.5 mm, the screw is rotating at 6.0 rpm, and a syringe with 1.6 cm inner diameter is used?

  a. 3 ml/hr
  b. 36 ml/hr
  c. 50 ml/hr
  d. 60 ml/hr

7. The upstream occlusion sensor of an infusion pump can detect the following condition:

  a. when the solution bag is emptied
  b. when there is a kink in the infusion line between the pump and the patient
  c. when there is a leak in the IV line
  d. when there is a large air bubble in the IV line

8. A slow flow rate maintained after a programmed volume of fluid is infused to the patient is to prevent

  a. under infusion
  b. over infusion
  c. blood clot at the catheter tip
  d. free flow of IV solution into patient

9. Which of the following factors determine the flow rate of a peristaltic infusion pump?

     i   the rotation speed of the stepper motor
     ii   the material of the IV tubing under the pumping mechanism
     iii   the height of the solution bag above the patient's venipuncture site

  a. i only
  b. ii only
  c. iii only
  d. I & ii only

10. Which of the following factors determine the flow rate of a piston cylinder pump?

     i   the rotation speed of the stepper motor

    ii   the stroke volume of the cylinder
    iii the inner diameter of the IV line

  a. i only
  b. ii only
  c. iii only
  d. I & ii only

11. How many drops of IV solution will be observed in the drip chamber in one minute if an infusion pump is programmed at a flow rate of 120 ml/hr, if the infusion line has a 60 drops/ml drip chamber?

12. What is a typical sensor used to measure air-in-line in an infusion pump?

  a. infrared light source and sensor
  b. ultrasound transmitter and detector
  c. electromagnetic force sensor
  d. heated thermistor

13. What is a typical sensor used to to measure flow rate in an infusion pump?

  a. infrared light source and sensor
  b. ultrasound transmitter and detector
  c. electromagnetic force sensor
  d. heated thermistor

14. The flow rate of an infusion pump is set at 120 ml/hr. If 19 ml of solution is collected in 10 minutes, what is the flow rate error?

Chapter 24

1. Explain why the ESU current frequency cannot be lower than 100 kHz.

2. Which of the following ESU current density in tissue is consider the threshold to produce tissue response?

  a. 20 μA/cm^2
  b. 10 mA/ cm^2
  c. 50 mA/ cm^2
  d. 100 mA/ cm^2

3. Which of the following is not a tissue effect created by electrosurgical current?

  a. cut
  b. fulguration

    c. blended
    d. desiccation

4.   In electrosurgery, which tissue effect is the primary mechanism of monopolar coagulation?

    a. cut
    b. fulguration
    c. blended
    d. desiccation

5.   Which tissue effect is primarily achieved in bipolar electrosurgical current?

    a. cut
    b. fulguration
    c. blended
    d. desiccation

6.   Which electrosurgical mode of operation creates the highest output power to the tissue?

    a. monopolar cut
    b. monopolar coagulation
    c. monopolar blended
    d. bipolar coagulation

7.   Which electrosurgical mode of operation creates the highest output voltage across the active and return electrodes?

    a. monopolar cut
    b. monopolar coagulation
    c. monopolar blended
    d. bipolar coagulation

8.   What is a typical test load used in ESU output verification?

9.   For ESU with dynamic control, the output power

    a. decrease with increase in the patient load
    b. increase with decrease in the patient load
    c. decrease with decrease in patient load
    d. stay almost constant with changes in the patient load

10. The ESU output isolation test measures the

a. power frequency leakage current
b. high frequency leakage current
c. decrease in output power
d. immunity to electromagnetic interference

11. Which of the following is not an essential characteristic of an ESU return electrode?

a. durable
b. large surface area
c. flexible
d. low electrode-skin resistance

12. An ESU with return electrode quality monitor will sound an alarm when

   i.   the return electrodes are in poor contact with the patient
   ii.   there is a short circuit between the dual conductive electrode pads
   iii.   the return electrode is disconnected from the ESU

a. i only
b. i or ii only
c. i or iii only
d. i, ii, or iii

13. Which of the following is not a hazard in electrosurgery

a. patient burn when return electrode is not properly attached to patient
b. adverse health effect from inhalation of ESU smoke and plume
c. micro-shock when ESU current passes through the heart
d. fire caused by ignition of flammable materials by ESU sparks

14. What is the surgical effect when the ESU is set in monopolar blended mode?

   i.   cut
   ii.   coagulation
   iii.   fulguration
   iv.   desiccation

a. i & ii
b. i & iii
c. ii & iii
d. ii & iv

Chapter 25

1. In the volume-time curve obtained from forced expiratory measurement of a patient, a low gradient magnitude indicates:

    a. restrictive lung disease
    b. obstructive lung disease
    c. pulmonary fibrosis
    d. reduced alveolar elasticity

2. The volume of air inhales in a normal breathing is called

    a. tidal volume
    b. total lung capacity
    c. inspiration volume
    d. vital capacity

3. Work of breathing in pulmonary function analysis can be obtained from the

    a. area under the flow-volume curve obtained from forced expiratory measurement
    b. area under the volume-time curve obtained from forced expiratory measurement
    c. area enclosed by the pressure-volume loop in pulmonary respiration
    d. area enclosed by the flow-volume loop in pulmonary respiration

4. In pulmonary function analysis, the residual volume can be measured by:

    a. the nitrogen washout method
    b. helium dilution method
    c. force expiratory measurement
    d. single-breath technique using a gas mixture of CO and helium

5. In pulmonary function analysis, the anatomical dead space can be measured by:

    a. the nitrogen washout method
    b. helium dilution method
    c. force expiratory measurement
    d. single-breath technique using a gas mixture of CO and helium

6. A critical requirement of a spirometer is low flow resistance.

    a. false
    b. true

7. Which of the following is not a condition of a gas volume adjusted to BTPS?

   i.   at room temperature
   ii. at atmospheric pressure
   iii. with saturated vapor pressure

   a. i only
   b. ii only
   c. iii only
   d. all are conditions of BTPS

8. What is a typical test load used in ESU output verification?

9. Explain the principle of respiration monitoring using heated thermistor method.

10. Respiration monitoring using the impedance pneumographic method

    a. measures the changes in electrical impedance of a heated thermistor placed on the chest the patient
    b. measures the change in electrical impedance of the inhaled and exhaled gas of the patient
    c. measures the biopotential across a pair of electrodes placed across the chest of the patient
    d. measures the changes in electrical impedance between a pair of electrodes placed across the chest of the patient

11. States methods to prevent patient cross contamination in pulmonary function testing.

12. In beside respiration monitoring using the impedance pneumographic method, to prevent electric shock

    i.   a high frequency current is applied to the patient
    ii. an isolation transformer is used to reduce the leakage current flowing from the monitor to the patient
    iii. a very low voltage is applied to the patient body

    a. i only
    b. i and ii only
    c. i and iii only
    d. i, ii, and iii

Chapter 26

1.   A biphasic cuirass ventilator creates a positive pressure in the patient's airway during the inspiratory phase of ventilation.

     a. true
     b. false

2.   Airway access using an endotracheal tube is considered invasive ventilation.

     a. true
     b. false

3.   The ventilator mode that allows the patient to breathe spontaneously under an elevated pressure is

     a. SIMV
     b. PEEP
     c. CPAP
     d. CMV

4.   The primary function of a nebulizer is to

     a. add humidity to the inspired air
     b. add humidity to the expired air
     c. prevent water condensation forming in the breathing circuit
     d. supply medication into the lungs

5.   Which of the following are functions of the exhalation valve of a mechanical ventilator?

     i.   prevent excessive pressure to build up in the patient's breathing circuit
     ii   create a positive pressure in the patient's breathing circuit in PEEP ventilation mode
     iii. maintain the positive pressure during the inspiratory phase of ventilation

     a. i only
     b. iii only
     c. i and ii only
     d. ii and iii only

6.   What is of the function of heated bacteria filter in a mechanical ventilator?

     a. kill bacteria in the patient's exhale gas
     b. block bacterial in the patient's exhale gas from entering the ventilator

   c. reheat the exhale gas to body temperature before releasing to the atmosphere

   d. warm the patient's exhale gas to ensure accurate measurement of the exhaled gas flow

7. The pneumatic component in the ventilator to prevent reverse gas flow is a

   a. check valve
   b. regulator
   c. gas blender
   d. air filter

8. A kink or obstruction in the patient breathing circuit can be detected by

   a. check valve
   b. regulator
   c. pressure transducer
   d. flow transducer

9. A small gas leak in the patient breathing circuit can be detected by

   a. check valve
   b. regulator
   c. pressure transducers
   d. flow transducers

10. Intrinsic PEEP is caused by

   a. low lung compliance
   b. atrophy of the diaphragm muscles
   c. high airway resistance
   d. fluid in the lungs

11. High frequency active ventilation uses a negative pressure to actively remove gas from the lungs during the expiratory phase of ventilation.

   a. true
   b. false

12. Which of the following are potential risk of a patient under prolonged period of invasive mechanical ventilation?

   i. pneumonia
   ii. diaphragm muscle atrophy iii. a very low voltage is applied to the patient body
   iii. trauma in the airway

    a. i only
    b. i and ii only
    c. i and iii only
    d. i, ii, and iii

Chapter 27

1. Which of the following regarding ultrasound is correct?

    a. it is a mechanical longitudinal wave
    b. it's repetition frequency is below the lower limit of the human audible range
    c. its propagation speed is the same in all media
    d. it can propagate in vacuum

2. The sound frequency detected by the receiver when it is moving toward a stationary sound source is higher than the source frequency.

    a. true
    b. false

3. Given the velocity of sound is air is 330 m/sec, the Doppler shift when the source is moving toward the sound receiver at 5 m/sec in air is

    a. + 20 m/sec
    b. – 20 m/sec
    c. + 5 m/sec
    d. – 5 m/sec

4. In the set up in Figure 27-2, D = 0.5 cm, Θ = 600. If the time difference for a sound pulse to travel from A to B and from B to A is 2.2 ns, what is the fluid flow velocity v? Given the speed of sound propagation in the fluid is 1500 m/sec. 0.1 m/sec and the?

    a. 50 cm/sec
    b. 100 cm/sec
    c. 150 cm/sec
    d. 200 cm/sec

5. For the ultrasound Doppler blood flowmeter shown in Figure 27.3, if Θ = Φ = 600, fs = 5 MHz, C = 1500 m/s and the Doppler shift is – 2.5 kHz, what is the velocity of blood flow?

    a. 50 cm/sec
    b. 75 cm/sec
    c. 100 cm/sec

d. 150 cm/sec

6.   In Doppler blood flow detection, the Doppler shift is due to reflection from moving blood cells.

   a. true
   b. false

7.   The error(s) to prevent accurate measurement of the flow velocity in a Doppler blood flowmeter is due to:

   i.   movement of blood vessel
   ii.   movement of blood vessel wall due to the pulsatile blood pressure
   iii. non-uniform blood flow velocity across the diameter of the blood vessel

   a. i only
   b. i and ii only
   c. i and iii only
   d. i, ii, and iii

8.   Theoretically, the Doppler shift detected when the probe of the Doppler blood flow detector is zero.

   a. true
   b. false

9.   The magnitude of the Doppler shift detected when the transmitter and receiver are in line with the flow of blood is

   a. lowest
   b. highest
   c. zero
   d. cannot be determined

10. The function of applying aqueous gel between the ultrasound probe and tissue is to

   a. dissipate heat to prevent skin burn
   b. reduce resistance when moving the probe on the skin surface
   c. reduce reflection loss of the ultrasound from the probe into the tissue
   d. amplify the ultrasound wave

Chapter 28

1. Which of the following physiological parameters is not usually monitored by fetal monitor?

   a. maternal temperature
   b. intrauterrine pressure
   c. fetal heart rate
   d. uterine activities

2. Continuous fetal heart rate can only be acquired invasively.

   a. true
   b. false

3. Interpretation of fetal heart rate traces includes analyzing its

   i.   baseline
   ii. variability
   iii. change of pattern over time

   a. i only
   b. i and ii only
   c. i and iii only
   d. i, ii, and iii

4. Uterine activity during labour refers to

   i.   frequency of contraction of the uterus
   ii. intensity of contraction of the uterus
   iii. the level of pain during intense labour

   a. i only
   b. i and ii only
   c. i and iii only
   d. i, ii, and iii

5. Fetal heart rate can be obtained from

   i.   direct ECG
   ii. abdominal ECG
   iii. phono ECG

   a. i only
   b. i and ii only

c. i and iii only

d. i, ii, and iii

6. In abdominal ECG acquisition, the maternal ECG can be removed from the fetal ECG by carefully positioning the electrodes.

a. true

b. false

7. In fetal heart rate acquisition using Doppler ultrasound, a commonly used sound frequency is:

a. 500 kHz

b. 2 MHz

c. 5 MHz

d. 10 MHz plus higher harmonics

8. In monitoring uterine activity, which of the following is an advantage of using a TOCO transducer over an intrauterine transducer?

a. TOCO transducer is less accurate

b. TOCO transducer is more accurate

c. TOCO transducer is non-invasive

d. TOCO transducer ensure continuous and reliable measurement

9. Which of the following may pose infection hazard in fetal monitoring?

a. phone heart rate

b. uterine activity using TOCO transducer

c. direct fetal ECG

d. abdominal ECG

10. Electronic fetal monitoring is recommended for all labour conditions as it can reduce complications during labour.

a. true

b. false

Chapter 29

1. The noise level inside the chamber of an infant incubator should be

a. less than 40 dBA

b. less than 50 dBA

c. greater than 50 dBA

d. greater than 40 dBA

2. Infant incubators are used to treat jaundice babies.

   a. true
   b. false

3. Which of the following is not a function of an infant incubator?

   a. provide an elevated oxygen level concentration for the infant
   b. provide a controlled environment for the infant
   c. facilitate feeding of the infant
   d. maintain a set steady temperature inside the chamber

4. The main function of the proportional heating system in an infant incubator is to

   a. maintain the set temperature
   b. reduce temperature fluctuation
   c. prevent overheating
   d. sense temperature inside the incubator chamber

5. Preferred characteristics of a transport infant incubator include:

   i.   light weight
   ii. small size
   iii. adequate portable power supply

   a. i only
   b. i and ii only
   c. i and iii only
   d. i, ii, and iii

6. Which region of the visible light contains the effective phototherapy spectrum?

   a. near infrared radiation
   b. yellow light
   c. blue light
   d. ultraviolet radiation

7. The primary source of heating in an infant warmer is

   a. near infrared radiation
   b. yellow light
   c. blue light
   d. ultraviolet radiation

8. A layer of Plexiglas between the phototherapy light source and the infant is to absorb:

    a. near infrared radiation
    b. yellow light
    c. blue light
    d. ultraviolet radiation

9. During observation of an infant under phototherapy treatment, blue light is turned on while white light is turned off.

    a. true
    b. false

10. To prevent overheating, the radiant heater of an infant warmer when set to skin temperature mode will turn off when the skin sensor is disconnected from the warmer.

    a. true
    b. false

11. Increasing the distance between the infant and the phototherapy light source may reduce the effectiveness of phototherapy.

    a. true
    b. false

12. Which of the following is not a component of an infant resuscitator in the labor and delivery area

    a. radiant warmer
    b. oxygen supply
    c. Apgar timer
    d. respiration monitor

Chapter 30

1. In general, which of the following provides a more accurate measurement of core body temperature

    a. pulmonary artery
    b. oral cavity
    c. skin surface
    d. rectum

2. The transducer element of a YSI400 temperature probe is a.

    a. RTD
    b. thermistor
    c. thermocouple
    d. IC temperature sensor

3. The resistance value measured across the terminals of a YSI400 temperature
    probe at 37 °C is 1,355 kΩ.

    a. true
    b. false

4. The main function of the proportional heating system in an infant incubator is
    to

    a. maintain the set temperature
    b. reduce temperature fluctuation
    c. prevent overheating
    d. sense temperature inside the incubator chamber

5. Which of the following is (are) correct for YSI400 and YSI700 temperature
    probes?

    i.   a YSI700 probe has more temperature sensing elements
    ii. the resistance-temperature characteristic of YSI400 is linear while
         YSI700 is non-linear
    iii. both have negative temperature coefficient

    a. i only
    b. i and ii only
    c. i and iii only
    d. i, ii, and iii

6. IR thermometer detect the infrared radiation from the patient to determine the
    patient's body temperature.

    a. true
    b. false

7. Which of the following body temperature sensor does not require good ther-
    mal contact with the patient?

    a. electronic thermometer using an RTD
    b. electronic thermometer using a thermistor
    c. liquid-in-glass thermometer

d. tympanic thermometer

8. Which of the following body temperature thermometer has the fastest respond time:

   a. electronic thermometer using an RTD
   b. electronic thermometer using a thermistor
   c. liquid-in-glass thermometer
   d. tympanic thermometer

9. In IR thermometry, which of the following is not an important factor to ensure accurate temperature measurement.

   i. field of view
   ii. distance from the object
   iii. emissivity of the object

   a. i only
   b. ii only
   c. iii only
   d. i, and iii only

10. The Wien's displacement law states that the wavelength corresponding to the maximum energy intensity in the radiated energy spectrum is proportional to the temperature of the heated object.

   a. true
   b. false

11. According to the Wien's displacement law, the temperature in °C of an object corresponding to a peak wavelength of 9.26 μm is

   a. 35
   b. 37
   c. 39
   d. 41

12. Which of the following contributes to errors in body temperature measurement using a tympanic thermometer.

   i. non straight ear canal
   ii. object blocking the optical pathway
   iii. too wide field of view

   a. i only
   b. i and ii only

c. i and iii only
d. i, ii, and iii

Chapter 31

1.   Which of the following is a SaO_2 level of a healthy adult?

a. 50%
b. 74%
c. 89%
d. 97%

2.   Which of the following is a PaO_2 level of a healthy adult?

a. 50 mmHg
b. 70 mmHg
c. 80 mmHg
d. 90 mmHg

3.   Red light is attenuated more than infrared light when it travels through a blood sample.

a. true
b. false

4.   The principle of transcutaneous oximetry is based on the Beer-Lambert Law.

a. true
b. false

5.   The main cause to prevent a pulse oximeter to determine the oxygen saturation in blood similar to a co-oximeter is:

a. The absorption of red and infrared light by oxyhemoglobin and deoxyhemoglobin are different
b. The absorption of red and infrared light cannot be isolated from each other
c. diameter of capillaries vary with time
d. there are other tissues attenuating the red and infrared light from the pulse oximeter probe

6.   Which of the following pulse oximeter probe is a reflecting probe which detects the scattered light from the tissue?

a. finger probe
b. earlobe probe

c. forehead probe
d. neonatal toe probe

7. The RED LED and IR LED in a pulse oximeter probe will never be tuned on at the same time.

    a. true
    b. false

8. Which of the following is/are true regarding a transcutaneous oxygen probe

    i.   the sensor is a Clark cell
    ii. a heater inside the probe to increase the temperature
    iii. a membrane at the surface to react with oxygen from the tissue

    a. i only
    b. i and ii only
    c. i and iii only
    d. i, ii, and iii

9. A defective temperature sensor under-detects the temperature of a tcpO2 probe. What is the effect on the monitor output?

    a. over estimation of the actual level
    b. under estimation of the actual level
    c. continue to provide accurate measurement
    d. produce a blank display

10. An oxygen analyzer measures

    a. FiO2
    b. SaO2
    c. SpO2
    d. PaO2

11. In pulse oximetry, too much noise at the probe optical detector will produce a lower than actual reading.

    a. true
    b. false

12. Which is the following is not decreasing the signal to noise ration (SNR) in pulse oximetry using a reusable finger probe?

    a. nail polish on the finger
    b. low perfusion

c. high blood pressure
d. strong ambient light

Chapter 32

1. End tidal carbon dioxide monitoring is used as a noninvasive method to esti-
    mate the carbon dioxide partial pressure of arterial blood in the pulmonary
    artery.

    a. true
    b. false

2. Which of the following is a $EtCO_2$ level of a healthy adult?

    a. 40 mmHg
    b. 70 mmHg
    c. 80 mmHg
    d. 90 mmHg

3. Which of the following patient ventilation parameters can be obtained from a
    capnogram?

    i. CO_2 level in the inspired air
    ii. CO_2 level in the expired air
    iii. respiration rate

    a. i only
    b. i and ii only
    c. i and iii only
    d. i, ii, and iii

4. The principle of end-tidal CO_2 monitoring is based on the Beer-Lambert Law.

    a. true
    b. false

5. In the $EtCO_2$ monitor, the center wavelength of the filter in the sensor to mea-
    sure the CO_2 absorption is:

    a. 1.2 µm
    b. 3.7 µm
    c. 4.2 µm
    d. 9.8 µm

6. Which of the following is not a critical component in a mainstream capnograph sensor?

    a. a Red LED
    b. an Infrared LED
    c. a heater
    d. a temperature sensor

7. In a side stream end-tidal carbon dioxide monitor, the optical bench is attached to the patient's breathing circuit.

    a. true
    b. false

8. The main function of heating the expired air passing through the sensor is to

    i.   prevent water condensation on the detection window
    ii. kill bacteria in the sensor
    iii. stabilize the CO_2 concentration in the expired air

    a. i only
    b. i and ii only
    c. i and iii only
    d. i, ii, and iii

9. The main reason of not using side stream sensing in end-tidal carbon dioxide monitoring on neonates is

    a. high risk of infection
    b. large error due to low tidal volume
    c. large error due to small dead space volume
    d. excessive condensation due to high humidity in expired air

10. Which of the following will cause error in end-tidal CO_2 monitoring?

    i.   fluctuation of gas pressure inside the sensor
    ii. fluctuation of gas temperature inside the sensor
    iii. fluctuation of anesthetic gas concentration in patient breathing circuit

    a. i only
    b. i and ii only
    c. i and iii only
    d. i, ii, and iii

Chapter 33

1. State the three stages of inhalation anesthesia

2. Which of the following is not a function of an anesthesia machine

    i. dispense a controlled mixture of anesthetic gases
    ii. assist patient's ventilation
    iii. monitor the patient's condition

  a. i only
  b. i and ii only
  c. i and iii only
  d. i, ii, and iii

3. Name the 3 basic subsystem of an anesthesia machine.

4. Which of the following is not a volatile anesthetic agent.

  a. desflurane
  b. nitrous oxide
  c. halothane
  d. sevoflurane

5. The anesthesia machine subsystem designed to protect the operating room personnel is

  a. gas supply and control
  b. breathing and ventilation
  c. scavenging
  d. monitoring

6. The component in an anesthesia machine to prevent back flow of gas is a

  a. regulator
  b. check valve
  c. flush valve
  d. shut of valve

7. The component in an anesthesia machine to reduce gas pressure is a

  a. regulator
  b. check valve
  c. flush valve
  d. shut of valve

8. In general anesthesia, which gas is never absence in the patient inspire air?

   a. oxygen
   b. nitrous oxide
   c. nitrogen
   d. helium

9. The pin-indexed safety system is for correct connection of

   a. a gas cylinder to the anesthesia machine
   b. a gas from piped-wall outlet to the anesthesia machine
   c. a patient breathing circuit to the anesthesia machine
   d. a scavenging circuit component to exhaust

10. The diameter-indexed safety system is for correct connection of

    a. a gas cylinder to the anesthesia machine
    b. a gas from piped-wall outlet to the anesthesia machine
    c. a patient breathing circuit to the anesthesia machine
    d. a scavenging circuit component to exhaust

11. Which of the following is not a feature to prevent wrong administration of anesthetic agent to the patient?

    a. Agent specific keyed filling spouts
    b. Colour coded vaporizer and filling bottle
    c. Interlock to prevent turning on more than one anesthetic agent
    d. Touch coded control knob on the vaporizer for each agent

12. Which of the following component in an anesthesia machine prevents barotrauma of the patient's lungs under anesthesia

    a. carbon dioxide absorber
    b. check valve
    c. APL valve
    d. pressure regulator

13. When the carbon dioxide absorber is reaching its end-of-life, it will change its color from pink to white.

    a. true
    b. false

14. Which is the following is not decreasing the signal to noise ration (SNR) in pulse oximetry using a reusable finger probe?

  a. nail polish on the finger
  b. low perfusion
  c. high blood pressure
  d. strong ambient light

Chapter 34

1. State the two physical principles of dialysis

2. Which of the following is not a function of an artificial kidney?

  a. remove waste products from body fluid
  b. maintain acid-base balance
  c. regulate external overtaken products
  d. control composition of urine

3. The parameter to measure the performance of water removal of an artificial kidney is call plasma clearance.

  a. true
  b. false

4. Which of the following is not a process to treat raw water for use in dialysis?

  a. Cartridge filter to remove particles
  b. Water softener to remove ions
  c. Reverse osmosis to remove smaller particles and remaining chemicals
  d. Chlorine and chloramines to kill bacteria and micro-organism

5. Which is the following is used to eliminate spore-forming bacteria inside a dialysis machine?

  a. ultra-violet irradiation
  b. ethylene oxide gas
  c. chemical such as formaldehyde
  d. heated solution above 85 °C

6. What is the volume of water removed from a patient in a 3 hour dialysis treatment if the dialyser Kuf = 2.5 ml/hr/mmHg, at a trans-membranous pressure setting = 200 mmHg?

  a. 1.0 L
  b. 1.5 L
  c. 2.0 L
  d. 2.5 L

7. The purpose of monitoring the conductivity of dialysate during a dialysis process is to

   a. monitor dialysate concentration
   b. maintain blood pH
   c. prevent blood clot in the dialyser
   d. detect amount of dissolved air in the extracorporeal blood circuit

8. Which of the following component(s) is(are) used to reduce air getting into the blood returning to the patient?

      i.   air/foam detector in extra-corporeal blood circuit
      ii. de-aeration pump in the dialysate circuit
      iii. heat-exchanger in the dialysate circuit

   a. i only
   b. i and ii only
   c. i and iii only
   d. i, ii, and iii

9. In peritoneal dialysis, dialysate is continuously being circulated into the peritoneal membrane of the patient.

   a. true
   b. false

10. A common complication in a patient on peritoneal dialysis is

    a. gas embolism
    b. peritonitis
    c. excessive removal of water from patient
    d. failure of semi-permeable membrane leading to dialysate mixing with blood

11. To remove potassium ions from the patient, the dialysate concentration of potassium ions needs to be higher than that in patient's blood.

    a. true
    b. false

12. Which of the following is not a common vascular access in hemodialysis?

    a. arteriovenous shunt
    b. arteriovenous graft
    c. arteriovenous fistula
    d. arteriovenous cannula

13. To remove water from the patient during hemodialysis, the pressure in the blood compartment must be

    a. higher than the osmotic pressure across the semipermeable membrane of the dialyser
    b. lower than the osmotic pressure across the semipermeable membrane of the dialyser
    c. greater than zero but lower than the osmotic pressure across the semipermeable membrane of the dialyser
    d. equal to the pressure of the dialysate

14. Blood clotting in the dialyser hollow fibers during hemodialysis is usually reduced by

    a. running higher blood flow
    b. running higher dialysate flow
    c. infusing heparin into the arterial blood
    d. infusing heparin into dialysate

Chapter 35

1. Laser is the acronym of

2. Which of the following is not a characteristic of laser beam from the lasing cavity?

    a. monochromatic
    b. coherent
    c. collimated
    d. high intensity

3. The energy density of a laser beam is highest at its focal spot.

    a. true
    b. false

4. CO_2 laser is considered a general surgical laser. The reason(s) is(are):

    i. it is highly absorbed by water
    ii. soft tissue are composed mainly of water
    iii. it is in the infrared radiation band

    a. i only
    b. i and ii only
    c. i and iii only

d. i, ii, and iii

5. Which laser is often used in retinal vascular procedures?

   a. Carbon dioxide laser
   b. Excimer laser
   c. Argon laser
   d. Nd:YAG laser

6. Which is the best transverse electromagnetic mode of the laser beam when used in surgery?

   a. TEM00
   b. TEM01
   c. TEM11
   d. TEM10

7. Extensive cooling is often built into a surgical laser because conventional laser generation is highly inefficient.

   a. true
   b. false

8. Laser tissue effect depends on

   i.   type of laser
   ii. type of tissue
   iii. exposure time

   a. i only
   b. i and ii only
   c. i and iii only
   d. i, ii, and iii

9. Which of the following surgical lasers cannot be transported by using optical fiber?

   a. Carbon dioxide laser
   b. Excimer laser
   c. Argon laser
   d. Nd:YAG laser

10. A Class 2 laser in use will not pose any risk to a bystander.

   a. true
   b. false

11. Which of the following measures are used to protect workers in an operating room where surgical lasers are used.

    i.   eye protection
    ii. physical distancing
    iii. smoke evacuation

  a. i only
  b. i and ii only
  c. i and iii only
  d. i, ii, and iii

12. According to ANSI Z136.3-2011, surgical lasers are

    i.   Class 4 lasers
    ii. Class 3B lasers
    iii. Class 3R lasers

  a. i only
  b. i and ii only
  c. i and iii only
  d. i, ii, and iii

Chapter 36

1. A light pipe and viewing optics are required in an endoscopic surgery .

  a. true
  b. false

2. What is the function of an insufflator in endoscopic procedures?

  a. illuminate the surgical field
  b. remove smoke in electrosurgical procedure
  c. supply medical air to the surgical site
  d. maintain a pneumoperitoneum

3. State the disadvantages of laparoscopic surgery.

4. Explain the main difference between a videoscope and fiberscope.

5. Which of the following is not an advantage of LED light sources used in endoscopy compare to conventional ones?

  a. better energy efficient

b. lower cost
c. generate less heat
d. last much longer

6. What is the preferred method to change the output intensity of an endoscopic light source without affecting its color temperature?

a. changing supply voltage to the light source
b. changing supply current to the light source
c. changing both the supply voltage and current to the light source
d. changing the output opening aperture of the light source

7. Which of the following is a rigid endoscope?

a. otoscope
b. colonoscope
c. bronchoscopes
d. gastroscope

8. Compared to open procedures, which of the following is(are) advantage(s) of laparoscopic surgeries?

i. less breeding
ii. faster recovery
iii. lower risk of infection

a. i only
b. i and ii only
c. i and iii only
d. i, ii, and iii

9. Major limitation(s) of laparoscopic surgery include

i. poor depth perception
ii. poor image resolution
iii. poor tactile feedback

a. i only
b. i and ii only
c. i and iii only
d. i, ii, and iii

10. Heat related injury related to endoscopic illumination can be reduced by

i. presence of infrared filter at the output of the endoscopic light source
ii. avoid prolonged illumination at the same tissue location

   iii. avoid unnecessary high illumination intensity

a. i only
b. i and ii only
c. i and iii only
d. i, ii, and iii

11. Three-dimensional endoscopy will

a. increase image resolution
b. improve tactile feedback
c. restore depth perception
d. increase image field of view

12. Which of the following processing is used on flexible endoscopes to prevent cross infection?

a. high level disinfection using glutaraldehyde
b. ultrasound cleaning
c. disinfection using alcohol
d. sterilization with pressurized steam

Chapter 37

1.   Which of the following is(are) differences between an HLM and ECMO unit?

   i.   HLM has a blood warmer which is absent in an ECOM units
   ii.   HLM has a built-in defibrillator to stop the patient's heart which is absent in an ECMO units
   iii. The procedures using ECMO units are often longer than those using HLMs

a. i only
b. ii only
c. iii only
d. i, ii and iii

2.   ECCO$_2$R procedures are procedures using CPB to remove CO$_2$ from patients

a. true
b. false

3.   About 97% of the oxygen transported from the lungs to tissues is carried by dissolved oxygen in the blood plasma.

a. true
b. false

4. A gas mixture at 780 mmHg has an oxygen concentration of 50%. What is the oxygen partial pressure in the gas mixture?

    a. 380 mmHg
    b. 390 mmHg
    c. 760 mmHg
    d. 780 mmHg

5. Oxygen in arterial blood is primarily stored in

    a. blood plasma
    b. dissolved gas in blood
    c. oxygenated hemoglobin
    d. carbamino-hemoglobin

6. In a membrane oxygenator, blood flows over a solid surface in an oxygen-enriched compartment to allow oxygen to diffuse into the blood.

    a. true
    b. false

7. The purpose of the blood defoamer in a CPB unit is

    a. to prevent air blockage in the hollow fibers of the membrane oxygenator
    b. to prevent air embolism when blood is returned to the patient
    c. to prevent blockage of the extracorporeal blood circuit
    d. to increase overall system gas transport efficiency

8. Which of the following is(are) desired characteristics of an oxygenator?

    i.   low blood pressure drop
    ii. high efficiency heat exchange
    iii. good O2 and CO2 transfer performance

    a. i only
    b. i and ii only
    c. i and iii only
    d. i, ii, and iii

9. The functions of the cold cardioplegia solution infused into the heart and coronary arteries include:

    i.   lowers the metabolic rate of the heart muscle
    ii. causes cardiac arrest
    iii. reduces myocardial oxygen consumption

a. i only
b. i and ii only
c. i and iii only
d. i, ii, and iii

10. The additional volume of fluid from the cardioplegia solution contributes to hemodilution during CPB. Which device can be added to the extracorporeal blood circuit to remove this extra fluid?

a. arterial defoamer
b. hemoconcentrator
c. cardiotomy suction pump
d. cardioplegic pump

11. Deep hypothermia using cold cardioplegia at temperature less than 28°C is used to create temporary cardiac arrest during a CPB procedure.

a. true
b. false

12. An advantage of a centrifugal pump over a roller arterial blood pump is

a. it produces higher blood flow
b. it does not require a one-way valve to prevent back flow
c. it causes more hemolysis to blood cells
d. it is able to maintain sterility during operation

13. Embolism caused by air or particulates is one of the most severe hazards during a CBP procedure. Which of the following are deployed prevent this hazard?

    i.   add bubble traps to the blood circuit
    ii. add filters to the blood circuit
    iii. ensure tight and proper connections of blood and fluid tubinb

a. i only
b. i and ii only
c. i and iii only
d. i, ii, and iii

14. To minimize patient's blood loss during a long CBP procedure, blood from the

surgical field is returned to the venous reservoir via

a. aortic root suction pump
b. ventricular suction pump
c. cardiotomy suction pump
d. cardioplegic pump

Chapter 38

1. The speed of sound is highest in a high temperature vacuum environment.

a. true
b. false

2. Which of the following is not a cause of hearing impairment?

a. aging
b. obesity
c. acoustic trauma
d. prolonged exposure to high level noise

3. The intensity of sound is proportional to the square of the sound pressure

a. true
b. false

4. A sound level 10 times the threshold of hearing is

a. 1 dB
b. 10 dB
c. 20 dB
d. 100 dB

5. The SPL in μPa at the threshold of hearing is

a. -10
b. 0
c. +10
d. + 20

6. The frequency response of the overall auditory system is independent of the size and shape of the ear canal

a. true
b. false

7. What is the function of the large size of the tympanic membrane and the lever mechanism of the ossicles?

    i.   amplify the sound vibration
    ii. transmit the sound vibration to the inner ear
    iii. change the frequency of sound vibration

    a. i only
    b. i and ii only
    c. i and iii only
    d. i, ii, and iii

8. Which of the following is not true regarding the cochlea?

    i.   high-frequency sounds stimulate the hair cells near the apex of the cochlea
    ii. vibration causes hair cells to bend and stimulate the neuron endings creating biopotential impulses
    iii. higher intensity sounds produce higher nerve impulses due to greater vibration intensity

    a. i only
    b. i and ii only
    c. i and iii only
    d. i, ii, and iii

9. Which weighting scale is used in human hearing measurement at normal sound level?

    a. dB
    b. dBA
    c. dBC
    d. dBX

10. Conductive hearing loss happens when sounds cannot get through

    i.   the inner ear
    ii. the outer ear
    iii. the middle ear

    a. i only
    b. i or ii only
    c. ii or iii only
    d. i, ii, or iii

11. Which of the following devices is used to assess middle ear problem?

    a. Audiometer
    b. Tympanometer
    c. Otoacoustic emission detector
    d. Auditory brain stem response unit

12. A tympanogram is a plot of sound impedance against the pressure inside the ear canal.

    a. true
    b. false

13. Which of the following is(are) challenge(s) in SOAE measurement?

    i. SOAEs are very small and easily corrupted by noise
    ii. it needs the cooperation of the subject to obtain accurate measurement
    iii. it requires an appropriate stimulation to trigger the SOAE

    a. i only
    b. i and ii only
    c. ii or iii only
    d. i, ii, or iii

14. Which of the following statements is not correct regarding cochlear implants?

    a. it replaces the function of damaged hair cells inside the patient's cochlear
    b. it requires healthy auditory nerves, otherwise it will not work
    c. it can improve the hearing of people who have severe sensorineural hearing loss
    d. patients who are born deaf are not going to benefit from cochlear implant

ANSWERS TO REVIEW QUESTIONS

Chapter 1

1. To reduce the risk of micro shock on patients

2. C

3. A

4. C

5. C

6. To study the response between the host and implant material during and after the implantation process

7. C

8. B

9. D

10. The time period when the sodium-potassium pump repolarizes the cell to its resting state is called the refractory period. During the refractory period, the cell is not responsive to any stimulation.

11. A

12. The risk index of a hazardous situation is a measure of its risk severity based on its frequency of occurrence and the degree of harm it may cause

13. C

Chapter 2

1. A

2. A

3. $\Delta Z = \sqrt{(c\Delta A)^2 + (m\Delta B)^2} = \sqrt{(0.5 \times 0.3)^2 + (2 \times 0.8)^2} = 1.61$

4. $\dfrac{\Delta Z}{Z} = \sqrt{\left(\dfrac{\Delta x}{x}\right)^2 + \left(-\dfrac{\Delta y}{y}\right)^2 + \left(2\dfrac{\Delta w}{w}\right)^2 + \left(\dfrac{\cos\theta\Delta\theta}{2-2\sin\theta}\right)^2}$

5. D

6. +C/(A+B)

7. C – D

8. D [since the input is outside of the range specified in the specification]

9. C

10. B

11. C

12. D

Chapter 3

1. A

2. D

3. B

4. C

5. D

6. A

7. B

8. $V_0 = V_E \ast \Delta Z/Z$

9. * $\Delta Z/\Delta m = 0.5\ \Omega\ /kg$, $V_0 = \frac{1}{2} * V_E * \Delta Z/Z = \frac{1}{2} * V_E * 0.5* \Delta m/Z$. $S = V_0/\Delta m$
 $= \frac{1}{2} * 0.5*V_E/Z = 0.25*10/1000\ V/kg = 2.5\ mV/kg$

10. 15 m

Chapter 4

1. D

2. 10.3 m [from Table 4-1, PSTP = 1.36×760 cm = 1033.6 cm = 10.3 m]

3. 14.7 psi

4. 984 cmH_2O [using Equation 4.3, $P_b = P_a - \rho gh = 1034$ cm − 50 cm = 984 cmH_2O]

5. * 16.0/10.6 kPa [$P_s = 120 \times 133$ Pa = 15960 Pa = 16.0 kPa, $P_d = 80 \times 133$ Pa = 10640 Pa = 10.6 kPa]

6. A

7. C

8. 2.5 [G.F. $= \dfrac{\Delta R/R}{\Delta L/L} = 1.50/100/6000 \times 10^{-6} = 2.50$]

9. 4.0 N [given G.F. $= \dfrac{\Delta R/R}{\Delta L/L}$

 Axial strain = $(102.5 - 100.0) / 100 / 2.50 = 1.0 \times 10^{-2}$
 $\Delta L = 1.0 \times 10^{-2} \times 1.0$ mm = 1.0×10^{-2} mm
 $F = 2.00$ N $* 1.0 \times 10^{-2}$ mm / 5.0 μm = 4.0 N]

Chapter 5

1. C

2. 34.4 °C

3. 307.6 K

4. A

5. 50 ^{0}C

6. 13 ^{0}C

7. 2.6 °C [= (0.5 + 0.5)/0.385]

8. R = 1390 Ω

9. -0.02 °C [temp coef = -1390×4000/(37+173)^2 = -57.9 Ω/K; T = 1/-57.9 = -0.02 K]

10. C

11. B

12. D

Chapter 6

1. 270*33/100 = 79^0

2. C

3. C

4. A

5. A

6. $v = \dfrac{dx}{dt}$, a = $\dfrac{dv}{dt} = \dfrac{d^2x}{dt^2}$

7. v = dD/dt = 4/(6-1) = 0.8 m/s

8. a = dv/dt = 0

9. a = dv/dt = 8/(6-1) = 1.6 m/s^2

10. m = F/a = 1.2/1.6 = 0.75 kg or 750 g.

Chapter 7

1. 30 cm/s [Q = v × A, 4500/60 cm^3/s = v × 2.5 cm^2, v = 4500/60/2.5 = 30 cm/s]

2. 47 cm/s [Cross sectional area proportional to D^2, Cross sectional area become
    2.5 × (1-0.2)^2 = 1.6 cm^2, v = 4500/60/1.6 = 47 cm/s]

3. D

4. D

5. A

6. C

7. A

8. 26 ml/s [Use Bernoulli's equation or Equation 7.9 in test, use 1000 kg/m^3]

9. 130 Pa [Use Poiseuille's Equation 7.4, $\Delta P = 8v\eta L/D^2$]

10. D

11. B

12. A

Chapter 8

1. A

2. 3.0×10^{17} Hz [f $= c//\lambda = 3 \times 10^8$ ms^{-1}/1.0 nm $= 3.0 \times 10^{17}$ Hz]

3. 2.7×10^{-16} J [E = hf $= 6.625 \times 10^{-34} \times 3.0 \times 10^{17} = 2.7 \times 10^{-16}$ J]

4. Radiometry is the measurement of quantities associated with radiant energy, while photometry is limited to the human visible spectrum.

5. C

6. A black body is an entity that absorbs all incident radiation, with no transmission or reflection.

7. D

8. B

9. C

10. A

11. A

12. C

13. A

14. C

Chapter 9

1. A

2. C

3. 0.78 V

   [using $E^0_{cell} = E^0_{Cathode} + (-E^0_{Anode})$, $E^0_{cell} = 0.34$ V + (-0.44 V) = 0.78 V]

4. A

5. hydrogen (0 V), Ag/AgCl (0.2 V), or Calomel (0.24 V)

6. 1.65 × 10-9 M

   [using the Nernst equation $E = +0.80 - \dfrac{0.0592}{1} \log \dfrac{1}{[Ag^+]} =$
   $+0.80 - 0.0592\log \dfrac{[Cl^-]}{K_{sp}}$,

   log{[Cl⁻]/(1.6 × 10⁻¹⁰)} = - (0.74 – 0.80)/0.0592 = + 1.014, therefore [Cl⁻] = 10.3 × 1.6 × 10⁻¹⁰ = 1.65 × 10⁻⁹ M]

7. D

8. B

9. B

10. D

11. A

12. D

Chapter 10

1. A

2. • absence of distortion or electrical noise
    • immunity to external interference
    • inertness of the electrode in the presence of tissue and bodily fluid
    • absence of interference with or influence on the tissue
    • absence of interference with or influence on the movement of the subject
    • ease of making contact with the biological source
    • invariance of the contact even during long periods of time
    • absence of discomfort to the subject
    • repeatability of results

3. C

4. see Figure 10.4

5. B

6. B

7. C

8. A

9. B

10. D

Chapter 11

1. • Very large input impedance
    • Very low output impedance
    • Constant differential gain with zero nonlinearity
    • High common mode rejection (that is, very small common mode gain)
    • Very wide bandwidth with no phase distortion
    • Low DC offset voltage or drift
    • Low input bias current and offset current
    • Low noise

2. 2.5×10^{-4} [Ac = 1.0 mV/4.0 V = 2.5×10^{-4}]

3. 4.0×10^3 [Ad = 8.0 V/(3.0-5.0) mV = -4.0×10^3]

4. 140 dB [CMRdB = 20 logCMRR = 20 log | Ad/Ac | = 20 log 1.6 × 10^7 = 140 dB]

5. A

6. C

7. 0.5 V [Vo = Ad × Vi = 1000 × 0.5 mV = 0.5 V = 500 mV]

8. 0.0125 [Ac = Ad/CMRR = 1000/80000 = 0.0125]

9. [60 Hz common mode signal at both IA input = 10 mV; Vo = 10 × 0.0125 = 0.125 mV

   60 Hz differential signal across the IA input = 10 × [2000/2040 − 2000/2050] mV = 4.78 × 10^{-2} mV; Vo = 4.78 × 10^{-2} × 1000 mV = 47.8 mV
   Total 60 Hz signal amplitude = 47.8 + 0.125 mV = 48 mV
   Note that this is almost 10% of the ECG magnitude at the IA output]

10. D

11. D

Chapter 12

1. C

2. D

3. B

4. D

5. 2.4 mA [I = 120 V/50 kΩ = 2.4 mA

6. > 1mA but < 5 mA; patient will feel the electrical current but below safe current limit [refer to Table 12-1]

7. > 20 µA; patient will suffer from ventricular fibrillation [refer to Table 12-2]

8. A

9. B

10. C

11. C

12. D

Chapter 13

1. B

2. B

3. A

4. 24,883,200 LCD elements [3840×2160×3 = 24,883,200]

5. D

6. A

7. 2.0 mV [5 mm/mv = 10 mm/V; V = 10/5 = 2 mV]

8. 60 bpm [d = 25 mm, Period = 25/25 = 1.0 s, HR = 60 bpm]

9. - 5.0% [(19-20)/20 ×100% = - 5.0%]

10. C

11. 0.075 Hz [use RC = (t1-t2)/ln(V2/V1) = (1-2)/ln(2.5/4) = 2.128, f_L = 1/(2πRC) = 0.075 Hz

12. 0.055 Hz [use f_L = 0.11/t = 0.11/2.0 = 0.055 Hz]

Chapter 14

1. B

2. A

3. D

4. C

5. A

6. • Easily adaptable and configurable to all site-specific requirements (i.e., different number of monitored bedside system, distances, etc.)
   • Easy and fast to move information throughout the network
   • Allow modifications or changes to system without loss of required function of the rest of the system
   • Disconnect and reconnect equipment to system without disruption
   • Scalable and allow variability of monitoring parameters
   • Compatible with other network equipment and system (i.e., adheres to industry standards)

7. D

8. A

9. D

10. D

11. A

12. D

Chapter 15

1. D

2. B

3. C

4. A

5. A

6. C

7. B

8. D

9. D

10. B

11. A

12. C

13. • Power frequency interference via conduction or capacitive coupling
    • Radiated electromagnetic interference (e.g., ESU, cell phones)
    • Interfering signals from other equipment connecting to the patient
    • Magnetic field induction

14. • Skeletal muscle contraction
    • Breathing action
    • Patient movement
    • Involuntary muscle contraction

Chapter 16

1. True

2. False

3. True

4. True

5. False

6. False

7. A

8. D

9. C

10. D

11. B

12. D

13. A

Chapter 17

1. D

2. B

3. D

4. Spatial recruitment – number of motor units activated, temporal recruitment – the firing rate of individual motor units. The more motor units recruited, the higher the firing rate, the higher strength developed by the muscle

5. A

6. A

7. C

8. B

Chapter 18

1. False

2. D

3. A

4. A

5. B

6. offset pressure = 1020 x 9.8 x 0.2 = +14 mmHg; BP reading = 145/115 mmHg.

7. C

8. D

9. B

10. transducer sensitivity S = 5 μV/V/mmHg. Output voltage = S x 5.0 x 100 = 2500 μVac = 2.5 mVac

Chapter 19

1.   A

2.   B

3.   C

4.   False

5.   True

6.   B

7.   C

8.   False

9.   A

10. C

Chapter 20

1.   A

2.   C

3.   B

4.   C

5.   B

6.   A

7.   B

8.   B

9.   C

10. B

11. C

12. A

Chapter 21

1. D

2. A

3. C

4. C

5. True

6. D

7. C

8. B

9. C

10. C

11. C

12. C

Chapter 22

1. B

2. B

3. A

4. C

5. D

6. B

7. D

8. C

9. B

10. C

11. B

12. B

13. use $E = 1/2CV^2 = 0.5 \times 64 \times 10^{-6} \times 3000^2 = 288$ J

14. use $E = V \times I \times t$
    $E = 200$ J $\times 10 = 2000$ J,
    $I \times t = 2000$ J $/ 12$ V $= 167$ A.s $= 167000$ mAs $= 167000/60/60$ mAh $= 463$ mAh

Chapter 23

1. A

2. Solution spike and drip chamber, regulating clamp, occlusion clamp, Y-connection site, luer lock connector.

3. B

4. B

5. D

6. B

7. A

8. C

9. A

10. D

11. 120 ml/hr = 120 x 60 drops/hr = 120 drops/min.

12. B

13. A

14. 19 ml/10 min = 114 ml/hr. % Error = (114-120)/120 x 100% = - 5%

Chapter 24

1. Frequency of current less than 100 kHz can stimulate muscle and nerve tissues

2. C

3. C

4. B

5. D

6. A

7. B

8. 500 Ω non-inductive resistance

9. D

10. B

11. A

12. D

13. C

14. A

Chapter 25

1. B

2. A

3. C

4. B

5. A

6. True

7. A

8. 500 Ω non-inductive resistance

9. It measures the change in resistance of the thermistor when it is heated or cooled by the breathing air from the patient.

10. D

11. Use bacterial filters, use single use disposable patient air circuits, frequent cleaning and disinfection of equipment.

12. A

Chapter 26

1. False

2. True

3. C

4. D

5. D

6. B

7. A

8. C

9. D

10. C

11. True

12. D

Chapter 27

1. A

2. True

3. C

4. B

5. B

6. True

7. D

8. True

9. B

10. C

Chapter 28

1. A

2. False

3. D

4. B

5. D

6. False

7. B

8. C

9. C

10. False

Chapter 29

1. A

2. False

3. C

4. B

5. D

6. C

7. A

8. D

9. False

10. True

11. True

12. D

Chapter 30

1. A

2. B

3. False

4. B

5. C

6. True

7. D

8. D

9.   B

10. False

11. C

12. D

Chapter 31

1.   D

2.   D

3.   False

4.   False

5.   C

6.   C

7.   True

8.   B

9.   A

10. A

11. True

12. C

Chapter 32

1.   False

2.   A

3.   D

4.   True

5. C

6. A

7. False

8. A

9. B

10. D

Chapter 33

1. induction, maintenance, and recovery

2. D

3. gas supply and control, breathing and ventilation, scavenging

4. B

5. C

6. B

7. A

8. A

9. A

10. B

11. D

12. C

13. False

14. C

Chapter 34

1. diffusion and osmosis

2. D

3. False

4. D

5. C

6. B

7. A

8. B

9. False

10. B

11. False

12. D

13. A

14. C

Chapter 35

1. light amplification by stimulated emission of radiation

2. D

3. A

4. B

5. C

6. A

7. True

8. D

9. A

10. True

11. C

12. A

Chapter 36

1. True

2. D

3. Limited dexterity and range of motion at the surgical site, loss of touch sensation and poor tactile feedback, procedures are not as intuitive due to the unconventional maneuvering mechanism of specialized instruments, poor depth perception from viewing anatomy on two-dimensional display monitor, cannot see surrounding and behind camera anatomy, may miss lesions or injuries outside the field of view, longer procedure time.

4. The image of a fiberscope is transmitted to the eyepiece by a flexible fiber optic bundle along the insertion tube, while a videoscope has a camera chip mounted at the tip of the insertion tube.

5. B

6. D

7. A

8. D

9. C

10. D

11. C

12. A

Chapter 37

1. C

2. True

3. False

4. B

5. C

6. False

7. B

8. D

9. D

10. B

11. True

12. D

13. D

14. C

Chapter 38

1. False

2. B

3. True

4. C

5. C

6. False

7.   B

8.   A

9.   B

10. C

11. B

12. False

13. B

14. D

INDEX

%O2Hb (see also fractional oxygen satura-
    tion), 586, 590
%SaO2, 269, 547, 583-600
    fractional, 583, 586, 590
    functional, 586-588, 590
%SpO2, 585, 589, 592, 596
%SvO2, 585, 711
    10-20 system, EEG, 318, 320, 321
    inion of, 321
    left auricular point of, 320
    nasion of, 321
    right auricular point of, 320
12-lead ECG, 291-306

A

AAMI Standards BP22 and BP23, 368
Ablation,
    ESU, 475, 493
    laser, 663
Abnormal fetal heart rate,
    bradycardia in, 547
    tachycardia in, 547
    variation in, 548
ABR (see auditory brainstem response)
Abrasion, laser 661
Absolute pressure, 69, 762
Absorbance, 586
Absorptivity, 586, 588
Acceleration, 103, 104
    Angular, 104
Access point, 760
Accuracy, 40
ACI (see activated clotting time)
Activated clotting time (ACL), 710
Action potential, 13-15, 180, 286
    muscle, 286

nerve, 337, 344
    single cell, 15, 180
Activated carbon, 649
Active breathing, 497, 516
Active cancellation, 208
Active electrode, 337, 338
Active matrix liquid crystal display
    (AMLCD), 249
Actuator, 56
Adaptive filtering, 597
ADC (see analog to digital conversion)
Adolf Fick, 385
AED (see automatic external defibrillator)
AGC (see automatic gain control),
AHA (see American Heart Association)
Airway resistance, 498-500, 511, 529, 530,
    532
AK (see artificial kidney),
Albert Einstein, 658
Alkaline cell, 175
Alveolar gas, 517, 704
Alveolar pressure, 497, 518
Alveolar ventilation, 504, 602
Alveoli, 354, 602, 704
Ambulatory ECG, 289
American Heart Association, 374, 442
American Society of Anesthesiologists, 584
AMLCD (see active matrix liquid crystal
    display)
Amperometry, 169
Amplification, 60, 199
Amplification factor, 60, 245, 739
Amplitude linearity, 60
Amplitude-Zone TimeEpoch-Coding
    (AZTEC), 274
Analgesic, 463, 609, 625
Analog modulation, 758

Analog to digital conversion, 30
Anatomic dead space, 500, 502
Anesthesia, depth of, 315, 610, 624
Anesthesia machine, 608-627
    adjustable pressure-limiting valve of, 619, 633
    agent monitor of, 624
    bellow of, 618
    breathing and ventilation subsystem of, 617-621
    breathing bag of, 618, 619
    check valve of, 613
    circle system of, 618
    CO_2 absorber of, 602, 624, 625
    CO_2 absorption canister of, 619
    color-coded gas cylinder of, 623
    color-coded gas hoses of, 623
    continuous-flow rebreathing, 612
    flow control valve of, 613
    flowmeter of, 613
    functional block diagram of, 612
    gas cylinder of, 612, 614, 624
    gas supply and control subsystem of, 613-617
    hanger yoke of, 612, 614
    hazard related to, 623
    manual breathing mode of, 618, 619
    oxygen failure protection detector of, 614
    oxygen flush valve of, 614
    oxygen ratio monitor of, 622, 623
    pop-off valve of, 618
    pressure-limiting valve of, 618, 620, 621, 623
    scavenging subsystem of, 621
    shutoff valve of, 613, 614
    T-piece design of, 620
    touch-coded control knob of, 622
    vaporizer interlock of, 616, 623
    vaporizer of, 613-617, 624
    ventilator mode of, 618-620
Anesthetic agent, 332, 609-625
    occupational hazard of, 621
Anesthetic gas, 610, 613-622, 625
    waste, 613, 618
Angular motion transducer, 104, 105
Anion, 160, 161
Anode, 143, 154, 156, 169, 170, 172, 175, 176, 244, 252, 593, 494
ANSI Z136.3-1988 Standards, 672, 675

Antepartum monitoring, 547
Antibradycardia pacing, 430
Anti-siphon valve (see syringe pump, anti-siphon vaive)
Antitachycardia pacing, 430
APL valve (see adjustable pressure-limiting valve of anesthesia machine)
Apnea, 313, 508, 521, 523
Appendectomy, 681
ARCnet network protocol, 278
Argon laser, 660, 666
Argon-enhanced ESU, 475
Arrhythmia, 273, 274, 383, 407, 428, 442
Arrhythmia detection, 166, 274, 275, 289
Arrhythmia detection algorithm, template
    cross-correlation, 274
    template matching, 274
    waveform feature extraction, 274
Arterial blood pressure monitor,
    amplifier of, 364
    bandwidth of, 364
    blood clot in, 359, 367
    continuous flush valve of, 358
    catheter of, 359, 367
    catheter error in, 346
    common problems of, 366-369
    display of, 365
    extension tube of, 362, 363, 367-369
    filter of, 364
    functional block diagram of, 362
    heparinized saline in, 358, 368
    patient port of, 359-361
    offset in, 359-361
    rapid flush valve of, 359
    setup of, 359
    setup error in, 366
    signal isolation of, 365
    signal processing of, 365
    spectral analysis of, 364, 365
    transducer of, 362-364
    transducer port of, 359, 360
    zero offset of, 364
    zero port of, 361
    zeroing process in, 359-361
Arterial blood sample, 585, 590
Arterial hypoxemia, 497
Arterial line, 358-361, 365, 366
Arterial oxygen saturation, 516, 521
Arterial tonometry, 380, 381
Arthroscopy, 681

Artificial kidney, 635-639
  blood compartment of, 629, 640, 644
  dialysate compartment of, 629, 640
  semipermeable membrane of, 629-631,
    634, 635, 637-639
Asepsis,
  medical, 765
  surgical, 765, 768
ATP (see standard atmospheric pressure)
Atrial fibrillation, 429, 440
Atrial flutter, 429, 440
Arteriovenous graft, 641
Arteriovenous shunt, 640, 641
Atrioventricular (AV) node, 284
Attenuation, 50, 723
Audiology equipment, 729-748
Audiometer,
  ear cups of, 735
  earphones of, 735
  insert earphones of, 732
  bone conductors (vibrators) of, 732, 734
  microphones of, 734
  loudspeakers of, 732, 733, 735
  types (1, 2, 3, 4) of, 733
Audiometric booth (see also sound booth),
    732, 73, 747
Audiometry,
  bone conduction threshold of, 733
  high frequency, 731
  pure tone, 731, 732
  speech, 731
  speech recognition threshold of, 733
  standard hearing threshold of, 733
Auditory ossicles, 726
Auditory brainstem response unit, 743
Augmented limb lead, 293
Auscultatory gap, 354
Autoclave, 768
Automatic external defibrillator (AED), 442
Automatic gain control, 689
Automatic NIBP monitor, 376
Arteriovenous fistula, 641, 651
AV node (see atrioventricular node)
Aversion response (see also laser safety haz-
    ard), 673
Axial strain, 69
Axonal velocity, 324
AZTEC (see Amplitude-Zone Time-Epoch-
    Coding)

B

BAHA (see bone-anchored hearing aid),
    744
Balanced bridge, 62
Band gap energy, 133
Bandwidth, 15, 49, 192, 266, 754
ECG, 16, 289, 305, 306
  blood pressure waveform, 264
  data transmission, 146, 254, 278, 759-761
  infrared thermometry, 576
  spontaneous acoustic emission, 740
Bang-bang, 44
Barb electrode, 412
Barometer, 68
Bassinet, 558, 561
Batteries, 172-176
  active capacity of, 174
  available compartment capacity of, 173,
    174
  capacity of, 173, 174
  dead compartment capacity of, 173, 174
  empty compartment capacity of,173, 174
  internal resistance of, 173, 174, 175
  nominal voltage of, 173, 175
  open circuit voltage of, 175
  operating voltage of, 175
  primary (see also primary cells), 172
  secondary (see also secondary cells), 172,
    174
  self-discharging of,174, 175
  shelf life of, 175
  thermal rrunaway of, 175
BCV (see biphasic cuirass ventilator)
Bedside monitor, 270, 271
  alarm of, 270
  common features of, 270
  freeze capability of, 270
  modular, 270
  preconfigured, 270, 271
  recorder of, 268, 271
  trending capability of, 270
Beer-Lambert Law, 585, 586, 588, 603
Bellow, 71, 72, 476, 506, 618
Bernoulli's equation, 114, 115
Bilirubin, 554, 778
Bimetallic sensor, 82, 83
Biocompatibility, 25, 26, 28
Biopotential,
  origins of, 12-15, 179-181

Biopotential amplifiers, 191-98
    noise in, 199-215
Biopotential electrodes, 178-187
    half-cell potential of, 182, 185
    nonpolarized, 182-185
    nonreversible, 183
    perfectly nonpolarized, 183-184
    perfectly polarized,182-184
    polarized,182-185
    reversible, 184
Biopotential signal, 14, 192, 199
Biosensors, 152, 170
BiPAP (see mechanical ventilator, bilevel
    positive airway pressure)
Biphasic cuirass ventilator (BCV), 515
Biphasic truncated exponential waveform,
    436
Bipolar electrode, 338, 479
Bipolar lead, 410
BIS (see bispectral index),
Bispectral index, 315, 323
Biventricular pacing, 411
Black body radiator, 129-131
Blink reflect (see laser safety hazard)
Blood compatibility, 27
Blood flow, 112, 116, 354, 356, 385, 387,
    398
Blood flowmeter, 536-544
Blood gas, 170, 497
    PCO_2 of, 168, 169, 705, 706
    PO_2 of, 168, 169, 595
Blood gas analyzer, 584, 589, 705, 711
Blood pH, 17, 409. 497
Blood pressure,
    arterial, 17, 354, 355
    venous, 17, 354, 355
    ventricular, 355
Blood pressure catheter error, 357, 368
    air bubble in, 367, 368
    blood clot in, 367, 368
    end pressure in, 367
    impact artifact in, 367
    leak in, 346, 367
    pinching in, 367
    whipping in, 367
    frequency response of, 368
Blood pressure transducer,
    calibration of, 368
    central floating block of, 362
    diaphragm of, 362, 363

excitation voltage of, 363
    piezoresistive strain gauge of, 362
    pressure dome of, 362
    resistive strain gauge of, 362, 363
    sensitivity of, 343
    strain wires of, 343
Bone vibrator, 732
Bony labyrinth, 726
    perilymph of, 726, 727
    semicircular canal of, 726, 727, 730, 745
    vestibule of, 626, 727
Body temperature, sea level pressure, and
    gas saturated with water vapor (see
    BTPS)
Body temperature monitors, 567-580
Blood compatibility, 27
Bouncing ball oscilloscope, 246
Bourdon tube, 70
BPEG (see British pacing and
    Electrophysiology Group)
Brain death, 315
Breakdown, 43
British pacing and Electrophysiology
    Group, 414
Bronchoscope, 682
BTPS, 507, 511, 522, 528
Bundle branches, 284
Bundle of His, 284, 406
Buried channel MOS capacitor, 137

C

Calibration, 37
Calibration factor, 368
Calomel reference electrode, 163, 164
Cannula,
    aortic, 710
    arterial, 713
    cardioplegia, 710
    laparoscopic 681,
    venous, 641, 711, 713
Capacitive coupling, 201, 696
CAPD (see continuous ambulatory peri-
    toneal dialysis)
Capnography (see also end-tidal CO_2 moni-
    tor), 602
Carboxyhemoglobin, 585
Cardiac activity, 286, 301
Cardiac arrhythmia, 274
Cardiac care units, 266

Cardiac cycle, 286, 355
Cardiac defibrillation, 442
Cardiac defibrillator,
  biphasic waveform of, 431, 436
  charge control of, 433, 438
  charge relay of, 434, 438
  charging circuit of, 433-435, 438
  common problems of, 443-445
  contactor of, 438
  current monitor of, 438, 439
  damped sinusoidal waveform of, 430,
    434, 438
  defibrillator paddles of, 302, 438
  discharge buttons of, 433, 434, 438
  discharge control of, 433, 441
  discharge relay of, 434, 438
  energy delivered in, 422, 435
  energy dumping in, 436, 439
  energy storage capacitor of, 433, 434,
    438
  energy stored in, 434-436
  functional block diagram of, 438-440
  inrush current of, 435
  inverter of, 439
  monophasic waveform of, 443
  output isolation of, 439
  patient load of, 443
  power supply of, 433, 439
  quality assurance of, 442
  triphasic defibrillation waveform, 431
  truncated exponential waveform, 436
  voltage monitor of, 438, 439
  wave-shaping inductor of, 434
  waveform shaping circuit of, 433, 438
Cardiac index, 386
Cardiac output, 17, 385
Cardiac output monitor, 385-402
  functional block diagram of, 396
Cardiac output monitoring,
  diffusible indicator of, 388, 389
  indicator (see also tracer) of, 386, 388
  pulse contour method of, 397, 398
  non-diffusible indicator of, 388
  thermodilution method of, 390, 398
  tracer (see also indicator) of, 388
Cardiac vector, 286, 291, 293, 301
Cardioplegia solution, 710, 711, 714
Cardiopulmonary bypass, 702-716
Cardiopulmonary system, 704
Cardiotocography, 546

Cardiotomy reservoir, 712, 713
Cardiovascular system, 112, 354-357
Cardioversion, 430, 440, 441
  synchronization circuit of, 441
Carotid artery, 543, 579
Carotid artery occlusion, 537
Cartridge filter, 649
Carrier sensed with multiple access and collision detection (CSMA/CD), 278
Cathode, 154
Cathode ray tube, 244-247
  anode of, 244
  brightness of, 245
  cathode of, 244
  contrast ratio of, 246
  deflection plate of, 246
  electron beam of, 244, 245
  electron gun of, 244, 245
  phosphor screen of, 244, 245
  refreshing rate of, 246
  resolution of, 246, 247
  triggering of, 245
  viewing angle, 253
Cations, 154
Cauterization (see also coagulation), 475,
  468
CCD (see charge-coupled device),
CCD Array, 138, 141, 142
CCD Pixel, 138
CCPD (see continuous cycler-assisted peri-
  toneal dialysis)
CCU (see cardiac care unit),
Celsius scale, 81
Cell diagram, 154
Cell membrane, 12, 179
Cellulose acetate, 173, 638
Central chart recorder (see central recorder)
Central nervous system, 314
Central processing unit, 31, 379
Central recorder, 268, 271
Central station, 271, 272
  basic capabilities of, 272
Cerebellum, 314, 315, 727
Cerebral cortex, 213-315, 609
  ridges and valleys of, 314
Cerebral oximetry, 590
Cerebral spinal fluid, 314, 747
Cerebrum, 314
CFU (see colony forming unit)
Charge-coupled device, 137-141

stop region of, 137
trapped charge of, 137
Chemical disinfection, 651, 697
Chest lead, 298
Cholecystectomy, 664, 681
Cholesterol esterase, 171
Cidex, 696
Cholesterol oxidase, 170
Clark oxygen electrode, 159
Clearance, 562, 636, 638, 639
plasma, 632, 633, 636
CMG (see common mode gain)
CMRdB, 198
CMRR (see common mode rejection ratio)
CMV (see continuous mandatory ventilation) CNS (see central nervous system)
CO (see cardiac output)
CO2 monitor (see also end-tidal CO2 monitor), 601-608
Coagulants, 644
Coagulation, 475, 478-481, 486-489, 660, 661, 714
Coagulation mode, 478
Cochlear, 724
frequency-sensitive locations of, 727, hair cells of, 726, 727
Cochlear implant, 745-747
electrode array of, 745, 747
Stimulation signal of, 746
transmitter and receiver of, 746
Cognitive limitation, 18
Coherent, 656
Collimated, 656
Collodion, 317
Colonoscope, 683, 684
Colony forming unit (CFU), 650
Color temperature, 129-131, 687, 688
Common aorta, 332, 334, 336
Common mode gain, 193
Common mode rejection ratio, 346
Common mode signal, 197-199, 304
Compressed gas cylinders, 593
Compressed spectral analysis (see electroencephalograph)
Conduction disorders, 408
Conductive hearing loss, 733, 736, 744
Conductive interference, 211-215
Conductor loop, 210, 211
Constraints in biomedical signal measurements, 23-25
Contact temperature sensor, 568
Continuous ambulatory peritoneal dialysis, 647
Continuous cycler-assisted peritoneal dialysis, 648
Continuous mandatory ventilation, 520
Continuous paper feed recorders, 238-241
building blocks of, 240
Continuous positive-airway pressure, 520
Continuous renal replacement therapy, 648
Continuous temperature monitors, 569-572
Control inputs, 23
Controlled mandatory ventilation, 520
Convection, 554, 568
Conversion factors, of pressure units, 69
Co-oximeter, 585, 588-590
Core temperature, 569, 573, 574
Corner frequency (see cutoff frequency)
Cortex, 313
cerebral, 313, 314, 609
premotor, 314
primary motor, 314
somatosensory, 314
CPAP (see continuous positive-airway pressure) CPB (see cardiopulmonary bypass),
Critical care ventilators, 517, 520, 528, 529
Cross talk, 423
CRRT (see continuous renal replacement therapy)
CRT (see cathode ray tube)
Cryogenic bulk central supply system, 762
CSA (see compress spectral analysis of electroencephalograph)
CSF (see cerebral spinal fluid)
Cuprophane, 638
Cutoff frequency, 48, 50, 256, 257, 289, 364, 367
Culture
cell, 29
organ, 29
tissue, 28, 29
CV (see controlled mandatory ventilation)
Cystoscope, 684

D

DAC (see digital to analog conversion)
Daniell cell, 154-156

Dark current, 135, 136, 141
Data management, 308, 609
Data transfer rate, 278, 280
Datapoint, 278
DC defibrillator, 430, 438
DC offset, 185, 193, 322
Dead space air, 502, 504, 602
    anatomical, 502
    functional, 502
    physiological, 502, 503, 508
Dead zone, 43
Defibrillation (see cardiac defibrillation)
Defibrillator protection, 213, 214, 302, 303
Demodulation (see modulation and demod-
    ulation)
Deoxygenated blood, 354, 386
Deoxyhemoglobin, 585, 587-589
Depolarization, 14, 286, 302, 407, 411
Desflurane, 610, 616, 625
Desiccation, 475-479, 481
Desired signal (see also desired input), 192,
    199, 596
Dextrose, 390, 392, 393, 449, 647
DG (see differential gain)
Diagnostic device, 7, 8, 16, 18, 29, 284
Diagnostic ECG, 288, 289, 306
    bandwidth of, 289, 305
Dial-type temperature gauge, 83
Dialysate, 629, 633-653
Dialysate delivery circuit, 643, 645
    sorbent regenerative, 643
    acetate base, 542
    batch processing of, 542
    bicarbonate base, 542
    composition of, 542, 643
    single-pass, 643
    single-pass recirculation, 643
Dialysis equipment, 628-653
Dialyzer (see also artificial kidney), 635-639
    clotting properties of, 638
    coiled tube, 637
    high efficiency, 638, 642, 652
    high flux, 637, 638
    high permeability, 637
    hollow fiber, 637, 638, 645
    mass transfer area coefficient of, 639
    parallel plate, 637, 638
    priming volume of, 638
    reuse of, 651, 652
    transmembrane pressure of, 636, 645

    ultrafiltration coefficient (KUf) of, 636,
        638
Diameter-indexed safety system, 623, 763
Diaphragm, 44, 70, 71, 75, 362
Diaphragm (in breathing), 424, 497, 498,
    516, 531, 532
Diaphragm pump, 450, 458, 464, 459
Diastolic blood pressure, 354, 356, 365, 372,
    377
DICOM (see Digital Imaging and
    Communications in Medicine)
Dicrotic notch, 356, 357
Differential amplifier, 193-197
Differential gain, 193-197
Differential mode signal, 197
Differential pressure flow transducer, 506
Diffusion, 13, 27, 179, 503, 592, 663, 664
Diffusion, in CPB, 704-706
Diffusion, in dialysis, 630, 634, 642
Diffusion, in ventilation, 519, 520, 530
Diffusion capacity, 504
Diffusion force, 13
Digital Imaging and Communications in
    Medicine (DICOM), 281, 691
Digital Processing, 31
Digital to analog conversion, 30
Digital display interface, 254
Direct Fick method, 386, 387, 400
Direct sequence spread spectrum (see
    DSSS),
Disinfectants, 765, 766
    high level, 762
    intermediate level, 767
    low level, 767
Displacement current, 201.202
Displacement transducer, 103-111
    Capacitive, 107, 108
    inductive, 105-107
    resistive, 104, 105
Display system,
    first-order high pass filter of, 257-259
    frequency response of, 256-258
    lower cutoff frequency of, 257-259
    paper speed of, 254-256
    performance characteristics of, 255, 256
    resolution of, 256
    sensitivity of, 255
    step response of, 257-259
    transfer function of, 256, 257
    upper cutoff frequency of, 256

DisplayPort, 244, 245
Disposable blood pressure transducer, 363, 368
DISS (see diameter-indexed safety system)
Dissipation constant, 84, 90
Distillation, 649
Distortion product otoacoustic emission, 741
Doppler blood flowmeter, 542-544
  audio frequency amplifier of, 542
  audio speaker of, 542
  blood vessel-wall motion in, 542
  demodulator of, 542
  functional block diagram of, 543
  integrator of, 542
  RF oscillator of, 542
  ultrasound receiver of, 542
  ultrasound transmitter of, 542
  zero crossing detector of, 542
DLCO (see also diffusion capacity), 504
Doppler effect, 379, 538, 540
Doppler flowmeter (see also Doppler blood flowmeter), 540
Doppler shift, 379, 538, 540, 542, 543, 549
Doppler ultrasound blood pressure monitor, 379, 380
  cuff pressure of, 380
Doppler shift in, 380
Double insulation, 232, 234
Double layer capacitance, 185
DPOAE (see distorted product otoacoustic emission)
DSSS (direct sequence spread spectrum), 760, 761
Dual chamber pacemaker, 411, 416, 420, 423
Duty cycle.
  electrosurgery, 478-489, 524, 489
  laser, 671
DVI (see digital visual display)

E

EAdi (see electrical activity of the diaphragm under mechanical ventilator)
Ear,
  external, 724, 725
  Inner, 726, 727
  Middle, 725, 726

ossicles of, (see auditory ossicles)
Ear drum (see tympanic membrane),
Ear thermometer (see tympanic thermometer)
EECO2R (see extracorporeal CO_2 removal)
ECG (see electrocardiograph and electrocardiogram)
ECG data management system, 308
ECG lead configuration, 291-297
  I of, 293
  II of, 293
  III of, 293
  aVF of, 293
  aVL of, 293
  aVR of, 293
  V1 of, 295
  V2 of, 295
  V3 of, 295
  V4 of, 295
  V5 of, 295
  V6 of, 295
ECG monitor, 290, 402
ECG event recorder (see also loop recorder), 290
ECMO (see extracorporeal membrane oxygenation)
Ectopic focus, 288
Eddies, 113
EEG (see electroencephalography or electroencephalogram)
EEG electrode, 317-321
  cortical electrode of, 317, 318
  depth electrode of, 318-320
  earlobe electrode of, 317
  impedance of, 322
  invasive electrode of, 317
  nasopharyngeal electrode of, 317
  needle electrode of, 317, 330
  placement of, 317-321
  subdural electrode of, 318, 319
  surface electrode of, 316, 317
Effector limitation, 18
Einthoven's triangle, 293
Electric arc, 477
Electrical hazard, 218, 222-224, 231
Electrical safety, 215-236
Electrical shock, 217, 218, 223, 225, 232
Electrical stimulation, 337, 347, 406, 407, 408
Electrocardiogram, 16, 284, 286-288

axis of, 299, 300, 301
baseline wander of, 309
common problems of, 308-310
extreme (see also northern) axis deviation of, 299, 300, 301
left axis deviation of, 299, 300, 301
normal axis of, 299, 300, 301
northern (see also extreme) axis deviation of, 299, 300, 301
P wave of, 286, 287, 289
PR interval of, 286
power frequency interference of, 306, 309, 310
PQ interval of, 286
QRS complex of, 286
QT interval of, 287
QTc interval of, 287
R wave of, 16, 286
resting, 289
right axis deviation of, 299, 200, 301
stress test, 289
surface, 286
T wave of, 286
U wave of, 286, 287
Electrocardiograph (see also ECG), 17, 284-310
ambulatory, 289, 290
amplifier of, 291, 304
calibration pulse of, 304
defibrillator protection of, 213, 214, 302
diagnostic, 289
esophageal lead of, 298
filter of, 289, 305, 305
frequency bandwidth of, 305
full disclosure, 289, 305
functional block diagram of, 202-306
high resolution, 289
Holter, 290
leads of (see ECG lead)
lead selector of, 304
lead-off detector of, 303
monitoring, 289, 305
notch filter of, 306, 310
preamplifier of, 304
recorder or display of, 306
right-leg-driven circuit of, 204-205, 304
signal isolation of, 304, 305
signal processor of, 306
Electrochemistry, 152, 174
Electroconductive pathways, 428

Electrocorticography, 313
Electrocution, 217
Electrode double layer, 182-184, 186
Electrode gel, 186, 330, 350
Electrode-electrolyte interface, 185, 186
Electrodes (see also biopotential electrodes), 152-176
half-cell potential of, 182, 185
Electroencephalogram, 313
alpha wave of, 315, 322, 324
beta wave of, 315, 322-324
delta wave of, 313, 315, 316, 322, 324
theta wave of, 322, 324
Electroencephalograph, 312-332
analog to digital converter of, 326-328, 379
bandwidth of, 328
bipolar connection of, 326
chart speed of, 329
compress spectral analysis of, 323
electrode impedance tester of, 329
filter of, 328
functional block diagram of, 324-330
head box, 325
montage of, 325-327
problems of, 331
sensitivity of, 327, 328
signal isolation of, 328
transverse bipolar, 327, 329
troubleshooting of, 331
unipolar connection of, 326
Electroluminescent display, 251, 252
Electrolyte, 12, 154, 175, 182-186, 594, 629, 634, 714
Electrolyte gel, 218, 317, 443
Electrolytic cell, 152
Electromagnetic flowmeter, 121
Electromagnetic immunity, 494
Electromagnetic interference, 199, 493
artifacts due to, 199, 308, 330, 331
Electromagnetic radiation, 127-129
Electromagnetic spectrum, 127, 129
Electromagnetic wave, 11, 128
Electromechanical transducer, 239
Electromedical device, 200, 217, 332
Electromyography, 334-350
Electron current, 181
Electroneurophysiology, 312, 313, 335
Electronic thermometer, 541
Electrophysiology studies, 167

Electrosurgery,
    coagulation mode of, 478
    common problems of, 492-494
    current density, 476, 482, 483
    bipolar mode of, 475, 479
    blended mode of, 478, 479
    cut mode of, 478, 480
    fire and explosion in, 493
    fluid mode, 479
    laparoscopic mode, 479
    modes of, 478-480
    monopolar operation of, 479, 480
Electrosurgical unit (see also ESU), 474, 495
    active electrode of, 480, 481, 487, 491
    bipolar electrode of, 479
    burst repetition frequency of, 486
    capacitive leakage current of, 491
    functional block diagram of, 487, 487
    high frequency leakage test of, 491
    isolated output of, 440, 487
    muscle and nerve stimulation in, 493
    output characteristics of, 487-489
    output power of, 477, 487, 488-490
    output power verification of, 490
    patient load of, 488, 489, 492
    percentage isolation of, 491, 492
    power amplifier of, 486, 487
    quality assurance of, 490-492
    return electrode of, 475, 481-485
    return electrode monitoring of, 483,485
    step-up transformer of, 486
Embolism, 466, 471, 504, 693, 706, 707,
        711, 715
EMG (see electromyography or electromyo-
        gram)
EMG & EP studies, 334-350
    concentric needle electrode of, 338
    end plate noise in, 340
    end plate spike in, 340
    fasciculation in, 340
    filter of, 341, 345
    grounding electrode of, 336
    insertion activity of, 340
    interference pattern of, 341, 342
    motor response of, 342-345
    needle electrode of, 338-340
    recording electrode of, 36, 337, 347-350
    sensitivity of, 345, 350
    single-fiber needle electrode of, 338
    spectral analysis of, 346

    spontaneous activity of, 340, 341
    stimulating electrode of, 336
    surface electrode of, 336-338
    sweep speed of, 345
    voluntary effort of, 341-342
EMG motor response, 342-344
    latency of, 343, 344
    supramaximal stimulation of, 343
EMG spontaneous activity, 340, 342
    biphasic potential of, 340, 341
    monophasic potential of, 341
EMG voluntary effort, 341, 342
    full effort of, 342
    mild effort of, 342
    moderate effort of, 342
EMI (see electromagnetic interference)
Emissivity, definition of, 129, 574
Endocardial lead, 410
Endoscope, 683-688
    disinfection and sterilization of,652
Endoscopic electrosurgical procedure,
    electrical leakage in, 696
    burn in, 687
    RF leakage current in, 695
Endoscopic video system, 689-699
    display monitor of, 683
    image management system of, 683, 691
    image processor of, 688, 689
    light source of, 687, 688
    video camera of, 688, 689
Endoscopy,
    bleeding in, 694-696
    camera capsule, 693
    cancer detection in, 689, 693
    perforation in, 693-695
    problems of, 694-699
    three-dimensional, 693
Endotoxin unit (or EU), 650
Endotracheal tube, 515, 525, 53, 602, 618
End-tidal carbon dioxide level, 269
End-tidal carbon dioxide monitor, 601-606
    airway adaptor of, 604
    errors in, 605, 606
    functional block diagram of, 606
    mainstream, 603-605
    nitrous oxide in, 603, 606
    pressure fluctuation in, 606
    sidestream, 604, 605
    water trap of, 605, 606
Enflurane, 610, 611, 616

Enriched oxygen environment, 223, 225, 493

Enteral feeding, 449, 455

Enzyme sensors (see biosensors)

EP (see evoked potential)

Epicardial lead (see also myocardial lead), 410

Epilepsy, 313, 315, 318

Epoxy housing of cardiac pacemakers, 418

Equipment standardization, 419

Equipotential grounding, 231, 234

Error, 34
    absolute, 35
    gross, 34
    random, 34
    relative, 35
    systematic, 34

ERV (see expiratory reserve volume)

ESU (see also electrosurgical unit)
    active electrode, 475-481
    ball electrode of, 481
    flat blade electrode of, 482
    foot switch of, 481
    loop electrode of, 481
    multiple use, 481
    needle electrode of, 482
    single use, 481
    smoke plume from, 494

ESU output waveform, 479
    crest factor of, 480, 481

ESU pencil, 481, 482

ESU return electrode, 481-485
    burn at, 483, 486, 492
    conductive gel pad, 482
    electrode-skin contact at, 482, 484, 496
    skin effect in, 482
    tissue damage at, 482

EtCO2 (see end-tidal carbon dioxide)

Ether, 609, 610

Ethernet, 278, 279, 760
    bandwidth of, 278
    collision detection of, 278

Ethylene oxide, 414, 580, 696, 768

E-type thermocouple, 97

EU (see endotoxin unit)

Eustachian tube, 737, 738

Evoked potential, 263 – 280
    auditory,314, 699
    electrical signal, 314
    somatosensory, 314

    visual, 314

Evoked potential study, 313, 334

Expiratory reserve volume, 499

Explosion hazard, 217, 225, 233

External invasive pacemaker, 408, 422

External pacemaker, 406, 422

Extra low voltage, 233

Extracorporeal CO2 removal, 703

Extracorporeal membrane oxygenation, 703, 713, 714

Extravascular lung water (EVLW), 399

Eye protection, 672, 673, 675

F

Fast Ethernet, 278

Feedback process, 10, 11

FEF25-75% (see maximal midexpiratory flow)

Fahrenheit scale, 81

Fetal ECG, 549

Fetal heart rate monitoring, 548-550
    abdominal ECG method in, 549
    direct ECG method in, 549
    phono method in, 549
    ultrasound method in, 549

Fetal monitoring
    fetal heart sound in, 549
    fetal heart rate in, 548
    intrauterine pressure in, 551

Fetal monitors, 546-552

Fetoscope, 684

FEV1 (see forced expiratory volume)

FHR (see fetal heart rate)

FHSS (see frequency hopping spread spectrum)

Fiber, optical, 141-150
    acceptance angle of, 143
    cable loss of, 143
    cladding of, 143
    core of, 143
    chromatic dispersion of, 146
    coupler of, 147
    graded index, 144, 145
    interruption problem of, 147
    intensity sensor, 148
    interferometric sensor, 148
    losses of, 147
    material dispersion of, 146
    modal dispersion of, 144, 146

multi-mode, 144, 145, 146
numeric aperture of, 143
polarization sensor, 149
refractive index of, 143, 148
single mode, 144, 145, 146
splice of, 147
splitter of, 147
step index, 144
Fiber bundle, optical, 146
coherent, 146, 147
flexible, 146
non-coherent, 146
packing fraction of, 147
rigid, 146
Fiberscope, 684, 688, 689
Fibrillation, 428
Fick principle, 385, 386
Fick's Law, 13, 164
Field of view, 575, 682, 684, 692
Filters,
band pass, 48, 49
band reject, 48, 49
high pass, 48, 49
low pass, 48, 49
FiO2, 584, 585
Fire hazard, 222, 260, 671, 673, 674
Flammable anesthetic agent, 225
Flammable gas, 225
Flexible endoscope, 682, 684-687
air channel of, 685
angulation control of, 686
control head of, 684-686
optical fiber of, 684, 689, 697
insertion tube of, 684-689
instrument channel of, 684, 685, 696, 698
light guide tube (or universal cord) of, 684, 686
suction of, 686
water channel of, 685
Floating surface electrode, 185, 186
Flow transducer (see also transducer, flow), 112-125
Fluid-in-glass thermometer, 82
Force transducer (see also transducer, force), 67-78
Forced expiratory volume, 500
Forced vital capacity, 500
Formaldehyde, 651, 767
Fourier analysis, 753-757

Fourier series, 44, 46, 753
Four-wire RTD, 89, 90
FOV (see field of view)
Fowler's method (see nitrogen washout method)
Fractional oxygen saturation, 586
Frequency domain, 44, 46-48, 322, 739, 754
Frequency hopping spread spectrum, 761
Frequency modulation, 761
Frequency response, 59-61, 236, 256, 257, 362, 367, 368, 725, 730, 734, 735
Frequency spectrum, 42, 302, 709-713
Frontal lobe, 314, 315
Frontal plane leads, 293, 295, 299
FRV (see functional residual volume)
Fuel cells, 172-176, 594
Fulguration, 477, 478
Full bridge, 64, 65
Functional building blocks, 29
Functional oxygen saturation, 586-588
Functional residual volume, 500
Fundamental frequency, 753
FVC (see forced vital capacity)

G

Galvanic cell, 152-158, 172, 594
Galvanic oxygen cell, 170
Gamma curve calibration, 690
Gas cylinder,
color-coded, 623, 764
dimension of, 764
stem of, 724, 762
yoke of, 624, 762
Gas Law, 70, 507
Gauge pressure, 69, 762
Gel-filled electrode, 186
General anesthesia, 602, 609, 610
Generalized epilepsy, 315
Gezo Jako, 656
GFCI (see ground fault circuit interrupter)
Glucose oxidase, 171
Glutaraldehyde, 696, 699
Grand mal seizures, 315
Gravity flow infusion, 450
Gravity flow intravenous infusion set, 453
Graybody,
definition of, 575
near, 575
Ground fault circuit interrupter, 231, 234

Ground reference, 230
Grounded power system, 225
Guarding shield, 204

H

Half bridge, 64, 65
Half-cell potential, 155
   electrode, 182
Half-reaction, 154-157
Hall effect sensor, 108-110
Hall voltage, 108, 109
Halothane, 610, 611, 616
Hand-switched ESU pencil, 481
Harmonics, 212, 753, 754
HDMI (see high definition multimedia
      interface)
Health care network standards, 281
Health Level 7 (HL7), 267, 281
Hearing aids, 744
   bone-anchored, 744
Heart block,
   1st degree, 407
   2nd degree, 407
   3rd degree, 407
Heart rate, 257, 548
Heart-lung (bypass) machine, 702, 709-713
Heart,
   aortic valve of the, 355, 356
   left atrium of the, 353-356
   left ventricle of the, 355-357
   mitral valve of the, 353
   pulmonary valve of the, 353
   right atrium of the, 353-357
   right ventricle of the, 353, 354
   tricuspid valve of the, 353
Heat disinfection, 650
Heat loss,
   conduction, 554
   convection, 554
   evaporation, 554
   radiation, 554
Heat sensitive paper (see thermal paper)
Heated thermistor, 408
Helical coil bimetallic strip, 83
Helium dilution method, 501
Hemoconcentrator, 712
Heliox, 611, 613
Hemodialysis system, 634-646
   air/foam detector of, 644

arterial blood line of, 643, 644
arterial drip chamber of, 644
blood leak detector of, 645
cleaning and disinfection of, 650-652
contamination of, 651, 652
deaeration chamber of, 645
deaeration pump of, 645
dialysate delivery circuit of, 645
extracorporeal blood circuit of, 643-645
fluid flow diagram of, 643
functional block diagram of, 634
heat exchanger of, 645
heater compartment of, 645, 646
metering chamber of, 645
metering pump of, 645
patient interface of, 639-641
pressure sensor of, 645
roller pump of, 645
ultrafiltration control of, 645
venous blood line of, 643
water treatment in, 648-651
Hemoglobin, 584, 585
Hemolysis (of blood cells), 28, 449, 651,
      652, 712, 715
Hemostatic, 475, 480, 481
Heparin, 28, 643-647, 710, 715, 716
Heparinized saline, 358, 368
High definition multimedia interface
      (HDMI), 254, 255
High frequency harmonics, 200
High-frequency ventilator, 517
High frequency ventilation, 530
HIS (see hospital information system)
Histocompatibility, 26
HL7 (see Health Level Seven)
Holter ECG (see also ambulatory ECG),
      290, 420
Homogeneous circuit,
   the law of, 96
Hospital information system, 267, 272, 276,
      556, 692
Hot film anemometer, 125
Hot wire anemometer, 124, 125
HR (see heart rate)
Hub, network, 280
Human error, 22, 23, 34, 467, 652
Human factors, 18-21, 23
Human-machine interface, 17-22
Humidifier, 525, 527, 693
Hydrogen electrode, 155, 156, 161, 162

Hydrogen reference electrode, 161
Hydrogen-oxygen fuel cell, 176
Hyfrecator, 475
Hypercapnia, 586
Hyperthermia, 558, 563, 564
Hypothalamus, 568, 575
Hypothermia, 16, 471, 598, 693, 699, 711,
Hypoxia, 547, 563, 584, 596, 611, 622
Hysteresis, 42, 44, 417

I

I:E ratio, 522, 523, 529, 532
IBP (see invasive blood pressure)
IC (see inspiratory capacity)
IC temperature sensor (see integrated circuit
    temperature sensors)
ICD (see implantable cardiac defibrillator)
ICHD code, 414
ICU (see intensive care unit)
Ideal electrode,
    characteristics of, 181
IEC (see International Electrotechnical
    Commission)
IEEE 802.11 Standards, 760
IEEE 802.3 Standards, 278, 760
IEEE 802.5 Standards, 278
Illuminance, 130
Impedance matching, 195
Implantable cardioverter defibrillator, 409,
    418, 430, 431
Implantable pacemaker, 420-425
    ADL rate of, 417, 418
    AOO, 414, 415
    battery monitor of, 420
    battery of, 413, 418-420
    DDD, 415-417
    DDDR, 416, 417
    functional block diagram of, 320-422
    hysteresis of, 42, 44, 417
    lead impedance of, 417
    lower rate of, 417, 418
    output circuit of, 421
    performance parameters of, 417, 418
    pulse width of, 413, 418, 423
    rate limit of, 417, 418, 423
    refractory period of, 417, 418, 422, 424
    sensing amplifiers of, 421
    sensitivity of, 409, 417, 418, 421, 424
    upper sensor rate of, 417, 418

VVI, 414-416, 420
IMV (see intermittent mandatory ventila-
    tion)
In vivo test, 29
    blood contact method of, 29
    tissue culture method of, 29
    functional, 29
    nonfunctional, 29
Incompressible fluid, 113, 116, 117
Incus, 726
Indicator dilution method, 381, 388-390
    indicator recirculation in, 389, 401
Indicator electrode, 160, 161, 164, 166, 169
Indocyanine green, 389, 395, 398
Induced current, 201-203, 309
Infant incubator, 533-558
    access door of, 555, 563
    access port of, 524
    cuffed ports of, 556
    disinfection or sterilization of, 563
    functional components of, 555
    modes of temperature control, 555
    oxygen control of, 556
    plastic hood of, 555
    proportional heating control of, 555
    relative humidity of, 556, 557
    water reservoir of, 556, 557, 563
Infant resuscitator, 561
Infant warmer, 554
Infection,
    community-based, 769
    nosocomial (hospital acquired), 769
Infection prevention and control (IPAC),
    765, 769, 771
Inferior vena cava, 354, 450, 713
Infrared spectroscopy, 603
Infrared thermometry, 573-580
    narrowband, 576
    two color, 576
    wideband, 576
Infusion controller, 450, 457, 458
Infusion devices, 448-471
Infusion pump, 458-471
    air-in-line detector of, 471
    anti-siphon valve of, 451, 466
    anti-reflux valve of, 452, 454
    back pressure of, 469
    bolus infusion of, 470
    bolus of, 463, 464, 469-471
    dose error reduction system of, 466, 467,

471
downstream occlusion of, 465, 469
flow accuracy of, 469-471
flow pattern of, 463, 464, 469, 470
flow rate of, 452, 455-459, 462-471
fluid depletion alarm of, 466
fluid viscosity of, 469
functional block diagram of, 467, 468
occlusion pressure alarm of, 465, 469
runaway prevention of, 466
upstream occlusion of, 466
Inkjet printer, 238, 240, 241
print cartridge, 243
print head of, 243
Input guarding, 204, 206
Input range, 41
In-service training, 444
Inspiration, 497
Inspiratory capacity, 498
Inspiratory reserve volume, 498
Intrinsic clearance (see mass transfer coefficient)
Instrumentation amplifier, 192-197
bandwidth of, 192, 193
characteristics of, 193
common mode rejection ratio (CMRR) of, 193, 194
input impedance of, 192-196
output impedance of, 192, 193
Insufflator, 681, 683, 692, 693, 698, 699
Integrated circuit temperature sensors, 101
current type, 101
voltage type, 101
Intensive care unit, 266
Intercostal muscle, 487, 498, 516
Interfering input, 58
Interfering signal (see also interference input), 25, 199, 309
Intermediate metals, the law of, 96, 98
Intermediate temperature, the law of, 96, 98, 99
Intermittent mandatory ventilation, 520
Intermittent positive pressure breathing, 517
Intermittent temperature measurement, 57, 568, 573, 575
International Practical Temperature Scale (IPTS), 81
Intracardiac blood pressure, 395
Intracranial pressure, 353, 532
Intrapartum monitoring, 547

Intrapleural pressure, 499
Intrathoracic pressure, 498
Intrathoracic blood volume (ITBV), 399
Intravenous access, 450
Intravenous infusion, purpose of, 449
Intravenous infusion set, 451-455
drip chamber of, 452, 455
drop nozzle of, 452
flexible PVC tubing of, 452
luer lock connector of, 452, 455
occlusion clamp of, 453, 455
primary solution bag of, 452
regulating (roller) clamp of, 452
secondary solution bag of, 452
solution bag of, 451
solution bag spike of, 451
Y-injection site of, 452
Invasive blood pressure monitor, 352-369
calibration factor of, 368, 369
transducer of, 362-364
zeroing process of, 359-361, 366
Invasive method, 58
Ion selective electrodes, 164-170
Ionic current, 181
Ionophores, 165
IPAC (see infection prevention and control)
IPPB (see intermittent positive pressure breathing)
IPTS (see International Practical Temperature Scale)
IR thermometry (see infrared thermometry)
Irradiance, 128, 558, 695
Iron lung, 575
IRV (see inspiratory reserve volume)
ISM band, 759, 760
Isoflurane, 611, 616
Isolated output, 440, 441, 487
Isolated power system, 224-226, 228
Isolation barrier, 31, 229, 230
Isolation impedance, 230
Isolation transformer, 225, 229, 230
ISO OSI communication model, 281
Isothermal block, 99
IV (see intravenous)
IV line (see intravenous infusion set)

J

Jaundice, 558
Jet nebulizer, 526

J-type thermocouple, 34
Jugular vein, 641, 713

K

Keep vein open, 465
Kelvin temperature scale, 68, 81
Kidney function, 631-634
KoA (see mass transfer area coefficient of dialyzer)
KoV (see mass transfer coefficient)
Korotkoff, N S, 372
Korotkoff sounds, 373-736
KUf (see Ultrafiltration coefficient of dialyzer)
KVO (see keep vein open)

L

Labor, 546-548
Laminar flow, 113-117
LAN (see local area network), 276, 760
Laparoscopic procedure, 147, 493, 681, 690, 694
Laparoscopy, 681, 682
Laser, surgical, 655-677
    aiming beam of, 662, 667, 673
    applications of, 658, 659
    Argon, 660, 664-666
    Carbon dioxide (CO_2), 656, 657, 662, 663
    characteristics of, 656
Class 1, 672
Class 1M, 672
Class 2, 673
Class 3R, 673
Class 3B, 673
Class 4, 673
    classification of, 672, 673
    collimated beam of, 658, 672
    continuous, 660-662
    cooling system of, 662, 663, 667, 668
    dye, 666
    diode (see semiconductor laser)
    excimer, 662, 666
    excitation source of, 657
    exposure time of, 659, 660
    focal spot of, 660
    functional components of, 667
    helium neon (HeNe), 662, 665

holmium YAG (Ho:YAG), 663
Krypton, 663, 664
KTP/532, 664, 665
lasing medium of, 657, 665
ND:YAG, 662, 665
optical cavity of, 657, 664
photostimulation effect of, 659
plume, 674, 676
power density of, 659, 660, 671, 672, 674
pulsed, 659, 663
population inversion of, 656, 664
problems, 676, 677
reflective mirror of, 657
resonator of, 657, 667
ruby, 664, 665
semiconductor, 666
surgical applications of, 661,, 663, 667
surgical effect of, 659, 670, 671, 677
thermal effect of, 658, 659
tissue effect of, 659, 660,
transport media of, 668
transverse electromagnetic mode of, 667, 668
Laser action, 657
Laser beam,
    diffused reflection of,673
    specular reflection of, 673
    scattered reflection of, 673
Laser delivery system, 669-671, 676
    articulation arm of, 669
    contact laser probe of, 676
    first surface mirror of, 669, 676
    optical fiber of, 671, 676
    photonic bandgap reflector of, 669
    total internal reflection in, 669
Laser plume, 674, 676
Laser power meter, 676
Laser printer, 240-243
    cleaning blade of, 242
    developing cylinder of, 241
    erase lamp of, 242
    laser beam of, 241
    paper speed of, 241
    photosensitive drum of, 241
    primary corona wire of, 241
    scanning mirror of, 241
    toner of, 241, 261
    transfer corona wire of, 241
Laser procedure,
    access control in, 674, 676

safety zone in, 636
warning sign in, 675
Laser safety committee, 672, 675
Laser safety hazard,
aversion response, 673
blink reflex, 673
burn, 671-673, 676
explosion, 672
eye damage, 673
eye injury, 695
fire, 671-674
scotoma damage. 672
Laser safety officer, 672, 675
Laser safety program, 675
Laser smoke evacuator, 674
Laser tissue effect, 659-661
ablation in, 663
coagulation in, 660, 661
cutting in, 661
necrosis zone of, 659
vaporization zone of, 660
vaporization in, 660, 661
zone of injury in, 660
Lateral strain, 73
LCD (see liquid crystal display)
Lead acid cell, 172
Lead error, 87-89
Leakage current, 221-224, 229, 231, 233-236
capacitive, 221, 222
earth, 222
enclosure, 222
measurement of, 233-236
patient, 222, 224
resistive, 221, 222
touch, 222
Let go current, 219
Light pipe, 576, 658, 681
Light polarization, 247
Light sources, endoscopes, 687-688
aperture of, 688
automatic brightness control of, 688
color temperature of, 688
infrared filter of, 687, 695
LED, 687, 688
life span of, 688
mercury vapor, 687
metal halide, 687
output intensity, 688
quartz halogen, 687, 688
xenon, 687, 688

Lightning surge, 213, 281
LIM (see line isolation monitor)
Limb electrodes, 293, 296, 299
Limb lead, 293, 295, 298, 299
Line isolation monitor, 288
Linear accelerator, 8, 688
Linear displacement transducer, 104
Linear peristaltic infusion pump, 459
Linear variable differential transformer, 107
Linearity, 41, 59
Liquid crystal, 247-249
Liquid crystal display, 247-250
active matrix, 249
addressing electrode of, 248-251
backlit of, 249, 250
brightness of, 245, 248, 250-253, 255
color filter of, 249, 250
contrast ratio, 246, 250-253, 255
organic, 244,252
refreshing rate of, 246, 249, 250-253, 255
resolution of, 246, 247, 249, 250, 253-256
twisted nematic, 248
viewing angle, 250, 253
Liquid junction potential, 154
Lithium-iodine battery, 418, 420
Lithium-ion battery, 172, 175
Lithium-manganese battery, 444
Load cell, 76
Local area network, 276
Longitudinal mechanical wave, 720
Loop recorder (see also ECG event recorder), 290, 420
LOWN waveform (see also monophasic damped sinusoidal waveform), 430
Luer lock connector, 451, 452, 455, 416
Luigi Galvani, 13, 178
Luminance, 253
Luminous density, 130
Luminous emittance, 130
Luminous energy, 129, 130
Luminous flux density, 130
Luminous flux, 129, 130
Luminous intensity, 130
Lung capacities, 498, 499
Lung compliance, 498, 499, 529, 530
LVDT (see linear variable differential transformer)

M

MAC (see minimum alveolar concentration)
MAC address (see media access control address)
MBC (see maximal breathing capacity)
MVV (see maximal voluntary ventilation)
Macroshock, 220-223, 232-234
    characteristics of, 221, 222
Magnetic field interference, 211
Magnetic flux, 106, 109, 210, 232
Malleus, 726
Manometer, 68-70
Manual gravity flow infusion, 450-457
Mass flow rate, 113
Mass transfer coefficient (or intrinsic clearance or KoV), 638, 639
Maternal ECG, 547, 549
Maternal heart sound, 549
Maternal monitoring, 547
Maximum breathing capacity, 504
Maximal midexpiratory flow, 500
Maximal voluntary ventilation, 504
MDS (see monophasic damped sinusoidal)
Mean blood pressure, 357, 365
Mechanical stylus recorder, 238-240
    galvanometer of, 239, 240
    servomotor drive of, 239, 240
Mechanical ventilation, 515, 531
Mechanical ventilation breaths
    assist control,
    mandatory, 519, 522
    patient initiated mandatory, 519
    sigh, 522
    spontaneous, 519
    ventilator initiated mandatory, 519
Mechanical ventilator, 514-553
    air compressor of, 523, 526, 529
    air supply line of, 526
    air leak alarm of, 528
    airway pressure release ventilation (APRV), 529
    airway resistance compensation (ARC), 530
    apnea submode, 521
    assist control mode of, 520
    automatic tube compensation (ATC), 530
    bilevel positive airway pressure (BiPAP), 517
    bacteria filter of, 524, 527
    breath types of (see mechanical ventilation breaths)
    breathing gas of, 524-528
    check valve of, 524, 526-528
    continuous mandatory ventilation of, 520
    controlled mandatory ventilation of, 520
    continuous positive-airway pressure of, 520
    collection vial of, 527
    disconnection alarm of, 528
    electrical activity of the diaphragm (EAdi), 531
    exhalation valve of, 527, 528
    expiratory limb of, 527, 528
    expired gas of, 522, 524, 527, 528
    flow cycled, 518
    flow control of, 526, 527
    flow sensor of, 525, 526, 528
    functional block diagram of, 524
    high frequency (see high frequency ventilation)
    high pressure alarm of, 528
    inspiratory limb of, 527
    inspired gas of, 522, 524-528
    intermittent mandatory ventilation of, 520
    loss of gas supplies of, 529
    loss of power of, 528, 529
    mandatory minute volume (MMV), 529
    NAVA (neurally adjusted ventilatory assist), 531
    oxygen supply line of, 526, 528
    oxygen/air blender of, 526
    patient circuit of, 528, 529, 531
    patient initiated mandatory breath of, 519, 523
    positive end-expiratory pressure of, 520
    pressure-regulated volume-control (PRVC), 529
    pressure support of, 521
    pneumatic system of, 524, 525, 527
    power-up self test of, 529
    pressure cycled, 517
    pressure sensor of, 526, 528
    safety backup of, 525
    safety features of, 528, 529
    sigh in, 522, 523
    synchronized intermittent mandatory ventilation of, 520
    time cycled, 488

types of, 517-519
user interface of, 525
ventilator-initiated mandatory breath of, 519
volume cycled, 517
volume support (VS), 530
waveform of, 530
work of breathing (WOB), 521, 523
Y-connection of, 527
Mechanics of breathing, 497, 498
Media access control address, 280
Medical air, 762, 764
Medical device,
classification of, 7-9
definition of, 6, 7
front-end of, 346, 364, 573
Medical gas supply system, 762-764
wall outlet of, 762, 763
Medical laser, 665, 666
Medical telemetry, 273, 758-761
Membrane indicator electrode, 164
Membrane potential, 13, 14, 166
Membrane resistance, 179
Membranous labyrinth, 726, 745
endolymph of, 726, 727
Metal indicator electrode, 160
Metal oxide semiconductor, 137
Methemoglobin, 585, 587, 591
Microshock, 31, 217, 220-229, 231, 233, 234
characteristics of, 221
Microsoft Windows, 279
Midaxillary line, 361
Middle ear analyzer, 733, 736-739
Minimally invasive surgery (MIS), 690
Minimum alveolar concentration, 610
Minute volume, 504, 552, 529
MIS (see minimally invasive surgery)
Mixed venous oxygen contents, 387
Modes of ventilation, 419-522
Modulation, 759, 761
Monitoring device, 8
Monitoring ECG, 289
Monochromatic, 146, 656, 659
Monophasic damped sinusoidal waveform (see also MDS and LOWN), 430, 434
Monophasic truncated exponential waveform (MTE), 430, 435, 436
Monopolar operation, 479, 480
Montage, 325-327, 330
electrode selector matrix of, 326

referential, 327
transverse bipolar, 327
Motion transducer (see also transducer, motion), 103-111
MOS (see metal oxide semiconductor)
Motor unit action potential, 335, 338
MTE waveform (see monophasic truncated exponential waveform)
MUAP (see motor unit action potential)
Multiparameter monitor, 267
Multiple sclerosis, 315
Multi-programmable pacemaker, 409
Muscle sensing, 424
Muscle stimulation, 424
Myocardial lead, 410
Myocardial damage, 431
Myosignal (or myoelectric signal), 347

N

NASPE (see also NBG), 414
Natural pacemaker, 284, 406, 408
NAVA (see neurally adjusted ventilatory assist)
NBG (see also NASPE), 414, 415
Near graybody, 575
Nebulizer, 525-527
Needle electrode, 15, 25, 58, 187, 317, 330, 337-340, 350
Negative pressure ventilator, 515
Nephrophane, 638
Nernst Equation, 158-161, 163
Nerve conduction velocity, 335, 347-350
Nerve potential, 17
Network, patient monitoring, 276-281
bridge, 280
hub, 280
interface card, 280
repeater, 280
router, 280
switch, 280
Network connection components, 279
Network interconnection devices, 280
Network model, 279
client-server, 279
host-terminal, 279
peer-to-peer, 279
Network operating system, 279
Network protocols, 278
token ring, 278

ethernet, 278
Network topologies, 276, 277
  bus or tree, 276, 277
  ring, 276, 277
  star, 276, 277
Neuromuscular transmission time, 348
Neurally adjusted ventilation assist (see NAVA in mechanical ventilator)
NIBP (see noninvasive blood pressure)
NIC (see network interface card)
Nickel metal hydride, 444
Nickel cadmium, 444
Nitrogen concentration curve, 502
Nitrogen washout method (Fowler's), 502
Nonfade display, 246
  erase bar of, 246
  waveform parade of, 246
Non-inductive resistor, 490
Noninvasive blood pressure measurement,
  auscultatory method of, 372, 373-376
  oscillometric method of, 372, 376-379
Noninvasive blood pressure monitor, 371-382
Noninvasive method, 58
Nonisolated output, 449, 441
Nonlinearity, 41, 101, 193
Non-thrombogenic surface, 28
North American Society of Pacing and Electrophysiology (see NASPE)
NOS (see network operating system)
Nosocomial infection, 580, 626, 696, 769
Notch filter, 49, 396, 310, 328
Novell Netware, 279

O

OAE (see otoacoustic emission)
Obstructive lung disease, 498, 500
Occipital lobe, 314, 315, 320
Ocular exposure, 672
OFDMSS (see orthogonal frequency division sion
  multiplexing spread spectrum)
OLED (see organic light emitting diode display)
Operating range, 60, 84, 91, 101
Operational amplifier, 192
Optical densitometer, 395
Optical intensity ratio, 589, 596, 597
Optical isolator, 290, 305, 308

Optical path length, 148, 586, 588, 589, 597, 598
Optical transducer (see also transducer, optical), 56, 126-150
Organic light emitting diode display (OLED), 252, 253
Orifice plate, 118-120
Orthogonal frequency division multiplexing spread spectrum, 761
Oscillometric NIBP monitor, 371-379
  amplifier of, 379
  analog to digital converter of, 379
  central processing unit of, 379
  display of, 379
  functional block diagram of, 378
  oscillometric filter of, 379
  overpressure switch of, 379
  pressure sensor of, 377, 379
  printer of, 379
  pump and solenoid valve of, 378
  watchdog timer of, 379
Osmosis, 630, 631
Osmotic pressure, 631
Otoacoustic emission detector, 739-742
  distortion product, 741
  spontaneous, 740
  transient-evoked, 741, 742
Oxygen analyzer, 387, 593-596, 622
Oxygen concentrator, 764
Oxygen content, 591
Oxygen cost of breathing, 500
Oxygen failure protection detector, 614
Oxygen-hemoglobin dissociation curve, 704, 705
Oxygen permeable membrane, 169, 595
Oxygen saturation, 17, 547, 584-589
  functional, 586-588
  fractional, 586
Oxygen sensor, 170, 171, 525, 556, 557, 593, 619
  galvanic, 594
  polarographic, 594
Oxygenated blood, 354, 386, 704, 706, 709
Oxygenator, 706-716
  bubble, 703, 706,
  film, 706, 707
  hollow fiber, 707
  membrane of, 703, 706, 707-709
Oxygenated blood, 354, 386, 704, 706, 709
Oxyhemoglobin, 584, 585-589, 591

P

Pacemaker analyzer, 413
Pacemaker programmer, 409
Pacemaker, cardiac, 405-455
  asynchronous mode, 409
  battery of, 409, 413, 418, 420
  demand mode, 409, 422
  escape interval of, 417
  external noninvasive, 408, 422
  external invasive, 408, 422, 423
  hysteresis of, 417
  implantable, 406, 408, 414, 417, 420
  lead system of, 408-412
  magnet mode of, 416
  modes of, 409, 414-417
  pacing rate of, 409, 416-418
  problems with, 423-425
  pulse amplitude of, 409, 418, 421
  pulse duration of, 409, 413, 418
  pulse generator of, 408
  rate-modulated, 409
  sensing threshold, 413
  sensitivity of, 417
  stimulation threshold, 412
  transcutaneous, 408, 422
Pacing current, 410, 424
Packet transmission, 254
PACS (see picture archiving and communication system)
PaCO$_2$, 585
PaO$_2$, 584
Paper chart assembly, 239
Paper chart recorder, 238-244
  continuous paper feed, 238-240
  inkjet, 243
  paper supply mechanism, 239
  single page feed, 240-244
  thermal dot array, 239, 240
  thermal stylus of, 239
Paradoxical sleep, 316
Parenteral nutrition, 449
Particle drift, 13, 179
Passive breathing, 497
Passive electrode, 475, 481
Pasteurization, 766
Patient breathing circuit,
  expiratory limb of, 527, 618
  inspiratory limb of, 527, 618, 619
Patient interface, 30, 639-641

Patient leakage current, 222, 224, 236
Patient-initiated breath, 519, 523
Paul M.Zoll, 406
PAWP (see pulmonary arterial wedge pressure),
PBW (see pulsed biphasic waveform),
pCO$_2$ electrode, 152, 168
PCW (see pulmonary capillary pressure)
PEEP (see positive end-expiratory pressure)
Peracetic acid, 697
Perception, 19
Perception of pain, 219
Percutaneous venous cannula, 641
Periodical signal, 44
Peripheral temperature, 569, 573
Peristaltic pumping mechanism, 450, 462, 470
  protruding finger of, 459, 461, 462, 464, 470
  rotating cam shaft of, 461
Peritoneal cavity, 646, 653, 681, 682
Peritoneal dialysis, 596, 608-610
  continuous ambulatory, 610
  continuous cycler-assisted, 610
  disposal bag of, 609
  drain bag of, 609
  heater compartment of, 609
  supply reservoir of, 609
  volume control compartment of, 609
  Continuous laser, 621, 622
Peritoneal membrane, 596, 608
Peritonitis, 614
Permeability, 89
  cell membrane, 14, 592, 597
  gas, 665
  solenoid magnetic, 102
Permittivity, dielectric, 77, 107
Personal protection equipment (PPE), 765, 769
Petit mal seizures, 295
Petrolate-in-glass thermometer, 78
pH electrode, 155-157
Phase distortion, 55, 57, 176, 177
Philip Drinker, 486
Phlebostatic axis, 361
Photocathode, 129
Photocoagulator, 621
Photoconductive sensor, 132
Photodetector, 122, 124
Photodiode, 129-132

photoconductive mode of, 132
photovoltaic mode of, 132
Photodynamic drug, 620
Photoelectric effect, 129
Photoelectric tube, 115
Photoemissive sensor, 129
Photometry, 123-126
Photon detector, 124
Photon emissions, 618
Photoresistive sensor, 127
Phototherapy light, 523, 527-529
    dehydration in, 527
    far-infrared radiation in, 527
    fluorescent tube of, 527-529
    height adjustable stand of, 528
    hyperthermia in, 527
    light sources of, 527, 529
    observation timer of, 528
    ultraviolet radiation in, 527
    UV filter of, 528
Phototransistor, 132, 133, 215
Physiological and tissue effects of risk current, 200
Physiological monitor, 250
    arrhythmia detection, 258-259
    alarm function of, 251
    analyze function of, 251
    condition function of, 251
    display function of, 251
    monitored parameters of, 253
    network, 260-265
    record function of, 251
    sense function of, 251
    telemetry, 256-258
Physiological monitoring, 249-266
Physiological parameters, 17, 253
Physiological signals, 16, 17
    characteristics of, 17
Picture archiving and communication system, 651
Piezoelectric constant, 77
Piezoelectric pressure transducer, 76-78
Piggyback infusion, 452, 454
PIM (see patient-initiated mandatory breath)
Pin-indexed safety system, 624, 762
Pinna, 724, 725
Piped-in wall outlets, 762
Piston cylinder pumping mechanism, 458, 459, 462, 464, 470
    cam of, 459

input port of, 459
output port of, 459
stepper motor of, 459
stroke volume of, 459
valves of, 459
Pitot tube, 119, 129
pK electrode, 165-167
Placental blood flow, 547
Planck's constant, 127
Planck's law, 547
Plasma display, 250, 251
Plexiglas, 555, 558, 559, 564
Plume, 494, 674, 676
P-N junction, 101, 134, 137
Pneumoperitoneum, 681, 692, 698
Pneumothorax, 624, 532
Pneumotachometer, 119
pO2 electrode, 169, 170
Poiseuille's law, 114-117
Poisson's ratio, 73
Polarization sensor, 149
Polarographic cell, 169, 594
Polysomnography (PSG), 313
Polytetrafluroethylene (PTFE), 640, 641
Pons and medulla, 497
Poor perfusion, 597
Population inversion, 656, 664
Portable ventilator, 517, 524
Position transducers, 103-111
Positive end-expiratory pressure, 520
Positive pressure ventilator (see also mechanical ventilator), 515, 517
Potentiometry, 169
PPE (see personal protection equipment)
Power line filter, 212, 213
Power line interference, 199, 200
Power line noise, 199
Precision in measurement, 40
Precordial lead (see also chest lead), 293-295
Premature ventricular contraction, 274, 288, 306
Pressure and force transducer, 67-79
Pressure support, 521
Pressure swing absorption, 764
Pressure transducer, 67-78
Prevost and Batelli, 428
Primary cells (see also batteries, primary), 172
Primary reservoir of medical gas supply sys-

tems, 762
Protheses, 346
Protective eyewear, 673, 674, 676
Pseudo-periodical, 47
PSG (see polysomnography)
Pulmonary arterial wedge pressure, 395
Pulmonary artery catheter, 390, 393, 397, 399
Pulmonary function lab, 497, 500
Pulmonary vein, 354
Pulse oximeter, 583-599
  dark signal of, 592
  functional block diagram of, 591, 592
  LED (light-emitting diode) of, 591, 592
  photodetector of, 592
  plethysmograph of, 592, 598
  reflecting probe of, 590, 591
  sensor probe of, 591
  signal to noise ratio (SNR), 597
  timing control circuit of, 592
  transmitting probe of, 590, 591
Pulsed biphasic waveform, 431
Pulse contour method, in cardiac output monitoring, 397, 398
Pulsed laser, 661
Pumping mechanism, of infusion pumps, 458-465
Purkinje fiber, 287, 406
P-Vectorcardiogram, 302
PVC (see premature ventricular contraction)
PvCO2, 585
PvO2, 585
Pyroelectric sensor, 132, 576, 579
  ferroelectric material of, 132, 576
Pyrometry, 573

Q

qEEG (see quantitative EEG)
QRS -Vectorcardiogram, 302
QT interval, 287
Quantitative EEG, 323
Quantum dot, 250
Quantum efficiency, 141
Quantum event, 127, 128, 131

R

Radiance, 128
Radiant density, 128

Radiant energy, 128, 129, 132, 575, 577
Radiant flux, 128
Radiant flux density, 128
Radiant intensity, 128
Radio frequency, 272, 280, 331, 424
Radiometry, 127, 128, 130
Radiopaque, 393, 418
Radiant emittance, 128
Rate-modulated pacemaker (see also rate responsive pacemaker), 409
Rate-responsive pacemaker, 409
Redox reaction, 152, 153, 182
Reference electrode, 152, 161-164
Refractory period, 14, 180, 417, 422
REM (see return electrode monitoring)
Remote monitoring station, 254
Refraction, of light, 142, 143
Renal deficiency, 633
Renal dialysis, 629
  concentration gradient in, 630, 631, 634, 635
  double-needle technique in, 641
  mass diffusion rate of, 630
  mechanism of, 634
  rate of diffusion in, 630, 634
  single-needle technique in, 641
Repeaters, 254, 280
Repolarization, 14, 180, 286, 287
  membrane, 13, 14
Reproducibility, 40
REQM (see return electrode quality monitoring)
Reserve supply, of medical gas, 762
Residual volume, 498, 500, 502
Resistance temperature device, 85-91
  characteristics of, 95, 87, 101
Resistivity, 72, 74, 131, 316, 476
Resolution, definition, 40
Respiration monitor, 507-511
  amplitude modulated signal in, 509
  functional block diagram of, 481
  lead selector of, 480-481
Respiration monitoring, 478
  heated thermistor method of, 478-479
  impedance pneumographic method of 500, 511
  muscle and nerve stimulation in, 509
Respiration rate, 17, 498, 508, 522, 530, 620
Respiratory arrest, 516, 609
Response time, 42, 100, 135, 245, 469

Resting potential, 14, 179
    cell's, 14
Restrictive lung disease, 498, 500
Resuscitator, infant (see infant resuscitator)
Retrolental fibroplasia, 563
Return electrode monitor, 483-485
Return electrode quality monitoring, 485-485
Return electrode,
    of pacemaker, 410
    of ESU, 481-483
Reusable pressure transducer, blood, 363, 368
Reverse bias current, 134, 135
Reverse osmosis, 649
Reynolds number, 116, 122
RF (see radio frequency)
RF current density, 476, 781
Right ventricle, 354
Right leg-driven circuit, 204-211, 304
Rigid endoscope, 684, 685
    eyepiece of, 684
    instrument channel of, 684
    light cable of, 684
    light source of, 684
    optical fiber of, 684
    rod lens of, 684
Risk classification, 9, 10
Risk current, 31, 200, 220, 223, 233
RLD circuit (see right leg-driven circuit)
Roller clamp (see also regulating clamp of IV set), 452, 453, 455, 456, 466
Roller pump (see also rotary peristaltic infusion pump), 432, 607, 670
Rotameter, 120
Rotary peristaltic, 459
Router, 280
Routing protocol, 281
Routing table, 281
RTD (see resistance temperature device)
RV (see residual volume)

S

SA node (see sinoatrial node)
Saline, 358, 362, 365, 390, 397, 449
Salt bridge, 154, 155, 163, 164
Scalp electrode, 187, 317, 318-322, 551
    electrode placement of, 318-322
Scavenging system, 613, 621, 622

Active, 621
    passive exhaust, 621
    vacuum, 621
    wall suction of, 621
Scotoma, 672
Screw pump (see also syringe infusion pump), 458
Sealed lead acid batteries, 444
Secondary burn, 440, 491, 696
Secondary cells (see also batteries, secondary), 172, 174
Secondary reservoir, 762
Seebeck coefficient, 94, 95
Seebeck voltage, 94
Seldinger technique, 358, 393, 413, 450, 455
Selective membrane, 165, 171, 635
    active permeability of, 635
    passive permeability of, 635
Selective radiator, 129-131
Self-heating error, 89-91
Semicircular canal, 726, 727, 730
Semipermeable membrane, 13, 593, 594, 629-631, 634, 635, 637-639
    permeability of, 630, 635, 637
Sensing threshold, 413
Sensitivity, 40
Sensitivity drift, 41, 368
Sensor, 56
Sensor-indicated interval, 414-416
Sensorineural hearing loss, 733, 740, 744
Sensory limitation, 18
Sensory nerve action potential, 344, 345
Servoflurane, 616
Shell temperature (see also peripheral temperature), 569
Shielded lead, 203, 304
Shock prevention methods, summary of, 234
Sigh, 522, 523
Signal averaging, 289, 346, 740, 741,
Signal conditioning, 31, 61, 62
Signal isolation, 30, 229-231
Signal to noise ratio, 133, 401, 597, 740
Silver/silver chloride electrode, 184, 185
    electrical equivalent circuit of, 184, 185
Silver/silver chloride reference electrode, 162
SIMV (see synchronous Intermittent mandatory ventilation),
Single forced expiration, 500

Single page feed recorder, 240-244
Single-cell membrane potential, 13, 179
Sinoatrial (SA) node, 284, 406-408, 428
Sinus bradycardia, 408
Sinus rhythm, 407, 428
Sinus tachycardia, 408
Sinusoidal signal, 44
Skin electrode (see also surface electrode), 184
Skin preparation, 203, 206, 310
Skin temperature, 555, 558, 561, 563
Sleep disorder, 313, 315, 316
Snell's Law, 142
SNR (see signal to noise ratio)
Sodium hypochlorite, 651
Sodium-potassium pump, 14, 179
Solar cell, 136
Solenoid, 104, 106
Somatosensory, 314, 315
Sound,
    acoustic admittance of, 721, 737
    acoustic impedance of, 721
    attenuation of, 723
    equal loudness contour of, 728, 729
    frequency of, 720, 721
    incident reflection coefficient of, 723
    incident transmission coefficient of, 723
    intensity of, 722, 723
    power of, 722
    pressure level of, 722
    propagation speed of, 720, 721
    reflection of, 681, 693
    refraction of, 723, 724
    specific acoustic impedance of, 721
    transmission of, 723
    velocity of, 721, 722
    wavelength of, 720
Sound booth (see also audiometric booth), 747
Spark-gap generator, 475
Specifications, 39
Spectral irradiance, 558
Sphygmomanometer, 373
    cuff of, 373-376
    hand pump of, 373, 376
    mercury manometer of, 373
    pressure measurement device of, 373
    rubber bladder of, 352, 376
    valve of, 373, 376
Spiral electrode, 412, 549

Spirometer, 8, 120, 504-507, 511
    bellow of, 506
    flow transducer of, 506, 507
    flow-sensing, 505
    functional block diagram of, 505
    volume transducer of, 506
    volume-sensing, 506
    water-sealed inverted bell, 506
SPL (see sound pressure level)
Spontaneous breath, 516, 517, 519-521, 523, 530
Spore-forming bacteria, 651
Spread spectrum technology, 759-761
Stability, 84, 91, 101, 186
Standard atmospheric pressure, 68
Standard Electrode potential, 155-157, 162
Standard oxidation potential, 156
Standard reduction potential, 156, 162, 164
Standard reference electrode, 155
Stapes, 762, 730
Statistical control, 42
Steady state characteristics, 43
Stephen Hales, 353
Sterilant, 766
Sterilization,
    Chemical, 766, 767
    dry heat, 766
    moist heat, 768
Steam, 766
Stethoscope, 373, 380, 549
Stimulated emission, 656, 664
Storage oscilloscope, 246, 247
STP (see standard atmospheric pressure)
Strain, 28
Strain gauge, 72-76
    bonded, 74
    diaphragm, 75, 76
    gauge factor of, 74, 75
    metal wire, 75
    piezoresistive, 75
    unbonded, 74
Stray capacitance, 222
Streamline flow, 113
Strength duration curve, 335, 413
Stress, 28
Stress test, 288, 289
Stroke volume,
    cardiac, 385, 398, 399, 532
    piston pump, 459,
Stroke volume variation (SVV), 399

S-type thermocouple, 97
Subsystems, 10, 11
Superior vena cava, 398, 406
Surface electrode (see also skin electrode),
    185
Surge protector, 213
SV (see stroke volume),
Swan-Ganz catheter, 392-396, 400
    balloon of, 393, 395
    distal lumen of, 393, 395
    injectate orifice of, 393
    injectate port of, 393, 395, 396
    thermistor of, 393, 395
Switches, network, 280
Switching transient, 200, 213
Synchronous cardioversion (see also car-
    dioversion), 441, 444
Synchronized Intermittent mandatory venti-
    lation, 520
Syringe pump, 458, 462-466, 470, 471
    anti-siphon valve of, 451, 466
    stiction of, 471
System boundary, 11
System input, 10, 11
System output, 10, 11
System,
    closed, 9
    open, 9
Systems Approach, 9-13, 19, 24
Systolic blood pressure, 354, 356, 373

T

TCP/IP, 278
TcO2 monitor (see transcutaneous oxygen
    monitor)
TD (see thermal dilution)
Teflon encapsulated, 82
Telemetry, 272, 273, 758-761
    bandwidth requirement of, 759
    channel overcrowding of, 759
    data rate of, 759, 760
    EM interference and immunity of, 759
    power requirement of, 759
    primary user in, 759
    receiver of, 273
    reliability of, 273, 759
    secondary user in, 759
    transmission range of, 759
    transmitter of, 272, 273, 758

TEM (see transverse electromagnetic mode
    of laser)
Temperature coefficient (see also tempera-
    ture sensitivity), 85-87, 92
Temperature monitor, bedside continuous,
    569-573
    excitation circuit of, 572
    functional block diagram of, 572, 573
    transducer element, 570
Temperature scale, 81
Temperature sensitivity (see also tempera-
    ture coefficient), 87
Temperature sensor (see temperature trans-
    ducer)
Temperature transducer, 81-101
    bimetallic, 82, 83
    characteristics comparison, 101
    electrical, 84-101
    empirical laws of thermocouples, 96
    fluid-in-glass, 82, 83
    infrared (see infrared thermometry)
    integrated circuit, 101
    lead errors, 87-89
    nonelectric, 82-84
    resistance temperature device, 85-91
    self-heating errors, 89
    thermistors, 91-94
    thermocouples, 94-100
    YSI 400 series thermistors, 91, 570
    YSI 700 series thermistors, 93, 95
Temporal lobe, 724, 726
Temporal artery thermometer, 579
Temporary cardiac pacing, 422
TEOAE (see transient-evoked optoacoustic
    emission)
TFT (see thin film transistor)
Therapeutic devices, 8, 19, 30
Thermal detector, 128
Thermal dilution curve, 391, 400
Thermal dilution method, 390-397
    bolus of injectate of, 401
    catheter dead space of, 400
    frequency of measurement of, 401
    injectate temperature of, 391-393, 401
    injectate volume of, 401
    injectate warming of, 400, 401
    injection rate of, 400
    injection timing of, 400
    intravenous administration of, 401
    problems and hazards, 399-402

recirculation in, 389, 390, 401
thermistor position of, 401
Thermal dot array recorder, 239, 240
print head of, 243
vertical resolution of, 39, 256
Thermal event, 127, 128, 131
Thermal flowmeter, 123, 124
Thermal paper, 238, 239
Thermionic electrons, 133
Thermistor, 84, 91-94, 101, 124, 393, 401,
508, 570, 577, 579
Thermistor linearization, 93, 94
Thermocouple, 94-100
cold junction of, 91, 128, 172, 577
exposed, 100
grounded, 100
hot junction of, 99, 100
thermoelectric sensitivity of, 96, 97
ungrounded, 100
Thermometry, 568, 569
Thermopile, 132, 576, 577
Thin film transistor, 249
Thomas Seebeck, 94
Three-wire RTD, 88
Threshold drift, 423
Threshold of perception, 219, 221
Thunderbolt, 255
Tidal volume, 498, 518
Time constant, 84, 257
Time domain, 322, 324, 754
Time-varying signal, 43, 44, 61
Tissue effect,
60 Hz electric current, 218, 219
Laser, 659-661
RF current, 676-678
Tissue-electrode interface, 181, 182, 184
Titanium housing, 418
TLC (see total lung capacity)
TOCO transducer, 550, 551
Token Ring,
bandwidth of, 278
characteristics of, 278
mutli-station access unit of, 278
Total lung capacity, 499
Touch current, 222
Trabeculae, 413
Tracheostomy tube, 21, 515, 524
Transcutaneous oxygen monitor, 584, 594,
596, 598
Transcutaneous pacemaker, 408, 422

Transducer, 55-
active, 57
definition of, 56
direct mode of, 57
electrochemical, 151-177
excitation, 62-65
flow, 112-125
force, 67-78
indirect mode of, 57
motion, 103-111
optical, 126-150
position, 103-111
passive, 57
pressure, 67-78
temperature, 80-101
Transduction element, 58
Transfer factor (see also diffusion capacity),
504
Transfer function, 67-68
Transient characteristics, 42, 43
Transit time flowmeter, 539-540
Transmission link, 276, 279
Transpulmonary dilution, 397, 399
Transthoracic pacemaker (see also transcutaneous pacemaker), 422
Trocar, 681
T-type thermocouple, 97
Turbine flowmeter, 120
Turbulent flow, 113, 114
TV (see tidal volume)
T Vectorcardiogram, 302
Twisted copper wire, 280
Tympanic membrane, 569, 575-580, 724-
726, 736-738
Tympanic thermometer, 573, 575-579
detector of, 576
errors in, 576, 578
filter of, 576
FOV of, 576, 577, 580, 584
functional block diagram of, 576
multiple scanning of, 577
offset of, 577, 580
optical lens of, 576
optical light pipe of, 576
shutter mechanism of, 577
signal processor of, 578
thermistor of, 577, 579
Tympanogram, 736-738
admittance of, 737, 739
types (A, B, C) of, 737, 738

Tympanometer (see middle ear analyzer)

U

U.S. Food and Drug Administration, 406
UA (see uterine activity)
UHF, 759, 760
Ultrafiltration coefficient (see also KUf), 636, 638
Ultrafiltration, 630, 631-, 634, 636, 638, 642, 645, 713
Ultrasound,
 frequency of, 537
 heating effect of, 537
 propagation speed (or velocity) of, 537
 receiver, 74, 118, 122, 123, 380, 467, 468, 537, 538-542, 549
 transmitter, 74, 118, 122, 123, 380, 467, 539-542, 549
Ultrasound blood flow detector, 536-544
 Doppler flowmeter, 540-544
 transit time flowmeter, 539, 540
Ultrasonic cleaner, 696
Ultrasound flowmeter, 122, 123
Ultrasound vortex flowmeter, 122, 123
Unipolar lead, 193, 410, 424
UNIX, 279
User fatigue, 20
User interface, 18, 30, 32, 471, 525
Uterine activity (UA), 547, 551
 amplitude of, 548
 duration of, 548
 frequency of, 548
 methods of monitoring, 550-552
 resting tone of, 548
 rhythm of, 548
Uterine activity monitoring,
 external pressure transducer method in, 550, 551
 intrauterine pressure method in, 551
Uterine contraction, 546, 550
Ultrasound transit time flowmeter (see also transit time flowmeter)r, 539 540

V

Valinomycin, 165, 166
Vaporizer, anesthetic agent, 613-617, 624
 agent chamber of, 616, 617
 concentration control of, 616

control valve of, 616, 617
 electronic controlled, 616, 617
 filling spout of, 624
 flow control of, 616
 measured flow, 616, 617
 mixing chamber of, 616, 617
 reservoir of, 616
 vaporizing chamber of, 616
 variable-bypass, 616, 617
Vascular access, 450, 455, 635, 639-641
 site survival rate of, 641
Vascular restriction, 536
VC (see vital capacity)
Vectorcardiogram, 301-302,
 QRS, 301, 302
 T, 302
Venous access, 422, 450, 643
Ventilation parameters, 522, 523
Ventilator (see mechanical ventilators)
Ventricular fibrillation, 217, 288, 409, 428
Venturi tube, 117, 118, 120
Veress needle, 681
Vestibule, 726, 727
Vestibulocochlear nerve, 724, 726, 730
 vestibular branch of, 724, 727
 cochlear branch of, 724
VF (see ventricular fibrillation)
VGA (see video graphic display)
VHF, 759
Video graphic display, 239
 brightness of, 245, 248, 250, 252, 253, 688
 contrast ratio, 246, 250-253, 255
 refreshing rate of, 246, 249, 250-253, 255
 resolution of, 246, 247, 250, 253-256
 viewing angle, 250, 253
Video processor, 688, 689
Video signal interface, 253-255
Videoscope, 687, 689
 CCD camera of, 688
 mosaic color filter of, 688
 rotating color wheel of, 689
VM (see ventilator-initiated mandatory breath)
Viscosity, 115, 122, 123, 449, 463, 469
 coefficient of, 115
Visking, 638
Viscoelasticity, 8
Visual display monitor, 243, 244
Visual Display technology, 243-253

brightness of, 253
comparison of, 253
contrast ratio of, 253
refreshing rate of, 253
resolution of, 253
viewing angle, 253
Vital capacity, 498, 500,
Vital sign, 276, 547
Voltage limiting device, 213, 214
Voltaic cell, 152
Volume flow rate, 116-122, 385, 452, 462,
463, 526
Volume of dead space air, 502
Volume to be infused (VTBI), 465
Volumetric infusion pump, 458, 467
VTBI (see volume to be infused of infusion
pump),

W

Wall outlets, piped gas, 524, 526, 528, 613,
624, 692, 763
Walton Lillehei, 406, 703
WAN (see wide area network),
Warmer, infant (see infant warmer)
Waste anesthetic gas, 613, 618, 621
Water softener, 649
Water treatment, 648-650
Waterborne bacteria, 650
Weaning, 520, 522
Wheatstone bridge, 62, 63, 571
excitation voltage of, 62
White balance, 689
Wide area network (WAN), 260, 264, 279-
281, 308
backbone of, 280
Wien's displacement law, 574
Wilson Greatbatch, 406
Wilson network, 206, 207
Wireless link, 280
Wireless medical telemetry system (see also
medical telemetry), 759
WLAN, 760
WMTS (see wireless medical telemetry sys-
tem)
Work of breathing, 500, 517, 520
WOB (see also work of breathing in
mechanical ventilator), 500, 521, 530

X

Xenon-neon plasma, 250
Xenon lamp, 688
X-ray fluoroscopy, 393

Y

Yellow Spring Instrument, 91, 570
YSI (see Yellow Spring Instrument)
YSI 400 probe, 570, 571
characteristics of, 570
YSI 400 series thermistor, 91
YSI 700 probe, 570, 571
characteristics of, 570
YSI 700 series thermistor, 93, 95, 570-572

Z

Zero drift, 40
Zero offset, 40, 364
Zeroing, in blood pressure measurement,
385-361, 363, 366
Zinc-air cell, 174
Zinc-carbon cell, 172, 174